METEOROLOGICAL MONOGRAPHS

VOLUME 23 JUNE 1990 NUMBER 45

ATMOSPHERIC PROCESSES OVER COMPLEX TERRAIN

Robert M. Banta, G. Berri, William Blumen, David J. Carruthers, G. A. Dalu, Dale R. Durran, Joseph Egger, J. R. Garratt, Steven R. Hanna, J. C. R. Hunt, Robert N. Meroney, W. Miller, William D. Neff, M. Nicolini, Jan Paegle, Roger A. Pielke, Ronald B. Smith, David G. Strimaitis, T. Vukicevic, C. David Whiteman
Contributing Authors

William Blumen
Editor

American Meteorological Society
45 Beacon Street, Boston, Massachusetts 02108

ISBN 1-878220-01-2
ISSN 0065-9401

Library of Congress catalog card number 90-80548

Published by the American Meteorological Society
45 Beacon St., Boston, MA 02108

Richard E. Hallgren, Executive Director
Kenneth C. Spengler, Executive Director Emeritus
Evelyn Mazur, Assistant Executive Director
Arlyn S. Powell, Jr., Publications Manager
Jon Feld, Publications Production Manager

Editorial services for this book were contributed by Pamela Jones.

We wish to thank Keith Seitter and Linda Esche.

Printed in the United States of America
by Lancaster Press, Lancaster, Pennsylvania

TABLE OF CONTENTS

Preface xii

List of Contributors xiii

Chapter 1. Mountain Meteorology
—WILLIAM BLUMEN 1

ABSTRACT 1

1.1 Introduction 1

1.2 Some historical footnotes 1

1.2.1 Surface winds 1

1.2.2 Observations and observers 2

1.2.3 The discovery of atmospheric waves 3

1.3 Current directions 3

Acknowledgments 4

Chapter 2. Observations of Thermally Developed Wind Systems in Mountainous Terrain
—C. DAVID WHITEMAN 5

ABSTRACT 5

2.1 Introduction to diurnal mountain winds 5

2.1.1 Summary of recent field experiments 6

2.2 Along-valley wind systems 7

2.2.1 Climatology 7

2.2.2 Basic physics 9

2.2.2.1 TOPOGRAPHIC AMPLIFICATION FACTOR 9

2.2.2.2 EQUATIONS FOR THE VALLEY WIND SYSTEM 13

2.2.3 Radiation and surface energy budgets 14

2.2.3.1 RADIATION BUDGET 15

2.2.3.2 SURFACE ENERGY BUDGET 17

2.2.4 Atmospheric budgets of mass, heat, momentum, and moisture 21

2.2.4.1 CONSERVATION OF ATMOSPHERIC MASS 22

2.2.4.2 THERMAL ENERGY BUDGET 23

2.2.4.3 MOMENTUM BUDGET 25

2.2.4.4 HUMIDITY BUDGET 25

2.3 Slope wind systems 25

2.3.1 Simple slope flows 25

2.3.2 Slope flows on valley sidewalls 26

2.4 Morning transition 28

2.5 Evening transition 32

2.6 The diurnal cycle 34

2.7 Other phenomena 35

2.7.1 Influence of external winds 35

2.7.2 Maloja winds 36

2.7.3 Jets at valley exits 36

2.7.4 Antiwind systems 38

2.7.5 Tributary flows 38

2.8 Future research 39

Acknowledgments 42

Chapter 3. Thermally Forced Flows: Theory
—JOSEPH EGGER 43
ABSTRACT 43
3.1 Introduction 43
3.2 The problem 43
3.3 Circulation in a cavity with differentially heated sidewalls 44
3.4 Sloping boundaries 48
3.5 Circulation in valley cross-sections 51
3.6 Toward three-dimensional modeling 53
3.7 Three-dimensional valley flow 55
3.8 Outlook 57

Chapter 4. Mountain Waves and Downslope Winds
—DALE R. DURRAN 59
ABSTRACT 59
4.1 Introduction 59
4.2 Fundamentals of internal gravity waves 59
4.3 Application to small-amplitude mountain waves 62
4.3.1 Sinusoidal ridges; constant wind speed and stability 62
4.3.2 Isolated mountain; constant wind speed and stability 63
4.3.3 Vertical variations in wind speed and stability 65
4.4 Downslope windstorms 66
4.4.1 Three explanations for the production of severe downslope winds 66
4.4.2 A comparison of the hydraulic and the vertically propagating wave theories 69
4.4.3 A comparison of the hydraulic and the wave-breaking mechanisms 71
4.4.4 Forecasting downslope winds 74
4.4.5 Gustiness near the surface in downslope winds 76
4.5 Flow over isolated mountains 78
Acknowledgments 81

Chapter 5. Fluid Mechanics of Airflow over Hills: Turbulence, Fluxes, and Waves in the Boundary Layer
—D. J. CARRUTHERS and J. C. R. HUNT 83
ABSTRACT 83
5.1 Introduction 83
5.1.1 Overview 83
5.1.2 Classification of flow regimes 84
5.1.2.1 STRATIFICATION EFFECTS 84
5.1.2.2 ROTATION EFFECTS 85
5.1.2.3 ROUGHNESS CHANGES 85
5.2 Modeling of turbulent airflow over hills 85
5.2.1 Linear analysis 86
5.2.1.1 GENERAL EQUATIONS 86
5.2.1.2 INNER REGION SOLUTION 86
5.2.1.3 MIDDLE LAYER SOLUTION 87
5.2.1.4 UPPER LAYER SOLUTION 87
5.2.1.5 SURFACE ROUGHNESS EFFECTS 88
5.3 Effects of stratification 88
5.3.1 Uniform stratification 89
5.3.2 More complex stability profiles 89

5.3.3 *Elevated inversion above neutral boundary layer* ... 90
5.3.4 *Strong stratification; large aspect ratios* ... 91
5.4 Turbulence ... 91
5.5 Numerical models and flow over complex terrain ... 93
5.5.1 *Isolated hills* ... 93
5.5.2 *Complex terrain* ... 94
5.6 Dispersion and deposition over complex terrain ... 95
5.6.1 *Overview and key processes* ... 95
5.6.2 *Localized sources near hills* ... 96
5.6.2.1 IDEALIZED HILL SHAPE [EPA-CTDM MODEL (PAINE ET AL. 1987)] ... 98
5.6.2.2 FOURIER ANALYSIS OF HILL SHAPES ... 98
5.6.3 *Dispersion and deposition over terrain for well-mixed scalars* ... 98
5.6.4 *Temperature and humidity fields* ... 102
5.7 Discussion ... 103
Acknowledgments ... 103
APPENDIX: Why Can't Stably Stratified Air Rise over High Ground?
—RONALD B. SMITH ... 105
5.A.1 Introduction ... 105
5.A.2 Combining the hydrostatic and Bernoulli equations ... 105
5.A.3 Application ... 107
5.A.4 Conclusion ... 107
Acknowledgments ... 107

Chapter 6. Rugged Terrain Effects on Diffusion
—STEVEN R. HANNA and DAVID G. STRIMAITIS ... 109
6.1 Introduction ... 109
6.1.1 *Problems* ... 109
6.1.2 *Overview of history of research* ... 110
6.2 Summary of EPA models and evaluations ... 111
6.2.1 *Model descriptions* ... 111
6.2.2 *Evaluations of regulatory models* ... 113
6.2.2.1 EPA EVALUATION AT LUKE MILL AND CINDER CONE BUTTE ... 113
6.2.2.2 EVALUATION OF COMPLEX I AND RTDM AT WIDOWS CREEK ... 114
6.3 Theories and experiments regarding diffusion over slopes and valleys ... 114
6.3.1 *Diffusion models for slope flows* ... 115
6.3.2 *Diffusion models for narrow valleys* ... 116
6.3.3 *DOE ASCOT experiments* ... 118
6.4 EPA Complex Terrain Model Development (CTMD) Program ... 121
6.4.1 *Objectives* ... 121
6.4.2 *Fluid modeling* ... 123
6.4.3 *Field experiments* ... 125
6.4.3.1 CINDER CONE BUTTE ... 126
6.4.3.2 HOGBACK RIDGE ... 127
6.4.3.3 TRACY POWER PLANT ... 127
6.4.4 *Assumptions contained in the Complex Terrain Dispersion Model (CTDM)* ... 131
6.4.4.1 DISPERSION PARAMETERS ... 133
6.4.4.2 CONCENTRATION EQUATION FOR LIFT (FLOW ABOVE H_d) ... 133
6.4.4.3 CONCENTRATION EQUATION FOR WRAP (FLOW BELOW H_d) ... 135
6.4.5 *Evaluation of CTDM* ... 135

6.5 Mesoscale flow models that include diffusion algorithms ... 137
6.5.1 General principles ... 137
6.5.1.1 APPROACH 1—LAGRANGIAN PARTICLE DIFFUSION ... 138
6.5.1.2 APPROACH 2—USE OF DIFFUSION EQUATION ... 139
6.5.2 Evaluation of mesoscale grid-based diffusion models in rugged terrain ... 141
6.6 Summary of findings and recommendations ... 141
Acknowledgments ... 142

Chapter 7. Fluid Dynamics of Flow over Hills/Mountains—Insights Obtained through Physical Modeling —ROBERT N. MERONEY ... 145

ABSTRACT ... 145
7.1 Introduction ... 145
7.1.1 Advantages and disadvantages of fluid modeling ... 145
7.1.2 Historical perspectives ... 146
7.2 Similarity considerations ... 146
7.2.1 Similitude parameters ... 147
7.2.2 Partial simulation of complex terrain flows ... 147
7.2.3 Performance envelopes for fluid modeling ... 148
7.2.3.1 NEUTRAL AIRFLOW MODELS ... 149
7.2.3.2 VALLEY DRAINAGE FLOWS ... 150
7.2.3.3 VERIFICATION EVIDENCE ... 150
7.3 Facilities for fluid modeling of complex terrain meteorology ... 150
7.3.1 Wind tunnels ... 151
7.3.2 Drainage flow facilities ... 151
7.3.3 Water channels and rotating tanks ... 151
7.3.4 Instrumentation ... 152
7.4 Neutral flow over hills, ramps, and escarpments ... 152
7.4.1 Idealized two-dimensional terrain flow studies ... 152
7.4.1.1 EFFECTS OF RIDGE SHAPE ... 153
7.4.1.2 EFFECTS OF TURBULENCE ... 155
7.4.1.3 EFFECTS OF SURFACE ROUGHNESS ... 155
7.4.2 Idealized three-dimensional terrain flow studies ... 156
7.4.3 Field/laboratory comparisons ... 157
7.4.3.1 RAKAIA RIVER GORGE, NEW ZEALAND ... 158
7.4.3.2 GEBBIES PASS, BANKS PENINSULA, NEW ZEALAND ... 158
7.4.3.3 KAHUKU POINT, OAHU, HAWAII ... 159
7.4.3.4 ASKERVEIN HILL PROJECT, OUTER HEBRIDES, SCOTLAND ... 159
7.4.4 Conclusions from neutral airflow terrain studies ... 160
7.5 Stratified flow over hills and ramps ... 161
7.5.1 Idealized two-dimensional flow domains for waves and blocking ... 161
7.5.2 Downslope winds, valley flows induced by crosswinds ... 162
7.5.3 Idealized three-dimensional terrain studies ... 163
7.5.4 Field/laboratory comparisons ... 165
7.5.5 Conclusions from stratified airflow terrain studies ... 166
7.6 Drainage flow phenomena ... 167
7.7 Diffusion phenomena in complex terrain ... 168
7.8 Summary ... 171

Chapter 8. Remote Sensing of Atmospheric Processes over Complex Terrain
—W. D. Neff 173
ABSTRACT 173
8.1 Introduction 173
8.2 Remote sensing techniques 174
8.2.1 Scattering mechanisms 174
8.2.2 The role of turbulence microstructure in remote sensing 175
8.2.2.1 Statically unstable conditions 176
8.2.2.2 Statically stable conditions 176
8.2.3 Sampling geometries 177
8.2.3.1 Fixed-beam systems 177
8.2.3.2 Scanning systems 180
8.2.4 Instruments 181
8.2.4.1 Sodars 181
8.2.4.2 Radars 182
8.2.4.3 Aerosol-mapping lidars 183
8.2.4.4 Doppler lidars 184
8.2.4.5 Optical crosswind sensors 185
8.3 Major complex terrain field studies using remote sensors 185
8.3.1 Atmospheric Studies in Complex Terrain (ASCOT) Program 185
8.3.2 EPA Complex Terrain Model Development (CTMD) Program 187
8.3.3 Urban and regional air quality studies 187
8.4 Application of remote and in-situ instrumentation to complex terrain studies—case studies 188
8.4.1 Sodar observations of simple drainage flows 188
8.4.2 Sodar observation of complex drainages and their interaction with ambient flows 191
8.4.2.1 The role of differential acceleration of air masses in echo creation 191
8.4.2.2 Entrainment by the external wind—turbulence and instability 193
8.4.3 Remote sensor observation of waves in complex terrain 194
8.4.3.1 Long-period oscillations observed in complex terrain flows 195
8.4.3.2 Scale analysis 196
8.4.3.3 Surface wind analysis 196
8.4.3.4 Waves associated with flow over ridges and mountains 199
8.4.4 Volume flux in simple drainage flows 200
8.4.5 Main canyon mass fluxes and merging 203
8.4.5.1 Lidar data analysis and correction 203
8.4.5.2 Volume flux growth and the role of tributaries 205
8.4.5.3 The merging of drainage flows from valleys with different physical characteristics 205
8.4.5.4 Comparison of lidar observations with other measurement methods in complex terrain 208
8.4.6 The role of remote and in-situ sensors in the study of elevated plumes during the 1984 CTMD experiment 209
8.4.6.1 Background 209
8.4.6.2 Lidar aerosol plume mapping 209
8.4.6.3 Complex terrain processes affecting plume transport and dispersion 209
8.4.6.4 Interpretation of elevated plumes 211
8.4.7 Interpretation of ground-based plumes during the 1980 ASCOT experiment 212
8.4.8 Remote-sensor observation of day/night transitions in complex terrain 212
8.4.8.1 Background 212

8.4.8.2 SODAR OBSERVATIONS OF NOCTURNAL INVERSION DESTRUCTION 213
8.4.8.3 LIDAR OBSERVATIONS OF MORNING FLOW REVERSALS 214
8.4.8.4 THE EVENING TRANSITION TO NOCTURNAL DRAINAGE IN A CONFINED VALLEY 215
8.4.8.5 THE EVENING TRANSITION AND THE EMERGENCE OF DRAINAGE FLOWS ONTO PLAINS 216
8.5 The use of remote sensors in large-scale complex terrain flows 220
8.5.1 Large-scale drainage flows 221
8.5.2 Denver Brown Cloud Study 222
8.5.2.1 DOPPLER LIDAR OBSERVATIONS 223
8.5.2.2 RASS OBSERVATIONS 224
8.6 Summary and future possibilities 226
Acknowledgments 228

Chapter 9. The Role of Mountain Flows in Making Clouds
—ROBERT M. BANTA 229

ABSTRACT 229
9.1 Introduction 229
9.2 Condensation and stability effects 230
9.2.1 Condensation 230
9.2.2 Lifting profiles 231
9.2.3 Stability effects on cloud forms 233
9.2.4 Interactional effects 234
9.3 Fog, stable clouds, and unstable snow clouds 236
9.3.1 Valley fog 236
9.3.2 Stable rain clouds 236
9.3.2.1 OROGRAPHIC STRATUS 238
9.3.2.1.1 Simple orographic flow 238
9.3.2.1.2 Microphysics 238
9.3.2.1.3 Blocking 240
9.3.2.1.4 Mountain-wave effects 241
9.3.2.1.5 Radiation 241
9.3.2.1.6 Other upslope flows 242
9.3.2.2 INTERACTION WITH LARGER-SCALE PROCESSES 243
9.3.3 Stable snow clouds 244
9.3.3.1 OROGRAPHIC STRATUS 244
9.3.3.2 INTERACTION WITH LARGER-SCALE PROCESSES 247
9.3.4 Unstable snow clouds 248
9.4 Unstable rain clouds 248
9.4.1 Initiation mechanisms 249
9.4.1.1 DIRECT LIFTING TO THE LFC 250
9.4.1.1.1 Potential instability release 250
9.4.1.1.2 Flash flooding 250
9.4.1.2 THERMALLY GENERATED MOUNTAIN CIRCULATIONS 254
9.4.1.2.1 Heat flux and soil moisture effects 256
9.4.1.2.2 Regional differences 256
9.4.1.2.3 Spatial and temporal distribution 257
9.4.1.2.4 Isolated peaks and small ranges 258
9.4.1.2.5 Larger ranges of mountains 262

9.4.1.2.6 Ambient wind effects on thermally forced mechanisms 263
9.4.1.2.7 Discussion 264
9.4.1.3 OBSTACLE (AERODYNAMIC) EFFECTS 265
9.4.1.3.1 Blocking 266
9.4.1.3.2 Flow deflection 266
9.4.1.3.3 Gravity-wave effects 267
9.4.2 Conditional instability 268
9.4.3 Moisture sources 268
9.4.4 Structures and transitions 269
9.4.4.1 DRY CIRCULATIONS 270
9.4.4.2 CUMULUS AND CUMULUS CONGESTUS 275
9.4.4.3 CUMULONIMBUS 276
9.4.5 Roles for large-scale forcing 277
9.4.6 Discussion 281
9.5 Conclusions 282

Chapter 10. Predictability of Flows over Complex Terrain
—J. PAEGLE, ROGER A. PIELKE, G. A. DALU, W. MILLER, J. R. GARRATT, T. VUKICEVIC, G. BERRI and M. NICOLINI 285
ABSTRACT 285
10.1 Overview of concepts of chaos and relation to predictability 285
10.2 Predictability studies on the synoptic and global scales 286
10.2.1 Predictability theory for large scales 286
10.2.2 Global experiments 287
10.2.3 Predictability experiments in limited area domains 288
10.3 Predictability studies of turbulence and clouds 289
10.4 Analysis of limited domain predictability 290
10.4.1 Theoretical considerations 290
10.4.2 Predictability of highly forced and dissipated flows 292
10.4.3 Predictability on small scales 294
10.4.4 Spread of lateral boundary errors 295
10.5 Contrast between average and realization parameterization in mesoscale models 297
10.6 Conclusions 298
Acknowledgments 299

References 301

PREFACE

A workshop on atmospheric processes over complex terrain, sponsored by the American Meteorological Society, was convened in Park City, Utah from 24–28 October 1988. The overall objective of the workshop was one of interaction and synthesis—interaction among atmospheric scientists carrying out research on a variety of orographic flow problems, and a synthesis of their results and points of view into an assessment of the current status of topical research problems. The final day of the workshop was devoted to an open discussion on the research directions that could be anticipated in the next decade because of new and planned instrumentation and observational networks, the recent emphasis on development of mesoscale numerical models, and continual theoretical investigations of thermally forced flows, orographic waves, and stratified turbulence.

The Park City Workshop was planned by a program committee consisting of Bob Banta, Jan Paegle, Bill Blumen, Bill Clements, Sumner Barr, and Mike Fosberg. The workshop could not have taken place without generous grants from the following organizations, with the responsible individuals also indicated: the U.S. Forest Service (M. Fosberg), the U.S. Department of Energy (A. Patrinos), the U.S. Army Research Office (W. Bach), and the U.S.A.F. Office of Scientific Research (Lt Cols J. Koermer and J. Stobie). We also thank both the Rocky Mountain Forest and Range Experiment Station of the U.S. Forest Service (D. Fox) for arranging for the federal funds from all sources to be transferred to the American Meteorological Society, and Joe Egger for contributing his illustrations of atmospheric processes over complex terrain to the workshop.

This monograph represents an outgrowth of the Park City Workshop. The authors have contributed chapters based on their lecture material. Workshop discussions indicated interest in both the remote sensing and predictability of orographic flows. These chapters were solicited following the workshop in order to provide a more balanced view of current progress and future directions in research on atmospheric processes over complex terrain. The timely publication of this monograph has been aided by additional grants from the U.S. Forest Service (M. Fosberg), the U.S. Department of Energy (A. Patrinos and H. Moses), the U.S. Army Research Office (W. Bach), the U.S.A.F. Office of Scientific Research (J. Stobie), and the Environmental Research Laboratories of the National Oceanic and Atmospheric Administration (NOAA/ERL).

Others have also provided important and timely contributions. Merlyn Holmes provided editorial assistance and Lynda McGinley was invaluable in keeping all of the editorial activities running smoothly. We thank both of them.

ROBERT M. BANTA
WILLIAM BLUMEN

List of Contributors

Chapter 1 William Blumen
University of Colorado
Campus Box 391
Boulder, CO 80309

Chapter 2 C. David Whiteman
Battelle Pacific NW Labs
2012 Hoxie Street
Richland, WA 99352

Chapter 3 Dr. Joseph Egger
Meteorological Inst.
University of Munich
Theresienstrasse 37
D-8000 Munchen 2
Fed. Rep. of Germany

Chapter 4 Dale R. Durran
Dept. of Atmos. Sci. AK-40
Univ. of Washington
Seattle, WA 98195

Chapter 5 Dr. David Carruthers
DAMTP
Cambridge University
Silver Street
England, CB39EW

Chapter 5A Dr. Ronald B. Smith
Dept. of Geology & Geophysics
Yale University
New Haven, CT 06520

Chapter 6 Dr. Steven R. Hanna
Sigma Research Corp.
234 Littleton Rd., Suite 2E
Westford, MA 01886

Chapter 7 Dr. Robert N. Meroney
Engineering Res. Ctr.
Dept. of Civil Engineering
Colorado State University
Fort Collins, CO 80523

Chapter 8 Dr. Bill Neff
NOAA/ERL, R/E/WP7
325 Broadway
Boulder, CO 80303

Chapter 9 Robert M. Banta
NOAA/ERL, R/E/WP2
325 Broadway
Boulder, CO 80303

Chapter 10 Dr. Roger A. Pielke
Dept. of Atmos. Sci.
Colorado State Univ.
Fort Collins, CO 80523

Jan Paegle
Dept. of Meteorology
Univ. of Utah
819 Wm. Browning Bldg.
Salt Lake City, UT 84112

ATMOSPHERIC PROCESSES OVER COMPLEX TERRAIN

CHAPTER 1

Mountain Meteorology

WILLIAM BLUMEN

Astrophysical, Planetary, and Atmospheric Sciences Department, University of Colorado, Boulder, Colorado

ABSTRACT

Some historical footnotes in the field of mountain meteorology are presented. This background provides a brief exposure to the early foundation for current research activities in the collection of meteorological data over complex terrain, physical modeling, and the development of diagnostic and theoretical models of thermal circulations and orographic wave motions. Nascent areas of research, including boundary-layer turbulence and the prediction of weather regimes and pollutant dispersal are assessed in light of the observational and theoretical limitations to be surmounted.

1.1. Introduction

The earth's orography is revealed by almost imperceptible bulges and depressions on a world globe. This is understandable when you consider that the highest mountain barriers only extend the earth's radius by about one-tenth of one percent from its sea-level value. Depressions in the earth's crust are even less noticeable. Nevertheless, the presence of mountains with their endless varieties of passes, valleys, and slopes provides a breeding ground for a myriad of extraordinary meteorological phenomena.

1.2. Some historical footnotes

1.2.1. Surface winds

The history of mountain climate has been etched into the landscape in one form or another. For example, the distortions of trees as shown in Fig. 1.1 are clear indicators of the wind direction over many years. Clues are also obtained from eroded rock surfaces that face the wind. Written history also provides evidence of unusual past events. To Strabo, the Greek historian and geographer, a wind descending through the Rhone Valley onto the Crau Plain was one of the great marvels that ravaged the South of France two thousand years ago. He writes (Jones 1917), "The Black North, a violent and chilly wind, descends upon this plain with exceptional severity; at any rate, it is said that some of the stones are swept and rolled along, and that by the blasts the people are dashed from their vehicles and stripped of both weapons and clothing." This north wind, the mistral, has hardly subsided. It rages frequently through the Rhone Valley and explodes into the Mediterranean basin often at speeds up to 100 mph. Ship logs tell of the havoc inflicted upon vessels caught in the path of the mistral and other winds that have emerged from coastal mountains.

Unusually large wind speeds are common in mountainous terrain and the surrounding foothills, even when the air is not funneled through a narrow gap or valley by the natural barriers on either side. Winds sweeping down the slopes of mountain ranges may match or even exceed the speed of the mistral. Quite often these blasts will cause thermometers to soar, producing temperature increases of 10°C or more in a matter of minutes. This combination of downslope winds and warmth has inspired a long list of names that either describe the phenomenon or its locale. The oroshi of Japan is "a wind blowing strong down the slope of a mountain." The foehn of the Alps has been traced to the Latin word *favonius,* a warm westerly wind; while the chinook, which appears on the eastern slope of the Rockies, is associated with the name of an Indian tribe that lived near the mouth of the Columbia River in the northwestern United States. The "snow eater" is an apt description of the chinook, since many inches of snow may be cleared away by this warm, dry wind.

Although in some cases strong downslope winds can cause events ranging from destruction and (foehn) sickness to a temporary disruption of daily life, such winds usually represent an intrusion upon the average climatic variability of a locale. The more prevalent gentle breezes, typically less than 10 m s^{-1}, have become embodied in the history of mountain culture. For the most part, these milder winds are of thermal origin. The pressure gradient force that provides the driving mechanism arises from the difference between the temperature of the air above a heated (or cooled) slope and that of the free air at the same level above a nearby valley or level plain. As simple as this widely accepted physical concept appears, the historical development leading to its delineation as the primary controlling mechanism for slope and valley winds has not been without controversy. Vergeiner and Dreiseitl (1987) aptly comment that the nineteenth century disputes over valley wind theory (Hawkes 1947) "seem so picturesque, laborious, and irrelevant today."

FIG. 1.1. A typical timberline tree at about 11 000 ft on a ridge near Arapahoe Glacier, Colorado (photo by Ives 1964). Coring showed this tree to be about 300 years old.

In contrast to strong winds that sweep out and disperse pollutants, low intensity orographic winds can transport harmful gases and particulate matter *en masse.* Since the weaker wind systems are pervasive over much of the year, the problem of anthropogenic pollution is exacerbated over mountainous terrain. This is not a new problem. For example, damage to crops in the State of Washington during the early 1930s from sulfur fumes from a Canadian smelter became an international controversy (Hewson and Gill 1944). Further, the contribution to smog conditions in the Los Angeles basin and the San Gorgonio Pass from orographic influences have been under study since at least the 1940s.

1.2.2. Observations and observers

The following accounts have been extracted from the daily journal of meteorological observations made on the summit of Pike's Peak, Colorado, between 1874 and 1888 (Annals of the Astronomical Observatory of Harvard College 1889).

November 30, 1875—Observer shot a large mountain lion this morning.

June 16, 1876—At 5:30 P.M. as the observer was sitting on a rock near the summit, a blinding flash of lightning darted from a cloud seemingly not more than five hundred feet northeast of him, accompanied by a sharp, quick, deafening report, and at the same time he felt the electricity dart through his entire person, jerking his extremities together as though by a most violent convulsion, and leaving lightning sensations in them for a quarter of an hour afterwards.

May 11, 1881—A hurricane struck the summit during the night, wind attaining a maximum velocity of one hundred and twelve miles per hour; at 12:15 A.M. the anemometer cups were blown from their socket and carried away. From this time until 2:30 A.M. the wind increased in violence and the estimated velocity must have been one hundred and fifty miles per hour.

July 1, 1882—At 4:31 P.M., during a heavy fall of hail, a bolt of lightning struck the station building near the southeast corner, having followed the course of the telegraph wire a distance of several rods. The fluid passed through the outer partition walls and entered the office in the southeast corner near the stove, tearing up the floor, melting and tearing off the zinc sheeting around the stove, jumped to the self-register, which it demolished, also the regulator clock on the wall, burned up completely the office wires, and, passing out the north window to the roof, burned out the dial of the anemometer. The explosion was terrific, breaking out every light in the office windows, and, bruising both observers.

The fact that many meteorological observations have been accumulated at Pike's Peak and at other mountain observatories[1] must be attributed to the tenacity of observers, who were exposed to perilous conditions. However, individuals confined to fixed mountain observing sites are not the only ones exposed to the hazards of weather observing. For instance, meteorological observations have been brought back from climbing assaults on Mounts Everest (Longstaff 1923) and Kilimanjaro (Hunt 1947). To the climbers involved in these expeditions, the foremost consideration is climbing the mountain—all other tasks are subordinate. The difficulty of recording observations during such ventures is dramatized in a report of the 1924 Everest expedition (Sommervell 1926). Although daily observations along this route were taken during May and early June, it should be underscored that "the greatest credit is due to those members of the expedition who summoned up the strength to swing a thermometer for five minutes, or to go out in a blizzard to see what was last night's minimum."

Fortunately, heroic deeds are not always a by-product of acquiring meteorological information above mountainous terrain. Besides outfitting climbers with instruments, a cog railroad at Mount Washington, New Hampshire, and a cable car reaching the Zugspitze Peak in Germany have been used as platforms for meteorological measurements; however, the accumulation of data through the efforts of climbing expeditions and by the use of moving surface vehicles have been sporadic endeavors. The quest for information near the surface of the earth is, for the most part, aided by routine measurements made on a day-to-day basis at ground level meteorological stations situated in virtually all land areas of the globe.

[1] Stone (1934) presents a succinct, highly informative summary of the history of meteorological observations obtained at North American and Pacific island mountain observatories.

1.2.3. The discovery of atmospheric waves

The realization that gliders or sailplanes could be used as atmospheric probes developed during the decade following the late 1920s, when the art of soaring was enjoying wide popularity in Europe. Evidence of pronounced vertical undulations of the airflow in the lee of mountain ridges and isolated peaks began to emerge when pilots were able to exploit these currents to rise to great heights and stay aloft for long periods of time. In fact, altitude records were frequently being surpassed during these years.

As Queney et al. (1960) point out, "Theoretical workers had not been idle," as a body of descriptive information about the characteristics of these undulations was being established. It is further noted that Queney, in his paper "Influence du Relief sur les Éléments Météorologiques," did in fact provide the theoretical explanation of the so-called mountain lee wave in 1936 *before* this phenomenon was properly depicted by observational means. Since Queney's pioneering work the study of lee waves has flourished. Not only has the theory been honed, for example, by Scorer (1949) and Long (1953b) among others, but field programs to investigate lee-wave characteristics have been carried out over many of the prominent peaks and ridges that characterize the earth's orography. The comprehensive monograph by Queney et al. and the extended reviews by Nicholls (1973) and Smith (1979) provide evidence of the worldwide scope of these studies.

It is now well established that lee waves are simply internal gravity waves maintained in an essentially steady state in a stably stratified airstream. It is also evident that, similar to surface water waves, lee waves may overturn and be associated with extremely turbulent conditions. Striking evidence for highly unusual and hazardous conditions was uncovered during the comprehensive field investigation of airflow over the Sierra Nevada range in the vicinity of Bishop, California (Holmboe and Klieforth 1957). Both gliders and powered aircraft were used during the 1951–52 and 1955 field programs to gather meteorological data. Vertical currents exceeding 10 m s^{-1} were encounted within the fully developed "Bishop wave." However, in one reported incident, the mountain induced updrafts in this locale were found to be of such intensity that the pilot of a P-38 aircraft was able to feather the propellers and soar like a glider for over an hour on the cushion of rising air. These updrafts were estimated to be about 40 m s^{-1}!

1.3. Current directions

Orographic flows encompass all scales of motion, and disturbances forced by orography may even extend to thermospheric altitudes. Both observational analyses and numerical model experiments have revealed that the planetary general circulation including the energy, momentum, heat, and moisture balances is significantly affected by orography. There continues to be widespread interest in the scientific exploration of phenomena either produced or affected by local inhomogeneities of the earth's terrain. The contributors to this volume consider those orographic responses that are principally confined to the troposphere and are of sufficiently small horizontal scale ($L < 10^5$ m) that they may be completely unobserved by the present synoptic-scale observational network.

The two most important mechanisms that force orographic flows are thermal and mechanical in nature. *Thermal circulations* are intimately related to differential heating and cooling associated with diurnal insolation. *Wave motions* arise because a resting state of stable stratification is perturbed by the low-level undulations of an incident airstream that is constrained to follow the topographic surface profile. The interactions between thermal and mechanical forcing give rise to convective and stratiform clouds, and to both stably stratified turbulence and small-scale instability that characterize the orographic boundary-layer regime. This wide range of physical phenomena are not necessarily restricted to orographic flows. Yet the fact that excitation, maintenance, and dissipation of these flows are intimately related to complex terrain features does not mitigate the problems of data collection, analysis, and prediction.

In situ measurements by ground-based measuring systems and by balloons and aircraft have traditionally been the mainstays of field programs. These traditional methods of acquiring data will undoubtedly continue for the foreseeable future, but will be increasingly supplemented by physical model experiments on the laboratory scale, and by remote sensing of thermal and motion fields and cloud features that characterize the real atmosphere. Expansion of these facilities will continue to depend on technological advances that are difficult to predict. Among the more promising observational tools is Doppler lidar, which provides real-time two-dimensional wind vectors over 10 s of kilometers with a resolution of about one-third of a kilometer (e.g., Bilbro et al. 1984). This range and resolution provides data for relatively detailed analyses of orographic waves and convective motions, and its usefulness for depicting fields adjacent to complex terrain has recently been demonstrated by Post and Neff (1986).

There has been notable progress, in recent years, both in the analysis and the modeling of thermal circulations, drainage flows, mountain waves, and related downslope windstorms. Yet the problems of data paucity and the sensitivity of physical responses to terrain features continue to pose formidable barriers to comparable progress in nascent research directed toward basic turbulent processes, and to the development of models for the reliable prediction of weather regimes and transport of pollutants over complex terrain.

The authors of the following chapters have provided historical perspectives and assessments of current knowledge that are intended to provide the fulcrum for the development of research goals in the decades ahead.

Acknowledgments. Financial support for this research has been provided by the National Science Foundation under NSF Grant ATM 86-17636. I also wish to thank C. David Whiteman and Joseph Egger for their welcome responses to my request for historical references.

CHAPTER 2

Observations of Thermally Developed Wind Systems in Mountainous Terrain

C. David Whiteman

Pacific Northwest Laboratory, Richland, Washington

ABSTRACT

Slope and valley wind systems are local thermally driven circulations that form frequently in complex terrain areas. Recent research has focused on the temperature structure along the slope and valley axes that leads to the wind systems. Two new tools being used in these analyses include the topographic amplification factor, which quantifies the role of the topography in producing bulk temperature gradients along a valley's axis, and atmospheric heat budgets, which identify key physical processes leading to changes in temperature structure. Both tools are in an early stage of development, are being applied primarily to steady-state nighttime periods, and are leading to new concepts and understanding.

Recent climatological evidence in Austria's Inn Valley and in several Colorado valleys supports the concept that valley winds are driven by horizontal pressure gradients that are built up hydrostatically by the changing temperature structure along a valley's length. Topographic amplification factors appear to be useful in assessing the strength of valley wind systems. Several components of valley atmospheric heat budgets have proven difficult to measure, and large imbalances are being experienced. Several recent experiments, in a range of climatological regimes, suggest that measured nighttime surface sensible heat fluxes are too small to result in balances. This may be caused by measurement errors or by nonrepresentative measurements. The advective and radiative flux divergence components are also uncertain.

A simple conceptual model of diurnal wind and temperature structure evolution in deep valleys is presented. During the morning transition period, upslope flows form over heated valley sidewalls and compensatory subsidence over the valley center produces warming that eventually reverses the down-valley winds. The key role of vertical motions in transferring energy through the valley atmosphere during the morning transition period has been demonstrated by field and modeling studies.

The evening transition period has received little observational attention, and the key physical processes are not yet well known. Investigation of slope wind systems has focused mostly on the nighttime flows. Flows on the sides of isolated mountains are reasonably well understood when external flows are weak, but slope flows on valley sidewalls are complicated by the continued evolution of temperature structure within the valley and the strong influence of the overlying along-valley flows.

Recent experiments have shown that thermally driven flows within the topography may be influenced in subtle ways by the overlying circulations. This influence is nearly always present to some extent, but has not yet been systematically investigated. Recent research on strong winds that issue from a valley's exit at night and on tributary flows is briefly summarized, and some comments are made on Maloja winds and antiwind systems. The chapter ends with a summary of topics needing further research.

2.1. Introduction to diurnal mountain winds

Two classifications of diurnal mountain wind systems are generally recognized (Fig. 2.1). *Slope winds* blow parallel to the inclination of the sidewalls and are called upslope and downslope winds. The slope winds are produced by buoyancy forces induced by temperature differences between the air adjacent to the slope and the ambient air outside the slope boundary layer. Typically, slope winds blow up the slope by day and down the slope by night. *Valley winds* blow parallel to the longitudinal axis of a valley. These winds are produced by horizontal pressure gradients that develop as a result of temperature differences that form along the valley axis or temperature differences between the air in the valley and the air at the same height over the adjacent plain. Valley winds typically blow up-valley during daytime and down-valley during nighttime, although their onset can be substantially delayed in valleys where large atmospheric volumes are involved. A variety of names have been applied to these wind systems, with usage varying somewhat from country to country. Alternative terminology for the diurnal mountain winds is listed in Table 2.1.

Study of the pure thermally developed winds is complicated by the influence of other wind systems that develop on different scales, of regional pressure gradients superimposed on the topography, and of mechanical effects induced by the topography on the wind systems themselves or on overlying wind systems. In this chapter, we emphasize thermally developed wind systems, and these complications will be considered as modifying influences.

Increases in understanding of slope and valley wind systems in the last decade have come from the combined efforts of observationalists, theoreticians, and modelers. This chapter deals primarily with observations of ther-

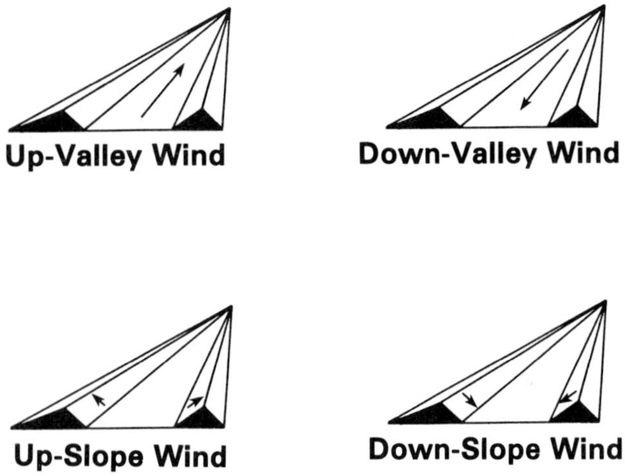

FIG. 2.1. Wind system terminology.

TABLE 2.1. Alternative names for thermally driven wind systems in valleys.

	Wind			
	Downslope	Down-valley	Upslope	Up-valley
<u>Generic terms</u>				
Thermal winds	×	×	×	×
Topographic winds	×	×	×	×
Valley winds	×	×	×	×
Mountain winds	×	×	×	×
Thermally driven winds	×	×	×	×
<u>Valley winds</u>		×		×
Valley wind (German, Talwind)				×
Mountain wind (German, Bergwind)		×		
In-valley wind (German, Taleinwind)				×
Out-valley wind (German, Talauswind)		×		
Ascending valley wind				×
Descending valley wind		×		
<u>Slope winds</u>				
Katabatic wind	×			
Anabatic wind			×	
<u>Mixed terms</u>				
Gravitation wind	×	×		
Drainage wind	×	×		
Night wind	×	×		
Day wind			×	×

mally developed wind systems, and the reader is assumed to have some general knowledge concerning the structure of valley and slope wind systems and wind system research through the 1970s. Prior reviews of these topics are listed in Table 2.2. Insights into theory and modeling are presented by Egger, Chapter 3.

In this chapter we summarize recent observations of along-valley and along-slope wind systems and their interactions, new findings on the morning and evening transition periods when the wind systems are reversed, and provide an overall view of the diurnal evolution of temperature and wind structure in deep valleys. The paper concludes with several peripheral topics, and points out the needs for future research.

2.1.1. Summary of recent field experiments

A partial list of complex terrain meteorology field experiments conducted in the last 10 years is provided in Tables 2.3 and 2.4. Most experiments have taken place during short campaigns that lasted typically 1 or 2 weeks during summer or fall and were intended to investigate well-developed circulations in simple topography during clear, undisturbed weather conditions. Relatively few experiments were conducted in winter or in large valley-ridge complexes, and few experiments were focused on developing long-term climatologies of local circulations. Finally, most field studies have been concerned with mean flows, so that little information is yet available on turbulent fluctuations or wave motions.

Manins and Sawford (1979a) summarized data on downslope winds by stating that "the data are almost all confined to wind fields and only rarely is information about their spatial and temporal variation included. Practically no temperature data are available." Their statement could also be fairly applied to valley wind systems in general. In the last 10 years, however, new instruments have become more widely used, including commercial tethered balloon systems, portable upper-air sounding systems, instrumented aircraft (including motorgliders), atmospheric tracer systems, and remote sensors such as Doppler sodars and Doppler lidars. Temperature data are now collected routinely, along with wind data. A current review of remote sensing technology as applied to complex terrain meteorology is presented by Neff, Chapter 8.

In many cases, different instrument systems have been used by American and European experimenters. Motorgliders, for instance, have been used extensively in the European experiments but have never been used in U.S. experiments. Tracer experiments and new remote sensing

TABLE 2.2. Previous reviews of mountain meteorology literature.

Author	Description
Hawkes (1947)	Good English summary of historical Alpine work.
Defant (1949)	Summary of Wagner's theory and interrelationship between slope and valley wind systems (in German).
Defant (1951)	English summary of portions of Defant (1949) and summary of other thermally developed circulations, including sea breezes, glacier winds, etc.
Brooks (1951)	Summary of mountain meteorology research, focusing on work published in English.
Geiger (1965)	Microclimatology text.
Yoshino (1975)	Updated microclimatology text.
Barr et al. (1977)	Summary of research needs in complex terrain meteorology focused on energy development activities.
Smith (1979)	Influence of mountains on the earth's atmosphere, focus on large scales of motion.
Atkinson (1981)	Summary of observational and theoretical work on slope and valley wind systems.
Orgill (1981)	Broad summary of mountain effects and results of a search for U.S. experimental areas.
Barry (1981)	Mountain meteorology textbook, focus on broad range of mountain effects.
Whiteman and Dreiseitl (1984)	Translation of classical German and French papers on theory of slope and valley flows. Summary of recent European field experiments.
Pielke (1984)	Modeling of mesoscale atmospheric processes.

tools, on the other hand, have been used predominantly in the United States. Studies have focused on a range of phenomena on different spatial scales, investigating katabatic flows on isolated hillsides, locally developed circulations in small well-defined valleys, and the meteorology of large valley complexes, especially in the Alps. The major experiments have utilized increasing resources and become logistically quite complicated, involving many collaborators. The result has been intensive experiments and large databases for a small number of valleys. These databases have been useful for the development and testing of dynamic models.

There is an increasing recognition among researchers that the valley meteorology problem is a continuum problem. While the physical processes affecting the circulations can be identified, the relative importance of the different processes varies from valley to valley, from time to time in the same valley, and even from segment to segment along a valley's length. The continuum in topographic complexity and scale, above-valley flows, climate, valley energy budgets, and even the scales of the local circulations, ensures that generalizations will be difficult. New tools and techniques are increasingly applied to address these continuum problems, however. Combined efforts by modelers and observationalists to design appropriate field experiments, and systematic model simulations in which different boundary conditions are imposed (e.g., terrain, surface energy budgets, and external ambient wind fields) seem particularly promising.

2.2. Along-valley wind systems

According to Wagner (1932a, 1938), along-valley wind systems are the result of a greater diurnal temperature range in a vertical column within the valley than in a similar column with its base at the same elevation outside the valley. The differing diurnal temperature ranges produce a thermally developed, diurnally varying pressure gradient that drives the valley wind system (Fig. 2.2). Recent climatological evidence that supports Wagner's theory is presented next, followed by a discussion of the basic physical processes that produce the along-valley wind system, with emphasis placed on the important role of topography.

2.2.1. Climatology

A key question regarding thermally developed circulations in complex terrain is the frequency with which such circulations appear in seasonal or long-term averages. Perhaps the most detailed published climatology of a single valley is for Austria's Inn Valley. There, Nickus and Vergeiner (1984) investigated the diurnal course of horizontal pressure gradients between the valley (Innsbruck) and the adjacent plain (Munich) as a function of season on sunny days. They found (Fig. 2.3) a regular diurnal reversal of the valley-plain pressure gradient in all seasons except winter, with pressure gradients supporting up-valley winds during daytime and down-valley winds during nighttime, in conformance with theory. The peak daytime pressure gradient typically occurred at 1500 UTC in all seasons. The thermal forcing of the pressure gradients is clearly seen in Fig. 2.4. At all altitudes within the valley the diurnal temperature range was larger than the corresponding temperature range over the plain. Further, the diurnal temperature range increased with up-valley distance (Table 2.5). At Landeck, 156 km up the valley, the diurnal temperature range had attained 3.6 times the range over the adjacent plain just beyond the valley's mouth at Rosenheim. The Inn Valley's wind system, as measured at

TABLE 2.3. Major valley meteorology field experiments in the last decade.

Experiment	Experimental dates	Key references
HangWindExperiment Innsbruck (HAWEI). Experiments focused on the slopes of a major, deep Alpine valley in the interior of the Alps. Inn Valley, Austria.	29 Sept–16 Oct 1978	Freytag and Hennemuth (1979) Vergeiner (1983)
Das Mesoskalige KlimaProgramm im Oberrheintal (MESOKLIP). Experiments in a wide, shallow valley flowing north from the Alps. Rhine Valley, FRG.	17–28 Sept 1979	Fiedler and Prenosil (1980)
DISKUS. Experiments in an idealized, medium-sized tributary or end valley of the Alps. Dischma Valley, Switzerland.	6–15 Aug 1980	Freytag and Hennemuth (1981) Freytag and Hennemuth (1982) Reiter et al. (1981)
Mesoskaliges Experiment im Raum Kufstein/Rosenheim (MERKUR). Experiments in a mesoscale region centered on a major, deep interior Alpine valley. Inn Valley, Austria/FRG.	23 Mar–5 Apr 1982	Freytag and Hennemuth (1983) Reiter et al. (1982) Reiter et al. (1984)
Atmospheric Studies in Complex Terrain (ASCOT). Experiments (1979 and 1980) in a valley basin on east side of coastal mountains in California. Experiments in 1981 in a V-shaped California valley draining west from coastal range. Experiments in Colorado (1982 and 1984) in an idealized semiarid end valley of the Rocky Mountains. Anderson Creek valley, CA; Big Sulfur Creek valley, CA; Brush Creek valley, CO.	16–28 Jul 1979 11–25 Sept 1980 12–24 Aug 1981 26 Jul–8 Aug 1982 17 Sept–6 Oct 1984	Dickerson and Gudiksen (1984) Orgill and Schreck (1985) Clements et al. (1989a) Special 1989 ASCOT JAM issues (June and July 1989). ASCOT report series

TABLE 2.4. Other valley meteorology field experiments in the last decade.

Experimental program	Key references
Latrobe Valley, Australia 1975, 1978	Manins and Sawford (1979a, 1979b, 1982)
NSF/CSU program of T. B. McKee 1975–present	Whiteman (1982) Bader et al. (1987)
South Park, CO 1977	Banta (1984, 1985, 1986) Banta and Cotton (1981)
Innsbruck Research Program 1979–1983	Vergeiner (1983) Vergeiner and Dreiseitl (1987)
Pajarito Mountain, NM 1979	Clements and Nappo (1983)
Rattlesnake Mountain, WA 1980, 1981	Horst and Doran (1982, 1983, 1986, 1988) Doran and Horst (1983)
LOWEX III 1981	Müller et al. (1982) Reiter et al. (1983)
Japanese valleys	Sato and Kondo (1988) Kondo and Sato (1988)
First Himalaya Soaring Expedition 1985	Neininger and Reinhardt (1986)
U. of Wyoming 1985	Kelly (1988)
ROMPEX 1985	Reiter et al. (1987)
Aare Valley, Switzerland 1985–1986	Filliger et al. (1987)
Touchet Valley, WA 1986	Doran et al. (1989)
Frijoles Canyon, NM 1987	Stone and Hoard (1989)
ASCOT Colorado valleys 1988	yet unpublished

the Innsbruck Meteorological Institute and on a 200-m-high isolated hill (Berg Isel) above the floor of the valley near Innsbruck, was consistent with the pressure gradients (Figs. 2.5 and 2.6). Wind system onset and cessation times varied throughout the year in agreement with the supposed thermal forcing. Up-valley winds were strongest in mid-afternoon, attaining mean values of nearly 4 m s^{-1} in

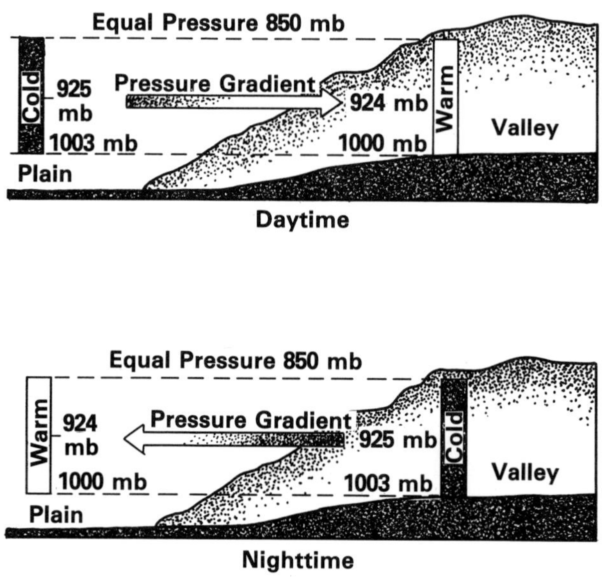

FIG. 2.2. Illustration of the thermal forcing of valley-plain pressure gradients leading to the development of an along-valley wind system. (Adapted from Hawkes 1947.)

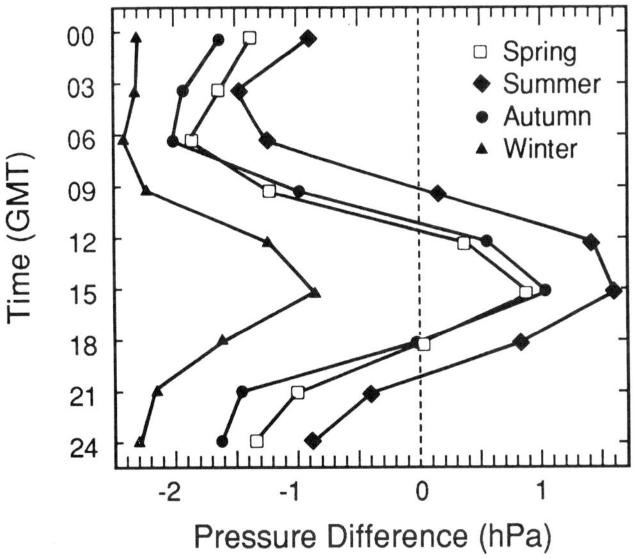

FIG. 2.3. Daily march of horizontal pressure gradient at 550 m MSL between a plain station (Munich) and a deep valley station (Innsbruck), sunny days. (Nickus and Vergeiner 1984.)

summer and fall. Down-valley winds attained mean values larger than 7 m s^{-1} in winter at Berg Isel and persisted nearly the entire day, in agreement with the sign of the calculated pressure gradients. Further, the development

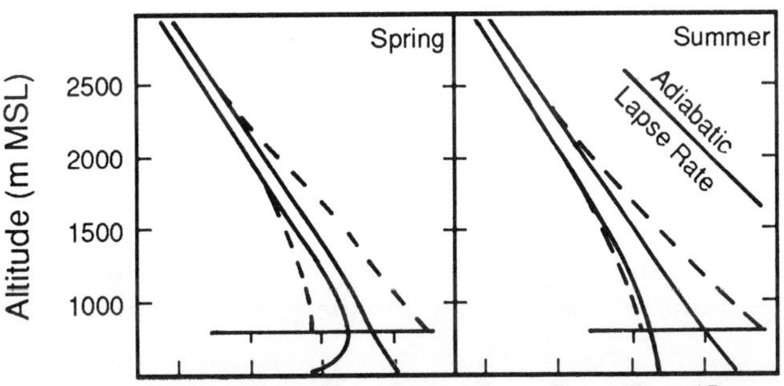

FIG. 2.4. Schematic vertical temperature profiles for a valley site (dashed line, Landeck, 821 m MSL in the Inn Valley) and for a site on the adjacent plain (solid curves, Munich, 529 m MSL) at 0600 and 1500 UTC, for all seasons. (Adapted from Nickus and Vergeiner 1984.)

of the local wind system depended on the synoptic weather type (Table 2.6). To investigate this dependence, Vergeiner (1983) stratified days according to the direction of the prevailing synoptic flow above the valley and determined whether the local wind system was evident on these days. At the Innsbruck Meteorological Institute and on the Berg Isel, valley winds were evident on 29% and 40% of the days of the year, respectively. In a smaller tributary to the Inn Valley (the Wipp valley) at Zenzenhof, valley winds occurred on 53% of the days. Near the valley exit at Kufstein, 60 km below Innsbruck, valley winds occurred on 32% of the days. Valley winds were most frequent on days with high pressure and weak synoptic winds.

The frequency of occurrence of valley wind days appears to be higher in continental areas of the western United States than in the Inn Valley, although direct comparison between published climatologies is difficult because of the differing definitions of valley wind days. In California's Anderson Creek valley, a relatively dry valley in the California coastal range that is subject to frequent marine intrusions, Gudiksen and Walton (1981) found strong monthly variations in drainage flow frequencies, with frequencies below 20% in February and March, 85% in August, and averaging 44% for the 10-month period investigated (Fig. 2.7). Colorado's semiarid Brush Creek valley, in the central Rocky Mountains, has well-defined valley wind circulations on more than 40% of the days in all months (Fig. 2.8), and in many months has valley winds on more than 60% of the days (Gudiksen 1989).

Climatological evidence thus supports Wagner's concept that valley wind circulations are driven by horizontal pressure gradients that are built up hydrostatically between valley and plain. (Nonhydrostratic flows may be significant in some cases, see Paegle et al., Chapter 10.) The pressure gradients produce a wind directed into the valley during the day when the valley column is warmer than the plain column, and directed out of the valley during the night when the valley column is colder. Pressure differences of several hPa in a distance of, say, 100 km (Fig. 2.2) may be considered typical of a very deep valley, with correspondingly smaller gradients in more shallow valleys. These thermally developed pressure gradients are comparable to typical synoptic scale pressure gradients, providing an explanation for the high climatological frequency of occurrence of the local circulations.

2.2.2. *Basic physics*

2.2.2.1. TOPOGRAPHIC AMPLIFICATION FACTOR

The diurnal reversal of the along-valley wind system arises from the larger diurnal temperature range in the valley atmosphere compared to the plains atmosphere.

TABLE 2.5. Daily ranges of valley mean temperature in the Inn Valley (Vergeiner and Dreiseitl 1987).

Station	Elevation (m)	Up-Valley distance (km)	Temperature range (1500–0600 UTC, °C)	Ratio with Rosenheim
Rosenheim	455	−17	1.7	1.0
Kufstein	508	15	3.0	1.8
Innsbruck	579	83	5.2	3.0
Landeck	821	156	6.1	3.6

This occurs as a consequence of the first and second laws of thermodynamics, expressed as

$$Q = \rho c_p V(T/\theta) d\theta. \qquad (2.1)$$

Following this equation, a given increment of heat Q added to or subtracted from an atmospheric volume will produce a potential temperature change $d\theta$ in proportion to air density ρ, specific heat c_p, and volume $V(T/\theta$, the ratio of actual temperature to potential temperature, is near unity and is defined as $(p/p_0)^{R/c_p}$, where p is atmospheric pressure, p_0 is atmospheric pressure at sea level, and R is the gas constant). The smaller the volume, the larger the potential temperature change for the same heat increment.

Applying this concept to the valley atmosphere, suppose that solar radiation enters a valley through the horizontal area at the top of the valley at ridgetop level and that, over the adjacent plain, solar radiation streams across an equivalent area at the same altitude. Assume further that the insolation, when received at the ground and converted to sensible heat flux, will heat the air below the horizontal areas, and we consider the elevations of the valley floor and the plain to be equal. In the case of the valley, the same energy is used to heat a smaller atmospheric volume than over the plain because the sloping valley sidewalls enclose less volume. The energy increment added to the valley atmosphere will thus result in a larger temperature change in the valley atmosphere than over the plain. Similarly, at night, when energy is lost through the equal horizontal areas at the tops of the volumes, the loss of energy is applied to a smaller volume within the valley, so that the valley atmosphere cools more strongly.

This concept, first proposed by Wagner (1932b), reinvestigated by Neininger (1982), and recently extended by Steinacker (1984) to account for realistic topography, can be quantified by defining a *topographic amplification factor* (TAF),

$$\tau = \frac{\left[\dfrac{A_{xy}(h)}{V_{\text{valley}}}\right]}{\left[\dfrac{A_{xy}(h)}{V_{\text{plain}}}\right]} \qquad (2.2)$$

where $A_{xy}(h)$ is the horizontal area through which energy enters the tops of the volumes at height $z = h$, where h is the height above the valley floor or plain. For an actual valley, a planimeter can be used with a topographic map to determine $A_{xy}(h)$ and the relationship between $A_{xy}(z)$ and z, from which the underlying valley volume can be estimated. Here V_{plain} is simply the product $hA_{xy}(h)$. This form of the TAF definition emphasizes volumetric comparisons between a valley and the adjacent plain. Yet an-

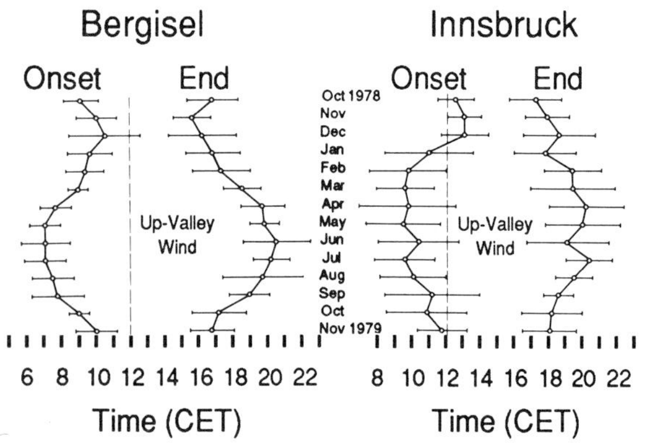

FIG. 2.5. Onset and cessation times for the down-valley wind at Innsbruck, Austria, and on a 200 m hill (Berg Isel) on the valley floor near Innsbruck as a function of month of year. Standard deviations are given as horizontal lines. (Dreiseitl et al. 1980.)

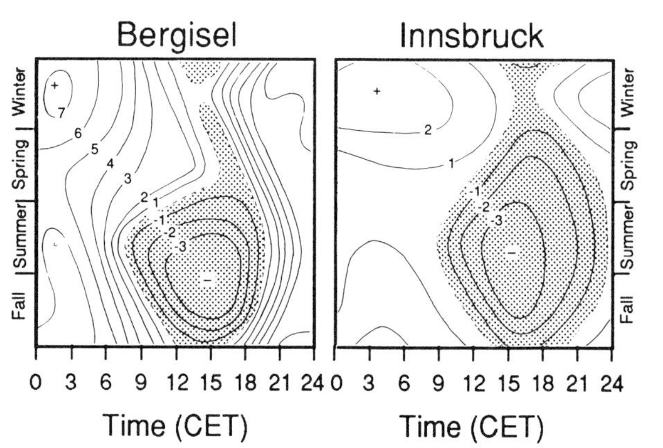

FIG. 2.6. Daily and yearly course of the along-valley wind speed components at Berg Isel and Innsbruck. Negative wind speeds are up-valley components. (Dreiseitl et al. 1980.)

TABLE 2.6. Number of valley wind days in the Inn Valley as a function of weather type, and the annual frequency of valley wind days (Vergeiner 1983).

Location	N, NE, E	SE, S, SW	W	NW	H	V	Total (year)
Kufstein	22	38	11	15	51	27	32%
Berg Isel	27	20	32	33	65	36	40%
Meteor. Inst.	24	15	24	22	43	28	29%
Zenzenhof	41	35	56	45	77	46	53%

Cardinal directions indicate direction of upper winds.
H = high pressure and weak gradients aloft.
V = variable upper winds.

other form of the TAF definition can be used to calculate topographic amplification factors for valley cross-sections. In that case, for a simple, unit-thickness, vertical valley cross-section, the TAF can be defined as

$$\tau = \frac{\left[\dfrac{W}{A_{yz_{\mathrm{valley}}}}\right]}{\left[\dfrac{W}{A_{yz_{\mathrm{plain}}}}\right]} \qquad (2.3)$$

where W is the width at the top of the two cross sections and A_{yz} is the area of the vertical cross-section. This formula is used to illustrate the calculation of τ for several idealized valley cross-sections of depth D, as shown in Fig. 2.9a. In the top part of the figure are cross sections for U- and V-shaped valleys, as well as for a valley with convex sidewalls. For illustration, the figure is drawn for valleys that are twice as wide as they are deep. The lower part of the figure shows cross sections for valleys with similar sidewall shapes, but with a horizontal valley floor of width L. The denominator of (2.3), the area to volume ratio for the plain, is simply $1/D$. Evaluation of (2.3) for the plain, and for the U-shaped, V-shaped, and convex valleys results in topographic amplification factors of 1, 1.27, 2, and 4.66, respectively, for the cross sections. For valley cross-sections with the same sidewall shapes, but with wider valley floors (Fig. 2.9b), the topographic amplification factors are reduced, approaching unity as the valley floor widens. Clearly, the valley with convex sidewalls, having less volume for the same area at its top, will have a substantially larger diurnal temperature range, if the energy change is confined within the volume. We tend to think of valleys as V-shaped or U-shaped because we typically view valleys from their floors, but valley drainage

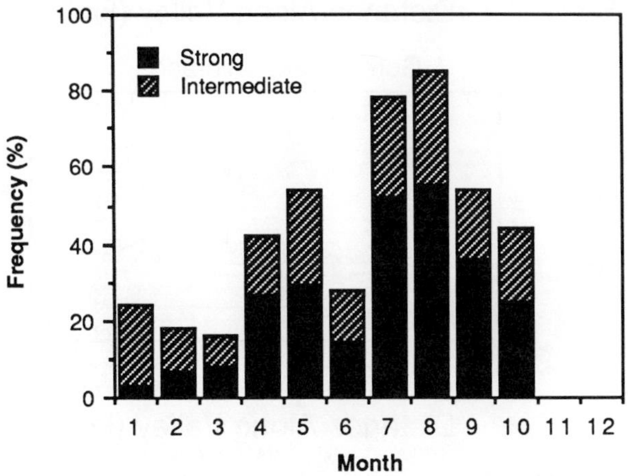

FIG. 2.7. Frequency of weak and strong drainage flows in California's Anderson Creek valley as a function of month of year. (Adapted from Gudiksen and Walton 1981.)

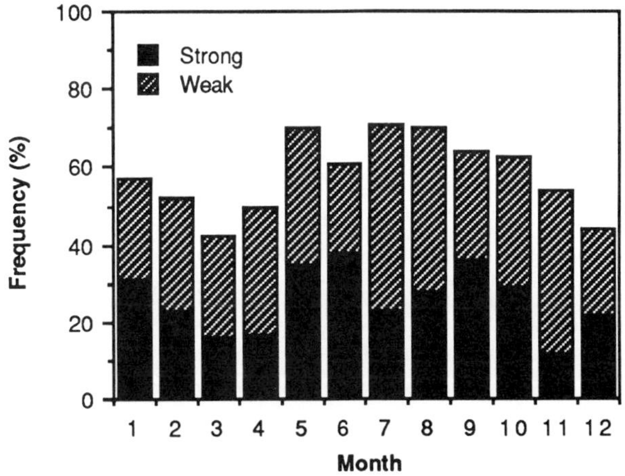

FIG. 2.8. Frequency of weak and strong drainage flows in Colorado's Brush Creek valley as a function of month of year. (Adapted from Gudiksen 1989.)

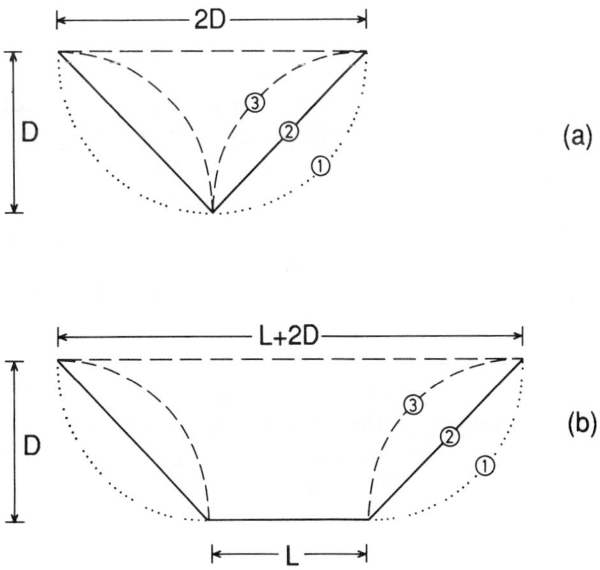

FIG. 2.9. Calculation of topographic amplification factors for idealized valleys of various cross sectional shape with (a) no horizontal valley floor, and (b) horizontal valley floor. (Müller and Whiteman 1988.)

areas typically have a pronounced convex shape in their upper sections.

According to Steinacker (1984), Wagner (1932b) originally made TAF calculations of diurnal temperature ranges and horizontal pressure gradients for a U-shaped section of the Inn Valley. These calculations gave horizontal pressure gradients that were somewhat smaller than observed. Subsequent calculations by Steinacker, including the Inn Valley and its tributaries, showed a substantially larger TAF, because of the preponderance of convex sidewalls in the tributaries and upper levels of the main valley. Steinacker's calculations have spawned a number of calculations by others, and reassessments of previous work, such as that of Whiteman and McKee (1982), in which calculations of TAFs were originally done using geometrically idealized cross sections. Böhm (1985) incorporated Steinacker's method in a version of the Whiteman and McKee inversion destruction model for Swiss topography. Müller and Whiteman (1988) used Steinacker's method of calculating TAFs to simulate the breakup of a valley temperature inversion in Switzerland's Dischma Valley. A key aspect of Steinacker's work that has received less attention is the effect of stability on the development of horizontal pressure gradients. According to Steinacker, stability may be effective in confining energy input to certain altitudes in the valley. If the energy is confined to the upper levels of a valley where large atmospheric volumes are present, the temperature change at that level will be small. On the other hand, if energy input is confined to lower altitudes in the valley where atmospheric volumes are small, temperature changes will be large. Through the hydrostatic equation, pressure depends on the mass of the column of air above the point of interest which, in turn, depends on the temperature distribution in the column. Small temperature changes at the upper altitudes of the valley will have less effect on the vertical pressure distribution than large temperature changes at the lower altitudes, even though the same amounts of total energy input may be involved in producing the respective temperature changes.

McKee and O'Neal (1989) carried Steinacker's approach a step further. Rather than considering that the valley-plain pressure gradient drives the valley winds, they postulated that valley winds can be driven by along-valley pressure gradients that develop within a valley due to variations of valley geometry along the valley's length. One can investigate the effect of the topography in developing such gradients by making TAF calculations with (2.3) at a number of points along a valley's axis, as calculated in Table 2.5 for Austria's Inn Valley. Stronger nighttime cooling at the head of a valley than at its mouth would result in a pressure gradient that would drive down-valley circulations. Calculations of TAF ratios along a valley's axis would allow one to determine whether the valley has a vigorous along-valley circulation or whether cold air would stagnate and pool in one or more segments of a valley. McKee and O'Neal made such calculations for several valleys in Colorado where wind data were available, and verified that the geometric calculations agreed with existing wind data (Fig. 2.10). This initial success has implications for future dynamic modeling of valley wind systems; it suggests that the TAF concept may be used to compare the model terrain to the actual terrain to verify that thermodynamic effects in the modeled terrain are realistic relative to the actual terrain being modeled.

The topographic amplification factor is purely a geometrical factor describing the valley topography. The influence of this factor on the valley's thermal structure and wind systems may be strengthened or weakened by other

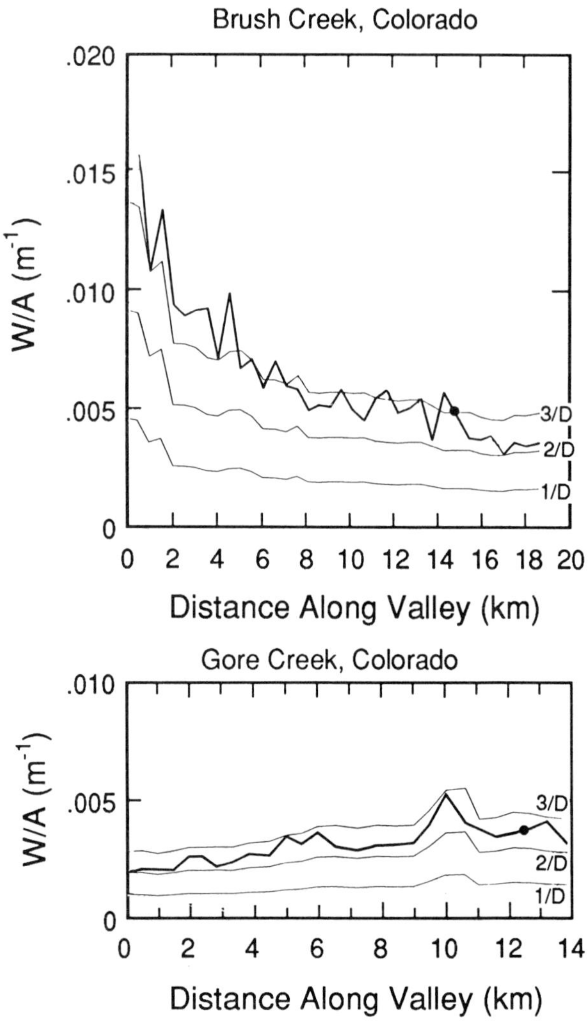

FIG. 2.10. Width to cross sectional area ratios for a draining valley (Brush Creek valley) and a pooling valley (Gore Creek valley). In the draining valley the width to area ratio decreases with down-valley distance; in the pooling valley the ratio increases. (McKee and O'Neal 1989.)

processes. First, since energy gain and loss from the valley volume occur primarily at the valley surfaces (i.e., valley floor and sidewalls), an efficient means of transport of energy from the sidewalls to the entire valley volume is required in order to drive the along-valley wind system. The valley volume is enclosed by topography so that radiation, conduction, and convection from the surfaces are relatively more effective in transporting energy than would be the case over a flat plain. The upslope and downslope flows, as we will see shortly, are especially adept at efficiently transporting energy to and from the entire valley volume. Second, if the energy gains and losses are to be effective in driving local wind systems in the valley, the energy gains and losses must be confined to the valley, rather than dissipated into the air above the valley. Observations show that one of the effects of valley geometry, at least under weak-to-moderate gradient wind conditions, is to limit the dynamic exchange between the air above the valley and the air within the valley (e.g., see Maki and Harimaya 1988). Finally, the effect of the geometrical factor can be strengthened or weakened by plain-valley or along-valley gradients of energy gain or loss. For example, gradients in albedo or soil moisture along a valley's axis may produce along-valley sensible heat flux gradients that may modify the topographically developed differences in diurnal temperature range between valley and plain or along a valley's axis.

The TAF concept provides an explanation for strong along-valley wind systems that develop in valleys like the Inn Valley having a weak valley floor slope ($\approx$0.001 between Innsbruck and the Alpine foreland). Following this concept, a valley cut into a mountain range surrounded by a plain and having an absolutely horizontal valley floor (Fig. 2.11) could still have a strong valley wind system because the topographic amplification factor produces a greater diurnal temperature range in the valley atmosphere than above the surrounding plain.

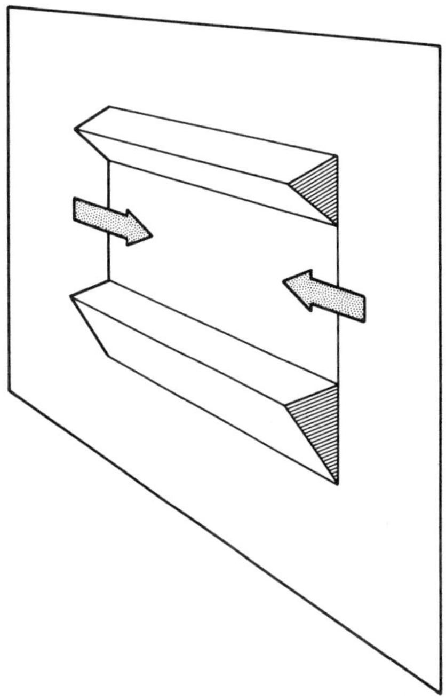

FIG. 2.11. A valley with a horizontal valley floor will still generate along-valley wind systems because the topographic amplification factor will produce a higher daily temperature range in the valley than over the surrounding plain.

2.2.2.2. EQUATIONS FOR THE VALLEY WIND SYSTEM

The following simplified set of equations can be used to illustrate the development of an along-valley wind system, the key physical processes involved, and the relationships between conservation equations:

$$\frac{\partial\theta}{\partial t} + u\frac{\partial\theta}{\partial x} + v\frac{\partial\theta}{\partial y} + w\frac{\partial\theta}{\partial z} = -\frac{\partial}{\partial x}(\overline{u'\theta'}) - \frac{\partial}{\partial y}(\overline{v'\theta'}) - \frac{\partial}{\partial z}(\overline{w'\theta'}) + \frac{\partial Q^*}{\partial z} \quad (2.4)$$

$$\frac{\partial u}{\partial t} + u\frac{\partial u}{\partial x} + v\frac{\partial u}{\partial y} + w\frac{\partial u}{\partial z} = -\alpha\frac{\partial p}{\partial x} + F \quad (2.5)$$

$$\frac{dp}{p} = -\frac{gdz}{RT} \quad (2.6)$$

$$\frac{\partial u}{\partial x} + \frac{\partial v}{\partial y} + \frac{\partial w}{\partial z} = 0 \quad (2.7)$$

$$p\alpha = RT, \quad (2.8)$$

where θ is potential temperature, u, v, and w are velocity components in orthogonal directions x, y, and z, t is time, Q^* is net all-wave radiation flux, α is specific volume, p is pressure, F is friction, g is gravitational acceleration, R is the gas constant, and T is temperature. The dependent variables have been averaged over some appropriate time scale, and overbars have been dropped except for the Reynolds terms. Two temperature variables have been used in the equations to facilitate discussions to follow. The defining equation for potential temperature is

$$\theta = T\left(\frac{1000}{p}\right)^{\frac{R}{c_p}}, \quad (2.9)$$

where the units of p are hectopascals. The equations, in order, are the thermodynamic energy equation, the x-momentum equation, the hydrostatic equation, the mass continuity equation, and the equation of state.

A somewhat heuristic or simplified view of the development of the along-valley circulation comes from considering (2.4) through (2.8) in turn for two unit vertical cross-sections in a valley at different down-valley distances. The atmosphere is initially considered stationary.

Different heating rates may be produced in the two cross sections by sensible and radiative flux convergence following (2.4). The amount of heating produced in the cross section depends on the topographic amplification factor, as we have described. Integrating (2.6) downward from a common pressure surface, we can calculate the pressure distribution in the column as a function of height. The cooler column will have a higher pressure at each height, so that an along-valley pressure gradient will be developed [$\partial p/\partial x$ in Eq. (2.5)], producing an acceleration that generates an along-valley wind. This acceleration is opposed by friction. Motions are constrained to occur so that, following (2.7), atmospheric mass is conserved. Once motion is initiated, advective terms become important in the thermodynamic energy and momentum equations. These terms do not generate cooling or momentum, but act to redistribute these quantities.

2.2.3. *Radiation and surface energy budgets*

A full understanding of the mechanisms by which heat is transferred to and from the valley atmosphere to generate along-valley flows requires an investigation of the budgets of mass, heat, and momentum for valley volumes using conservation equations (2.4), (2.5), and (2.7). Present experimental efforts along these lines are discussed in section 2.2.4. First, however, it will be convenient to review what is known concerning the radiation and energy budgets of the valley surfaces, since these constitute the lower boundary condition for the atmospheric volume, and it is largely through this lower surface that the valley atmosphere is warmed and cooled. Thus, we begin with a review of radiation and energy budget fundamentals at surfaces, illustrating the concepts with recent observational data from valleys.

Net all-wave radiation Q^* is the sum of net shortwave (K^*) and net longwave (L^*) radiation, such that

$$Q^* = K^* + L^*, \tag{2.10}$$

where net shortwave and longwave radiation are given as the sums of the appropriate incoming and outgoing radiation streams (Fig. 2.12a),

$$K^* = K{\downarrow} + K{\uparrow}, \tag{2.11}$$

$$L^* = L{\downarrow} + L{\uparrow} \tag{2.12}$$

and incoming shortwave radiation can be written as the sum of the direct (S) and diffuse (D) shortwave radiative components

$$K{\downarrow} = S + D. \tag{2.13}$$

The sign convention to be used here is that downward-directed radiation is positive and upward-directed radiation is negative. By combining (2.10) through (2.13) we can express the radiation budget as

$$Q^* = S + D + K{\uparrow} + L{\downarrow} + L{\uparrow}. \tag{2.14}$$

This expression allows one to calculate net all-wave radiation at a surface at a given instant in time. Measurements on clear days show that, typically, net all-wave radiation is a large positive value during the day and a small negative value during the night. This points out the general lack of a radiative balance at a given instant in time, or diurnally. Deficits or excesses in the radiation budget must be made up through fluxes of energy from nonradiative sources (Fig. 2.12b), following the expression for the energy balance of a surface:

$$Q^* + Q_G + Q_H + Q_E = 0, \tag{2.15}$$

where Q^* is net all-wave radiation, Q_G is ground heat

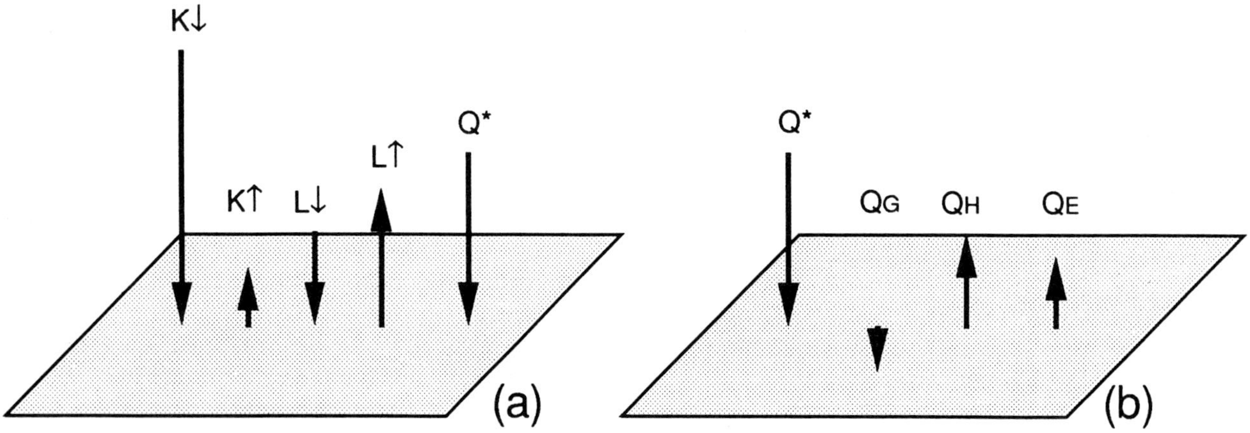

FIG. 2.12. Components of (a) the radiation budget, and (b) the surface energy budget on a horizontal surface. Here $K{\downarrow}$ = incoming solar radiation, $K{\uparrow}$ = reflected solar radiation, $L{\downarrow}$ – incoming longwave radiation, $L{\uparrow}$ = outgoing longwave radiation, Q^* = net all-wave radiation, Q_G = ground heat flux, Q_H = sensible heat flux, and Q_E = latent heat flux. The components shown are scaled to represent midday values of the radiation and surface energy budgets for a dry continental valley in the fall.

flux, Q_H is surface sensible heat flux (over flat terrain usually considered to be given by the single term $\rho c_p \overline{w'\theta'}$), and Q_E is latent heat flux. In this equation we use the convention that all fluxes toward the surface, whether from the atmosphere or the ground, are considered positive.

Following this expression, excess radiative energy (e.g., during daytime) may be used to heat the ground, warm the atmosphere, or evaporate water. Radiative energy deficits during nighttime may be balanced by an upward flux of heat from the ground, a downward flux of heat from the atmosphere, or condensation of water on the surface. Note that (2.15) states only that the sum of the terms is zero; it gives no information on how a change in one term will be partitioned into the other terms. Such partitioning may involve differing signs among the three remaining terms.

Thermally developed valley wind systems are driven by heat transfer to and from the valley atmosphere, so that our primary interest in the radiation and surface heat budgets is the estimation of sensible heat flux. Direct measurements of sensible (and latent) heat flux are notoriously difficult to make on heterogeneous sloping surfaces. Estimation of sensible heat flux often involves measurement of the individual terms in (2.15), under the overall constraint that the surface energy budget must be balanced. Ground heat flux measurements are also problematical in complex terrain, although, like sensible and latent heat fluxes, several different measurement methods are available (Kimball et al. 1976). Definitive net radiation measurements are largely unavailable for mountains of the western United States, even though rather extensive measurements of solar, thermal, and net radiation were reported in the Alps in the 1950s (Sauberer and Dirmhirm 1958), and the required equipment is simple and reliable. If the measurements of individual terms in (2.15) were straightforward and error-free, there would still be the question of the representativeness of the point measurements for estimating sensible heat flux over the valley as a whole. An initial approach to estimating variability of energy budget terms in complex terrain has been to consider that the surface energy budget is driven primarily by net radiative gains during daytime and losses during nighttime, and to estimate the spatial variability of such gains and losses by investigating the contributions to net radiation from individual terms in the radiative energy budget (2.14). In the next section an initial step in this approach is illustrated for Colorado's Brush Creek valley.

2.2.3.1. Radiation budget

As part of the 1984 ASCOT experiment (Table 2.3), radiation budget component measurements were made at five sites in Colorado's 650-m-deep Brush Creek valley. (See Fig. 2.13 for the site locations and a topographic map of the experimental area.) In the dry, high-altitude, Brush Creek valley, observations (Whiteman et al. 1989a,b) showed a large direct-beam solar radiation component in the radiation budget, so that slope aspect and inclination angles of the valley surfaces had a strong effect on incoming solar radiation. Surface properties, including vegetation cover and soil moisture, were important in explaining differences in reflected radiation and outgoing longwave radiation between sites. These differences were especially significant between the dry, barren, southwest-facing sidewall and the opposing, relatively moist, brushy, northeast-facing sidewall. Daily totals of radiation and surface energy budget parameters varied significantly between sites depending on the length of the daytime period, and thus on local sunrise and sunset times at individual stations. Large daytime net radiation gains and nighttime losses were experienced on the mesa-shaped ridgetop of the valley, supporting the concept of the ridgetops being elevated heating surfaces during daytime and cooling surfaces during nighttime. Observations averaged over a 15-day period of variable weather illustrated the general effect

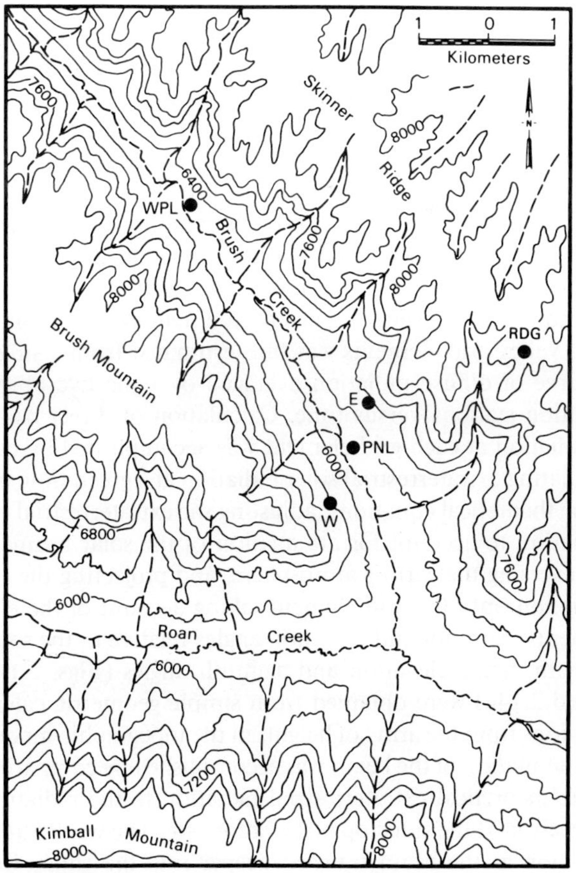

FIG. 2.13. Locations of radiation and surface energy budget measurement stations in the Brush Creek valley, September–October 1984.

of cloudy weather in reducing contrasts in radiation climate among sites.

The spatial variations in climatic contrasts experienced within this dry, high-elevation, continental valley can be illustrated with the help of a digital terrain model. The reader interested in reviewing recent progress in the modeling of solar and longwave radiation fields in complicated terrain areas is referred to prior work by Marks and Dozier (1979), Dozier (1980), Isard (1983), and Olyphant (1984, 1986a,b). For the Brush Creek valley, modeled spatial patterns of global radiation on 25 September 1984 will be used to illustrate variations in instantaneous and daily total surface microclimates. The date in question is near the autumnal equinox when the sun rises due east and sets due west with a 12-h daylight period.

A digital terrain model of the Brush Creek valley region was produced using a 30-m grid interval. A small subdomain was chosen for modeling, representing a segment of the Brush Creek valley running up-valley from the Brush–Roan Valley confluence for 8 km and bounded by the valley ridgetops (Fig. 2.14a). Using this model, combined with theoretical equations for the sun's movement through the sky on the date of interest (Whiteman and Allwine 1986), a solar shading algorithm was used to simulate the propagation of shadows throughout the valley. Sunrise and sunset times were then determined at each of the grid elements in the valley by running the shading algorithm with sun positions determined at 10-min intervals throughout the day. Calculated sunrise and sunset times are illustrated in Figs. 2.14b and 2.14c.

Global radiation is the sum of direct-beam radiation (S) and diffuse radiation (D). Observations in the valley on 25 September showed that diffuse radiation was generally small and, while varying in time, did not vary greatly among the five sites at a given time. Diffuse radiation at all valley grid elements was thus estimated from a single curve of diffuse radiation averaged over the five observation stations versus time. Calculation of direct beam radiation at each grid element was accomplished by calculating extraterrestrial solar radiation at a given time using theoretical equations, adjusting the extraterrestrial radiation to account for attenuation of the solar beam in traversing the earth's atmosphere, and projecting the radiation onto the grid element taking account of the elements' azimuth and elevation angles relative to the solar beam. Slope elevation and azimuth angles (Figs. 2.14d and 2.14e) were obtained from simple geometric calculations from the array of heights in the topographic model. Attenuation of the beam was accomplished by comparing the theoretical extraterrestrial radiation to the radiation received at the ridgetop observation site. The weak attenuation as the beam traveled deeper into the vallcy was determined by developing an empirical attenuation versus altitude relationship from measurements at the five sites.

The resulting calculations of global radiation are shown in Fig. 2.14f for 0730 MST 25 September. At this time, the east sidewall was in shadow and the maximum global radiation was received on steeply inclined east-facing surfaces on the west sidewall. Since all surfaces receive diffuse radiation, the global radiation was positive even on the shaded east sidewall. At 1600 MST (Fig. 2.14g) the west sidewall was in shadow and strong insolation was received on the east sidewall. Strong contrasts in insolation were present between the east and west sidewalls, as expected, but strong contrasts were also present between adjacent terrain elements on the east sidewall on account of shadows and differences in aspect and inclination angles. By integrating such instantaneous global radiation fields generated at 10-min intervals through the course of a day, a daily total global radiation field was determined, as shown in Fig. 2.14h. Daily totals varied over an order of magnitude at different locations in the valley. Ridgetop sites were favored by the long period of direct beam radiation relative to shaded locations deeper in the valley. Inclined south- and southwest-facing slopes received daily totals exceeding 26 MJ m^{-2}, while north-facing slopes in box canyon tributaries of the west sidewall received daily totals less than 3 MJ m^{-2}. Daily totals were generally much higher on the valley's east sidewall than on the opposing sidewall, but interesting differences occurred within the finestructure of the topography on each sidewall. Thus, the field of daily total global radiation on the west sidewall contains a mosaic of microclimatic regimes, with pronounced differences occurring over very small spatial scales.

The amount of incoming longwave and diffuse radiation received on a surface in complex terrain depends on the fraction of the sky visible in the viewing hemisphere of the surface, called the sky view factor. For a simple valley cross-section with a horizontal valley floor and two linear sidewalls of different inclination angles the smallest sky view factors on the slopes will occur at their bases (Petkovsek 1978c). On the valley floor, view factors will increase as the valley center is approached. Figure 2.15 provides an example of sky view factor calculations for Colorado's Brush Creek valley, computed from a 30-m resolution digital topographic model using the formula

$$\beta = \frac{1}{n} \sum_{i=1}^{n} (1 - \sin\phi(\theta_i)) \tag{2.16}$$

where $n = 2\pi/\Delta\theta$, $\Delta\theta = 10°$ is the azimuth angle increment, and $\phi(\theta_i)$ is the mean elevation angle of the horizon over the azimuth angle increment $\Delta\theta$ centered around azimuth angle $\theta_i = (i - 0.5)\Delta\theta$. In the figurc the lightest shade of gray represents a view factor of 1 and the darkest shade of gray represents the lowest calculated sky view factor of 0.47. The mesa tops on either side of the Brush

Creek valley have high sky view factors, but the view factors drop rapidly with depth in the valley. The smallest view factors are at the base of cliffs in the box canyons that feed Brush Creek along its length. The valley center has noticeably higher sky view factors than at the lower edges of the sidewalls.

The distribution of sky view factors in real valleys causes complicated spatial variations in the downward longwave and diffuse solar radiation components. Locations with high sky view factors are exposed to the cold radiating sky and receive relatively small downward longwave radiative fluxes. As one descends deeper and deeper into a valley the effect of the relatively warm radiating sidewalls (compared to the cold radiating sky) increases the downward longwave irradiance. Other things being equal, this results in a decrease in nighttime radiative loss with distance below the ridgetops. Given equivalent outgoing longwave, soil, and latent heat fluxes, downward sensible heat flux from the near-ground layer of air should be enhanced [Eq. (2.15)] by this sky view factor effect, with the result that the upper slopes and ridgetops should prove to be more effective generators of cold air at night than the lower slopes.

Actual calculation of the downward irradiance at a surface element in a valley must be accomplished with the radiative transfer equation, and will depend on the sky view factor, atmospheric profiles of radiatively active gases (water, carbon dioxide, ozone) and the distribution of radiating temperature on the sidewalls. McKee and Whiteman (1977) made such calculations for a circular basin with a flat valley floor of 1 km diameter and 31° sidewall slopes, for isothermal sidewalls at temperatures that varied from 20°C colder to 20°C warmer than the base of the temperature sounding. Calculations were performed using average summer and winter temperature and humidity soundings at Grand Junction, Colorado. At the valley floor center the decrease of net radiative loss from values calculated over a flat plain having the same atmospheric profiles can vary from 8% to 24% depending on the radiating temperature of the sidewall. The largest relative decreases occur for dry winter soundings.

2.2.3.2. Surface energy budget

Few observations of sensible heat flux are yet available in valleys because of difficulties inherent in making the measurements. Several sets of measurements are now available, however, to illustrate the essential differences between levels of sensible heat flux in different climate settings.

Observations on the floor of Switzerland's Dischma Valley (Fig. 2.16) on the clear day of 6 August 1980 show that, in daytime, much of the available energy is used for evapotranspiration (Halbsguth et al. 1984). Sensible heat flux, while somewhat larger than ground heat flux, is only about 40% of the latent heat flux. Observations on this day followed a period of heavy rain, suggesting that the results may not be entirely typical. On the clear day of 7 September 1980 on the floor of Austria's Inn Valley (Fig. 2.17), daytime net radiation was much weaker than experienced a month earlier in the Dischma Valley. Only a small fraction of the net radiation was used to heat the ground, and most of the available energy was used for evapotranspiration. Sensible heat flux was about a third of the daytime integrated latent heat flux (Vergeiner and Dreiseitl 1987).

A comparison can be made between the more moist Alpine setting and Colorado's semiarid Brush Creek valley. Brush Creek valley observations (Fig. 2.13) were made at 5 sites—two on the valley floor (PNL and WPL sites), one on the ridgetop above the valley (RDG), and one on each of the opposing valley sidewalls (E and W). The Brush Creek dataset (Fig. 2.18) is characterized by large values of net radiation for the date in question (7 weeks later in the year than the Dischma Valley data) and conversion of a high fraction of the available energy into sensible heat flux. Peak sensible heat flux values at some of the sites are larger than the net radiation peak in the Inn Valley. The Colorado valley is thus characterized by very high rates of energy transfer to the overlying valley atmosphere during daytime (Whiteman et al. 1989b)—a feature of the climate that strongly affects the development of local wind systems.

Strong contrasts in instantaneous latent and sensible heat fluxes occurred between the opposing northeast- and southwest-facing sidewalls of the Brush Creek valley as solar insolation varied through the course of the day and as shadows propagated across the valley. This differential heating and moistening of the air above the opposing slopes produced cross-valley circulations that resulted in redistribution of moisture and heat throughout the valley volume. The cross-valley circulations strongly affected the morning breakup of artificial tracer plumes introduced into the valley (Whiteman 1989; Bader and Whiteman 1989; Gudiksen and Shearer 1989).

The ridgetop site, with a nearly unobstructed view of the sky and the longest daytime period, received the highest daily total of net radiation and the highest sensible heat flux total. The dry southwest-facing slope produced a nearly equivalent daily total sensible heat flux, despite the later sunrise and earlier sunset at this site, because of the dry soil, lack of vegetation, and intense afternoon radiation on the sloping surface. One of the valley floor sites, located in a wheatgrass meadow, produced a daily total latent heat flux over four times larger than the dry southwest-facing sidewall. Mean daytime Bowen ratios (sensible/latent heat flux ratio) varied from 0.86 at the valley floor meadow site to 7.60 on the southwest-facing

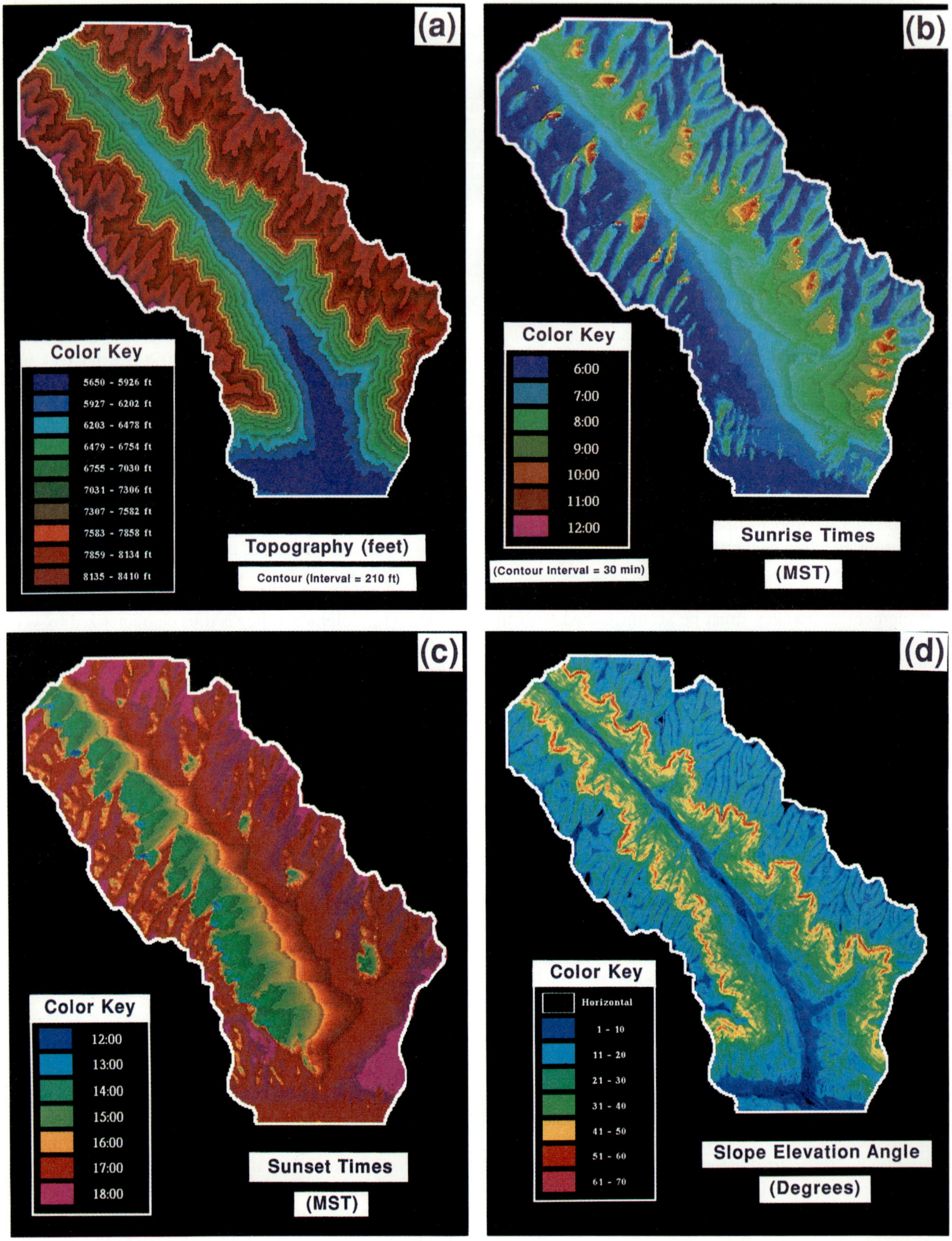

FIG. 2.14. Illustration of topographic and microclimatic variations in Colorado's Brush Creek valley on 25 September 1984. (a) Topography, (b) sunrise time, (c) sunset time, (d) slope elevation angle.

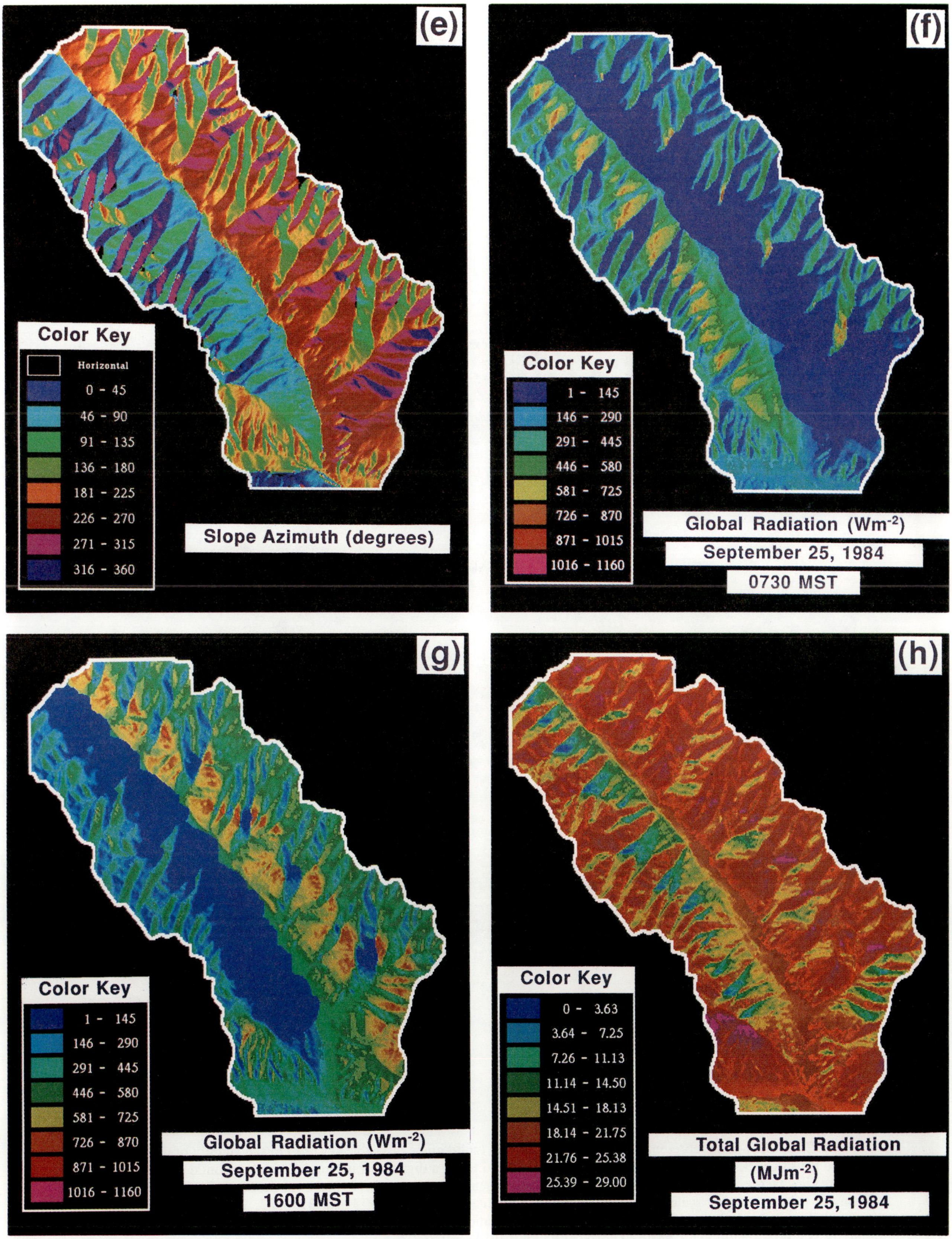

FIG. 2.14, continued. (e) slope azimuth angle, (f) global radiation at 0730 MST, (g) global radiation at 1600 MST, (h) daily total global radiation.

FIG. 2.15. Gray-scale representation of sky view factors (solid angle of visible sky/2π) in the vicinity of the mouth of the Brush Creek valley. Sky view factors were calculated using a digital terrain model with horizontal resolution of 30 m. View factors in the scene range from 1.00 at the ridgetops (white) to 0.47 (black) at the base of cliffs within box canyon tributaries; gray scales represent intermediate values (*Weatherwise*, 39(6), 319–322, 1986.) Reprinted with permission of the Helen Dwight Reid educational foundation. Published by Heldreff Publications, 4000 Albemarle St., N.W., Washington D.C. 20016 (Thorpe and Orgill 1986). ©1986.

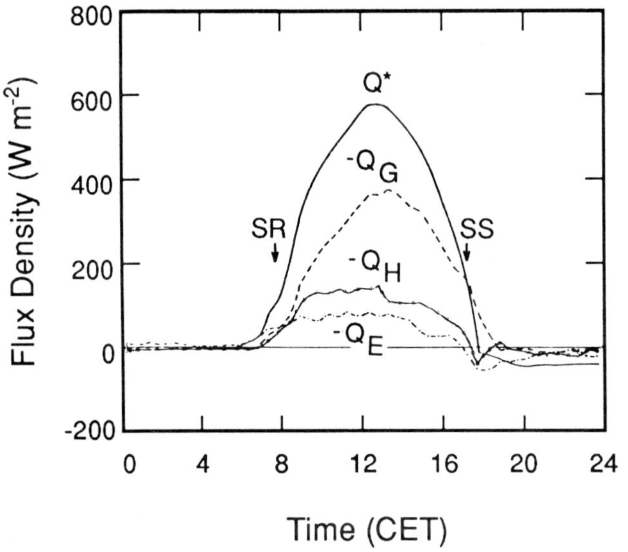

FIG. 2.16. Surface energy budget for the Dürrboden site in Switzerland's Dischma Valley during the DISKUS experiment, 6 August 1980. (Halbsguth et al. 1984.)

sidewall. Thus, we see that significant spatial differences in instantaneous and daily total sensible heat fluxes occur within complex terrain areas.

Daily total sensible heat fluxes in the valley as a whole were much larger than required to destroy typical nocturnal temperature inversions, and the excess was available on clear fall days to grow deep convective boundary layers over the region. The times of reversal of the slope wind systems at the individual energy budget sites were closely related to the time of sign reversal of sensible heat flux.

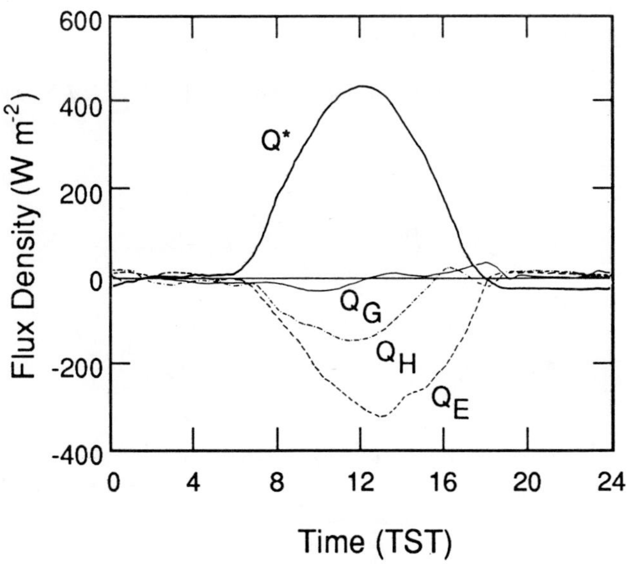

FIG. 2.17. Radiation and surface energy budgets for a site on the floor of the Inn Valley, 7 September 1980. (Vergeiner and Dreiseitl 1987.)

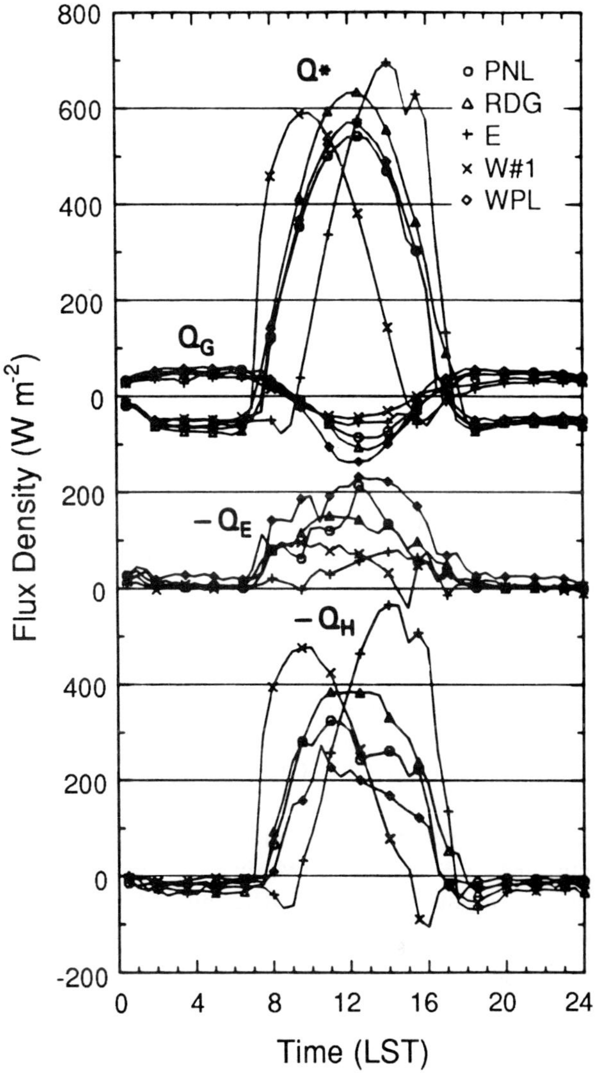

FIG. 2.18. Surface energy budgets at five sites in the Brush Creek valley, Colorado, during the ASCOT experiment, 25 September 1984. (Whiteman et al. 1988b.)

Thus, the evolution of the wind systems at individual sites is strongly dependent on sensible heat input into individual topographic elements within the valley.

2.2.4. *Atmospheric budgets of mass, heat, momentum, and moisture*

Several recent major field studies have investigated atmospheric budgets of mass, heat, momentum, and moisture in valley circulations. Initial studies have generally investigated individual budgets, although in actual atmospheric flows all budgets are inextricably coupled and must be simultaneously balanced. The initial observational studies are stimulating efforts to make such budget calculations from dynamic models. Present models have not been entirely successful in simulating realistic wind and, especially, temperature fields in actual topography so that improvements may be necessary before the model budgets can be considered representative of actual valleys.

In this section, volumetric budget calculations of Vergeiner et al. (1987) for Colorado's Brush Creek valley illustrate the coupling between the mass, momentum, and thermal energy budgets. Work by other investigators is summarized to illustrate the range of results obtained for the individual budgets. Results obtained thus far should be considered preliminary, since such calculations are in an early stage and assumptions, methods, and data requirements are still being questioned.

Vergeiner et al. (1987) evaluated the coupled atmospheric energy, momentum, and mass budgets for Colorado's Brush Creek valley for the 30–31 August 1982 ASCOT experiment. In this experiment, a down-valley flow was present through the entire depth of the valley, with peak winds of 7 m s^{-1} at the 100-m level. Multiple tethered balloon soundings observed the wind and temperature distribution along the valley axis. Computations were performed for the steady-state nighttime period, for which atmospheric variables were assumed to be uniform in the cross-valley direction. Brush Creek valley width and the observed wind and temperature fields were fit with analytical functions that defined their variations in the along-valley and vertical directions. Individual terms in the thermal energy, continuity, and x-momentum equations were then calculated on a two-dimensional $x - z$ grid. The vertical motion field was determined from the continuity equation and the pressure field was calculated hydrostatically from the temperature field. The friction term was calculated as a residual in the momentum equation, and the diabatic terms were calculated as a residual in the thermal energy equation. The following momentum and thermodynamic energy equations illustrate the results obtained from these initial Brush Creek valley investigations. Under steady-state conditions, the terms have the signs indicated.

$$\frac{\partial u}{\partial t} = -u\frac{\partial u}{\partial x} \quad - \quad w\frac{\partial u}{\partial z} \quad - \quad \alpha\frac{\partial p}{\partial x} + \text{Friction}$$

$$0 = \quad + \quad \begin{matrix} -\text{ (above jet)} \\ +\text{ (below jet)} \end{matrix} \quad + \quad - \qquad (2.17)$$

$$\frac{\partial \theta}{\partial t} = -u\frac{\partial \theta}{\partial x} - w\frac{\partial \theta}{\partial z} + D$$

$$0 = \underbrace{\quad - \qquad + \quad}_{} \quad -$$

$$0 = \qquad + \qquad\qquad -. \qquad (2.18)$$

The valley vertical motion field was found to be quite sensitive to small changes in the fit of the analytical function to the along-valley wind field. Along-valley mass flux divergence produced subsidence and warming throughout the valley atmosphere. The subsidence caused slower moving air to subside into the valley from above but, for the lower levels of the valley below the jet, subsidence caused advection of faster moving air. Evaluation of the terms of the momentum equation showed a rough balance between the pressure gradient force and friction in the lower half of the valley atmosphere, but total advection balanced the pressure gradient force in the upper valley atmosphere where friction became small.

The change of temperature of the valley volume during the 2200–0500 LST period was at the rate of 0.4 K h^{-1}. Strong subsidence produced warming of 2.5 K h^{-1} that was only partly counteracted by down-valley cold air advection (-0.5 K h^{-1}), so that strong diabatic cooling was required to balance the thermal energy budget. The cooling rate required from sensible and radiative heat flux divergence (ca. -2.0 K h^{-1} for steady-state conditions) seemed somewhat too large, although measurements were not available in the 1982 ASCOT experiment. The imbalance could have been caused by the horizontal homogeneity assumption used in calculating down-valley volume fluxes, which would cause subsidence warming to be overestimated. This assumption for the Brush Creek flow had been originally suggested by Whiteman and Barr (1986), but has since been shown to produce significant errors. This will be discussed further in section 2.2.4.1.

2.2.4.1. Conservation of atmospheric mass

Atmospheric mass conservation is often expressed in terms of volume fluxes rather than mass fluxes. The two fluxes are simply related by atmospheric density, which is near 1 kg m^{-3}. Mass conservation is illustrated for a closed valley and for a valley segment in Fig. 2.19. For a homogeneous atmosphere in a closed valley, conservation of mass results in the expression

$$\bar{u}A_{yz} = \bar{w}A_{xy} \quad [\mathrm{m^3\,s^{-1}}] \qquad (2.19)$$

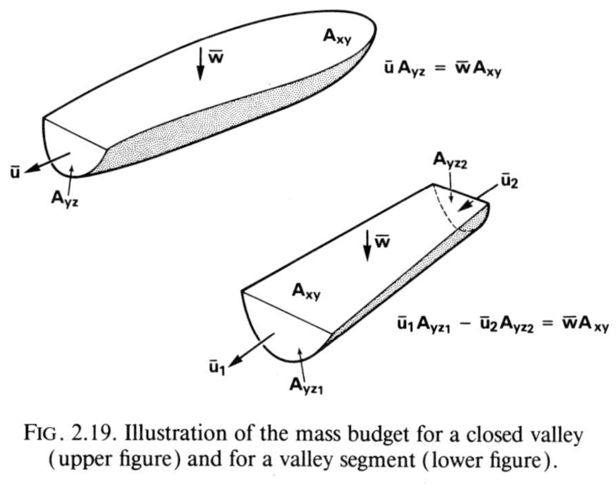

FIG. 2.19. Illustration of the mass budget for a closed valley (upper figure) and for a valley segment (lower figure).

where $\bar{u}$ is the mean along-valley wind speed averaged on a vertical $y - z$ plane and $\bar{w}$ is the mean vertical velocity averaged over the top of the valley atmosphere.

Most calculations of valley volume fluxes have been made for nighttime down-valley flows. The earliest such published calculations were apparently made by Ekhart (1944) for the Salzach Valley. Rao (1968, 1970) used such calculations to determine vertical motions from the mass continuity equation by assuming that measured mid-valley wind profiles were representative of winds on the entire cross section, from sidewall to sidewall. He found that mean down-valley wind speeds, as averaged over vertical cross-sections, increased with down-valley distance in three Vermont valleys during summer but, in the limit of long distances, tended to approach a constant value. Sinking motions at the top of the valley atmosphere on the order of several centimeters per second were required to balance the along-valley mass flux divergence. Whiteman and Barr (1986) made similar calculations for Colorado's well-drained Brush Creek valley in summer and found volume flux divergences that required subsidence rates of 15 cm s^{-1} or less in various segments of the valley. They stated that subsidence rates were, no doubt, overestimated by the horizontal homogeneity assumption. Since this work was published, lidar measurements in the Brush Creek valley (Post and Neff 1986) have provided new information on the cross-valley structure of the down-valley wind system. The horizontal homogeneity assumption in this narrow valley produces volume flux overestimates and mean vertical velocity overestimates of about 43% (Clements et al. 1989b; King 1989).

Calculations in Austria's Inn Valley, made by assuming that actual volume fluxes are 85% of those obtained with the horizontal homogeneity assumption, similarly show nocturnal along-valley volume flux divergence (Freytag 1987). Surprisingly, daytime measurements in the up-valley wind regime also show along-valley volume flux divergence. This suggests that net downward vertical velocities are the general rule in the segment of the Inn Valley investigated, except for possible rising motions that compensate the downslope flow during the brief evening transition. Freytag suggests that the daytime results, which were taken in a segment near the exit of the Inn Valley, may be a feature of the flow in that segment only, caused by a number of large tributaries just up-valley from the segment.

The presence of nighttime along-valley volume flux divergence requires compensatory subsidence over the valley. Such subsidence brings potentially warmer air into the valley during nighttime, warming the valley atmosphere. If the warming is too strong, the driving force for the along-valley circulation will disappear. Thus, the steady-state equilibrium is thought to be maintained by continuous diabatic cooling of the valley atmosphere. This cooling is facilitated by the topographic amplification factor, and the dynamic equilibrium is established through the combined action of the continuity equation, thermodynamic energy equation, and the momentum equation, involving friction at the valley floor and sidewalls.

The required net nocturnal subsidence is quite unlikely to occur uniformly over the valley cross-section since precooled air would be thermodynamically preferred, and such air is available over the slopes and tributaries so that sinking rates may be enhanced there. During daytime, sinking motions over the valley center compensate the upslope flows to warm the valley atmosphere. During nighttime, however, the evidence is not clear. Are rising motions in the valley center required to compensate the downslope flows and produce cooling throughout the valley atmosphere, or do other terms in the thermodynamic energy equation (2.4) play a relatively more important role than the vertical advection term?

2.2.4.2. Thermal Energy Budget

Atmospheric heat budgets have been computed in recent years for valley and basin atmospheres using data collected with motorgliders, tethered balloons, rawinsondes, Doppler sodars, and lidars. Most budgets have focused on the nighttime period when the wind and temperature structure change only slowly with time and the cooling is more or less confined to the valley; however, success at balancing such budgets has been mixed. The problem appears to be that the nighttime fluxes are relatively small, and measurement errors, particularly in the advective and sensible heat flux terms, can be relatively large.

Several conclusions can be derived from the nighttime thermal energy budget computations. In a well-drained valley, once the down-valley flow sets up and attains a reasonably steady state, the rate of cooling of the valley atmosphere becomes small. This is, no doubt, because air carried down the valley must be replaced by potentially warmer air from above. An approximate balance is apparently maintained between the inflow of this warmer air and the limited rate of cooling produced by the diabatic processes (sensible and radiative flux divergences). In closed basins, on the other hand, air can be cooled in place, with little warm air entrained from above, so that moderate rates of cooling can continue throughout the night. All studies to date have found radiative flux divergence to be small relative to sensible heat flux divergence. This conclusion, however, comes from model calculations of radiative flux divergence for plane parallel atmospheres, and does not properly account for within-valley processes such as radiation from the sidewalls. Advective terms are particularly important in well-drained valleys. The choice of coordinate system affects the calculated values of the individual advective terms so that it is necessary to look at the sum of the advective terms to compare with other terms in the heat budget. Typically, the individual terms are large numbers of different sign and with significant errors, so that the sum is somewhat uncertain. Calculations suggest, however, that the advective sum is about the same size as the sensible heat flux divergence term. Thus, local closure of the heat budget within a small valley segment is not assured. Larger scale processes produced by the entire valley drainage area, including tributaries, affect the heat budget of a valley segment. This suggests that successful numerical modeling of the down-valley circulation system will require large three-dimensional model domains. Further, since important physical processes occur within the shallow downslope flows on the sidewalls, fine vertical grid resolution may be necessary.

We illustrate several key findings with reported heat budget computations. Heat budget studies have been published for Austria's Inn Valley (Freytag 1985), Switzerland's Dischma Valley (Hennemuth 1985), Colorado's Brush Creek valley (Horst et al. 1987), Japan's Akaigawa Basin (Maki et al. 1986), and Japan's Aizu Basin (Kondo et al. 1989). The Inn Valley is over 300 km long, 800 to 1500 m deep, and 10 km wide at the ridgetops. The valley structure is quite complicated, including major tributary valleys and subsidiary mountain ranges. The Dischma Valley is a small Alpine end valley of 15-km length, 1000-m depth, and 4-km width. The Brush Creek valley is a 25-km long, 650-m-deep, 4-km-wide (ridge-to-ridge) end valley. The Akaigawa Basin is a 10-km wide near-circular basin with a rather flat 5-km-wide floor, and with surrounding peaks of 250-m height. The Aizu Basin is a large elongated basin of 130-km length with a mean depth of about 1000 m. References to studies of these valleys and basins are made with the understanding that the results come from the respective authors' works.

In the well-drained Brush Creek valley, atmospheric cooling rates are relatively small. During the bulk of the nighttime period (2300–0600 MST), they are near 0.2 K h^{-1} when averaged over the valley depth. In closed basins where advection into or out of the basin is precluded, larger rates of cooling can be expected. For example, in the Akaigawa Basin, tethered balloon profiles showed mean nighttime cooling rates after 2100 h of about 0.5 K h^{-1}.

Observations in the Aizu Basin, the Akaigawa Basin, and the Brush Creek valley show that nighttime ground heat fluxes can largely counter longwave net radiation losses so that nighttime sensible heat fluxes are small (0 to 35 W m^{-2}). This conclusion is also supported by the Alpine data of Halbsguth et al. (1984) and Vergeiner and Dreiseitl (1987) in Figs. 2.15 and 2.16 where sensible heat fluxes are near zero or, even, upward into the atmosphere on the mornings of 6 August and 7 September. Evening observations in both valleys show sensible heat fluxes toward the ground at rates near 15 W m^{-2}. Observations in the Brush Creek valley and the Aizu Basin, however, indicate that sensible heat fluxes are generally smaller on the valley floor than on the sidewalls. This could be caused by several factors. First, the valley floor is typically more moist than the sidewalls, so that latent and ground heat fluxes may be relatively more important than radiative and sensible heat fluxes. Second, net radiative losses at lower altitudes in the valley are decreased somewhat by the sky view factor effect described previously. Third, downslope flows on the sidewalls generate shear and turbulence in a shallow layer that would enhance the turbulent sensible heat flux on the sidewalls relative to the valley floor.

In the Aizu and Akaigawa basins, continuous nocturnal cooling occurred with near-zero sensible heat fluxes at the basin floors, suggesting that the cooling was caused by advection of air cooled on the surrounding sidewalls where sensible heat fluxes were apparently larger, although unmeasured. Advective terms are strong, as well, in well-drained valleys such as the Inn and Brush Creek valleys. The calculation of cooling contributed by along-valley advection is prone to experimental errors because of weak along-valley potential temperature gradients. In the 1984 Brush Creek valley experiments, the gradients ($\partial\theta/\partial s$) were on the order of 0.2 K km^{-1}. In the Inn Valley, the nighttime gradients were near 0.1 K km^{-1}, although aircraft measurements during daytime showed gradients of only 0.03 K km^{-1}.

Daytime energy budgets are not discussed in detail because generalizations are not yet warranted from the small amount of research material so far available. Several findings seem important, though. The Inn Valley research of Freytag (1985) shows that the heat budget can be complicated in large valleys by the large phase shift (up to 5 h) between the diurnal heating cycle and the along-valley wind system. Further, the Dischma Valley and Aizu Basin research has shown that daytime heating may not be confined to the valley atmosphere, so that computations must be extended above the valley and must be corrected for above-valley advective effects. Daytime heating rates in the moist Dischma Valley were small (0.5 K h^{-1}), and the role of advection was rather inconsequential relative to sensible heat flux convergence. Further, the distribution of surface cover appeared to produce spatial variations in sensible heat flux. Regions of heat surpluses and deficits within the valley were expected to transfer heat from the sunny to the shaded sidewall, from the end to the entrance of the valley and, after noon, from the valley to the surrounding region. Finally, in the Aizu Basin, the observed increase in energy storage over the center of the basin appeared to be caused by subsidence warming as a compensation for upslope flows that formed over the valley sidewalls.

2.2.4.3. MOMENTUM BUDGET

An atmospheric momentum budget was completed for a 30-h period during March 1982 for the Inn Valley using data from the MERKUR experiment (Freytag 1986). Pressure gradient and friction were the most important terms in the u-momentum equation. The pressure gradient decreased linearly from the surface to the 1000-m ridgetop, while friction decreased linearly from the surface to 400 m. Diurnal accelerations produced by the pressure gradient term reached 3 m s^{-1} h^{-1} during daytime and −4 m s^{-1} h^{-1} at night. The friction term had the same shape and roughly the same amplitude as the pressure gradient term, but lagged by 1 to 2 h. Actual accelerations and decelerations of the along-valley winds lagged the pressure-gradient zero crossings by 5 to 6 h. The 1- 2-h lag of the friction term is a measure of the reaction time of the wind systems to the forcing pressure gradient. For comparison, calculations by Hennemuth (1986) in the smaller Dischma Valley resulted in a reaction time of only 30 min, and Vergeiner et al. (1987) estimated a value of 8 min for the Brush Creek valley. Freytag (1986) found that the momentum advection terms played a role in damping the wind field of the Inn Valley. Horizontal advection was usually advection of air with less momentum. The role of vertical advection was small, but was most significant in the upper levels of the valley's atmosphere where sinking motions imported slower moving air from above.

2.2.4.4. HUMIDITY BUDGET

A valley moisture budget was completed for two daylight periods in August 1980 in Switzerland's Dischma Valley as part of the DISKUS experiment (Hennemuth and Neureither 1986). This appears to be the only valley humidity budget accomplished to date. Evaporation was the source of water vapor in the valley atmosphere during daylight hours. Vertical turbulent transport accounted for only a minor export of humidity from the valley. The observed morning humidity decrease and afternoon increase were attributed to advection in the local wind systems, with the drainage winds exporting humidity in the morning and the up-valley winds importing humidity in the afternoon. Humidity transport between valleys was thought to be important.

2.3. Slope wind systems

Slope wind systems are driven by buoyancy forces and are formed on any slope where a layer of air adjacent to the slope is either cooled or warmed relative to air at the same level in the ambient environment away from the slope. Slope flows occur quite frequently over sloping terrain of different scales, since thermal boundary layers form daily over most land areas. Slope flows are known on various scales, occurring over large sloping land areas such as Antarctica, where katabatic or downslope winds may reach gale force on the periphery of the continent. We confine our discussion to slope flows occurring on the slopes of isolated mountains and on valley sidewalls. Such slope flows are subject to local terrain, vegetation, and soil moisture influences, are intermittent in time, and may change over short distances (Vergeiner 1982).

2.3.1. Simple slope flows

Mahrt (1982) investigated modeling and observational studies of downslope flows and pointed out, on the basis of a careful scale analysis of the momentum equation, the many different types of downslope flow regimes that are possible. A useful summary of the application of the thermodynamic energy equation in the valley and slope flow layers has been recently provided by Brehm (1986).

The system of equations generally used for slope flows on simple slopes is written in a natural coordinate system oriented along (s) and perpendicular (n) to a slope of constant inclination angle α. The system of equations includes the s- and n-momentum equations, the thermodynamic energy equation, and the continuity equation, respectively, as described by Horst and Doran (1986):

$$\frac{\partial u}{\partial t} + u\frac{\partial u}{\partial s} + w\frac{\partial u}{\partial n} = -\frac{1}{\rho_0}\frac{\partial(p-p_a)}{\partial s} - g\frac{d}{\theta}\sin\alpha - \frac{\partial\overline{u'w'}}{\partial n} \quad (2.20)$$

$$\frac{\partial w}{\partial t} + u\frac{\partial w}{\partial s} + w\frac{\partial w}{\partial n} = -\frac{1}{\rho_0}\frac{\partial(p-p_a)}{\partial n} + g\frac{d}{\theta_0}\cos\alpha \approx 0 \quad (2.21)$$

$$\frac{\partial \theta}{\partial t} + u\frac{\partial \theta}{\partial s} + w\frac{\partial \theta}{\partial n} = -\frac{1}{\rho_0 c_p}\frac{\partial R}{\partial n} - \frac{\partial\overline{w'\theta'}}{\partial n} \quad (2.22)$$

$$\frac{\partial u}{\partial s} + \frac{\partial w}{\partial n} = 0. \quad (2.23)$$

Symbols u and w represent flows down and perpendicular to the slope, and R is the upward flux of net all-wave radiation. The potential temperature stratification is given by

$$\theta = \theta_0 + \gamma z + d(s, n, t), \quad (2.24)$$

where γ is the ambient potential temperature gradient, d is the potential temperature excess near the slope (this will be negative for downslope flows), and θ_0 and ρ_0 are reference values for potential temperature and density in the absence of surface cooling. The ambient pressure p_a is determined from the formula

$$\frac{\partial p_a}{\partial z} = -\rho_a g, \tag{2.25}$$

where ambient air density is a function of ambient potential temperature, as defined by $\theta_a = \theta_0 + \gamma z$. Equation (2.21) assumes that accelerations perpendicular to the slope are negligible so that the slope flow is in hydrostatic balance normal to the slope.

Following this system of equations, the driving force for the downslope flow is the buoyancy force generated by cooling of the layer adjacent to the slope, as represented by the second term on the right-hand side of (2.20). This force is balanced by the divergence of the turbulent momentum flux (third term), and the adverse pressure gradient associated with the change of depth of the slope flow layer with downslope distance (first term). The temperature deficit develops as a result of radiative and turbulent sensible heat flux divergences, following (2.22).

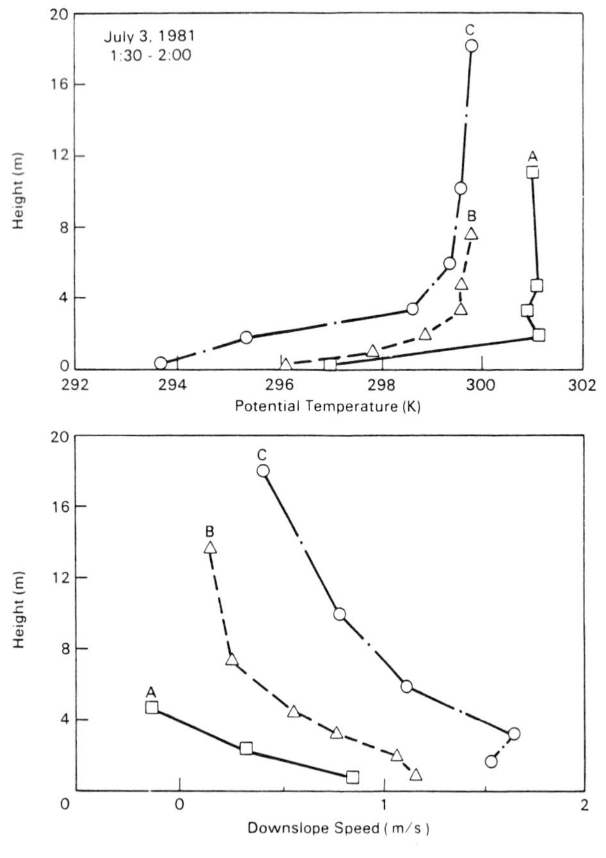

FIG. 2.20. Temperature and wind profiles over the slope of Washington's Rattlesnake Mountain illustrating the evolution of profiles with downslope distance. The upper slope of the mountain is 21° until just below tower B, where it changes to 8°. Ambient winds had upslope and cross-slope components of 1 to 2 m s^{-1}. Reprinted by permission of Kluwer Academic Publishers. (Horst and Doran 1986.)

A large number of slope flow models have been developed over the years, but there have been few datasets that could be used for comparison with simulations. Two datasets collected recently on simple slopes have alleviated this problem somewhat, although the datasets cannot be considered comprehensive because of the lack of surface energy budget and radiation measurements. Horst and Doran (1982, 1986, 1988) have collected slope flow observations on the side of Rattlesnake Mountain, Washington, while Clements and Nappo (1983) developed a similar dataset on the side of Pajarito Mountain, New Mexico. At the Rattlesnake Mountain site, downslope winds did not form regularly from night to night, being quite subject to the prevailing winds and, apparently, to variations in the surface energy balance. Figure 2.20 shows observations of wind and temperature structure as a function of downslope distance on a good drainage night. The topmost tower on the slope, located 69 m below the ridge crest, was designated tower A. Tower B was located 151 m below the ridgecrest and tower C was located 229 m below the ridgecrest.

The shallow inversion over the slope increased in depth with downslope distance, while the drainage flow became stronger and deeper and the height of the peak speed increased. Inversion strength did not change significantly along the slope's length. According to Horst and Doran (1986) the development of the drainage flow with downslope distance was predicted well by Manins and Sawford's (1979b) drainage flow model, except that the inversion strength in the downslope flowing layer did not decrease with downslope distance.

2.3.2. *Slope flows on valley sidewalls*

Downslope flows on valley sidewalls can be expected to differ from flows on simple slopes because of two main factors—one dealing with atmospheric stability and one dealing with ambient winds. The formation and growth of a surface-based inversion over the valley floor ensures that the temperature structure of the ambient environment will undergo temporal changes as the inversion deepens. Potential temperature gradients in the valley inversion typically become weaker with altitude, so that downslope increases in ambient stability are expected. These stability variations in the ambient atmosphere can be expected to organize the slope flow structure in ways that differ from the slope flow structure over simple slopes where ambient stability may vary weakly with altitude. Second, along-valley circulations can be expected to influence the structure and evolution of the shallow downslope flow layer over the slope. Many investigators have previously remarked on the difficulty of finding pure downslope flows on valley sidewalls because of the overriding influence of the down-valley flows (e.g., Hennemuth and Schmidt 1985).

A recent experiment (Doran et al. 1989) investigated the development of downslope flow on a sidewall of the 225-m deep Touchet Valley in Washington. This valley had a uniform sidewall angle of 23 degrees. Three towers, designated A, B, and C, were installed on the fall line of this slope at positions 75 m, 125 m, and 175 m below the ridge crest, respectively. The development of wind and temperature profiles above the slope was monitored by instruments placed on the towers, while changes in the ambient (valley) environment were monitored with tethered balloon and Doppler sodar observations over the valley floor. Results are summarized in Table 2.7. The variation of inversion depth, inversion strength, and buoyancy deficit with distance along the sidewall are summarized in the table. The first column gives predictions of the Manins and Sawford (1979b) model for a simple slope (i.e., a slope on the side of an isolated mountain) in which ambient inversion strength does not vary along the slope. The second column summarizes Horst and Doran's (1982) measurements over the simple slope of Rattlesnake Mountain, and the final column summarizes Doran et al.'s (1989) observations over the sidewalls of the Touchet Valley. The notation A > B for inversion depth, for example, means that inversion depth is greater at tower A than at tower B. Simple slope predictions agree well with observations except for inversion strength, as mentioned previously. Observations on the valley sidewall, however, are at odds with simple slope observations and predictions. These results have been attributed by Doran et al. (1989) to increases in the stability of the ambient environment with downslope distance and to the effects of ambient (down-valley) winds on the downslope layer. Variations in surface energy budget along the slope may also be a contributing factor.

The effects of ambient stability on downslope drainage flows are further illustrated in Fig. 2.21. Here, a succession of pictures shows the dispersal of a smoke plume released into a shallow downslope flow on the upper east sidewall of Colorado's Brush Creek valley, as described by Thorp and Orgill (1986). The upper half of the valley atmosphere on this night was nearly isothermal, with a strong inversion

TABLE 2.7. Relative values of inversion depths, inversion strengths, and buoyancy deficits from predictions for a simple slope, measured on a simple slope, and measured on a valley sidewall (Doran et al. 1989).

	Model prediction	Simple slope	Valley sidewall
Inversion depth	A < B < C	A < B < C	A > B ~ C
Inversion strength	A > B > C	A ~ B ~ C	A ~ B > C
Buoyancy deficit	A < B < C	A < B < C	A > B > C
Katabatic wind	A < B < C	A < B < C	B > A ~ C

in the lowest 150 m of the valley. The smoke tracer cascaded over a cliff at the head of the box canyon tributary, was carried down the slope of the tributary and, as the slope flow encountered strong ambient stability, the plume became detached from the sidewall and left the sidewall to attain its buoyancy equilibrium level in the valley inversion. It turned the corner of the tributary and was then transported in the Brush Valley down-valley flow. This sequence of events illustrates several points. The flow of the oil fog plume down the cliff requires that a sufficient buoyancy deficit exist in the flow as it approaches the cliff. Further, an accurate model of the slope flow must incorporate the cooling of the parcel during its downslope transport caused by downward sensible heat flux to the underlying ground, warming caused by subsidence, entrainment of ambient air into the slope flow layer, and the effects of the ambient stability. These factors, and the slope angles, may change frequently along the parcel path so that the parcels' equilibrium is continually disturbed. The adequate incorporation of such disparate processes represents a significant challenge to the development of a realistic model.

Daytime, upslope flows on valley sidewalls have not yet been investigated in detail. In particular, the variation in structure with upslope distance has received little observational attention. Vergeiner predicted (Vergeiner 1982; Vergeiner and Dreiseitl 1987), on the basis of a simple slope layer energy budget, that flow in the layer will be quite sensitive to ambient temperature structure, and that upslope volume flux may decrease or increase with upslope distance depending on the presence or absence of strong stability layers in the ambient environment. Actual observations at a particular location on a slope may be quite unrepresentative of the mean conditions over the slope because of the known inhomogeneity and nonstationarity of these flows in real topography. Dynamic modeling of idealized sidewall topography supports Vergeiner's view of the nonstationarity of the flows as well as the sensitivity of the flows to ambient stratification (Bader and McKee 1985). On the other hand, it may not be necessary to know the detailed turbulent structure of the upslope flows to determine bulk valley warming or bulk changes in the along-valley flows. Such changes are caused by the total energy input to the valley volume and its distribution through the valley atmosphere by the integrated slope flows. Brehm (1986) has recently developed a modeling approach that couples the structure and evolution of sidewall slope flows with the developing structure of the ambient valley temperature structure. This approach has been extended to three dimensions by Egger, Chapter 3.

February observations of temperature structure evolution over a deep snow cover in Colorado's Yampa Valley have suggested that upslope flows are present even in win-

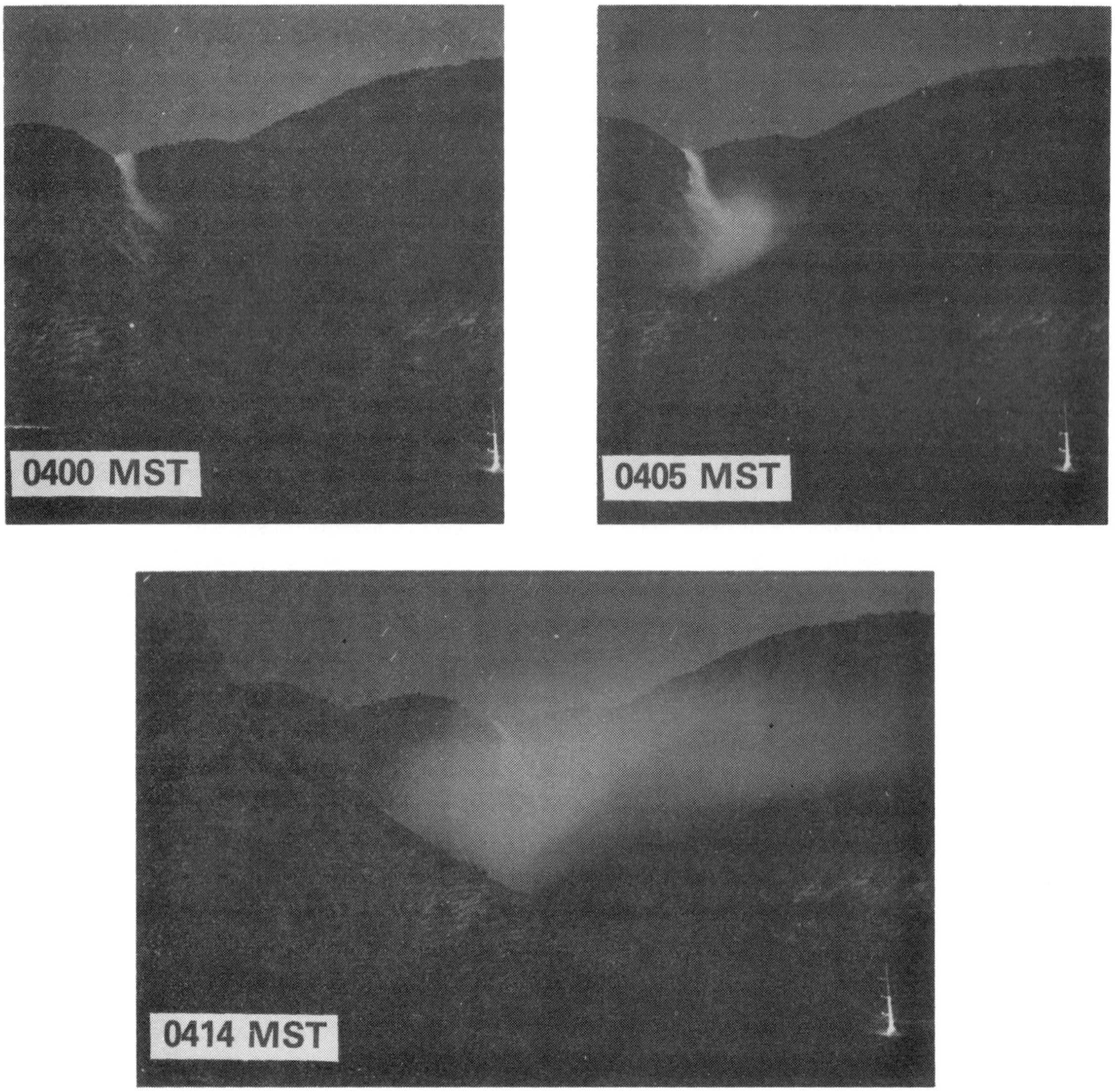

FIG. 2.21. A smoke plume released on the mesalike ridgetop of Colorado's Brush Creek valley was carried over a cliff into a box canyon tributary and down the sidewall until separating from the sidewall and flowing out into the main valley as an elevated plume. (Photographs provided by J. M. Thorp and M. M. Orgill, Battelle NW Laboratories, Richland, Washington.)

ter. Effective heat transfer to the air by the evergreen forests is thought to be the means by which such flows are developed, as illustrated in Fig. 2.22. Special opportunities to compute valley atmospheric energy budgets are present in such conditions that have, as yet, been unrealized. Ground and latent heat fluxes at the surface [see Eq. (2.15)] may be negligible in such conditions because of the insulating effect of the snow cover and the subfreezing temperatures. Further, on days when air temperatures rise above freezing, the direction of the sensible heat flux will be affected by the inability of the snow surface temperature to rise above 0°C.

The superposition of the valley and slope wind systems results in a diurnal turning of the wind over the sidewalls, as shown in Fig. 2.23, so that, regardless of valley orientation, winds on the left sidewall of a river valley (taking the river's point of view) turn counterclockwise with time, while winds on the right sidewall turn clockwise with time. This turning is illustrated with data for the opposite sidewalls of Colorado's Brush Creek valley in Fig. 2.24.

2.4. Morning transition

Early observations of valley temperature structure evolution (Ekhart 1948, 1949; Machalek 1974) showed that a typical feature of the morning transition period was the descent into the valley of the top of the nocturnal temperature inversion. Pibal wind observations collected by other observers (Ayer 1961; Davidson and Rao 1963) showed a similar feature in the wind observations. After

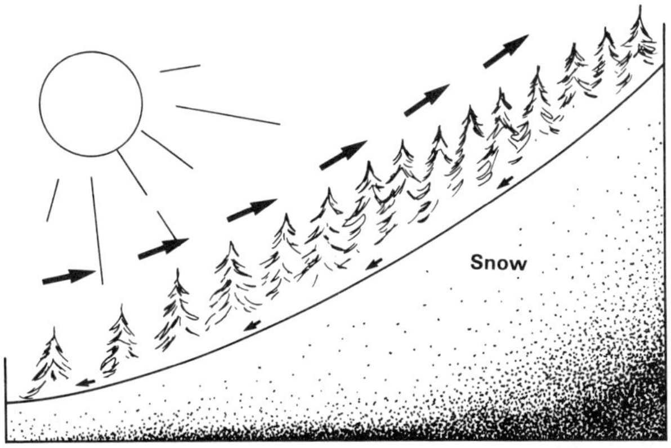

FIG. 2.22. Upslope flow can form over a forested slope in winter even when the ground is covered with snow. (Adapted from Hawkes 1947.)

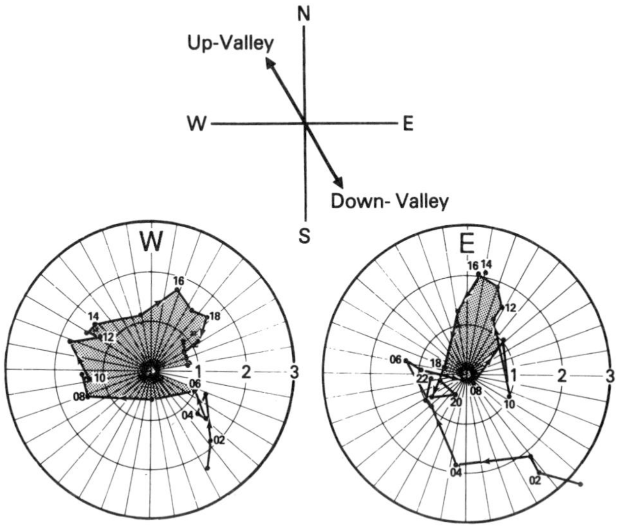

FIG. 2.24. Wind hodographs for the opposing sidewalls of Colorado's Brush Creek valley, 25 September 1984. (Whiteman et al. 1989b.)

sunrise, the winds within a valley would continue to blow down the valley, but a wind reversal (to up-valley) would first occur aloft and the wind reversal point would descend into the valley with time.

Interpretations of the cause of the descent phenomena were hampered by a lack of good data; especially critical was the need for concurrent wind and temperature structure data in a number of valleys. Initial hypotheses attributed the descent of the wind reversal point and the (assumed) top of the temperature inversion to "turbulent erosion" by strong winds aloft (Davidson and Rao 1963) or the draining of the nocturnal valley cold air pool down the valley's axis after sunrise (Ayer 1961).

Analysis of observations taken in the late 1970s in a number of Colorado valleys (Whiteman 1980, 1982) has provided an explanation for the inversion descent phenomenon that emphasizes the role of subsidence in the valley atmosphere in effectively transferring heat received at the slopes to the entire valley cross-section.

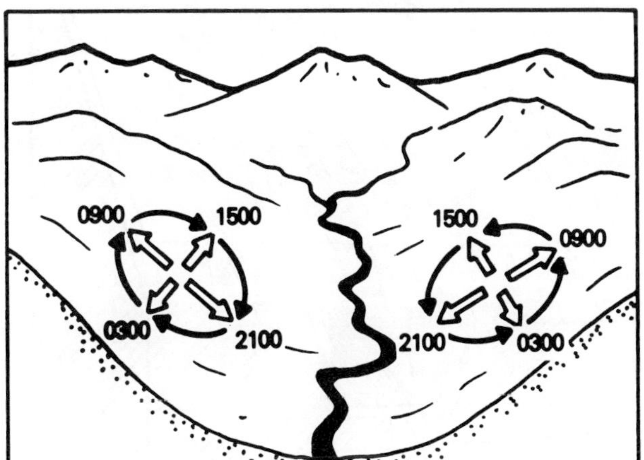

FIG. 2.23. Winds on the left bank of a river (i.e., right side of the figure) turn counterclockwise with time; winds on the right bank turn clockwise. (Adapted from Hawkes 1947.)

Temperature inversions in the Colorado valleys typically developed to the full depths of the valleys and were destroyed after sunrise following one of three patterns of temperature structure evolution (Fig. 2.25). The first pattern, observed in the widest valley studied, approximates inversion destruction over flat terrain, in which the nocturnal inversion is destroyed after sunrise by the upward growth from the ground of a warming convective boundary layer (CBL). The second pattern, observed in snow-covered valleys, differs significantly from the first. Here the growth of the CBL, which begins after sunrise, is arrested once the CBL has attained a depth of 25 to 50 m. The inversion is then destroyed as the top of the nocturnal inversion descends into the valley. Successive profiles of the valley atmosphere show a warming consistent with a simple subsidence of the previous profiles. The third pattern of temperature structure evolution was observed in all of the valleys when snow cover was not present and describes the majority of case studies observed in field experiments. Following this pattern, inversions are destroyed by two processes: the continuous upward growth from the valley floor of a warming CBL and the continuous descent of the top of the nocturnal temperature inversion. Warming of the elevated inversion layer above the CBL is consistent with a simple subsidence of the previous profiles. In the Colorado valleys studied, the time required to break an inversion and establish an unstable atmosphere within the valley was typically $3\frac{1}{2}$ to 5 h after sunrise. Temperature structure evolution during clear, undisturbed weather differed little from day to day and from season to season. Further, the valley inversions

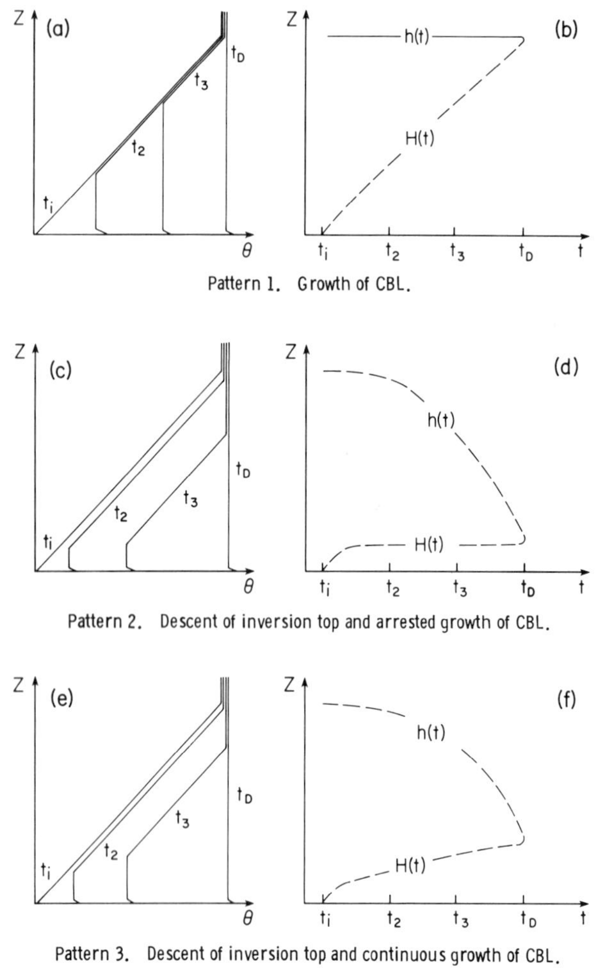

FIG. 2.25. Summary of the three different types of inversion breakup observed in Colorado valleys. Pattern 1 is the pattern typically seen over flat plains and occurs in very wide valleys, pattern 2 occurs in snow-covered valleys, and pattern 3 occurs in most non-snow-covered valleys. (Whiteman 1982.)

showed less variability from day to day and from season to season than inversions over the adjacent plains (Whiteman 1982). This suggests that, at least for a narrow range of weak to moderate synoptic flow conditions investigated, deep valley topography produces more consistent inversions by protecting them from winds aloft.

The common element of all three patterns of temperature structure evolution is the development of a CBL over the valley floor after it is illuminated by direct sunlight. Observations of the sidewall of one of the valleys indicate that a CBL begins to grow above the sidewall after direct sunlight illuminates it. The temperature structure of the sidewall CBL is similar to that of the CBL over the valley floor, but upslope winds blow at speeds up to 3 m s^{-1} in the sidewall CBL.

The Colorado observations have indicated that five different temperature structure layers may be present during the inversion destruction period. Above the valley floor CBL and the sidewall CBL just mentioned is the stable core of the potential temperature inversion—the remnants of the nocturnal inversion. A near-neutral stability layer above the stable core appears to be part of a large-scale convective boundary layer that forms over the entire western slope of the Rocky Mountains. Above this layer is the free atmosphere. Each of these temperature structure layers, identified primarily by their potential temperature structure, can also be identified by the winds that prevail within them, as illustrated in Fig. 2.26 for a time midway through the morning transition period. During inversion destruction, the CBLs over the valley floor and sidewalls contain winds that blow up the floor of the valley and up the slopes. The CBL, or neutral layer, above the valley inversion has winds that blow up the inclined western slope of the Rocky Mountains during the day. Winds in the stable core typically continue to blow down-valley after sunrise until the stable core is nearly destroyed. Winds in the stable free atmosphere may blow from any direction with speeds determined by synoptic-scale pressure gradients. Despite some variability in the strength and timing of reversal of the winds, the temperature structure evolves uniformly from day to day in individual valleys.

A hypothesis was developed to explain the observed temperature structure evolution (Fig. 2.27). Initially, just before sunrise, a deep temperature inversion fills the valley. After sunrise, solar radiation received on the valley floor and sidewalls is partly converted to sensible heat flux. Sensible heat flux causes a CBL to develop over the

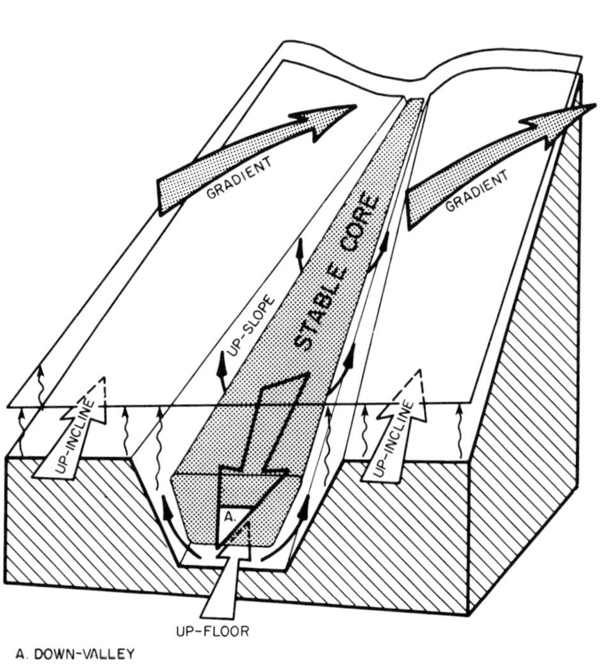

FIG. 2.26. View of the interacting wind systems during the inversion breakup period. (Whiteman and McKee 1982.)

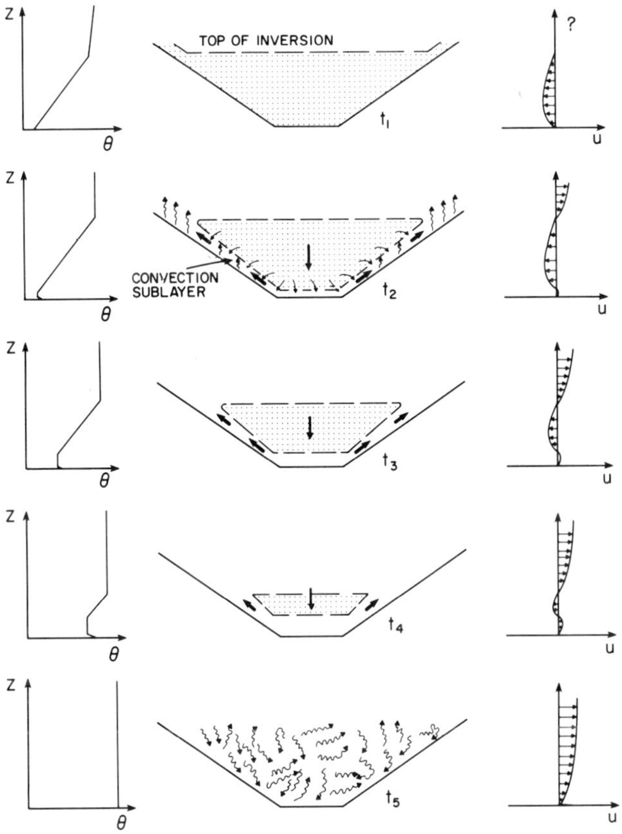

FIG. 2.27. Inversion breakup hypothesis. Stable core air is entrained into the convection sublayer and removed from the valley in upslope flows, producing a compensating subsidence over the valley center. In the center of the diagram cross sections of a valley are shown at times t_i, t_2, t_3, t_4, and t_D. On the left are corresponding potential temperature profiles as taken from the valley center. On the right are corresponding up-valley wind components as a function of height. At sunrise, t_i, an inversion is present in the valley. At t_2, a time after sunlight has illuminated the valley floor and slopes, a growing CBL is present over the valley surfaces. Mass and heat are entrained into the CBLs from the stable core above and carried up the sidewalls in the upslope flows. This results in a sinking of the stable core and growth of the CBLs (t_3 and t_4) until the inversion is broken (t_D) and a turbulent well-mixed neutral atmosphere prevails through the valley depth. Down-valley winds continue to blow in the stable core during the inversion breakup period. Winds in the CBL below, and in the region above the stable core, often blow up-valley during this same period. (Whiteman 1982.)

surfaces. Mass is entrained into the CBLs from the stable core above and carried from the valley in the upslope flows that develop in the CBLs over the sidewalls. This removal of mass from the base and sides of the stable core causes the elevated inversion to sink into the valley and to warm adiabatically as a result of subsidence. Incidentally, the subsidence (or, equivalently, the mass flux divergence from the CBLs) decreases the rate of growth of the CBLs over the valley floor and sidewalls relative to normal CBL growth rates over flat plains.

The physical concepts applied here are quite closely related to concepts discussed in earlier sections of this chapter. The valley atmosphere is driven by incoming solar radiation. Some of the radiation, when received on the valley surfaces, is converted to sensible heat flux. The topographic amplification factor plays an important role in determining the rate of warming of the valley atmosphere for a given energy input. Vertical convergence of sensible heat flux over the sidewalls warms the inclined CBL. An upslope flow quickly develops in this layer. Transport of mass by upslope flows requires compensatory subsiding motions over the valley's center. This subsidence in the stable valley atmosphere produces warming and provides a very effective mechanism for transporting the energy received at the valley sidewalls throughout the entire valley atmosphere. The warming of the valley atmosphere, enhanced by the topographic amplification factor, decreases the nighttime along-valley pressure gradient and eventually causes the down-valley winds to reverse. When the elevated remnant of the nocturnal inversion is finally destroyed, continued upslope flow will fail to produce compensatory warming in the valley center because the vertical potential temperature gradient vanishes. Much of the energy is then transported from the valley and goes to develop CBLs above the mountain range.

A thermodynamic model based on a simplified energy budget of the valley atmosphere and incorporating a topographic amplification factor (Whiteman and McKee 1982) can adequately describe the evolving temperature structure during the inversion breakup period in typical valleys. The model, however, requires the specification of a free parameter indicating the fraction of the sensible heat flux to be used for CBL growth, with the remaining fraction used for transporting mass up the sidewall and producing compensatory subsidence and warming over the valley center. Further research is necessary to gain a better understanding of the factors that affect the partitioning of energy in actual valleys, and thus affect the pattern of inversion destruction to be followed (Fig. 2.25). Observations show that high rates of input of sensible heat flux into the valley atmosphere favor the growth of convective boundary layers and pattern 1 inversion destruction. This pattern is also favored in wide valleys [e.g., the wide South Park valley investigated by Banta and Cotton (1981)], where compensatory subsidence is weak because the upslope flows carry small amounts of volume relative to the total atmospheric volume in the cross section. Pattern 3 is the most common pattern of temperature inversion destruction in valleys since inversions are typically destroyed by a combination of CBL growth and subsidence warming, but changes in the surface energy budget in a given valley from day to day (say, following a rain) have an effect on the relative importance of the CBL growth and subsidence warming mechanisms (Whiteman 1982).

Recent work has clarified important aspects of wind and temperature structure evolution during the morning transition period. The relatively minor role played by turbulent erosion at the top of the temperature inversion has been documented in two-dimensional dynamic model simulations by Bader and McKee (1983). Further, their simulations have shown that variations in albedo on a valley cross-section can make important changes in the inversion destruction pattern followed. Gravity waves play an important role in transferring heat across the valley atmosphere when differential heating occurs on the opposite sidewalls (Bader and McKee 1985). Further, the dynamic model simulations show that the upslope flows are not necessarily continuous along the slope and are quite nonstationary. The noncontinuous slope flows were predicted independently by Vergeiner (1982) based on an energy budget equation that he developed for the slope flow layer. Small-scale topographic features on the slopes as well as ambient atmospheric layers of intense stability may cause mass to leave the slope and converge into the valley atmosphere, carrying its heat content, and producing organized large-scale cross-valley circulations. Brehm (1986) has carried Vergeiner's energy budget approach further and developed a model of temperature structure evolution that explicitly includes the evolution of the slope flow layer and allows the along-slope volume flux to converge or diverge depending on ambient atmospheric stability. His comparison of the model with data suggests that the direct heating of the valley atmosphere through mass leaving the slope is small compared to subsidence warming.

Dispersion of elevated nocturnal pollutant plumes in valleys is strongly affected by inversion breakup processes, and high ground-level pollution concentrations have been frequently observed during the morning transition period (e.g., Hewson and Gill 1944). When valleys are oriented in a north–south direction, solar insolation can be quite different on the opposing sidewalls during the morning transition period. If the valley is narrow this can produce significant cross-valley pressure gradients and cross-valley flows (Urfer-Henneberger 1964, 1970). These cross-valley flows may carry elevated air pollution plumes toward the more strongly heated sidewall. There, high concentrations may be experienced as the plume is fumigated into the growing CBL and carried upslope. These physical processes were found to be important contributors to dispersion of elevated tracer plumes in Colorado's northwest–southeast oriented Brush Creek valley in summer and fall experiments (Whiteman 1989; Gudiksen and Shearer 1989). Summer and winter cross-valley advection of elevated plumes in the Brush Creek valley have been simulated recently with a dynamic model (Bader and Whiteman 1989). Summer simulations show that a cross-valley flow toward the heated sidewall occurs in the lowest altitudes of the valley, but that this may be accompanied by a weak cross-valley flow toward the shaded sidewall in the upper altitudes of the valley, so that the plume's altitude can have an important bearing on the direction of plume transport and ultimate pollutant concentration distribution. Winter simulations suggest that cross-valley advection during the morning transition period will be weak, since winter insolation falls more equally on the two sidewalls of the northwest–southeast oriented valley.

Observations of inversion destruction in other valleys and basins throughout the world (Brehm and Freytag 1982; Böhm 1985; Brehm 1986; Neininger and Reinhardt 1986; Müller and Whiteman 1988) have tended to confirm the important role of vertical motions in transferring heat to the valley atmosphere. Very recent results in Japanese basins (Kondo et al. 1989) suggest that subsidence that compensates for upslope flow above the sidewalls plays an important role in the heating of basins as well.

2.5. Evening transition

The evening transition period has received much less observational and theoretical attention than the morning transition period. This is true despite early observations (Hewson and Gill 1944) of air pollution episodes on the valley floor during the evening transition period. The evening transition period is generally perceived to be a difficult observational, theoretical, and modeling problem because of the nonstationary nature of the transition period and the combination of important physical processes acting on multiple time and space scales.

The evening transition period is generally defined as the time period through which daytime up-valley flows are reversed to down-valley flows and attain a reasonably steady state, characteristic of nighttime. Key physical processes during the transition period include the reversal of the sensible heat flux and the development of downslope flows. The spatial pattern of sensible heat flux reversal and the near-coincident slope flow reversal can be quite complicated in a real valley. Sensible heat flux reversal is closely related to the time of local sunset, so that the progression of shadows in a valley can sometimes give an indication of the spatial field of negative sensible heat flux. Downslope flows begin locally after the sensible heat flux reverses (Whiteman et al. 1989b). The air cooled above the sidewalls flows downslope so that the cooling is effectively confined within the valley. Compensatory rising motions over the valley center, which arise as a response to the downslope flows over the valley sidewalls, are thought to contribute to the cooling there. From this, one obtains a picture of nonhomogeneous, nonstationary processes producing cooling within the bulk of the valley atmosphere. One could speculate about the difficulty of obtaining representative measurements of turbulence, etc., during this time period. This cooling of the valley atmosphere eventually becomes sufficient to reverse the temperature contrast between valley and plain (or, more properly, along the valley axis) so that the along-valley pressure gradient reverses sign and down-valley winds are initiated.

In addition to the sensible heat flux divergence that produces cooling in a shallow layer above the slopes and the rising motions that produce cooling over the valley center, cooling of the valley atmosphere is also produced by radiative flux divergence and, on a given valley cross-section, may be produced by horizontal down-valley cold air advection. Advective effects could be especially significant if along-valley gradients in sensible heat flux were present along the length of the valley at a given time. This must be a quite frequent occurrence, since such gradients could be produced by differences in shading caused by changes of direction of the valley axis, as well as along-valley differences in soil moisture, vegetation cover, soil conductivity, or other factors affecting the surface energy budget.

Over flat plains, cooling typically begins in a shallow layer near the ground that deepens slowly through the course of the night. In valleys, on the other hand, several mechanisms combine to produce deep cooling through the entire valley atmosphere rather than confining the cooling to a shallow layer at the valley's floor. Further, the topographic amplification factor enhances the total cooling within the valley relative to the plain, since energy losses out the top of the valley come from a smaller volume of air and thus produce larger temperature decreases. An example of such cooling and the corresponding wind structure evolution over a 4-h evening transition period is shown for Colorado's 700-m-deep Eagle Valley in Fig. 2.28. Initial soundings showed that the daytime winds blew up valley, as expected. A shallow down-valley wind layer formed first near the ground and rapidly grew in depth. A shallow transition layer wind is evident at the top of the down-valley wind layer and appears to be at about the same height as a sharp inversion layer. The evolution of peculiar small-scale features in the potential temperature profiles appears to be related to the convergence of downslope flows that leave the sidewalls. The upward propagation of these features suggests that rising motions occur over the valley center as a compensation for downslope flows over the sidewalls, as shown in Fig. 2.29. Further observations will be necessary before this interpretation can be tested.

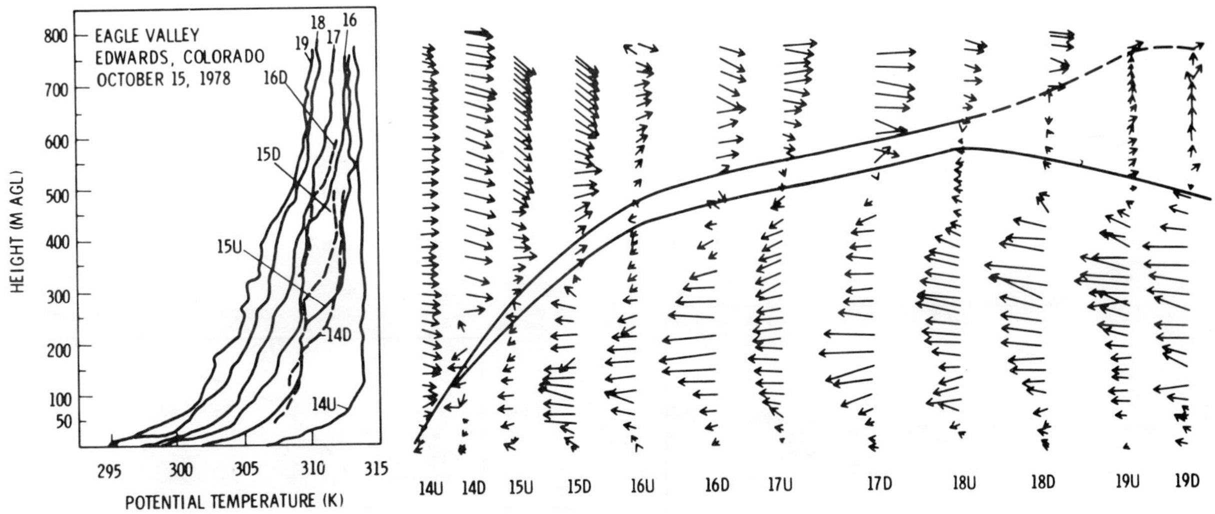

#	TIME (MDT)	#	TIME (MDT)	#	TIME (MDT)
14U	1802-1820	16U	1931-1947	18U	2100-2118
14D	1820-1836	16D	1947-2001	18D	2118-2133
15U	1846-1903	17U	2014-2031	19U	2143-2202
15D	1903-1922	17D	2031-2045	19D	2202-2216

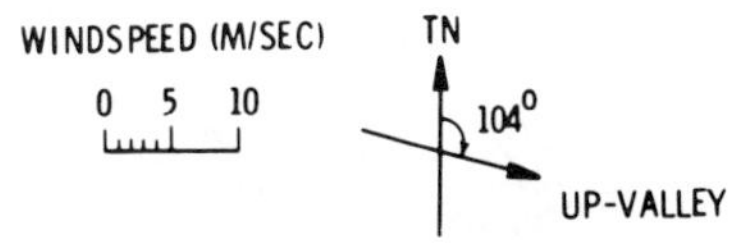

FIG. 2.28. Temperature and wind structure evolution during the evening transition in the 700-m-deep Eagle Valley on 15 October 1978, illustrating the rapid development of the down-valley wind system and the growth of the temperature inversion. Note that small-scale features in the temperature structure are maintained from sounding to sounding. In the figure, vector winds are plotted as a function of height; vector lengths correspond to wind speeds (see wind speed legend). The up-valley direction is given for reference. The time intervals over which the up (U) and down (D) soundings were conducted are given in the table. The solid lines in the wind profiles delineate the boundaries of the wind transition layer. (Whiteman 1986.)

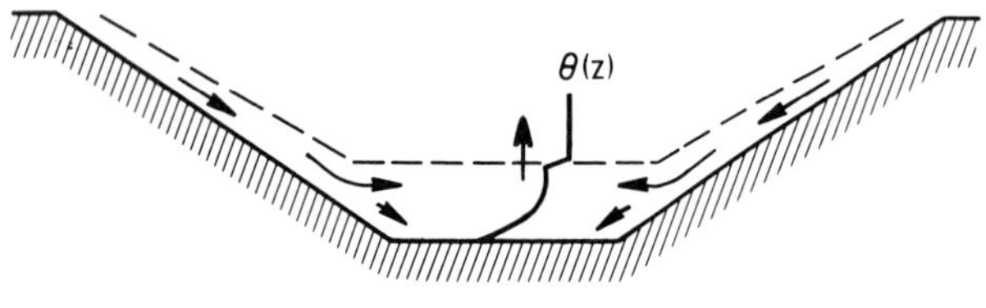

FIG. 2.29. Postulated flow in relationship to temperature inversion features in the previous diagram. Slope flows appear to leave the slope and converge toward the valley center producing compensating rising motions that cool the atmosphere over the valley center.

The rapid development of valley temperature inversions during the evening transition period has been confirmed in recent fall experiments in Washington's 225-m-deep Touchet Valley (Doran et al. 1989). On clear nights in the Touchet Valley, inversions attain the full valley depth within 3 h of the start of inversion formation (Fig. 2.30a); inversion growth is accompanied by an initial rapid growth in inversion strength (potential temperature difference between the ground and the top of the inversion) followed by the gradual attainment of an equilibrium strength of 7 or 8 K (Fig. 2.30b). Loss of energy storage

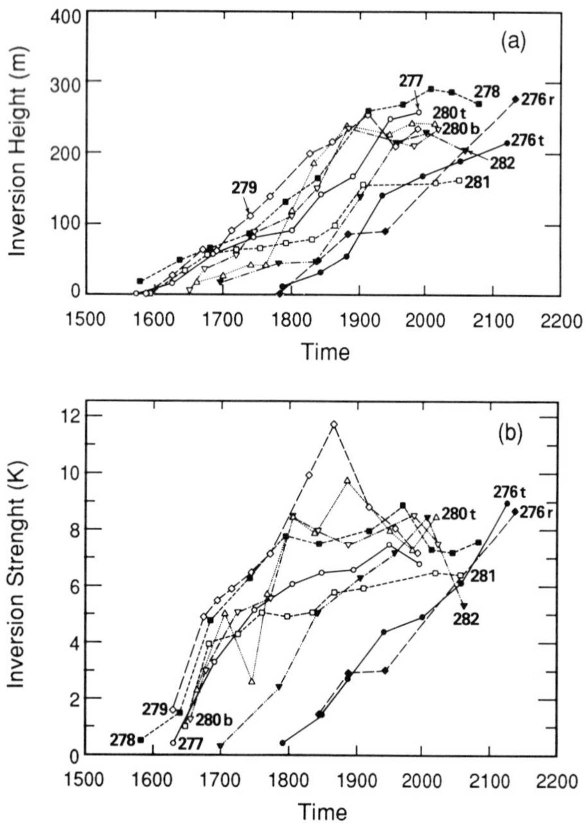

FIG. 2.30. Growth of (a) temperature inversion depth and (b) inversion strength with time in Washington's 225-m-deep Touchet Valley during clear October nights in 1986. Dates given are Julian dates; 4 October is equal to Julian date 276. Dual tethersondes were used at different valley locations on Julian dates 276 and 280. (Doran et al. 1989.)

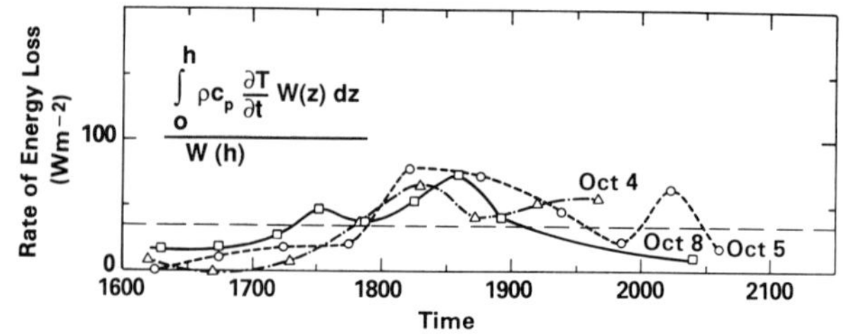

FIG. 2.31. Rate of loss of energy from a unit vertical cross-section of Washington's Touchet Valley in 1986 as normalized by the width of the cross section. The horizontal line provides a rough estimate of the rate of loss of energy that could be explained by diabatic processes in the valley cross-section.

in the valley atmosphere follows the pattern shown in Fig. 2.31, in which valley energy loss (normalized by the cross sectional area of the valley's lid) reaches a peak of 60 to 70 W m^{-2} during the 1730 to 1930 PDT period when the down-valley winds become strong. Estimates of sensible and radiative flux divergences from the valley cross-section suggest that this rapid cooling cannot be explained solely by diabatic processes occurring in the cross section, suggesting that down-valley cold air advection may play an important role during the evening transition period.

2.6. The diurnal cycle

The normal diurnal cycle and the relationships between wind and temperature structure are illustrated in Fig. 2.32 for a typical 500-m-deep Colorado valley on the western slope of the Rocky Mountains. The figure presents typical wind and temperature structure evolution over the valley center during the course of a day. The evolution takes place during the following four distinct regimes:

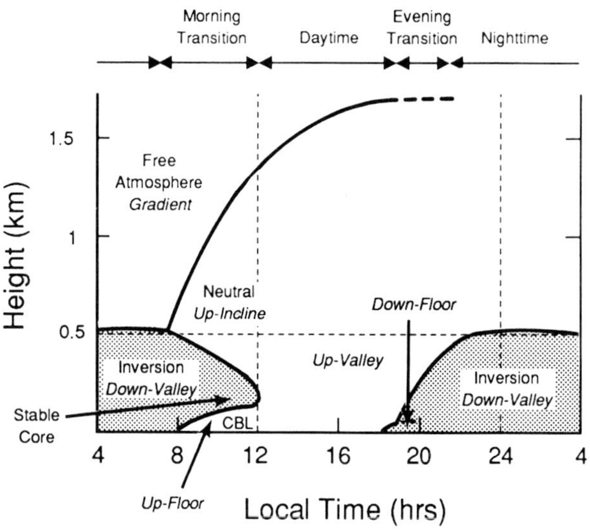

FIG. 2.32. Typical diurnal evolution of temperature (bold) and wind (italics) structure layers in a 500-m-deep valley.

• *Nighttime:* Observations taken over the valley center during the nighttime show that a temperature inversion is generally present, with depth approximately equal to the valley depth. Downslope flows occur on the sidewalls during this period. Down-valley winds prevail within the inversion, but larger-scale pressure gradients produce gradient winds that may blow from any direction in the slightly stable atmosphere above the valley. During this period the valley flows are decoupled from the flows above the valley by the presence of the surrounding ridges and by the stable temperature structure within the valley that suppresses vertical exchange of momentum.

• *Morning transition:* At or shortly after sunrise, convective boundary layers begin to grow above heated surfaces. Over the ridges, a CBL (designated the "neutral layer" in Fig. 2.32 to distinguish it from other convective boundary layers) is produced. Winds in this layer blow up the inclined western slope of the Rocky Mountains. During the morning transition period (and the daytime period, as well) the neutral layer grows into the slightly stable free atmosphere above, which represents the remnant of the previous day's mixing layer. The rate of growth of the neutral layer is high if sensible heat flux is strong since the stability of the ambient atmosphere into which it grows is weak. Over the valley floor and sidewalls, however, convective boundary layers grow slowly upward into the remnants of the strong nocturnal inversion, while the top of the inversion sinks deeper into the valley. CBL growth is slowed by the divergence of mass from the slope boundary layer or, equivalently, by subsidence of the ambient environment. The important role of the upslope flows in heating the valley atmosphere during this period has been previously mentioned. By midday the remnant of the nocturnal inversion within the valley (stable core) is finally destroyed by CBL growth and inversion sinking. The valley atmosphere becomes well mixed, and potential temperature becomes constant with height, although the valley atmosphere continues to warm with time.

• *Daytime:* After the remnants of the nocturnal inversion are destroyed, convection occurs above all of the valley surfaces—valley floor, sidewalls, and ridgetops. Thermally developed winds within the valley blow up the valley and up the slopes, but winds gradually change with height to the wind directions prevailing above the valley. During this period the valley atmosphere may become closely coupled with the atmosphere above the valley by enhanced vertical transfer of momentum. Winds in the valley can become turbulent and upper winds can be channeled into the valley, enhancing or countering the up-valley winds that develop locally. This daytime overpowering of the thermally developed winds is a common feature of meteorology in the western United States and in other areas where deep daytime convective layers may develop. Strong gusty daytime winds may be quite important for air pollution dispersion, prediction of forest fire behavior, etc. In the absence of such winds, the thermally developed daytime up-valley and nighttime down-valley wind systems have about the same strength. This is evidently because the topographic amplification factor becomes ineffective in creating strong along-valley temperature gradients after the valley inversion is destroyed since the incoming energy can no longer be confined within the valley.

• *Evening transition:* The evening transition period begins in late afternoon after cooling over shaded surfaces initiates downslope flows on the sidewalls and valley floor. Once the downslope flows begin, a valley temperature inversion containing down-valley winds grows through the valley depth. Compensatory rising motions over the valley center act to cool the valley atmosphere through its entire depth.

2.7. Other phenomena

2.7.1. *Influence of external winds*

Because most valley meteorology field experiments to date have focused on periods when upper winds were weak, little evidence has been accumulated on the effects of variable ambient winds on the development of thermally forced local wind systems. Some influence on the local wind system is nearly always present and is more or less manifest, depending on atmospheric stability and the strength and direction of the upper winds or, equivalently, on the above-valley pressure gradients. These pressure gradients may counteract or be superimposed on locally developed pressure gradients. Many mechanisms can be proposed to explain the influence of upper winds on local circulations, but the effects of topography may be difficult to quantify and present observational tools are not readily available for investigating the details of interactions at valley ridgetop levels.

Germany's 35-km-wide, shallow, flat-bottomed, Rhine Valley is quite susceptible to influence from prevailing flows above the valley, but the means by which this influence is felt is somewhat surprising. Under stable conditions, winds within the valley typically blow parallel to the north–south valley's axis, and the direction depends strongly on the direction of the geostrophic wind above the valley at 850 hPa, as shown in Fig. 2.33. Note in the figure that geostrophic winds from the west are most frequently associated with a surface wind from the south and that, when geostrophic winds blow from the south, small changes in direction can result in surface winds in the valley from either south or north. The explanation for this is related to the pressure pattern associated with the geostrophic wind and its superposition on the local pressure pattern in the valley. The geostrophic wind blows parallel to isobars, with higher pressure on the right. Thus,

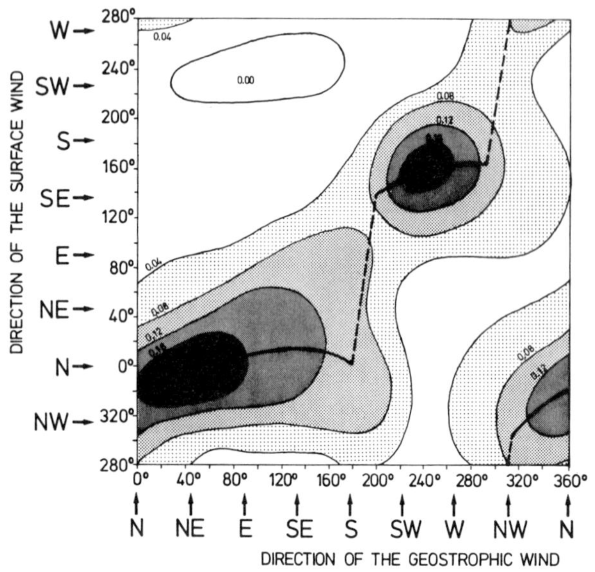

FIG. 2.33. Relative frequency of surface wind directions in the Rhine Valley at Mannheim, Federal Republic of Germany, as a function of geostrophic wind direction. The heavy solid line connects the most frequent wind directions. Data are for the period 1969–74, for stable stratifications only. (Wippermann and Gross 1981.)

westerly geostrophic winds above the valley are associated with a high pressure to the south and a low pressure to the north. The superposition of this large-scale pressure pattern on the valley forces surface winds in the valley to blow from the south—from high to low pressure along the valley channel. This extreme sensitivity of the valley wind system to the large-scale pressure pattern is thought to be a feature of shallow valleys where thermally developed wind systems are weak.

Even in deep valleys, however, such as the valleys in California and Colorado investigated by the ASCOT program, important aspects of the structure of valley circulations can depend on rather subtle differences in ambient atmospheric conditions (Barr and Orgill 1989). In the 1984 ASCOT experiment in Colorado's Brush Creek valley, down-valley drainage flow depth was closely related to the strength of the above-valley ambient flows (Fig. 2.34). When ambient winds were strong, drainage flows were shallow; when ambient winds were weak, drainage flows tended to fill the entire valley depth. Such observations suggest that the degree of influence of ambient flows on valley circulation patterns varies over a wide range, so that it may be inappropriate to speak of the influence as a dichotomous process in which ambient flows are considered either coupled or decoupled. In the 1980 ASCOT experiment in California's Anderson Creek valley, the influence of ambient flows varied from night to night. Strong ambient flows intruded into the valley on two successive nights, disrupting well-established drainage flows (Fig. 2.35).

2.7.2. Maloja winds

Occasionally, observations are made of deviant wind systems operating in apparent violation of valley wind theory. The best-known example is the Maloja wind in the upper Engadine Valley of Switzerland near the Maloja Pass. The wind system in the upper section of the valley blows up-valley during nighttime and down-valley during daytime, in apparent contradiction with theory. The accepted explanation is that the wind system from the steep Bergell Valley crosses the Maloja Pass and intrudes into the shallow upper Engadine Valley. The ridgelines of the Bergell extend beyond the pass into the upper Engadine, suggesting that the intrusion may be thermally driven rather than momentum driven. Calculations of topographic amplification factors may thus help to clarify the mechanism for this wind system (Fig. 2.36).

Two valleys draining Mount Rainier, Washington, produce winds of the Maloja type (Buettner and Thyer 1966). These are apparently the only U.S. valleys where such wind systems have been documented, but one could expect more Maloja-type winds to be discovered as further wind systems in the western United States are investigated.

2.7.3. Jets at valley exits

The down-valley wind often attains its maximum strength at the valley's exit, or a short distance beyond. Several factors may be responsible for this phenomenon. First, enhanced horizontal pressure gradients occur near the exit because pressure differences there are manifested over shorter length scales. Second, at the valley exit, down-valley accelerations occurring along the valley's length have had the maximum time to act. Finally, a conversion of potential energy to kinetic energy may occur at the valley's exit as the cold valley air sinks and fans out over the plain. These factors would appear to be maximized for deep valleys ending abruptly at the plain, rather than for valleys that become progressively shallower with down-valley distance.

Pamperin and Stilke (1985) report on the strong wind that issues from the mouth of Austria's Inn Valley. The development with time of the nocturnal wind speed profiles over Thalreit, Federal Republic of Germany (located on the southern German plain less than 10 km from the mouth of the Inn Valley) on the night of 25–26 March 1982 is shown in Fig. 2.37. The winds issued directly from the valley exit, with a clear jet profile beginning as early as 2015 UTC. The jet became stronger with time, attaining peak speeds of 14 m s^{-1} by sunrise, with the strongest winds at about 200 m AGL. On the same night, as part of the MERKUR experiment, wind observations were taken at Niederbreitenbach, 20 km up-valley from the valley exit. Along-valley wind components at the two stations are shown in time–height sections in Fig. 2.38.

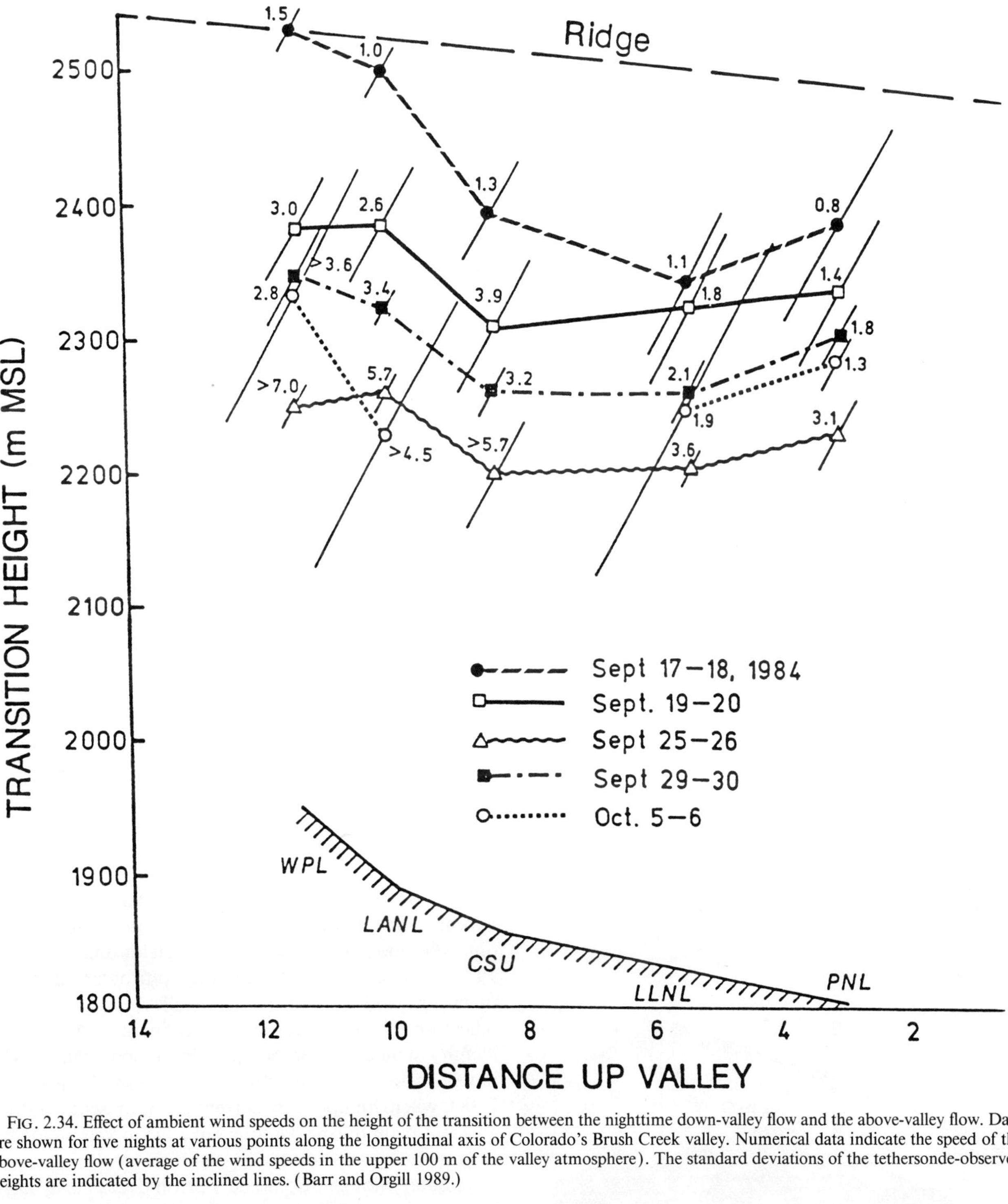

FIG. 2.34. Effect of ambient wind speeds on the height of the transition between the nighttime down-valley flow and the above-valley flow. Data are shown for five nights at various points along the longitudinal axis of Colorado's Brush Creek valley. Numerical data indicate the speed of the above-valley flow (average of the wind speeds in the upper 100 m of the valley atmosphere). The standard deviations of the tethersonde-observed heights are indicated by the inclined lines. (Barr and Orgill 1989.)

Down-valley wind speeds reached 7 m s^{-1} at a height of 625 m inside the valley just before sunrise, but attained speeds of 13 m s^{-1} at a height of 200 m beyond the valley exit. Similar jet profiles have been observed at the mouths of other Alpine valleys—at the south of Lago Maggiore, Italy, near Salzburg, Austria, in the Loisach Valley below Garmisch, Federal Republic of Germany, and in the Beierbach Valley, Austria. The author has seen similar nighttime winds at Paonia, Colorado, where the valley of the North Fork of the Gunnison River fans out into a wide plain.

Elevated jets are sometimes found in valleys where a constriction in the lower altitudes of the valley causes pooling of air in the lowest layers, but a significant wid-

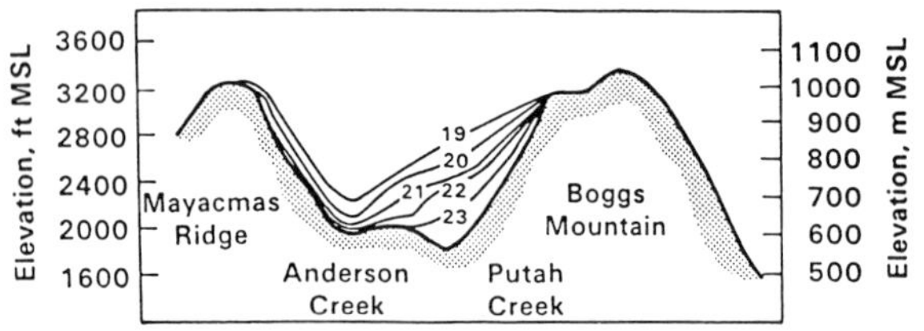

FIG. 2.35. Incursion of northeasterly ambient winds into California's Anderson Creek and Putah Creek valleys on 22 September 1980. The labeled lines show the boundaries between the drainage and external flows at the times indicated. The topographic cross-section runs from northeast (right) to southwest (left). (Barr and Orgill 1989.)

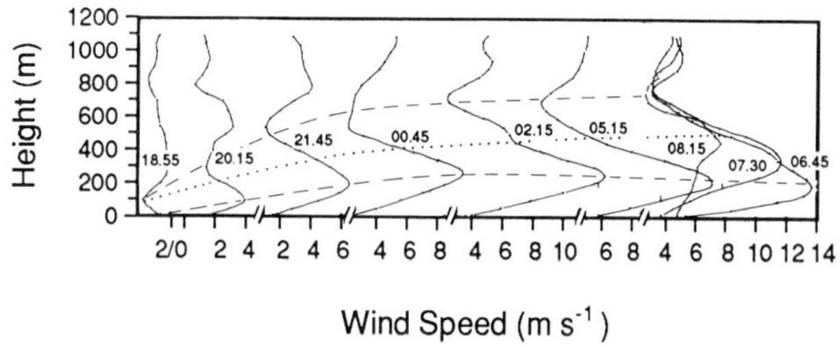

FIG. 2.37. Nocturnal development of jet profiles of down-valley winds at the exit of the Inn Valley on 25–26 March 1982. Wind speeds attain nearly 14 m s^{-1} by sunrise. (Pamperin and Stilke 1985.)

ening of the valley occurs at upper levels, as in Colorado's Gore Valley, near Vail (Fig. 2.39).

2.7.4. *Antiwind systems*

The different streams of the valley wind system (up-valley and down-valley currents) are thought to represent the lower branch of a circulation that exists between mountain and plain. Thus, for example, the up-valley current during daytime is compensated by a flow from mountain to plain above the valley ridgetops. These return circulations are rarely observed directly, except in a statistical sense (Ekhart 1944), because the compensation current is not confined to a channel like the lower branch of the circulation. Thus it is much weaker than the lower branch and is weaker than typical prevailing synoptic winds. In some cases, however, and especially in a series of experiments conducted in valleys extending radially outward from the volcanic cone of Mount Rainier, Washington, the return circulations have been observed in the upper altitudes of the valleys below ridgetop level (Buettner and Thyer 1966). In the years since these observations were taken there have been few reports of other antiwind systems.

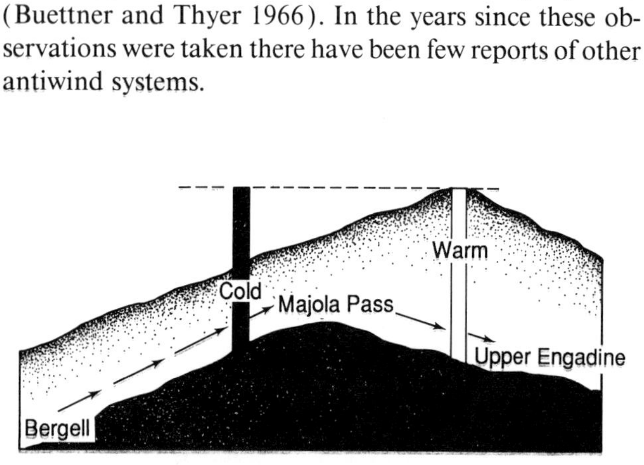

FIG. 2.36. Daytime winds in the Upper Engadine Valley blow down-valley, in apparent contradiction to theory. This wind system feature is an example of a wind system from one valley (the Bergell) streaming over a pass (Maloja Pass) into the upper reaches of another valley (Upper Engadine). The extension of the Bergell ridgeline into the Upper Engadine may produce along-valley topographic amplification factor gradients that lead to horizontal temperature gradients and the unusual winds. (Adapted from Hawkes 1947.)

2.7.5. *Tributary flows*

No effort has yet been reported in the literature to evaluate the details of tributary/main valley atmospheric interactions at valley junctions. Temperature measurements at the Ziller Valley/Inn Valley juncture show unusual warming in the main valley during daytime (Brehm 1986), which Freytag (1987) has suggested represents enhanced sinking motions compensating for the up-tributary flows (Fig. 2.40). The details of interactions at such junctures would certainly depend on the relative topography of the two valleys, the local topography at their juncture, and the detailed wind and temperature structure of the individual valley flows. It would thus be difficult to gain general knowledge from the investigation of individual cases. Nonetheless, the bulk characteristics of the individual flows have important applications for air pollution dispersion. For example, the inflow from a tributary represents, in many cases, the import of clean air into a polluted main valley atmosphere.

During nighttime, drainage currents into the broad, shallow Rhine Valley from small tributaries (Fig. 2.41) appear to maintain their identity for some distance out onto the floor of the main valley (Heldt and Höschele 1989). There is no corresponding up-tributary flow apparent in wind roses in the main valley during daytime, reflecting the influence of stability in maintaining the identity of nighttime flows. This effect is apparent, as well, in wind observations near the exit of the Inn Valley (Fig. 2.38), where an up-valley current is not present over the adjacent plain during daytime but a down-valley current is strongly in evidence during nighttime.

Tributary flows in Colorado's Brush Creek valley have been studied recently by a number of investigators (e.g., Coulter et al. 1989; Porch et al. 1989a,b). Data collected in the valley on 6 October 1984 by Neff and King (1987) suggested that shallow drainage flows in the upper drainage areas of the valley are susceptible to disruption by strong above-valley winds. Drainage flow in a tributary canyon decreased suddenly when winds above the valley accelerated to 7 m s^{-1} at 0400 MST. This was followed a half-

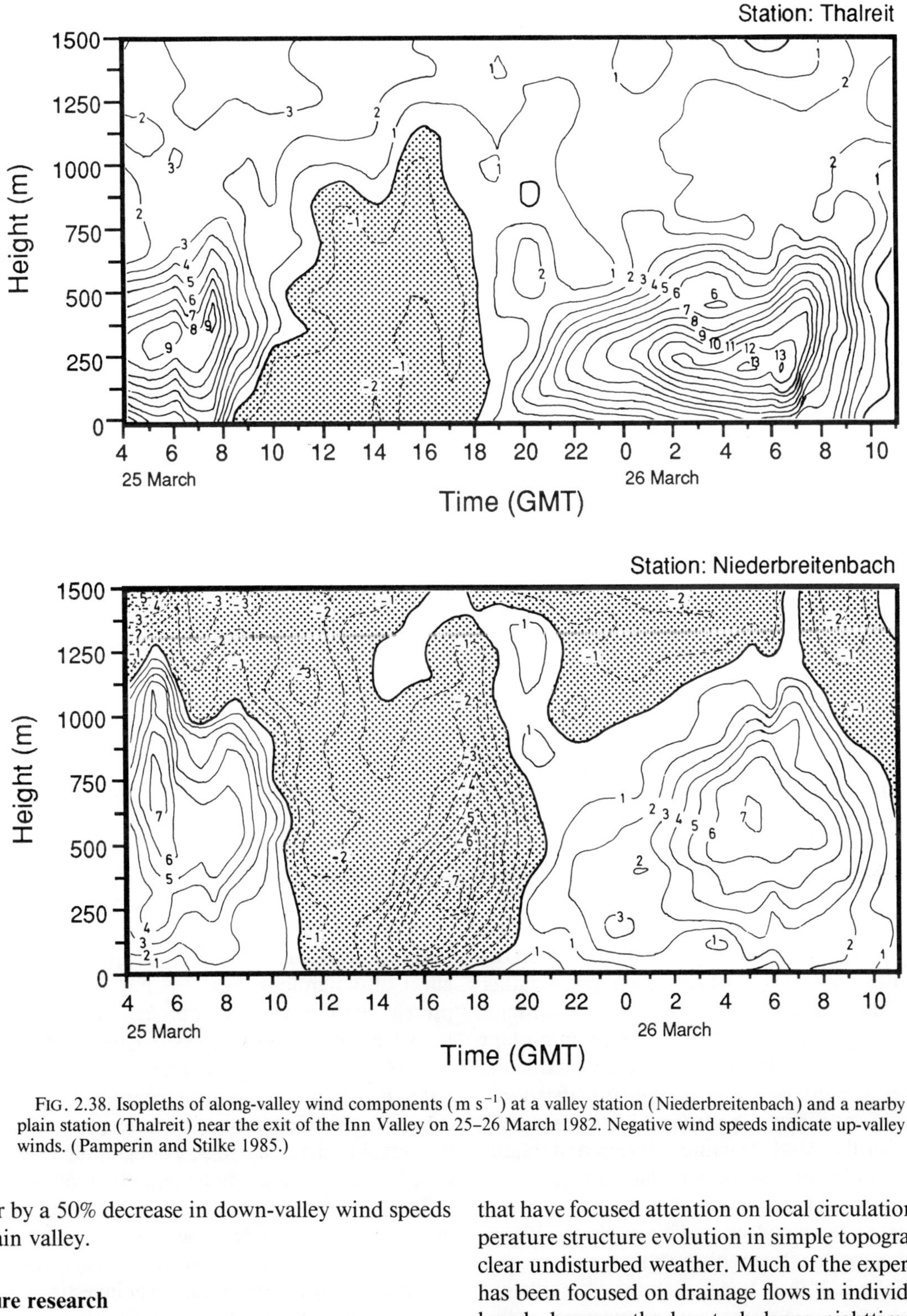

FIG. 2.38. Isopleths of along-valley wind components (m s^{-1}) at a valley station (Niederbreitenbach) and a nearby plain station (Thalreit) near the exit of the Inn Valley on 25–26 March 1982. Negative wind speeds indicate up-valley winds. (Pamperin and Stilke 1985.)

hour later by a 50% decrease in down-valley wind speeds in the main valley.

2.8. Future research

Complex terrain meteorology experiments conducted in the last decade have improved our understanding of physical processes leading to local thermally developed wind systems. The experiments have benefited from the availability of new research tools, including remote sensors, and have largely been conducted in short campaigns that have focused attention on local circulations and temperature structure evolution in simple topography during clear undisturbed weather. Much of the experimentation has been focused on drainage flows in individual valleys, largely because the low-turbulence nighttime flows can be investigated with relatively inexpensive tethered balloon measurement systems. A few experiments have investigated downslope flows on the slopes of isolated mountains, using instrumented towers. These studies have given way to investigations of nocturnal down-valley flows in simple valleys. The logistical difficulty of instrumenting

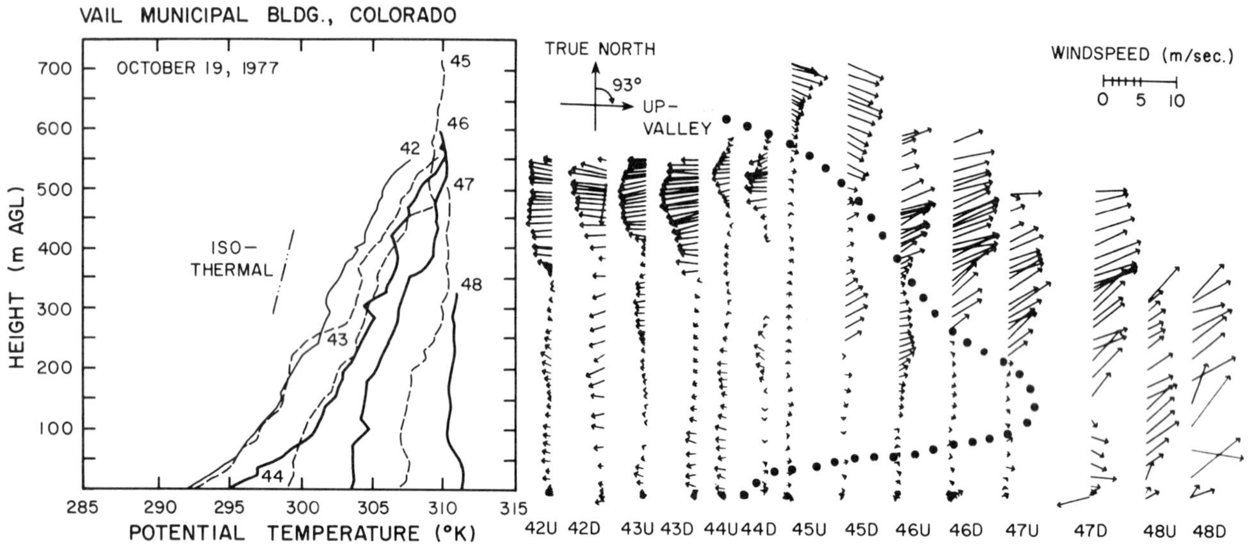

SOUNDING	TIMES
42U	6.91–7.34
42D	7.34–7.60
43U	7.75–8.22
43D	8.22–8.58
44U	8.62–8.97
44D	8.97–9.31
45U	9.43–9.81
45D	9.81–10.09
46U	10.23–10.64
46D	10.64–10.99
47U	11.00–11.23
47D	11.23–11.44
48U	11.57–11.71
48D	11.71–11.80

FIG. 2.39. Nocturnal down-valley jet above a 400-m-deep stagnant pool in the Gore Valley at Vail, Colorado, near sunrise on 19 October 1977. The jet winds in the 700-m-deep valley occur at an altitude corresponding to a sudden widening of the valley above a downstream constriction. Profile sounding times are given in decimal hours.

the slopes of deep valleys has resulted in too little emphasis on interactions between the slope and valley wind circulations. Important gains in the understanding of the morning transition period have been made in the last decade through a combined observation/modeling approach, but the evening transition period and the daytime period have received much less attention.

Most experiments have been conducted in summer and fall, and few long-term climatological data have been developed to ascertain the influence of synoptic weather events on local thermally developed circulations. Major winter experiments are unknown, although important gains in understanding may come from such experiments because of certain surface energy budget simplifications.

The focus of most past experiments has been on the development of the wind structure. Increased emphasis now seems desirable on several boundary conditions affecting wind system development. The roles of topography, the surface energy budget, and interactions with external circulations are topics that should now receive increased emphasis. If these topics are pursued, a different experimental design may become more appropriate. Experiments, rather than focusing intensive efforts on a single valley, may need to become more *extensive,* focusing on differences between multiple valleys. In this way, for example, the influence of differing boundary conditions can be investigated and the general applicability of detailed knowledge gained in a single valley can be better determined.

The role of topography in the development of thermally developed circulations has recently been investigated using the topographic amplification factor (TAF) concept. The great practical value of this purely geometric or topographic method of estimating the strengths of wind circulations in valleys is a strong incentive for evaluating, refining, and extending the TAF approach, although the effect of dynamic processes is not accounted for in the present thermodynamic approach. Testing of our basic knowledge of physical processes causing complex terrain circulations, as well as the TAF concept, would be facilitated by extending experiments into valleys, basins, and tributaries of different size, shape, length, width, and other topographic characteristics. Topographic effects could be tested by a combined modeling/observation approach.

The topographic amplification factor can be viewed as one of a number of possible integral geometrical or topographical measures (in this case, area to volume ratios) that are relevant to meteorological investigations in complex terrain. Other such measures should prove useful in improving our understanding of complex terrain meteorology. Petkovsek (1978b, 1980), for example, has defined several topographical parameters related to the buildup and outflow of cold air pools in basins, as well as an integral measure of the susceptibility of a basin to

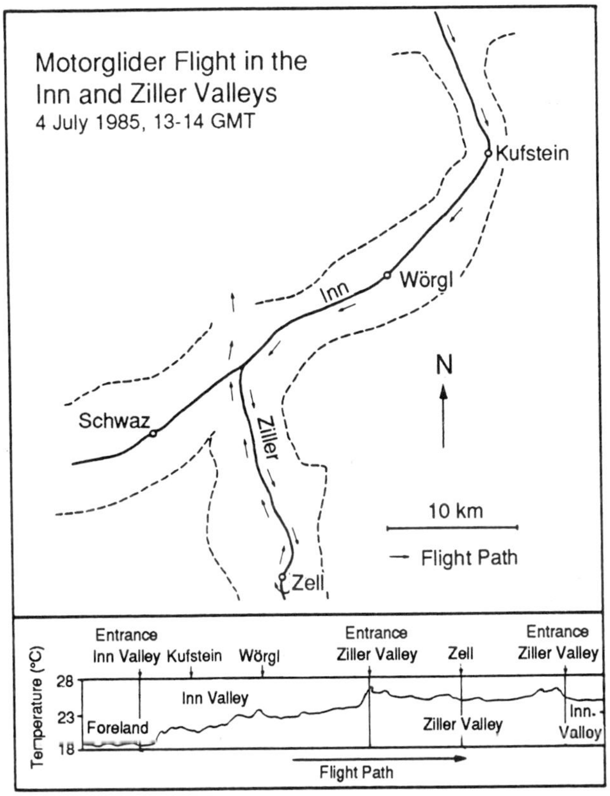

FIG. 2.40. Motorglider temperature measurements over the foreland, in the main Inn Valley, and in a major tributary valley (Ziller Valley) on 4 July 1985. Temperature increases with distance up the Inn Valley and a warm zone is observed at the exit of the Ziller Valley. The aircraft flew approximately 750 m above the ground along the flight path shown. (Brehm 1986.)

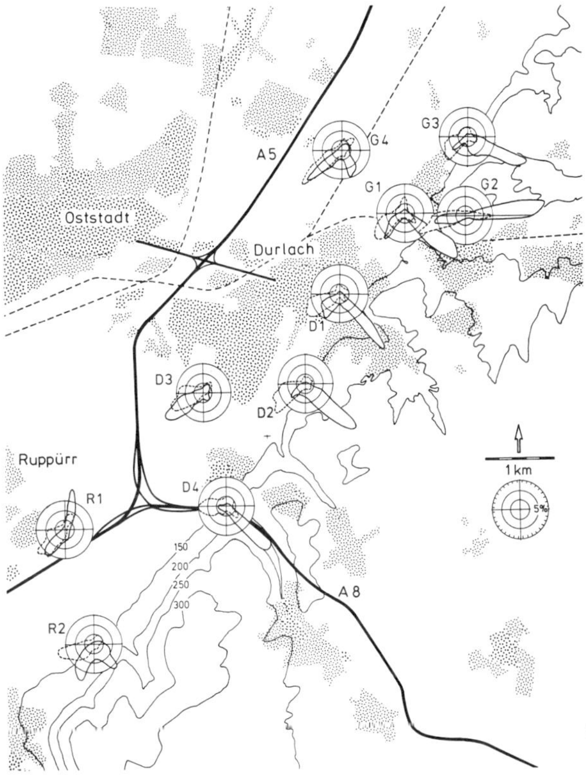

FIG. 2.41. Wind directions and speeds at the east edge of the wide, shallow, Rhine Valley near Karlsruhe, Federal Republic of Germany at 0000–0300 UTC 30 May 1979. During this moderately cloudy period, regional flows were from the southwest. Flows issuing from the tributaries maintain their identity for some distance from the tributary exits. Daytime flows (not shown) do not show a corresponding effect on the wind roses on the main valley floor. (Heldt and Höschele 1989.)

influence by upper winds. Neff and King (1987, 1988, 1989) have used other topographic parameters, such as the cross sectional area of constrictions, to investigate the effects of basin topography on local wind systems in the Geysers region in California. In another approach, Pielke and Kennedy (1980) computed the characteristic spectrum of terrain scales on a cross section of the Blue Ridge Mountains to determine the appropriate grid size for an atmospheric numerical model. Nevertheless, only a few efforts have so far been made to systematically classify terrain in ways that are relevant for meteorological investigations, although terrain has been classified for military, geomorphic, geologic, and other purposes (e.g., see Carr and Van Lopik 1962; Orgill 1981). The condensation of detailed topographical information into a few simple integral parameters that are relevant to the physics of the development of local circulations is an interesting new approach that will be facilitated by the increased availability of digital topographic databases and graphical presentation capabilities (McEwen et al. 1983; Elassal and Caruso 1983).

The role of the valley radiation and surface energy budgets in driving topographic circulations should be investigated further. Techniques and equipment should be developed so that such measurements can become a regular feature of complex terrain field experiments. Techniques are needed to make such measurements over slopes, to choose representative locations for such measurements, and to estimate spatial variations over complex terrain areas from small numbers of point measurements or from remotely sensed data.

External winds may produce more or less strong influences on the development of valley circulations. A series of carefully designed experiments could be used to determine the mechanisms by which these influences are felt, and climatological studies would prove helpful in quantifying their effects on valley meteorology over time.

Most valley meteorology investigations to date have

focused on mean flows. Turbulence and wave motions, while frequently present in the valley atmosphere, have received too little attention, especially in light of their effect on air pollution dispersion. These topics are discussed in more detail by Carruthers and Hunt, Chapter 5. Finally, laboratory modeling appears to provide a means for extending investigations of thermally developed circulations in complex terrain. This topic is discussed further by Meroney, Chapter 7.

Acknowledgments. Special thanks are due to Dr. Thomas B. McKee of Colorado State University for interesting and enlightening technical discussions during his sabbatical leave at Pacific Northwest Laboratory in early 1988. Harlan P. Foote, L. Guy McWethy, and John M. Hubbe participated in digital terrain modeling work leading to Fig. 2.14. Drs. J. Christopher Doran, David C. Bader, Robert M. Banta, Ignaz Vergeiner, and Joseph Egger are thanked for their reviews of the manuscript. The North Atlantic Treaty Organization is acknowledged for providing a one-year fellowship that allowed the author to review recent European work during a visit to the Institute for Meteorology and Geophysics at the University of Innsbruck, Austria, in 1983, and the Meteorological Institute at the University of Munich, Federal Republic of Germany, in 1985.

This work was supported by the U.S. Department of Energy under Contract DE-AC06-76RLO 1830 at the Pacific Northwest Laboratory. The Pacific Northwest Laboratory is operated by Battelle Memorial Institute for the U.S. Department of Energy.

CHAPTER 3

Thermally Forced Flows: Theory

JOSEPH EGGER

Meteorologisches Institut der Universität München, Munich, Federal Republic of Germany

ABSTRACT

Conceptual models of thermally induced valley winds are described. First, an account is given of Gill's work on two-dimensional flow in a cavity with differentially heated sidewalls. Second, we focus on models of flow along heated (cooled) slopes. Third, we discuss in detail the work by Vergeiner and Brehm on cross-valley circulations. Finally, highly simplified models of three-dimensional valley flows are discussed and a novel approach to this problem is presented.

3.1. Introduction

As in many other fields of dynamic meteorology, theoretical work on thermally induced flows in complex terrain proceeded mainly along two avenues: explicit numerical simulation and conceptual modeling. Conceptual models are designed to help us understand observed flow phenomena. The mathematical structure of a conceptual model should be relatively simple so that a good insight into the model's reactions and responses becomes feasible. This simplicity is achieved by introducing assumptions and parameterizations beyond what is thought to be acceptable in simulation models intended to simulate or predict observed flows in as much detail as possible. A clear distinction between simulation models and conceptual models is sometimes difficult—and perhaps not even necessary—since both types of models should complement each other. Results from conceptual models should help us to better understand and also to improve simulation models, while simulation models can be used to calibrate conceptual models (an example will be given in section 3.4).

Conceptual thinking about valley flows has a considerable history. In particular, the "Austrian school," represented by Wagner (1932c), Ekhart (1939), Jelinek (1937) and others, arrived relatively early at a qualitative understanding of valley winds but failed to cast these ideas in a quantitative form. Later, conceptual models played a central role in research on thermally forced valley wind systems. The complexity of the mountainous terrain made it difficult to use simulation models and only in recent years have explicit simulations of valley flows been attempted (e.g., Bader et al. 1987; Yamada and Bunker 1988). The main emphasis of this chapter will be on conceptual models of thermally forced valley winds. Results from simulation models will be described when necessary.

We begin by presenting a conceptual model of two-dimensional flow in a cavity with differentially heated sidewalls (Gill 1966). Gill's work provides a good starting point since much of the subsequent work relies on the concepts and techniques he used. Slope wind models are discussed in section 3.3. A more advanced two-dimensional valley wind model that relies on what has been learned from these slope wind models, but that is similar to Gill's model in many other respects, is presented in section 3.4. Examples of highly simplified solutions to the problem of three-dimensional valley flows are provided in section 3.6. The chapter ends with the presentation of a novel conceptual model of three-dimensional valley flow that relies on the modeling efforts described in sections 3.3–3.6. A list of symbols is provided in the Appendix.

3.2. The problem

Consider the valley presented in Fig. 3.1. The valley axis coincides with the x-axis. The valley's length is L and its width $B(x, z)$ may vary with distance x to the mouth and height z. For convenience we specify a constant crest height $z = H$ and a constant width $B(x, H) = B_0$ on top. Assume now a heating or cooling at or near the slopes and at the valley bottom; the heating induces flows in the valley. In principle, we expect along-valley (down-valley) and upslope (downslope) flow if there is heating (cooling). (Related observations are described in Whiteman, Chapter 2.) Despite the simplicity of this pattern the interplay of slope winds and flow along the valley's axis is only partially understood. Moreover, there are many factors; e.g., the state of the atmosphere outside the valley, the valley geometry, or the daily march of the sun that considerably complicate the problem.

For a description of valley winds we rely on the Boussinesq equations of hydrostatic motion

$$\frac{d\mathbf{v}}{dt} = -\frac{1}{\rho_0}\nabla p + \mathbf{F}, \tag{3.1}$$

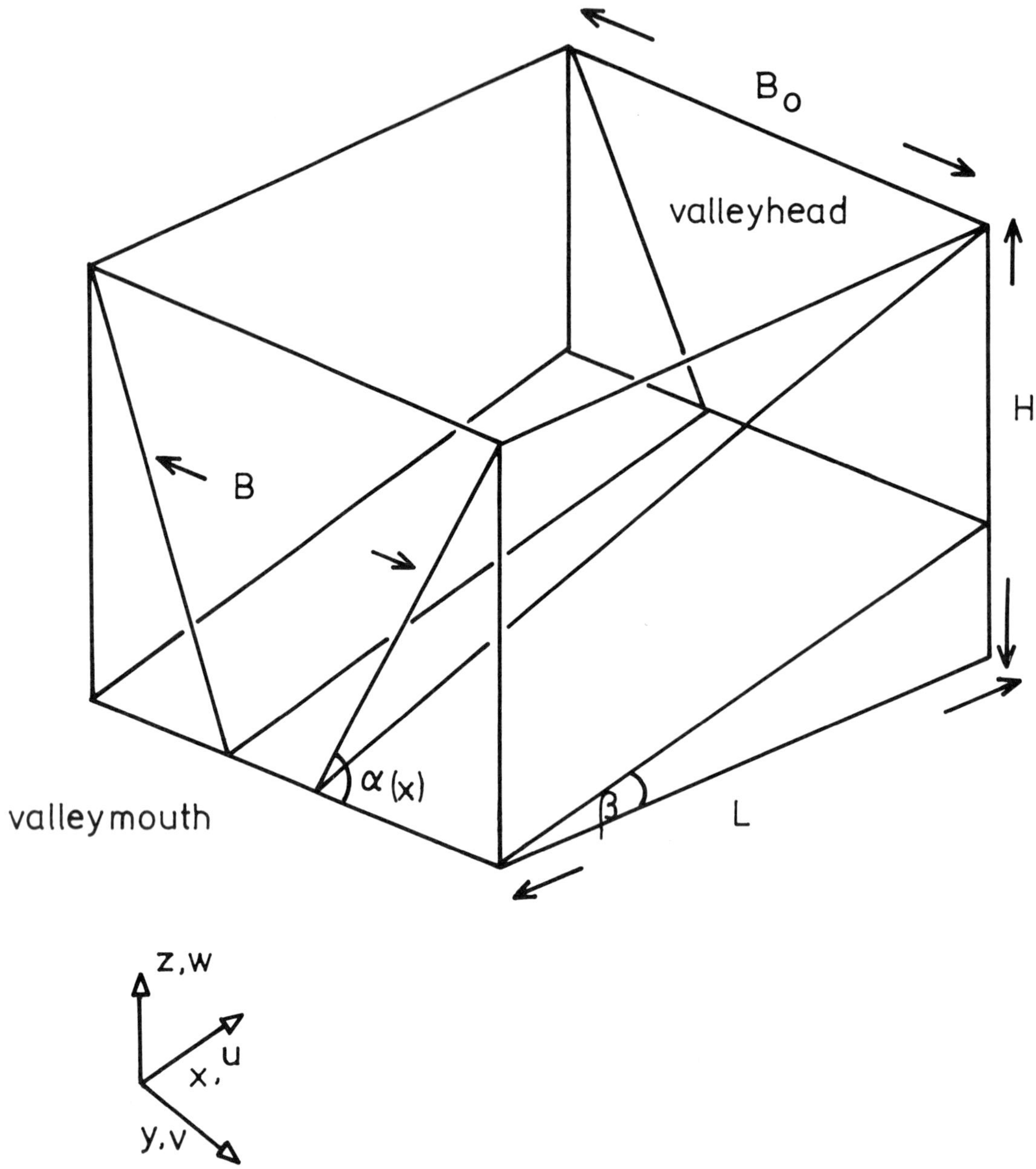

FIG. 3.1. Geometry of the valley and flow variables.

$$\frac{d\theta}{dt} = A, \qquad (3.2)$$

$$\frac{\partial p}{\partial z} = -g\rho_0(\theta - \theta_0)/\theta_0, \qquad (3.3)$$

$$\nabla \cdot \mathbf{v} + \frac{\partial w}{\partial z} = 0, \qquad (3.4)$$

where $\mathbf{v} = (u, v)$ is horizontal velocity, $\rho(\theta_0)$ a constant reference density (potential temperature), A (K s^{-1}) is heating, **F** represents frictional force, w is vertical velocity, and p is pressure. No-slip conditions are posed at the ground. The problem is, then, to find solutions to (3.1)–(3.4) for a heating specified in the valley depicted in Fig. 3.1. In a simulation model the approach is to discretize (3.1)–(3.4) in the valley domain and to numerically resolve the flow in as much detail as possible. Conceptual methods are described in section 3.3.

3.3. Circulation in a cavity with differentially heated sidewalls

Let us break down the valley flow problem posed in the previous section into simpler parts that may be amenable to analytic treatment. One obvious simplification is to reduce the dimensions of the problem by looking at cross sections of the valley under the assumption that variations along the axis can be neglected. A further simplification is to replace the slopes by vertical walls and to close the flow domain on top with a lid; i.e., to consider two-dimensional convective motion in a rectangular cavity. To simplify matters further we assume that the vertical walls are maintained at different temperature but exclude heat flow through the horizontal boundaries. An approximate steady-state solution to this problem has been given by Gill (1966). Although Gill did not consider valley flows when dealing with this situation, the techniques he employed will prove useful in the following analysis. Moreover, the case studied by Gill corresponds to the situation found in a valley early in the morning when there is no heating of the valley bottom, but where one slope receives sunlight while the opposite slope is in the shade.

The geometry and the detailed boundary conditions of the problem are summarized in Fig. 3.2. We adopt the convention that the left-hand wall is hotter than the right-hand wall so that $\Delta\theta = \theta_l - \theta_r > 0$, where θ_l and θ_r are the imposed potential temperatures of the walls. A streamfunction ψ is introduced with $w = \partial\psi/\partial y$, and $v = -\partial\psi/\partial z$. Gill subdivided the flow domain into boundary layers at the vertical walls and a core region in the interior of the valley atmosphere. He assumed that the potential temperature θ_v and the streamfunction ψ_v in the interior depend on z only. This assumption excludes vertical motion outside the boundary layers.

It is convenient to introduce a deviation $\delta\theta(y, z) = \theta - \theta_v$ of potential temperature in the boundary layer. The approximate boundary-layer equations used by Gill are

$$v_s(z)\frac{\partial}{\partial y}\delta\theta + w\frac{d\theta_s}{dz} = K_H\frac{\partial^2}{\partial y^2}\delta\theta, \qquad (3.5)$$

$$0 = K_M\frac{\partial^2}{\partial y^2}w + g\frac{\delta\theta}{\theta_0}, \qquad (3.6)$$

where $v_s(z)$ is a mean cross-valley velocity and $d\theta_s/dz$ stands for a mean stability in the boundary layer. Constant eddy diffusivities for momentum and heat are denoted,

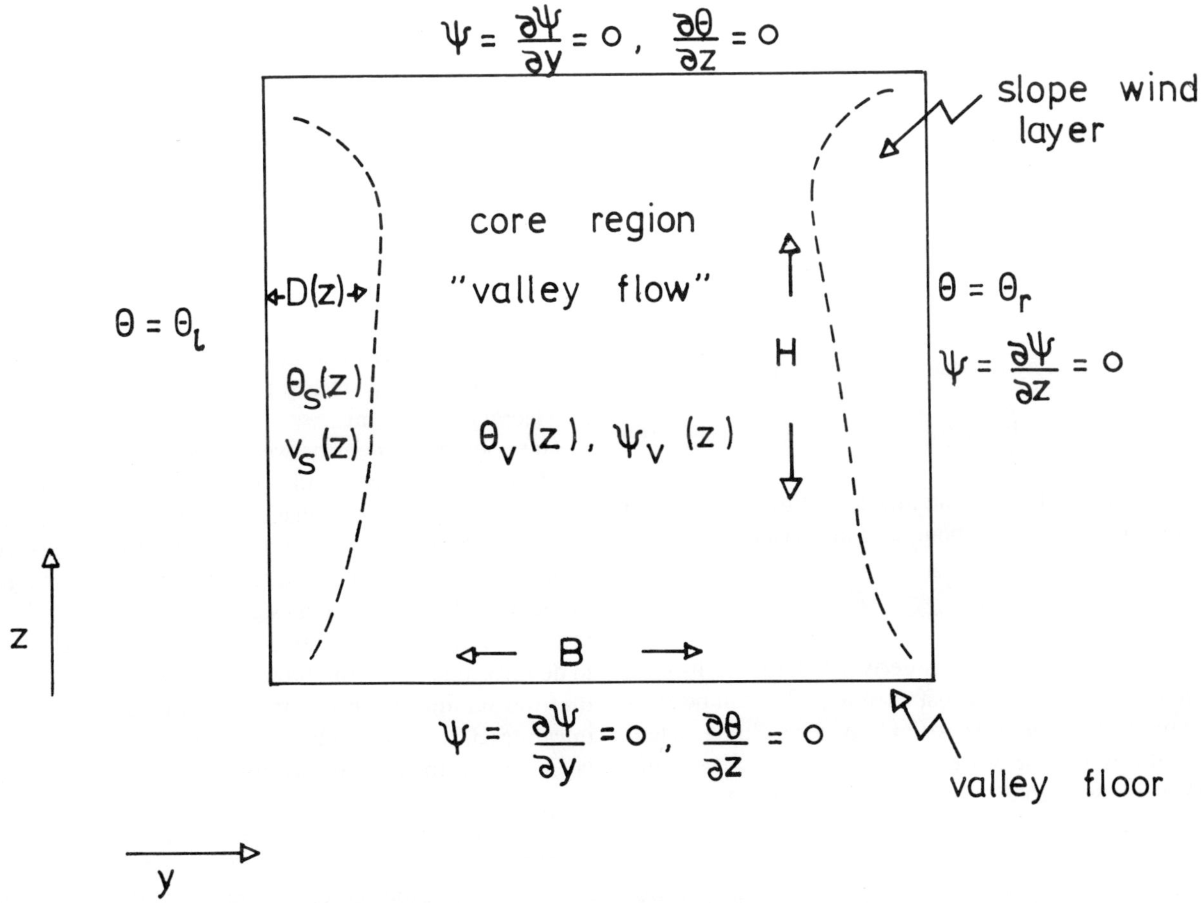

FIG. 3.2. Cavity with differentially heated sidewalls as considered by Gill (1966). Notation and boundary conditions. The dashed line separates the boundary layer at the walls from the core region. See text for further details.

respectively, by K_M and K_H. Gill considered molecular viscosity and conductivity when solving this problem. For our purposes it is preferable to think in terms of turbulent boundary layers. As can be seen from (3.5), horizontal advection of excess temperature is included in the boundary-layer equation; however, advective transport of momentum is excluded. Note that Gill did not include the boundary layers on top and at the bottom.

Let Ψ scale the streamfunction in the boundary. The terms on the left-hand side of (3.5) are of order $\Psi\Delta\theta/DH$, where D is the characteristic thickness of the boundary layer and we anticipate that $d\theta_s/dz \sim \Delta\theta/H$ (see Fig. 3.3). The diffusive term on the right is $\sim K_H\Delta\theta/D^2$. Correspondingly, the factor $\Delta\theta$ drops out and balance requires

$$\Psi \sim K_H H/D. \qquad (3.7)$$

From (3.6) and (3.7), we obtain

$$D^4 \sim K_H K_M H \frac{\theta_0}{g\Delta\theta}. \qquad (3.8)$$

As one would expect, the boundary-layer thickness increases with increasing diffusivities. Somewhat surprisingly, however, D decreases with increasing $\Delta\theta$. According to (3.6) an increase of $\Delta\theta$ can be balanced by a decrease of D, while Ψ is not directly affected by the choice of $\Delta\theta$.

Since v_s and θ_s do not depend on y one obtains solutions to (3.5) and (3.6) of the form

$$\begin{pmatrix} w \\ \delta\theta \end{pmatrix} = \sum_{n=1}^{4} \begin{pmatrix} w_n \\ \delta\theta_n \end{pmatrix} \exp(-\lambda_n y), \qquad (3.9)$$

which must satisfy the boundary conditions $\delta\theta = \theta_l$ for $y = 0$ (left wall), $\delta\theta \rightarrow 0$ for large y, $w = 0$ at $y = 0$, and for large y. The coefficients w_n, $\delta\theta_n$, λ_n in (3.9) depend on v_s and $d\theta_s/dz$. The boundary-layer flow and the interior flow are matched by imposing continuity through

$$\int_0^\infty w dy = \psi_v(z). \qquad (3.10)$$

Further integration of the first law [Eq. (2.6) of Gill 1966] over the left-hand boundary layer yields

$$\frac{d}{dz}\int_0^\infty w\delta\theta dy + \psi_v \frac{d\theta_v}{dz} = -K_n \frac{\partial}{\partial y}\delta\theta|_{y=0}. \qquad (3.11)$$

A corresponding equation is derived for the right-hand boundary layer. Since the first term in (3.11) can be evaluated on the basis of (3.9), we end up with a closed system of equations for the variables $\psi_v(z)$, $\theta_v(z)$, $\theta_s(z)$, and $v_s(z)$. (Gill was able to solve this nonlinear problem.) Results are presented in Figs. 3.3 and 3.4. The potential temperature $\theta_v(z)$ in the interior increases from θ_r at the bottom to $\theta_v = \theta_l$ at the top (Fig. 3.3). The valley atmosphere is stably stratified, characterized by a Brunt–Väisälä frequency $N_v{}^2 \sim g\Delta\theta/\theta_0 H$. The thickness of the boundary

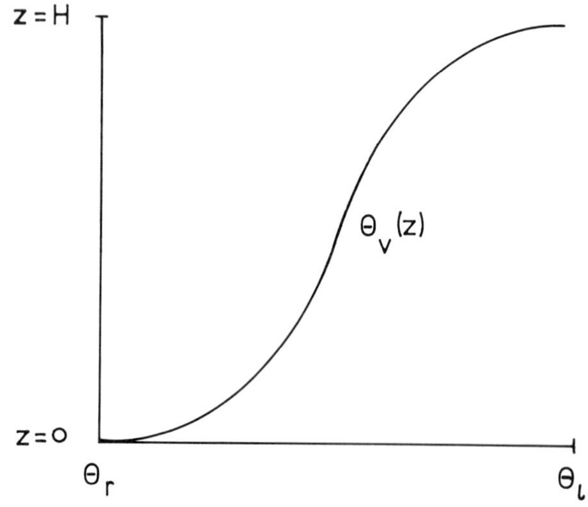

FIG. 3.3. Potential temperature θ_v in the core as derived by Gill (1966). Adapted from Fig. 3 of Gill (1966).

layer in the warm layer increases somewhat with z (Fig. 3.4). The vertical motion in the boundary layer has a maximum at a distance of approximately $2D$ from the wall. Note the weak downward motion for $y > 5D$. There is strong inflow toward the heated boundary layer near the lower boundary and strong outflow on top. In the interior the streamlines are strictly horizontal. The temperature in the hot boundary layer falls off from the maximum at the wall and negative values of the excess temperature occur for $y > 2.5D$. The stratification in the boundary layer is almost neutral within a distance $y < 2D$ from the wall. Altogether we have a direct circulation with strong flow along the boundaries and a stably stratified core with almost no motion. Elder (1965) conducted laboratory measurements of the flow in a cavity with differentially heated sidewalls. Gill (1966) compared his solutions to Elder's observations and obtained quite satisfactory agreement.

The good agreement between observations and Gill's solution suggests that similar techniques should be used when tackling a more realistic version of the valley flow problem. There is, however, a technical point that deserves attention. Gill (1966) assumes in (3.5) that the stability in the boundary layer differs significantly from that in the interior so that a new variable $d\theta_s/dz$ must be introduced to describe this deviation; however, in (3.11) it is assumed that the stability is $d\theta_v/dz$ in the boundary layer. If we integrate (3.5) horizontally over the depth of the boundary layer and subtract the result from (3.11) we obtain

$$\frac{d}{dz}\int_0^\infty w\delta\theta dy + \psi_v\left(\frac{d\theta_v}{dz} - \frac{d\theta_s}{dz}\right) - v_s\frac{\Delta\theta}{2} = 0. \qquad (3.12)$$

If we had $\theta_s = \theta_v$ the vertical divergence of the horizontally integrated excess heat flow in the boundary would have to be balanced by a fictitious transport of excess temper-

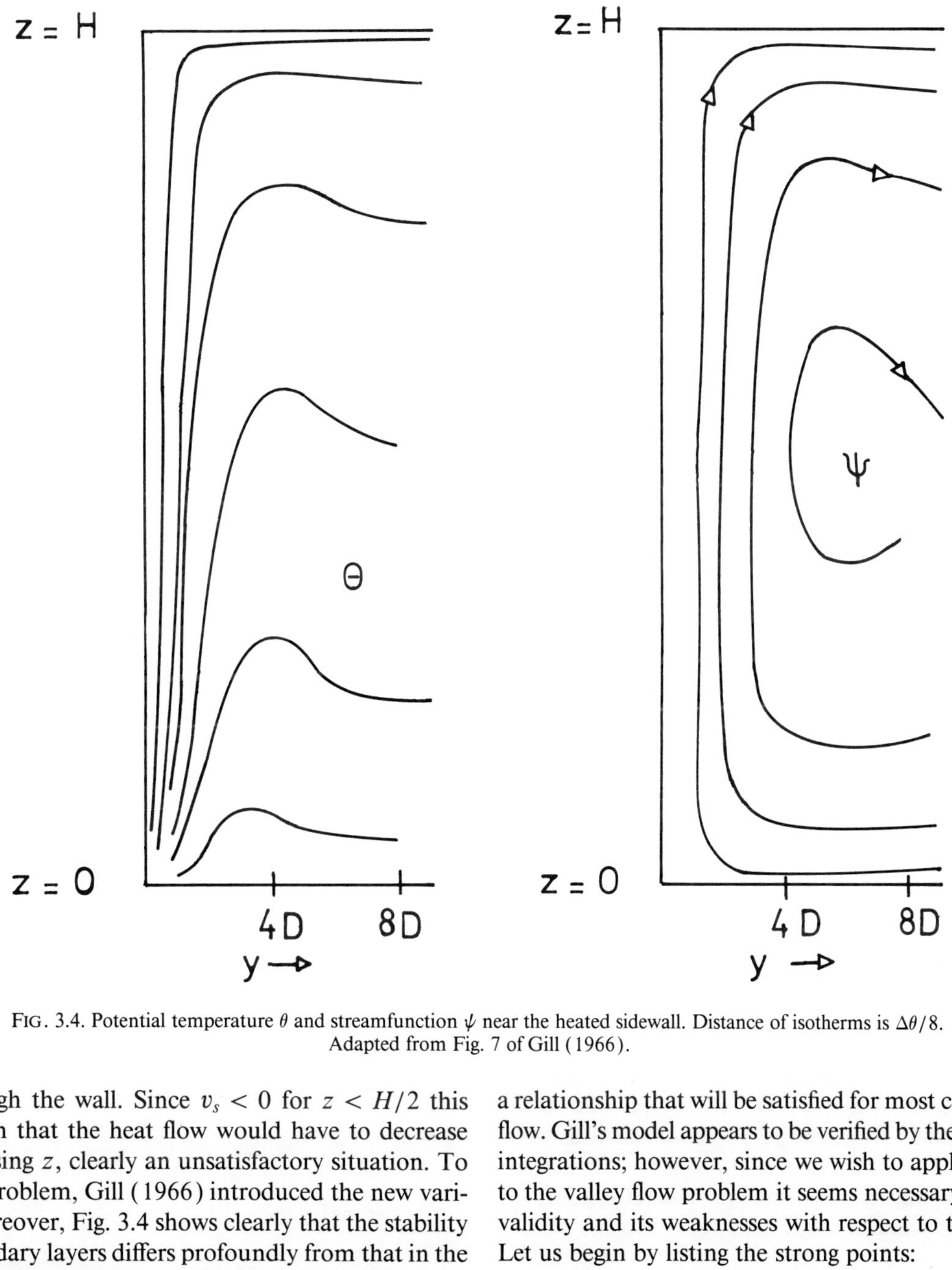

FIG. 3.4. Potential temperature θ and streamfunction ψ near the heated sidewall. Distance of isotherms is $\Delta\theta/8$. Adapted from Fig. 7 of Gill (1966).

ature through the wall. Since $v_s < 0$ for $z < H/2$ this would mean that the heat flow would have to decrease with increasing z, clearly an unsatisfactory situation. To avoid this problem, Gill (1966) introduced the new variable θ_S. Moreover, Fig. 3.4 shows clearly that the stability in the boundary layers differs profoundly from that in the core regions. Nevertheless, one may need to look for alternatives when dealing with the valley flow problem. Data do not suggest a marked difference in the stability of the slope wind layer versus that in the interior. Correspondingly, one would not prescribe constant temperatures at the slope of a valley wind model.

Patterson and Imberger (1980) presented numerical solutions to the cavity flow problem for an aspect ratio $H/B = 1$ in addition to an extensive scale analysis. They found that flows such as those described by Gill arise for

$$g\frac{\Delta\theta}{\theta_0} > \frac{K_M^{\ 3}}{H^3K_H}, \quad (3.13)$$

a relationship that will be satisfied for most cases of valley flow. Gill's model appears to be verified by these numerical integrations; however, since we wish to apply this model to the valley flow problem it seems necessary to assess its validity and its weaknesses with respect to this problem. Let us begin by listing the strong points:

(i) A solution is offered for the complete flow problem. In particular the stability of the core region is not prescribed but is part of the solution.

(ii) The basic characteristics of the boundary-layer flow agree with what is observed in slope wind layers (see Fig. 2.20 of Whiteman, Chapter 2).

The situation found in valleys, on the other hand, differs in many respects from that considered by Gill (1966):

1) In most situations of interest there is a mean heating or cooling of the valley. In particular heat flow at the lower boundary must be admitted. All this implies that there must be vertical motion in the core region.

2) Real valleys do have sloping sidewalls where the surface temperature varies along the slope.

3) In most situations the upper boundary of the valley atmosphere cannot be treated as a rigid lid.

4) Valley flow is three-dimensional.

5) A K-parameterization with K constant, as assumed in analogy to the laws of molecular conduction, may provide too crude a treatment of turbulent processes.

3.4. Sloping boundaries

The profiles of wind and temperature in the sidewall boundary layers as obtained by Gill (1966) look reasonably realistic when compared to ensembles of observed slope wind profiles. One may ask why we should not be satisfied with vertical sidewalls in a conceptual valley flow model. As will become evident later, however, there is no along-valley wind in a valley with vertical sidewalls. Since the along-valley flow is an important part of the valley flow system, we must allow for sloping boundaries in a conceptual model of valley flow if we wish to deal with the three-dimensional problem.

Prandtl (1942) provided a simple solution to the problem of buoyant flow along a slope of angle α. He assumed constant stability $N_v{}^2$ and a fixed difference $\Delta\theta_l$ of temperature between the fluid near the sloping wall and that in the interior at the same level. Under these conditions a one-dimensional problem can be posed with flow equations:

$$\hat{w}N_v{}^2 \sin\alpha/g = \frac{\partial}{\partial\hat{y}}\left[K_H \frac{\partial}{\partial\hat{y}}\left(\frac{\delta\theta}{\theta_0}\right)\right], \qquad (3.14)$$

$$0 = g\sin\alpha\frac{\delta\theta}{\theta_0} + \frac{\partial}{\partial\hat{y}}\left(K_M\frac{\partial\hat{w}}{\partial\hat{y}}\right), \qquad (3.15)$$

where the $(\hat{y}, \hat{z})$ axes are, respectively, normal and parallel to the slope, and $\hat{w}$ is the flow velocity parallel to the slope. Note that (3.14) and (3.15) coincide with (3.5) and (3.6) for $\alpha = \pi/2$ except that here $d\theta_s/dz = d\theta_v/dz$ and horizontal advection is omitted. Accordingly, the solution to (3.15) and (3.16) must be of the form (3.9). It reads

$$\delta\theta(\hat{y}) = \Delta\theta_l \exp(-\hat{y}/D)\cos(\hat{y}/D), \qquad (3.16)$$

$$\hat{w}(\hat{y}) = \Delta\theta_l\left[\frac{K_H g^2}{(\theta_0{}^2 K_M N_v{}^2)}\right]^{1/2} \exp(-\hat{y}/D)\sin(\hat{y}/D), \qquad (3.17)$$

where the characteristic depth is now

$$D = \left[\frac{4K_M K_H}{(N_v{}^2\sin^2\alpha)}\right]^{1/4}. \qquad (3.18)$$

Since $N_v{}^2 = g\Delta\theta/H\theta_0$ in Gill's solution (Fig. 3.3), (3.8) and (3.18) essentially coincide.

Prandtl (1942) considered laminar viscous flow when examining this problem. However, Defant (1949) has shown that Prandtl's laminar solution fits observed profiles of slope winds quite well if the free parameters in (3.16) and (3.17) are chosen such that the height and intensity of the slope wind maximum, as given by (3.17), coincide with the corresponding observations. The theoretical profiles for intercomparison with observations are shown in Fig. 3.5. The agreement is surprisingly good for the night cases and acceptable during the day.

Egger (1981a) and Kondo (1984) extended the Prandtl problem by allowing for variations of the surface excess temperature along the slope, thus rendering the problem two-dimensional. Not surprisingly they found that the flow induced by a localized excess temperature turns away from the slope at the upslope end of the heating. Off the slope there is descent and an almost horizontal return flow toward the lower end of the heated strip (Fig. 3.6).

Despite the good agreement between observations and (3.17) it remains a major shortcoming of the Prandtl solution that eddy diffusivities must be tuned. One way to prevent this dilemma is to use more advanced boundary-layer models where the eddy diffusivities depend on stability and on the slope wind flow. The so called (q^2, l) models appear to be the most advanced version of such models (e.g., Yamada 1983), where the turbulent kinetic energy q^2 and a mixing length scale l are prognostic variables in addition to wind and temperature. Such models do not fall into the category of conceptual models because they demand a fairly good numerical resolution of the boundary layer. Nevertheless, we must discuss some of the results obtained with these models since this information is needed in what follows. Rao and Snodgrass (1980), for example, carried out extensive integrations with a model based on the time-dependent versions of

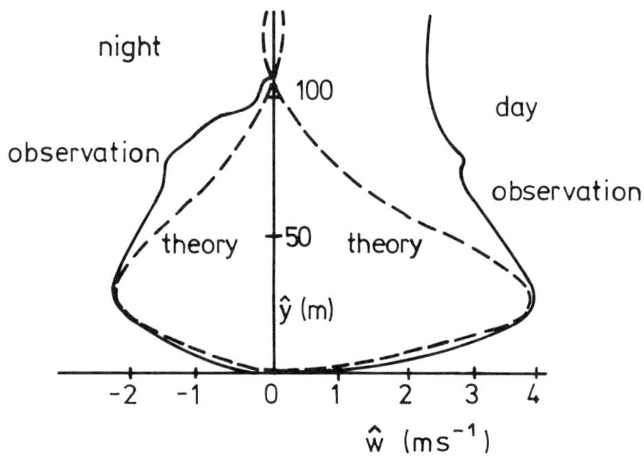

FIG. 3.5. Profiles of slope winds as observed (bold) and as obtained from (3.17) (dashed). Adapted from Defant (1949).

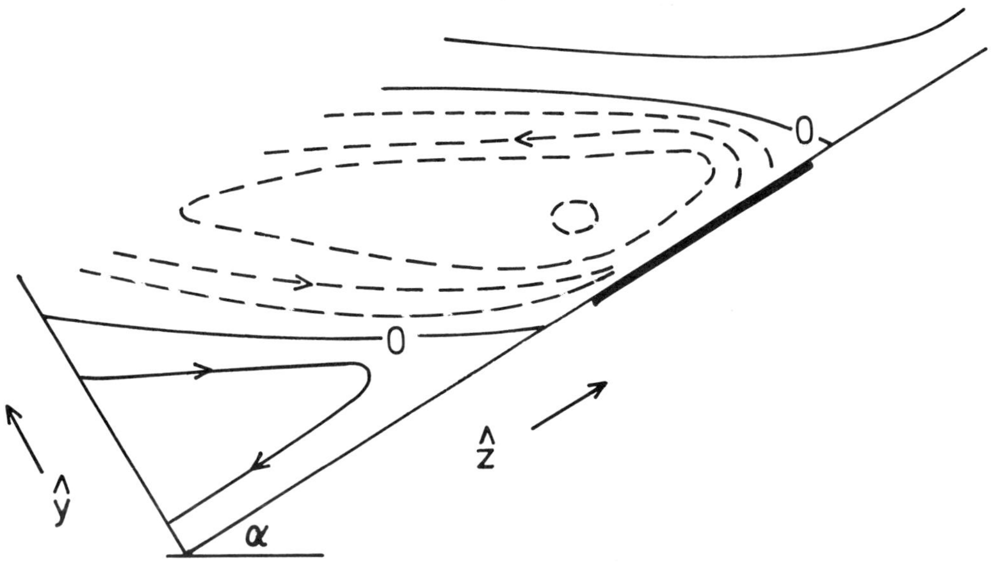

FIG. 3.6. Circulation induced by excess heating along the bold line at a slope. Nondimensional units; see Egger (1981a).

(3.14) and (3.15) but with eddy diffusivities specified according to

$$K_{M,H} \approx l_{M,H} q, \tag{3.19}$$

where a diagnostic relationship links the mixing lengths $l_{M,H}$ to y and stability. Moreover, a prognostic equation is provided for q^2 that is not given here [see Rao and Snodgrass, 1980, Eq. (3b)]. Nappo and Rao (1987) ran a two-dimensional version of this model to study nocturnal boundary-layer flow. Results are shown in Fig. 3.7 where we give the depth D of the boundary layer, the vertically integrated mean wind W [see (3.20)–(3.22) for definitions], and the entrainment factor E [to be discussed later; see (3.25)] as a function of the distance to the upper end of a slope. It is seen that the depth D increases downslope; however, there appears to be an equilibrium farther downslope, at least if the stability $d\theta_v/dz$ is strong enough. The wind speed is closely linked to D. The entrainment factor relates the entrainment on top of the slope wind layer to the speed of the slope wind. It is seen from Fig. 3.7 that E decreases with increasing stability and may even vanish. Results from such models have been compared extensively to observed slope wind profiles. A main problem in model verification is to find datasets where model assumptions are satisfied; i.e., the influence of ambient winds must be small. By and large a satisfactory agreement of observations and model results has been found under favorable conditions.

Let us turn from these high-resolution models to the so-called hydraulic slope wind models where the detailed vertical structure of the flow is replaced by a vertically integrated flowing layer that is subject to heating or cool-

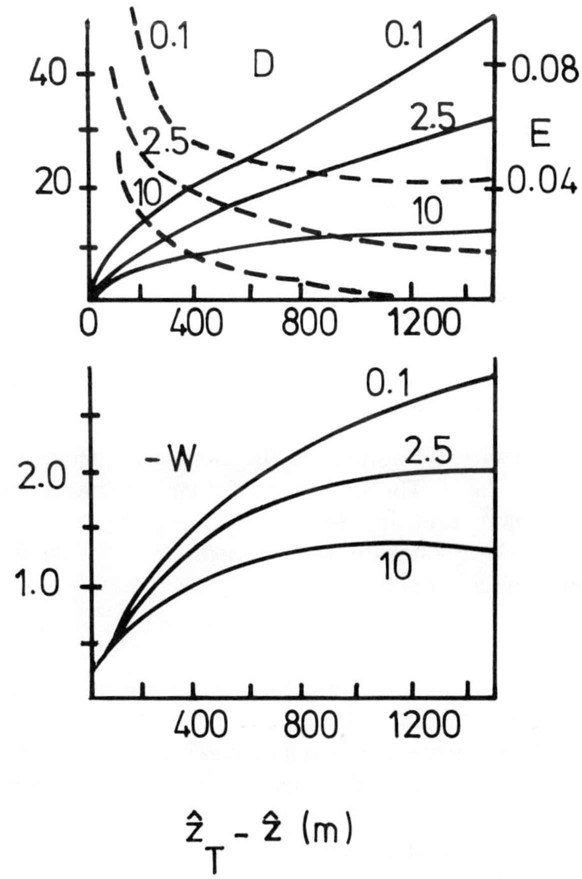

FIG. 3.7. Depth D (m) (bold), entrainment factor E (dashed) and vertically averaged steady-state flow speed W (m s^{-1}) of downslope winds as a function of distance ($\hat{z}_T - \hat{z}$) to the top for various stabilities $d\theta_v/dz$ (K/km), as obtained by Nappo and Rao (1987) through numerical integrations of a two-dimensional slope wind model.

ing, surface stress, and interfacial entrainment. In such a model one need not worry about the specification of the eddy diffusivities since the vertical integration eliminates the mixing terms in (3.14) and (3.15).

We follow the customary approach to define scales of velocity W along the slope, boundary-layer thickness D, and potential excess temperature T by

$$WD = \int_0^S \hat{w} d\hat{y}, \tag{3.20}$$

$$W^2 D = \int_0^S \hat{w}^2 d\hat{y}, \tag{3.21}$$

$$WTD = \int_0^S \hat{w}\delta\theta d\hat{y} \tag{3.22}$$

(e.g., Manins and Sawford 1979b) where $S > D$ is a distance to the slope large enough so that $\hat{w} \sim 0$ for $\hat{y} > S$.

An integration of the complete equation of motion over the depth of the boundary layer yields

$$\frac{d}{d\hat{z}} w^2 D = -\frac{d}{d\hat{z}}\left(\frac{1}{2} S_1 g T D^2 \cos\alpha/\theta_0\right) + S_2 g T D \sin\alpha/\theta_0 - C_D W^2, \tag{3.23}$$

where S_1 and S_2 are form factors (see Manins and Sawford 1979b for details) and steady-state conditions have been assumed. The term on the left-hand side represents the divergence of momentum transport along the slope. The first term on the right describes the pressure gradient force followed by the buoyancy term and the surface stress term where a quadratic drag formulation is assumed.

The equation of continuity reads

$$\frac{d}{d\hat{z}} WD = -\hat{v}_D, \tag{3.24}$$

with velocity $\hat{v}_D$ normal to the slope well above the boundary layer. The specification of $\hat{v}_D$ is a crucial part of hydraulic modeling. Manins and Sawford (1979b) in their work on katabatic winds propose to adopt the entrainment assumption

$$\hat{v}_D = EW \tag{3.25}$$

made by Ellison and Turner (1959) in interpreting their laboratory study of turbulent entrainment into inclined wall plumes. Note that $W < 0$ for downslope flow so that we have flow toward the slope and a corresponding increase of the volume flux in the slope wind layer down the slope. The entrainment parameter E is assumed to be a function of the Richardson number $\mathrm{Ri} = gTH\cos\alpha/W^2\theta_0$. Note that Gill (1966) did not have to introduce an entrainment assumption. Inflow to and outflow from the boundary layer is computed explicitly via (3.5) and (3.10) in Gill's model. Of course, there is no entrainment in (3.14) and (3.15) since Prandtl's problem is one-dimensional.

Finally, the first law needs to be integrated vertically as well:

$$WD\frac{d\theta_v}{d\hat{z}}\sin\alpha + \frac{d}{d\hat{z}} WTD = AD, \tag{3.26}$$

where the term AD represents the cooling (or heating) of the slope wind layer. When writing (3.26) we discarded an additional term introduced by Manins and Sawford (1979b), which is of minor importance in all cases of interest here.

Manins and Sawford (1979b) were able to find an analytical solution to (2.22)–(2.25) for neutral stratification, where Ri is constant along the slope and

$$D \sim (\hat{z}_T - \hat{z}), \quad W \sim -(\hat{z}_T - \hat{z})^{1/3}, \quad T \sim -(\hat{z}_T - \hat{z})^{-1/3}. \tag{3.27}$$

The solution depicts a boundary layer of growing thickness with distance to the upper end of the slope at $z = z_T$, where there is an infinitely large temperature deficit in the slope wind layer; however, this singularity can be circumvented by prescribing appropriate boundary values at a slope point $\hat{z} < \hat{z}_T$. Manins and Sawford (1979b) integrated (3.23)–(3.26) numerically for $N_v > 0$. We show W and the volume flux WD in Fig. 3.8 as a function of $(\hat{z}_T - \hat{z})$ for a slope $\alpha = 5°$ and the set of parameters

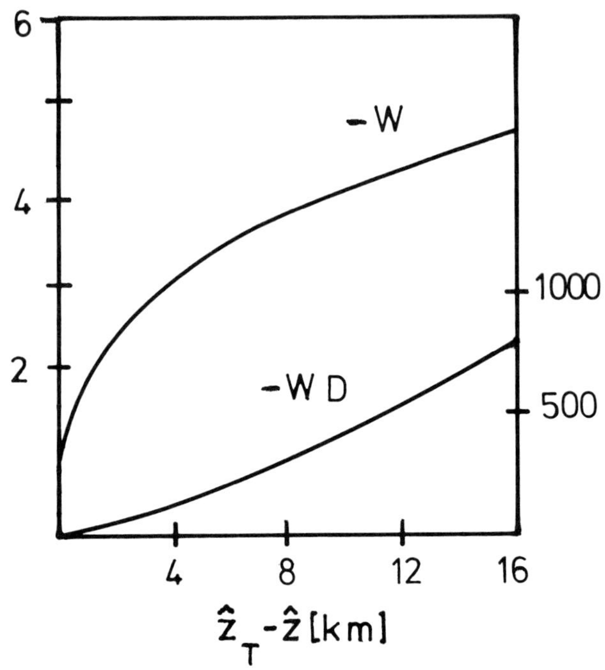

FIG. 3.8. Vertically averaged flow speed W (m s^{-1}) and volume flux WD (m^2 s^{-1}) of downslope flow in steady state as a function of downslope distance $\hat{z}_T - \hat{z}$. Adapted from Fig. 2 of Manins and Sawford (1979b).

listed in Manins and Sawford (1979b). In particular the atmosphere is stably stratified with $N_v = 10^{-2}\ s^{-1}$. The downslope variation of W is described reasonably well by (3.27) although this agreement is found to be sensitive to the specification of the functional dependence of E on Ri. As can be seen from Fig. 2 of Manins and Sawford (1979b), however, this sensitivity becomes important only for long slopes. The volume flux increases like $(\hat{z}_T - \hat{z})^{4/3}$.

There is a problem with (3.25) that limits the applicability of this set of equations. Assume a katabatic wind entering a region downslope where the stability N_v is much larger than upslope. In that case we anticipate that WD will decrease but the downslope flow will continue; however, (3.24) and (3.25) exclude such a response. Once there is downslope flow the slope wind layer must grow according to the entrainment assumption (3.25). This suggests that (3.25) is not suited for valley flow problems where N_v depends on height. Vergeiner (1982) and Brehm (1986) proposed a solution to this problem. Their work on two-dimensional valley flow will be discussed in section 3.5.

3.5. Circulation in valley cross-sections

Vergeiner (1982) proceeded essentially as Gill (1966) in proposing to treat the valley atmosphere separately from the slope wind layer. Vergeiner (1982) realized, however, that vertical motion w_v must be admitted in the core region since there is a net heating or cooling in the slope wind layer that must be partially transferred to the interior through vertical motions. Vergeiner (1982) argued that it should be sufficient for a valley flow model to rely on continuity and the first law. Vergeiner invoked a closure condition based on observations of convective plumes to relate the volume flux in the slope wind layer to the heat transport. Brehm (1986) followed Vergeiner's lead but introduced a new closure condition derived from results obtained with a slope wind model. In what follows we will describe Brehm's approach.

In Fig. 3.9 we consider a situation similar to that given in Fig. 3.2. In particular we assume a separation of the slope wind layers and the interior of the valley; however, we prescribe a flux of sensible heat Q at the slopes as well as at the valley floor instead of fixing the surface temperature. The first law, when integrated horizontally over the left-hand slope wind layer [see also (3.26)] becomes:

$$\frac{\partial}{\partial t}(\theta_v + T_l)D'\gamma + (WD)_l\frac{\partial \theta_v}{\partial z} + \frac{\partial}{\partial z}(WTD)_l + \epsilon(v_l + |v_l|)\frac{T_l}{2} = Q_l/\sin\alpha, \quad (3.28)$$

where $D' = D/\sin\alpha$ and the switch-on parameters γ and ϵ equal 0 or 1. We admit in (3.28) a transport of excess

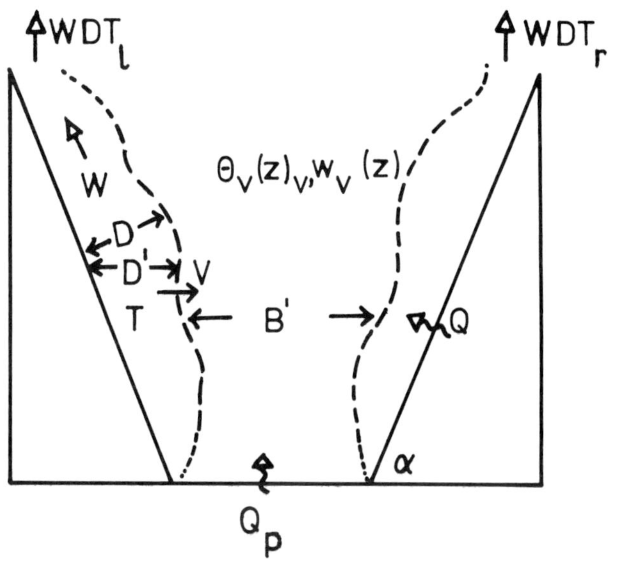

FIG. 3.9. Schematic of the valley flow model designed by Brehm (1986). See text for further details.

temperature out of the boundary layer and choose an upstream formulation for this term to exclude an unrealistic transport of T into the slope wind layer. Since $z = \hat{z}\sin\alpha$, (3.26) and (3.28) become identical for $\gamma = \epsilon = 0$. The corresponding equation for the interior is

$$\frac{\partial}{\partial t}\theta_v B' + w_v B'\frac{\partial \theta_v}{\partial z} - \epsilon(v_l + |v_l|)T_l/2 + \epsilon(v_r - |v_r|)T_r/2 = 0. \quad (3.29)$$

The valley atmosphere experiences no external heating except at the ground, where a heat flux Q_p is prescribed. Continuity requires

$$(WD)_l + (WD)_r + w_v B' = 0. \quad (3.30)$$

During the day $WD > 0$ in both slope wind layers so that we expect downward motion in the interior. (It is the opposite during the night.) By adding (3.28) for both slopes and (3.29), we obtain with (3.30):

$$\frac{\partial}{\partial t}(\theta_v B' + \gamma(\theta_v + T_l)D'_l + \gamma(\theta_v + T_r)D'_r) + \frac{\partial}{\partial z}(WDT)_{l+r} = Q_{l+r}/\sin\alpha. \quad (3.31)$$

We follow Brehm who assumed steady-state conditions for the boundary layers. As can be seen from (3.31) this implies $D' \ll B'$, an assumption that appears to be justified for most valleys. Nevertheless, we must realize that we violate the thermal energy balance of the valley atmosphere by omitting the first term in (3.28). Brehm (1986) closed the system by linking the volume flux WD in the boundary layer to the vertically integrated transport of temperature WDT. The closure (3.32) is derived from numerical experiments with a one-dimensional slope wind layer model, similar to that developed by Rao and Snodgrass (1981). He integrated this model for many different sets of slope angle α, stability N_v, and heating rate Q until steady-state conditions prevailed, to arrive at an empirical relation

$$WD = P(WTD)^i, \tag{3.32}$$

where $i \sim 0.66$ and P depends on stability, slope angle, and roughness of the slope. Note that (3.32) disagrees with (3.27) where $i = \frac{4}{3}$. Equation (3.32) is valid for an infinitely long slope wind layer without entrainment, however, while (3.27) reflects the entrainment condition (3.25). The quantity P is displayed in Fig. 3.10 as a function of the vertical gradient of the temperature in the core for two slope angles. Here P decreases with increasing stability and/or slope angle. Since (3.32) implies

$$T^i = (WD)^{1-i}/P, \tag{3.33}$$

a certain volume flux goes with large excess temperatures if the stability is large or if the slope is steep. Finally, we must require

$$\frac{\partial}{\partial z}(WD) = \pm v, \tag{3.34}$$

where the positive sign must be chosen for the right-hand slope. A separate entrainment equation is not needed in this model because it is sufficient to impose continuity. Brehm omitted all terms involving the meridional transports in (3.28) and (3.29) ($\epsilon = 0$) so that (3.34) is not needed either in his model. In any case (3.28)–(3.34) can be stepped forward in time under the assumption of steady state in (3.28). If WD, T, w_v, v and θ_v are given at a certain time, we can predict θ_v from (3.29). Next we derive WD and v from (3.28) with $\gamma = 0$, (3.32) and (3.34), and obtain the excess temperature T from (3.33). Finally, w_v follows from (3.30) where we can either replace B' with B or use one more of Brehm's empirical relations to derive D from WD. Then we can again step θ_v forward in time. The model equations are discretized in the z-direction. A distinction between slope wind layer and valley atmosphere cannot be made above the valley. Brehm assumed a layer above the cross section that is heated by the heat flow WDT_r and WDT_l on top of the slopes. Finally he prescribed a simple surface heating at the slopes and at the bottom of the valley that varies sinusoidally in time. Convective adjustment is imposed in the core region.

In Fig. 3.11 we show results from a model run where a stable stratification is prescribed initially in a valley of 800 m depth. Near the top there is a neutral layer. The potential temperature near the valley floor increases mainly because of the surface heating. Convective adjustment leads to neutral stratification near the valley floor. This process erodes the stable core of the valley atmosphere from below (see also Figs. 2.25 and 2.27 of Whiteman, Chapter 2). After two hours we have a slope wind layer whose depth D increases up the slope in the lowest 100 m where stratification is neutral; however, D decreases rapidly when the slope winds enter the stably

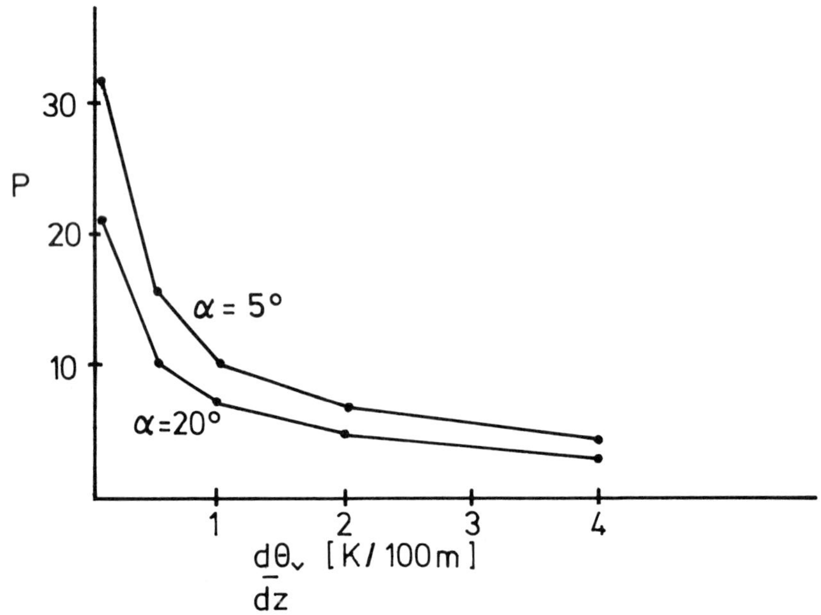

FIG. 3.10. Factor $P\,(\mathrm{m}^{0.66}\,\mathrm{s}^{-0.33}\,\mathrm{K}^{-0.66})$ in (3.32) as a function of $d\theta_v/dz$ and for two slope angles. The figure is based on Table B1 of Brehm (1986) (roughness length $z_0 = 0.03$ m).

stratified region, and settles at an equilibrium depth of about 25 m above $z = 250$ m. At $t = 4$ h the near-neutral domain is already quite deep and correspondingly the slope wind layer is deeper and has its maximum depth higher up. Nevertheless we still find a decrease of D with z in the stably stratified region. The profiles of the slope wind velocity W match closely those of D and we find upward motion along the slope at all times. Correspondingly [see (3.30)] there is downward motion in the interior. This leads to the warming of the valley atmosphere above the well-mixed layer (Fig. 3.11). Note, however, that the only mechanism of heat transfer from the slope wind layer to the valley atmosphere is through vertical advection of potential temperature as expressed by the second term in (3.29). In particular this transfer vanishes if $N_v = 0$.

It appears that Brehm's model is able to handle situations quite well when N_v depends on height, and most of the difficulties with Gill's model have been overcome. There is vertical motion in the core and we can prescribe sloping sidewalls. Since we use vertically integrated boundary-layer equations, the K-parameterization is not a problem, nor is the treatment of entrainment, as was the case with (3.25). On top the valley is essentially open. Of course, Brehm's model does have its shortcomings. The steady-state assumption for the slope wind layer leads to an immediate adjustment of the boundary layer to any changes of θ_v or Q. As can be seen from (3.31) this assumption is tantamount to storing all the thermal energy made available to the valley in the interior. Thus the valley atmosphere becomes too warm; however, the effect is small. Moreover, since Brehm chose $\epsilon = 0$ excess temperature is not transported out of the slope wind layers. This process could play a role at $t = 2$ h, say, in Fig. 3.11, where $\partial/\partial z(DW) < 0$ for $100 < z < 250$ m. The corresponding warming of this layer might slow the growth of the well-mixed layer. Moreover, there is no balance of momentum. Nocturnal drainage flows have not been considered by Brehm, but his approach can be extended to nighttime cases, as will be shown in section 3.6.

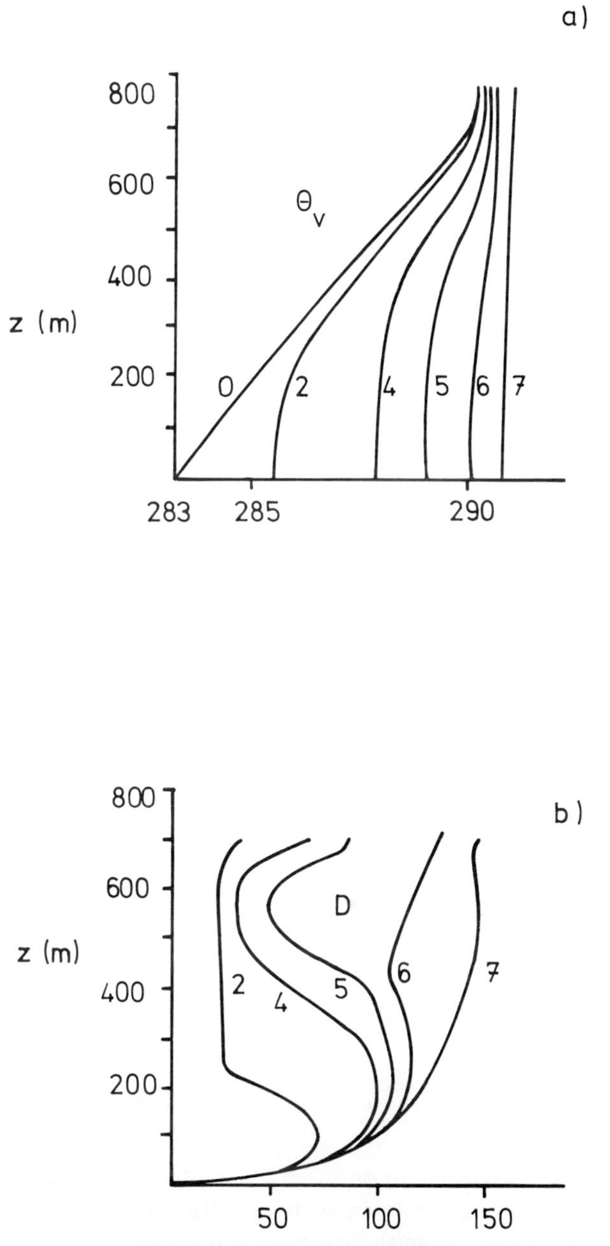

FIG. 3.11. Numerical integration of Brehm's model started at $t = 0$ h with a stably stratified state of rest. (a) Potential temperature θ_v (K) at various times, and (b) depth D (m) of the slope wind layer at various times. Adapted from Figs. 6.9 and 6.12 of Brehm (1986).

3.6. Toward three-dimensional modeling

The problem of three-dimensional valley flows must still be examined. While elaborate simulation models have been used to reproduce valley flow and dispersion observations even in quite complicated terrain (e.g., Yamada and Bunker 1988), conceptual modeling has so far attempted to deal with this problem at a relatively elementary level. The idea is to prescribe the topographic amplification factor for a valley as an additional source of heating. Since the concept of the topographic amplification factor is thoroughly discussed by Whiteman, Chapter 2, we can be brief here. (See subsection 2.2.2.1 of Chapter 2, where a definition of the topographic amplification factor is given.) The idea is simple: the slope wind layers interact with the interior of the valley and contribute to a heating (or cooling) of the valley atmosphere in addition to the energy provided through the heat flux at the valley floor. In contrast, on the plain it is just the energy from the surface that heats the atmosphere. Correspondingly $\theta_v > \theta_{\text{plain}}$ during the day and the related gradient of the surface pressure will induce a circulation between the valley and the plain. This circulation will be stronger the larger the topographic amplification factor.

The topographic amplification factor is unity for a valley with vertical sidewalls as considered by Gill. Correspondingly, we would not expect to find along-valley winds in a three-dimensional version of Gill's problem. The amplification factor exceeds unity in valleys with sloping sidewalls. In highly simplified models of three-dimensional-valley flows one simply assumes that the effect of the slope wind layers on θ_v need not be calculated explicitly as in Brehm's model but can be prescribed as an additional heating or cooling (see Vergeiner 1987; Egger 1987).

To demonstrate this approach we consider the valley depicted in Fig. 3.1. An extremely coarse numerical resolution of the valley flow in the vertical is adapted by defining a mean temperature $\theta_v(x)$ and flow $u_v(x)$ at height $z = H/2$ only, and w_v just at $z = H$. We do not specify a $B(z, y)$ profile of the valley but assume that the topographic amplification factor can be taken into account by prescribing an extra heating θ^*/τ in the first law

$$\frac{\partial \theta'_v}{\partial t} + \epsilon u_v \frac{\partial \theta'_v}{\partial x} + \frac{w_v}{2}\frac{d\bar{\theta}_v}{dz} = (\theta^* - \theta'_v)/\tau, \quad (3.35)$$

where $\theta'_v = \theta_v - \bar{\theta}_v$ is the deviation from a prescribed mean temperature $\bar{\theta}_v$ in the valley with fixed Brunt–Väisälä frequency N_v, and τ is a relaxation time that characterizes the speed of heat transfer from the slope wind layers to the core. In (3.35) it is assumed that there is no vertical transport of θ'_v on top of the valley. Note in particular that there is no separate equation for the boundary layers since the impact of the slope wind layers is represented by the heating in (3.35). The equation of motion is

$$\frac{\partial u_v}{\partial t} + \epsilon u_v \frac{\partial u_v}{\partial x} = \nu \frac{\partial \theta_v}{\partial x} - u_v/\tau, \quad (3.36)$$

where the first term on the right-hand side involves the hydrostatic assumption with $\nu = Hg/2\theta_0$, and the assumption of a vanishing deviation pressure at $z = H$. In principle, we should have chosen in (3.36) a frictional time scale that differs from the relaxation time τ for the heating; however, it is argued in Egger (1987) that both time scales are of the same order of magnitude. Continuity requires

$$\frac{\partial u_v}{\partial x} + \frac{w_v}{H} = 0, \quad (3.37)$$

where we neglect any effects of the valley's shape [as we did in (3.35) and (3.36) except for the heating]. In the plain we assume $\theta = \bar{\theta}_v$ and, of course, $u_v = 0$. Thus $\theta'_v = 0$ is the boundary condition to be posed at the mouth of the valley. Consider the case $L \to \infty$. If we assume periodic heating $\theta^* = \hat{\theta}^* \exp(i\omega t)$, $\omega = 2\pi/\text{day}$, and linearity ($\epsilon = 0$), the solution is

$$\theta'_v = \text{Re}\{\hat{\theta}^*[1 - \exp(-\mu x)] \exp(i\omega t)/(i\omega\tau + 1)\} \quad (3.38)$$

$$u_v = \text{Re}\{\nu\hat{\theta}^*\mu \exp(-\mu x + i\omega t)/(i\omega\tau + 1)^2\} \quad (3.39)$$

with

$$\mu = 2(i\omega\tau + 1)/(\tau H N_v). \quad (3.40)$$

The response to the jump in heating at the mouth has a length scale $\Lambda = \tau N_v H$ that varies between a few kilometers for daytime conditions where $N_v \sim 0$ and, say, $\Lambda = 70$ km during the night where we insert $\tau \sim 2$ h and $H = 1000$ m. The response decays like $\exp(-2x/\Lambda)$ with distance from the mouth. In particular, $\theta'_v = \theta^*$ for $x \gg \Lambda$. The wind speed is proportional to μ; i.e., winds should be strongest under near neutral conditions. Of course, there is symmetry between day and night if $\epsilon = 0$.

In Fig. 3.12 we show u_v as obtained from (3.35)–(3.37) through a numerical integration with $\epsilon = 0$ (dashed) and with the boundary condition $u_v = 0$ at $x = L$. The variables are staggered in a grid, where u_v is predicted also at $x = 0$ and $u_v = 0$ at the head, while θ'_v is defined in between momentum grid points. In the nonlinear situation ($\epsilon = 1$) we choose an upstream formulation for the transport terms in (3.35) and (3.36). Inflow is strong at noon and maximum flow speeds are seen at the mouth. Of course, the linear solution (dashed) decays with increasing x, where $\Lambda \sim 50$ km in this case. At night, most of the characteristics of the linear solution carry over to the nonlinear regime. The down-valley wind profile at midnight (bold) mirrors almost exactly the linear solution at noon. If N_v is reduced to 10^{-3} s^{-1} (dots), the decay with x is more rapid—in agreement with (3.39). Note, however, that the flow speed at the mouth is hardly affected by the reduction of N_v in contrast to (3.39). Additional tests showed that it is the advection of air from the plain that causes this strong deviation from linear theory. During the day nonlinear effects are more pronounced. In particular, the reduction of N_v to 10^{-3} s^{-1} has almost no impact at noon. It is the along-valley wind that transports deviations of the temperature into the valley so that Λ is no longer the relevant parameter. The corresponding length of penetration is approximately $tu_v \sim 100$ km for $t = 6$ h. Even so, winds vanish for, say, $x > 150$ km. This clearly contradicts the observations in the Inn Valley, Austria where vigorous valley winds are found far from the mouth. To induce such flows in the model we must assume that the amplification factor increases with distance to the mouth. If that is done the wind regime extends throughout the valley (see also Fig. 2.10 in Whiteman, Chapter 2). At sunrise, we have down-valley flow.

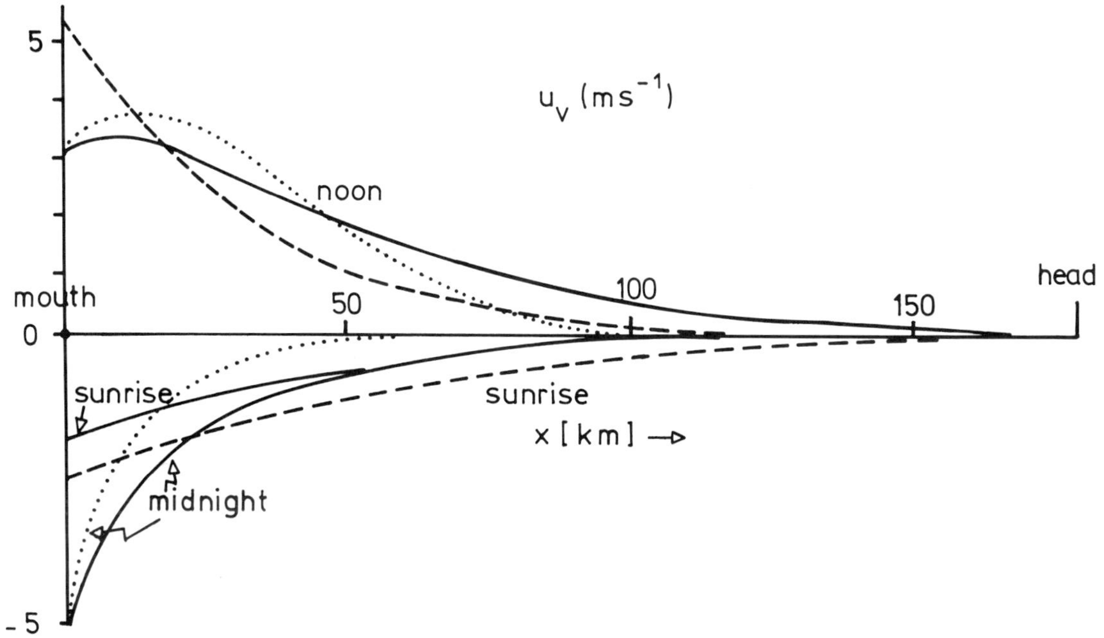

FIG. 3.12. Wind u_v along the valley axis as obtained in integrations of (3.35)–(3.37) with periodic forcing. Bold: nonlinear solution with $N_v = 1.2 \times 10^{-2}$ s^{-1}. Dashed: linear solution for $N_v = 1.2 \times 10^{-2}$ s^{-1}. Dotted: nonlinear solution for $N_v = 10^{-3}$ s^{-1}; $\tau = 2$ h; $\theta^* = 2.5$ K.

We infer from Fig. 3.12 that useful results can be obtained by prescribing the topographic amplification factor as an additional heat source. Egger (1987) extended this approach by considering a valley cut into an escarpment and opening into a plain. Air motion above the escarpment and the plain was treated explicitly, but with extremely coarse resolution in the vertical. By and large, one finds that (3.35) and (3.36) provide good guidance to the results obtained with this more elaborate model; therefore, a detailed discussion is not necessary. Vergeiner (1987) considers a flow domain that is subdivided into a foreland, a valley, and a basin in the interior of the mountain massif, which is meant to represent the side valleys and tributaries of the main valley. Vergeiner (1987) does not include the return flow above the mountains in his model. Nevertheless he was able to reproduce basic features of the observations in the Inn Valley, Austria.

It is a major shortcoming of the models discussed in this chapter that the Brunt–Väisälä frequency must be prescribed. As can be seen from Fig. 3.11 the stability of the valley atmosphere has a pronounced impact on the flow. A satisfactory conceptual model of three-dimensional valley flow should predict the stratification in the valley.

3.7. Three-dimensional valley flow

The results presented in Fig. 3.12, and Vergeiner's (1987) work, suggest that we may treat the mouth of the valley as a boundary between the foreland and the valley. The same appears to be advisable with respect to the top of the valley. If we can specify the boundary conditions properly, we can circumvent the enormously complex problem of how to determine the circulation above the mountains. Correspondingly, we propose here an extension of Brehm's model to three dimensions where the computational domain has its boundary at $x = 0$ and $z = H_0$. As in Fig. 3.9, we assume a separation of valley atmosphere and slope wind layers. We assume that the cross-valley circulation can be modeled as in section 3.4. Thus we use (3.28) with $\epsilon = 1$ in steady-state form ($\gamma = 0$) to evaluate the volume flux WD in each boundary layer as a function of z and x. Two types of vertical motion must be distinguished in the valley: w_v as obtained from (3.30) gives the vertical motion induced by the slope winds. Since, however, we now allow for along-valley winds, the divergence of volume transports along the valley may lead to a vertical motion $\tilde{w}$ through

$$\frac{\partial}{\partial x}(u_v B) + \frac{\partial}{\partial z}(\tilde{w} B) = 0, \tag{3.41}$$

so that the total vertical motion in the valley is now $w_v + \tilde{w}$. Then, the equations of the model are (3.28) in steady-state form, (3.30), (3.32) and (3.34), and

$$\frac{\partial u_v}{\partial t} + \frac{1}{B}\frac{\partial}{\partial x}(u_v{}^2 B) + \frac{1}{B}\frac{\partial}{\partial z}(\tilde{w} B u_v) = -\frac{1}{\rho_0}\frac{\partial p}{\partial x} + F_u, \tag{3.42}$$

$$\frac{\partial \theta_v}{\partial t} + \frac{1}{B}\frac{\partial}{\partial x}(u_v\theta_v B) + \frac{1}{B}\frac{\partial}{\partial z}(\tilde{w}\theta_v B) + w_v\frac{\partial \theta_v}{\partial z}$$

$$- \epsilon(v_l + |v_l|)T_l/2B + \epsilon(v_r - |v_r|)T_r/2B = 0, \quad (3.43)$$

where we assume $B' \sim B$. The model equations are discretized in the vertical with mesh size Dz. A simple quadratic law is used to parameterize the surface stress so that

$$F_u Dz = -C_D|\mathbf{v}|u_v \quad (3.44)$$

in the lowest layer of the model of height Dz, but $F_u = 0$ in all other layers. For the sake of simplicity C_D is assumed constant. The pressure is derived from θ through (3.3). Note that w_v is not included in the transport terms of (3.42). It would be inconsistent to include the impact of a cross-valley circulation on u_v, which by itself does not satisfy the momentum equation.

The grid of the model in the horizontal is the same as for (3.35) and (3.36), so that u_v is also defined at the mouth. To evaluate the pressure gradient in (3.42) at $x = 0$ we must know the potential temperature θ_p in the foreland. We solve

$$\frac{\partial \theta_p}{\partial t} = Q_p/Dz \quad (3.45)$$

in the lowest layer, where Q_p is a prescribed heat flux, and impose convective adjustment at all levels as in the valley proper. It is more difficult to handle the situation at $z = H$. We normally have nonvanishing heat transports WDT in both boundary layers at $z = H$ for daytime conditions. It is difficult, if not impossible, to say which part of that flux is available to heat the valley atmosphere above the valley. In many cases even light ambient winds may carry the excess temperature away. We consider two extremes here. In one set of experiments we transfer all the heat flux at the upper boundary to the uppermost layer of the model. This is tantamount to assuming that there is an extremely stable layer above $z = H$. The other choice is to simply let the heat flux leave the valley atmosphere. Note that this problem does not occur in a nighttime simulation, since $WDT = 0$ at $z = H$ in that case. It is important, however, to know the stability N_v at $z = H$ during the night, since the outflow at the mouth requires corresponding downward flow at $z = H$, which carries potential temperature downward. Finally we must ensure that the total amount of heating is the same in the valley and in the foreland; i.e., we require

$$B_0 Q_p = B(x, 0)Q_p + 2QH/\sin\alpha. \quad (3.46)$$

We consider a valley where $\alpha = 20°$, $\beta = 0$, $L = 110$ km, $H = 1000$ m, and the valley width is 7.8 km at the bottom. We prescribe a stable stratification at $t = 0$. The heating is turned on at $t = 0$ h and does not vary in time. In Fig. 3.13 we show the along-valley wind at $t = 2$ h (dashed) and at $t = 6$ h (bold). At $t = 2$ h the along-valley wind is quite weak and restricted to a small part of the valley close to the mouth. At $t = 6$ h, however, the along-valley wind regime essentially fills the valley. Wind intensities are strongest below $z = H/2$. The cross-valley circulation follows essentially the pattern given in Fig. 3.11 with a slow erosion of the stable layers in the core. The stratification is almost neutral at $t = 6$ h. The air is warmest near the head and relatively cool at the mouth. The vertical motion is dominated by the downward motion linked to the cross-valley circulation. We find $|w_v| \sim 0.1$ m s^{-1} while the ascending motion induced by the along-valley winds is <0.05 m s^{-1} at $t = 6$ h. Unlike (3.35) the thermal energy equation (3.43) of the model does not contain a damping term. Moreover, the damping by surface friction

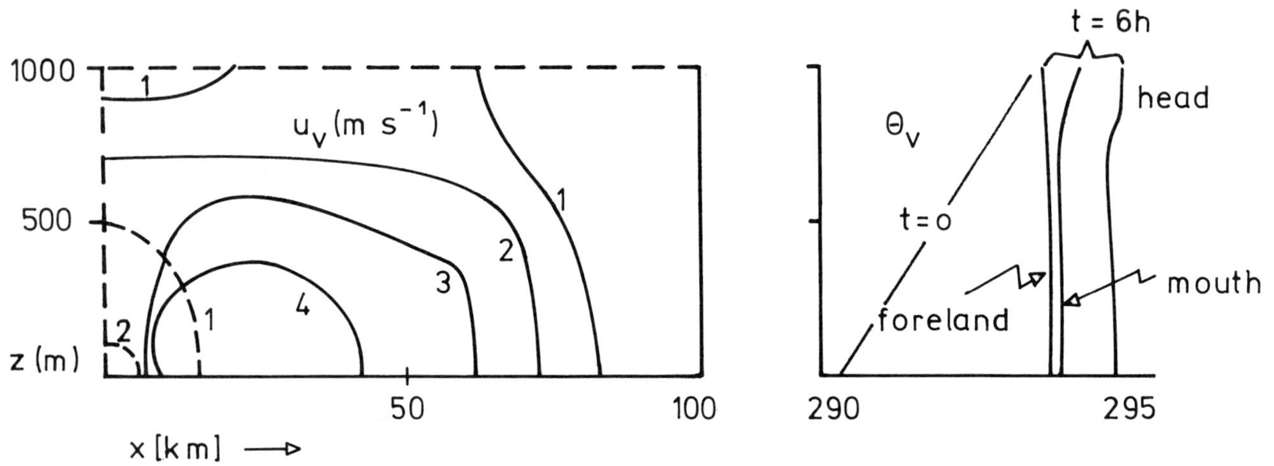

FIG. 3.13. Along-valley wind u_v (m s^{-1}) and profiles of θ_v (K) as obtained in numerical integrations of the three-dimensional model for a daytime situation. The wind is given at $t = 2$ h (dashed) and $t = 6$ h (bold). The heating is switched on at $t = 0$ h, and $Q_p = 0.1$ (K m s^{-1}). Vertical resolution $Dz = 100$ m; horizontal resolution $Dx = 10$ km.

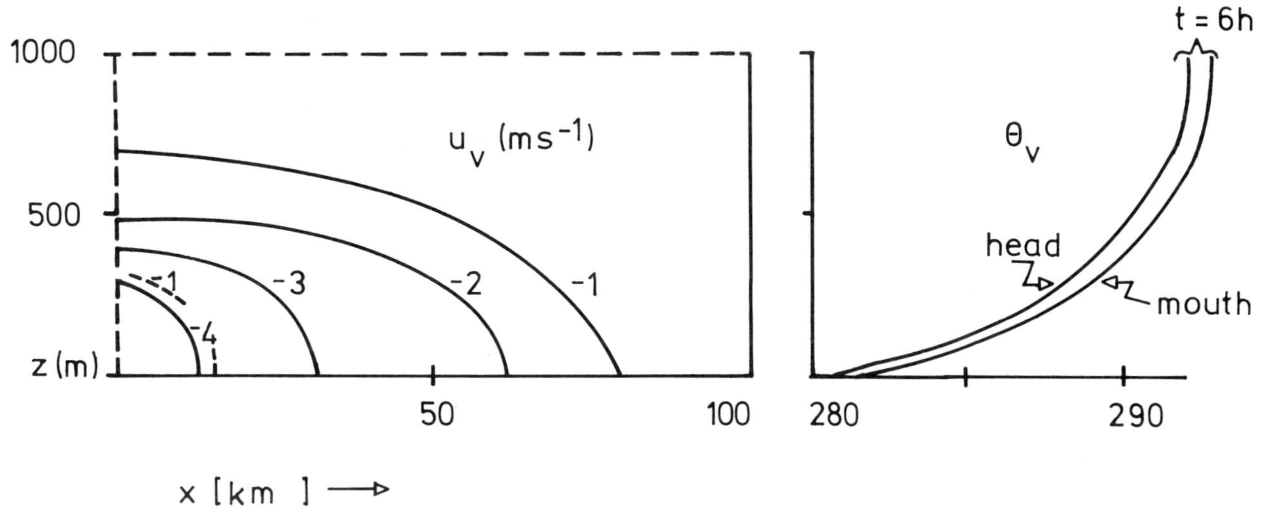

FIG. 3.14. As Fig. 3.13, but for a nighttime case.

is relatively weak. If we solve (3.35)–(3.37) in time with $\tau \sim \infty$ and a switch-on of the heating at $t = 0$ h, as in the experiments presented in this section, we find that the along-valley wind regime penetrates into the valley with speed of propagation $N_vH/2$. Since $N_vH/2 \sim 5$ m s^{-1} at $t = 0$ the linear theory predicts that the inflow regime reaches the head after about 6 hours—in agreement with Fig. 3.13. We assume in this experiment that the heat flow on top of the slope wind layers is injected into the valley atmosphere. If we assume that this flow is lost, the along-valley wind is less vigorous with maximum speed of 2.5 m s^{-1} at $t = 6$ h. Moreover, the along-valley wind regime extends only 60 km into the valley; i.e., our results are highly sensitive to the specification of the boundary conditions on top of the slope wind layer.

We are now in a position to test many conjectures about three-dimensional valley flows. For example, one expects to see weaker along-valley flow in a valley that is wider but has the same slope. After all, the topographic amplification factor is smaller for such a valley. The maximum along-valley flow speed at $t = 6$ h is just 3 m s^{-1} if the width is 15 km at the bottom. The inflow becomes more intense if the valley bottom is inclined.

The three-dimensional model can be used to study nighttime situations as well. Since Brehm (1986) restricted his attention to daytime conditions a relationship in analogy to (3.32) is not available at present; however, we can at least guess what such a relation should be. Nappo and Rao (1987) provide an estimate of P in (3.32) as a function of stability for slopes with an angle of $\alpha = 20°$ (their Fig. 8). We run the valley flow model with this estimated value of P, for a cooling identical in intensity to the heating prescribed in the foregoing experiments. The valley width B_0 on top is 15 km as in the daytime experiment presented in Fig. 3.13. Since the slope is now steeper the topographic amplification factor is smaller. In (3.42) and (3.43) a diffusive term is added on the right-hand side. The resulting down-valley flow is shown in Fig. 3.14. By and large the outflow profiles in Fig. 3.14 can be obtained from Fig. 3.13 by a reversal of sign. This suggests that linear processes are dominant; however, the depth of the layer of intense outflow is smaller than that of inflow. The total vertical motion is downward. The temperature profiles in the valley show a strong inversion near the ground. This inversion would be restricted to the lowest layer of depth $Dz = 100$ m if we had not added a diffusive term to (3.43) in order to smooth this extreme profile.

3.8. Outlook

It appears that conceptual modeling of valley wind systems has reached a stage where we can now reproduce and understand at least the basic features of thermally induced flows in valleys. The interaction of slope wind layers and the core of the valley atmosphere is the key to the problem. Since Brehm's approach allows us to describe this interaction we were able to present a three-dimensional model that captures the essence of the observations. It must be stressed, however, that the results presented in section 3.5 are preliminary in the sense that not all cases of interest have yet been considered; e.g., periodic forcing. Moreover, it is unclear to what extent we can take into account the momentum balance of slope wind layers through (3.32) (see Mahrt 1982).

In any case, we have taken the first steps toward an understanding of the observations. As is clear from the presentation by Whiteman, Chapter 2, valley flows in the atmosphere rarely occur under such idealized circumstances as assumed here. There is nearly always flow above the mountains linked to a large-scale pressure field.

Therefore we have mixing of momentum on top of the valley, a process that will be difficult to describe. Moreover, the imposed pressure gradient force will accelerate (or decelerate) the flow in the valley. Actual valleys tend to have a much more complicated geometry than assumed here. There are bifurcations, tributaries, and bends that may complicate the situation considerably. The impact of moisture is another factor that needs to be considered. After all, much of the solar energy absorbed at the slopes is transferred to the atmosphere in the form of latent heat, which influences the circulation and valley winds in a manner that is totally different from what has been described here. Very little work has been done on these problems. We anticipate, however, that conceptual modeling will continue to play a major role in clarifying the physical processes involved.

APPENDIX

List of symbols

g	gravity
i	exponent in parameterization; (3.32)
l	mixing length scale; (3.19)
p	pressure
q	turbulent kinetic energy; (3.19)
u	along-valley velocity
u_v	along-valley velocity in valley
v	cross-valley velocity
$\hat{v}_D$	velocity normal to slope on top of boundary layer; (3.24)
v_s	cross-valley velocity in Gill's model; (3.5)
v_l, v_r	cross-valley velocity on top of left (right) slope wind layer; (3.28)
$\hat{w}$	velocity parallel to slope
w_v	vertical velocity in valley as induced by slope winds; (3.30)
$\tilde{w}$	vertical velocity in valley linked to divergence of horizontal mass flux; (3.41)
x	along-valley coordinate
y	cross-valley coordinate
$\hat{y}$	coordinate normal to slope
z	vertical coordinate
$\hat{z}$	coordinate parallel to slope
$\hat{z}_T$	upper end of slope
A	heating rate; (3.2)
B	valley width
B'	valley width (core only)
B_0	valley width on top
C_D	drag coefficient
D	characteristic thickness of boundary layer; (3.7) boundary layer thickness; (3.20)
D'	$D/\sin\alpha$
Dz	vertical mesh size
$\mathbf{F}$	frictional force (3.1)
H	valley depth
K_H, K_M	eddy diffusivity for heat and momentum
N_v	Brunt–Väisälä frequency in valley
P	factor in parameterization; (3.32)
$Q(Q_p)$	heat flux at ground (in plain)
Ri	Richardson number; $\mathrm{Ri} = gth\cos\alpha/W^2\theta_0$
S_1, S_2	form factors; (3.23)
T	mean temperature deviation in slope wind layer; (3.22)
W	mean wind parallel to slope; (3.20)
α	slope angle
β	angle of valley bottom
γ, ϵ	switch-on parameters
Λ	length scale
μ	complex spatial decay rate; (3.40)
ρ_0	mean density
ψ	streamfunction
τ	relaxation time
θ	potential temperature
θ_v	potential temperature in valley
θ_p	potential temperature in foreland (3.45)
θ_s	potential temperature in slope wind layer
θ^*	heating (3.35)
$\hat{\theta}^*$	Fourier coefficient of heating
$\delta\theta$	potential temperature excess in slope layer
θ_l, θ_r	potential temperature at left (right) sidewall

CHAPTER 4

Mountain Waves and Downslope Winds

DALE R. DURRAN

Department of Atmospheric Sciences, University of Washington, Seattle, Washington

ABSTRACT

In an effort to enhance the reader's physical understanding, this chapter begins with a discussion of fundamental concepts including parcel oscillations in a stable atmosphere, wave propagation, and the fluid motions associated with vertically propagating internal gravity waves. This is followed by a brief review of the theory of small-amplitude mountain waves, beginning with the case of air flowing over a series of sinusoidal ridges in a basic state with uniform wind speed and stability. The linear theory is then extended to cover isolated mountains and situations with vertical variations in the basic-state wind speed and stability.

The effects of nonlinearity on the flow are introduced in a discussion of downslope windstorms. Three different theoretical models of the downslope windstorm are compared. The first argues that the development of strong downslope winds is analogous to the transition from subcritical to supercritical flow in a hydraulic jump. The second proposes that downslope winds are produced by the superposition of partially reflected waves generated at atmospheric interfaces where there is a rapid change in the static stability, such as the tropopause. The third suggests that strong winds are produced by processes that become active when vertically propagating waves become unstable and "break" in the upper troposphere.

The problem of forecasting downslope winds is discussed. Recent observational and theoretical investigations of the gust structure embedded in the downslope wind current are reviewed. The paper concludes with a selective overview of recent work on the flow around isolated three-dimensional mountains.

4.1. Introduction

A disturbance is created when stably stratified air is forced to rise over a topographic barrier. The energy associated with that disturbance is usually carried away from the mountain by gravity waves. Gravity waves forced by mountains are specifically referred to as *mountain waves.* Mountain waves exert a drag on the upper levels of the atmosphere; indeed, the cumulative worldwide effect of mountain-wave drag is believed to have a significant influence on the strength of the mean zonal circulation near the polar jet streams. Large-amplitude mountain waves can be associated with regions of clear-air turbulence that pose a hazard to aviation. Mountain waves can also be associated with strong surface winds that blow down the mountain along its lee slope; wind gusts in excess of 50 m s^{-1} have been measured in extreme events.

This chapter reviews some of the fundamental aspects of mountain waves and examines recent progress in our understanding of downslope windstorms and the airflow around an isolated mountain. A discussion of the basic physical properties of internal gravity waves is provided in section 4.2, and is intended to improve the reader's physical intuition about the motions in these waves. Although the material in section 4.2 is certainly elementary, it is not often covered in textbooks [Gill (1982) is the exception]. The theory of section 4.2 is applied to linear mountain waves in section 4.3. Detailed expositions of linear mountain-wave theory have been provided by a number of previous authors; therefore, section 4.3 is limited to a brief review. Downslope windstorms are discussed in section 4.4; this material draws on a number of recent studies that appear to have clarified our understanding of the dynamics governing these events. Recent progress in our understanding of the airflow over an isolated mountain is described in section 4.5.

4.2. Fundamentals of internal gravity waves

While most people have some intuitive familiarity with external gravity waves through their contact with surface water waves, many meteorologists are not familiar with the atmospheric motions that occur in an internal gravity wave. Therefore, in an attempt to gain a more thorough physical understanding of mountain waves, let us review some fundamental aspects of gravity waves. For simplicity, the influence of the Coriolis force will be neglected and the waves will be assumed to lie in the two-dimensional x–z plane.

The basic forces that give rise to internal gravity waves are buoyancy restoring forces. If an air parcel is displaced vertically in a stably stratified flow, the buoyancy difference between the parcel and its environment will produce a restoring force, accelerating the parcel back toward its equilibrium position. In the absence of pressure gradient forces, the parcel will oscillate along a vertical axis through its equilibrium position at the Brunt–Väisälä frequency

$$N = \left(\frac{g}{\theta_0}\frac{d\theta_0}{dz}\right)^{1/2}, \qquad (4.1)$$

where $\theta_0(z)$ is the vertically varying mean-state potential temperature. [The derivation of this result may be found in Holton (1979) p. 50, or Gill (1982) p. 129.] If pressure gradient forces are again neglected and the parcel is constrained to oscillate along a path tilted off the vertical axis by an angle ϕ (one might imagine that it is attached to a slanted rod by a frictionless coupling), buoyancy forces will produce an identical type of oscillatory motion at the reduced frequency $N\cos\phi$. The factor $\cos\phi$ arises from the reduction in the component of gravitational acceleration, and the reduction in the apparent stratification, along the slanting path [see Holton (1979) p. 160, or Gill (1982) p. 132]. Thus, buoyancy driven oscillations can be accommodated at any frequency less than N by selecting a suitable angle for the sloping trajectory. However, it is not possible to support buoyancy driven parcel oscillations at frequencies greater than the Brunt–Väisälä frequency.

The pressure gradients and buoyancy forces in an internal gravity wave play the role of the slanting rod with the frictionless coupling in the preceding illustration—they act in concert to keep air parcels oscillating along a path slanted off-vertical at the angle that matches the frequency of the wave to the resonant frequency $N\cos\phi$. In order to see how this works, consider small-amplitude perturbations about a basic state at rest in a Boussinesq atmosphere, which satisfy the linearized governing equations

$$\frac{\partial u}{\partial t} + \frac{\partial P}{\partial x} = 0, \qquad (4.2)$$

$$\frac{\partial w}{\partial t} + \frac{\partial P}{\partial z} = b, \qquad (4.3)$$

$$\frac{\partial b}{\partial t} + N^2 w = 0, \qquad (4.4)$$

$$\frac{\partial u}{\partial x} + \frac{\partial w}{\partial z} = 0, \qquad (4.5)$$

where $b = g(\theta - \theta_0)/\theta_s$, $P = c_p\theta_s(\pi - \pi_0)$, and $N^2 = (g/\theta_s)d\theta_0/dz$ [the Boussinesq version of Eq. (4.1)]. Here π is the Exner function pressure $(p/p_s)^{R/c_p}$; $\pi_0(z)$ and $\theta_0(z)$ are the hydrostatically balanced basic-state pressure and potential temperature; p_s and θ_s are constant reference values, and b represents buoyancy. Equations (4.2)–(4.5) are formally equivalent to the linearized Boussinesq equations for an incompressible fluid (for which $b = -g(\rho - \rho_0)/\rho_s$, $P = (p - p_0)/\rho_s$, and $N^2 = -(g/\rho_s)d\rho_0/dz$). These equations reduce to a single equation for w:

$$\frac{\partial^2}{\partial t^2}\left[\frac{\partial^2}{\partial x^2} + \frac{\partial^2}{\partial z^2}\right]w + N^2\frac{\partial^2 w}{\partial x^2} = 0. \qquad (4.6)$$

A solution to (4.6), describing the vertical velocity perturbations in an internal gravity wave, is

$$w = w_0 \cos(kx + mz - \nu t). \qquad (4.7)$$

Given the preceding expression for w, (4.5) implies

$$u = -\frac{m}{k} w_0 \cos(kx + mz - \nu t), \qquad (4.8)$$

and (4.4) requires

$$b = \frac{N^2}{\nu} w_0 \sin(kx + mz - \nu t). \qquad (4.9)$$

Finally, (4.2) and (4.8) imply

$$P = -\frac{\nu m}{k^2} w_0 \cos(kx + mz - \nu t). \qquad (4.10)$$

A schematic diagram is plotted in Fig. 4.1, representing the perturbation fields (4.7)–(4.10) at a single instant in time. All perturbation quantities are constant along the straight lines $kx + mz$ = constant; these are lines of constant phase, or "wave fronts." Equations (4.7) and (4.8) imply that $u/w = -m/k$; thus *all air parcel motions are parallel to the wave fronts.* The angle between the slanting air parcel trajectories and the vertical (ϕ) is determined by the dispersion relation

$$\nu^2 = \frac{N^2 k^2}{k^2 + m^2} = N^2 \cos^2\phi. \qquad (4.11)$$

It follows that the air parcels in an internal gravity wave oscillate along trajectories exactly as if they were attached to a rod slanting off the vertical at the angle $\phi = \cos^{-1}(\nu/N)$.

The case illustrated in Fig. 4.1 is one in which $k < 0$ and $m < 0$ (ν is assumed greater than zero by convention, in order to eliminate redundant solutions). The slanting solid lines in Fig. 4.1 represent those lines of constant phase where the perturbation velocities reach their extrema; the direction of the velocity along each of these lines is indicated by open arrows. These lines are also the location of the extrema in the pressure field, and are labeled as required by (4.10). The perturbation velocities and pressure are zero along the dashed wave fronts. Equation (4.9) shows that, when $k < 0$, $m < 0$, the extrema in the buoyancy field lead the extrema in w by 90°. As a consequence, the buoyancy perturbations achieve their extremum along the dashed wave fronts. These dashed lines have been labeled "most buoyant" or "least buoyant" using (4.9). Now consider how the situation in Fig. 4.1 will evolve with time. According to (4.4), buoyancy per-

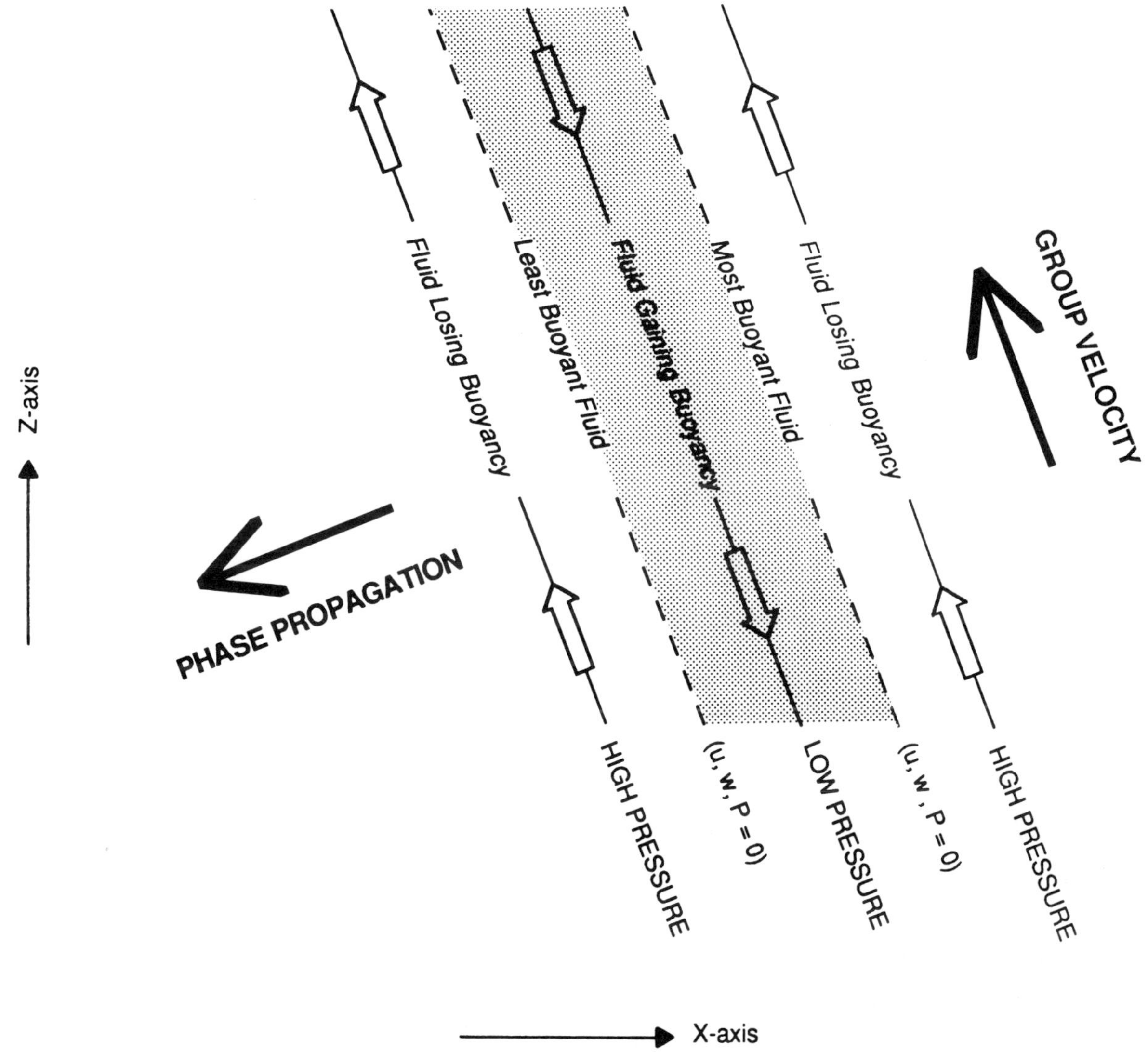

FIG. 4.1. The instantaneous distribution of velocity, pressure, and buoyancy perturbations in an internal gravity wave. This is a view in the x–z plane. The phase of the wave is constant along the slanting, dashed, and solid lines. Velocity and pressure perturbations have extrema along the solid lines; buoyancy perturbations are zero along the solid lines. Buoyancy perturbations have extrema, and velocity and pressure perturbations are zero along the dashed lines. Small arrows indicate the perturbation velocities, which are always parallel to the lines of constant phase. Large heavy arrows indicate the direction of phase propagation and the group velocity.

turbations develop through the action of vertical advection on the mean-state stratification. Since $w < 0$ everywhere within the shaded region in Fig. 4.1, the air within the shaded region is becoming more buoyant; therefore, the line of most buoyant fluid must be moving into the shaded region. Since the wave maintains itself as a coherent structure, the phase lines associated with the other perturbation quantities also propagate down and to the left, as indicated in Fig. 4.1. Note, for example, how the phase lines in the horizontal perturbation velocity field will shift in response to the horizontal pressure gradient. As evident in Fig. 4.1, *the lines of constant phase, or wave fronts, move perpendicular to the actual air-parcel trajectories.*

The total kinetic and potential energy (per unit mass) associated with the wave perturbations may be expressed as

$$E = \frac{1}{2}\left(u^2 + w^2 + \frac{b^2}{N^2}\right). \tag{4.12}$$

The governing equations (4.2)–(4.5) require that E satisfy

$$\frac{\partial E}{\partial t} + \frac{\partial Pu}{\partial x} + \frac{\partial Pw}{\partial z} = 0. \tag{4.13}$$

Thus, the waves transport energy in the direction of the energy flux vector $P\mathbf{V}$. Examination of the correlation between P and $\mathbf{V}$ in Fig. 4.1 shows that energy is transported parallel to the air parcel trajectories and the wave fronts in the direction indicated by the "group velocity" vector. The group velocity, defined as $\mathbf{c}_g = (\partial\nu/\partial k, \partial\nu/\partial m)$, points in the direction of energy propagation because it satisfies the relation $\widehat{P\mathbf{V}} = \hat{E}\mathbf{c}_g$, where $(\hat{\ })$ denotes an average over one wavelength. In summary, the preceding analysis of the structure of an internal gravity wave shows that *the phase speed is perpendicular to the energy flux (or group velocity) and the energy flux is directed parallel to the actual air parcel trajectories.*

4.3. Application to small-amplitude mountain waves

Several authors have provided thorough discussions of the linear theory of mountain waves in previous reviews or textbooks (Queney et al. 1960; Smith 1979; Gill 1982; Durran 1986b). Therefore, an extended exposition of linear mountain-wave theory will not be presented in the following section. Our focus will be on the application of the theory of section 4.2 to the basic case of a uniform airstream flowing over an infinite series of sinusoidal mountains. The generalization of this basic case to situations with more complex atmospheric structure and isolated topography will be summarized, but not discussed in detail.

4.3.1. Sinusoidal ridges; constant wind speed and stability

The most fundamental properties of small-amplitude mountain waves can be profitably examined by considering the steady-state, two-dimensional airflow over an infinite series of periodic ridges of the form

$$h(x) = h_m \cos kx. \tag{4.14}$$

The two-dimensional assumption is appropriate if the mountains extend indefinitely in the direction parallel to the ridge. As further simplifications, the Rossby number governing the flow (u_0k/f) is assumed large, so Coriolis accelerations may be neglected, and the atmosphere is assumed inviscid and Boussinesq. Then, if the mean horizontal wind speed u_0 is a constant, the dynamics of the problem are governed by equations identical to (4.2)–(4.5), except that each time derivative $\partial/\partial t$ is replaced by an advection term $u_0\partial/\partial x$. After two integrations with respect to x (the integration constants are zero because w has no mean or linear trend), the equation corresponding to (4.6) reduces to

$$\frac{\partial^2 w}{\partial x^2} + \frac{\partial^2 w}{\partial z^2} + \frac{N^2}{u_0^2} w = 0. \tag{4.15}$$

Solutions to (4.15) may be written in the form

$$w = \Re\{Ae^{i(kx+mz)} + Be^{i(kx-mz)}\}, \tag{4.16}$$

where $\Re$ denotes the real part, A and B are complex coefficients, and

$$m = (N^2/u_0^2 - k^2)^{1/2}. \tag{4.17}$$

In (4.16), redundant solutions are eliminated by choosing the positive root for m and requiring $k > 0$.

The coefficients A and B are determined by the boundary conditions. At the lower boundary, the velocity normal to the topography must vanish. This condition can be approximated, to the same order of accuracy as the linearized equations, by

$$w(x, 0) = u_0 \frac{\partial h}{\partial x} = -u_0 k h_m \sin kx. \tag{4.18}$$

Substitution of (4.16) into (4.18) yields the conditions

$$\Re\{A\} + \Re\{B\} = 0; \quad \Im\{A\} + \Im\{B\} = u_0 h_m k, \tag{4.19}$$

where $\Im$ denotes the imaginary part. A second boundary condition is imposed as $z \to \infty$. The nature of this boundary condition depends on the vertical structure of the solution, which depends on the relative magnitudes of u_0k and N. When $u_0k > N$, the solution (4.16) may be written

$$w = \Re\{Ae^{-\mu z}e^{ikx} + Be^{\mu z}e^{ikx}\}, \tag{4.20}$$

where, according to (4.17), μ is a real number equal to $(k^2 - N^2/u_0^2)^{1/2}$. In this Boussinesq system, it is physically unreasonable for the wave amplitude to grow exponentially without bound as $z \to \infty$, so the upper boundary condition requires that $B = 0$. Then from (4.19), $\Re\{A\} = 0$ and $\Im\{A\} = u_0 h_m k$; the complete solution appears in (4.23).

Now consider the case where $u_0k < N$, then m is real and both A and B are coefficients of waves that are bounded as $z \to \infty$. Instead of a simple condition on the boundedness of the solution, the appropriate upper boundary condition for this linear steady-state problem is the "radiation condition," which requires that all waves at an arbitrarily great height above the mountain must be transporting energy away from the mountain. In general, the imposition of a radiation boundary condition as $z \to \infty$ is justified whenever all processes that could generate waves (such as nonlinear wave–wave interactions or reflections from regions where there is a rapid change in atmospheric structure) are contained within some finite height of the topography. In order to impose the radiation condition, it is necessary to determine the direction of the energy transport by each wave in (4.16). One approach, due to Eliassen and Palm (1960), is to determine the vertical energy flux $\widehat{Pw}$. The perturbation pressure associated with each wave may be evaluated from the horizontal momentum equation, the continuity equation, and (4.16). In the wave with coefficient A (the upstream tilting wave, whose vertical velocity will be denoted w_a),

$$P_a = \frac{u_0 m}{k} \Re\{Ae^{i(kx+mz)}\}. \tag{4.21}$$

In the downstream tilting wave

$$P_b = -\frac{u_0 m}{k} \Re\{Be^{i(kx-mz)}\}. \tag{4.22}$$

It follows that $\widehat{P_a w_a} > 0$, so the upstream tilting wave is associated with upward energy flux. Similarly, $\widehat{P_b w_b} < 0$, and the downstream tilting wave is associated with downward energy flux. The radiation boundary condition is thus imposed by requiring $B = 0$. Then using (4.19), $\Re\{A\} = 0$ and $\Im\{A\} = u_0 h_m k$.

In summary, the perturbation vertical velocity field in waves forced by the sinusoidal terrain profile (4.14) can be written

$$w(x, z) = \begin{cases} -u_0 h_m k e^{-\mu z} \sin kx, & u_0 k > N; \\ -u_0 h_m k \sin(kx + mz), & u_0 k < N. \end{cases} \tag{4.23}$$

The difference between these two wave structures is illustrated in Fig. 4.2. In the case $u_0 k > N$ (Fig. 4.2a), the waves decay exponentially with height (evanescent waves) and the wave crests are aligned vertically. In the case $u_0 k < N$ (Fig. 4.2b), the waves propagate vertically without loss of amplitude and the wave crests tilt upstream with height. The waves decay away from the forcing when the intrinsic frequency exceeds the Brunt–Väisälä frequency ($u_0 k > N$) because, as discussed in section 4.2, there is no way for buoyancy restoring forces to support the oscillation. On the other hand, when the intrinsic frequency is less than the Brunt–Väisälä frequency, vertical propagation occurs because buoyancy restoring forces can support air-parcel oscillations along a path slanted off the vertical at an angle $\phi = \cos^{-1}(u_0 k/N)$. In steady mountain waves ϕ is the angle at which lines of constant phase tilt off the vertical (see the dashed line in Fig. 4.2b). A schematic diagram of the perturbation velocity, pressure, and density fields in Fig. 4.2b would be identical to that in Fig. 4.1. However, since there is a mean flow in the mountain-wave problem, the actual air-parcel trajectories do not follow the slanted paths in Fig. 4.1, but rather the wavy lines shown in Fig. 4.2.

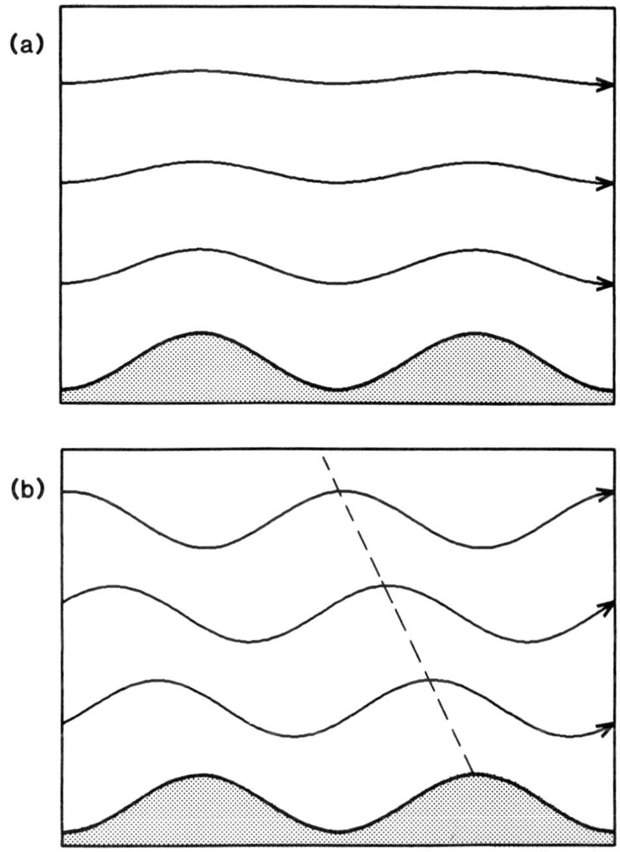

FIG. 4.2. Streamlines in steady airflow over an infinite series of sinusoidal ridges when (a) $u_0 k > N$, or (b) $u_0 k < N$. The dashed line (b) shows the upstream tilt of the lines of constant phase. Unless otherwise stated, the airflow in this and all subsequent figures is from left to right.

4.3.2. *Isolated mountain; constant wind speed and stability*

The mountain-wave solution (4.23) applies only to an airstream with constant basic-state wind speed and stability flowing across an endless series of sinusoidal ridges. If more realistic terrain profiles and atmospheric structures are considered, other linear solutions can be obtained that strongly resemble observed mountain waves. In the remainder of this section, we will describe how the wave response is influenced by isolated topography and vertical variations in atmospheric wind speed and stability. In order to focus on the physical results, the mathematical derivations will be omitted; they can be found in Smith (1979) or Queney et al. (1960).

Suppose that the mountain contour consists of a single ridge, so that the terrain elevation drops to some reference level at all distances sufficiently far upstream and downstream. Just as Fourier series can be used to represent a wide variety of periodic functions with an infinite sum of sines and cosines, the isolated mountain can generally be constructed from periodic functions by the use of Fourier transforms. The Fourier transform (F) of a real function ϕ and its inverse (F^{-1}) may be defined:

$$\begin{aligned} \tilde{\phi}(k) &= F[\phi(x)] = \frac{1}{\pi}\int_{-\infty}^{\infty} \phi(x)e^{-ikx}dx, \\ \phi(x) &= F^{-1}[\tilde{\phi}(k)] = \Re\int_0^{\infty} \tilde{\phi}(k)e^{ikx}dk. \end{aligned} \tag{4.24}$$

The Fourier transform is particularly useful in this application because it has the property that $F(\partial^n\phi/\partial x^n) = (ik)^n\tilde{\phi}$.

When N and u_0 are constant, each component of the Fourier transformed vertical velocity $\tilde{w}(k, z)$ must satisfy the Fourier transform of the governing equation (4.15),

$$\frac{\partial^2\tilde{w}}{\partial z^2} + \left(\frac{N^2}{u_0{}^2} - k^2\right)\tilde{w} = 0. \qquad (4.25)$$

The solution to (4.25), subject to appropriate upper and lower boundary conditions, is

$$\tilde{w}(k, z) = iku_0\tilde{h}(k)\exp[i(N^2/u_0{}^2 - k^2)^{1/2}z], \quad k > 0. \qquad (4.26)$$

Equation (4.26) is just the complex analog of (4.23); therefore, each component $\tilde{w}(k, z)$ of the transformed vertical velocity is identical to the velocity forced by an infinite series of sinusoidal ridges having wavenumber k and amplitude $\tilde{h}(k)$. Thus, the solutions obtained in section 4.3.1. are also applicable to the case of isolated topography. The only complication arises from the requirement that after $\tilde{w}(k, z)$ is determined, the total vertical velocity $w(x, z)$ must be obtained by computing an inverse Fourier transform. According to (4.26), the relative weight attached to each individual wavenumber in the composite solution is determined by Fourier transform of the terrain profile.

Queney (1948) calculated solutions for the waves generated in a fluid, with constant N and u_0, flowing over the "Witch of Agnesi" terrain profile:

$$h(x) = \frac{h_m a^2}{a^2 + x^2}, \qquad (4.27)$$

where h_m is the maximum mountain height and a determines the mountain width. For a very narrow mountain, $u_0a^{-1} \gg N$; the profile is dominated by wavenumbers greater than N/u_0, and the mountain primarily forces evanescent waves (Fig. 4.3a; cf. Fig. 4.2a). For a wide mountain, $u_0a^{-1} \ll N$; the dominant wavenumbers are less than N/u_0, and the waves propagate vertically with lines of constant phase tilting upstream (Fig. 4.3b; cf. Fig. 4.2b). The wide mountain limit, being equivalent to the hydrostatic limit, eliminates the dependence of vertical structure on the horizontal wavenumber in (4.26). As a result, the mountain profile is reproduced at every level that is an integral multiple of $2\pi u_0/N$. This result is independent of the shape of the mountain contour. It should be emphasized that if an isolated mountain is sufficiently wide that it only forces hydrostatic waves (but not so wide that Coriolis forces become important), there will only be one wave crest in the air flowing over the mountain. Additional crests do not appear downstream from the mountain unless nonhydrostatic effects are significant. The most obvious way that nonhydrostatic effects manifest themselves is in the development of trapped lee waves.

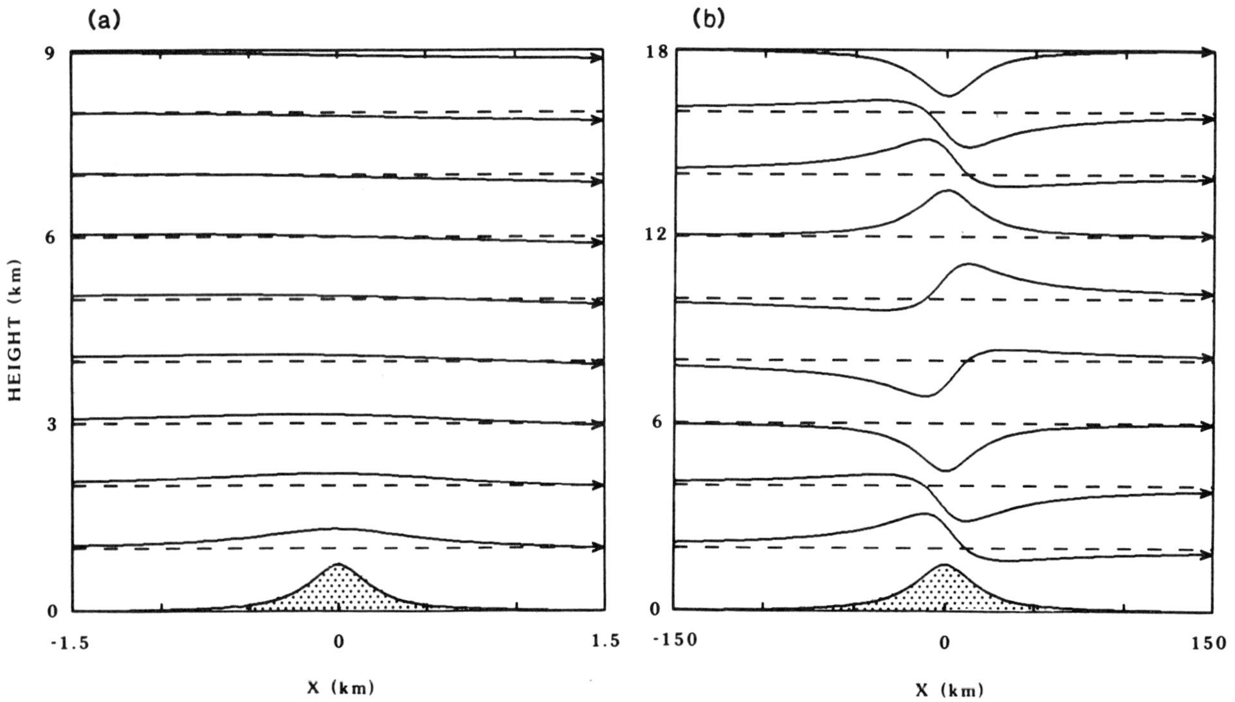

FIG. 4.3. Streamlines in steady airflow over an isolated bell-shaped ridge when (a) $u_0a^{-1} \gg N$, or (b) $u_0a^{-1} \ll N$.

4.3.3. *Vertical variations in wind speed and stability*

Now consider the effects of vertical variations in the mean atmospheric structure. The most important variations are those that occur in u_0 and N. Vertical variations in u_0 and N can support a new qualitatively different type of wave—the trapped lee wave. The airflow in a series of trapped lee waves (also known as resonant lee waves) is shown in Fig. 4.4; most of the wave activity is confined to the lower troposphere on the lee side of the mountain. Scorer (1949) used linear theory to show that trapped waves arise as a result of vertical variations in $N(z)$ and $u_0(z)$. These variations introduce an additional term in the governing equation, so that (4.15) becomes

$$\frac{\partial^2 w}{\partial x^2} + \frac{\partial^2 w}{\partial z^2} + l^2 w = 0, \tag{4.28}$$

where l is the Scorer parameter, defined as

$$l^2 = \frac{N^2}{u_0^2} - \frac{1}{u_0}\frac{d^2 u_0}{dz^2}. \tag{4.29}$$

Scorer considered the case where l was constant within each of two layers. (Note that a discontinuity in l does not require a discontinuity in either θ_0 or u_0.) Scorer showed that a necessary condition for the existence of trapped waves in the two-layer problem is that

$$l_L^2 - l_U^2 > \frac{\pi^2}{4H^2}. \tag{4.30}$$

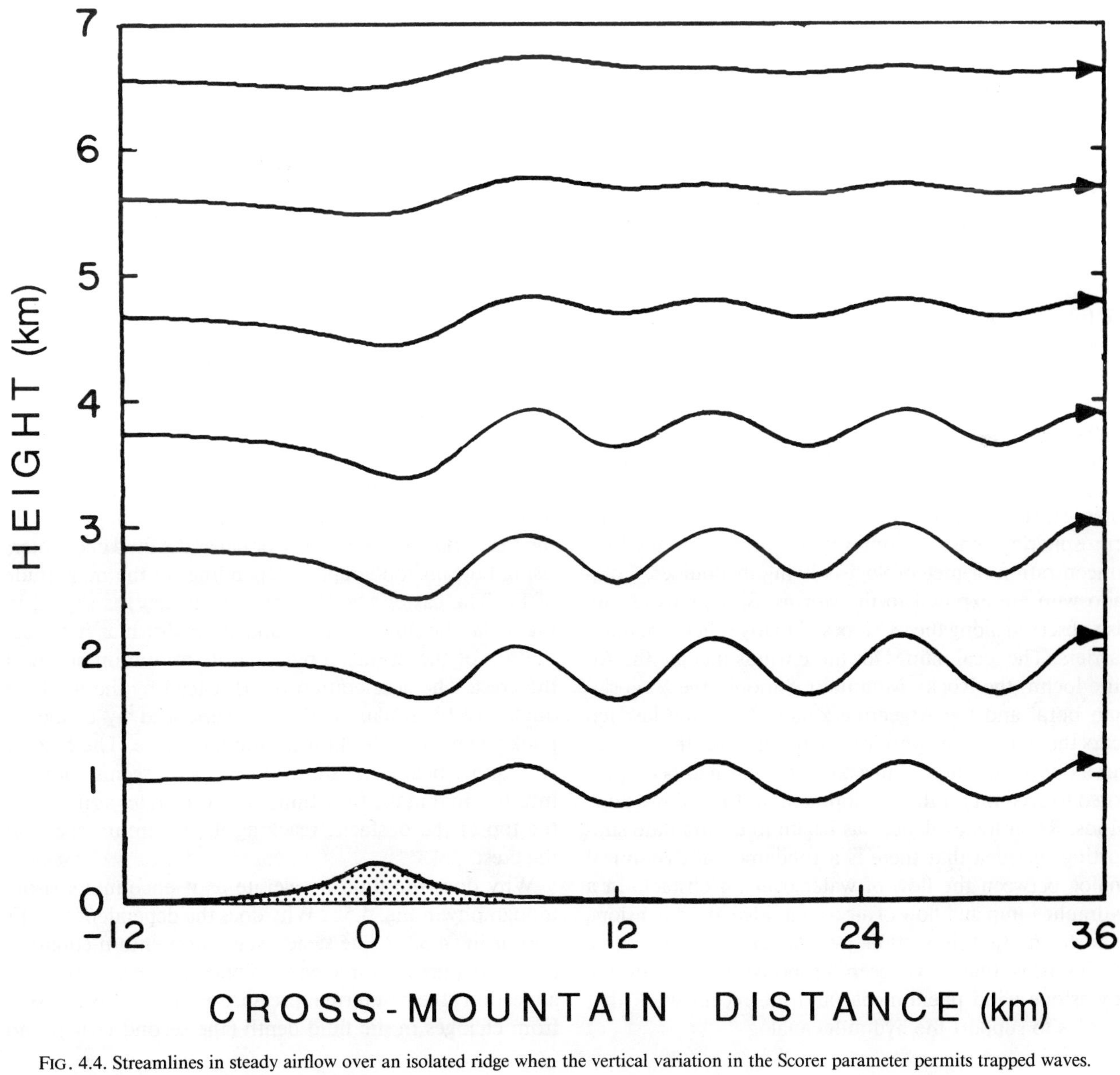

FIG. 4.4. Streamlines in steady airflow over an isolated ridge when the vertical variation in the Scorer parameter permits trapped waves.

In the preceding paragraphs, l_U and l_L are the Scorer parameters in the upper and lower layers, and H is the depth of the lower layer. As implied in Eq. (4.30), the difference in wave propagation characteristics in the two layers must exceed a certain threshold before the waves can be "trapped."

Earlier in this section it was emphasized that vertically propagating waves are permitted when $u_0 k < N$. The appropriate generalization of this result to an environment with vertical wind shear is the condition that $k < l$. The horizontal wavenumber of any resonant lee wave in the two-layer system satisfies $l_L > k > l_U$, implying that the wave propagates vertically in the lower layer and decays exponentially with height in the upper layer. As shown in Fig. 4.4, trapped waves have no tilt, even though they can propagate vertically in the lower layer. The reason for this is that wave energy is repeatedly reflected, without loss of amplitude, from the upper layer and the flat ground downstream from the mountain. As a result, the downstream disturbance is the superposition of equal-amplitude upward and downward propagating waves, a combination that has no tilt.

Vertical variations in l not only provide a way of supporting trapped waves, they can also modify the amplitude of long-wavelength hydrostatic waves that propagate vertically through both layers. The influence of rapid changes in l on vertically propagating waves will be discussed in the following section on downslope windstorms.

4.4. Downslope windstorms

Every few years the eastern slope of the Colorado Front Range (part of the Rocky Mountains) experiences a damaging windstorm, with peak gusts as high as 60 m s^{-1}. These storms have received considerable attention in the atmospheric science community due, in part, to the high concentration of meteorologists living in Boulder, Colorado who are exposed to the storms. Similar winds are also observed along the lee slopes of many other mountain barriers. The local names for these winds include the Alpine foehn, the Rocky Mountain chinook, the Yugoslavian bora, and the Argentine zonda. Over the last ten years there has been considerable debate over the relative merits of three different mechanisms that had been proposed to account for the development of strong downslope winds. Recently, evidence has begun to accumulate supporting the idea that there is a fundamental dynamical analog between the flow of water over an obstacle in a hydraulic jump and flow of air over a ridge in a downslope windstorm. In this section we will examine the three mechanisms that have been proposed to explain the downslope wind phenomena, and the recent work that appears to support the hydraulic analog.

4.4.1. *Three explanations for the production of severe downslope winds*

The oldest of the three conceptual models was proposed by Long (1953a) over 35 years ago. Long suggested that there is a fundamental similarity between downslope windstorms and hydraulic jumps—essentially the same theory that is currently enjoying a resurgence in popularity. In order to better understand Long's hypothesis, let us consider the dynamics of a homogeneous fluid flowing over a ridgelike obstacle. Assume that the flow is in hydrostatic balance and bounded by a free surface. Then, if there are no variations in the coordinate direction parallel to the ridge axis, the steady-state behavior of the system is governed by the shallow-water momentum and continuity equations:

$$u\frac{\partial u}{\partial x} + g\frac{\partial D}{\partial x} + g\frac{\partial h}{\partial x} = 0, \tag{4.31}$$

$$\frac{\partial uD}{\partial x} = 0. \tag{4.32}$$

Here x is the coordinate direction oriented perpendicular to the ridgeline, u is the velocity in the x direction, D is the thickness of the fluid, and h is the height of the obstacle. Using (4.32) to substitute for $\partial u/\partial x$ in (4.31), one obtains:

$$(1 - \mathrm{Fr}^{-2})\frac{\partial(D+h)}{\partial x} = \frac{\partial h}{\partial x}, \tag{4.33}$$

where the Froude number Fr, defined as

$$\mathrm{Fr}^2 = \frac{u^2}{gD}, \tag{4.34}$$

is the ratio of the fluid velocity to the speed of propagation of linear shallow-water gravity waves. According to (4.33), the free surface can either rise or fall as the fluid encounters rising bottom topography, depending on the magnitude of Fr. The case Fr > 1 (supercritical flow) is shown in Fig. 4.5a; the fluid thickens and slows down as it crosses the top of the obstacle, reaching its minimum speed at the crest. The accelerations experienced by the fluid are qualitatively similar to those experienced by a hockey puck traversing a frictionless mound of ice. The case Fr < 1 (subcritical flow) shown in Fig. 4.5b seems counterintuitive in that the fluid thins and accelerates as it crosses the top of the obstacle, reaching its maximum speed at the crest.

Why does the fluid accelerate as it encounters rising topography in Fig. 4.5b? Why does the dependence on Fr appear in (4.33)? The steady-state momentum equation (4.31) requires a three-way balance between nonlinear advection (the first term), pressure gradient forces arising from changes in the fluid depth (the second term), and

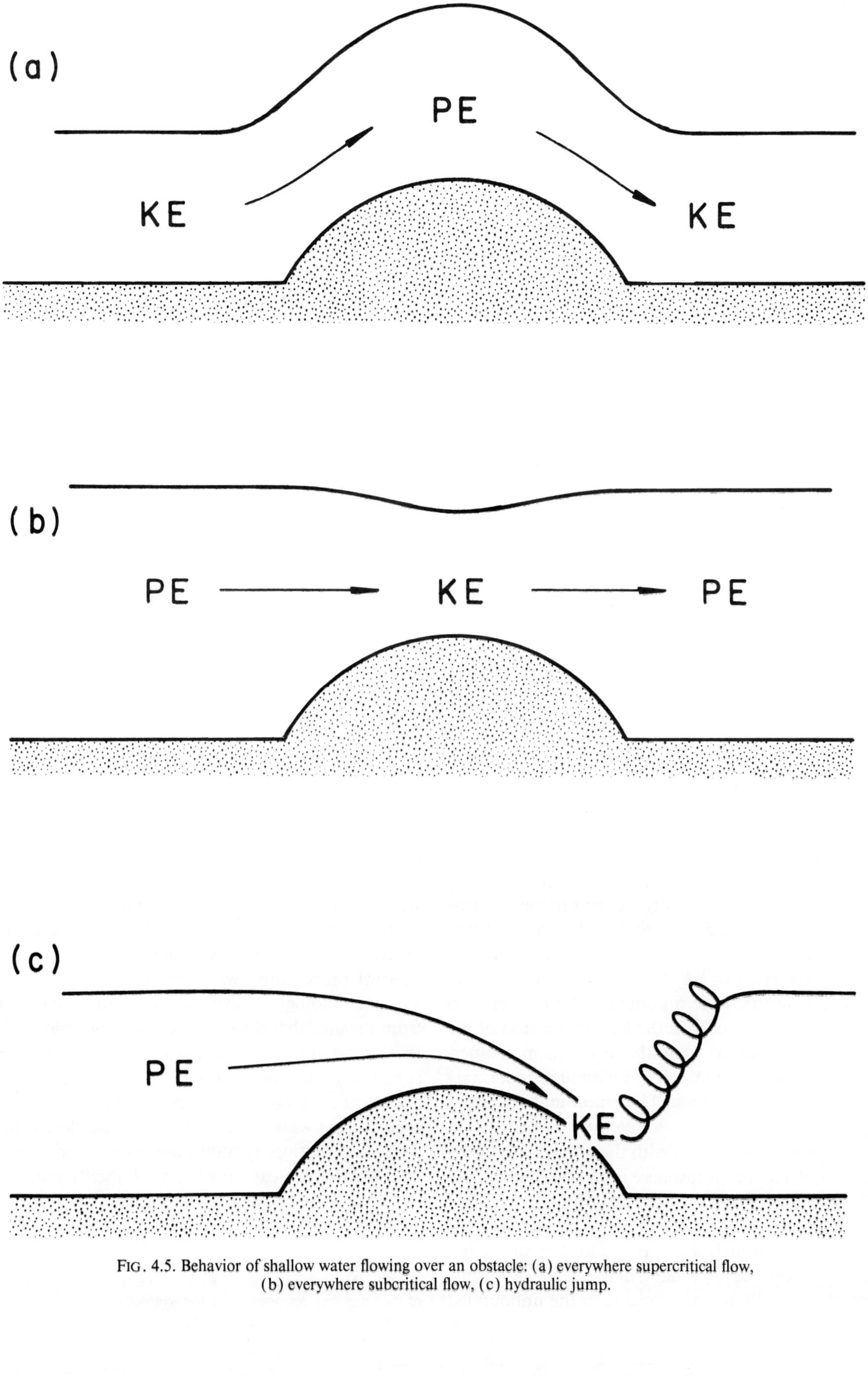

FIG. 4.5. Behavior of shallow water flowing over an obstacle: (a) everywhere supercritical flow, (b) everywhere subcritical flow, (c) hydraulic jump.

the component of gravity parallel to the sloping topography (the third term). Using (4.32) one can show that

$$\left(u\frac{\partial u}{\partial x}\right)\Big/\left(g\frac{\partial D}{\partial x}\right)=\left(u\frac{\partial u}{\partial x}\right)\Big/\left(\frac{-gD}{u}\frac{\partial u}{\partial x}\right)=-\mathrm{Fr}^2. \tag{4.35}$$

Thus, advection always opposes the pressure gradient due to changes in fluid depth. Furthermore, *Fr^2 is the ratio of the magnitude of nonlinear advection to the magnitude of the pressure gradient generated by changes in the fluid depth.* In supercritical flow (Fr > 1) nonlinear advection dominates the pressure gradient term and the three-way balance in (4.31) is satisfied when fluid parcels are accelerated in the same direction as the gravitational force. Consequently, as a fluid parcel ascends the obstacle, it slows, converting kinetic energy (KE) to potential energy (PE); after passing the crest, it reaccelerates as PE is converted back to KE (Fig. 4.5a). On the other hand, in subcritical flow (Fr < 1) the pressure gradient term dominates advection and (4.31) is balanced when fluid parcels accelerate in the direction opposite the gravitational force. Then, as shown in Fig. 4.5b, a fluid parcel ascending the obstacle accelerates as the free surface drops and PE is converted to KE; after passing the crest it decelerates as KE is converted back to PE. The disturbance centered over the obstacle in Fig. 4.5b is a stationary surface gravity wave.

The flow regime that Long proposed as a model for downslope windstorms is shown in Fig. 4.5c. If there is a sufficient acceleration in the stationary gravity wave; i.e., a sufficient increase in velocity and decrease in thickness as the fluid ascends toward the crest, a transition from subcritical to supercritical flow occurs at the top of the obstacle (Fig. 4.5c). Since the flow along the lee slope is supercritical, the fluid continues to accelerate as it falls down the mountain; it eventually recovers to the ambient downstream conditions in a turbulent hydraulic jump. Very high velocities are produced along the lee slope because PE is converted to KE during the entire time that a fluid parcel traverses the mountain. The deceleration that would otherwise occur in the lee-side portion of the gravity wave is disrupted when the flow becomes supercritical. Further discussion of the hydraulic model may be found in Long (1954) and Houghton and Kasahara (1968).

One uncertainty associated with the application of the hydraulic model to the atmosphere is that the atmosphere is not bounded by a free surface. The presence of a free surface (or in different formulations, a rigid lid) prevents energy transport through the upper boundary of the hydraulic layer; however, internal gravity waves can transport energy vertically to great heights in the unbounded continuously stratified atmosphere. In at least some instances, it appears that this energy can be removed in the upper atmosphere without significantly affecting the low-level wave structure. Thus, the hydraulic model may be too limited for application to the atmosphere. Furthermore, as indicated by the variation in spacing between streamlines in Figs. 4.2b and 4.3b, the surface wind speed distribution in vertically propagating internal gravity waves is asymmetric about the crest, with a maximum along the lee slope. The preceding observations suggest an alternative to hydraulic theory, which better accounts for the possibility of vertical energy transport, namely that downslope windstorms are produced by large-amplitude vertically propagating mountain waves. Eliassen and Palm (1960) showed that when an upward propagating linear gravity wave encounters a region in which the Scorer parameter changes rapidly, part of its energy can be reflected back into a downward propagating wave. Klemp and Lilly (1975) extended the results of Eliassen and Palm, and Blumen (1965), to the case of small-amplitude hydrostatic waves in a multilayer atmosphere with constant stability and wind shear in each layer. Klemp and Lilly suggested downslope windstorms occur when the atmosphere is tuned so that the partial reflections at each interface produce an optimal superposition of upward and downward propagating waves. They found that, in practice, the most important tuning requirement is that the tropopause be located one-half vertical wavelength above the ground.

The third explanation for the development of strong downslope winds was proposed on the basis of simulations performed with a comprehensive numerical model. In a series of papers Clark and Peltier (1977, 1984), Peltier and Clark (1979, 1983), and Clark and Farley (1984) found significant increases in the lee-slope surface winds occurred after vertically propagating waves became statically unstable and "broke." The wave-breaking region was characterized by strong mixing and a local reversal of the cross-mountain flow. They proposed that this "wave-induced critical layer" acts as a boundary, reflecting upward propagating waves back toward the mountain. The terminology "wave-induced critical layer" derives from the idea that the wave overturning region influences the underlying flow in a manner similar to a turbulent (i.e., low Richardson number[1]) mean-state critical layer. A critical layer (or critical level) occurs where the phase speed of the wave is equal to the mean flow velocity. In the case of stationary mountain waves, a critical level occurs where the mean flow is zero. If the Richardson number at the critical layer is less than $\frac{1}{4}$, nonlinear waves encountering a critical layer are believed to reflect from

[1] The Richardson number, $\mathrm{Ri} = N^2/(U_z)^2$, is a measure of the ratio of the buoyancy restoring force to the shear.

that layer without significant loss in amplitude. Using linear theory to describe the reflection process, Peltier and Clark suggested that if the depth of the cavity between the self-induced critical layer and the mountain is suitably tuned, the reflections at the critical layer should produce a resonant wave that amplifies linearly with time and ultimately produces very strong surface winds. Additional numerical studies by many different researchers have repeatedly verified the importance of wave breaking in the development of strong surface winds in both the 11 January 1972 Boulder windstorm and in cases with constant upstream values of N and u_0.

4.4.2. *A comparison of the hydraulic and the vertically propagating wave theories*

The linear theory of vertically propagating waves predicts that some enhancement of the lee-side surface winds will be produced by arbitrarily small mountains and that the strength of the horizontal wind perturbation will scale linearly with the height of the mountain. In contrast, hydraulic theory suggests that strong downslope winds will only occur if the mountain height exceeds the threshold required to force a transition to supercritical flow. This fundamental difference in amplitude dependence can be used to test each theory. Consider first the simple case when the upstream profiles of u_0 and N are constant with height. Then, within the accuracy of the Boussinesq approximation, the finite-amplitude response can be calculated analytically from Long's equation (Long 1953b). As shown by Lilly and Klemp (1979), these finite-amplitude solutions are both qualitatively and quantitatively similar to the solutions obtained from linear theory. Thus, on the basis of the constant N and u_0 case, one might suppose that linear theory provides a satisfactory description of finite-amplitude mountain waves. However, as noted by Smith (1977), the assumption of constant u_0 and N places a special constraint on the nonlinear interactions possible in steady flow, and caution should be exercised in generalizing from the constant u_0 and N case to more realistic atmospheric profiles.

A slightly more complex atmospheric structure was studied by Durran (1986a), who used a numerical model to examine the effects of nonlinearity on the partial reflection of vertically propagating mountain waves in a mean state with constant wind speed and a two-layer stability structure where the Brunt–Väisälä frequency was constant within each layer. In each case, a measure of the amplitude of the wave response was obtained by calculating the pressure drag exerted on the mountain by the flow

$$\int_{-\infty}^{\infty} p \frac{\partial h}{\partial x} dx. \tag{4.36}$$

The drag was normalized by $\pi \rho N_L u_0 h_m^2/4$, the drag produced by linear hydrostatic waves on a Witch of Agnesi mountain [see Eq. (4.27)] in a single-layer atmosphere with constant wind speed and stability equal to that in the lowest layer. This normalization emphasizes the effects of atmospheric structure and nonlinearity on the drag. According to the linear theory of hydrostatic mountain waves, when the thickness of the lower layer is one-half vertical wavelength ($\pi u_0/N_L$), the normalized drag (D_n) is N_U/N_L. When the thickness of the lower layer is one-quarter vertical wavelength, the effect of the layering reverses and $D_n = N_L/N_U$.

Durran considered the four basic configurations shown in Fig. 4.6. In Figs. 4.6a,b the more stable layer ($N = 0.02$ s^{-1}) is on the bottom; in Figs. 4.6c,d the less stable layer ($N = 0.01$ s^{-1}) is on the bottom; in all cases $u_0 = 20$ m s^{-1}. The depth of the lowest layer is set to one-quarter or one-half vertical wavelength, so that linear theory predicts values for D_n of 2.0 in the cases in panels 4.6a,d and 0.5 for those in panels 4.6b,c. Numerical simulations conducted with a maximum value of $Nh_m/u_0 = 0.001$ confirmed that the model was capable of reproducing the linear result in circumstances where nonlinear effects are genuinely insignificant. When $(Nh_m/u_0)_{max}$ was increased to 0.6, nonlinear processes had a significant influence on the solution. The amplification predicted in the linear limit in panels 4.6a,d disappeared in the finite-amplitude simulations as D_n dropped to values less than or equal to 1. The most dramatic differences between linear theory and the finite-amplitude simulations appeared in Fig. 4.6b, where D_n increased from a linear value of 0.5 to a nonlinear value of 2.9. Although there are quantitative differences, the waves in Figs. 4.6a,c,d are qualitatively similar to the vertically propagating waves described by linear theory. On the other hand, the situation in panel 4.6b is very different: it is the only finite-amplitude case associated with high drag (and high surface winds) and the lower-layer flow qualitatively resembles a hydraulic jump.

A detailed look at the effects of nonlinearity on the drag and surface wind speed in a two-layer atmosphere similar to that shown in Fig. 4.6b is provided in Fig. 4.7. The normalized drag D_n is relatively independent of mountain height for $N_L h_m/u_0 < 0.2$ and $N_L h_m/u_0 > 0.6$. In between these two limits, there is a rapid transition between low and high drag regimes. Linear theory predicts that the mountain drag should vary as h_m^2, and this amplitude dependence is removed when the drag is normalized. Thus, the variation of D_n with $N_L h_m/u_0$ is due exclusively to nonlinear effects. The amplitude dependence revealed in Fig. 4.7 is qualitatively similar to the increase in drag that occurs in hydraulic theory as the height of the obstacle becomes sufficient to force a transition from subcritical to supercritical flow. The finite-amplitude results shown

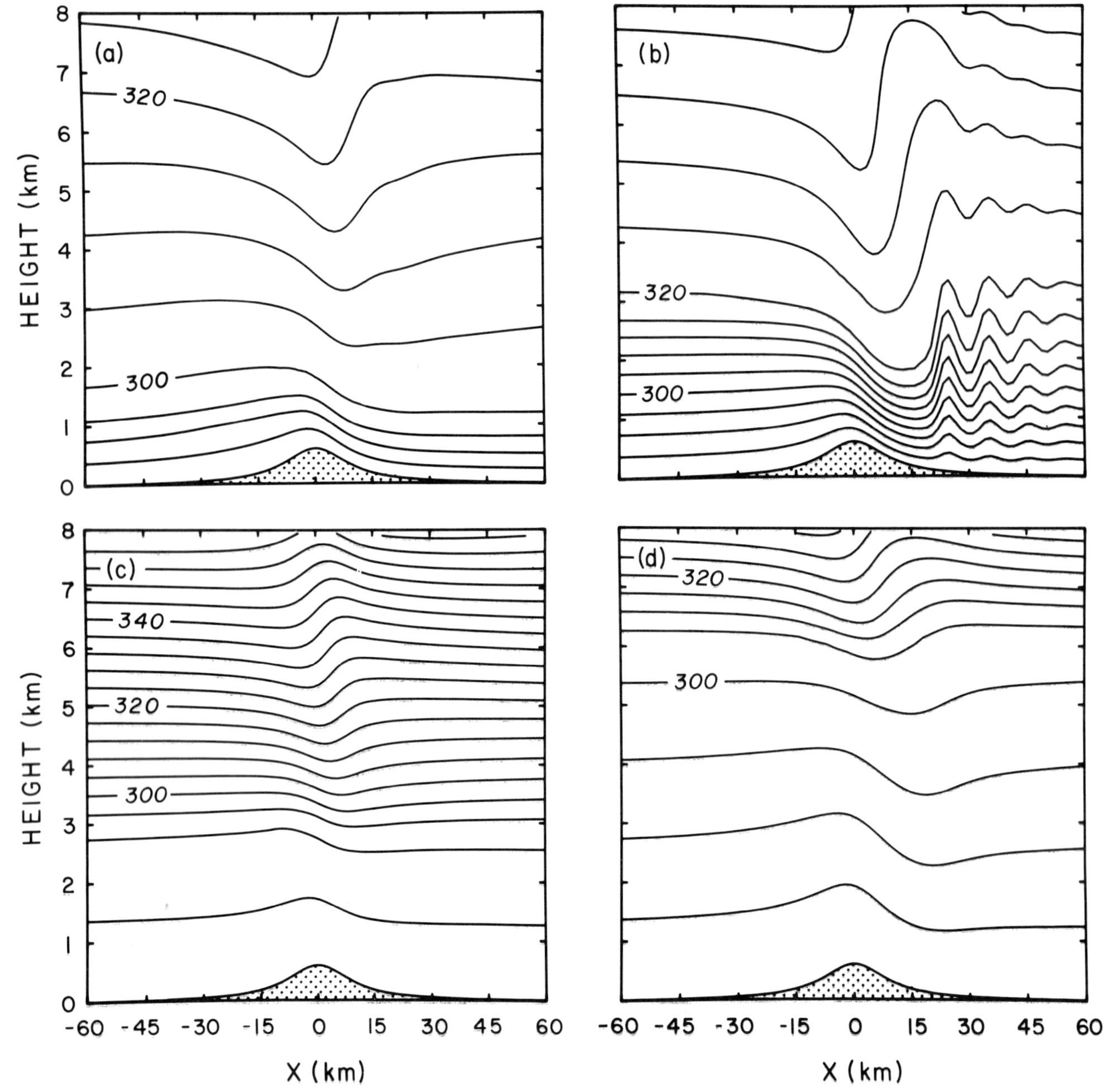

FIG. 4.6. Isentropes in a two-layer atmosphere flowing over an isolated mountain at a nondimensional time $u_0t/a = 20$. In each case, the maximum value of Nh_m/u_0 is 0.6. (a) interface at $\frac{1}{4}$ vertical wavelength, $N_L = 2N_U$; (b) interface at $\frac{1}{2}$ wavelength, $N_L = 2N_U$; (c) interface at $\frac{1}{4}$ wavelength, $N_L = \frac{1}{2}N_U$; (d) interface at $\frac{1}{2}$ wavelength, $N_L = \frac{1}{2}N_U$. (From Durran 1986a.)

in Fig. 4.7 deviate significantly from linear theory, even when the mountain is very small. For example, when $N_L h_m/u_0 = 0.3$, D_n is more than double its true linear value; however, as shown by Lilly and Klemp (1979), if N and u_0 were everywhere constant and Nh_m/u_0 was 0.3, there would be very little quantitative difference between the linear drag and the drag in the finite-amplitude solution. Thus, the sensitivity of the solution to nonlinear processes is greatly enhanced by the presence of the interface.

Further similarities between the two-layer solutions and hydraulic theory are shown in Fig. 4.8, in which the mountain height is fixed and the depth of the lowest, most stable layer is varied. As the depth of the lower layer increases, the flow in the lower layer first assumes a form similar to everywhere supercritical flow (Fig. 4.8a), then it resembles a propagating jump (Fig. 4.8b), a stationary jump (Fig. 4.8c), and finally, everywhere subcritical flow (Fig. 4.8d). The same dependence of the flow on fluid depth is observed in hydraulic theory (Houghton and Ka-

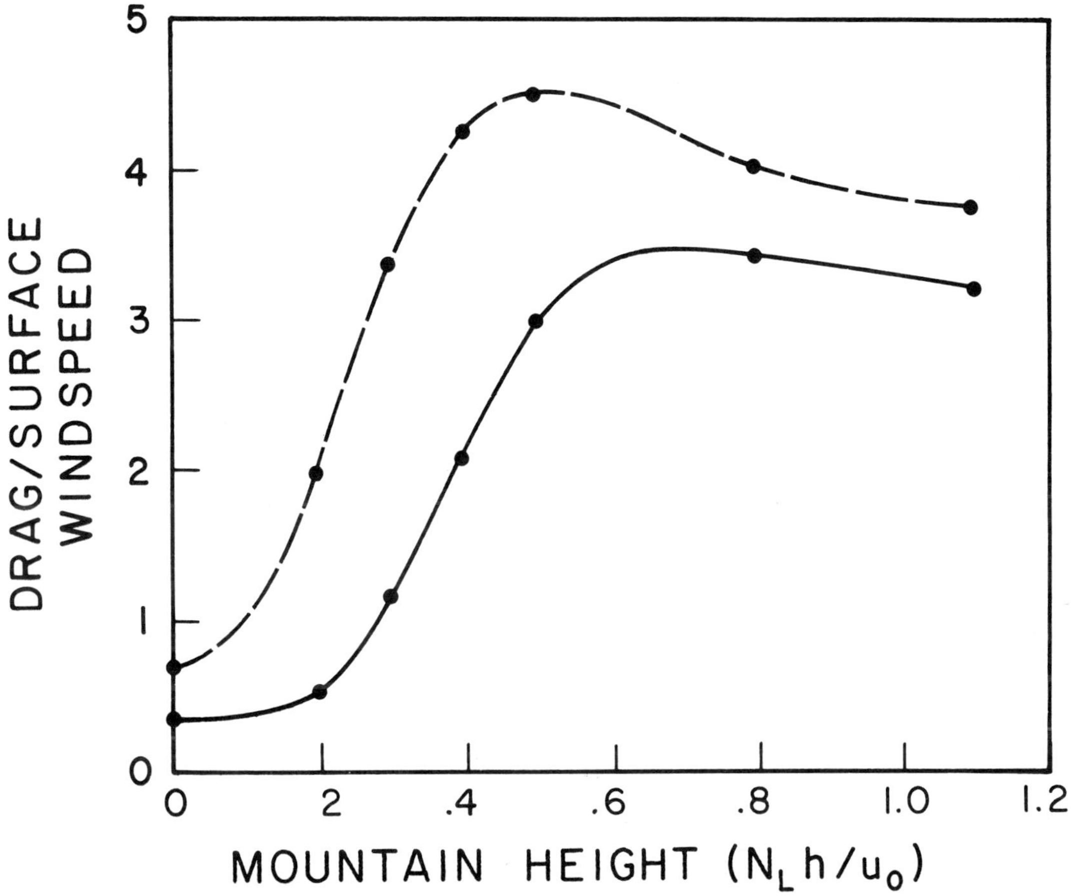

FIG. 4.7. Pressure drag (solid) and maximum surface wind speed perturbation (dashed) as a function of nondimensional mountain height $N_L h_m / u_0$ for a two-layer atmosphere with $N_L = 2.5N_U$. The drag is normalized by $\pi \rho N u_0 h_m^2/4$ and the wind speed by $Nh_m/2$ where ρ, u_0, and N are evaluated at the surface. (From Durran 1986a.)

sahara 1968). In summary, Durran's (1986a) results suggest that the partial reflection of vertically propagating waves from a layer interface is strongly dependent on wave amplitude. Furthermore, in those cases associated with high surface wind speeds and high drags, the influence of the interface on the nonlinear solution is qualitatively similar to that of a free surface in hydraulic theory.

4.4.3. *A comparison of the hydraulic and the wave-breaking mechanisms*

Evidence from observations (Smith 1987) and numerical models (Clark and Peltier 1977, and many others) suggests that wave breaking plays an important role in many downslope windstorms. Clark and Peltier (1977) suggested that after the isentropes in a mountain wave overturn, a resonant cavity forms between the terrain and the wave-breaking region, and the wave within that cavity amplifies dramatically. Peltier and Clark's (1983) mathematical model of wave amplification in this resonant cavity is based on linear theory and the physical assumption that the wave-overturning region behaves like a critical level where the Richardson number is less than $\frac{1}{4}$. Smith (1985a), and Smith and Sun (1987), have offered another mathematical model of the wave-breaking amplification process, beginning with the same fundamental assumption used by Clark and Peltier: that the breaking region imposes an effective upper boundary condition that traps energy within the underlying flow. Smith's mathematical treatment of the upper boundary is based on the identification of a hypothetical "dividing streamline" separating the laminar flow beneath the wave-breaking region from the turbulent flow above. Smith obtained an upper boundary condition stating that the cross-mountain wind speed is constant along the dividing streamline, by postulating that 1) the turbulent flow in the breaking region is well mixed, and 2) pressure perturbations at the top of the well-mixed region are negligible. Note that the presence of Clark and Peltier's low-Richardson-number critical layer is implicitly used in the

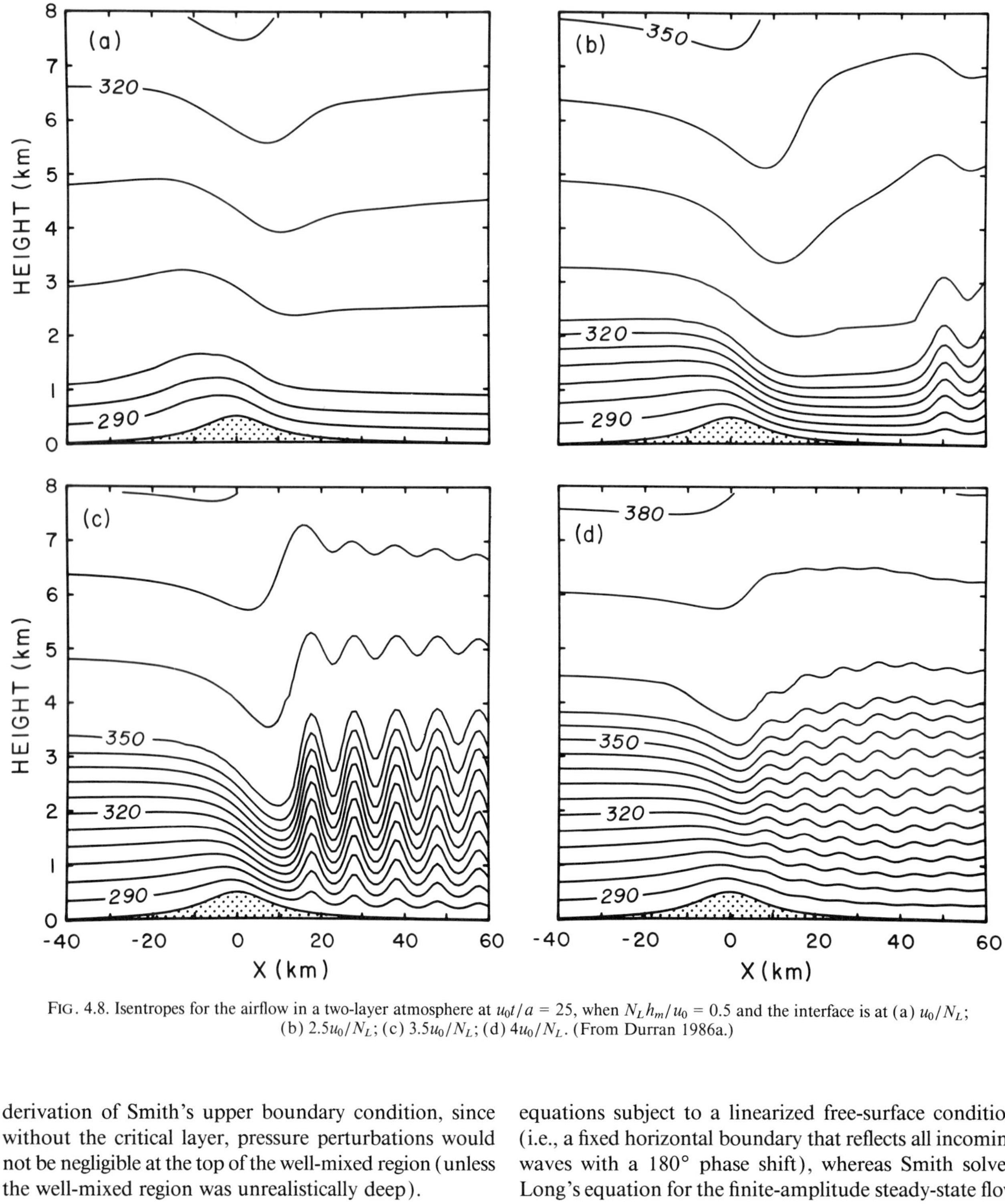

FIG. 4.8. Isentropes for the airflow in a two-layer atmosphere at $u_0t/a = 25$, when $N_L h_m/u_0 = 0.5$ and the interface is at (a) u_0/N_L; (b) $2.5u_0/N_L$; (c) $3.5u_0/N_L$; (d) $4u_0/N_L$. (From Durran 1986a.)

derivation of Smith's upper boundary condition, since without the critical layer, pressure perturbations would not be negligible at the top of the well-mixed region (unless the well-mixed region was unrealistically deep).

Although the theories of Peltier and Clark, and of Smith, might both be considered theoretical descriptions of the flow beneath a low-Richardson-number ($\mathrm{Ri} < \frac{1}{4}$) critical layer, significant differences in the mathematical formulation of each theory lead to very different results. Peltier and Clark (1983) solved the time-dependent linear wave equations subject to a linearized free-surface condition (i.e., a fixed horizontal boundary that reflects all incoming waves with a 180° phase shift), whereas Smith solved Long's equation for the finite-amplitude steady-state flow beneath a deformable upper boundary along which the wind speed was constant. Peltier and Clark's results imply that a low-Richardson-number critical layer will produce amplification only when it is positioned approximately $\frac{1}{4} + n/2$, $n = 0, 1, \ldots$ vertical wavelengths above the topography. On the other hand, Smith's results suggest amplification is possible over the entire range of critical-layer heights between $(\frac{1}{4} + n)$ and $(\frac{3}{4} + n)$ vertical wavelengths. Given the presence of a suitably located low-Richardson-number critical level, the amplification process proposed by Peltier and Clark is independent of mountain height. (Note, however, that when the critical level is "self-induced" by wave breaking, the development of the critical level, and hence the basic structure of the total flow, can be very sensitive to mountain height.) In contrast, the results of Smith are explicitly dependent on mountain height in a manner similar to conventional, shallow-water hydraulic theory.

Durran and Klemp (1987) and Bacmeister and Pierrehumbert (1988) used these differences to conduct a series of numerical experiments testing each theoretical model of the wave-breaking process. These investigators examined the response of air flowing over a mountain to changes in the mountain height and changes in the height of a critical layer in the mean flow. In most of the cases examined by Durran and Klemp, and Bacmeister and Pierrehumbert, a local region of turbulent flow forms over the mountain where the mountain wave interacts with the mean-state critical layer. (To be specific, Durran and Klemp found that significant regions of $\mathrm{Ri} < \frac{1}{4}$ developed above the mountain in all simulations with the critical layer located between $\frac{1}{4}$ and $\frac{3}{4}$ vertical wavelengths above the ground, whenever Nh/U exceeded 0.2.) Thus, although the wave-breaking theories of Peltier and Clark, and of Smith, were originally formulated to describe the flow field beneath a region of wave breaking, they should also apply to the cases with a mean-state critical layer studied by Durran and Klemp, and Bacmeister and Pierrehumbert.

In accordance with Smith's theory, Durran and Klemp, and Bacmeister and Pierrehumbert found a high amplitude response could be obtained if the critical layer was located anywhere between $\frac{1}{4}$ and $\frac{3}{4}$ vertical wavelengths above the ground. They also found a dependence on mountain height closely paralleling the predictions of Smith. A representative example of these simulations is shown in Fig. 4.9. In the case illustrated in Fig. 4.9, the mean stability is constant everywhere, and the wind speed is constant between the ground and 5 km, and zero in the region above 7 km. The two regions of constant wind speed are connected by a layer with linear wind shear between 5 and 7 km. The mean Richardson number in the shear layer is 1.1; however, in both cases a region of stagnant flow develops over the mountain in which the local Richardson number is less than $\frac{1}{4}$.

Following Peltier and Clark, the critical layer in Fig. 4.9 would be estimated to lie $\frac{7}{12}$ of a vertical wavelength above the ground. This is far from the resonant level of $\frac{9}{12}$ vertical wavelength, and thus should not permit the development of a large-amplitude response. Nevertheless, as shown in Fig. 4.9b, a large response does develop when $Nh_m/u_0 = 0.4$. Turning to Smith's theory, a reasonable choice for the height of the dividing streamline would be the middle of the shear layer, or $\frac{6}{12}$ of a vertical wavelength above the ground. Smith's theory predicts that, if the dividing streamline is fixed at this location and the mountain height is gradually increased, the flow should undergo a change from weak waves to a large-amplitude response

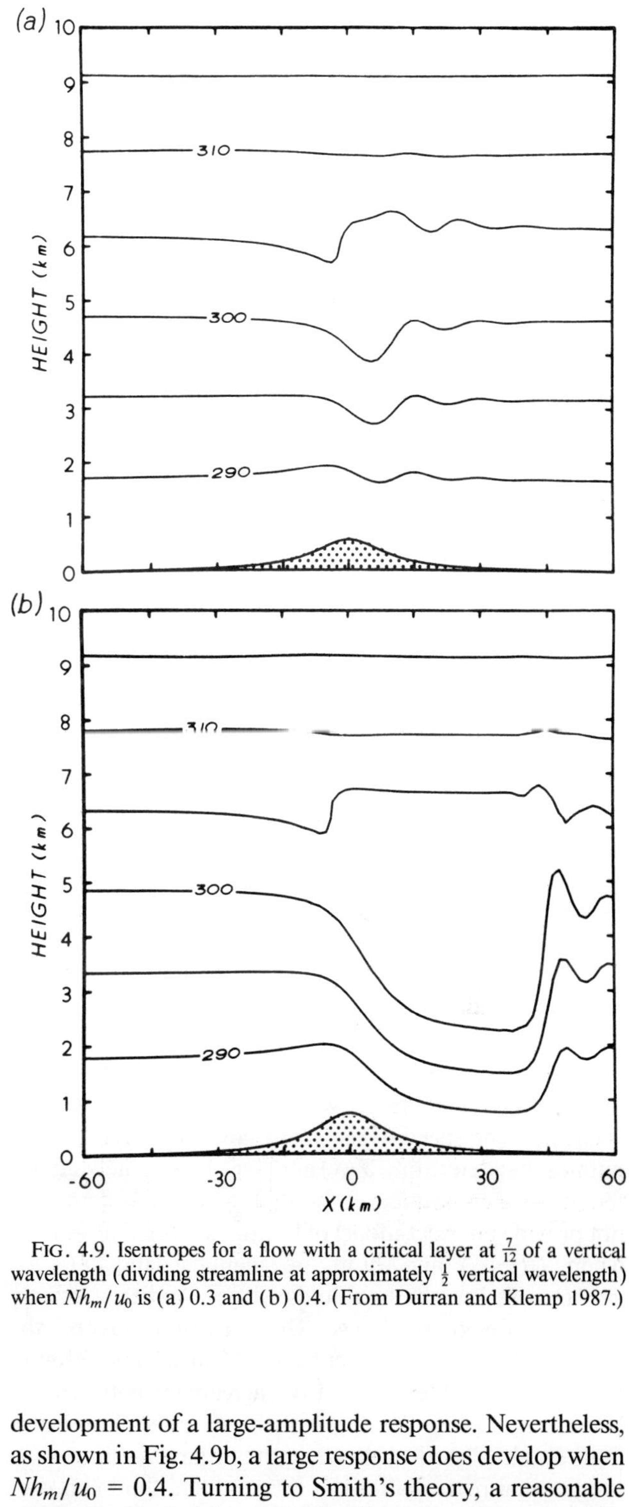

FIG. 4.9. Isentropes for a flow with a critical layer at $\frac{7}{12}$ of a vertical wavelength (dividing streamline at approximately $\frac{1}{2}$ vertical wavelength) when Nh_m/u_0 is (a) 0.3 and (b) 0.4. (From Durran and Klemp 1987.)

when Nh_m/u_0 exceeds a critical value between 0.3 and 0.4; a result that agrees with the numerical simulations shown in Figs. 4.9a,b. Other simulations, conducted by Durran and Klemp, and Bacmeister and Pierrehumbert over a wide variety of different situations, confirm the ability of Smith's theory to adequately describe topographically disturbed flow in an airstream with a mean critical layer. These simulations do not support the linear resonance mechanism of Clark and Peltier.

The most significant aspect of the recent work by Smith, Durran and Klemp, and Bacmeister and Pierrehumbert is not a demonstration of the importance of wave breaking, which had been previously provided by Clark and Peltier, but an elucidation of the connection between wave breaking and the hydraulic amplification mechanism. Durran and Klemp (1987) have shown that Smith's solution satisfies

$$(1 - \mathcal{F}^{-2})\frac{\partial(D+h)}{\partial x} = \frac{\partial h}{\partial x}, \tag{4.37}$$

where

$$\mathcal{F}^2 = \frac{1 - (N\delta_c/u_0)\sin(ND/u_0)}{1 - \cos(ND/u_0)}. \tag{4.38}$$

Here $D(x)$ is the total depth of the fluid between the dividing streamline and the topography, $\delta_c(x)$ is the displacement of the dividing streamline from its undisturbed height upstream, and N and u_0 represent the constant values of Brunt–Väisälä frequency and cross-mountain wind speed below the critical-level in the undisturbed upstream flow. The parameter $\mathcal{F}$ is analogous to the conventional shallow-water Froude number; in particular it renders (4.37) equivalent to (4.33), and both $\mathcal{F} = 1$ and $\mathrm{Fr} = 1$ imply that the phase speed of the most rapidly propagating linear gravity wave in each respective system is equal to the speed of the mean flow. However, the dynamics that determine $\mathcal{F}(x)$ and $\mathrm{Fr}(x)$ are somewhat different. As a consequence, the shallow-water system does not provide an exact model of Smith's stratified flow, even when $D(-\infty)$ and $h(x)$ are identical and $\mathcal{F}(-\infty) = \mathrm{Fr}(-\infty)$; i.e., the two fluids will not undergo exactly the same thickness change, $D(x)$, as they traverse the obstacle. However, as discussed by Durran and Klemp, there is reasonable quantitative agreement between the two systems when the depth of the stratified flow is less than one-half vertical wavelength. At larger values of D the internal perturbations in the stratified fluid can produce significant quantitative differences, but a qualitative similarity remains.

In order to use Smith's hydraulic theory to predict the steady-state flow over a mountain, it is necessary to specify the initial height of the dividing streamline, $D(-\infty)$. In a case with a mean-state critical layer, Durran and Klemp (1987) and Bacmeister and Pierrehumbert (1988) have shown that good results can be obtained by setting $D(-\infty)$ equal to the height of the critical layer. However, in the case of breaking waves and wave-induced critical layers, no height for the dividing streamline can be specified a priori. Consequently, the application of the hydraulic model to breaking waves is limited to a posteriori diagnosis; it cannot be used to predict the final steady state or the flow evolution. Nevertheless, the successful a posteriori application of hydraulic theory to the case of breaking waves allows one to construct a simple unified model for the creation of high lee-slope fluid velocities in a wide range of geophysical flows, including the flow of water over a rock, the flow of a multilayered airstream over a mountain (without wave breaking), the flow of a single-layered airstream over a mountain when breaking waves are present, and cross-mountain flow bounded above by a mean-state critical layer. That simple unified model consists of a two-part amplification process. In the first part of the amplification process, high fluid velocities are produced near the obstacle's crest as pressure gradient forces in a standing gravity wave dominate the force balance on fluid parcels in the upstream flow. Fluid near the crest experiences a net acceleration and a net elevation gain (as in subcritical shallow-water flow). The second part of the amplification process only occurs if the obstacle is sufficiently high, in which case the gravity-wave-induced acceleration becomes so great that it is impossible to develop a lee-side pressure gradient capable of decelerating the flow (as in supercritical shallow-water flow). In this circumstance, the flow undergoes a transition; the force balance on a fluid parcel is dominated by gravity, and very high velocities develop as the fluid accelerates down the lee slope like a hockey puck descending a mound of ice.

4.4.4. *Forecasting downslope winds*

As discussed in the preceding sections, there appears to be a fundamental dynamical similarity between the development of downslope winds in the atmosphere and the development of supercritical flow in a hydraulic jump. This is not a new idea, nor is it one that has provided much practical assistance to forecasters in the past. Indeed, one might suppose that if the simple hydraulic analog introduced by Long in the early 1950s had led to a set of highly successful forecasting guidelines, the alternative mechanisms proposed to explain the development of strong downslope winds would not have enjoyed much popularity.

One problem with any attempt to apply shallow-water hydraulic theory directly to the problem of downslope wind forecasting is that there appear to be at least three rather different circumstances in which the atmosphere can undergo a transition from subcritical to supercritical flow. These three circumstances are:

1) *Wave breaking:* as in an atmosphere with constant N and u_0 where the mountain is large enough to force breaking waves (Clark and Peltier 1977).

2) *Scorer-parameter layering:* as in an atmosphere with constant u_0, and a two-layer structure in N where the mountain is too small to force breaking waves (Durran 1986a).

3) *Capping by a mean-state critical layer:* as in an atmosphere with constant N and u_0 below a critical layer, where in the absence of the critical layer, the mountain is too small to force breaking waves (Smith 1985a).

Until more detailed guidance is developed, the forecaster might do well to examine the observed or forecast atmospheric sounding for factors that would promote the development of atmospheric hydraulic jumps through the action of one of the preceding three processes.

In a case where there is a deep cross-mountain flow and no mean-state critical layer (as in most of the Boulder, Colorado and Owens Valley, California windstorms), observational evidence suggests that conditions favorable for downslope winds occur when:

(i) The wind is directed across the mountain (roughly within 30° of perpendicular to the ridgeline) and the wind speed at mountaintop level exceeds a terrain dependent value of 7 to 15 m s^{-1}.

(ii) The upstream temperature profile exhibits an inversion or a layer of strong stability near mountaintop level (Colson 1954; Brinkmann 1974).

These factors would favor the development of strong winds by creating conditions similar to situation 2). In addition they promote the development of larger amplitude mountain waves and thereby increase the chances for breaking waves; i.e., situation 1). Breaking waves are also favored when the upper-tropospheric winds are not excessively strong. Thus, cross-mountain wind speeds of 30 to 40 m s^{-1} may be more likely to produce strong downslope winds than the higher values observed near the axis of the jet stream.

In cases where there is a mean-state critical layer in the flow above mountaintop level (e.g., many Yugoslavian boras and "canyon winds" on the western side of the Wasatch Front in Utah), criterion (i) still applies. There is likely to be an inversion near the critical level, but the critical level itself appears to be more important than the inversion in the generation of strong-downslope winds. However, even when a mean-state critical level is present it may not be the feature that is primarily responsible for the development of downslope winds. Klemp and Durran (1987) conducted numerical simulations suggesting that the 15 April 1982 bora, observed during ALPEX (the Alpine Experiment), was produced by simple wave breaking below a mean-state critical layer [i.e., situation 1)] and that the development of strong surface winds was not inhibited by the absence of the mean-state critical layer.

Additional factors that may influence the strength of a given wind event are the terrain shape, atmospheric humidity, and the time of day. Long mountain ridges with gentle windward slopes and steep lee slopes are most favorable for the generation of strong downslope winds (Smith 1977; Lilly and Klemp 1979; Hoinka 1985). High relative humidity in the lower troposphere appears to reduce the strength of windstorms (Barcilon et al. 1979; Durran and Klemp 1983). Climatological studies of windstorms in Boulder, Colorado show a distinct tendency for the highest winds to occur in the nighttime or the early morning (Whiteman and Whiteman 1974; Brinkmann 1974). This diurnal signal is probably due to a diurnal modulation of boundary-layer friction. Although they did not investigate diurnal effects, Richard et al. (1989) recently demonstrated the importance of surface friction by showing that its inclusion in a numerical model eliminates the unrealistic tendency of models with a free-slip lower boundary to form a layer of high surface wind that propagates indefinitely downstream (as in Figs. 4.8b and 4.9b).

In a case with no mean-state critical layer, the forecaster might also examine the relationship between the tropopause height and the vertical wavelength of hydrostatic mountain waves. Klemp and Lilly (1975) found the strongest downslope wind events occurred in Boulder when a one-half wavelength phase shift was present between the ground and the tropopause. The theoretical support for this relationship rests on the partial reflection mechanism of linear theory; however, as discussed in section 4.4.2, the application of this theory to finite-amplitude waves can produce misleading results. It is therefore not entirely clear why the phase-shift correlation is as high as it seems. Although the strongest Boulder windstorms seem to exhibit the half-wavelength phase shift, exclusive reliance on phase shift and wind speed/direction criteria can lead to the prediction of windstorms when none actually occur. Bower and Durran (1986) discuss a nonwindstorm event that did satisfy the half-wavelength phase shift criteria, along with appropriate conditions on the wind speed and direction. Unlike other windstorm cases, the lower tropospheric stability in the nonwindstorm case was very weak and there was no significant inversion near mountaintop level. On the basis of the climatologies of Brinkmann (1974) and Colson (1954) one might suppose that the absence of low-level stability was responsible for the failure of a downslope wind to develop in Bower and

Durran's "well-tuned" cross-mountain flow. Further support for the idea that lower tropospheric stability plays an important role in Boulder windstorms has been provided by numerical simulations of the 11 January 1972 windstorm. Durran (1986a) showed that the development of the windstorm (including wave breaking in the upper troposphere) was triggered by the formation of a hydraulic-jump-like disturbance in an elevated inversion near mountaintop level. As shown in Figs. 4.10a,b, no windstorm developed when the numerical simulation was repeated using a modified upstream sounding from which the elevated inversion had been removed. In summary, it is possible that the half-wavelength criteria may be a necessary condition for the strongest Boulder windstorms, but it does not appear to be a sufficient criteria for even a moderate storm.

4.4.5. Gustiness near the surface in downslope winds

In many downslope windstorms, the surface wind is very gusty. An example of this gustiness is shown in Fig. 4.11, the anemometer trace recorded at the National Center for Atmospheric Research during the onset of the 17 January 1982 Boulder windstorm. Gusts in excess of 210 km h^{-1} (58 m s^{-1}) with lulls of less than 15 km h^{-1} (4 m s^{-1}) are evident after the onset of this severe storm (time reads right to left in Fig. 4.11). Fluctuations with a period of slightly more than one minute appear to dominate the surface wind observations. In addition to the surface data, observations of the gust structure a few hundred meters above the surface have recently been obtained by Neiman et al. (1988) using a pulsed Doppler lidar; an example of their analysis appears in Fig. 4.12. These observations, taken during a moderate Boulder, Colorado windstorm, show a series of gusts forming at the upwind edge of the observation area and maintaining their coherence as they propagate downstream at the approximate speed of the mean winds. The dominant period of these fluctuations is approximately 4 minutes; the maximum and minimum velocities are roughly 34 and 16 m s^{-1}. The gustiness observed above the surface by Neiman et al. is less extreme than that recorded in the preceding surface anemometer trace. Some of the difference can be attributed to differences in the strength of the two windstorms; however, an additional enhancement in the surface gustiness is probably produced by surface friction.

Clark and Farley (1984) investigated gustiness in the 11 January 1972 windstorm using two- and three-dimensional numerical models. The two-dimensional simulations did not appear to produce gusty surface winds. The three-dimensional simulations were nominally similar to those in two-dimensions, because there were no variations in the topography or the mean atmospheric structure

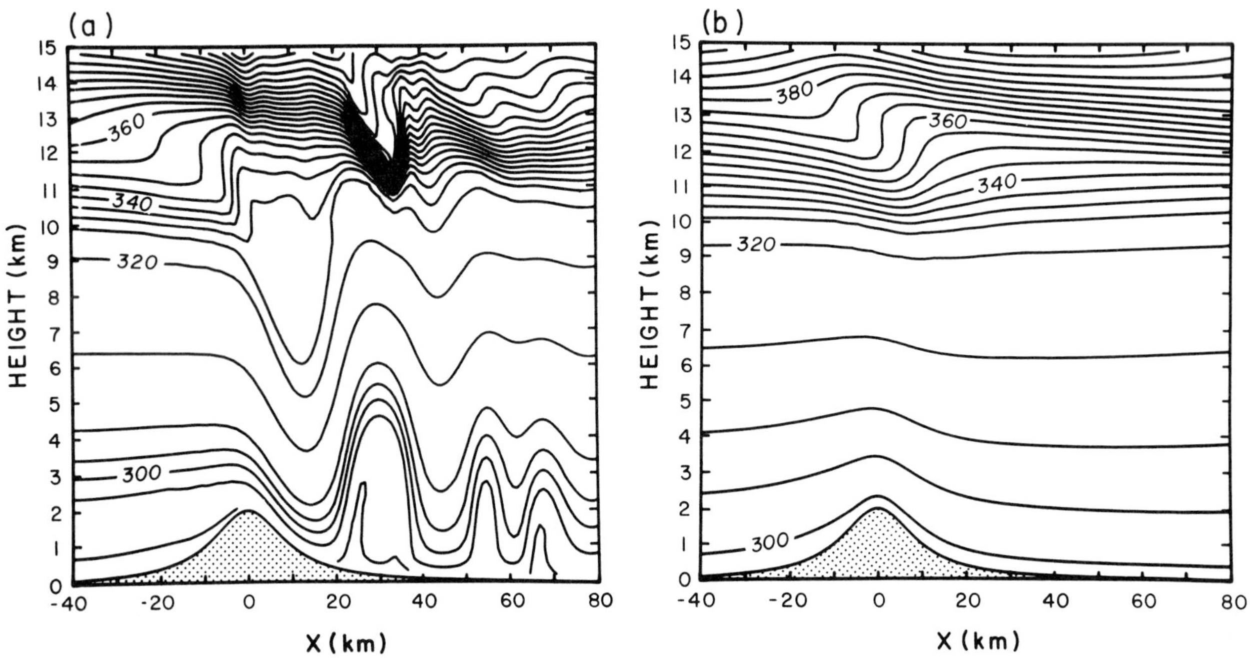

FIG. 4.10. (a) Isentropes form a simulation of the 11 January 1972 Boulder windstorm using the upstream conditions observed at Grand Junction, at a model time of 12 000 s. (b) As in (a), except that the upstream sounding has been modified to remove the elevated inversion. (From Durran 1986a.)

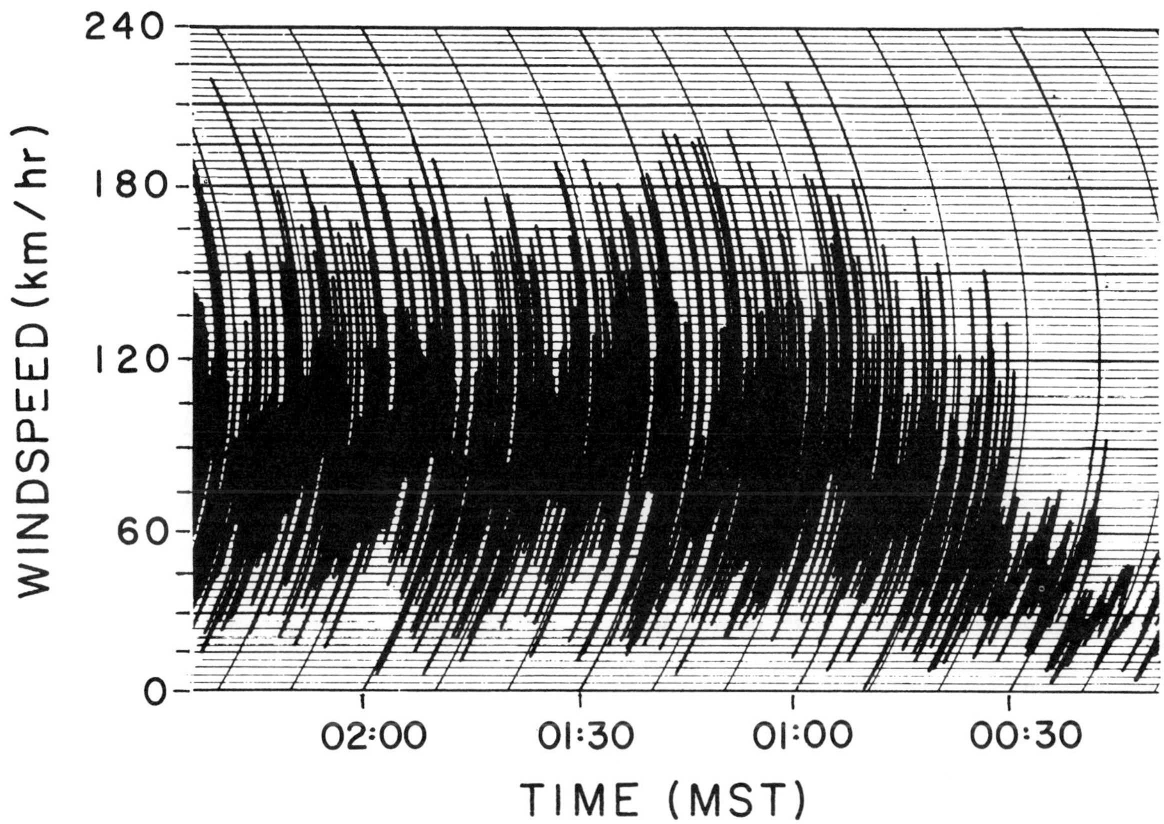

FIG. 4.11. Anemometer trace recorded at the National Center for Atmospheric Research during the onset of the 17 January 1982 Boulder windstorm. Time reads from right to left.

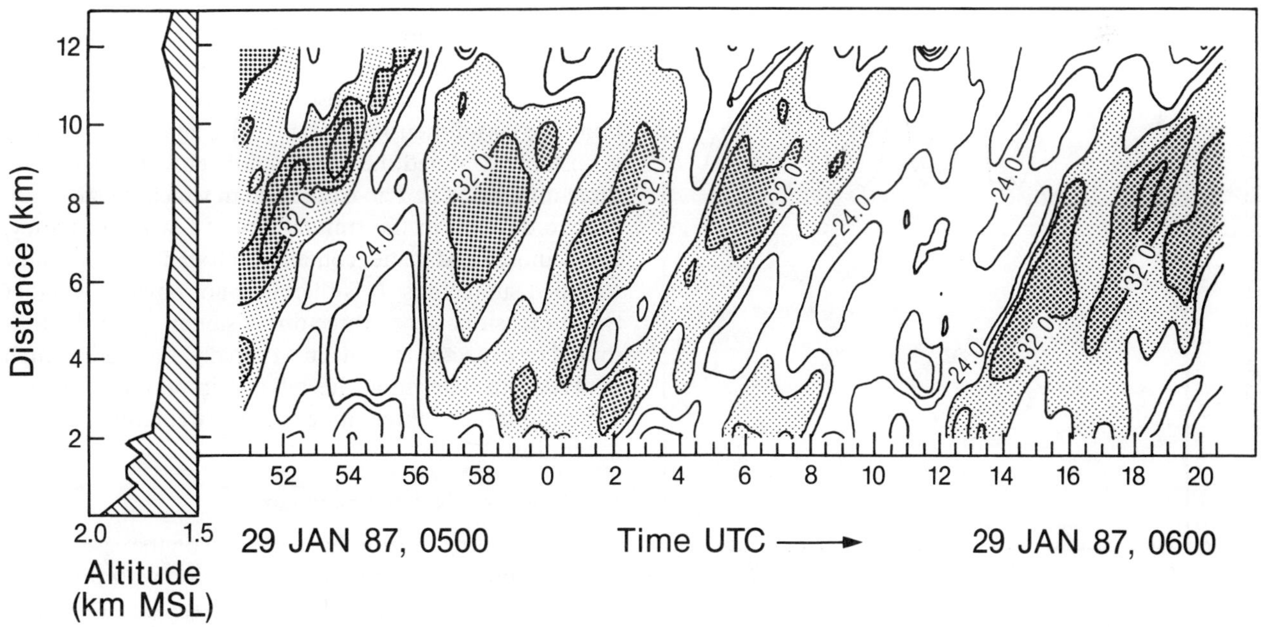

FIG. 4.12. Range–time display of Doppler-lidar-measured radial velocities ($m\ s^{-1}$) on 29 January 1987 in the vicinity of Boulder, Colorado. The lidar was located in the foothills and pointed eastward, out of the mountains at an elevation angle of $-2°$. The optical beam intersected the ground 12 km downrange. The terrain profile underneath the lidar beam is displayed along the vertical axis. (From Neiman et al. 1988.)

along the direction parallel to the ridge. Clark and Farley tested the stability of two-dimensional flow to three-dimensional perturbations by applying a small random three-dimensional forcing to the potential temperature field in the three-dimensional simulations. They found that the three-dimensional flow was stable with respect to these perturbations until the waves began to overturn. In the wave-breaking region, the random three-dimensional forcing triggered unstable disturbances that eventually penetrated to the surface and produced gusty winds along the lee slope. Clark and Farley suggested that this gustiness is produced by a competition between wave buildup via forced gravity wave dynamics and wave breakdown via convective instability. Figure 4.13 provides an example of the surface wind variation at one point on the lee slope in Clark and Farley's three-dimensional simulation. The dominant period of these fluctuations is approximately 8 minutes; the maximum and minimum velocities during the gusty period are approximately 70 and 50 m s^{-1}. Clark and Farley have hypothesized that better numerical resolution would reduce the period of the gusts in their simulation, which could bring it into closer agreement with the measurements of Nieman et al. (1988).

Given the crudeness with which the topography and boundary processes are represented in their model, and the fact that they were not attempting to simulate the case observed by Nieman et al., Clark and Farley's simulation appears to provide a qualitatively correct description of

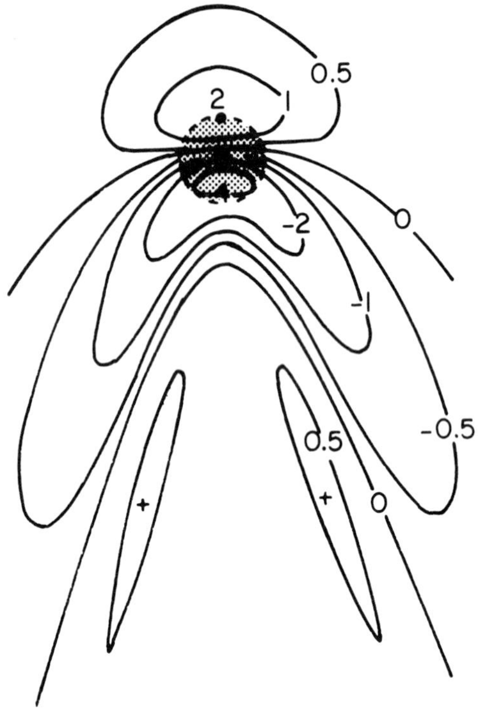

FIG. 4.14. Horizontal cross-section of the vertical displacement field in hydrostatic flow over a circular bell-shaped mountain of the form (4.39). The cross section is at an elevation of $\frac{1}{4}$ vertical wavelength. Shading indicates the topography within a radial distance a from the peak. (From Smith 1980.)

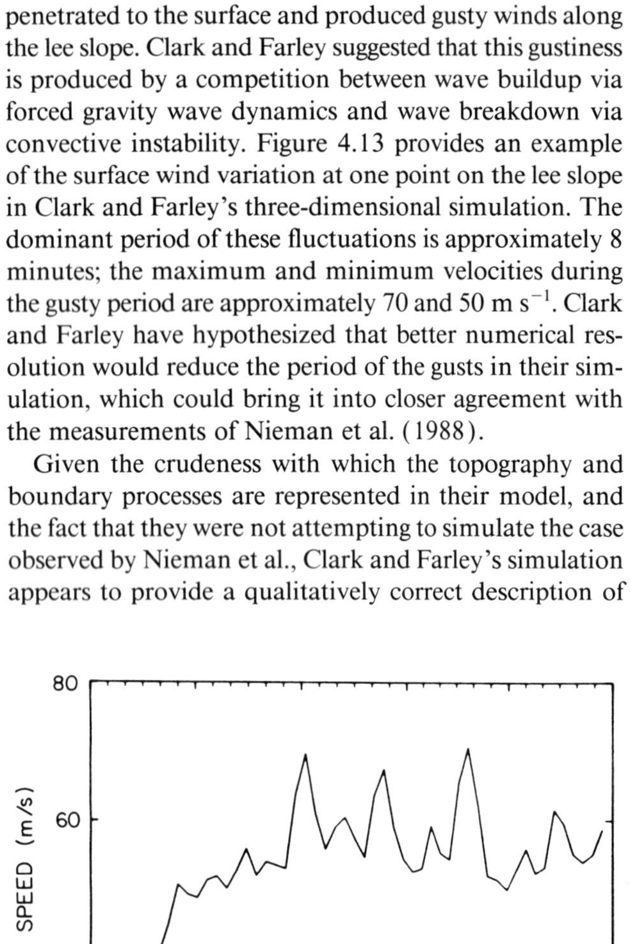

FIG. 4.13. Time series plot of the horizontal wind speed 9.5 km downwind of the ridge crest in a three-dimensional simulation of the 11 January 1972 Boulder, Colorado windstorm. (From Clark and Farley 1984.)

the gustiness in the flow; however, Scinocca and Peltier (1989) have recently found that realistic gust structures can be produced in high resolution simulations of strictly two-dimensional flow. Kuo (1963) has shown that small-amplitude perturbations in a stably stratified unidirectional shear flow grow most rapidly when the perturbations are two-dimensional, but that when the same flow is unstably stratified the most rapidly growing modes are three-dimensional. The basic state in which the gusts appear to develop in the simulations of Clark and Farley, and Scinocca and Peltier contains both stably and unstably stratified subregions. It is therefore not obvious whether two-dimensional or three-dimensional perturbations should be expected to grow most rapidly. Further research is needed to determine the relative importance of two-dimensional and three-dimensional circulations in gust generation.

4.5. Flow over isolated mountains

The preceding sections have focused on the case of flow over an infinitely long mountain barrier, with no variations in the direction parallel to the ridge axis. Most real mountains are highly three-dimensional, and unfortunately our current understanding of the flow around three-dimensional mountains is far from complete. Significant

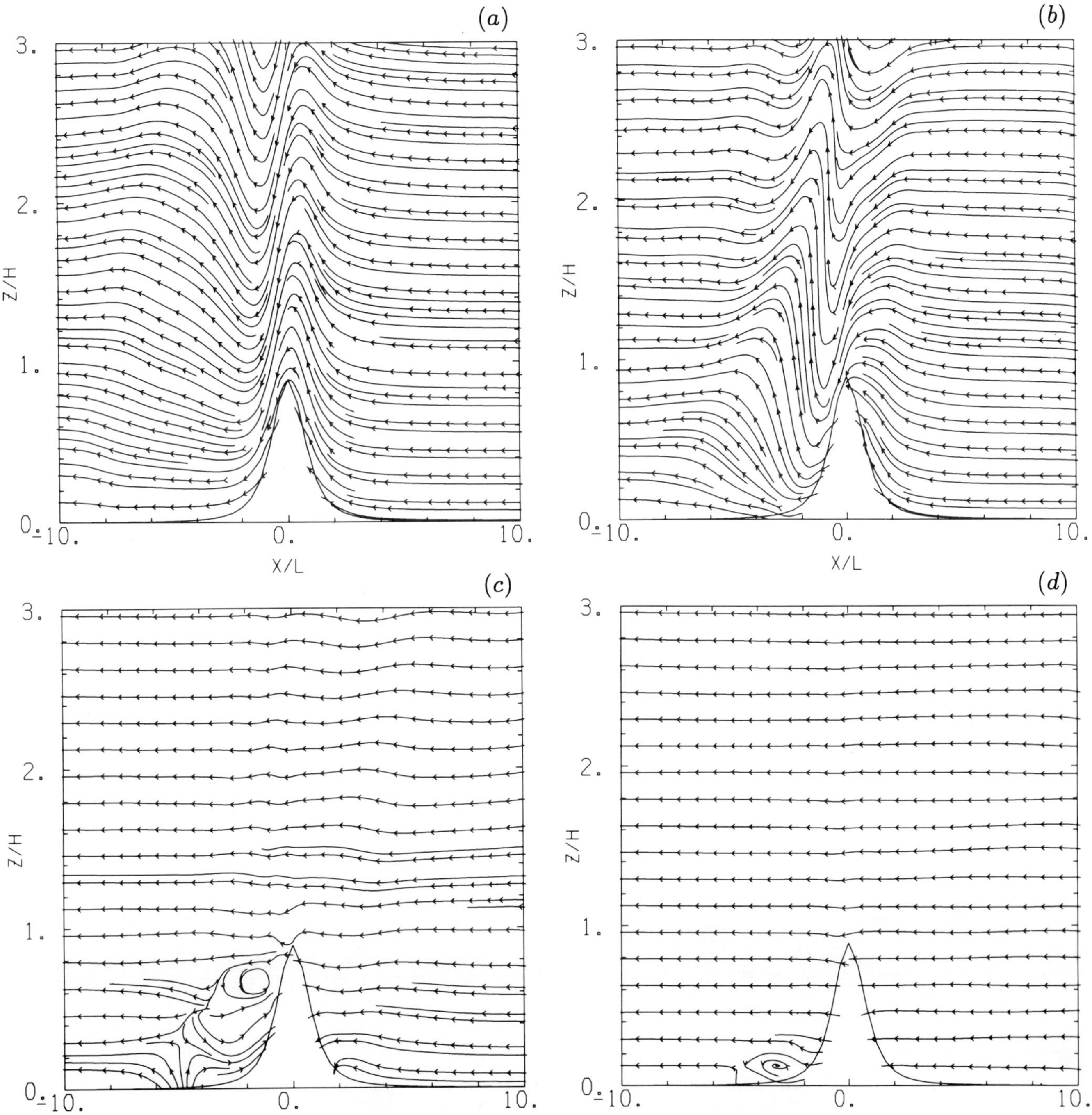

FIG. 4.15. Streamlines in the x–z plane through the center of a circular bell-shaped mountain, when the upstream value of Nh_m/u_0 is (a) 0.45, (b) 1.5, (c) 4.5, (d) 18. Airflow is from the right to the left. (From Smolarkiewicz and Rotunno 1989.)

progress has been made, however, in one case of relatively simple three-dimensional geometry—the isolated circular mountain. A discussion of the airflow over circular mountains and hills is presented by Carruthers and Hunt, Chapter 5, and in reviews by Smith (1979) and Queney et al. (1960). An analysis of the circumstances under which air diverts around a mountain, instead of flowing over it, appears in the Appendix to Chapter 5 (Smith). In the following section we will limit our attention to some recent results describing the dynamics of flow

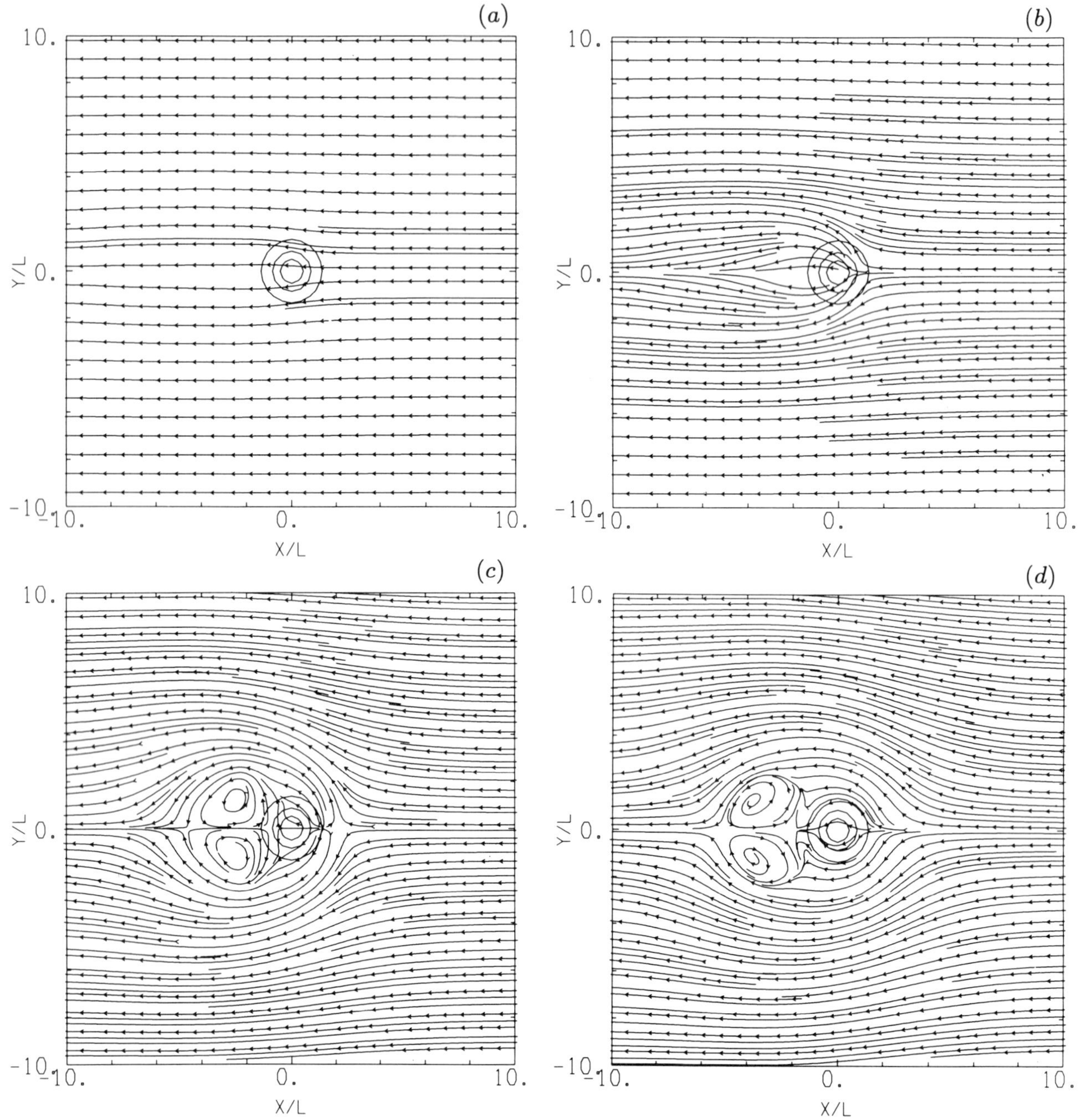

FIG. 4.16. As in Fig. 4.15, except that the streamlines are displayed in the x–y plane and represent the flow on the surface of the topography. (From Smolarkiewicz and Rotunno 1989.)

around a circular obstacle when the mean wind speed and stability are constant with height.

Smith (1980) obtained linear solutions for waves in an atmosphere with constant N and u_0 flowing over a bell-shaped circular mountain

$$h(x, y) = \frac{h_m}{[(x^2 + y^2)/a^2 + 1]^{3/2}}, \quad (4.39)$$

where h_m and a are the height and horizontal scale of the mountain. An example of Smith's solution for the wave-induced vertical displacement at a level one-quarter wavelength above the surface is shown in Fig. 4.14. The flow in Fig. 4.14 is hydrostatic ($Na/u_0 \gg 1$); yet unlike the hydrostatic waves forced by a two-dimensional mountain, the disturbance is not confined to the region directly over the mountain. Instead, the perturbations are concentrated along parabolas extending downstream from the mountain.

Smith's solutions are based on linearized equations and are formally valid only in the limit $Nh_m/u_0 \ll 1$. In the opposite limit, $Nh_m/u_0 \gg 1$ the behavior is similar to two-dimensional potential flow in a horizontal plane where all the fluid is diverted laterally around the mountain (Drazin 1961). Analytic models are not available to describe the flow when Nh_m/u_0 is of order unity. Smolarkiewicz and Rotunno (1989) bridged this theoretical gap by calculating a series of numerical solutions for the flow around the circular bell mountain (4.39); their results are illustrated in Figs. 4.15 and 4.16, which represent steady or quasi-steady solutions. Streamlines in a vertical cross-section through the center of the hill are shown in Fig. 4.15. The case $Nh_m/u_0 = 0.45$ (Fig. 4.15a) is in quantitative agreement with Smith's linear theory; qualitative agreement remains at $Nh_m/u_0 = 1.5$ (Fig. 4.15b). However, a major change in the flow regime occurs as Nh_m/u_0 is increased to 4.5 (Fig. 4.15c). The vertically propagating gravity waves evident at $Nh_m/u_0 = 1.5$ are virtually eliminated and closed circulations develop in the lee of the mountains. When $Nh_m/u_0 = 18$ (Fig. 4.15d), the flow is largely horizontal. Figure 4.16 shows a plot of the surface streamlines associated with each of the four cases in Fig. 4.15. Note that as the flow becomes more nonlinear, there is an increased lateral displacement of the flow around the mountain. Closed vortices appear in the lee of the mountain in the two simulations with the highest values of Nh_m/u_0.

Satellite imagery has revealed vortices downstream of isolated mountains in the atmosphere. Lee vortices have also been observed in laboratory experiments (Hunt and Snyder 1980). The development of these vortices has traditionally been explained by analogy with the flow of a homogeneous fluid past a cylinder, where vorticity is generated by surface friction in the viscous boundary layer and advected downstream after the boundary layer separates from the obstacle. This mechanism cannot be responsible for the development of the lee vortices in Smolarkiewicz and Rotunno's simulations, however, because those simulations were conducted without surface friction. Yet the vorticity apparent in Figs. 4.16c,d had to be created by some process because there was no vorticity in the upstream flow (N and u_0 were constant). Smolarkiewicz and Rotunno have suggested that the lee vortices were generated in their simulations through a three-step process. First, as the stratified flow is disturbed by the mountain, horizontal density gradients are produced. Second, these horizontal density gradients baroclinically generate horizontal vorticity. Third, the sharply descending flow in the lee of the mountain redistributes the horizontal vorticity into the vertical by tilting and stretching the vortex filaments. Since the depth of the surface boundary layer is often small compared with the height of a large mountain, it is likely that this inviscid baroclinic vorticity-generation mechanism is responsible for the generation of many of the lee vortices observed in the atmosphere. Smolarkiewicz et al. (1988) have successfully applied the numerical model to the simulation of airflow around the island of Hawaii.

Acknowledgments. This effort was supported by NSF Grant ATM-8796281.

CHAPTER 5

Fluid Mechanics of Airflow over Hills: Turbulence, Fluxes, and Waves in the Boundary Layer

D. J. CARRUTHERS and J. C. R. HUNT

Department of Applied Mathematics and Theoretical Physics, University of Cambridge, Cambridge, United Kingdom

ABSTRACT

Recent developments in the understanding of turbulent airflow over hills and complex terrain are described. We first review linear calculations for the mean flow over isolated hills including the effects of different wind and stratification profiles upstream, and variations in surface roughness. This approach is extended to model flow over complex terrain and is contrasted with other numerical developments. The turbulence structure is also examined with a simple linear approach that is used to calculate how length scales, variances, and spectra vary over hills. More sophisticated modeling is required near the top of the inner layer where rapid distortion, curvature, and nonlinear interaction all combine in determining the structure. Finally, we discuss the dispersion and deposition of particles over hills, considering both thin plumes and plumes far from their source that are well mixed in the boundary layer.

5.1. Introduction

5.1.1. Overview

Hills and mountains affect the flow over them in many different ways and over a great range of length and time scales. An overview is given of the main mechanisms in these flows, and we review in detail flows over mesoscale/microscale topography ($\leqslant$30 km). In particular we describe recent developments in understanding the changes in turbulence and diffusion and in analytical models of these flows, which enable them to be easily programmed and calculated on small computers. The review updates an earlier American Meteorological Society review by Hunt (1980). Other recent reviews include those of Finnigan (1988) and Taylor et al. (1987).

We show in Fig. 5.1 a schematic picture of the various processes occurring over hills and complex terrain. Considering first neutral flows, the airflow approaches the terrain with a logarithmic profile and it flows *over* the hills. Changes in the mean flow occur due to induced pressure gradients, shear-stress perturbations, and the curvature of the surface. In general the flow speeds up over the summits and slows down in the valleys. On the lee side, a separated wake region may form and vortex shedding may take place. The structure of turbulence is also affected over the hill; in general all three components increase in magnitude near the hill surface, where the turbulence adjusts to the local velocity gradients. Above the surface there are anisotropic distortions to the turbulent structure caused by changes in the mean flow.

The presence of stratification, both elevated and at the surface, changes the flow patterns considerably. With increasing stable stratification there is an increasing likelihood of flow around rather than over the hill and upstream blocking may occur. Gravity waves may be excited resulting in a transfer of energy and momentum from the lower flow. These waves generally induce an asymmetry in the flow over the hill—relative to neutral flow and upwind of the summit, the streamlines rise and slow down, and downwind of the summit the streamlines descend through lee waves. This can result in very low wind speeds near the surface upwind of the summit and strong downslope winds in the lee side. The intensity of turbulence (relative to the local mean velocity) is decreased, resulting in a change in the wind profile, with smaller mean gradients near the surface, but larger gradients above.

This flow pattern gradually changes as the stratification increases further. When the Froude number based on the hill height H, F_H is less than unity, the airflow below a critical level tends to move in horizontal planes around the hill and the flow stagnates in the wake. As F_H decreases, this critical level rises to the hilltop.

The changes in the mean flow and turbulence characteristics affect the transfer and fluxes of heat, moisture, and pollutants; however, one also must consider that surface vegetation over hills is often different from the form upwind of the hills and this will also affect these fluxes through both their effect on the turbulence structure and changes in the availability of moisture. Other effects of hills include their effect on clouds and precipitation (see Banta, Chapter 9) and the excitation of propagating gravity waves (see Durran, Chapter 4). When the speed of the approach wind or gradient wind is low enough, thermal winds (see Whiteman, Chapter 2) are induced by horizontal pressure gradients in the airflow caused by heating and cooling of sloping ground.

Before describing the important parameters of the problem and then presenting the feature of the flow in

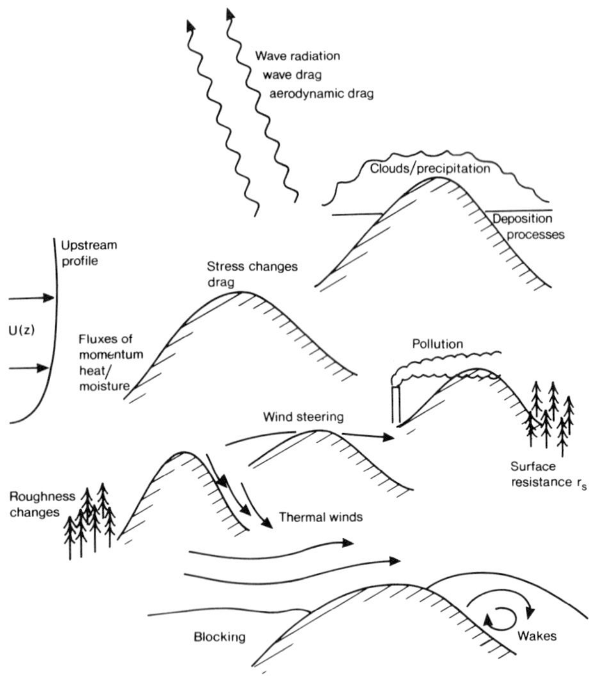

FIG. 5.1. Schematic picture of the various processes occurring over complex terrain.

detail, we note that understanding the airflow over hills is very important for a number of reasons, which include:

(i) Hills affect aerodynamic and wave drag; this has a strong impact on the general circulation in the troposphere, stratosphere, and mesosphere (Palmer et al. 1986).

(ii) Mesoscale and local meteorology is affected through lee-side cyclogenesis, severe downslope winds, and wind steering (Smith 1986).

(iii) The dispersion and deposition of pollution are both affected by changes in the mean streamline pattern, the turbulence, and the presence of orographic clouds (e.g., Egan 1984; Carruthers and Choularton 1984).

(iv) Correct location of wind-energy conversion systems is greatly helped by knowing how the maximum surface wind speed varies with hill shape and atmospheric stability (Meroney and Sandborn 1978).

(v) Making estimates of loads on structures over complex terrain.

(vi) Planning tree cultivation to avoid wind damage and to increase control of forest fires.

(vii) Soil erosion, dust and sand transport, the movement of sand dunes (Howard and Walmsley 1985), and snow drifting are all significantly affected by the presence of hills.

5.1.2. Classification of flow regimes

The different kinds of flow over hills can be characterized by a few major parameters (Hunt and Richards 1984; Snyder 1985). For the purposes of this discussion we assume that the wind approaches a single hill height H and length L with a speed U_0 (taken at a height about equal to L) and friction velocity u_*. The dynamical effects of the potential temperature or density gradient $\partial\theta/\partial Z$ are defined by the Brunt–Väisälä or buoyancy frequency $N^2 = (g/\theta)(d\theta/dZ)$. For most hills, the overall height-to-length ratio (H/L) is small and the following arguments are based on this assumption. The surface roughness length z_0 typically changes by a characteristic amount Δz_0 over hills. We refer to different ranges of these parameters throughout this chapter.

As the buoyancy frequency is approximately zero ($N^2 \approx 0$) in unstable boundary layers except near the surface ($Z \leqslant h/10$), where h is the boundary-layer depth, we shall only consider cases where $N^2 > 0$ and we define $N = \sqrt{N^2}$.

5.1.2.1. STRATIFICATION EFFECTS

It is assumed that the stratification is uniform with buoyancy frequency N; the flow types are shown in Fig. 5.2. Similar flows occur when the stratification is nonuniform.

- Neutral flows:

$$\frac{u_*}{NH} \gg 1; \quad \frac{U_0}{NL} = F_L \gg 1; \quad \frac{U_0}{NH} = F_H \gg 1$$

(e.g., $u_* = 0.3\ \mathrm{m\ s^{-1}}$, $N = 0.005\ \mathrm{s^{-1}}$, $H = 20$ m, $L = 100$ m) where the upwind friction velocity u_* is the longitudinal shear stress at the surface. Buoyancy has a negligible effect so that the flow passes over the topography with small changes in direction. The flow structure is independent of the wind speed, but depends on the hill shape and the form of the upwind profile, characterized by the ratio H/L and u_*/U_0.

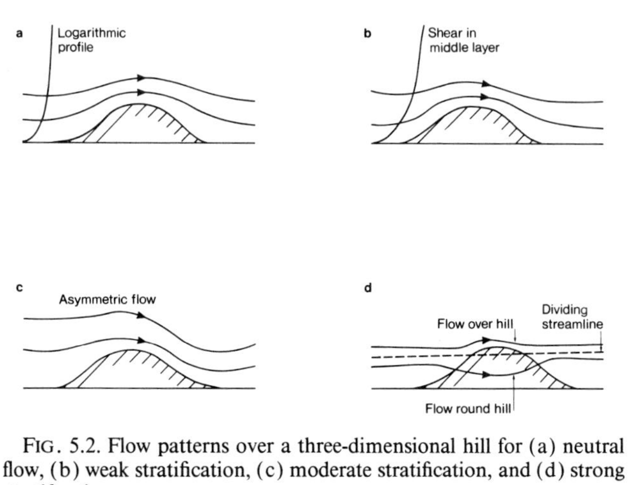

FIG. 5.2. Flow patterns over a three-dimensional hill for (a) neutral flow, (b) weak stratification, (c) moderate stratification, and (d) strong stratification.

- Weak stratification:

$$\frac{u_*}{NL} \leqslant 1; \quad \frac{U_0}{NL} \geqslant 1; \quad \frac{U_0}{NH} \geqslant 1$$

(e.g., $u_* = 0.2$ m s^{-1}, $N = 0.008$ s^{-1}, $L = 400$ m, $H = 100$ m, $U_0 = 10$ m s^{-1}). In this case the upwind profile is affected by stratification (so that U_0/u_* increases or decreases with $u_*/(NH)$ for stable or unstable stratification), and this influences the changes in the flow over the hill. Buoyancy forces are not important over the hill.

- Moderate stable stratification:

$$\frac{U_0}{NL} \leqslant 1; \quad \frac{U_0}{NH} > 1; \quad \Delta u \sim (H/L)\left(\frac{NL}{U}\right)U_0$$

(e.g., $U_0 = 10$ m s^{-1}, $N = 0.01$ s^{-1}, $L = 1000$ m, $H = 200$ m). Buoyancy forces have a significant effect on the mean flow over the hill as well as on the upstream velocity gradient; gravity waves are produced; the flow is highly asymmetrical.

- Strong stable stratification:

$$F_H = \frac{U_0}{NH} \leqslant 1, \quad \Delta u \sim U_0$$

(e.g., $U_0 = 5$ m s^{-1}, $N = 0.04$ s^{-1}, $H = 200$ m). Buoyancy forces are so strong that they effectively suppress vertical motion below a critical height $H_c = F_H H$. Above this height H_c, the flow is similar to moderately stratified flow over a cutoff hill of height $H - H_c$.

5.1.2.2. ROTATION EFFECTS

Just as with stratification, the Coriolis parameter f is important because of the effects on the profile, as well as on the mean acceleration over the hills.

- Weak effects:

$$\frac{u_*}{fL} \leqslant 1, \quad \frac{U_0}{fL} > 1.$$

Rotational effects change the upwind profile (in particular the variation of flow direction with height); they are unimportant over the hills.

- Moderate effects combined with strong stratification:

$$\frac{U_0}{fL} > 1, \quad \frac{U_0}{NH} \leqslant 1.$$

When the flow is strongly stratified, the horizontal distance over which hills affect the flow increases, which leads to significant deviations of the flow over a distance of order U/N.

- Strong effects:

$$\frac{U_0}{fL} \leqslant 1.$$

Coriolis effects are significant and change the flow pattern, especially in the wake.

5.1.2.3. ROUGHNESS CHANGES

$$\ln(\Delta z/z_0) \geqslant 1.$$

Whether the ground is sloping or not, the changes in the mean velocity near the surface caused by roughness changes are of order $\Delta u_r \approx u_* \ln(\Delta z_0/z_0)$. Over hills whose slopes are large enough (typically $H/L > \frac{1}{3}$) that the flow is close to separating, a significant change of roughness can change the velocity in the wake by an amount that is of order U_0, which is much greater than Δu_r.

This chapter is organized as follows. In section 5.2 we describe in some detail a linear perturbation analysis for turbulent flow over arbitrary terrain with low slope and where the incident flow is stratified. The effects of different kinds of stratification are discussed in section 5.3, the structure of turbulence in section 5.4, and numerical models in section 5.5. In section 5.6 we describe recent developments in the modeling of diffusion and deposition over hills.

5.2. Modeling of turbulent airflow over hills

A satisfactory description of turbulent airflow over hills involves, at the very least, an understanding of the following features:

(i) the dynamics in different regions—and the length scales of these regions;

(ii) the magnitudes of the perturbation in the velocity field, the perturbation pressure, and the shear-stress perturbation;

(iii) the location of the maximum and minimum wind speed; occurrence of wakes, regions of separation; and

(iv) the turbulence structure.

The basis of our discussion is the linear three-dimensional theory of Hunt et al. (1988a; hereafter HLR). This is a development of an earlier theory by Jackson and Hunt (1975) and originates with Townsend's (1966) analysis of roughness changes and the inviscid and laminar flow theories of perturbed shear flow and stratified flow.

The advantage of linear theory is that it enables formulae for the flow to be derived analytically. These describe a wide range of upwind conditions and shapes of hill so that for the environmental purposes of dispersion, wind energy siting, etc., calculations can be conducted over a wide parameter range. Of course linear theories *cannot* describe large nonlinear perturbations to the flow or nonlinear synergisms where two or more effects combine such as roughness change and separation.

As in most theories of perturbed flow, HLR divide the flow over the hills into two main regions (Fig. 5.3); an inner region of height l in which the changes in the shear stresses have an important effect on the mean flow and where the flow field is very different from that obtained from inviscid flow models, and an outer region in which the flow is inviscid but in which stratification is likely to be important. (It is only with very strong stratification and in thermally driven flows that stratification effects in the inner region are significant.) Both regions are further divided into two layers; the inner region has a very thin inner surface layer used to match the solution correctly at the surface and a shear-stress layer, while the outer region has a middle layer where the effects of upwind shear, particularly in stable flows, are important, and an outer layer where the flow is affected by pressure and buoyancy forces. In the theory of HLR it is assumed that the air flows *over* the hill, so that the Froude number $F_H > 1$; we also assume that the slope is small enough ($H/L \leqslant 0.3$) for the linear theory to be a reasonable approximation of two or more effects combined such as roughness and topographic effects.

5.2.1. *Linear analysis*

5.2.1.1. General equations

The linearized equations of motion for the inner region (where $Z \leqslant l$) for turbulent flow, where buoyancy forces are locally small, are for three-dimensional flow with no rotation effects,

$$U_0 \frac{\partial u}{\partial x} + w \frac{\partial U_0}{\partial Z} = -\frac{1}{\rho}\frac{\partial p}{\partial x} + \frac{\partial \tau_{zz}}{\partial Z}$$

$$U_0 \frac{\partial v}{\partial x} = -\frac{1}{\rho}\frac{\partial p}{\partial y} + \frac{\partial \tau_{yz}}{\partial Z}$$

$$\frac{\partial u}{\partial x} + \frac{\partial v}{\partial y} + \frac{\partial w}{\partial Z} = 0, \qquad (5.1)$$

where Z is the displaced vertical coordinate $Z = z - Hf(x, y)$; u, v, w represent the velocity perturbations, and $U_0 = U(Z)$ is the upwind velocity so that $U = U_0(Z) + u$. (Note that in section 5.1 U_0 represents a velocity scale.) The shear stresses are expressed using a mixing-length approximation with the assumption that the turbulence is in local equilibrium (Townsend 1965):

$$\tau_{xz} = 2\kappa u_* Z \frac{\partial U}{\partial Z}$$

$$\tau_{yz} = \kappa u_* Z \frac{\partial v}{\partial Z}\,. \qquad (5.2)$$

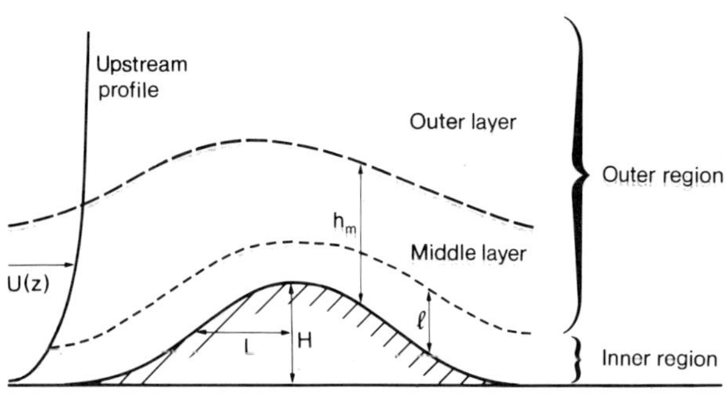

FIG. 5.3. Flow regions for the linear analysis.

In the outer region ($Z \geqslant l$) the perturbation shear stresses have a negligible effect on the perturbation mean flow (although the shear stresses play a significant part in establishing the mean flow in this region). In the lower part of the outer region (the middle layer where $Z \leqslant h_m$), provided the slope is small, the dynamical equations reduce to the Scorer equation (Scorer 1949):

$$\left[\frac{\partial^2}{\partial x^2}\left(\nabla^2 - \frac{1}{U_0}\frac{d^2U_0}{dZ^2}\right) + \frac{N^2}{U_0^2}\nabla_H^2\right] w = 0, \qquad (5.3)$$

where $\nabla_H^2 = \partial^2/\partial x^2 + \partial^2/\partial y^2$.

In the upper layer, where $Z > h_m$, the Scorer equation is written in Cartesian coordinates with $\partial/\partial z$ replacing $\partial/\partial Z$. These two layers must be carefully matched. The buoyancy term $(N^2/U_0^2)\nabla_H^2$ is considered in detail in Hunt and Richards (1984) and Hunt et al. (1988b), hereafter HRB.

5.2.1.2. Inner region solution

Solutions to the inner region equations (5.1) involve using an expansion procedure in terms of the small parameter $\delta = 1/\ln(l/z_0)$ and solving Fourier transformed versions of the equations. We substitute the following expansions for the velocity and shear-stress perturbations

$$u = -\frac{P_0}{U(l)}\,[u^{(0)}(x, y, Z) + \delta u^{(1)}(x, y, Z)\cdots]$$

$$v = -\frac{P_0}{U(l)}\,[\delta v^{(1)}(x, y, Z) + \delta^2 v^{(2)}(x, y, Z)\cdots]$$

$$\Delta w = -\frac{P_0}{U(l)}\,[\delta w^{(1)}(x, y, Z) + \delta^2 w^{(2)}(x, y, Z)\cdots]$$

$$\Delta\tau = -\frac{2P_0u_*^2}{U^{(2)}(l)}\,[\tau^{(1)}(x, y, Z) + \tau^{(2)}(x, y, Z)\cdots], \qquad (5.4)$$

where

$$w = f'(x, y)[U_0(Z) + u(x, y, Z)] + \Delta w(x, y, Z)$$

and

$$\tau_{x,z} = u_*^2 + \Delta\tau. \qquad (5.5)$$

The terrain height is given by $f(x, y)$ and P_0 is the pressure calculated in the outer layer. The inner-region height, l, is given by

$$l \ln(l/z_0) = 2\kappa^2 L, \tag{5.6}$$

so that typically $l \approx L/20$. Denoting Fourier transforms with a tilde; e.g.,

$$\tilde{u}(k_1, k_2, Z) = \int_{-\infty}^{\infty}\int_{-\infty}^{\infty} u(x, y, Z)e^{i(k_1x+k_2y)}dxdy, \tag{5.7}$$

then the solution for $\tilde{u}^{(0)}(k_1, k_2)$ in terms of Fourier transform of the terrain surface $\tilde{f}$ is

$$\tilde{u}^{(0)} = \frac{k_1^2\tilde{f}(k_1, k_2)}{k_{12}^2}\frac{U_0^2(h_m)}{U_0(l)} \times \left[1 + \frac{k_{12}^2/k_1^2 + 4K_0/(\alpha_u\sqrt{i}) - \ln(Z/l)}{\ln(l/z_0)}\right], \tag{5.8}$$

where $\beta = 2\kappa^2 \ln(l/z_0)$, $\alpha_u = 2\sqrt{2(k_1/\beta)}$, and K_0 is the modified Bessel function.

5.2.1.3. Middle layer solution

In the middle layer (defined by $l < z < h_m$) the shear term in Eq. (5.3) dominates the buoyancy term; i.e.,

$$\frac{1}{U_0}\frac{d^2U_0}{dZ^2} > \frac{N^2}{U_0^2}.$$

Data show that this layer exists (e.g., see Fig. 5.4 from the Boulder Atmospheric Observatory) and is frequently quite deep (Hunt et al. 1985; Kaimal and Gaynor 1983). If the boundary layer is approximately logarithmic, then the above inequality is satisfied if $h_m \sim L/[\ln(L/z_0)]^{\frac{1}{2}}$; however, in stable flows the layer may be deeper than this expression suggests. Equation (5.2) simplifies to

$$\frac{\partial^2 w}{\partial Z^2} - \frac{1}{U}\frac{d^2U}{dZ^2}w = 0, \tag{5.9}$$

which has the solution

$$w = [f'(x, y) - f(x, y)]U_0(Z) + \frac{B'(x, y)}{ZI(Z)/U_0(Z)}$$

where

$$I(Z) = 1 + \int_{z_0}^{Z} \frac{[U_0^2(Z)/Z]}{(U_0^2(Z') - U_0^2(Z))}dZ' \tag{5.10}$$

and

$$B(x, y) = P_0\sigma(x, y) - fU_0'(h_m)/J(h_m)$$

$$\sigma(x, y) = \frac{1}{2\pi}\int_{-\infty}^{\infty}\int_{-\infty}^{\infty}\frac{f'(x', y)(x - x')}{[(x - x')^2 + (y - y')^2]^{3/2}}dx'dy';$$

$$J(Z) = 1 + U_0U_0'\int_{z_0}^{Z} dZ'/U_0^2(Z).$$

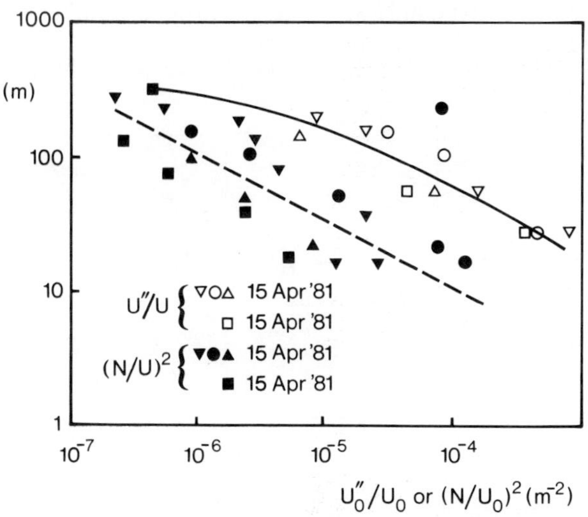

FIG. 5.4. Data from Hunt et al. (1988b) showing the magnitude of the shear and buoyancy terms in Scorer's equation (5.3).

5.2.1.4. Upper layer solution

In the middle layer the shear of the upwind profile is dominant and the vertical velocity increases with height; however, in the upper layer this is no longer the case. The velocity perturbation decays in a potential flow in neutral stratification, whereas it may contain propagating or trapped waves in stable flows (see Durran, Chapter 4). The solution in terms of Fourier transforms is

$$\tilde{w} = ik_1\tilde{f}e^{-mz}, \tag{5.11}$$

where the vertical wavenumber

$$m = (k_1^2 - S^2)^{1/2}\frac{k_{12}}{|k_1|}, \quad S_2 < |k_1|$$

$$m = -i(S^2 - k_1^2)^{1/2}\frac{k_{12}}{|k_1|}, \quad S^2 > k_1^2;\ k_1 > 0$$

$$m = i(S^2 - k_1^2)^{1/2}\frac{k_{12}}{|k_1|}, \quad S^2 > k_1^2;\ k_1 < 0. \tag{5.12}$$

This solution is derived making use of the upper radiative condition for $|k_1| < S$ and the condition that the vertical velocity $w \to 0$ for $z \to \infty$ for $|k_1| > S$. Solutions for $\tilde{u}$ and $\tilde{v}$ can be obtained from (5.11) by use of the continuity equation (5.1).

Solutions for the velocity field in the three regions are obtained by inverting the Fourier transforms. To a first approximation, [zeroth order in the small parameter $\delta = 1/\ln(l/z_0)$] the inner region horizontal velocity perturbation u is not affected by shear stresses in the shear-stress layer. In the next approximation, however, there is a balance between accelerations and the perturbations in the shear stress with u reaching a maximum value at $Z \approx l/3$; this is given approximately by

$$u_{\max} = (h/L)\frac{U^2(h_m)}{U(l)}(1 + 1.9\delta). \qquad (5.13)$$

In the previous linear model of Jackson and Hunt the maximum perturbation was at the top of the inner region $(Z = l)$.

Profiles of the speedup ΔS and change in wind speed u at 10 m over Askervein, a hill in the Hebrides of Scotland (Zeman and Jensen 1987), are shown in Fig. 5.5. The calculations show that the linear theory predicts the speedup quite accurately even for a hill with an aspect ratio as great as $H/L = 0.4$, and it correctly predicts the height of the maximum wind speed; however, in the wake the decrease in wind speed is considerably underestimated, for reasons explained in section 5.5.

5.2.1.5. Surface roughness effects

Previous theories of flow over changes in surface roughness (e.g., Townsend 1966; Panofsky and Townsend 1964) have been based on the assumption that throughout the internal boundary layer the advective velocity is the upwind velocity at the top of the inner region $U(l)$. This leads to significant errors near the surface where $\ln(l/z_0)$ is the same order as $\ln(z/l)$. These theories have only considered step changes in surface roughness rather than arbitrary roughness change.

Changes in a surface roughness using the flow structure of HLR have recently been considered by Walmsley et al. (1986) who considered the first-order solution and Xu and Hunt (1989) who extended this analysis to a second-order solution. Because this theory is also linear, the combined effect of changes in both terrain elevation and changes in surface properties can be calculated by *superposition.* In practice, these changes frequently occur together. In the surface roughness problem, the analysis is very similar to that already described except that the surface boundary conditions are changed. In this case the governing parameters of the perturbed flow is

$$m = \ln(z_1/z_0), \qquad (5.14)$$

where the upstream velocity profile is $U(z) = u_*/\kappa \ln(z/z_0)$ and z_1 is the local roughness. Note that m can be positive or negative for decreases or increases in roughness. The surface boundary conditions are that as $z \to z_1$

$$u_r = \frac{u_*}{\kappa} m, \qquad (5.15)$$

or $u_r^{(0)} = m/\kappa$ and $u_r^{(1)} = 0$ as $\eta \to \eta_1$ $(=z/z_1)$. The subscript r signifies that the perturbation is due to changes in surface roughness. Then the expressions for the first-order terms of the Fourier transforms of the changes of u_r and w_r due to the surface roughness changes are

$$\tilde{u}_r = U_0 A \tilde{m} K_0(2\sqrt{2(\mathrm{sgn}(k_1))}) \qquad (5.16)$$

$$\tilde{w}_r = ikU_0 A\tilde{m} \int_{z_1}^{Z} K_0(2\sqrt{i(\mathrm{sgn}(k_1)\eta})dZ) \qquad (5.17)$$

where

$$A = \frac{2u_*}{U_0}[\ln(l/z_0) - 2\gamma - \ln(i\ \mathrm{sgn}(k_1))]^{-1},$$

γ ($\approx$0.577) is the Euler number, and $\tilde{m}$ is the Fourier transform of m.

The use of Fourier transforms to analyze the flow perturbation over a sudden change in roughness is more accurate than methods based on a perturbed problem in the inner region (cf. Panofsky and Townsend 1964). The results agree closely with the field experiments of Bradley (1968) and with the results of finite-difference computations where the turbulent shear stresses are modeled using auxiliary "second-order" equations for the shear stresses and the rate of energy dissipation.

The linear theory that the two perturbations caused by surface slope and surface roughness can be superposed was shown to be valid for flow over a model hill ($H/L = 0.4$) in a wind tunnel by Britter et al. (1981), except in the wake where the linear theory is not valid anyway.

5.3. Effects of stratification

We first describe the general features of inviscid flow over hills and then show how these flows interact with the turbulent flow in the inner layer. There have been many theoretical and modeling studies of inviscid stably stratified flow over hills at moderate Froude number (e.g.,

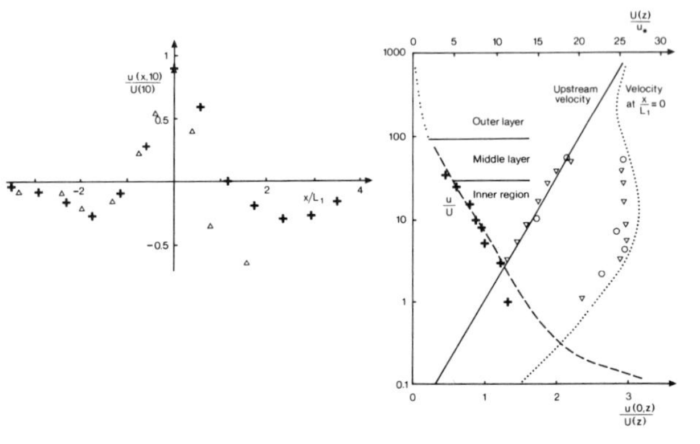

FIG. 5.5. (a) Horizontal velocity perturbations at 10 m over Askervein Hill, Δ measurements + calculations of the linear model; (b) u and ΔS above the summit. Curves, triangles, and circles are measurements, lines are calculations of the linear model.

Scorer 1949, 1953, 1954; Lilly and Klemp 1979; Klemp and Lilly 1978). [For further discussion of these flows see Durran, Chapter 4 and Smith (1979).] There have, however, been far fewer studies of how stable flow in the outer layer interacts with the turbulent flow in the inner region. In this section we consider these flows and describe developments of the linear model to include moderate stratification of different types. We include a uniformly stable atmosphere and an elevated inversion above a neutral boundary layer. Strong stratification cannot be treated by this modeling approach but is discussed in section 5.3.3.

5.3.1. *Uniform stratification*

In Fig. 5.6 we show the laminar flow of a neutral and a uniformly stratified atmosphere over a ridge. The figure displays the main features of the flow, which will determine the velocity field near the surface with a lull upwind of the summit and strong flows on the lee side in the uniformly stratified case.

The upper-layer velocity perturbation in two-dimensional flow over a hill as a function of the Froude number of the hill is presented in Fig. 5.7 from HRB. Initially, as the Froude number decreases (stratification increases) in nonhydrostatic flow, a secondary maximum develops downwind of the hill. This increases in magnitude and moves upwind as F_L further decreases, until eventually the maximum downwind of the hill dominates. In the hydrostatic limit, there is one maximum downwind of the hill summit. The intensity and position of the maximum velocity as a function of the Froude number are shown in Fig. 5.8.

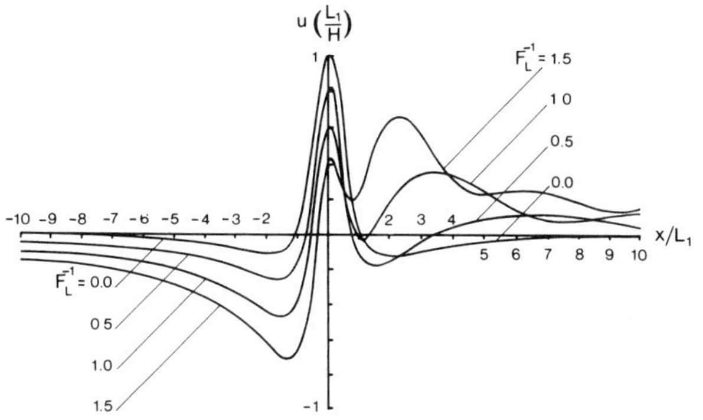

FIG. 5.7. Upper-layer horizontal velocity perturbations over two-dimensional ridge (from HRB).

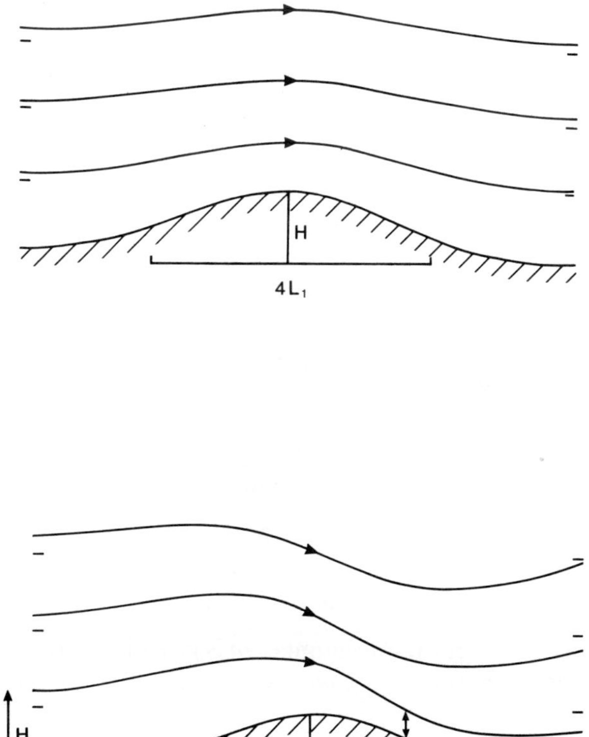

FIG. 5.6. Inviscid flow of a neutral and uniformly stable ($U_0/NL_1 = 2$) atmosphere over a ridge.

5.3.2. *More complex stability profiles*

The linear modeling approach of HRB enables treatment of the effect of a range of different atmospheric stability profiles on the turbulent flow near the surface provided an analytic solution to (5.2) can be obtained, and that the stability in the lower layer is not so large that the flow blocks, or, for the three-dimensional cases, flows around the hill. The profile types that have been considered using this approach are shown in Fig. 5.9. Cases (i) and (ii), neutral flow and uniform stratification, have already been discussed. Case (iii), a uniform boundary layer, is capped by a strong elevated inversion with the

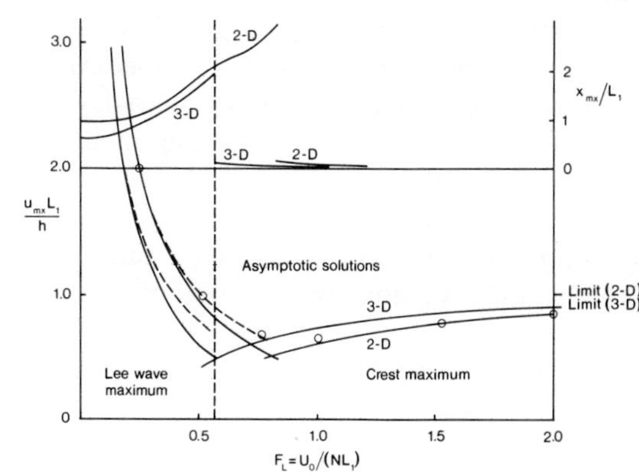

FIG. 5.8. Uniformly stratified flow. Intensity and position of maximum velocity as a function of Froude number over two- and three-dimensional hills.

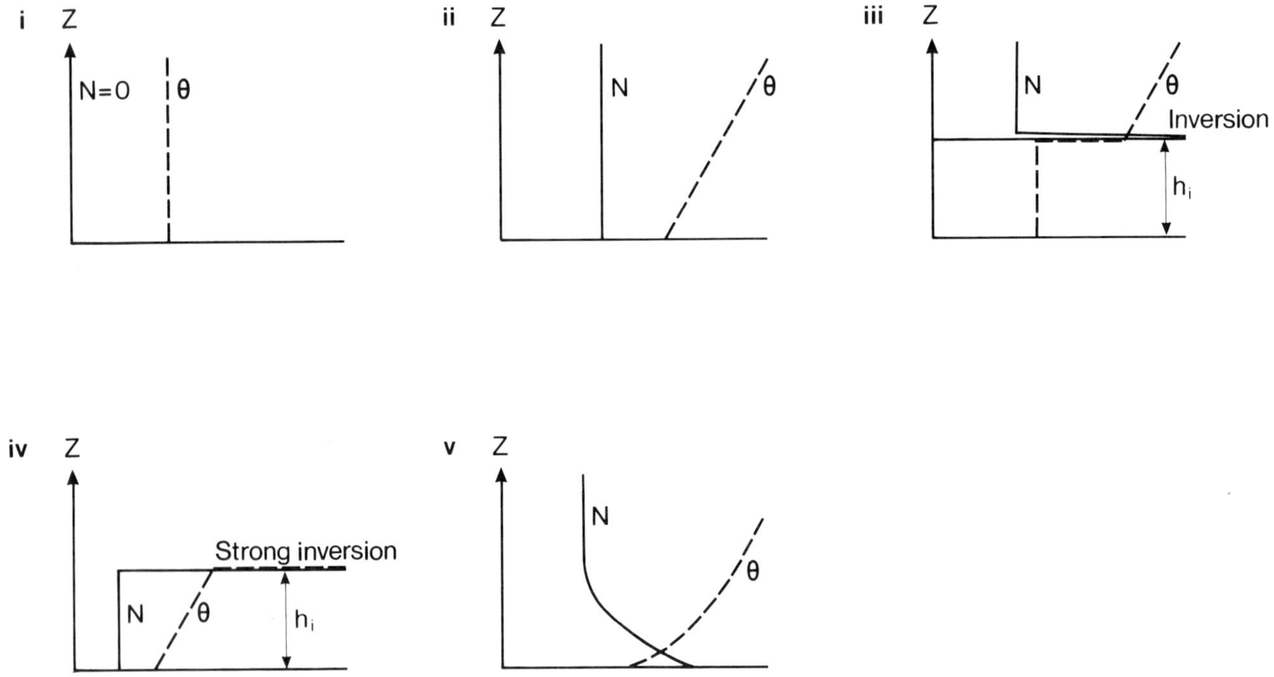

FIG. 5.9. Stability profiles for the upstream flow treated by the linear model.

free troposphere above; in case (iv) a stable boundary layer is capped by a very strong inversion; and in case (v) stratification decreases from a very high value near the surface to a uniform value above. In this chapter we consider case (iii). Details of cases (iv) and (v) may be found in Carruthers et al. (1988).

5.3.3. *Elevated inversion above neutral boundary layer*

This situation occurs very frequently, so that it is of considerable importance to understand how the inversion layer passes over complex terrain and how it affects the airflow below. Using a nonlinear analysis, Carruthers and Choularton (1982) considered a case where the inversion has finite width. Using a linear analysis HRB assumed a discontinuity in the temperature profile. This is frequently a good approximation because the inversion layer is very thin and is a small fraction of the boundary-layer depth. We shall follow the analysis of HRB; the stability profile is shown in Fig. 5.9.

Letting the upwind height of the inversion be h_i with temperature change across if $\Delta\theta_i$, then integrating Eq. (5.3) across the inversion layer, we obtain

$$\left[\frac{\partial w}{\partial z}\right]_{z=h_i} = -F_i^{-2}\frac{w}{h_i}(z=h_i) = \frac{-F_i^{-2}}{h_i}U_0(h_i)\frac{d(\Delta h_i)}{dx}, \qquad (5.18)$$

where $[\partial w/\partial z]_{z=h_i}$ is the change in $\partial w/\partial z$ at $z = h_i$, Δh_i the vertical displacement of the inversion, and F_i the Froude number of the inversion layer; i.e.,

$$F_i^{-2} = \frac{h_i}{U_0^2(h_i)}\frac{g\Delta\theta_i}{\theta_i}. \qquad (5.19)$$

The solutions for the Fourier transform of the vertical velocity are obtained by integrating (5.3) in both layers and determining the coefficients by satisfying (5.18) and the upper and lower boundary conditions. The solutions are as follows:

$z < h_i$:

$$\tilde{w} = ik_1U_0\left\{\left[\frac{\tilde{f}e^{k_{12}h_i} - \widetilde{\Delta h_i}}{2\sinh(k_{12}h_i)}\right]e^{-k_{12}z} - \left[\frac{\tilde{f}e^{-k_{12}h_i} - \widetilde{\Delta h_i}}{2\sinh(k_{12}h_i)}\right]e^{k_{12}z}\right\}, \qquad (5.20)$$

$z > h_i$:

$$ik_1\widetilde{\Delta h_i}e^{-m(z-h_i)}, \qquad (5.21)$$

where the vertical wavenumber m is given by (5.12) and the Fourier transform of the displacement of the inversion is

$$\Delta h_i = \frac{(k_{12}h_i)\tilde{f}(\operatorname{cotanh}(k_{12}h_i) - 1)}{mh_i + k_{12}h_i\operatorname{cotanh}(k_{12}h_i) - F_i^{-2}}e^{k_{12}h_i}. \qquad (5.22)$$

The most striking feature of the flow is the presence of waves trapped by the inversion layer when the denominator of the expression for $\widetilde{\Delta h_i}$ vanishes. These waves obtain large amplitudes when the wavelength of the resonant mode corresponds closely to the dominant mode in the Fourier transform of the hill. They are discussed in detail by, for example, Scorer (1949), Crapper (1959, 1962), and Carruthers and Choularton (1982). The waves occur downwind of the hill summit and can result in rotors with regions of reverse flow at the ground. The presence of an inversion also has a strong influence on the flow over the hill summit. The main features are shown in Fig. 5.10. In general, when $F_i \gg 1$ the inversion rises over the hill summit, and when $F_i \ll 1$ the flow dips near the summit; increasing the upper-layer stratification (decreasing F_L) increases the asymmetry of the flow with flow speed increasing on the lee side. When $F_i \approx 1$ or $F_l \approx 1$ large-amplitude motions may occur. Carruthers and Choularton (1982) calculated examples of the effect of elevated inversions on the flow in the inner region using the linear model of Jackson and Hunt. They showed that the presence of the inversion could change the speedup significantly from its value in neutral flow increasing or decreasing if dependent on F_i and F_L.

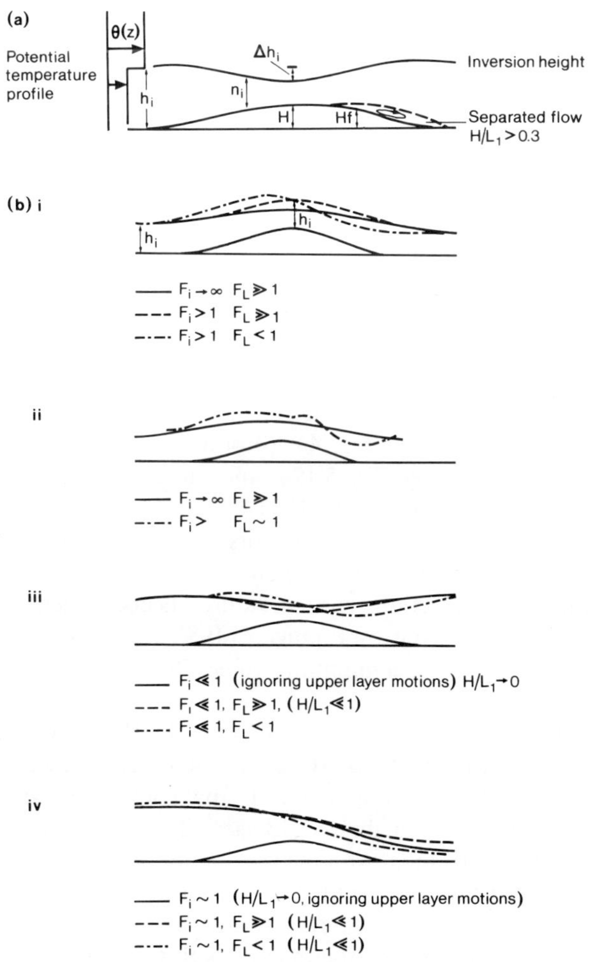

FIG. 5.10. Flow features associated with an inversion layer for different inversion strengths and upper level stratification (from HRB).

5.3.4. Strong stratification; large aspect ratios

When stable stratification in the boundary is so strong that the Froude number $F_H = U_0/NH \leqslant 1$, the perturbation to the flow is of the same order as the velocity U_0 of the incident flow. The linear approach used in section 5.2 is then unable to describe the flow observations. Hunt and Snyder (1980) have shown that in three-dimensional flows when $F_H \leqslant 1$ the flow divides into two regions (Fig. 5.11). At height $z < H_c$, the flow is around the hill, while for heights $z \gtrsim H_c$, the flow is over the hill (i.e., a hill of height $H - H_c$). Experiments over a wide range of hill shapes in the laboratory and in the field (Snyder 1988; Egan 1984) suggest the height of dividing streamline is given by

$$\frac{H_c}{H} \approx 1 - F_H \quad \text{for} \quad F_H < 1. \tag{5.23}$$

Simple dimensional and energy arguments (Sheppard 1956) give an identical result, but Smith (see the Appendix) has shown that Sheppard's analysis is not strictly correct since he ignored the pressure forces. Smith has also shown, however, that a new analysis leads to rather similar results for the position of the dividing streamline. In strongly stratified flow over a ridge there is blocking of the lower flow when $F_H < 1$. In laboratory experiments Brighton (1978) and Hunt and Snyder (1980) noted other features of strongly stratified flow over a hump including the tendency to form lee waves, hydraulic jumps, or trailing vortices on the lee side at approximately the height of the hill summit (see Fig. 1 in Hunt and Snyder). The recent theory of Newley et al. (1989) also predicted that the maximum vertical displacement would be in this summit region.

5.4. Turbulence

The study of turbulence over complex terrain has recently been the subject of much investigation; some aspects are covered in the review of Finnigan (1988). To analyze the changes of turbulence over hills, we divide the boundary layer into two regions which, following the approach of section 5.2, correspond broadly to the inner and outer regions of the mean flow (Britter et al. 1981). Near the surface ($Z \ll l$) where the eddy turnover time $T_L(Z) \sim Z/u_*$ is much smaller than the travel time $T(Z) \sim L/U(Z)$ for mean flow over the hill, the turbulence is close to local equilibrium. Therefore all the components are changed by an equal amount and their increases are proportional to the increase in shear stress $\Delta\overline{u_i^2}/\overline{u_i^2} = \Delta\tau/\tau$ for $i = 1$, 2, and 3, and τ is the upstream shear stress.

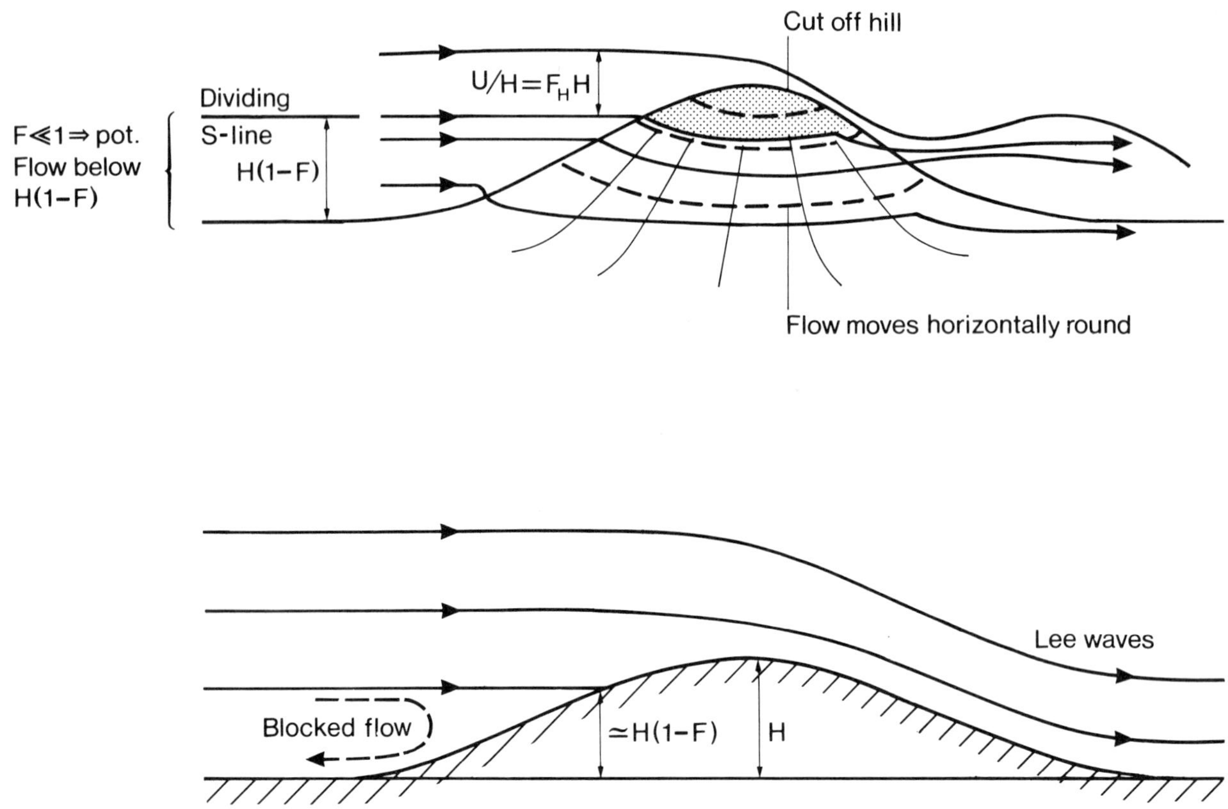

FIG. 5.11. Schematic picture of strongly stratified flow over a three-dimensional hill and a two-dimensional ridge.

The hypothesis is supported reasonably well by the measurements of Bradley (1980) and Mason and King (1985). Over the top of the hill the maximum of $\overline{\Delta u_i^2}/\overline{u_i^2}$ occurs at the surface where $\Delta\tau$ is maximum, but the perturbation turbulence variances $\overline{\Delta u_i^2}$ of all components decrease rapidly with height, as does the shear stress, to near zero at $Z \sim l/3$. In the outer region the eddy turnover time is greater than the travel time, so changes in the turbulent structure caused by the distorting effect of the mean flow are relatively "rapid" and can be modeled by rapid distortion theory. This implies that the initial state of the turbulence is important. For simplicity Britter et al. (1981) assumed the upstream turbulence was isotropic, but subsequently Newley (1985) and Zeman and Jensen (1987) allowed for the anisotropy of the upstream turbulence. They assumed it to be axisymmetric with $\overline{u_2^2} = \overline{u_3^2} < \overline{u_1^2}$ and obtained the following estimates for the changes in streamwise and vertical turbulence for flow over two-dimensional hills above the inner region:

$$\frac{\overline{\Delta u_1^2}}{\overline{u_1^2}} \approx -\frac{6}{5}\left(1 - \frac{1}{3A}\right)\Delta S, \tag{5.24}$$

$$\frac{\overline{\Delta u_3^2}}{\overline{u_3^2}} \approx \frac{6}{5}\left(1 - \frac{A}{3}\right)\Delta S, \tag{5.25}$$

where $A = \overline{u_1^2}/\overline{u_3^2} > 1$ is the upstream ratio of turbulence components. Note that the effect of anisotropy is to reduce both $\overline{\Delta u_1^2}$ and $\overline{\Delta u_3^2}$. The main feature of these results is that above the inner region, the three velocity components change differently. The changes agree in sign and approximate magnitude with the field measurements of Bradley (1980, their Fig. 4), Mason and King (1985), Mason (1986) (see Fig. 5.12), and the laboratory measurements of Britter et al. (1981).

In the region $Z \approx l$ the factors determining the turbulence structure include the effects of rapid distortion, nonlinear interaction, and curvature effects. A detailed description is given by Finnigan (1988). In this region $\overline{w^2}$ and τ have a local minimum value (Zeman and Jensen 1987; Newley 1985).

Panofsky et al. (1982), Gallagher et al. (1988), Mason and King (1985), and Mason (1986) have presented spectra of the longitudinal and normal components over hills, and over roughness changes. These spectra are broadly in agreement with the spectral form inferred from the aforementioned arguments. Near the ground the small-scale, high-frequency wavenumber ranges (where $K_1 \gtrsim l^{-1}$) of the perturbations of both the longitudinal and vertical spectra ($\Delta\Theta_{11}$, $\Delta\Theta_{33}$) are close to local equilibrium; therefore

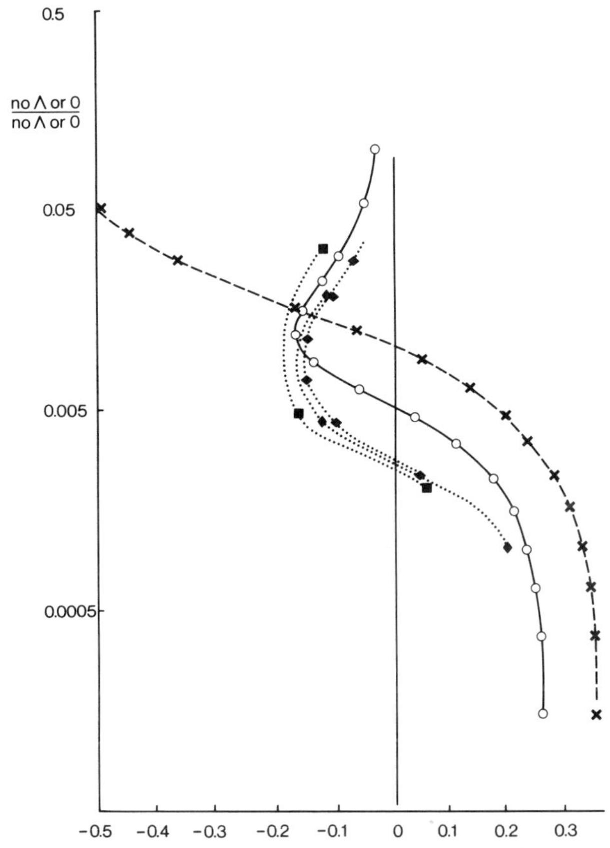

FIG. 5.12. Comparison of the perturbation shear-stress profile at the summit of an isolated hill with field data, from Newley (1985). Here (–○–) denotes second-order closure model, (–×–) mixing-length model, (–■–) Blashaval data (Mason and King 1985), (–♦–) Nyland Hill data (Mason 1986), and (–◆–) Askervein data (Zeman and Jensen 1987).

$$\Delta\Theta_{11} = \frac{3}{4}\,\Delta\Theta_{33} \approx \frac{3}{4}\left(\frac{\Delta\tau}{\tau}\right)\alpha u_*^2 K_1^{-5/3} l_\epsilon^{-2/3}, \quad (5.26)$$

where $l_\epsilon(Z)$ ($=\epsilon/\sigma_2^3 \approx \epsilon/\tau^{3/2}$) is the energy dissipation scale. This macro length scale is determined both by the blocking effect of the surface and by the mean shear. Physical arguments (Hunt et al. 1988c) and comparisons with computations using a second-order turbulent model (Newley 1985) show that these two effects can be combined to yield

$$l_\epsilon^{-1} = \frac{A_R}{Z} + A_S\frac{dU/dZ}{\sigma_w}, \quad (5.27)$$

where the coefficients A_R and A_S are 0.45 and 0.27, respectively. Note that over the summit where $\Delta\tau$ and $\partial U/\partial Z$ decrease rapidly with height, $\Delta\Theta_{11}$ and $\Delta\Theta_{33}$ also decrease from their maximum value at the surface. In the wake $\Delta\Theta_{11}$ and $\Delta\Theta_{33}$ follow $\Delta\tau$ and $\partial U/\partial Z$ in having maximum values *above* the surface at $Z \sim l$.

Since at the lowest frequencies in distorted turbulent flows, the spectra $\Delta\Theta_{11}$ and $\Delta\Theta_{33}$ are proportional to the square of the mean velocity of the relevant component (i.e., are quite strong), it follows that for $K_1 < L^{-1}$,

$$\Delta\Theta_{11}(K_1) \approx (u/U)^2\Theta_{11}(K_1) \quad \text{and}$$

$$\Delta\Theta_{33} \approx (w/U)^2\Theta_{11}(K_1). \quad (5.28)$$

Both these perturbation spectra are of the order of $(H/L)^2$. At intermediate wavenumbers $L^{-1} < K_1 < l^{-1}$, the spectra in the inner region are determined by the distortion of the eddies that are larger than the inner region and are centered at height $Z \sim K_1^{-1}$; therefore $\Delta\Theta_{11}$ is reduced but $\Delta\Theta_{33}$ is increased in proportion to the slope H/L. Simple analytical expressions are not available for this range. The above arguments suggest, however, that over a hill the changes in the horizontal spectra are highly variable in magnitude and sign, but less so for the vertical spectra.

Over a change in surface roughness or in the wake, all significant changes are confined to the inner region. Therefore the large-scale eddies above the inner region are not affected. In addition, *within* the inner region the scales that are much larger than the thicknesses of the inner region are not affected by the changes in the mean flow; i.e., for $K_1 \ll l^{-1}$, $\Delta\Theta_{ii} = 0$ (no summation). For scales much less than l, the turbulence is in local equilibrium and is determined by the local rate of dissipation ϵ, and described by formulae (5.26).

Since for a roughness change the thickness l of the inner region grows with distance downwind [$l\ln(l/z_0) \approx 0.3x$ (Townsend 1965)], the turbulent eddies are advected into the inner region and adjust to the local turbulence on a time scale $T(K_1)$ that depends on the eddy scale. Panofsky et al. (1982) and subsequently Hogstrom and Hogstrom (1985) showed that this adjustment could be modeled by a first-order process. For $K_1 \leqslant Z^{-1}$,

$$d\Delta\Theta_{ii}/dt = (\Delta\Theta_{eqii} - \Delta\Theta_{ii})/T(K_1), \quad (5.29)$$

where the change to the equilibrium value $\Delta\Theta_{11}$ is given by (5.26) and $T(K_1) = \epsilon^{-1/3}K_1^{-2/3}$. Thus at a height Z above the surface, it takes a distance of about $40Z$ for the turbulence to be equilibrium (to within 10%) for $K_1 \geqslant Z^{-1}$; i.e., for a wavenumber such that $K_1 Z \geqslant 1$, only the bottom half of the layer (height Z) is approximately in local equilibrium.

5.5 Numerical models and flow over complex terrain

5.5.1. *Isolated hills*

There have been many models developed of varying levels of sophistication to calculate the flow over isolated topography. Taylor (1977) presented a two-dimensional numerical model to compute the turbulent shear stress by means of a kinematic eddy viscosity proportional to $K^{1/2}l_m$, where K is the turbulent kinetic energy per unit mass and l_m is a mixing length. Here K must be computed from an auxiliary equation coupled to the momentum equation. This approach contains several assumptions including an estimate for the dissipation length l_ϵ. Taylor (1977) and Britter et al. (1981) assumed $l_m = l_\epsilon = 0.4Z$, and they found that the velocity defect on the lee side of the hill decreased too rapidly on the lee slope. In other words the extent of the wake was greatly underpredicted because of the neglect of the reduction of the mixing and dissipation length scale. As a simple model to allow for this effect, we have used (5.27) and assumed that the mixing length scale (l_m) is equal to the dissipation length scale (l_ϵ); i.e.,

$$l_m{}^{-1} = l_\epsilon^{-1} = \frac{A_R}{Z} + A_S \frac{\partial U / \partial Z}{\sigma w}. \qquad (5.30)$$

This model leads to a lower value of shear stress in the upper half of the inner region in the wake and hence to a larger wake defect that extends further downwind. In more complex models, the length scale is modeled by a second auxiliary equation (the k–ϵ method; e.g., Rodi 1980) or the equation is derived for the Reynolds stress (e.g., "second-order" models). These second-order models give improved representations of the turbulence in the upper part of the inner region and agree with results of rapid distortion theory in the outer region [as Fig. 5.12 from Newley (1985) illustrates]. This figure compares observations from the Scottish hills Blashaval, Nyland Hill, and Askervein with the second-order closure model of Newley and the linear model of HLR. A large number of laboratory studies of flow in complex terrain have also been performed (see Meroney, Chapter 7).

5.5.2. *Complex terrain*

The two main limitations of the linearized analytical models are that they exclude both large positive or negative changes in the mean flow and more complex models of turbulent shear stresses, in particular those involving auxiliary differential equations. Comparison between computation and linear models shows that the greatest errors in neutral flows occur on the lee side and in the wake, whereas over hills of slope even as high as $\frac{1}{3}$ the *increase* in wind speed over the summit is well estimated.

In fact, even nonlinear computational models can also be unsatisfactory in the lee side and wake region, unless they incorporate a turbulence model that allows for the reduction in the scale $L_x{}^{(w)}$ of the vertical turbulence, where $\partial U / \partial Z$ is a maximum in the upper half of the inner region. This reduction in $L_x{}^{(w)}$ leads to a reduction in l_ϵ, to a *lower* shear stress, and hence to a *larger* value of $\partial u / \partial Z$ and a more pronounced reduction in velocity near the surface. This sensitivity of turbulent boundary layers to adverse pressure gradients means that small changes in surface conditions produce large effects on the whole wake region, such as a reduction in roughness or a slight increase in stable stratification suppressing separation (Britter et al. 1981; Scorer 1955).

Because of the large amount of computer time required by nonlinear numerical models, they are impractical for very high resolution studies over complex terrain, although mesoscale models can be used to analyze many features of the flow (see Peagle and Pielke, Chapter 10). Using a numerical representation of the Fourier transform of the terrain and roughness, however, the HLR model can be used to construct a fast computer model for flow over complex terrain. Such models have been developed [Walmsley et al. (1986) and Geljaars et al. (1987) for neutral flow only, and Carruthers et al. (1988)] for different types of stratification. The later model (FLOWSTAR) was developed for Cambridge Environmental Research Consultants.

The method of calculation is to compute Fourier transforms of the velocity field following HLR; then the transform is inverted numerically to calculate the actual flow variables at a point. In contrast to numerical models that solve the equations of motion on a grid, there is no iteration involved and no doubt about the solution once the algorithms and their assumptions have been established. This makes the Fourier transform method quite appropriate for use on small personal computers. There are a number of conditions concerning the terrain data and meteorology that must be satisfied in order for the model to give useful results:

(i) The slopes of the hill are small (typically less than about $\frac{1}{4}$, although useful results are obtained for slopes as great as $\frac{1}{2}$).

(ii) The changes in the natural logarithm of the roughness length z_0 are also small (less than 1.0, although useful results are obtained for larger changes).

(iii) The profile of potential temperature in the atmosphere can be approximated by one of the three basic forms described earlier in this section.

(iv) The upwind velocity profile increases from the ground upward and does not have a strong elevated shear layer.

(v) The upwind conditions are varying slowly on a time scale comparable with that taken by a fluid particle to cross the flow region under consideration.

(vi) Rapid cooling or heating of the surface of the hillside are absent; this can induce significant motions that are not considered here (see Whiteman, Chapter 2 and Banta, Chapter 9).

Some examples of output from FLOWSTAR are shown in Figs. 5.13 and 5.14. The first set of calculations was

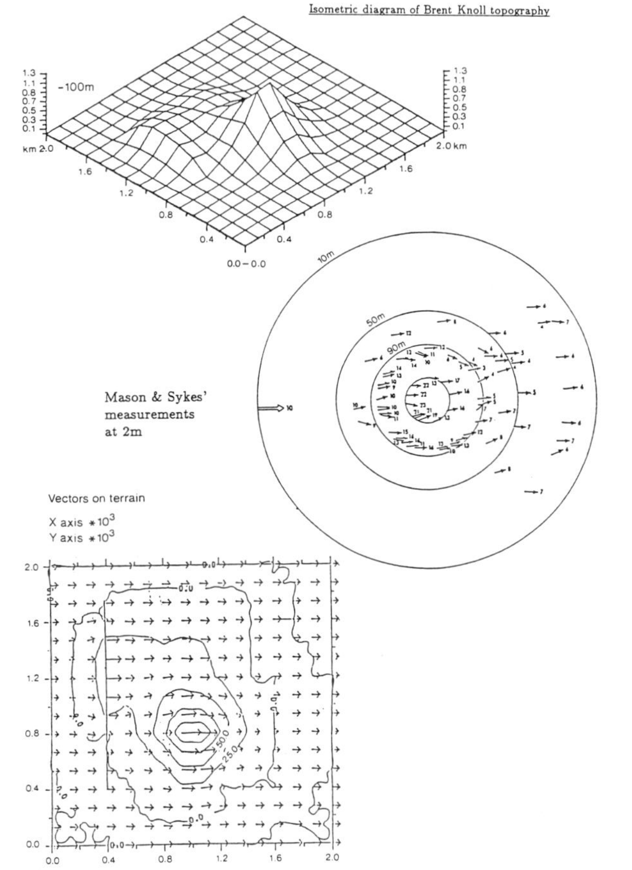

FIG. 5.13a. Brent Knoll FLOWSTAR calculations at 2 m in neutral flow S_0. Upwind speed at 2 m is 10 m s^{-1}: L = 290 m, H = 137 m, l = 14 m, and $h_m \approx$ 100 m, z_0 = 0.02 m. The maximum of 24 m s^{-1} compares well with observed values of approximately 23 m s^{-1}.

performed for Brent Knoll in England. In neutral flow, the speedup (ΔS) at the summit was calculated to be 1.4, which compares well with the observed value of $\Delta S \approx 1.3$ (Mason and Sykes 1979). The cross section of the flow shows that over the summit there is a low-level maximum at about 10 m above the surface; however, away from the summit there is a monotonic increase with height. The vector plot for flow over Brent Knoll with strong stratification (S_1) shows both that the maximum wind is downwind of the summit and that regions of considerable transverse and reverse flow exist.

The second set is for flow over a large ridge with dimensions similar to those of Great Dun Fell in Cumbria, a county in northwest England (Carruthers and Choularton 1982). This case with a strong elevated inversion results in stronger flow downwind at the surface than in neutral flow. There is strong shear across the inversion layer and wave activity above the inversion.

Using the results for the changes in mean velocity and shear stress, the analytical expression for the changes in turbulence spectra, reviewed in section 5.4, can be incorporated into the simple models such as FLOWSTAR to provide estimates for the changes in spectra over complex terrain.

5.6. Dispersion and deposition over complex terrain

5.6.1. Overview and key processes

We have seen how the presence of hills and variation in surface roughness change the airflow in the atmospheric boundary layer. These changes inevitably affect the dispersion of heat and airborne matter. Present research on atmospheric dispersion in complex terrain has focused on dispersion from localized sources of pollution situated within a few kilometers of isolated hills and in valleys, primarily with a view to estimating the increase in the surface concentrations compared with the levels found on flat, uniform terrain. (Increases by factor of 10 are quite possible!) [Recent reviews have been given by Egan (1984) and Hunt (1985)], and there are further discussions in Hanna and Strimaitis, Chapter 6, where a number of recent numerical models are compared with field measurements.) Even when sources of pollution are located hundreds or thousands of kilometers upwind, however, the presence of the complex terrain can still change the dispersion and deposition of the pollutants. The processes involved in these changes are beginning to be systematically investigated theoretically, computationally, and in field studies (e.g., Choularton et al. 1986; Padro 1987; Hunt et al. 1988). Unlike the localized sources near hills, these long-range dispersion/deposition problems over complex terrain have not been extensively investigated with physical modeling. The only relevant experiments known to the authors are by Dawkins and Davis (1981) and Verma and Cermak (1974), who studied evaporation over undulating surfaces.

The fact that deposition onto hills can be many times greater than deposition onto level terrain caused much concern in Europe following the release of radioactivity at Chernobyl in April 1986.

The ways in which the changes in airflow affect dispersion over hills are (following Hunt 1985):

(i) The streamlines of the mean flow (or mean streamlines) converge and diverge, leading to increases or decreases in concentration gradients, and decreases and increases in the cross-stream widths of "plumes."

(ii) The distant n between mean streamlines and the ground changes, leading to high surface concentrations, where streamlines from a source impinge onto a surface.

(iii) The mean velocity increases or decreases leading to reduction or increase in concentration at the source (other factors being equal).

(iv) The turbulence and turbulence gradients change leading to variations in the concentration gradients,

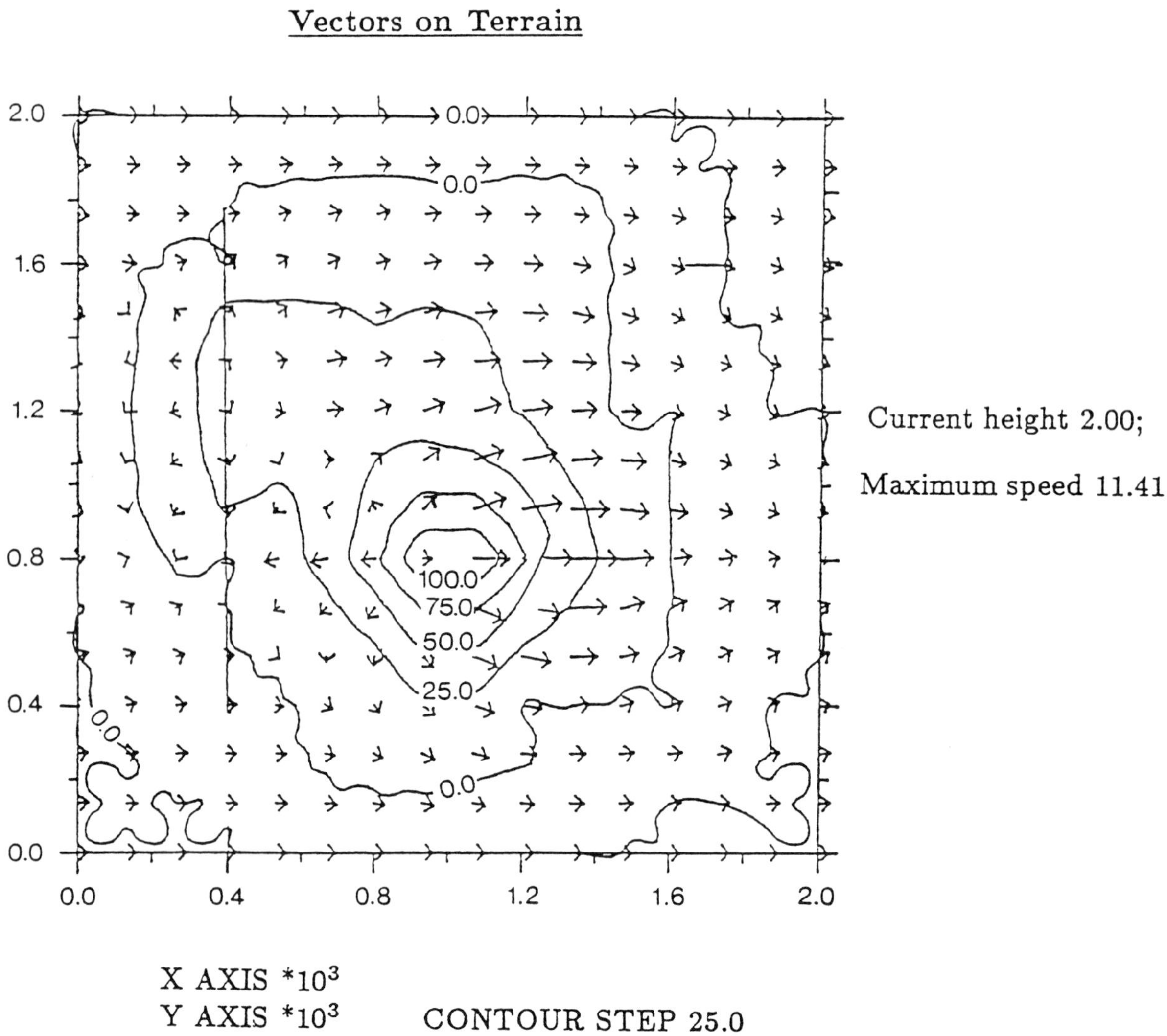

FIG. 5.13b. Brent Knoll—model calculation at $Z = 2$ m, strong uniform stratification $N = 0.08\ s^{-1}$, upwind speed at 2 m is 2.5 m s^{-1}.

changes in "plume widths," changes in the mean position of a plume, and changes in rates of deposition to the surface.

(v) Recirculating and reversing mean flow in valleys on hills with moderate or steep slopes leads to rapid rates of mixing; the fact that the flow *reverses* in wakes appears to be significant in determining the concentration distribution from sources within the wake region through the recirculation (Turfus 1986). For sources outside the wake, the recirculation leads to approximately constant wake concentration (Puttock and Hunt 1979).

5.6.2. *Localized sources near hills*

If the rate of deposition of a pollutant or other matter is slow (typically $\frac{1}{5}u_*C$ per unit area per unit time where C is the mean concentration), then the mean concentration distribution downwind of a localized source is *independent* of the deposition rate for a distance downwind x of the order of $5z_sU/u_*$, where z_s is the same height. Consequently, for typical source heights of 30–100 m, the effects of deposition are generally not important within 5 km, so that if the ground surface is impervious over this range, it is possible to estimate the various complex effects listed in section 5.6.1 on the dispersion of a plume from a localized source. It has been established both theoretically and by experiment (e.g., Pasquill and Smith 1983) that the mean concentration distribution downwind of a point source can be approximated by a Gaussian distribution with "reflection" terms to allow for the impervious surface and the presence of an elevated inversion; e.g.,

$$C(x, y, z) = \frac{Q}{\sqrt{2\pi}U_\psi\sigma_y}\exp\left(-\frac{(y-y_\psi)^2}{2\sigma_y^2}\right)G_z(z, z_\psi), \tag{5.31}$$

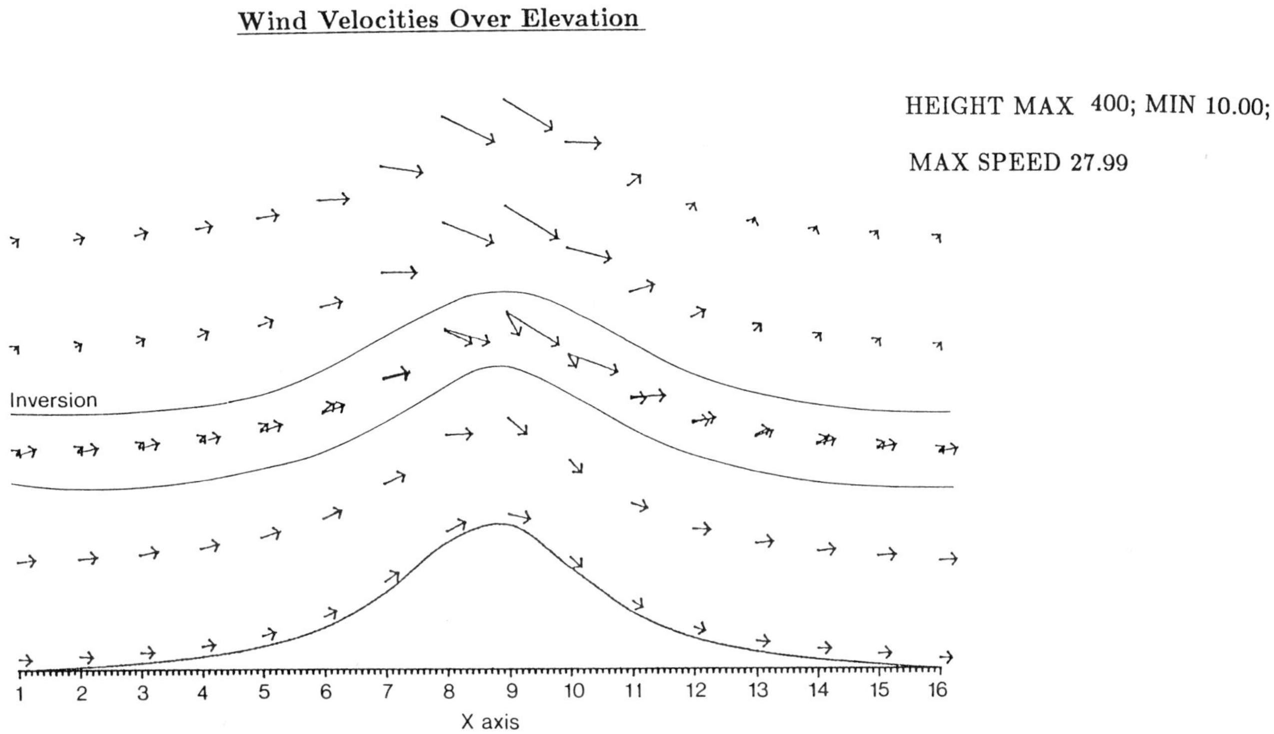

FIG. 5.14. Great Dun Fell. Calculation using FLOWSTAR, with elevated inversion. $F_i^{-2} = 0.77$, buoyancy frequency of upper layer = 0.02 s^{-1}. Height of inversion $h_i = 200$ m, $z_0 = 0.02$ m, $L = 2300$ m, $H = 665$ m, $h_m = 200$ m, and $l = 90$ m. Wind at 10, 100, 200, 300, and 400 m.

where U_ψ is the mean velocity along the streamline ψ, $y_\psi(x)$ is the y-coordinate of the mean streamline, and

$$G_z(z, z_\psi) = \frac{1}{\sqrt{2\pi}\,\sigma_z}\left[\exp\left(-\frac{(z - z_\psi)^2}{2\sigma_z^2}\right) + \exp\left(-\frac{(z - z_\psi)^2}{2\sigma_z^2}\right)\right]. \quad (5.32)$$

It is useful here to show how the concentration distribution is changed by the presence of hills. For sources outside recirculating flow regions, the center line of the plume and the location of the reflected plume are calculated from the mean flow; to calculate the horizontal and vertical plume widths (σ_y, σ_z), it is necessary to consider both the converging and diverging of the mean streamlines *and* the turbulence structure. Surprisingly, where the mean flow changes are more important, calculations of the variances σ_y and σ_z are insensitive to the model used for turbulent diffusion.

For idealized shapes, such as cylindrical or axisymmetrical hills, the model based on potential flow and eddy diffusion calculations describes the major features of dispersion from upwind sources (Hunt et al. 1979). For dispersion on the lee side and in the wakes of hills, eddy diffusion and random-flight methods have been used (Puttock and Hunt 1979; Turfus 1986, 1988; Thomson 1988).

Comparison between these kinds of models and laboratory and idealized field experiments have been made by Snyder and Hunt (1984), Egan (1984), Maryon et al. (1986), Castro and Snyder (1982) and many others. An important finding of the experimental results in stable conditions was the low-frequency variation of the flow. For example, the critical height H_c (see Fig. 5.11) could vary by 10–20 meters over a 20-minute period, and the wind direction could vary by 20° or so, so that in a 1-hour period a plume could be swept horizontally around a hillside below H_c and also dispersed over the top above H_c. The resulting distribution for a 1-hour period would not correspond to any calculation of the mean concentration based on meteorological statistics averaged over a 1-hour period (Egan 1984).

Comparison between the models and field measurements of computations from large sources, such as power stations, in large-scale complex terrain (with elevation changes of the order of 10^3 m) has been attempted as part of the EPA (Environmental Protection Agency) Complex Terrain Project (Strimaitis et al. 1986). The same broad features concerning plume impaction were found as in the smaller-scale experiments, but in the larger-scale flows thermal wind effects were also important.

It is equally necessary to develop simple computational models for dispersion over complex terrain for the airflow; i.e., models that do not require direct computations of the full flow and diffusion questions. Using the basic concepts of flow models for different flow regions and different regimes of flow depending on the atmospheric stability, algorithms based on Gaussian plume models have been developed for dispersion from elevated sources near hills. Two approaches have been used for these "simple models."

5.6.2.1. IDEALIZED HILL SHAPE [EPA-CTDM MODEL (PAINE ET AL. 1987)]

Each hill near a source is approximated by an ellipsoid; the flow is assumed to be given by potential neutral flow or stratified flow at strong stratification $F_H \lessdot 1$. No interaction is allowed for in the flow between the hills. Great emphasis is placed on modeling the flow above and below H_c, the dividing streamline. The model has been validated against three field experiments and some idealized laboratory experiments. The model has been written for use on a PC, and is available from the EPA (see Hanna and Strimaitis, Chapter 6 for further details).

5.6.2.2. FOURIER ANALYSIS OF HILL SHAPES

In FLOWSTAR (Carruthers et al. 1988) and the codes developed by Walmsley et al. (1986) flow over arbitrary terrain is analyzed provided that the slope is low and that $F_H \gtrsim 1$. Compared with the previous case, this kind of model is better at near neutral conditions than in very stable conditions. General purpose codes based on this approach are not yet available.

5.6.3. *Dispersion and deposition over terrain for well-mixed scalars*

Far downwind of a source (when $x > 10hU/u_*$) the vertical distribution of mean concentration in the atmospheric boundary layer depends on the rate of deposition to the surface per unit area (F_d). If F_d is zero, there is no variation in the vertical concentration (i.e., $\partial C/\partial Z = 0$ for $z_0 < z < h$). If F_d is nonzero, however, the variation in C depends on the rate at which turbulence transports matter to the surface and redistributes the concentration throughout the layer. The same processes also control the distribution of other scalar quantities in the atmosphere, such as water vapor, NO_x, etc.

Consequently, over complex terrain, where the turbulence and the mean flow change, the distribution of concentration (as explained in section 5.6.1) changes; the deposition rate must also change, since F_d is proportional to the local value of shear stresses that increases over hilltops. There are also other factors associated with hills that change F_d; e.g., the clouds, precipitation rate, and solar radiative flux to the surface. A two-dimensional model to describe how some of these processes and factors affect F_d and the distribution of concentration is described in Hunt et al. (1988d). We describe this method briefly. An alternative approach has been described by Padro (1987); however, he assumed that the surface completely absorbs the pollutant.

We assume that the undisturbed or upstream flux of the species is uniform over the inner region depth. Then we consider the normalized perturbation quantities for the stress and perturbations in concentration and flux, $\Delta\tau$, c, ϕ as

$$\tau = u_*^2(1 + \Delta\tau(x, Z))$$

$$\bar{C} = \frac{c_*}{\kappa}(C(Z) + c(x, Z))$$

$$\bar{F} = F + \Delta F = -u_* c_*(1 + \phi(x, Z)). \quad (5.33)$$

The equation describing the concentration is

$$U\frac{\partial c}{\partial x} + w\frac{\partial \bar{C}}{\partial Z} = \kappa\epsilon\frac{\partial \phi}{\partial Z} = -\frac{\partial F}{\partial Z}, \quad (5.34)$$

where the small parameter $\epsilon = u_*/U_0$ and the flux

$$\bar{F} = -2\kappa u_*(1 + \Delta\tau/2)\frac{\partial}{\partial Z}\left[(C + c)\frac{c_*}{\kappa}\right]. \quad (5.35)$$

Using (5.33) and (5.35) and substituting into (5.34), taking linear terms only and normalizing so that $\zeta = x/L$ and $\eta = Z/l$, we obtain an inhomogeneous equation describing the concentration perturbation c:

$$L(c) = (1 + \delta \ln\eta)\frac{\partial c}{\partial \zeta} - \frac{1}{2}\frac{\partial}{\partial \eta}\left(\eta\frac{\partial c}{\partial \zeta}\right)$$

$$= \underbrace{\frac{1}{\eta U(l)}\int_0^\eta \frac{\partial u}{\partial \zeta}\partial\eta}_{\text{streamline convergence/divergence}} + \underbrace{\frac{1}{4}\frac{\partial(\Delta\tau)}{\partial \eta}}_{\text{shear stress}}. \quad (5.36)$$

with the upper boundary condition that the perturbation in concentration becomes small for large η (i.e., $c \to 0$ and $\eta \to \infty$) and the surface condition that

$$\eta\frac{\partial c}{\partial \eta} \to \phi_0 + \Delta\tau_0/2 \quad \text{as} \quad \eta \to 0, \quad (5.37)$$

where ϕ_0 and $\Delta\tau_0$ are the surface flux perturbation and surface shear-stress perturbation.

Following Chamberlain (1966) and Hunt et al. (1988d), we assume that the flux of both gases, and wet and dry aerosol particles are governed by the empirical relationship

$$F = -V_g C. \quad (5.38)$$

At low wind speeds or in conditions when plants are not

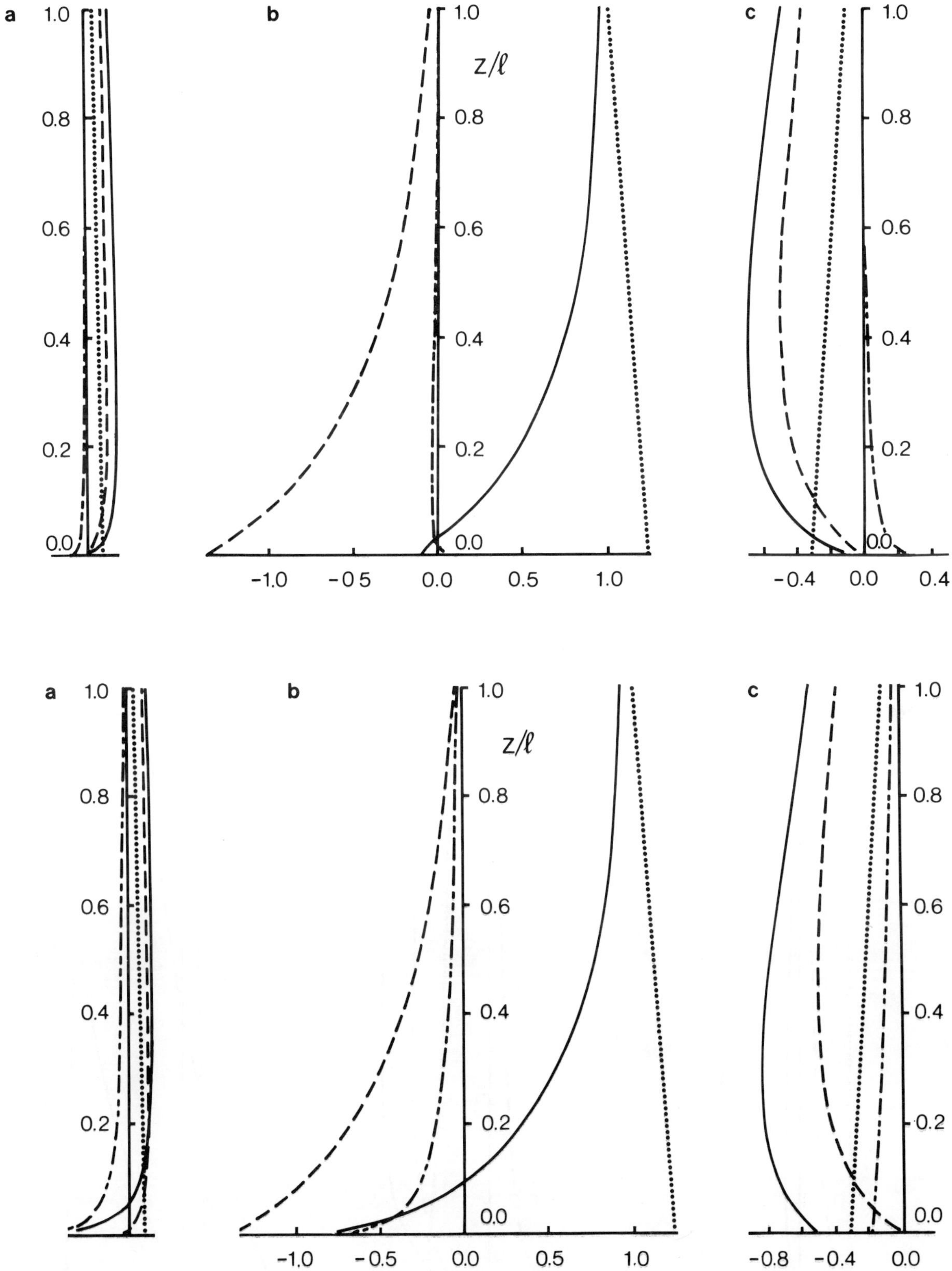

FIG. 5.15. The vertical profiles of concentration perturbation components at different positions over a hill (neutral conditions), normalized by $(c_* H/\kappa L)/U^2(l)$ (from Hunt et al. 1988d). Figure 5.15(i) $\chi = 0.0$; Fig. 4.15(ii) $\chi = 0.1$. Here (– —— –) represents the total perturbation $c\,(=c_u + c_{\Delta\tau} + c_s)$, (- - - - - -) the streamline convergence and divergence component c_u, (– — — –) the effect of shear stress component $c_{\Delta\tau}$, and (– — — - - -) the surface flux component c_s. (a) $x/L = -1.0$, (b) $x/L = 0.0$, (c) $x/L = 1.0$.

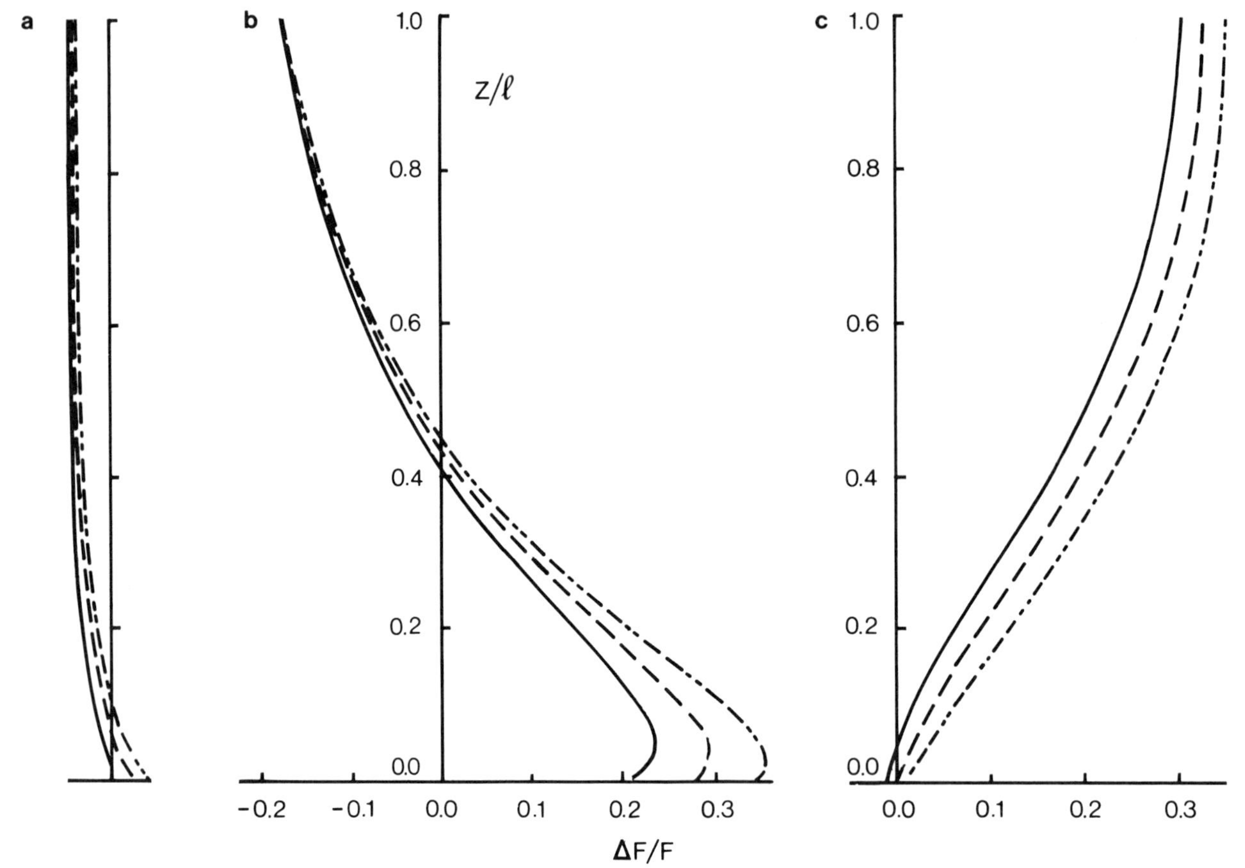

FIG. 5.16. The vertical flux perturbation profiles at the different hill positions (neutral conditions), normalized by upwind flux ($-u_* c_*$). Here (-——-) represents $\Delta F/F$ when $\chi = 0.0$, (-————-) $\chi = 0.1$, (-———---) $\chi = 0.2$. (a) $x/L = -1.0$, (b) $x/L = 0.0$, (c) $x/L = 1.0$. The scales for (a) are the same as in (b).

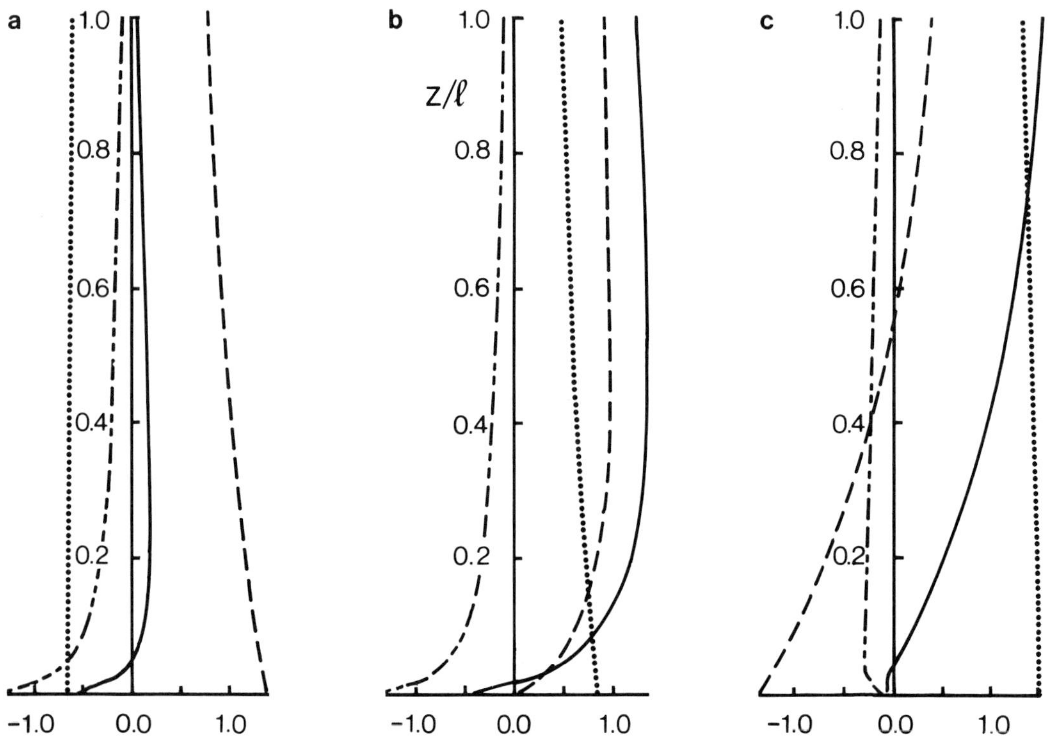

FIG. 5.17. The vertical profiles of the perturbation components at the different hills positions (uniform stratification, $F_L = 2.0$), normalized by $(c_* H/\kappa L)/U^2(l)$. Here (-——-) represents the total perturbation $c\,(=c_u + c_{\Delta\tau} + c_s)$, (-------) c_u, (-———-) $c_{\Delta\tau}$, (-———---) c_s. (a) $x/L = -1.0$, (b) $x/L = 0.0$, (c) $x/L = 1.0$.

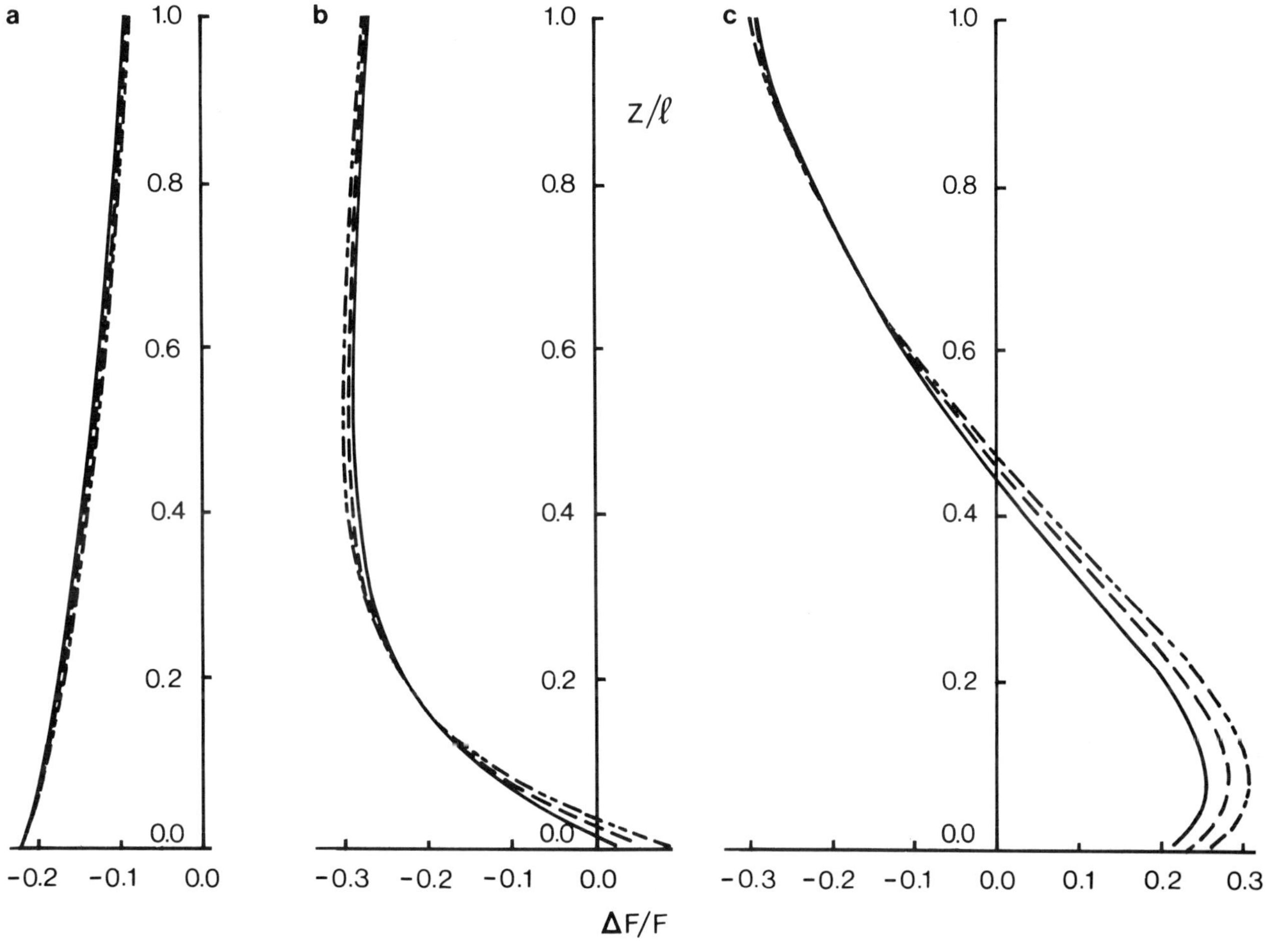

FIG. 5.18. The perturbation flux profiles at different positions of the hill (uniform stratification, $F_L = 2.0$), normalized by upwind flux $(-u_* c_*)$. Here (-——-) represents $\Delta F/F$ when $\chi = 0.0$, (-———-) $\chi = 0.1$, (-————---) $\chi = 0.2$. (a) $x/L = -1.0$, (b) $x/L = 0.0$, (c) $x/L = 1.0$.

absorbing gases, V_g is independent of u_*, but in many conditions V_g is largely controlled by turbulent eddies close to the vegetation, so that $V_g \propto u_*$. In this analysis we describe the concentration and flux changes over the hill, so that following (5.37) the flux and perturbation over the hill is

$$F + \Delta F = -(V_g C + V_g c + \Delta V_g C)$$

or

$$\Delta F = V_g c + \Delta V_g C. \qquad (5.39)$$

This shows that the surface flux changes as a result of changes in both deposition velocity and concentration.

Changes in the deposition velocity V_g over the hill are affected by two separate effects. These are first, the changes due to the surface shear stress, which are expressed as

$$V_{g1} = V_{g0} \kappa u_* (1 + \Delta\tau/2), \qquad (5.40)$$

where V_{g0} is a dimensionless constant of order unity. Substitution of (5.40) into (5.38) shows that

$$V_{g0} = 1/\ln(Z/z_0). \qquad (5.41)$$

Second, variations in V_g that are caused by changes in the nature of the surface (including changes in vegetation type, varying exposure to sunlight, or different surface wetness due to changes in rainfall) and by changes in processes of deposition [such as caused by higher humidities in the atmosphere over the hill, resulting in more wet deposition due to the formation of cap clouds (Carruthers and Choularton 1984)]. These effects can only be determined precisely by a detailed study for each case considered. As an example, we model the effects as a change in V_g, which varies in proportion to the height of the terrain and which might be especially relevant to enhanced deposition over hilltops by clouds (Dollard et al. 1983). In this case

$$V_{g2} = V_{g0} \kappa u_* \chi f(x, y), \qquad (5.42)$$

where the definition of the terrain shape function is such that $f/H \leqslant 1$ and χ is a dimensionless parameter, we will take values of $\chi = 0, 0.1$, and 0.2. Then using (5.40) and (5.42), the expression for the change in deposition velocity is

$$\Delta V_g = V_{g0}\kappa u_*[\Delta\tau + \chi f(x, y)/H], \quad (5.43)$$

and the lower-boundary condition becomes as $\eta \rightarrow \eta_1(Z \rightarrow z_1)$,

$$\eta \frac{\partial c}{\partial \eta} \rightarrow V_{g0}c\,(\eta = \eta_1) + \chi f(x, y). \quad (5.44)$$

Here $z_1 \approx 10z_0$ is approximately the height of the vegetation or roughness elements. The contributions from the three processes that change the concentration and flux as given by the two source terms on the right-hand side of (5.36) and the boundary condition (5.39) are solutions of

$$L(c_u) = \frac{1}{\eta U(l)} \int_0^\eta \frac{\partial u}{\partial \zeta}\,\partial\eta$$

$$\eta \frac{\partial c_u}{\partial \eta} \rightarrow 0 \quad \text{as} \quad Z \rightarrow z_1 \quad (5.45)$$

$$L(c_{\Delta\tau}) = \frac{1}{4}\frac{\partial(\Delta\tau)}{\partial\eta}$$

$$\eta \frac{\partial c_u}{\partial \eta} \rightarrow 0 \quad \text{as} \quad Z \rightarrow z_1 \quad (5.46)$$

$$L(c_s) = 0$$

$$\eta \frac{\partial c_s}{\partial \eta} \rightarrow V_{g0}(c_u + c_{(\Delta\tau)} + c_s) + \chi f(x, y). \quad (5.47)$$

Profiles of concentration and fluxes are shown in Figs. 5.15, 5.16, 5.17, and 5.18. In the first case the surface flux parameter $\chi = 0$ so that the contribution from changes in surface processes is very small. The convergence of the streamlines increases the concentration (c_u) while the increase in shear stress results in more rapid mixing to the surface and lower concentrations ($c_{\Delta\tau}$). In the case $\chi = 0.1$, $c_s < 0$ over the hill summit as the surface now absorbs the particles more readily in this region. The flux perturbation over the hill shows that these are greatest over the hill summit and increase with increasing χ. When the upstream flow is uniformly stratified (Figs. 5.17, 5.18), the perturbations in both concentration and flux are greatest on the lee side.

5.6.4. *Temperature and humidity fields*

Similar techniques can be used to calculate changes in humidity and temperature fields. In this case the fluxes of the sensible and latent heat are coupled and in order to obtain solutions, new concentration variables are defined that enable separate advective diffusion equations to be obtained with self-contained boundary conditions. These are (McNaughton 1978):

$$a = \rho c_p T + \rho\lambda Q$$

$$b = \nu\rho c_p(\bar{T} - T_1) - \rho\lambda(Q - Q_{\text{sat}}(\bar{T}_1)), \quad (5.48)$$

where $\bar{T}_1$ is the potential temperature at the undisturbed surface far from the hill, ν is the dimensionless slope with temperature of Q_{sat} the saturated vapor mixing ratio, Q

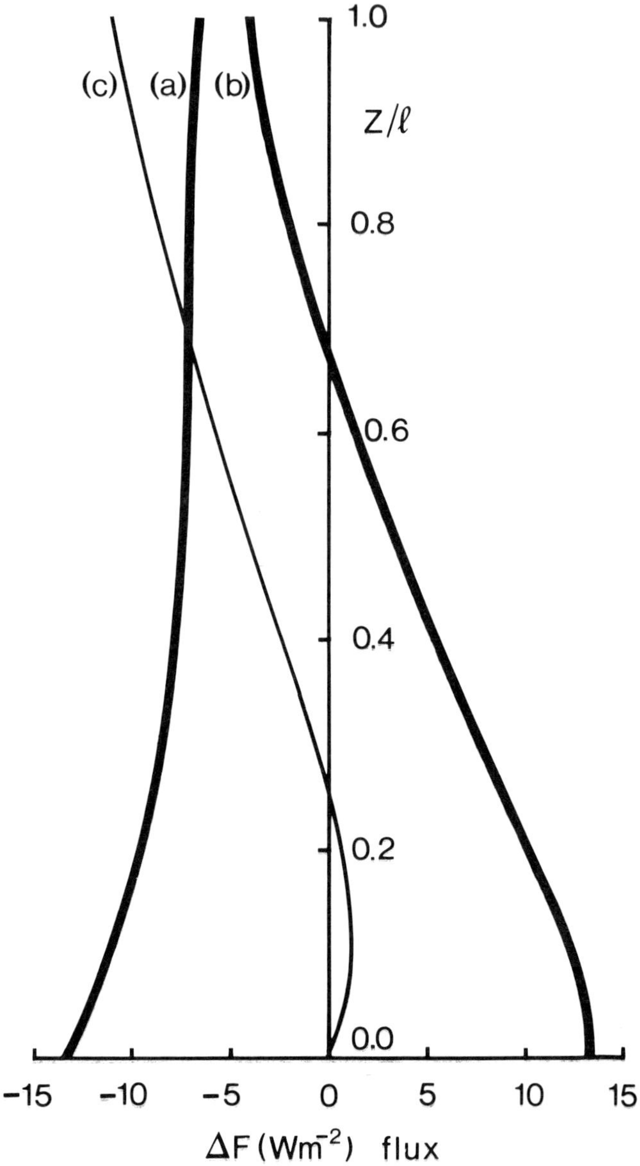

FIG. 5.19. Calculated heat fluxes over a hill summit using a linear model. (a) Latent flux, (b) sensible flux, (c) total flux: L = 500 m, H = 100 m, z_0 − 0.01 m, u_* = 0.38 m s^{-1}. The upwind conditions are H = 300 W m^{-2}, λE = 150 W m^{-2}, and the surface resistance r_s = 420 m s^{-1}.

is the water vapor mixing ratio, and T the potential temperature. The full solution is described in Hunt et al. (1989). An example of the results shows how at the summit of the hill the sensible heat flux increases, and the latent heat flux decreases (Fig. 5.19), resulting in an increase in the Bowen ratio, the ratio of the fluxes of sensible and latent heat at the surface.

5.7. Discussion

There has recently been much development in our understanding of the airflow over hills through the use of both theoretical and numerical modeling, laboratory experiments, and comprehensive field experiments. In this review we concentrated on situations in which the flow is *over* rather than *around* the hill(s) ($F_H > 1$). The emphasis of current research has now moved into the study of mean and turbulent structures over regions of complex terrain rather than over single isolated hills. Simple numerical models based on solutions of the linear equations and a mixing length representation for the turbulence can describe many aspects of these flows quite well; these include:

(i) The mean flow upwind and over summits when the stratification is not strong.

(ii) The turbulence structure, including velocity variances and length scales close to the hill surface ($Z \ll l$) upwind and over summits, and well above the inner region ($Z \gg l$). It is likely that such models will be used increasingly in the future as they require only limited computer resources (running satisfactorily on personal computers).

In weak or neutral stratification downwind of hills with large aspect ratio ($H/L > 0.3$) separation is likely to occur, so that the simplest mixing-length models are not sufficient; however, mixing-length models including the effect of local shear used in conjunction with numerical models predict separation very well. Second-order models also predict separation more precisely and give better estimates of the turbulence structure in the region $Z \approx l$ where rapid distortion, curvature effects, and nonlinear processes are all important. In moderate stratification ($F_H \approx 1$) high resolution numerical models are required to describe many aspects of the flow.

Acknowledgments. DJC was supported by a grant from the Natural Environmental Research Council during the course of this work. JCRH and DJC were partly supported by a grant to Cambridge Environmental Research Consultants Ltd. from the Central Electricity Generating Board. Our thanks to Naomi Coyle for typing the manuscript, to an anonymous reviewer for pointing out many improvements, and to many colleagues whose work is described here.

APPENDIX

Why Can't Stably Stratified Air Rise over High Ground?

RONALD B. SMITH

Department of Geology and Geophysics, Yale University, New Haven, Connecticut

5.A.1. Introduction

A central issue in complex terrain meteorology is the question of whether low-level air will split and flow around, or rise over a hill. Most of the chapters in this volume touch on this issue in one way or another, including discussions on mountain waves, orographic precipitation, and plume impaction. Although a complete understanding of airflow splitting requires a solution to the entire three-dimensional hill-induced flow field, some insight can be gained from a consideration of just the hydrostatic and Bernoulli equations, both well-established and well-understood principles of fluid mechanics. These equations provide constraints on how much hill-induced lifting can occur before flow stagnation will begin. Flow stagnation, when it occurs near the surface, will allow streamline splitting. Such an analytic approach was first suggested by Sheppard (1956). Below, we show serious deficiencies in Sheppard's method but indicate how it can be extended to provide a physically sound basis for understanding limits to airflow lifting over hills.

An equally important problem in stratified flow is the question of when gravity-wave breaking will begin aloft. Like flow splitting, wave breaking begins with the formation of a stagnation point. Whatever method is used to predict flow splitting should be equally useful for predicting wave breaking.

5.A.2. Combining the hydrostatic and Bernoulli equations

Consider a streamline in steady flow rising from an initial level z_0 to some new level $z = z_0 + \eta$ (Fig. 5.A.1). The vertical displacement of this particular streamline is considered to be part of a field of vertical displacement $\eta(x, y, z)$ induced by a hill.

Assuming the flow to be incompressible,

$$\rho = \rho_0 = \text{constant} \qquad (5.\text{A}.1)$$

along the streamline. Bernoulli's equation is

$$p + \tfrac{1}{2}\rho_0 U^2 + \rho_0 g z = C = \text{constant} \qquad (5.\text{A}.2)$$

along a streamline. Far upstream ($x = -\infty$), this is

$$C = p_0 + \tfrac{1}{2}\rho_0 U_0{}^2 + \rho_0 g z_0, \qquad (5.\text{A}.3)$$

where z_0 is the height of the streamline upstream. If the stratification $N^2 = (-g/\rho_0)(\partial\rho/\partial z)$ is independent of height

$$\rho_{-\infty}(z) = \rho_0\left(1 - \frac{N^2}{g}(z - z_0)\right). \qquad (5.\text{A}.4)$$

Using the hydrostatic equation and (5.A.4)

$$p_{-\infty}(z) = p_0 - p_0 g(z - z_0) + \rho_0 \frac{N^2}{g}\frac{(z - z_0)^2}{2}. \qquad (5.\text{A}.5)$$

Defining p^* as the difference between the pressure at some point on the streamline and the pressure at a point at the same elevation far upstream,

$$p^*(z) = p(z) - p_{-\infty}(z) \qquad (5.\text{A}.6)$$

and substituting (5.A.6) into (5.A.2) and (5.A.3) gives

$$p^* + p_{-\infty} + \tfrac{1}{2}\rho_0 U^2 + \rho_0 g z = p_0 + \tfrac{1}{2}\rho_0 U_0{}^2 + \rho_0 g z_0. \qquad (5.\text{A}.7)$$

Including (5.A.5) in (5.A.7) and setting $\eta = z - z_0$ gives

$$p^* + \left(p_0 - \rho_0 g\eta + \rho_0 \frac{N^2}{g}\frac{\eta^2}{2}\right) + \tfrac{1}{2}\rho_0 U^2 + \rho_0 g\eta = p_0 + \tfrac{1}{2}\rho_0 U_0^2. \qquad (5.\text{A}.8)$$

Simplifying (5.A.8)

$$U^2 = -\frac{2}{\rho_0}p^* - N^2\eta^2 + U_0^2. \qquad (5.\text{A}.9)$$

If we assume with Sheppard that

$$p^* = 0 \qquad (5.\text{A}.10)$$

then

$$u^2 = U_0^2 - N^2\eta^2. \qquad (5.\text{A}.11)$$

Having neglected p^*, (5.A.11) seems to offer a simple explanation for airflow speed variations in terms of kinetic and potential energy. The kinetic energy [i.e., U^2 in (5.A.11)] is decreased as the parcel increases its potential energy [i.e., η^2 in (5.A.11)].[1] Such a conceptual picture

[1] In this problem, potential energy appears as a term proportional to the square of the vertical displacement rather than the displacement to the first power, as in particle mechanics.

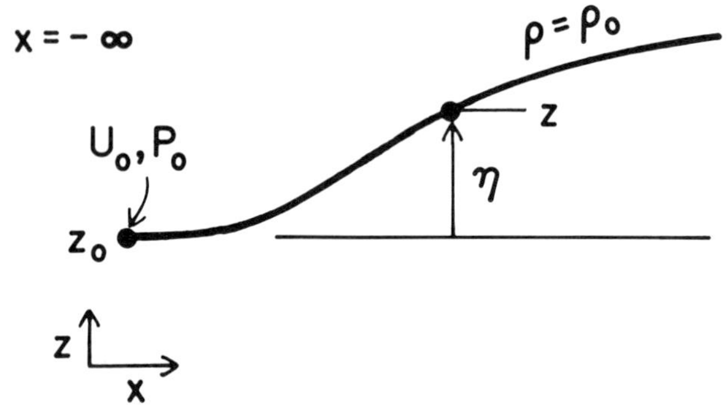

FIG. 5.A.1. A steady streamline along which Bernoulli's equation applies.

certainly applies to ideal problems in particle mechanics, but its application to a parcel in a continuous fluid requires the assumption (5.A.10), and may have no validity.

The predictions that follow from (5.A.11) are

- Parcels can lift an amount no greater than

$$\eta_{\max} = U_0/N \qquad (5.A.12)$$

without stagnating (i.e., $U = 0$). Thus hills higher than $h_{\max} = U_0/N$ cannot lift air over the top, independent of the hill shape.

- The minimum speed along the surface occurs at the mountaintop where $\eta = \max[h(x, y)]$.
- Aloft, parcels slow whenever they move higher or lower than their initial level. The fluid speed is everywhere

$$U \leqslant U_0. \qquad (5.A.13)$$

The first prediction (5.A.12) agrees, to a certain degree of accuracy, with laboratory and numerical experiments over a limited range of hill shapes. Hills significantly higher than $h_{\max} = U_0/N$ are observed to force low-level fluid to pass around, while only streamlines originating at higher elevations are able to lift over the top.

The second and third predictions of Sheppard's formula (5.A.11) are quite at odds with laboratory and numerical experiments, and even common experience. This leads us to question the validity of (5.A.11) and wonder whether the agreement with (5.A.12) might be fortuitous. We shall see below that this is the case.

The fundamental problem is that in stratified airflow over hills, p^* is not zero, nor is it small in any practical sense. Variations in pressure play an important and sometimes dominant role in airflow speed variations.

To go further, we need an expression for p^*. Using the hydrostatic equation in the disturbed region

$$p^* = g\int_z^{\infty} \rho' dz = \rho_0 N^2 \int_z^{\infty} \eta dz \qquad (5.A.14)$$

as the density anomaly at each level is proportional to the displacement $\eta(x, y, z)$. The difficulty in combining (5.A.9) with (5.A.14) is that the lower limit of (5.A.14) is evaluated at a constant level while (5.A.9) is true along a streamline. This problem can be overcome by writing the displacement field in Lagrangian or density coordinates; i.e., $\eta = \eta(x, y, z_0)$, where $z_0(\rho)$ is the height of each density surface upstream. Then as

$$dz = dz_0 + d\eta \qquad (5.A.15)$$

(5.A.14) becomes

$$p^* = \rho_0 N^2\left(\int_{z_0}^{\infty} \eta dz_0 + \int_{\eta}^{0} \eta d\eta\right) \qquad (5.A.16)$$

assuming decay aloft; i.e., $\eta \rightarrow 0$, as $z_0 \rightarrow \infty$. The decay assumption is reasonable as Boussinesq mountain-wave solutions in three dimensions decay with height due to wave dispersion. Upon defining

$$I_\eta = \int_{z_0}^{\infty} \eta dz_0 \qquad (5.A.17)$$

(5.A.16) becomes

$$p^* = \rho N^2(I_\eta - \tfrac{1}{2}\eta^2). \qquad (5.A.18)$$

Substituting (5.A.18) into (5.A.9)

$$U^2 = -2N^2 I_\eta + U_0^2. \qquad (5.A.19)$$

The cancellation of the η-squared term between (5.A.9) and (5.A.18) is significant as it removes from consideration the potential energy term; the basis of the Sheppard stagnation concept.

Equation (5.A.19) is [just as (5.A.11)] a diagnostic equation relating the speed and displacement fields. According to (5.A.19) it is not the local displacement "η" at a point, but the integral of the displacement "I_η" above a point, that controls the flow speed. Over some of the flow field, I_η can be considered as the product of the local displacement of the parcel "η" and "d," the height above the parcel where that displacement persists:

$$I_\eta \approx \eta d. \qquad (5.A.20)$$

The predictions of (5.A.19) are:

- Stagnation begins at a point where

$$I_\eta = U_0^2/2N^2. \qquad (5.A.21)$$

- On the lower boundary, stagnation begins on the windward slope. Due to the upstream phase tilt of the displacement field with height, points on the windward slope of the hill have the strongest region of positive displacements above them (Fig. 5.A.2).
- The critical hill height depends on the hill shape. For example, a two-dimensional flow over a ridge has a vertical displacement field that decays less rapidly that does an isolated hill. Thus, for the same surface elevation "η," the

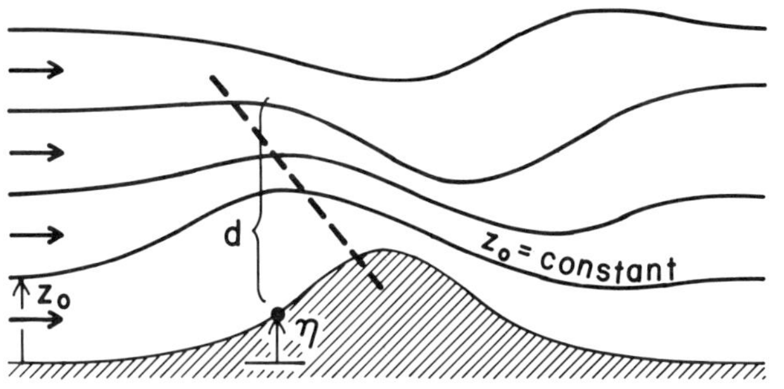

FIG. 5.A.2. A schematic illustration of the streamlines over a hill showing how the upwind phase tilt (heavy dashed line) of the vertical displacement pattern gives rise to large positive values of I_η on the windward slope. The quantity d is the height over which positive displacements exist above the points in question.

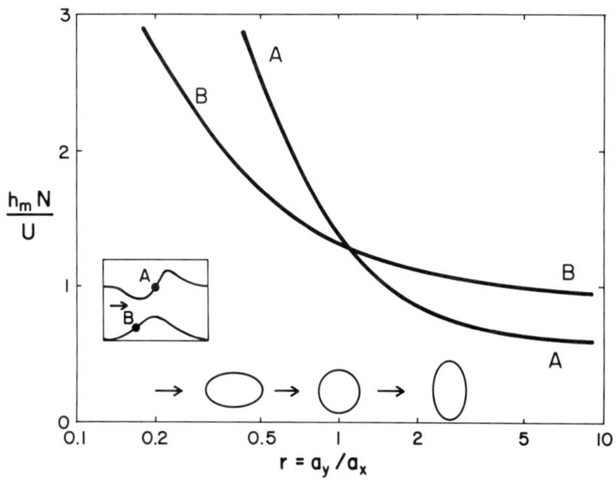

FIG. 5.A.3. The nondimensional hill height at which stagnation will begin [i.e., (5.A.21) is met] either aloft (curve A) or at the lower boundary (curve B). The curves are estimates based on linear theory calculations of $I_\eta(x, y, z_0)$. For a particular hill aspect ratio r, one may consider higher and higher hills by moving upwards on the diagram until one of the stagnation curves is reached (from Smith 1989).

quantity d will be greater for the two-dimensional flow and stagnation is more likely (see Fig. 5.A.3).

• Stagnation aloft begins at a point where fluid parcels have temporarily returned to their original undisturbed altitude. To see this, note from (5.A.19) that the slowest flow will occur at a local maximum of $I_\eta(x, y, z_0)$. Thus

$$\frac{\partial}{\partial z_0} I_\eta = \frac{\partial}{\partial z_0} \int_{z_0}^{\infty} \eta dz_0 = 0 \qquad (5.A.22)$$

and therefore

$$\eta = 0 \qquad (5.A.23)$$

at the point of minimum speed.

5.A.3. Application

A prediction of the critical hill height using (5.A.19) requires a method for estimating I_η. Lacking an exact nonlinear solution, we use a linear theory calculation (Smith 1989) to construct Fig. 5.A.3. For an isolated hill

$$h(x, y) = \frac{h_{\max}}{[1 + (x/a_x)^2 + (y/a_y)^2]^{3/2}}, \qquad (5.A.24)$$

two regions of large I_η (i.e., weak winds) exist in the flow field. One of these (point A), occurs aloft at a nondimensional height $zN/U = 4.0 \pm 0.7$ above the hilltop. The other (point B), occurs on the windward slope of the hill along $y = 0$ between $x/a_x = -1.0$ and -0.1. The hill height at which stagnation first begins at point A or B is shown in Fig. 5.A.3 as a function of hill (planform) aspect ratio $r = a_y/a_x$.

For $r > 1$; i.e., airflow across a ridge, stagnation aloft precedes stagnation at the surface. To the extent that the B curve is correct, however, stagnation at the surface is predicted at hN/U values between 1.0 and 1.3. This is essentially in agreement with observations (just as Sheppard's formula was) but with different terms operating. For $r < 1$, low-level stagnation (curve B) requires a higher hill. This prediction disagrees with Sheppard's formula.

5.A.4. Conclusion

The diagnostic equation put forward by Sheppard in 1956 appears now to be physically incorrect due to its emphasis on potential energy rather than pressure variation in controlling the wind speed. Its agreement with early laboratory and numerical model results may have been fortuitous. Predictions concerning other aspects of steady stratified flow are more obviously incorrect.

Sheppard's development can be extended and corrected by deriving a Lagrangian form of the hydrostatic equation. The resulting diagnostic equation provides a more consistent view of flow stagnation aloft, leading to wave breaking, and flow stagnation at the lower boundary, leading to flow splitting. Stagnation is associated with increasing pressure along a streamline caused by lifting and positive density anomalies aloft.

Acknowledgments. This work was partially supported by the National Science Foundation, Division of Atmospheric Science, Grant ATM-8605939.

CHAPTER 6

Rugged Terrain Effects on Diffusion

STEVEN R. HANNA AND DAVID G. STRIMAITIS

Sigma Research Corporation, Westford, Massachusetts

ABSTRACT

Recent research on diffusion in rugged terrain is reviewed, with emphasis on the Environmental Protection Agency's (EPA) Complex Terrain Model Development Program and the Department of Energy's (DOE) Atmospheric Studies in Complex Terrain Program. It is shown how fluid modeling studies, field experiments, and theoretical model development programs have led to practical diffusion models, such as the EPA's Complex Terrain Diffusion Model (CTDM). This model uses the concept of a dividing-streamline height to separate the flow below this height that cannot rise over the hill from the flow above this height that does lift over the hill; however, neither CTDM nor any of the other simpler models treat any terrain beyond the "first hill."

This chapter also covers statistical evaluation of rugged terrain diffusion models, special studies of flow and dispersion down slopes and valleys, and methods of including diffusion in mesoscale flow models for rugged terrain. It is seen that progress in most of these research areas is in a preliminary stage, and that field verification of fundamental concepts and models is minimal. The best diffusion models typically have an error of ±30% in simulating the maximum and mean concentrations, have a root mean square error that is the same order as the mean, and yield a correlation coefficient of about 0.50. The influences of external meteorology, such as subsidence, gravity waves, multivalley circulations, sea breezes, and pooling are also not well understood.

There are a few mesoscale wind field models available that can be used in conjunction with statistical particle dispersion models to estimate dispersion in complex terrain where the flow field is significantly altered by the terrain. Although these models have not been adequately tested, they are potentially very useful because they can be applied to terrain past the first hill.

6.1. Introduction

The perturbations to meteorological variables caused by rugged terrain are discussed in other chapters in this monograph. Various field experiments and modeling studies have led to a partial understanding of phenomena such as thermally developed wind systems in mountainous terrain (Whiteman, Chapter 2; Egger, Chapter 3), gravity waves over mountains (Durran, Chapter 4), and turbulence and waves in the boundary layer (Carruthers and Hunt, Chapter 5). Some understanding of turbulence over rugged terrain has also been acquired from these studies; however, our knowledge is still far from complete. A good understanding of wind, turbulence, and temperature structure is required in order to properly estimate diffusion in rugged terrain since plume diffusion is a one-to-one function of the meteorological parameters. If these parameters were known on a three-dimensional grid with a grid size of about 10 to 100 m and domain size about 10 to 100 km, and their variation with time steps of about 10 to 100 s were known, then a properly formulated, very detailed numerical diffusion model could be implemented. In any case, the accuracy of a diffusion model is critically dependent on the accuracy of the meteorological variables that are input to the model.

This chapter emphasizes recent developments in diffusion models for rugged terrain. Keeping in mind the uncertainties imposed by our imperfect knowledge of the input variables, each section describes the input data that are assumed to be available. The discussion is centered on four main topics:

1) Environmental Protection Agency (EPA) regulatory model descriptions and evaluations.
2) Atmospheric Studies in Complex Terrain (ASCOT) results.
3) Complex Terrain Dispersion Model (CTDM) development.
4) Mesoscale wind field/dispersion models.

Under each topic related field experiments and diffusion model efforts are briefly reviewed.

We first provide an overview of the important problems in rugged terrain diffusion modeling and discuss the major scientific programs of the past two decades.

6.1.1. Problems

Pollution problems are usually thought to be more serious in complex terrain because of the decrease in wind speed and the possibility of plume impaction on high terrain. There are, however, certain situations where air pollution problems are lessened due to the increased turbulence intensities, the increased plume rise caused by light winds, and the deflection of plumes around obstacles. Despite these exceptions, most scientists believe that the following air pollution situations can lead to increased concentrations in rugged terrain:

- *Plume impingement on high terrain*—The situation depicted in Fig. 6.1a has been observed to cause high pollutant concentrations in stable conditions when the plume trajectory cannot lift over the terrain obstacle. In this case the maximum concentration on the terrain equals the maximum concentration on the plume centerline.
- *Pooling in valleys*—Often the airflow in a valley is cut off from the airflow above the terrain. For example, wind speeds in the Potomac River Valley in western Maryland are observed to be a factor of 6 less than those on the surrounding ridges, and an inversion is frequently present at ridgetop (Hanna et al. 1984). Consequently, pollutants released into the air at the valley floor can accumulate or "pool" due to the limited dilution rate and the resistance to vertical diffusion (see Fig. 6.1b).
- *Drainage toward population centers*—Wind flow is known to be parallel to valley orientation in rugged terrain. At night, if synoptic winds are light and there is strong radiational cooling, persistent downslope drainage winds can develop such that pollutants may be carried downstream (see Fig. 6.1c) toward population centers, which are typically sited along drainage systems.
- *Persistence due to channeling*—Any synoptic wind direction with a westerly component will lead to a fixed westerly wind direction in the valley depicted in Fig. 6.1d. It is possible for a plume of pollutants to follow the same trajectory for hours or days at a time, possibly exceeding long-term standards (e.g., the 24 hour SO_2 standard) for some pollutants at fixed monitor locations in a valley environment. In fact, such persistent flow can last for several days, ceasing only with a change in the synoptic pattern.

6.1.2. Overview of history of research

The meteorological effects of rugged terrain have been known in qualitative terms for centuries. This popular knowledge plus the results of limited field experiments were summarized by Defant in 1951. The question of diffusion in rugged terrain received little study until the 1970s, when it became necessary to use diffusion models as a requirement of the U.S. Clean Air Act and its amendments. In the first applications of these models to the Four Corners Power Plant in the southwestern United States, it was assumed that the plume impacted nearby steep terrain as if the terrain did not affect the plume, and as if the receptor was mounted on a "pole" (Van der Hoven et al. 1972). Soon afterward, field experiments by Start et al. (1975) in Huntington Canyon, Utah contradicted the earlier approach when it was found that dispersion in complex terrain exceeded that in flat terrain by as much as an order of magnitude.

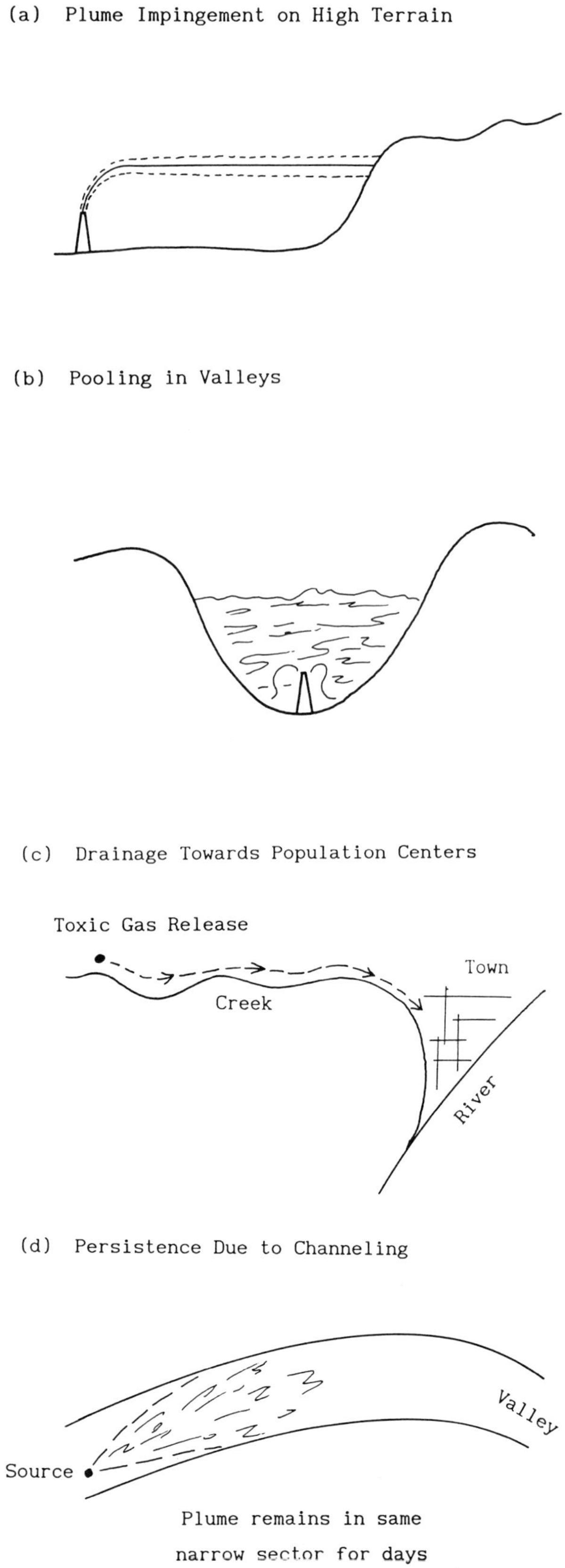

FIG. 6.1. Four types of scenarios that lead to high air pollution concentrations in rugged terrain.

Drazin (1961) and others had developed the concept that the deflection of the airflow over rugged terrain was a function of the terrain height, H, the wind speed, u, and the frequency characterizing the atmospheric stratification, $N = [(g/T)(d\theta/dz)]^{1/2}$. These parameters may be combined to give the Froude number, $\mathrm{Fr} = u/NH$. In the spirit of these findings, the EPA developed a series of simple dispersion models for regulatory applications in complex terrain (e.g., VALLEY, COMPLEX I, COMPLEX II), which assumed no vertical deflection of plumes in stable conditions, and "half-height" deflection in neutral and unstable conditions. These models will be discussed in section 6.2.

Egan (1975) described the status of research and applications regarding diffusion over rugged terrain in a review paper. This paper was used as a starting point for several new research initiatives in the United States as the EPA and the Department of Energy (DOE) began planning for field experiments and new modeling efforts to help solve their problems in this area. Each group sponsored a planning workshop, and extensive field experiments were begun in 1979 by the DOE and in 1980 by the EPA. These experiments are now completed and most of the data analyses and model development have been performed. Details of these programs are given in sections 6.3 and 6.4, and brief outlines are given below:

EPA Complex Terrain Model Development Program

- Objective—development of regulatory model for diffusion in rugged terrain in stable conditions.
- Field experiments—Cinder Cone Butte, Idaho (three-dimensional hill); Hogback, New Mexico (two-dimensional ridge); Tracy, Nevada (full scale power plant).
- Fluid modeling at EPA laboratory.
- Data analysis and CTDM model development.
- Modification of model for daytime condition.

Reference: Strimaitis et al. (1988).

Department of Energy (DOE) Atmospheric Studies in Complex Terrain (ASCOT) Program

- Objective—Development of transport and dispersion models for nocturnal drainage flows on slopes and in narrow valleys.
- Field experiments—Geysers, California; Brush Creek, Colorado; plus a number of more limited experiments at other sites.
- Data analysis and model development—Conducted at DOE's National Laboratories plus some NOAA laboratories and universities.

Reference: Clements et al. (1989a).

Besides these large agency programs, several other individual research studies are under way. In particular, many researchers are including diffusion in their three-dimensional wind flow models (e.g., Yamada 1985; McNider et al. 1981). Some of these studies will also be described in the following four sections. A list of symbols for this chapter is provided in the Appendix.

6.2. Summary of EPA models and evaluations

The primary purpose of this chapter is to describe recent significant research findings on diffusion in rugged terrain. The EPA regulatory diffusion models are generally considered to be representative of the state-of-the-art 10 years ago. The models are slightly behind the times because of the time required to thoroughly review and evaluate them prior to application to situations where billion-dollar decisions on pollution control equipment, fuel types, and new plant construction must be made. Most scientists and engineers working in industries, environmental consulting firms, and air pollution control agencies have extensive experience with these models. We provide a short overview of them here, since they are in such wide use and may be considered as bases for comparison of new analyses and models.

6.2.1. Model descriptions

Turner (1979) summarizes the approved EPA regulatory models as of that year. The only model approved for use at that time at sites where terrain heights exceeded stack height was the VALLEY model (Burt 1977). This algorithm may be used now only in the "screening" mode. It is based on the Gaussian plume formula, and is applied using a default wind speed of 2.5 m s^{-1}, stability class F, (i.e., very stable conditions), and assumes that pollutants are mixed uniformly in the lateral direction in a 22.5° sector. No lifting of the plume over the terrain is permitted. The EPA decided that these conditions represent a likely "worst-case" scenario in rugged terrain. In any given problem, a person would first apply the worst-case VALLEY model, and if the predicted concentrations were much below the standard, no further analysis would be necessary. If concentrations are not much below the standard or are above the standard, then further analysis using more refined models is necessary.

The COMPLEX I and COMPLEX II models (EPA 1983) are recommended for more refined calculation of diffusion in rugged terrain. They are based on the MPTER model (Pierce and Turner 1980) for nonrugged terrain. They are intended for application to a full year of hourly meteorological data, including wind speed and direction, mixing height, and cloudiness. The only difference between the two models is that the COMPLEX I model assumes that pollutants are uniformly mixed in the lateral direction in a 22.5° sector (as in VALLEY), while the COMPLEX II model assumes that there is a Gaussian distribution in the lateral direction, with standard deviation given by the Pasquill–Gifford–Turner σ_y curves (e.g., Gifford 1976). These two models have the following common characteristics:

- Treatment of plume rise and vertical dispersion (σ_z) is the same as in EPA's flat terrain models.
- In stable conditions the plume trajectory does not lift over the terrain, but continues at the same elevation until it impacts the terrain. (Actually the model does not strictly adhere to this assumption, since the plume centerline is never allowed to come closer than a distance of 10 m from the terrain; this value has not been justified on physical grounds.)
- In neutral and unstable conditions, the plume trajectory follows the "half-height" relation, in which the revised elevation of the plume centerline above the terrain is the larger of 1) 0.5 times the elevation over flat terrain, and 2) H_0 minus 0.5 times the hill height, where H_0 is the initial elevation of the plume centerline above the flat terrain. This concept is illustrated schematically in Fig. 6.2, reproduced from the review article by Egan (1975).

The EPA also supported the SHORTZ and LONGZ models for calculating diffusion over urban areas located in rugged terrain (Bjorklund and Bowers 1982). An advantage of these models is that they permit the use of observed turbulence intensities for the direct calculation of σ_y and σ_{zi}; however, they assume no deflection of the plume centerline over terrain obstacles.

In order to expand the list of useful models, the EPA solicited new models from developers and evaluated the resulting models that were submitted using field data. This evaluation is described in section 6.2.2. As a result, the Rough Terrain Dispersion Model (RTDM; Paine and Egan 1987) was added to the list of approved models for certain circumstances. Although the model can accept observations of turbulence intensities, the EPA has approved its use only with parameterized turbulence intensities. The main difference between RTDM and the other models is in its treatment of deflection of the plume centerline over terrain obstacles. The model has the following characteristics:

- First a dividing-streamline height, H_d, is calculated from $H_d = H(1 - \mathrm{Fr})$, where H is hill height and $\mathrm{Fr} = U/HN$ is the hill Froude number. (See the Appendix to Chapter 5 by Smith for a discussion of this assumption.) Then if the initial plume centerline height, H_0, is less than H_d, no lifting of the plume is assumed. If H_0 is greater than H_d, then the "half-height" assumption is used for plume lifting, with the correction that elevations are referred to H_d rather than ground level at the stack. Note that $H_0 > H_d$ for all neutral and unstable conditions, when N and H_d equal zero.
- Plume reflection from the ground is modified such that the plume crosswind averaged centerline concentration never increases with downwind distance as steep terrain is approached.

In order to define the Brunt–Väisälä frequency, N, observations of the vertical potential temperature gradient at plume height are useful. Otherwise, the model uses

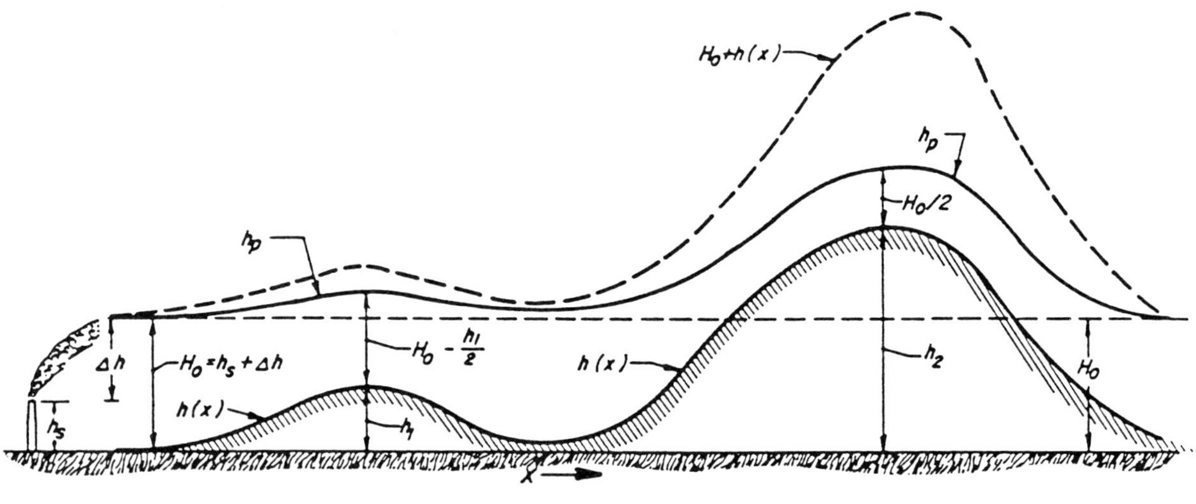

h_s is physical stack height
Δh is plume rise
H_0 is effective stack height (physical stack height plus plume rise)
$h(x)$ is terrain height above stack base elevation
h_p is plume centerline height above stack base elevation

FIG. 6.2. Plume–terrain interaction under stable conditions assuming plume path coefficient = 0.5 (from Egan 1975).

simple climatological default values for this gradient. Also, on-site wind information at an elevation close to stacktop is desirable. In practice, however, it is often difficult to find appropriate input data or (the opposite problem) to decide which wind data are more appropriate for input. For example, if data from towers on a ridge and in a valley are available, which should be used in the model?

The reader will note that there are many arbitrary decisions made in defining and applying the EPA-approved models. This is because the EPA is concerned with protecting the environment, and does not want to use any model assumptions that could potentially lead to underpredictions. When the models are run with one year of hourly meteorological input data, they should always err on the side of conservatism or overprediction, if anything. It is desirable that the models be scientifically plausible, but they should first be consistent and conservative.

6.2.2. *Evaluations of regulatory models*

The regulatory models described above, plus others, have been evaluated using data from a number of field studies. It is found that the best model typically has an error of ±30% in simulating the maximum and mean concentrations, shows a root mean square error as large as the mean, and yields a correlation coefficient of about 0.50. Specific results are given below.

6.2.2.1. EPA EVALUATION AT LUKE MILL AND CINDER CONE BUTTE

As part of the EPA model evaluation program, diffusion models for rugged terrain were solicited from model developers and were evaluated using field data from two sites. EPA models (COMPLEX I, COMPLEX II, SHORTZ, and COMPLEX / PFM) were included, as well as models submitted by industry (4141, RTDM, and PLUME5). All models are based on the Gaussian plume formula, but have different methods of accounting for diffusion coefficients and for plume deflection by terrain.

The following model performance measures were used in the analysis by Wackter and Londergan (1984a,b), where C_o is observed concentration and C_p is predicted concentration, and an overbar indicates an average:

- Highest and "highest second high" (HSH) concentrations for 3 and 24 h averages. (These values are prescribed by the national ambient air quality standards.)
- Average difference:

$$\bar{C}_o - \bar{C}_p.$$

- Root mean square error:

$$\text{rmse} = [\overline{(C_o - C_p)^2}]^{1/2}.$$

- Average absolute residual:

$$\overline{|C_o - C_p|}.$$

- Correlation coefficient:

$$R = \overline{(C_o - \bar{C}_o)(C_p - \bar{C}_p)} / \sigma_{C_p}\sigma_{C_o}.$$

Here 95% confidence intervals were calculated for each of these performance measures.

Field data from two rugged terrain sites were used in the evaluation. The first site was the Potomac River Valley in Luke, Maryland, where there were continuous SO_2 emissions from a 183 m tall stack, and there were 11 SO_2 monitors collecting hourly observations for a period of two years. On-site meteorological data were also taken (Hanna et al. 1984). The ridges surrounding the stack rose to elevations of 200 m above the stack at distances less than 2 km from the stack, and most of the SO_2 monitors were located on these ridgetops.

The second field site was Cinder Cone Butte, Idaho, the subject of an intensive one-month tracer study by the EPA during fall 1980. Most of these data were taken during stable conditions. Tracer gas was released upwind of the 100 m hill and hundreds of monitors were set out on the hill to detect the resulting concentrations. Detailed meteorological data were taken on a 150 m tower and on other shorter towers. A cross section of the hill, showing the placement of some of the equipment, is given in Fig. 6.3 (taken from Spangler 1986). This experiment will be described in more detail in section 6.3, since it played a large role in the development and testing of the CTDM model.

Wackter and Londergan (1984a,b) present several tables containing their summaries of the model evaluations performed with these two datasets. Because most SO_2 regulations concern the highest second-high (HSH) concentration for 3 and 24 h averages, these data were analyzed for the seven models applied to the Luke Mill dataset. The only model whose predictions are even close to the HSH observations is RTDM, which exhibits overpredictions ranging from 24% to 57%. The other models, however, exhibit overpredictions ranging from a factor of 4 to a factor of 22. The most severe overpredictions are given by the COMPLEX II model, probably because of its assumptions of a narrow plume width in stable conditions, its use of full reflection from the ground surface in conjunction with the partial plume height adjustment for terrain, and its underestimation of rate of plume spread in neutral conditions.

The performance measures calculated for eight models for the Cinder Cone Butte dataset emphasize the highest observed and predicted concentrations for each hour anywhere on the monitoring network (total of 104 hours). The following models are compared: COMPLEX I and II, 4141, RTDM, PLUME5, COMPLEX/PFM,

SHORTZ, and IMPACT. The RTDM model shows zero bias in this comparison, and also has the highest correlation coefficient (0.60). All models except RTDM and IMPACT show an overprediction of a factor of 2 or 3, in agreement with the results at Luke, Maryland that were reported in the previous paragraph. The IMPACT model shows an underprediction of a factor of 2. The rmse for RTDM is approximately equal to the mean, and the rmse values for the other models are about two to four times the mean.

As a result of the evaluation by Wackter and Londergan (1984a,b) reported above, the EPA decided to recommend RTDM for use in rugged terrain, with certain restrictions related to data input and with the recognition that it is a "third-level screening model." The highest class of regulatory model to be applied to rugged terrain, the "refined model," is an empty set at the moment, but it is expected that CTDM will be assigned that rating when used with adequate on-site data.

6.2.2.2. Evaluation of COMPLEX I and RTDM at Widows Creek

The TVA Widows Creek Steam Plant is located in the Tennessee River Valley in Alabama. SO_2 is continuously released from two 152 m stacks and one 305 m stack, and the terrain rises to an elevation of 250 m above stack base at a distance of 3 km from the plant. Figure 6.4 contains a map of the region, showing the topography in the area and the location of the meteorological towers. The stack is just to the north of the valley meteorological tower. The 14 SO_2 monitors are sited in both valley and mountaintop locations, at distances from 1.5 to 10 km from the stack. Because there are two meteorological towers, one near the stack and another on a neighboring plateau, the modeler faces the dilemma of deciding which data to use as input to his model. As shown in Fig. 6.5, the wind data from the two towers can differ markedly, due to the channeling effects of the valley (Hanna 1980). In the COMPLEX I and RTDM model evaluation study by Egan et al. (1985), a complex priority system for meteorological data input was devised, emphasizing wind data from the plateau tower. The vertical temperature gradient was assumed to be given by the difference between the temperatures at the 61 m levels of the plateau and valley towers.

The resulting observations and model predictions of the highest two concentrations (unpaired in time and space) for 1979 and 1980 are listed in Table 6.1 for averaging times of 1, 3, and 24 hours (Egan et al. 1985). The RTDM model produces results ranging from a 26% underprediction to a 64% overprediction, with a median overprediction of about 30%. In contrast, the COMPLEX I model typically overpredicts by a factor of 2 to 10. We stress, however, that the degree of overprediction or underprediction would change if a different priority scheme were set up for input of meteorological data.

6.3. Theories and experiments regarding diffusion over slopes and valleys

Most of the discussion in the last two sections has been concerned with whether elevated plumes will impact rugged terrain in stable conditions. This scenario is important because it usually defines the worst-case air pollution episode for SO_2 emissions from tall stacks at power plants, paper mills, or other large industries. Environmentalists are also interested in smaller emissions of air pollution from stacks closer to the ground, such as from mining,

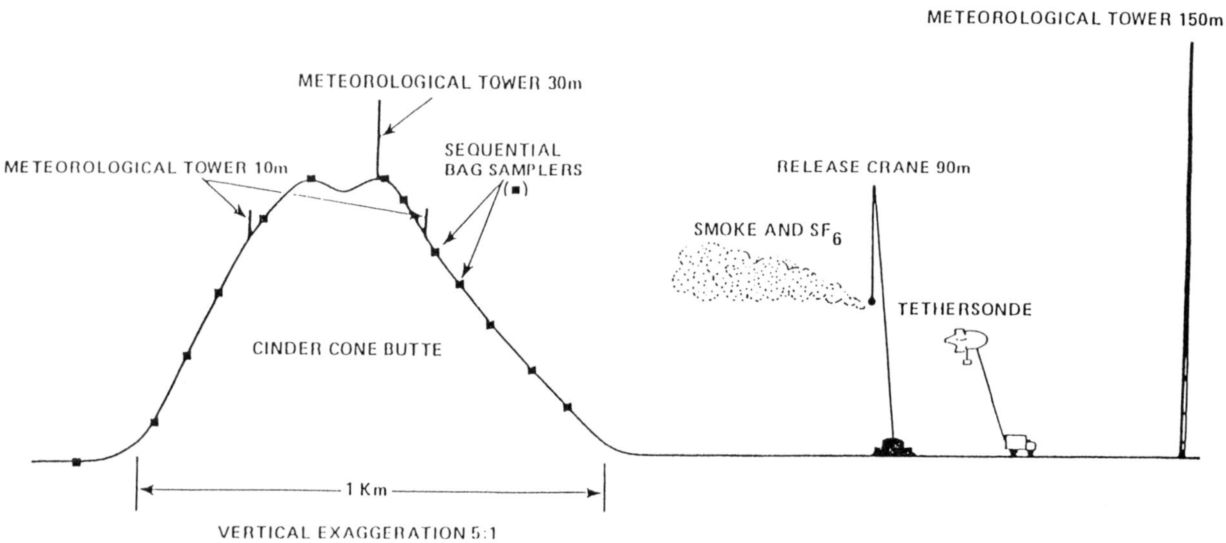

FIG. 6.3. Depiction of Cinder Cone Butte field study (from Spangler 1986).

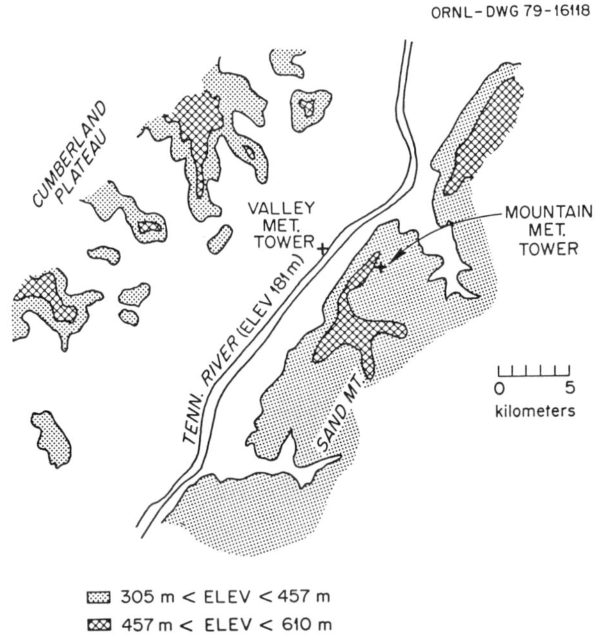

FIG. 6.4. Topographic map of area within 10 km of Widows Creek Steam Plant. Meteorological station locations are given (from Hanna 1980).

processing, and refining industries. In these scenarios, the questions of downslope drainage winds and valley stagnation become important. In this section, the principles of diffusion in stagnating valley flows and drainage flows are reviewed, and descriptions of field experiments are given. Emphasis is on the DOE's ASCOT program.

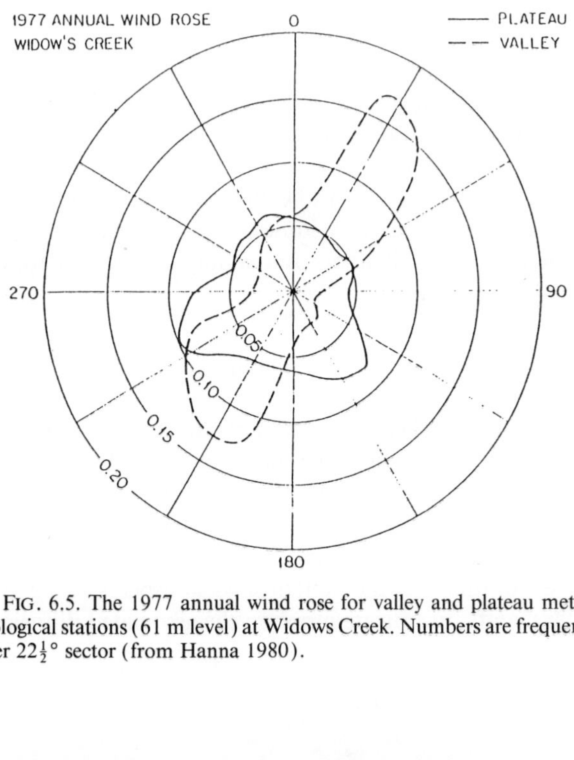

FIG. 6.5. The 1977 annual wind rose for valley and plateau meteorological stations (61 m level) at Widows Creek. Numbers are frequency per $22\frac{1}{2}°$ sector (from Hanna 1980).

TABLE 6.1. Summary of the highest two concentrations observed in 1979 and 1980 at Widows Creek; unpaired in time and space; full onsite data (from Egan et al. 1985).

	Top two concentrations individually listed		
	1 hour	3 hour	24 hour
1979 Observed values,	3482.9	2363.6	410.7
μg m^{-3}	3430.5	1545.0	385.2
1979 RTDM values,	5483.3	2486.6	427.8
μg m^{-3}	5325.5	1921.9	427.8
1979 COMPLEX I values,	23 474.2	14 986.5	2724.2
μg m^{-3}	22 962.2	14 623.0	2464.7
1980 Observed values,	5918.2	2444.1	577.4
μg m^{-3}	4608.9	2374.3	474.6
1980 RTDM values,	9454.8	3266.8	423.1
μg m^{-3}	7569.6	2712.6	352.8
1980 COMPLEX I values,	12 453.3	6871.0	1663.6
μg m^{-3}	11 473.5	6411.6	1469.9

6.3.1. *Diffusion models for slope flows*

Diffusion in a slope flow cannot be accurately estimated until the wind, temperature, and turbulence distribution in the flow are known. There are many models available for nocturnal slope or drainage flows, as mentioned by Sato and Kondo (1988). A few of these models are simple analytical expressions, but most require a computer to obtain a numerical solution to a set of governing equations. The characteristics of the nocturnal slope flow are determined by the following parameters:

- Geostrophic or free stream wind speed.
- Slope angle.
- Ground friction.
- Surface–air temperature difference or sensible heat flux.

Another name for the slope or drainage flow is the katabatic wind, which is used more to describe drainage flows over long, uniform slopes (see Whiteman, Chapter 2).

Generally the intensity of the slope flow increases as the slope angle increases, the ground friction decreases, and the temperature difference or the downward sensible heat flux increases. Briggs (1979) proposed an analytical formula for the thickness, h, of the drainage layer on a simple slope:

$$h = 0.05x \sin\beta \quad (\beta > 20°) \tag{6.1}$$

$$h = 0.037x(\beta 2\pi / 360°)^{2/3} \quad (\beta < 20°), \tag{6.2}$$

where x is distance along the slope and β is slope angle in degrees. For example, if $\beta = 20°$ and $x = 500$ m, then

$h = 9$ m. Briggs (1979) also suggested a formula for the wind speed, u, in this drainage layer:

$$u = 2.15(\sin\beta)^{2/9}(Q_H x)^{1/3}, \qquad (6.3)$$

where $Q_H = -g\overline{w'T'}/T'$ is the buoyant energy production (in units of $m^2\ s^{-3}$) associated with downward sensible heat flux. A typical value of Q_H is 0.001 $m^2\ s^{-3}$, which would give $u = 1.8\ m\ s^{-1}$ if $\beta = 27°$ and $x = 1000$ m.

This simple solution is impractical in most cases because the system is usually nonlinear, with the surface energy balance dependent on the slope flow. Furthermore, these flows are invariably three-dimensional in the real world. Observations (e.g., Stone and Hoard 1989) also show the presence of low-frequency fluctuations with periods in the range from 10 to 100 min, which are usually attributed to a reduction in temperature gradients due to adiabatic warming of the descending air.

Nappo et al. (1989) present a model for pollutant transport and diffusion in slope flows. Their predicted vertical wind profiles for slope angles of 2.5° and 20° and downslope distances of 0.5, 1, 2, and 3 km are illustrated in Fig. 6.6, where the along-slope component of the free wind speed is about 3 m s^{-1} above the surface layer. For this example the depth of the slope flow perturbation is about 50 m and the elevation of the slope "jet" is about 10 to 20 m.

The flow model is described by Nappo and Rao (1987). The paper by Nappo et al. (1989) adds an equation for pollutant conservation:

$$\frac{\partial C}{\partial t} + u\frac{\partial C}{\partial s} + w\frac{\partial C}{\partial n} = \frac{\partial}{\partial n}\left(K_n\frac{\partial C}{\partial n}\right), \qquad (6.4)$$

where the s-axis is along the fall-line of the slope and the n-axis is normal to the slope. Finite difference approximations to this equation are made at 31 × 31 points, with a domain of 3000 m in the s direction and 600 m in the n direction.

The eddy diffusivity, K_n, in Eq. (6.4) is calculated from

$$K_n = \ell(0.2e)^{1/2}, \qquad (6.5)$$

where e is the local turbulent kinetic energy and l is a turbulent mixing length given by

$$\ell = 0.4n/(0.74 + 4.7(n - z_o)/L). \qquad (6.6)$$

In this equation z_o is the surface roughness and L is the Monin–Obukhov length.

In practice, the pollutant plume is initialized using a Gaussian plume until the plume dimensions exceed the grid size of the model. One of their more interesting findings is that the subsidence in the slope flow causes plume centerlines to descend to the ground, as shown in Fig. 6.7. This phenomenon could lead to higher than expected ground level concentrations in slope situations. A maximum mean subsidence velocity of 3 cm s^{-1} is calculated

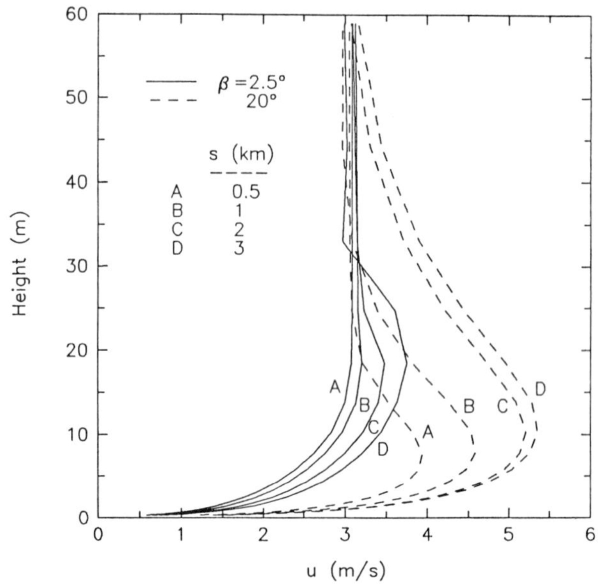

FIG. 6.6. Mean velocity profiles calculated for the katabatic flow over 2.5° and 20° slopes at several downslope distances using the model by Nappo et al. (1989). The ambient wind is 3 m s^{-1} downslope.

for the 20° slope, which will bring the centerline of a plume released at a height of 10 m to the ground after a time of 333 s.

The model predictions have been evaluated using observations obtained during the ASCOT Anderson Creek valley field study. Figure 6.8 contains a time series of observed and predicted concentrations of tracer gas at one of the sequential sampler sites. The tracer gas was released over a time period of one hour. Several such plots are given by Nappo et al. (1989), and this one is one of the better ones in terms of showing reasonable agreement between the observed and predicted time series. The maximum concentration is simulated within about 20%. The observed concentration decreases with time much less rapidly than the predicted concentration, in agreement with the results of other field experiments involving short-term releases. Apparently some tracer gas is slowed down by the roughness elements and is only very slowly transported downstream.

6.3.2. Diffusion models for narrow valleys

The boundary layer evolution of narrow mountain valleys during the early morning has been studied extensively by, for example, Whiteman (1982) and Bader and McKee (1985), and a detailed description of their work is given by Whiteman, Chapter 2. In this overview we will provide a brief summary of the meteorological aspects of narrow valleys, and concentrate more on the associated diffusion models.

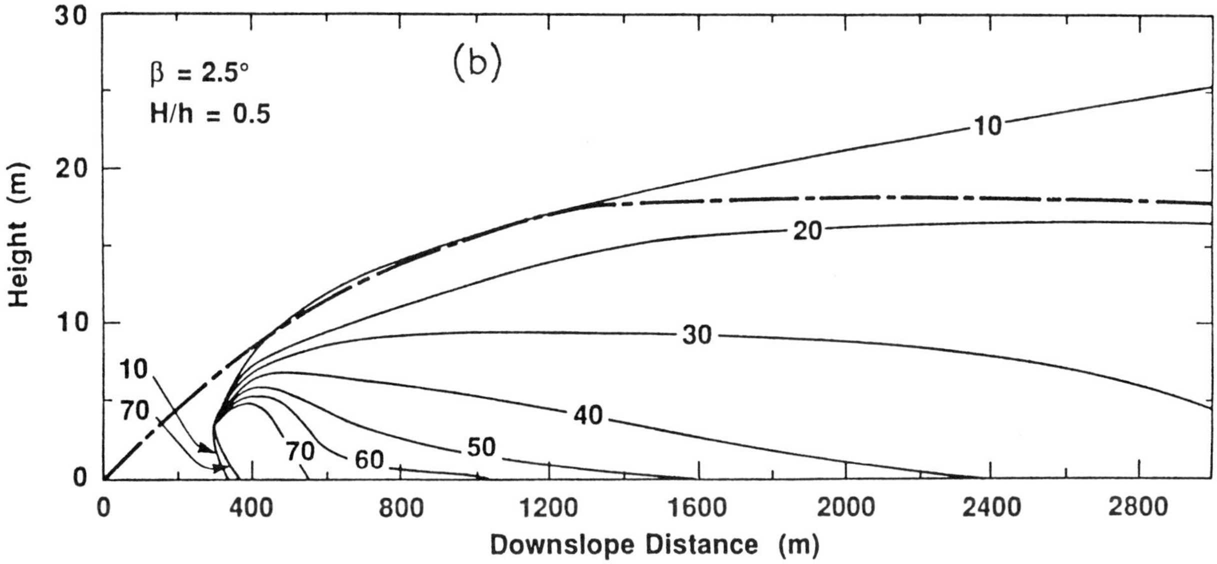

FIG. 6.7. Calculated concentrations above a 2.5° slope for a source with height equal to 0.5 times the local drainage depth, which is shown as a heavy broken line (from Nappo et al. 1989).

The conceptual model for the evolution of the early-morning boundary layer emphasizes the following principles:

- The nocturnal stable layer in a valley is destroyed by 1) the growth of the convective boundary layer over the valley floor and sidewalls, and 2) the subsidence of the stable air mass or "core" in the valley center as the upslope motions transport mass out of the valley.

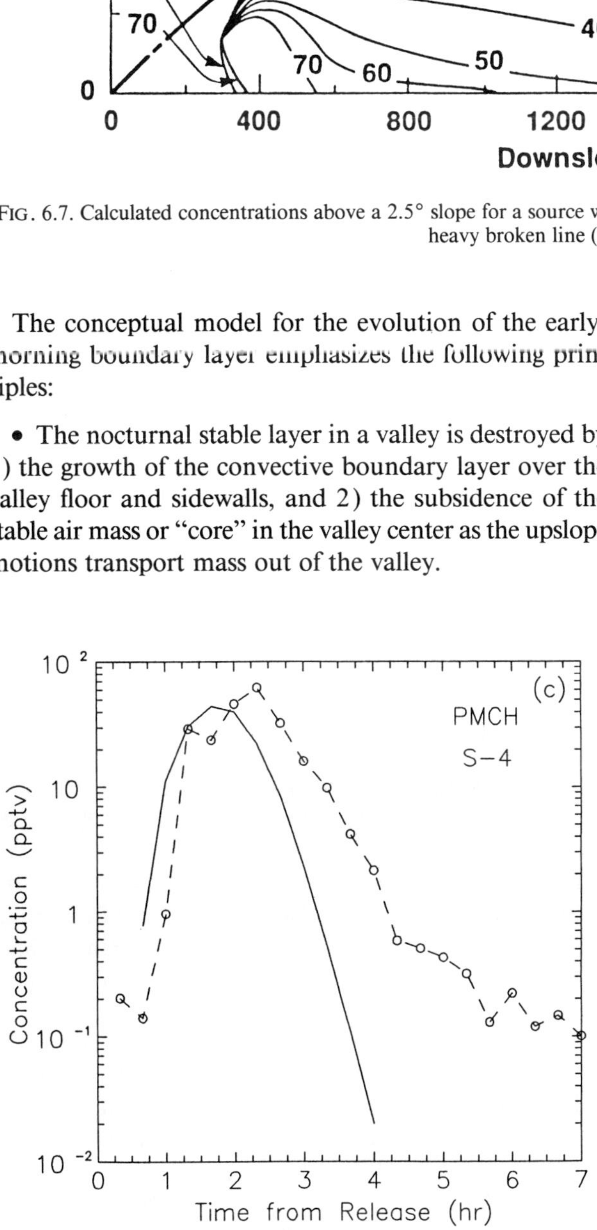

FIG. 6.8. Comparison of predicted (solid line) and observed (dashed line) tracer gas concentrations, as a function of time after release, at a sampler site in the ASCOT Anderson Creek valley experiment (from Nappo et al. 1989).

- Asymmetric heating of the valley sidewalls by the sun can skew the development of the boundary layer, with a tendency towards upslope motions on the heated sidewall and residual stability on the shaded sidewall. Lateral turbulence exchange processes tend to smooth out these differences.

A combined thermodynamic and diffusion model for valleys (VALMET) and its application to a tracer study is described by Bader and Whiteman (1989). This is a three-dimensional mesoscale model that carries the following variables: wind components, temperature, pressure, and subgrid scale turbulent kinetic energy (TKE), the product of TKE and the turbulent length scale, and any number of passive scalar concentrations. Turbulent fluxes are parameterized using K theory, where the Ks are functions of local stability, turbulent length scale, and TKE. The model includes a surface energy balance, and accounts for conditions on the valley walls at the boundary of the domain.

Bader and Whiteman (1989) simulate a tracer gas release, similar to that observed in the Brush Creek valley (Whiteman 1989), but they do not attempt to evaluate the model through quantitative comparisons of predicted and observed concentrations. The initial postulated potential temperature structure at sunrise over a symmetric valley, along with cross sections through two plumes are shown in Fig. 6.9. The mean flow in the valley is parallel to the valley orientation. At a time of 120 minutes after sunrise, the model predicts that the plume cross-sections will be modified to the shapes in Fig. 6.10, where both plumes have diffused appreciably, and the lower plume has shifted laterally toward the heated valley sidewall. This

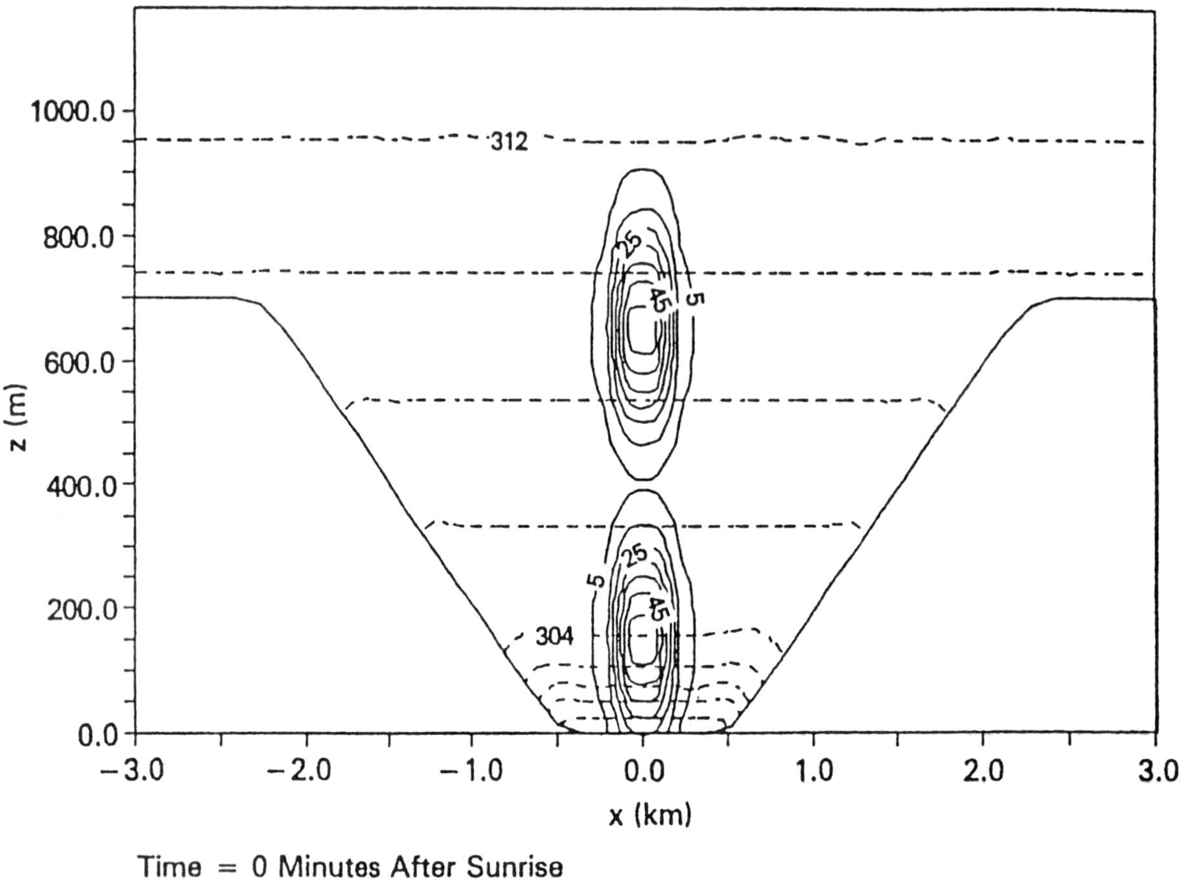

FIG. 6.9. Initial plume concentration (solid line) and potential temperature (dashed line, °K) profiles in a narrow valley, as assumed by Bader and Whiteman (1989) in their model simulations.

prediction of lateral shifting is validated qualitatively by the field observations from Brush Creek valley in Fig. 6.11 (from Whiteman 1989), which also suggests that the plume moves towards the heated sidewall, where maximum concentrations are achieved.

6.3.3. DOE ASCOT experiments

The principal reason for the initial interest of the U.S. Department of Energy (DOE) in diffusion in rugged terrain was the potential development of the vast fossil fuel energy resources of the western United States, where little information was available regarding transport and diffusion in the rugged terrain surrounding most energy development regions. In 1978 the ASCOT program was initiated, with the following goals (Clements et al. 1989a):

1) To improve fundamental knowledge of transport and diffusion processes in complex terrain.

2) To utilize this improved insight into the physics of terrain dominated flows to provide a methodology for performing air quality model assessments.

A group of scientists from national laboratories, universities, and other federal laboratories has participated in two major field studies and has conducted a wide variety of theoretical analyses. The focus of the program has been on nocturnal drainage flows. Twenty-six papers describing this work were published in the June and July 1989 issues of the *Journal of Applied Meteorology.*

The first set of major field studies took place in 1979–81 in the Geysers geothermal area of Northern California, with emphasis on understanding the fundamental characteristics of the nocturnal circulations (Gudiksen and Dickerson 1982). In 1982, 1984, and 1988 the ASCOT program moved to the then active oil shale region of western Colorado, with emphasis on the dynamics of deep and narrow valleys and the interactions among drainage flows as they converge. Papers in the cited issues of the *Journal of Applied Meteorology* are primarily concerned with the 1984 ASCOT field study in Brush Creek valley, Colorado (Clements et al. 1989a). A topographic map of the region is contained in Fig. 6.12, and the locations of the tethered instrumented balloon sites are also indicated.

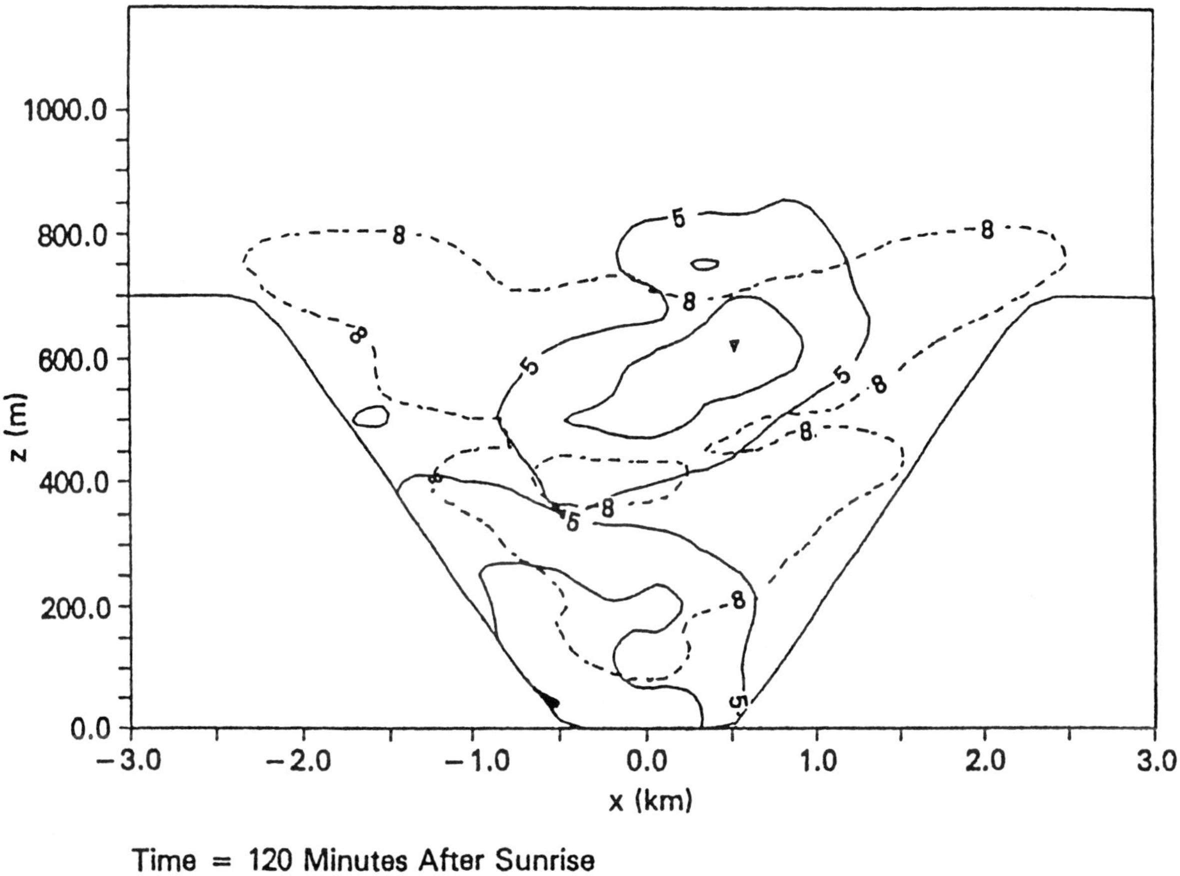

FIG. 6.10. Predicted plume concentrations (solid contours in same relative units as Fig. 6.9) and lapse rate (dashed contours in units of degrees Kelvin per kilometer) at a time 120 minutes after the initial conditions specified in Fig. 6.9. This time period includes the morning transition period (Bader and Whiteman 1989).

The valley floor has a gentle slope of 1.5° and the sidewalls have a relatively steep slope of 35° to 40°, rising to about 600 m above the valley floor.

Table 6.2 contains a list of the instruments that were used, showing that great diversity and numbers are possible when several groups come together for a cooperative program. The good spatial coverage permitted mass fluxes to be calculated with fair accuracy. The series of tracer studies employed oil fog and three gaseous tracers. The oil fog was released on the mesa top, to evaluate the entrainment of air into tributaries to Brush Creek. The transport and dilution of the oil fog was documented by photography. Two of the perfluorocarbon tracers were released within the valley, one near the ground surface and the other 200 m above the surface release site. The third tracer was released on the mesa near the oil-fog release site.

Gudiksen and Shearer (1989) present some of the results of their analyses of the tracer data. They found that the observed transport and diffusion within the valley at night was generally independent of the synoptic flow above the valley. Tracer gas was ventilated out of the valley in the morning as convection developed. The cross sections of down-valley flows observed by Doppler lidar in expt 4 during the night, during the morning transition period, and during the daytime are presented in Fig. 6.13. The strong drainage flow (6 m s^{-1}) at night near the valley floor can be seen in the top figure. In the morning transition period, this flow weakens and shifts toward the heated valley sidewall. During the day the flow field becomes well mixed and switches to an up-valley direction, with maximum flow speeds of 4 m s^{-1}. Corresponding tracer gas concentration cross-sections for these same periods are plotted in Fig. 6.14. The nocturnal distribution is contained near the valley floor and shows little cross-valley variation. During the morning transition period, the tracer gas distribution on the two sidewalls is noticeably different, as the plume moves toward the heated sidewall and is advected up to the top of the valley. The daytime distribution is much more well mixed over the cross section, with the maximum concentration dropping by an order of magnitude. Analysis of the observed vari-

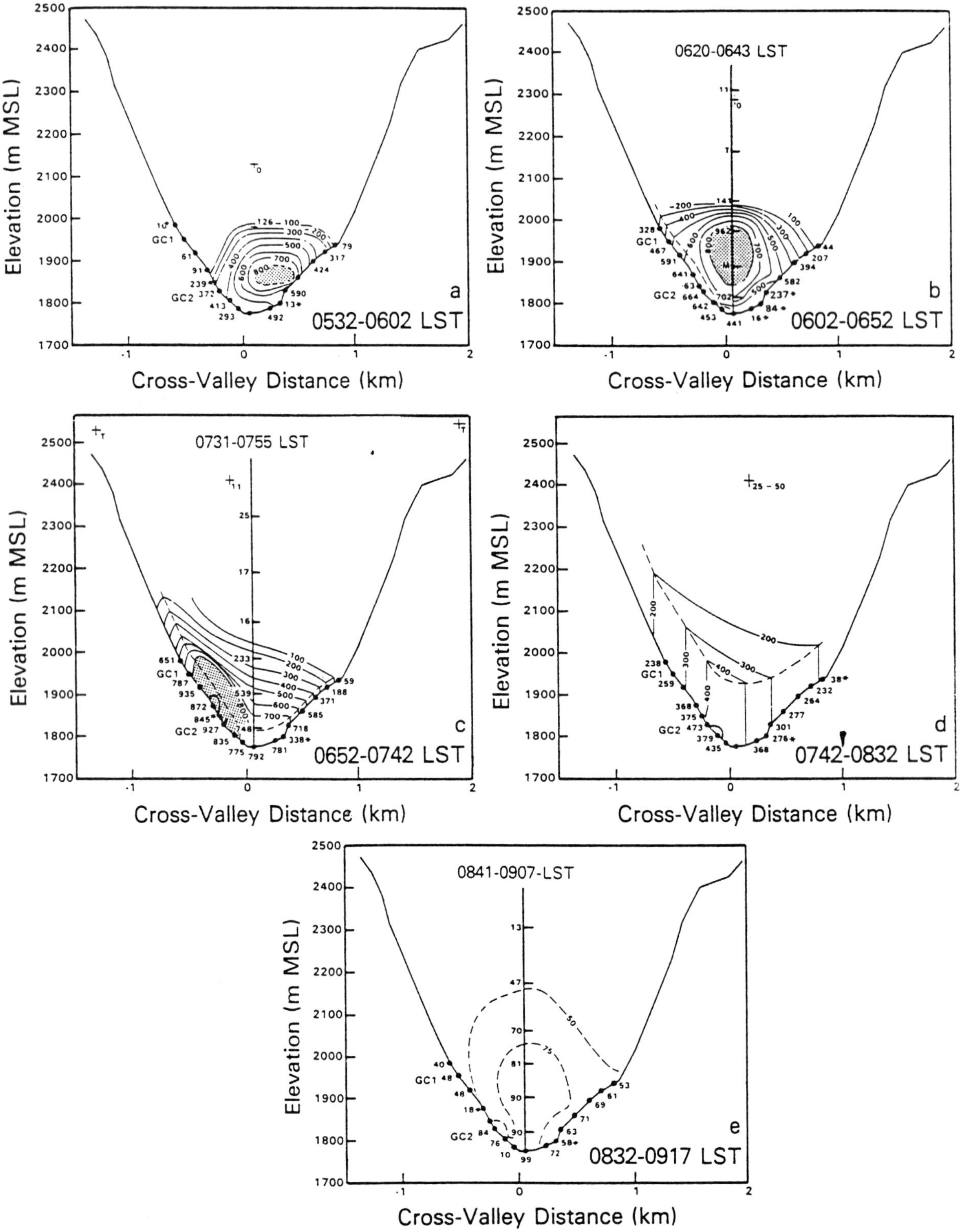

FIG. 6.11. Observed tracer gas concentrations (ppt) in Brush Creek valley during the morning transition period on 4 August 1982. The dashed line indicates convective boundary layer depth (from Whiteman 1989).

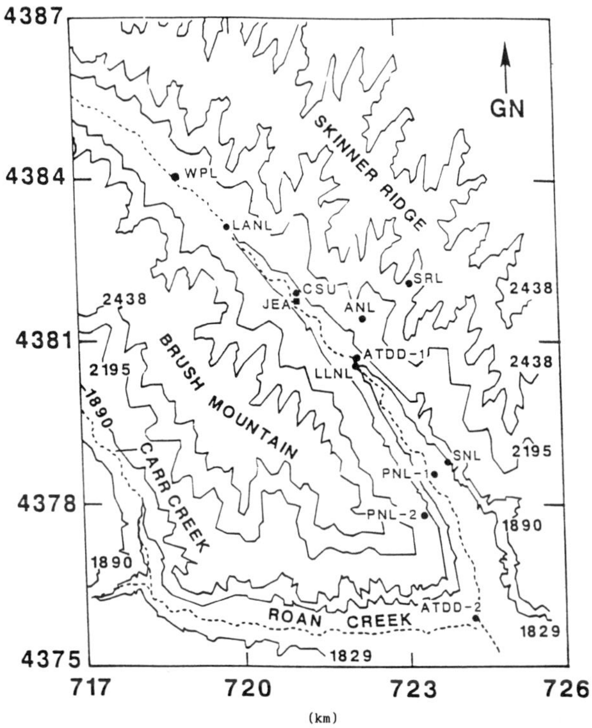

FIG. 6.12. Topographic map of Brush Creek valley field site, including locations of tethered balloons (from Clements et al. 1989a). Elevations are in meters.

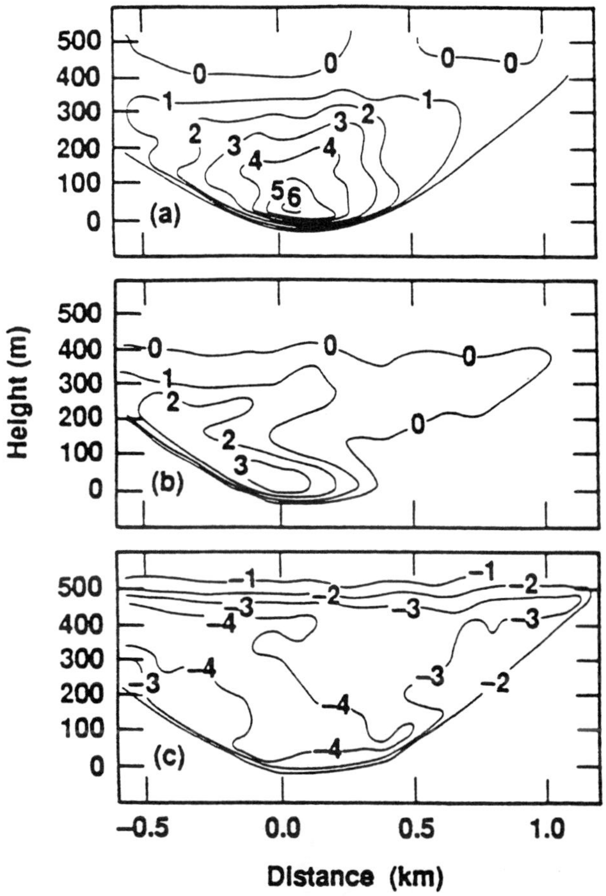

FIG. 6.13. Cross-valley views of the flows observed in expt 4 by the Doppler lidar within the Brush Creek valley. The data were obtained during the night (top), the morning transition period (middle), and the daytime (bottom). Positive values denote down-valley flow in meters per second (from Gudiksen and Shearer 1989).

ation of maximum concentration with time leads to the conclusion that for maximum solar heating cases, the concentrations are higher during the night but lower during the day, and that average concentrations start decreasing around sunrise. We note that these data have not yet been quantitatively compared with the predictions of the model by Bader and Whiteman (1989).

TABLE 6.2. Instruments used in the ASCOT 84 field study (from Clements et al. 1989a).

Instrument	Number
Tethered instrumented balloons	12
Doppler acoustic sounders	7
Vertical turbulence profiler	1
Instrumented towers	25
Optical anemometer paths	13
Surface energy budget stations	5
Dual Doppler lidar	1
Upper air stations	5
Tracers	4
Vertical tracer profiler	11
Surface tracer samples	90

6.4. EPA Complex Terrain Model Development (CTMD) Program

6.4.1. Objectives

Holzworth (1980) presents an overview of the objectives and design of the EPA's Complex Terrain Model Development (CTMD) Program at its inception. The goal of the program was to improve "mathematical modeling capabilities to describe concentrations at ground level (i.e., at the 2 m height) on elevated terrain due to an elevated plume from a large source (e.g., a power plant) in a stably stratified atmosphere interacting with a terrain obstacle . . ." In many regulatory assessments of air quality near sources of pollution in complex terrain, peak concentrations estimated by means of screening models result from direct impingement of a plume on nearby elevated terrain during periods of stably stratified flow. The model development effort during the CTMD program was directed at providing refined estimates of concentrations during

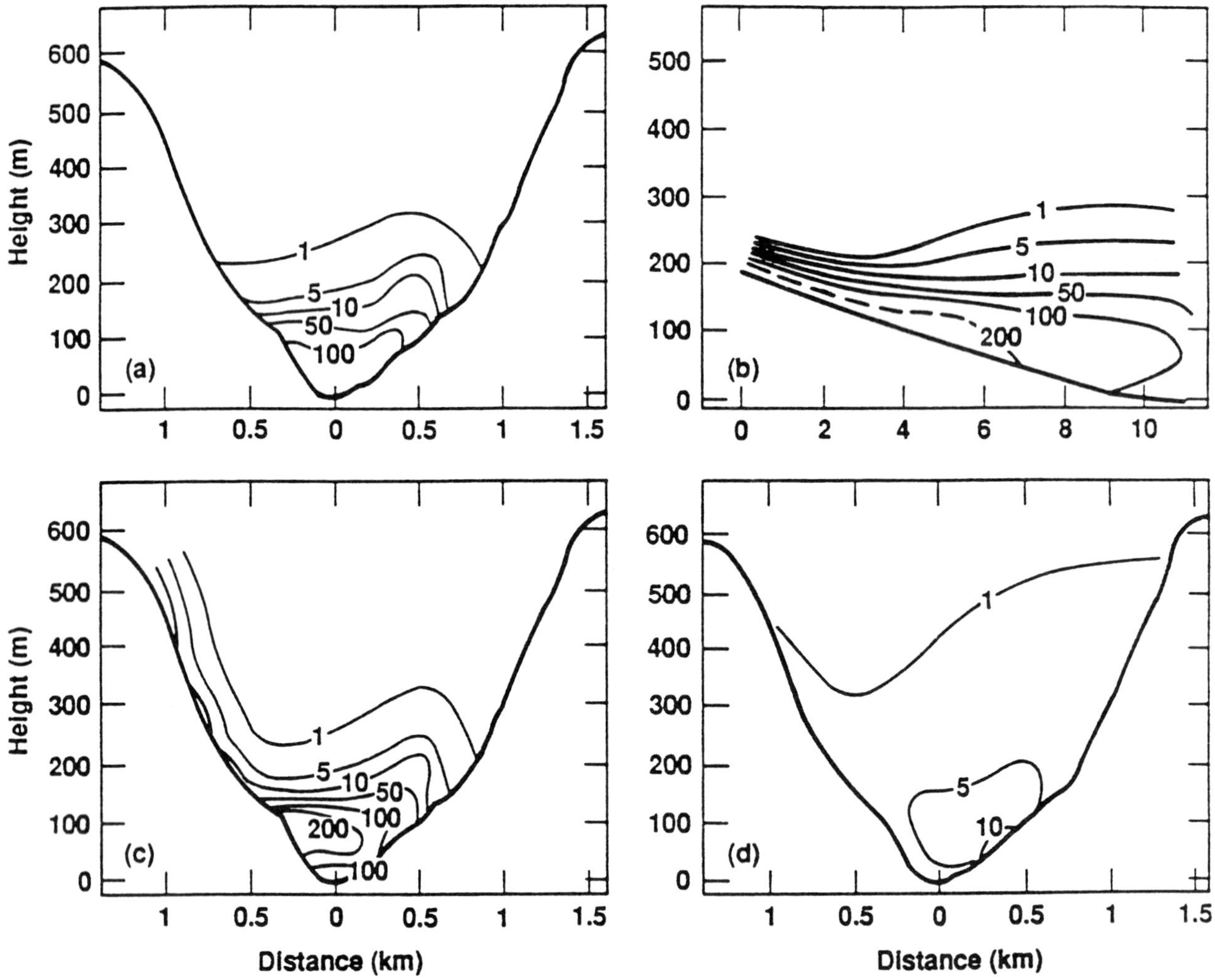

FIG. 6.14. Cross-valley distributions of tracer gas concentration observed during expt 4 in Brush Creek valley, concurrent with the wind data plotted in Fig. 6.13. The data were obtained during (a) and (b) the nighttime, (c) the morning transition period, and (d) the daytime (from Gudiksen and Shearer 1989). Here (b) shows the distribution along the valley axis.

this type of plume–terrain interaction. This focus explicitly relies on site-specific measurements of the wind field and/or detailed models for estimating the wind field just upwind of a terrain feature, as the program does not address how to model the flow from the source to the terrain feature.

The CTMD program included fluid modeling studies and field experiments to provide data on flow and dispersion around terrain features of varying complexity. The fluid modeling studies conducted by the EPA's Fluid Modeling Facility (FMF) explored the limits of the applicability of fluid modeling in simulating flow and dispersion over three-dimensional and nearly two-dimensional obstacles in neutral and stably stratified flow, and were instrumental in clarifying many phenomena predicted by theoretical arguments, or suggested by observations from the field experiments. The field experiments provided data on plume transport and diffusion, and the resultant field of ground level concentrations at three sites. The first two sites were selected to perform "controlled" experiments on flow and dispersion around an isolated three-dimensional and two-dimensional hill, in which the source position could be controlled, although the meteorology could not. The third site was selected to obtain data that would allow the modeling concepts to be refined and evaluated at a typically complex site, with two fixed source locations and many surrounding terrain features. The field experiments were generally performed during the hours between early evening and midmorning to obtain data for stably stratified to neutrally stratified flow regimes.

In thc following subsections, results obtained from the fluid modeling studies and field experiments are reviewed. Corresponding assumptions and features contained in the

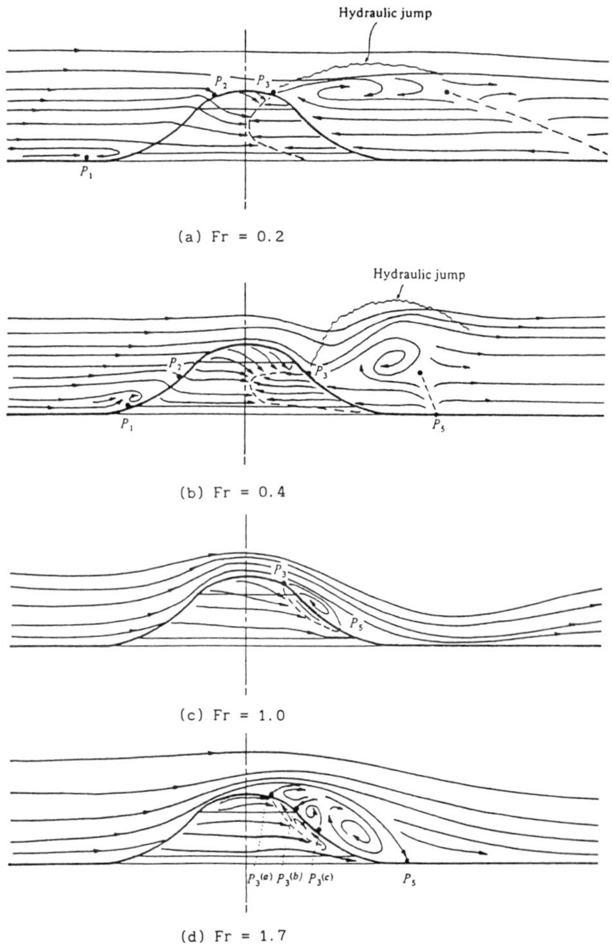

FIG. 6.15. Derived center-plane streamline patterns (Hunt and Snyder 1980). The symbols, p, denote stagnation points.

Complex Terrain Dispersion Model (CTDM) developed during the project are presented along with the results obtained from evaluating the performance of CTDM.

6.4.2. *Fluid modeling*

Studies conducted at the EPA FMF contributed greatly to the design of the program, as well as to the conceptual formulation of CTDM. A comprehensive summary of the FMF contributions to the CTMD program is contained in Snyder (1987). Other discussions of fluid modeling of flows around complex terrain are given by Meroney, Chapter 7.

Prior to the start of the program, studies were conducted at the facility to determine the structure of stably stratified flow over idealized three-dimensional hills. Flow over a bell-shaped hill was studied for neutral flow simulated in a wind tunnel, and stably stratified flow simulated in a tow tank. Hunt et al. (1978) and Hunt and Snyder (1980) discussed the nature of the observed flow, including lee-wave patterns, regions of separated flow, applicability of the Drazin (1961) model for strongly stratified flow, and the presence of a dividing streamline that separated a lower region in which the flow is deflected laterally around the hill, from an upper region in which the flow is deflected over the hill.

Conclusions from Hunt and Snyder (1980) include the following:

1) When the Froude number ($Fr = u/NH$) is much less than 1 (say <0.3) the flow is approximately horizontal except in narrow regions of depth $FrH = u/N$ near the top and bottom of the hill (Fig. 6.15a). Between these two thin layers, the flow that passes around the hill separates on the lee side of the hill into a horizontal, recirculating flow (the axis about which the circulation takes place is vertical).

2) As the Froude number increases, the region of flow over the top (u/N) increases and the region of nearly horizontal flow diminishes. More of the flow passes over the top of the hill, and the region of recirculation in the lee of the hill is reduced (Fig. 6.15b). Once the Froude number is on the order of unity, the region of horizontal flow is virtually absent, and lee-wave patterns can dominate the structure of the flow in the lee of the hill (Fig. 6.15c).

3) When the Froude number becomes greater than approximately 1.5 for this hill, the structure of the flow is qualitatively similar to that found in neutral flow (Fig. 6.15d).

By quantifying the depth of the upper layer, in which the flow travels over rather than around the hill, as u/N, Hunt and Snyder (1980) suggest a simple formula for the dividing streamline that establishes a criterion for determining when a plume may impinge on the surface of a hill. They point out that the dividing-streamline concept is only referenced to the bell-shaped hill, and a flow in which both u and N are constant with height. Snyder et al. (1980) later showed that the dividing-streamline relationship also applied to a cone-shaped hill and a hemisphere.

The utility of the dividing-streamline concept was extended when a series of tests were conducted with a scaled model of Cinder Cone Butte (CCB), Idaho, and density profiles in which the density gradient of the fluid decreases with vertical distance away from the base of the hill; i.e., the Brunt–Väisälä frequency, N, decreases with height above the surface. The purpose of these tests was to evaluate an integral formula for predicting the dividing-streamline height, which was to be used in planning the field experiments at Cinder Cone Butte. The integral formula that Hunt (1980) had proposed from a simple energy balance argument, and that had actually been published earlier by Sheppard (1956), strikes a balance between the kinetic energy of a parcel far upwind of the hill and the

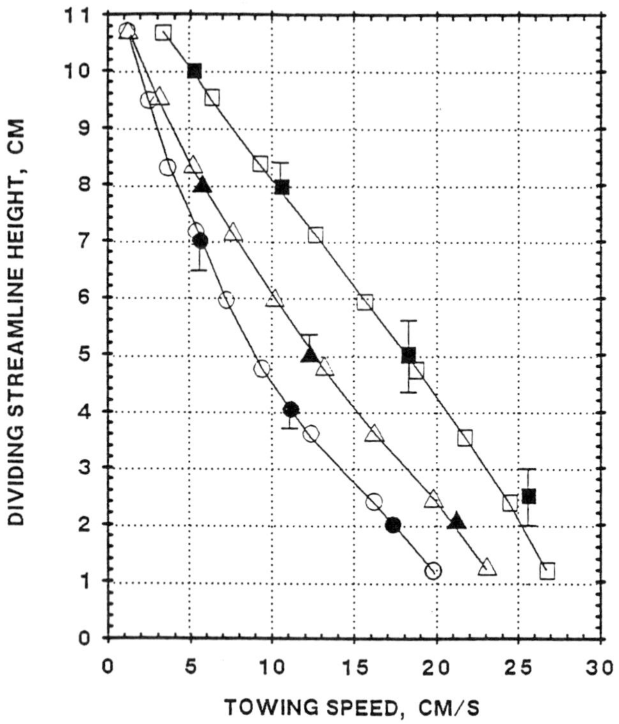

FIG. 6.16. Comparison of predicted dividing-streamline heights with observations as functions of towing speed. Open symbols denote predictions using integral formulas; closed symbols denote observations (from Snyder 1987).

potential energy acquired by that parcel when lifted to the top of the hill:

$$\frac{1}{2}u^2(H_d) = \int_{H_d}^{H} N^2(z)(H - z)dz, \quad (6.7)$$

where H_d is the dividing-streamline height, H the height at the top of the hill, and $N(z)$ the Brunt–Väisälä frequency at height z:

$$N^2 = -\frac{g}{\rho}\frac{\partial \rho}{\partial z} \approx \frac{g}{\theta}\frac{\partial \theta}{\partial t}. \quad (6.8)$$

The dividing-streamline height is that value which satisfies the equality of the left and right sides of Eq. (6.7). This balance between potential and kinetic energy implies that the speed along a streamline just above H_d nearly drops to zero at the crest of a hill, a result at odds with experience. In spite of this contradiction, Eq. (6.7) does reduce to the simpler relation $H_d = H(1 - \mathrm{Fr}) = H - u/N$ when both u and N are constant, which was found to work well for simple obstacles. A series of tows was made at different speeds with streamers of dye, and two layers of constant N. Three streamers spaced 0.064 H apart in the vertical served to identify the elevation of the observed dividing streamline of the flow. The excellent agreement that was attained with Eq. (6.7) is shown in Fig. 6.16.

Physical deficiencies in the energy-balance derivation for the dividing-streamline height have recently been addressed by Smith (Appendix to Chapter 5). In particular, Smith emphasizes the treatment of the pressure along a streamline to obtain the speed and finds that the potential energy term is replaced by the vertical integral of the displacement field induced by the hill. He concludes that the success of the energy-balance approach in locating H_d may be fortuitous.

Other tows that produced flows around and over the model of CCB indicated that concentrations on the surface of the hill below H_d are very sensitive to wind direction. Streamers would impinge on the hill only if they marked a stagnation streamline of the flow. A vortex street was also frequently established in the lee of the model hill. The alternating pattern in shedding these vortices introduced an oscillation in the flow around the hill that caused streamers marking the stagnation streamlines below H_d to meander across the hill. In these tow-tank simulations, long-term averaged concentrations at the surface of the hill would result from a time-averaged impact from a narrow plume sweeping back and forth across the hill.

The dividing-streamline concept was also tested for ridgelike obstacles, and for obstacles that include a range of slopes. "Infinitely" long ridgelike obstacles proved very troublesome with strong stratification because upwind "columnar" disturbances were generated that precluded establishing steady-state conditions. But when the "ridge" was reduced to a segment with finite length, the integral formula for H_d was found to be appropriate, so that crosswind aspect ratio was not an important factor. The summary report by Snyder (1987) provides a good overview of these studies.

Wind shear was found to be important in predicting the structure of stratified flow on the upwind face of steep terrain features (Snyder and Ogawa 1982; Snyder et al. 1985). The shear produced a gradient in the dynamic pressure on the face of steeper features that created a vortex that tended to "downwash" plumes on the leading face of the features. Features with low to moderate slope such as CCB showed no such vortex formation.

The conceptual model that emerged from these results suggested that plumes above and below the dividing-streamline height might be treated separately. In particular, a simple approximation for plumes above H_d would treat H_d as a solid, planar surface, and the flow in this region would be treated as though only the segment of the hill above H_d were important. This approach has been called the "cutoff hill" approach. It was tested by Snyder and Lawson (1987) using the bell-shaped hill of earlier studies.

The bell-shaped hill was towed in a tank of constant N at a speed that produced a dividing-streamline height of one-half the hill height. Concentrations were obtained by

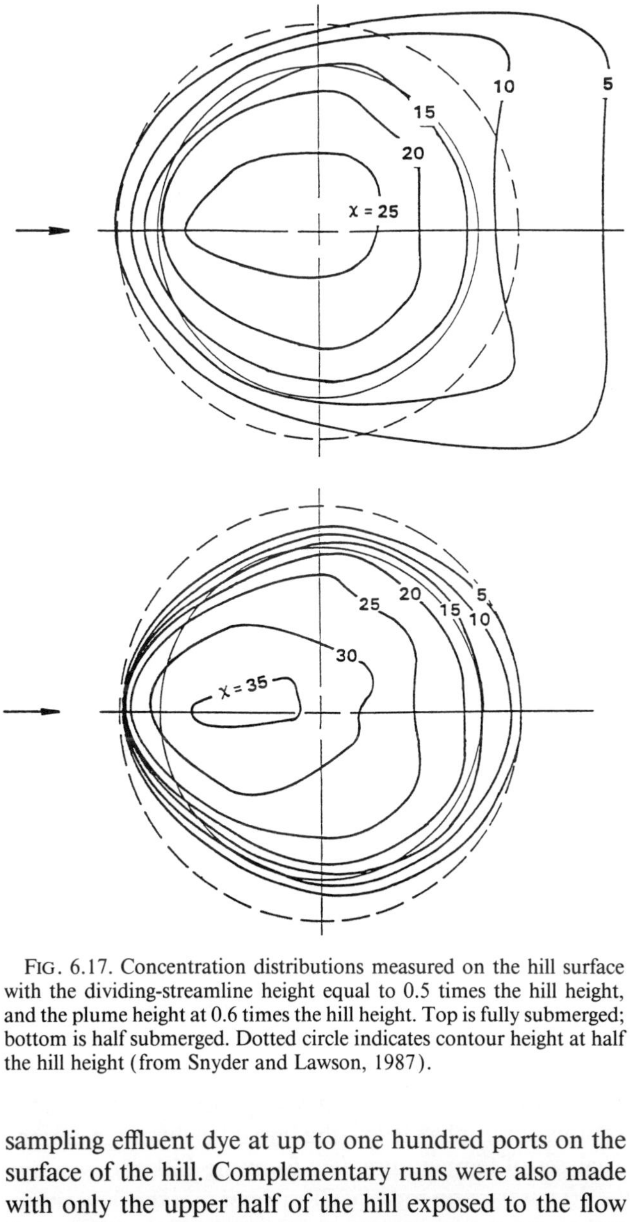

FIG. 6.17. Concentration distributions measured on the hill surface with the dividing-streamline height equal to 0.5 times the hill height, and the plume height at 0.6 times the hill height. Top is fully submerged; bottom is half submerged. Dotted circle indicates contour height at half the hill height (from Snyder and Lawson, 1987).

sampling effluent dye at up to one hundred ports on the surface of the hill. Complementary runs were also made with only the upper half of the hill exposed to the flow (i.e., the "cutoff" hill). It was found that the "cutoff" hill approximation is useful in predicting the locations and values of maximum concentrations, and also in predicting the areas of plume coverage of the upwind face of the hill. An example of concentration patterns measured on the full hill and the corresponding "cutoff" hill for one of the configurations is shown in Fig. 6.17. The "cutoff hill" model results were least similar to the full hill results for plume heights of order 0.8 H or greater. In these tows, the plume makes contact with the hill on the lower portion of the lee side. The differences in concentration patterns appear to result from the presence of a hydraulic jump in the full immersion tow that was absent in the half immersion tow.

Another study performed at the FMF investigated the magnitude of the presence of a hill on peak ground level concentrations in a simulated neutral atmospheric boundary layer. An axisymmetric "CCB" model was placed in the wind tunnel, and concentrations were obtained for a matrix of source positions both upwind and downwind of the hill. The peak concentration for each position was divided by the peak concentration for the same height of release without the hill to obtain the terrain amplification factor, A. The method was also applied to a two-dimensional ridge feature, and results were reported by Lawson and Snyder (1985) and Lawson et al. (1989). Figure 6.18 maps their results. For neutral flow, there exist "windows" of source positions that result in peak ground level concentrations that significantly exceed those expected in flat terrain. The greatest amplification factors are found for sources positioned in the lee of the features, where plumes can be brought quickly to the surface in the cavity flow region.

Note that amplification factors for sources placed upwind of the features are greater for the three-dimensional feature. The deformation of the flow over the three-dimensional feature is greater, and plumes are mixed more rapidly toward the surface.

6.4.3. *Field experiments*

A major component of the CTMD program was the design and execution of tracer and flow visualization experiments at three sites. These experiments progressed from the study of flow and dispersion over an isolated, three-dimensional hill (Cinder Cone Butte—CCB), to an isolated two-dimensional ridge (Hogback Ridge—HBR), to a typically complex river valley site (Tracy Power Plant—TPP). The intent of the program was to gain an understanding of flow and dispersion at sites with simple terrain first, and then to proceed to a "real-world" site.

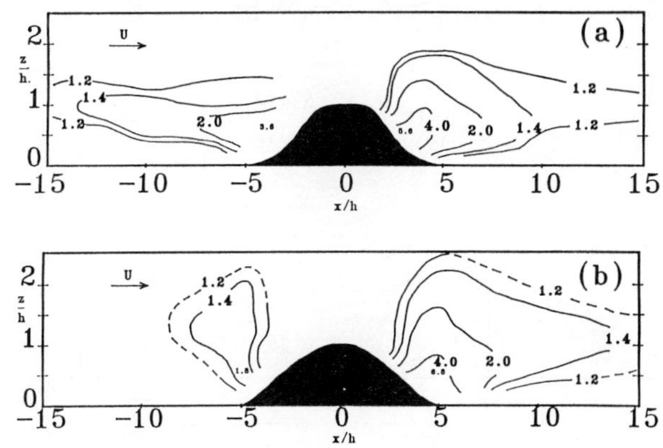

FIG. 6.18. Contour of constant terrain amplification factors over (a) axisymmetric hill, and (b) two-dimensional ridge. Note that vertical scale is exaggerated by a factor of 3 (from Snyder 1987). The distribution of amplification factors identifies the locations of a source that produces a peak ground level concentration in excess of that obtained in the absence of the hill.

The first two sites with simple terrain also needed to have a manageable scale to facilitate the logistics of obtaining detailed flow measurements and tracer concentration measurements.

6.4.3.1. Cinder Cone Butte

The first field program was named Small Hill Impaction Study #1 (SHIS #1), and a description of the study is contained in the "EPA Complex Terrain Model Development Program: First Milestone Report—1981" (Lavery et al. 1982). CCB rises approximately 100 m above the broad Snake River basin near Boise, Idaho. A cross section of the hill is given in Fig. 6.3. Although nearly circular in cross section at its base, it takes on a two-peak structure with a prominent draw at its top (Fig. 6.19). A 150 m tall meteorological tower was positioned approximately 2 km to the north of the hill to document the flow and temperature structure near CCB, while one 30 m tower on the north peak of CCB and four 10 m towers at elevations of approximately 65 m above the basin floor provided wind and temperature information over CCB. Two tracer gases were released from cranes positioned on roads approximately 600 to 1000 m "upwind" of the center of CCB, and these gases were sampled at approximately one hundred locations on CCB. Oil fog was released near one of the tracers to provide flow visualization. The oil-fog plume was sampled by lidar to obtain cross sections, and it was photographed from a number of locations on and around CCB with the aid of illumination from a revolving carbon arc-lamp placed approximately 3 km away.

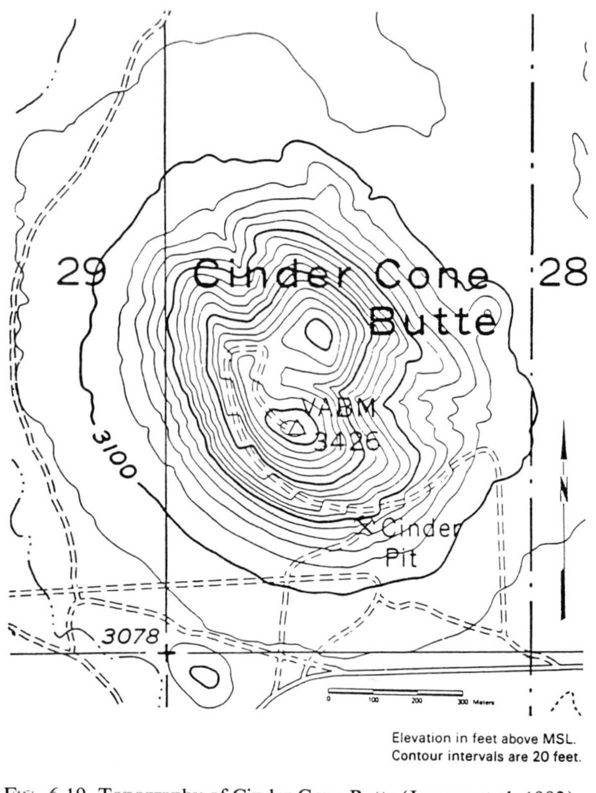

Fig. 6.19. Topography of Cinder Cone Butte (Lavery et al. 1982). Elevations are in feet MSL.

The phenomena observed at CCM largely confirmed features of the flow seen in the fluid modeling studies of neutral and stratified flow around a three-dimensional terrain feature. The dividing-streamline values computed from Eq. (6.7) using data obtained from the 150 m tower did differentiate between distinct patterns of flow over the hill for plume material above H_d, and impingement or flow around the hill for plume material below H_d. Plumes above H_d that rose over the top of CCB grew noticeably thinner in the vertical as they traveled over the crest, and exhibited a lee-wave dominated structure at times downwind of the crest. Plumes below H_d that passed around CCB tended to skirt the samplers, causing relatively small concentrations at the surface. Plumes below H_d that were carried toward CCB within a small angle of the direction to the center of the hill meandered back and forth across the face of CCB, producing large concentrations within the sampler array on the hill.

Further analyses of the data collected at CCB and reported in the "EPA Complex Terrain Model Development Program: Second Milestone Report—1982" (Strimaitis et al. 1983) showed that the relative locations of peak concentrations of the tracer gases depend on the ratio of the initial plume elevation to the dividing-streamline height, H_d. For plumes well above H_d, the larger concentrations are found either near the top of the hill, or on the lee side. Plumes below H_d produced peak concentrations on the windward side of the hill. In the latter cases, some streamline lift was usually apparent, but vertical mixing was sufficient at times to produce peak concentrations at samplers below the initial height of the plume. Analyses of temperature data from the 150 m tower and the 30 m tower on top of CCB suggest that the flow above H_d bears some resemblance to weakly stratified flow over the bell-shaped hill studied in the FMF tow tank.

A summary of the meteorological variables that are associated with the ten largest observed concentrations of each tracer scaled by the emission rate are reported in the "EPA Complex Terrain Model Development Program: Fifth Milestone Report—1985" (DiCristofaro et al. 1986). Half of the peak scaled SF_6 concentrations occurred when the plume height was approximately equal to H_d. The other five peaks include three hours in which the plume was above H_d, and two hours in which the plume was below H_d. In the SF_6 dataset, roughly 55% of the hours were associated with plumes above H_d, 18% with plumes approximately equal in height to H_d, and 27% with plumes below H_d.

The CF_3Br releases produced plumes that were predominantly above H_d (about 88% of the hours in the dataset), while CF_3Br plumes were near H_d in height for approximately 10% of the hours, and the plumes were below H_d for 2% of the hours. Only one of the hours associated with the ten largest scaled concentrations is for a plume approximately equal in elevation to H_d; the rest are made up of hours in which the plume was above H_d. For these ten hours, peak concentrations were found at elevations that either exceed the initial plume height of the plume, or occur in the lee of the hill below the height of the plume.

6.4.3.2. Hogback Ridge

The second field experiment was named Small Hill Impaction Study #2 (SHIS #2), and is described in the "EPA Complex Terrain Model Development Program: Third Milestone Report—1983" (Lavery et al. 1983). Hogback Ridge (HBR) is located on the Colorado Plateau near the western slopes of the San Juan Mountains, about 15 miles west of Farmington, New Mexico. The experiments were carried out along a 1.5 km section of HBR (Fig. 6.20) in which the ridge rises approximately 90 m above the upwind section of the plateau. They were conducted using methods similar to those used at CCB. The 150 m meteorological tower was positioned approximately 500 m upwind of the base of HBR; the 30 m tower was placed right at the base of HBR; and a 10 m tower was placed near the crest.

Two tracer gases were released from either a crane located within 100 m of the base of HBR or from the 150 m tower. Oil fog was routinely released along with one of the tracers to provide a visible plume for the photography program and the lidar-sampling program. Samplers were located along height contours, with four major rows along the windward face and crest of HBR, and two rows in the lee.

The structure of the flow at HBR was quite different from that observed at CCB during SHIS #1. The temperature gradient was weak above the stable surface layer. As a result, the oil-fog plume produced by the jet-fired fog generator typically exhibited significant buoyant rise and marked flow above H_d, regardless of its actual release height. Because the SF_6 tracer gas was introduced into this jet, all of the SF_6 concentration data pertain to plumes above H_d, and all of the flow visualization data based on observations of the oil-fog plume also pertain to the flow above H_d.

Data obtained from SHIS #2 indicate that the dividing-streamline height calculated by means of Eq. (6.7) with data from the tall meteorological tower located upwind of the hill does differentiate between flow that is readily able to go over the feature and flow that is blocked. Releases below H_d dispersed in the blocked flow and produced peak ground level concentrations on the windward side. Releases above H_d dispersed in a flow that went over the ridge, producing peak ground level concentrations near the crest or in the lee.

Plume material below H_d tended to travel along the ridge away from the sampler array when there was a substantial component of the flow along the ridge. When this component was small, however, some material eventually moved up the face of the ridge. (The winds along the face of HBR tended to be unsteady with occasional "surges" toward the hill that appeared to carry material over the ridge.) Other material drifted away from the sampler array after being mixed into the shallow drainage flow from the feature.

The "Fifth Milestone Report" (DiCristofaro et al. 1986) summarizes the meteorological regimes that are associated with the ten largest observed concentrations of each tracer scaled by the emission rate. Due to the buoyancy of the oil-fog plume containing SF_6, and weak temperature gradients, all of the SF_6 plumes are in the flow above H_d and produce peak concentrations at receptors near the crest of HBR, or in the lee. The ten largest scaled CF_3Br concentrations correspond to plumes below H_d. Meteorological data obtained from the 30 m tower at the base of HBR indicate that light and variable winds are associated with the largest scaled concentrations at HBR.

The two-dimensional character of the flow over the top of HBR, and the large number of hours in which good lidar cross-sections of the oil-fog plume are available allowed an assessment to be made of the structure of the flow above H_d. The analysis, described in the "Fifth Milestone Report" (DiCristofaro et al. 1986) shows that the "cutoff" hill approach is useful in modeling streamline elevations above HBR. This is demonstrated in Fig. 6.21 by comparing the observed streamline heights with potential flow calculations for three two-dimensional ridge shapes. Even without stratification and shear, the potential flow calculations for flow above H_d appear broadly applicable. When stratification is incorporated into a model for the height of streamlines over the crest of a two-dimensional, inverse polynomial hill, the observations are modeled fairly well, as illustrated by Fig. 6.22. The model used for the predictions in this figure was developed by Hunt and co-workers (e.g., Hunt et al. 1981). Virtually the same results were obtained using a formulation by Queney (1947).

6.4.3.3. Tracy Power Plant

Two studies were carried out in the Truckee River valley around the Tracy Power Plant located east of Reno, Nevada. The first was a feasibility and design study. The second was the full scale plume study (FSPS). This study

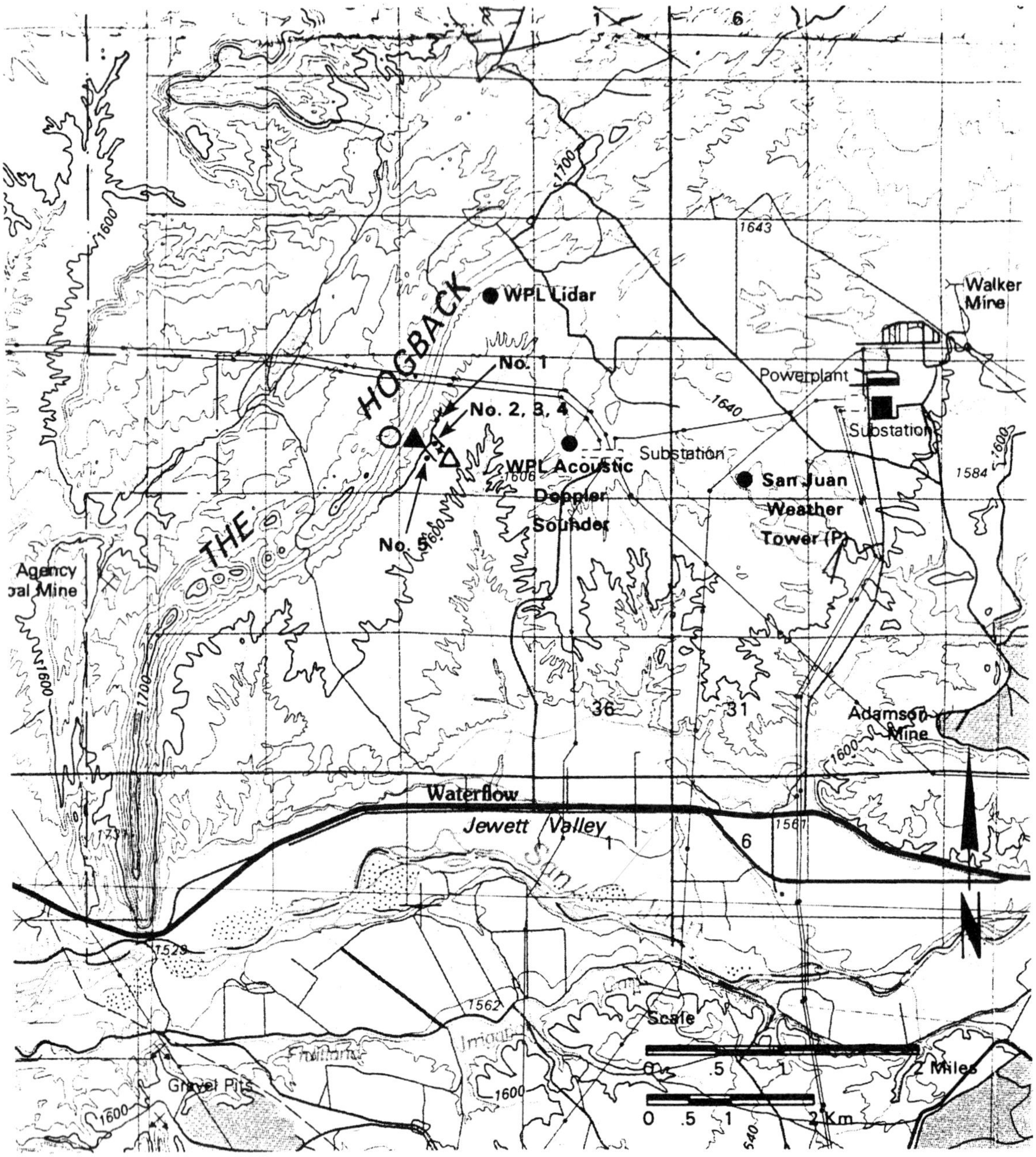

LEGEND

△ 500′ Tower A

▲ Tower B

○ Tower C

1 Tracer Release Pt. No. R-80

2 Tracer Release Pt. No. 203

• 3 Tracer Release Pt. No. 215

× 4 Tracer Release Pt. No. 216

5 Tracer Release Pt. No. 111

FIG. 6.20. SHIS #2 (Hogback Ridge) field experiment layout (Lavery et al. 1983). Elevations are in meters MSL.

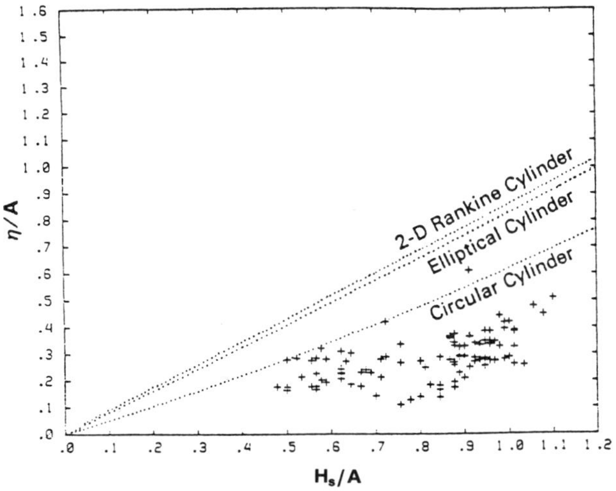

FIG. 6.21a. Height of the source streamline (η) above the creast of HBR for various source heights (H_s). Values are scaled by the height of HBR (A) (DiCristofaro et al. 1980).

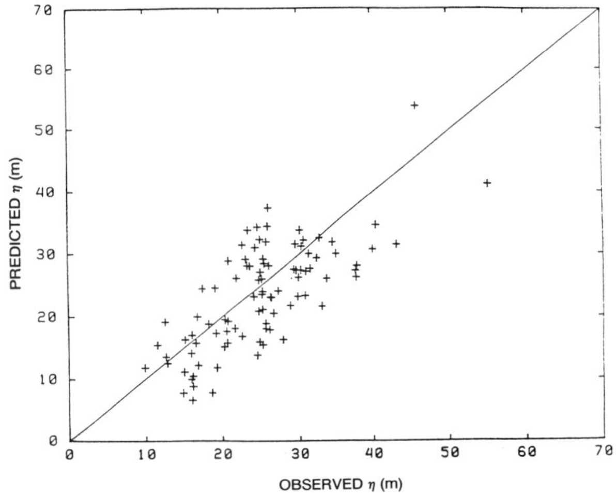

FIG. 6.22. Predicted versus observed plume centerline standoff distance (η) at the crest of HBR using the "outer-layer" solution of Hunt et al. (1981), which includes stratification but neglects shear (DiCristofaro et al. 1986).

differs from the earlier "small hill" studies in that the source locations are fixed, the topography is considerably more complex, and the vertical scale of the terrain is on the order of 300 to 900 m in height rather than 100 m in height.

The SF_6 tracer gas and oil fog were introduced to the effluent stream just upstream of a 91.4 m tall stack at the power plant. The CF_3Br tracer was released from one of three heights on a 150 m meteorological tower located 1.3 km to the east of the stack. As in the "small hill" experiments, the oil-fog plume was photographed throughout the night with the aid of two rotating carbon arc-lamps, and it was also sampled remotely by a ground-based lidar system. In addition, an airborne lidar system tracked the transport and development of the oil-fog plume throughout the study area during selected periods.

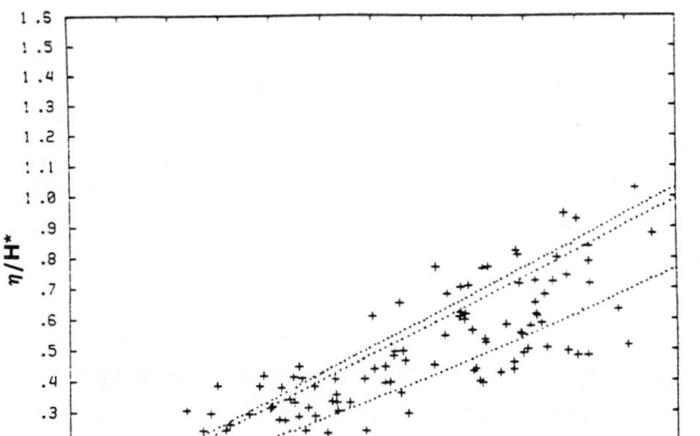

FIG. 6.21b. Height of the source streamline (η) above the crest of HBR for various effective source heights ($H_s - H_d$). Values are scaled by the height of the cutoff hill (H^*) (DiCristofaro et al. 1986).

The topography of the study area and terrain features of significance to the study are displayed in Fig. 6.23. The Truckee River flows from west to east, and drainage winds also tend to flow toward the east in the lower portion of the valley. Obviously, local drainage is downslope, and drainage winds fill the valley from all directions. The greatest potential for plume–terrain interactions during stably stratified flow exists for the terrain features identified as Beacon Hill, South Ridge, and Target Mountain. These features lie down-valley from the stack, and rise to elevations in excess of plume heights expected during nighttime flow conditions. Higher terrain to the north and south of these features constitutes the sidewalls of the valley that contain and channel the flow in the valley. A full description of the FSPS is contained in the "Fifth Milestone Report" (DiCristofaro et al. 1986).

The FSPS dataset contains a variety of meteorological regimes, including near-neutral periods, very stable periods with weak and variable winds, and periods in the morning in which inversion breakup occurs with fumigation of the elevated plume to the surface. Neff (1985) suggests that a typical FSPS test period contains the following sequence: the period from sunset to midnight is generally windy with a deep mixing layer and consequently, a weakly stable stratification. Between 0000 and 0800 LST the stability increases. The relative strengths of the inversion and the wind speed control whether waves or instabilities dominate the smaller scale motions. Several wind maxima may be present during the more stable conditions, while a single elevated jet forms under conditions with stronger winds. After 0800 LST, a convective

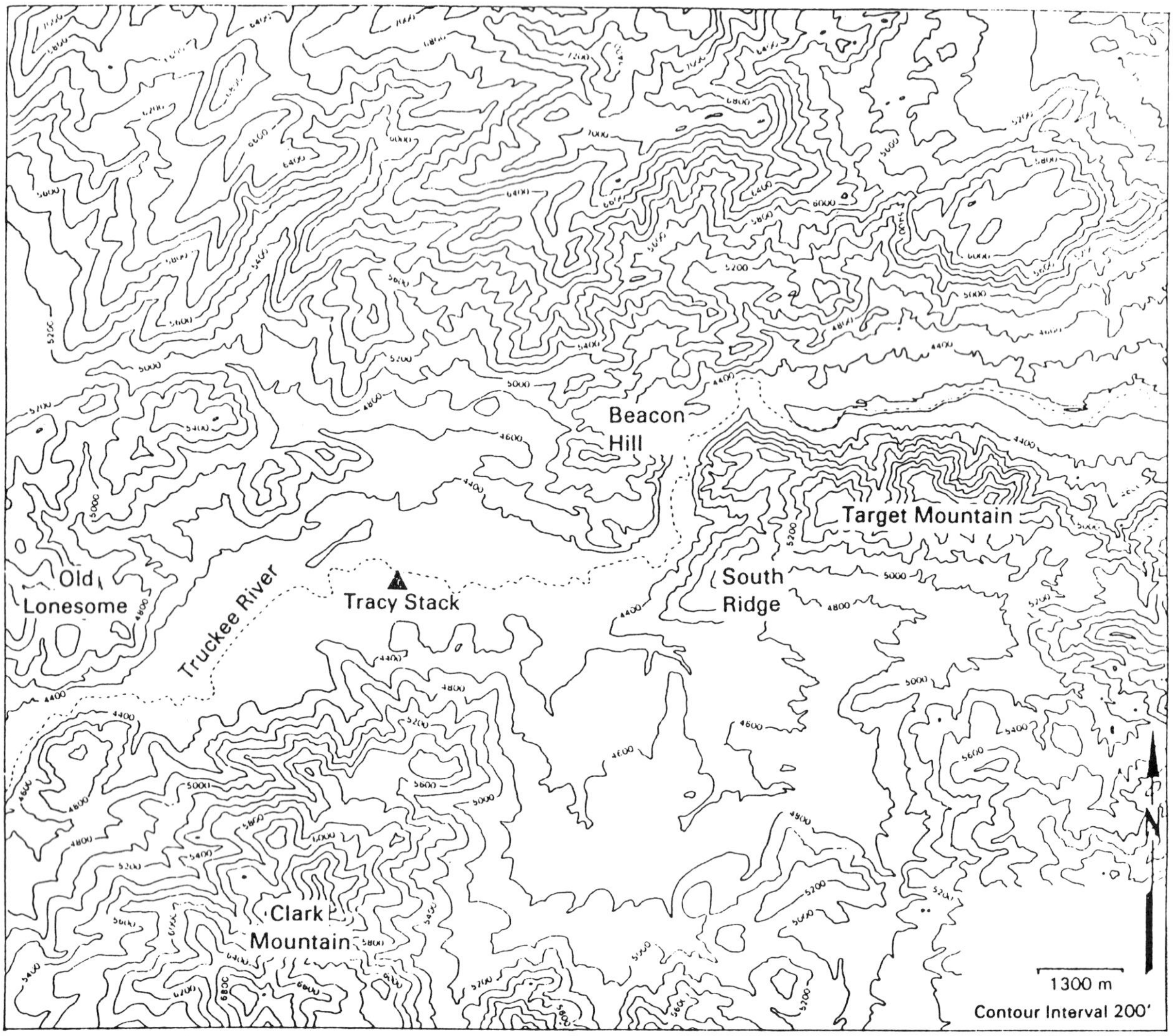

FIG. 6.23. Topography surrounding the Tracy Power Plant site (DiCristofaro et al. 1986). Elevations are in feet MSL.

boundary layer forms and grows while the elevated inversion layer descends. These layers generally meet by 1000 LST.

During strong layering events, the visible plume remained compact in the vertical, and marked time-dependent wavelike phenomena while over the river valley as well as over discrete terrain features. Of particular importance to the CTMD program was the manner in which the plume would interact locally with specific features like Beacon Hill. The channeling effects of the valley would force the flow toward Beacon Hill at times, and the plume would travel over or around it depending on the height of the dividing streamline calculated for the feature. The appearance of this interaction was strikingly similar to observations made at CCB during SHIS #1.

The "Fifth Milestone Report" (DiCristofaro et al. 1986) summarizes meteorological variables associated with the occurrence of the ten largest observed concentrations of each tracer scaled by the emission rate. The larger concentrations were measured on terrain features with values of H_d greater than or about equal to the plume elevation. Most of the peak SF_6 concentrations occurred at receptors whose elevations exceeded the top of the stack, but few of these receptors lie at elevations greater than the estimated height of the plume. Lavery et al. (1986) point out that the largest concentration was measured on Beacon Hill during an hour with strong stable stratification in which plume material (oil fog) was confined to a single sheet or layer. Other large concentrations occurred during events in which the tracer plumes were contained in a channeled flow into the gorge between Beacon Hill and Target Mountain. In essence, the role of H_d in defining flow regions, and the behavior of plumes both above and below H_d were found to be similar to what had been observed at the SHIS sites, and in the FMF studies.

6.4.4. *Assumptions contained in the Complex Terrain Dispersion Model (CTDM)*

The Complex Terrain Dispersion Model (CTDM) is a point-source Gaussian plume dispersion model designed to estimate hourly averaged concentrations of plume material at receptors near an isolated hill or near a well-defined segment of an array of hills. It incorporates those features of flow and dispersion observed in the FMF studies and in the field experiments that are of primary importance in modeling ground level concentrations arising from plume–terrain interactions during periods of near neutral to stably stratified flow.

The following restrictions are identified in the "User's Guide to the Complex Terrain Dispersion Model" (Paine et al. 1987):

- CTDM contains no wake algorithms for simulating the mixing and recirculation found in cavity zones in the lee of a hill. Therefore, sources within the lee of terrain features are not treated in the model, and estimates of concentrations at receptors in the lee will not be reliable when such zones are present.
- CTDM contains no global flow calculation that accounts for the presence of many hills. The path taken by a plume through an array of hills cannot be simulated by the model. It relies on measurements of the flow obtained in the neighborhood of the source to define the incident flow field for each of the terrain segments independently. If there is a strong channeling of the flow due to large-scale terrain features (e.g., a valley setting), then this influence is reflected in the modeling only insofar as it is contained in the measurements.
- All hills that are explicitly modeled are done so in isolation; any changes to the plume size caused by one hill are not carried forward to subsequent simulations downwind.
- For situations in which an individual hill is difficult to isolate from a complex terrain structure, caution must be used in interpreting the CTDM results.
- CTDM does not simulate calm meteorological periods.
- CTDM assumes that the meteorological data are representative of the entire averaging period, and apply to the entire spatial domain. Spatial variability resolved by an array of meteorological towers cannot be used directly in the model.
- CTDM is designed for neutral to stable flow conditions; therefore the model does not predict concentrations during unstable hours when any plume is within the convective mixed layer.

Many basic features in the formulation of the model follow concepts developed by Hunt et al. (1979) in modeling plumes embedded in small and large Froude number flows about a three-dimensional obstacle. In the absence of strong stratification, all streamlines in the flow pass over a hill. The centerline of a plume in this flow follows the streamline that passes through the source of that plume. As the plume grows in the vertical and horizontal directions (in the plane perpendicular to the flow), plume material diffuses across adjacent streamlines, eventually reaching the set of streamlines that marks the surface of the terrain. Distortions in the flow that are induced by the hill change the position and relative spacing of the streamlines from their initial distribution over level terrain, and therefore change the shape of the plume as it passes over the hill. Ground level concentrations (GLCs) of plume material also change, and this happens in two ways. The first and most obvious change in the GLCs is the shift in the distribution on the surface of the hill, arising from the change in the shape of the plume. The second change in the GLCs is the change in magnitude, arising from the effect of the distortion on the rate of diffusion of plume material across streamlines (Hunt and Mulhearn 1973). Typically, spacing between streamlines is reduced in the vertical and expanded in the horizontal, while the speed of the flow increases over the crest. These changes tend to increase the diffusion in the vertical and reduce it somewhat in the horizontal, thereby altering the magnitude of the GLCs on the hill.

When the flow is very stably stratified, H_d is nonzero, and a two-layer structure develops in which the flow in the layer between H_d primarily deflects around the hill, while the flow in the layer above H_d travels over the top of the hill. Above H_d, the flow is assumed to be similar to that just described although a cutoff hill must be used. Below H_d, the flow is approximated as an ideal, steady, two-dimensional flow. Within this flow, only one streamline at each elevation touches and follows the surface of a hill, and is referred to as the stagnation streamline. Plume material reaches the surface of the hill only if it reaches the stagnation streamline. If the plume centerline lies along the stagnation streamline and if it also lies below H_d, the center of the plume impinges on the hill. If it lies to one side of the stagnation streamline, however, the centerline will pass to one side of the hill.

In CTDM, the position of H_d and the stagnation streamline relative to the centerline of the plume dominates the degree to which a hill in stratified flow is able to alter the peak ground level concentration obtained in the absence of the hill. The H_d surface "slices" the plume

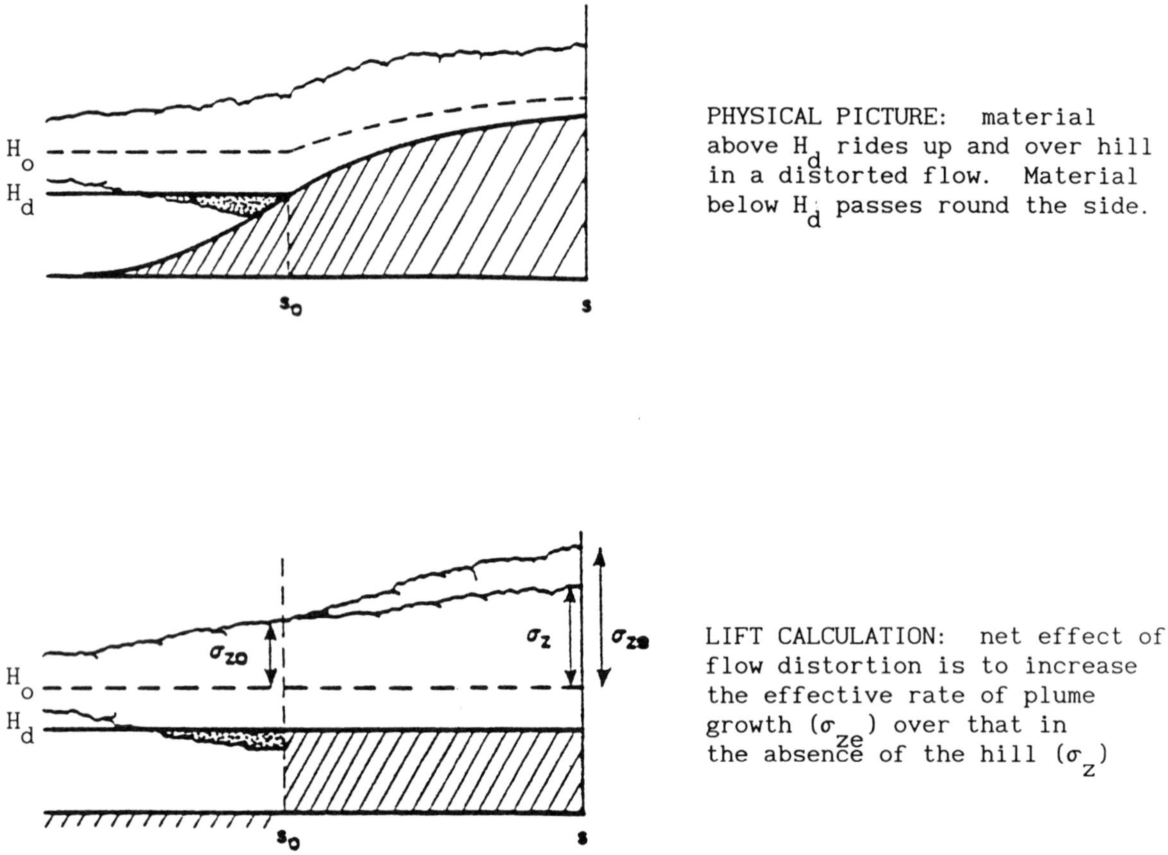

FIG. 6.24. Illustration of terrain effect on the vertical distribution of plume material above H_d as modeled in LIFT (Strimaitis et al. 1989). The initial height of the plume is denoted by H_0.

into two pieces as the hill is encountered. Plume material now residing below H_d is "sliced" once again by the plane of stagnation streamlines.

Figure 6.24 addresses plume material above H_d. The upper portion of the figure is a vertical cross-section of the plume as it travels along the surface. Material below H_d is excluded at s_o, and the remaining material is distorted in the flow and reflected from the surface of the hill. The size of the plume in the vertical at s depends on the amount of distortion in the shape of the plume as well as the amount of additional growth of the plume caused by changes to the rate of diffusion. The lower portion of Fig. 6.24 illustrates how this process is treated within CTDM. Reflection of plume material from the surface $z = 0$ is allowed over the distance from the source to s_o as if the terrain were perfectly flat. Beyond s_o, only the material above H_d is modeled. This material, which is no longer distributed in the vertical as a complete, reflected Gaussian distribution, is bounded by H_d, which is treated as a reflecting surface. Furthermore, the distortion in the flow (and the plume) beyond s_o is accounted for in the model by a coordinate transformation, leaving only the effect of the distortion on the diffusivity in what is termed the effective sigma-z (σ_{ze}). As illustrated, σ_{ze} exceeds σ_z (the plume size in the absence of the hill) because the diffusion in the vertical across streamlines is increased by the contraction in the vertical spacing of the streamlines.

Figure 6.25 addresses plume material below H_d. The diagram on the left illustrates the plume in horizontal cross-section as it splits and flows around a hill. In this case, plume material lies on both sides of the stagnation streamline before the hill is encountered, so plume material is found on both sides of the hill. Once the plume wraps around the leading edge of the hill, the stagnation streamline (which forms the boundary of the hill) becomes a reflecting surface in addition to the plane $z = 0$ and material cannot diffuse from the segment of the plume on one side of the hill to the segment on the other side of the hill. The diagram on the right side of Fig. 6.25 illustrates how the model treats this flow. The actual surface of the hill is replaced by a line in the plume that corresponds to the position of the stagnation streamline. All distortion is eliminated by this transformation, as was done for the flow above H_d. The additional effect of the distortion on lateral diffusion is ignored in the model. The lateral positions of all receptors below H_d essentially collapse onto the stagnation streamline, as depicted by

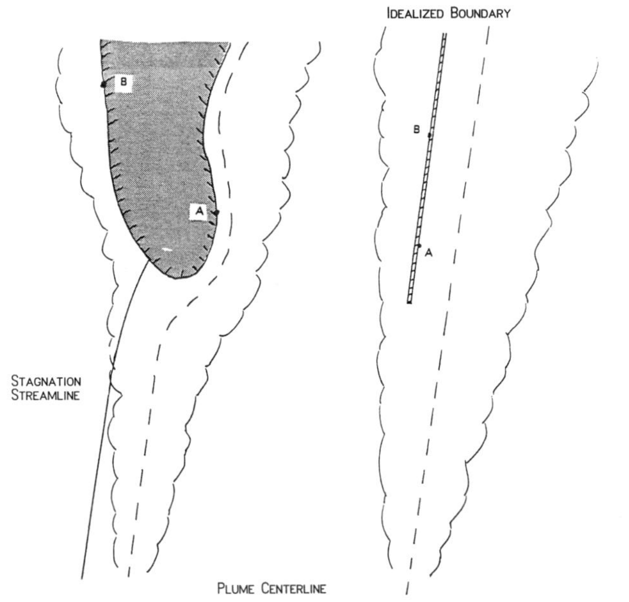

FIG. 6.25. Depiction of plume behavior in CTDM as it is deflected around a hill as seen from above: the stagnation streamline in the left figure separates flow going around the right and left sides of the hill. In CTDM, the hill is treated as being collapsed into an impenetrable wall that separates the flow and plume material going on either side of the hill (Strimaitis et al. 1988).

the points labeled A and B, so that each receptor is the same distance from the plume centerline; however, concentrations along one side of the line differ from those on the other side because diffusion through the line is not allowed.

An important aspect of this formulation of terrain influences in CTDM is that all surfaces from which reflection of plume material is allowed are explicitly maintained parallel to streamlines in the flow. Problems with reconciling simple plume height adjustments in the Gaussian plume equation and the image source method of simulating reflection from sloping terrain are thereby avoided.

6.4.4.1. DISPERSION PARAMETERS

The formulation for σ_z is described in Venkatram et al. (1984). It is based on the form of Taylor's (1921) theorem of diffusion for very short and very long times of travel:

$$\sigma_z = \sigma_w t; \quad t \ll T_L \tag{6.9}$$

$$\sigma_z = (2K_z t)^{\frac{1}{2}}; \quad t \gg T_L, \tag{6.10}$$

where σ_w is the standard deviation of the vertical velocity fluctuations, t the travel time, T_L the Lagrangian time scale, and K_z the eddy diffusivity. Using mixing-length arguments, K_z is defined as

$$K_z = \sigma_w \ell; \quad \ell = \sigma_w T_L \tag{6.11}$$

so that in the limit $t \gg T_L$, Eq. (6.10) becomes

$$\sigma_z = \sigma_w (2tT_L)^{\frac{1}{2}}. \tag{6.12}$$

An interpolation formula spans the gap between small and large times of travel:

$$\sigma_z = \sigma_w t \left[1 + \frac{t}{2T_L}\right]^{-\frac{1}{2}}. \tag{6.13}$$

In Eq. (6.13), σ_w is measured directly, but T_L must be related to other measurements before σ_z can be calculated in the model. This is accomplished by using the empirical flux-profile relationships of surface similarity theory (Businger 1973) to estimate the mixing length, ℓ. The derivation relies on the appropriateness of surface similarity theory and on the assumption that σ_w is proportional to the friction velocity u_*. If σ_w is due mainly to turbulent fluctuations, then the formulation is appropriate.

An equation similar to that used for σ_z is used to compute σ_y as a function of the time of travel and the turbulence velocity scale for lateral fluctuations, σ_v:

$$\sigma_y = \sigma_v t \left[1 + \frac{t}{2T'_L}\right]^{-\frac{1}{2}}. \tag{6.14}$$

The departure of this expression from that for σ_z arises in specifying the functional form of T'_L, which is the Lagrangian time scale of the transverse correlogram. In this case, T'_L cannot be derived from measurements of the mean flow and its statistics. Instead, T'_L is set equal to the time-of-travel required to cover a distance of 10 km. This choice reduces σ_y by about 18% from the value that would be obtained using a linear growth law at a distance of 10 km. All of the CTMD field experiments included sampler locations within 10 km of the source, so that the growth in σ_y beyond 10 km was not documented. Nearly linear growth in σ_y over the first 5 km was typical of many of the experiments.

With T_L and T'_L specified, the following equations are used to calculate the plume spread parameters σ_z and σ_y:

$$\sigma_z = \frac{\sigma_w t}{\left[1 + \sigma_w t\left(\frac{1}{.72z} + \frac{N}{.54\sigma_w}\right)\right]^{1/2}} \tag{6.15}$$

$$\sigma_y = \frac{\sigma_v t}{[1 + ut/20\,000]^{1/2}} \tag{6.16}$$

where N is the Brunt–Väisälä frequency.

6.4.4.2. CONCENTRATION EQUATION FOR LIFT (FLOW ABOVE H_d)

The concentration at a receptor located at (s, y_R, h_R) due to plume material released from $(0, y_s, z_s)$ is obtained from

$$C(s, y'_R, h'_R; s_o) \frac{Qe^{-0.5\left[\frac{y'_R - y_s}{\sigma_{ye}}\right]^2}}{4\pi u \sigma_{ye}\sigma_{ze}} F_z, \quad (6.17)$$

where

$$\begin{aligned} F_z = {} & \exp\left[-0.5\left(\frac{h'_R - z_s + H_d}{\sigma_{ze}}\right)^2\right] \\ & \times \{1 + \text{ERF}(A_z[(z_s - H_d)(\sigma_z^*/T_z)^2 + h'_R\sigma_{zo}^2])\} \\ & + \exp\left[-0.5\left(\frac{h'_R + z_s - H_d}{\sigma_{ze}}\right)^2\right] \\ & \times \{1 + \text{ERF}(A_z[(z_s - H_d)(\sigma_z^*/T_z)^2 - h'_R\sigma_{zo}^2])\} \\ & + \exp\left[-0.5\left(\frac{h'_R + z_s + H_d}{\sigma_{ze}}\right)^2\right] \\ & \times \{1 - \text{ERF}(A_z[(z_s + H_d)(\sigma_z^*/T_z)^2 - h'_R\sigma_{zo}^2])\} \\ & + \exp\left[-0.5\left(\frac{h'_R - z_s - H_d}{\sigma_{ze}}\right)^2\right] \\ & \times \{1 - \text{ERF}(A_z[(z_s + H_d)(\sigma_z^*/T_z)^2 + h'_R\sigma_{zo}^2])\} \end{aligned} \quad (6.18)$$

and

$$A_z = (\sqrt{2}\sigma_{zo}\sigma_{ze}\sigma_z^*/T_1)^{-1}. \quad (6.19)$$

The error function (ERF) is tabulated in many mathematical handbooks. The effective receptor position is denoted as (s, y'_R, h'_R); its relationship to the actual receptor position is described later.

Dispersion parameters in these equations are called "effective" plume spread parameters because they contain a terrain factor that alters the rate of diffusion from that prescribed for flat terrain (see Fig. 6.24). The effective plume spread, accounting for the effect of strain in the flow on the rate at which material diffuses across streamlines, is defined as

$$\begin{aligned} \sigma_{ze}^2 &= \sigma_{zo}^2 + (\sigma_z^*/T_z)^2 \\ \sigma_{ye}^2 &= \sigma_{yo}^2 + (\sigma_y^*/T_y)^2. \end{aligned} \quad (6.20)$$

The quantities σ_z^*/T_z and σ_y^*/T_y that appear in Eqs. (6.18) through (6.20) are obtained from an approximate relation derived from the theory of Hunt and Mulhearn (1973). For axisymmetric strain, the theory for a thin plume embedded in a straining flow shows that

$$\sigma^2(t) = \frac{2}{\beta^2(t)} \int_0^t \beta^2(t')K(t')dt', \quad (6.21)$$

where σ denotes either σ_y or σ_z, β is the strain function for the corresponding component, and K the diffusivity for that component. By assuming that the strain is weak over hills of low slope, this equation leads to

$$\begin{aligned} \left(\frac{\sigma_z^*}{T_z}\right)^2 &= \int_{t_o}^t 2K_z(t') \exp\{-2[T_h(z_m, t') - 1]\}dt' \\ \left(\frac{\sigma_y^*}{T_y}\right)^2 &= \int_{t_o}^t 2K_y(t') \exp\{-2[T_\ell(z_m, t') - 1]\}dt', \end{aligned} \quad (6.22)$$

where t_o is the time of transport to s_o, and T_h and T_ℓ are distortion factors for streamline spacing in the vertical and lateral directions across the flow evaluated at time t' along a streamline whose height above the surface when far upwind of the hill is z_m.

Equations (6.17) through (6.19) provide the framework for estimating concentrations due to plume material that travels up and over a hill. The error functions also arise from treating only the portion of plume material that lies above H_d at s_o. If H_d is zero, then Eq. (6.17) becomes

$$C(s, y'_r, h'_R; s_o) = \frac{Q\exp\left[-0.5\left(\frac{y'_R - y_s}{\sigma_{ye}}\right)^2\right]}{2\pi u \sigma_{ye}\sigma_{ze}} \times \left\{\exp\left[-0.5\left(\frac{h'_R - z_s}{\sigma_{ze}}\right)^2\right] + \exp\left[-0.5\left(\frac{h'_R + z_s}{\sigma_{ze}}\right)^2\right]\right\} \quad (6.23)$$

which, with the exception of alterations in the plume size and in the distance between the plume centerline and the receptor, is the familiar Gaussian plume equation for the concentration at an elevated receptor.

The receptor position is denoted as (h'_R, y'_R), where the primes indicate that the distance of the receptor from the centerline of the plume has been altered by the presence of the hill. Upwind of the hill, streamlines are straight and parallel. Over the hill, the spacing is variable and the streamlines are subject to deflections. Because concentrations are computed for a parallel flow in which the influence of terrain is manifested in the H_d-partition of plume material, in the altered rates of diffusion, and in the altered streamline that passes through the receptor, the chief task in computing (h'_R, y'_R) is in identifying the streamline that passes through the source, and the streamline that passes through the receptor. The position of these two streamlines in the undisturbed flow upwind of the hill defines the effective receptor location relative to the centerline of the plume. These streamlines are obtained from a flow model.

The flow model is formulated as a "backwards-looking" solution. It is designed to compute the deflection experienced by a streamline that passes through a given point over the hill. If the given point is a receptor, then the model will compute the deflection experienced by the streamline that passes through that receptor. Knowing the actual position of the receptor, the deflections allow h'_R and y'_R to be computed.

The flow algorithm uses the linearized equations of motion for steady-state Boussinesq flow (Smith 1980) and constant u and N. An approximate solution is obtained for a hill of Gaussian shape in cross section. Comparisons of streamline positions obtained from the algorithm with those measured over the crest of a bell-shaped (polynomial) hill in the FMF tow tank (Snyder et al. 1985) are illustrated in Fig. 6.26. These results indicate that the linear theory performs best for neutral (unstratified) flow. Lateral deflections near the surface of the hill are progressively underestimated as the Froude number approaches unity.

6.4.4.3. CONCENTRATION EQUATION FOR WRAP (FLOW BELOW H_d)

Concentrations due to plume material that lies below H_d at the front of a terrain feature (s_o) are obtained from

$$C(s, 0, z_R; s_o) = \frac{Q}{4\pi u \sigma_y \sigma_z} \exp\left[-0.5\left(\frac{d}{\sigma_y}\right)^2\right] \times \left[1 + \text{sign}(y_R)\,\text{ERF}\left(\frac{d\sigma_y^*}{\sqrt{2}\,\sigma_{yo}\sigma_y}\right)\right] \times \left\{B_1 \exp\left[-0.5\left(\frac{z_s - z_R}{\sigma_z}\right)^2\right] + B_2 \exp\left[-0.5\left(\frac{z_s + z_R}{\sigma_z}\right)^2\right]\right\}. \quad (6.24)$$

The factor sign(y_R) denotes the sign of the receptor position in the coordinate system with x-axis aligned with the flow, and it signals whether or not the plume centerline passes around the side of the hill where the receptor lies. The factors B_1 and B_2 are given by

$$B_1 = \text{ERF}\left(\frac{b_1 - b_2 - b_3}{b_o}\right) + \text{ERF}\left(\frac{b_1 + b_2 + b_3}{b_o}\right)$$

$$B_2 = \text{ERF}\left(\frac{b_1 - b_2 - b_3}{b_o}\right) + \text{ERF}\left(\frac{b_1 + b_2 + b_3}{b_o}\right), \quad (6.25)$$

where

$$b_0 = \sqrt{2}\,\sigma_z\sigma_{zo}\sigma_z^* \quad b_1 = H_d\sigma_z^2 \quad b_2 = z_R\sigma_{zo}^2 \quad b_3 = z_s\sigma_z^{*2}.$$

The subscript R denotes the receptor location, and the subscript s denotes the source location. Here the distance from the stagnation streamline associated with the mean wind direction to the centerline of the plume is denoted as d, the total spread of the plume is denoted as σ_z and σ_y, the amount of plume spread experienced over the distance to the stagnation point is denoted as σ_{zo} and σ_{yo}, and the rate of plume growth beyond the stagnation point to the receptor is denoted as σ_z^* and σ_y^*, where

$$\sigma_z^{*2} = \sigma_z^2 - \sigma_{zo}^2 \quad \text{and} \quad \sigma_y^{*2} = \sigma_y^2 - \sigma_{yo}^2. \quad (6.26)$$

The ellipse used to estimate the flow below H_d is taken from the horizontal cross-section of the hill at the minimum of the following two elevations: either the elevation of the centerline of the plume, or the elevation of H_d. In this way the shape of the hill selected is associated with the peak concentration of plume material found within the layer of fluid below H_d. Analytical solutions for two-dimensional potential flow around the ellipse are used to identify the streamline through the source and the stagnation streamline, and thereby obtain the offset distance d. Note that the stagnation streamline forms the boundary of the hill surface in the horizontal cross-section.

6.4.5. Evaluation of CTDM

The model evaluation portion of the CTMD program tested how well CTDM performed relative to RTDM (in both its "default" and "full on-site" modes) and COMPLEX I. Five datasets were used in the evaluation: CCB (SHIS #1), HBR (SHIS #2), Tracy (FSPS), the Westvaco pulp and paper mill dataset from Luke, Maryland (Wackter and Londergan 1984), and the TVA Widows Creek Steam Plant dataset (Egan et al. 1985). For each of the datasets, three sets of meteorological data were prepared, reflecting varying degrees of use of on-site data. The preferred dataset makes use of all available data. Two "degraded" datasets make use of data from only two measurement levels or one measurement level on the main tower(s). These degraded sets were designed to evaluate how well CTDM performs with limited one-site data. COMPLEX I and RTDM were run with a single level of meteorology near stacktop.

The results obtained for peak hourly averaged concentrations, paired in time, but not in space, exhibit the following characteristics (Strimaitis et al. 1988):

> The CCB evaluation results show a significant decline in CTDM performance toward underprediction for both SF_6 and CF_3Br as the meteorological input is degraded. CTDM results for the full meteorological dataset are quite good, however. COMPLEX I overpredicts for both tracers, but by less than a factor of 2. RTDM in on-site mode shows a modest underprediction tendency, while in default mode it overpredicts for SF_6 and underpredicts for CF_3Br.
>
> The HBR results indicate that CTDM overpredicts for SF_6 (except for the run with one tower level), but underpredicts for CF_3Br. COMPLEX I shows a large overprediction for SF_6 (released usually above H_d). These differ-

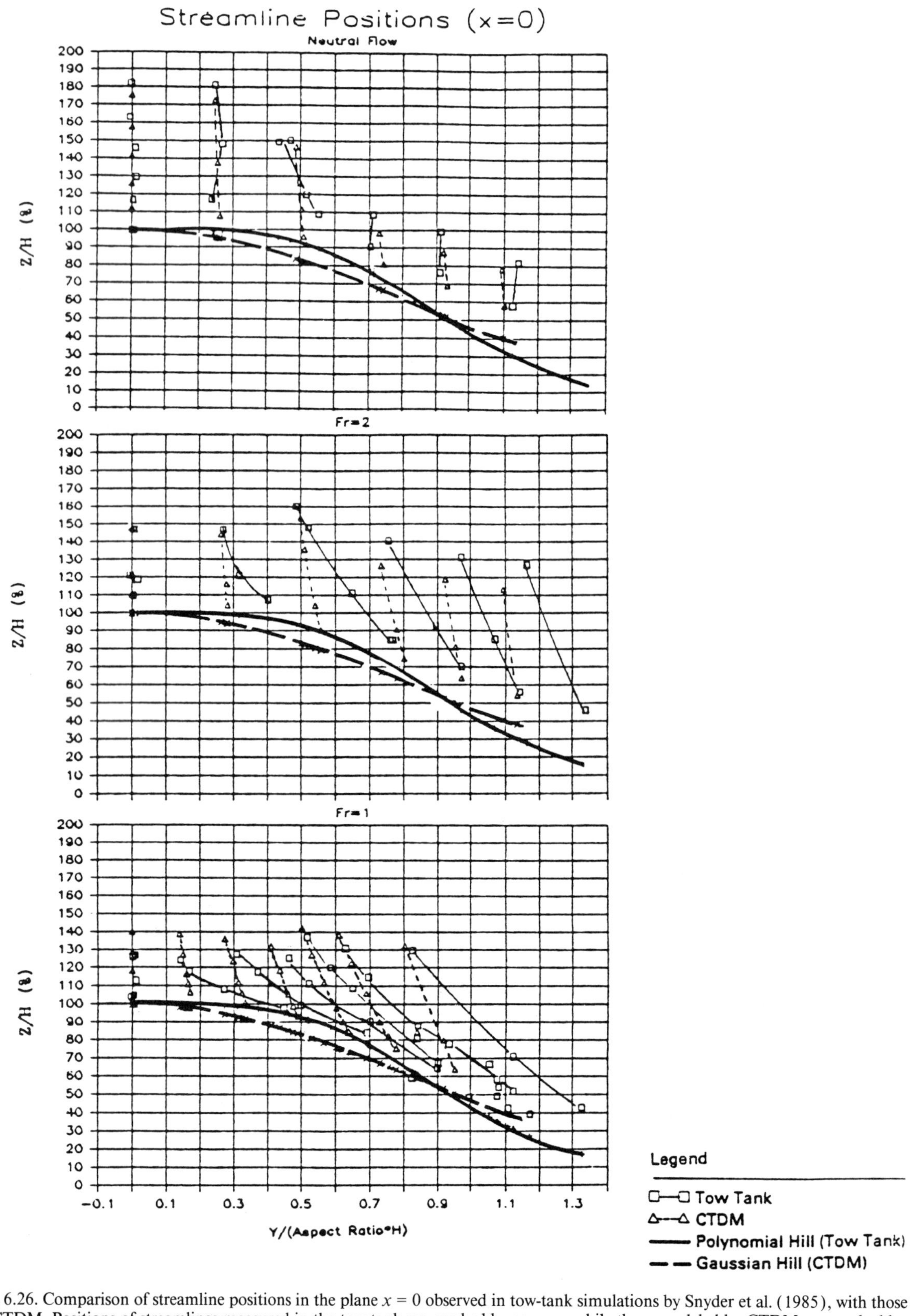

FIG. 6.26. Comparison of streamline positions in the plane $x = 0$ observed in tow-tank simulations by Snyder et al. (1985), with those obtained from CTDM. Positions of streamlines measured in the tow tank are marked by squares, while those modeled by CTDM are marked by triangles. Streamlines that originated at the same lateral position are joined by either a thin line (tow tank) or a thin dashed line (CTDM). Note that the polynomial hill used in the tow tank (heavy solid line) differs from the Gaussian hill used by CTDM (heavy dashed line) (Strimaitis et al. 1988).

ences reveal the inability of COMPLEX I to distinguish between flows passing over or around a terrain obstacle rather than impinging on the obstacle. RTDM results for the on-site mode are very good for both tracers, showing a modest tendency toward overprediction. The default mode exhibits a large tendency toward overprediction. Note that CTDM's relatively good performance with the HBR SF_6 dataset (not used in the model development) is consistent with its performance at other CTMD sites. This result suggests that CTDM's relatively good evaluation results at the CTMD sites are due to theoretically sound design rather than from "tuning" and calibration.

The FSPS (Tracy Power Plant) results are listed in Table 6.3, showing that CTDM has predicted-to-observed ratios close to 1.0 and a high number of cases with predictions within a factor of 2 of observations. COMPLEX I shows large overpredictions while RTDM (default) overpredictions are more modest. For this site, RTDM (on-site mode) shows a moderate underprediction tendency. Results for the models are consistent between the two tracers.

For the Westvaco Luke database, the results show good performance for both CTDM (all tower levels) and RTDM (on-site mode) (see Table 6.4). With fewer tower levels, CTDM overpredicts significantly, probably due to the use of unrepresentative or default $d\theta/dz$ values that are too high. Overpredictions are quite apparent for RTDM default ($d\theta/dz$ too high) and especially COMPLEX I ($d\theta/dz$ too high and no plume lifting over terrain in stable conditions).

As in the case for Westvaco, the CTDM results show more overpredictions at Widows Creek as the meteorological data input is degraded. The best results are attained by CTDM (all tower levels) and RTDM (on-site mode), while RTDM (default) and COMPLEX I overpredict for the same reasons as stated above for Westvaco.

When these modeling results were characterized in terms of a skill score, CTDM (with all tower data) and the full on-site mode of RTDM displayed a clear superiority in modeling peak hourly concentrations. This finding is attributed to the use of site-specific values of turbulence (to characterize plume growth), stratification, and the use of H_d values derived from full vertical profiles of wind speed and stratification.

6.5. Mesoscale flow models that include diffusion algorithms

6.5.1. *General principles*

Most of the information in the preceding sections is concerned with diffusion in rugged terrain at scales of a few kilometers or less. This "microscale" region is of interest because maximum ground level concentrations are likely to occur at those distances. Theoretical approaches and models at those ranges are based on relatively simple similarity concepts such as the Gaussian plume model or the critical dividing-streamline approach. These models usually assume that the flow streamlines are straightforward and put nearly all of their emphasis on the first terrain obstacle encountered by the plume. This set of models, however, does not adequately handle the time and space-varying meteorological fields that commonly occur on the mesoscale over rugged terrain. Here we assume that mesoscale is defined by distance scales ranging from

TABLE 6.3. Summary statistics for data paired in time (not in space) at Tracy Power Plant (from Strimaitis et al. 1988).

Data subset	Average hourly peak value	Ratio of PRE/OBS	rms errors	% Cases: 0.5 < PRE/OBS < 2.0
Tracer: SF_6 (highest value from each hour used), 111 hours				
Observed	1.96	—	—	—
CTDM, several tower levels	1.94	0.99	2.11	61
CTDM, two tower levels	2.07	1.07	2.24	49
CTDM, one tower level	2.41	1.24	3.08	38
COMPLEX I	6.14	3.13	7.27	19
RTDM, default mode	3.05	1.56	3.38	48
RTDM, full onsite mode	1.18	0.60	2.22	34
Tracer: CF3Br (highest value from each hour used), 111 hours				
Observed	2.84	—	—	—
CTDM, several tower levels	2.46	0.87	2.94	60
CTDM, two tower levels	3.02	1.06	4.01	53
CTDM, one tower level	3.04	1.07	4.99	31
COMPLEX I	8.54	3.01	9.14	23
RTDM, default mode	3.76	1.32	4.79	39
RTDM, full onsite mode	1.95	0.69	3.17	52

TABLE 6.4. Summary statistics for data paired in time (not in space) at Westvaco Luke (from Strimaitis et al. 1988). Units ($\mu s\ m^{-3}$); 1-hour averages.

	Tracer (SO_2) (Highest value from each hour used)			
Data subset	Average hourly peak value	Ratio of PRE/OBS	rms errors	% cases: 0.5 < PRE/OBS < 2.0
Observed	0.33	—	—	—
CTDM, several tower levels	0.29	0.88	0.83	14
CTDM, two tower levels	1.60	4.85	3.30	19
CTDM, one tower level	1.70	5.15	2.00	15
COMPLEX I	4.12	12.48	9.44	14
RTDM, default mode	0.98	2.97	2.46	16
RTDM, full onsite mode	0.21	0.64	0.82	15

about 10 km to about 1000 km. Distance scales larger than about 1000 km would be considered to be synoptic scales.

Paegle et al., Chapter 10, provide an extensive discussion of mesoscale meteorological modeling. There are several such models in existence, and most have the following characteristics:

- Horizontal domain size is on the order of 100 km and the vertical domain size is from the ground up to about 10 km.
- The horizontal grid contains 30 × 30 grid points, and there are about 10 levels in the vertical. With a time step of about 1 minute, these models therefore require the services of as fast a computer as possible with as much storage as possible.
- Given initial and boundary conditions, the model solves the equations of motion, continuity, and enthalpy for each grid point and each time step.

These models have not been thoroughly compared with detailed meteorological data, but can be said to produce wind, temperature, and turbulence distributions that are "not counterintuitive." Various well-known phenomena are simulated, such as channeling, downslope winds at night and upslope winds during the day, stagnation in isolated valleys, sea and land breeze cycles, and lee waves.

The diffusion of pollutants on the mesoscale can be simulated by assuming that the equation for the conservation of pollutant mass is independent of the other governing equations. Consequently, once the time-variable three-dimensional field of wind, temperature, and turbulence is known, then the pollutant mass conservation equation can be applied in the same manner as if these variables were observed in a large field experiment. In general the pollutant is assumed to move with the wind and diffusion at a rate proportional to the local turbulent velocities. Two approaches for calculating diffusion used in mesoscale models are described in subsections 6.5.1.1 and 6.5.1.2.

6.5.1.1. APPROACH 1—LAGRANGIAN PARTICLE DIFFUSION

Segal et al. (1988a), Yamada (1985), Gross et al. (1987), and Etling et al. (1986) use the Lagrangian particle diffusion approach in their mesoscale models. Segal et al. (1988a) briefly describe the equations that are discussed in more detail by Pielke et al. (1983). In this approach, a large number (about 1000) of "particles" are released at the source location in the model domain and their trajectories are calculated using a "random walk" method. After all the trajectories are calculated, the numbers of particles that are in each grid volume at each time are counted in order to estimate the concentration field as a function of time and space. The three-dimensional particle trajectory is given by the formula (from Segal et al. 1988a):

$$x_i(t + dt) = x_i(t) + (u_i(t) + u_i'(t))dt \qquad i = 1, 2, 3, \tag{6.27}$$

where $x_i(t)$ is the old x, y, and z position of the particle and $x_i(t + dt)$ is its new position after time interval dt was passed; u_i is the u, v, and w velocities computed by the mesoscale meteorological model, and u_i' is turbulent velocity fluctuations u', v', and w' parameterized statistically using the boundary-layer predictions of the mesoscale meteorological model.

The turbulent components are given by the relation (from Segal et al. 1986)

$$u_i'(t) = u_i'(t - dt)R_i(dt) + u_i'' + \delta_{i3}(1 - R_i(dt))w_d \qquad i = 1, 2, 3, \tag{6.28}$$

where $R_i(dt)$ is the Lagrangian autocorrelation function at time dt and u_i'' is a random component with a mean of zero and a variance defined by $(1 - R_i^2(dt))\sigma_i^2$. The variance σ_i^2 is the local turbulence variance of component i. The final term is a correction suggested by Legg and Raupach (1982) for cases where there is a vertical gradient in the turbulence variance; they suggest that the "drift velocity" w_d should be given by the formula

$$w_d = T_{Lz}\partial\sigma_w^2/\partial z, \tag{6.29}$$

where T_{Lz} is the Lagrangian integral time scale in the vertical. Also note that in the last term, δ_{i3} is the Kronecker delta operator.

The Lagrangian autocorrelation function is usually approximated as an exponential:

$$R_i(dt) = \exp(-dt/T_{Li}) \qquad i = 1, 2, 3, \tag{6.30}$$

where T_{Li} is the Lagrangian integral time scale in the i direction. Hanna et al. (1982) provide some suggestions for this time scale, which can be assumed to be a constant equal to about 10^3 or 10^4 seconds for the x and y directions, and approximately equals z/σ_w in the vertical direction.

An example of the particle distribution calculated by Segal et al. (1988a) for an instantaneous release is given in Fig. 6.27. The particles are seen to be transported and dispersed out of the valley. The particle distribution calculated by Yamada (1985) for a continuous release is given in Fig. 6.28. These distributions "look" correct when viewed by eye, but it remains to be seen whether they agree with field observations.

6.5.1.2. Approach 2—Use of diffusion equation

In a second approach, the diffusion equation (i.e., the equation for conservation of pollutant mass) can be written in the same format as the equations of motion and other thermodynamic equations used in the mesoscale meteorological model:

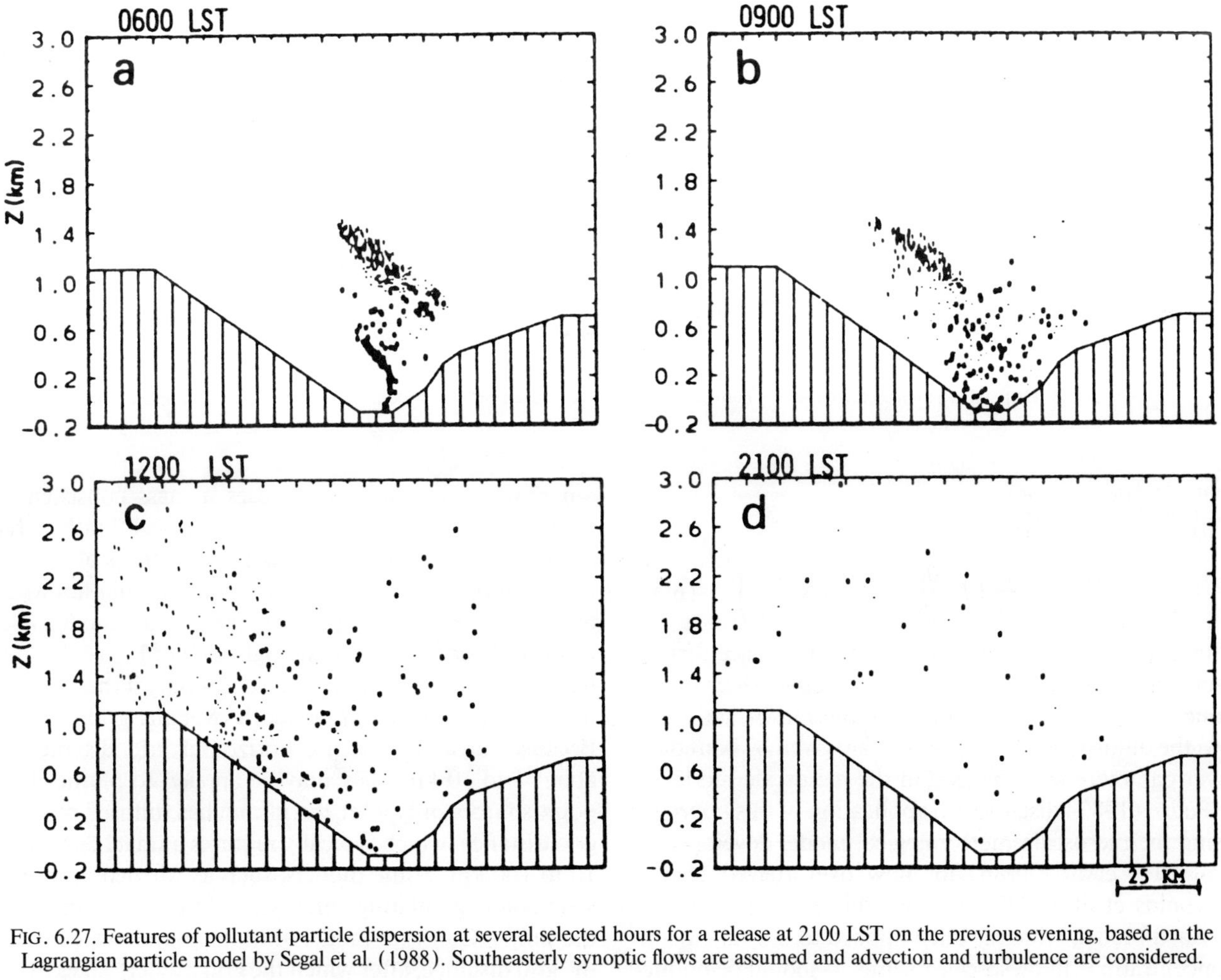

FIG. 6.27. Features of pollutant particle dispersion at several selected hours for a release at 2100 LST on the previous evening, based on the Lagrangian particle model by Segal et al. (1988). Southeasterly synoptic flows are assumed and advection and turbulence are considered.

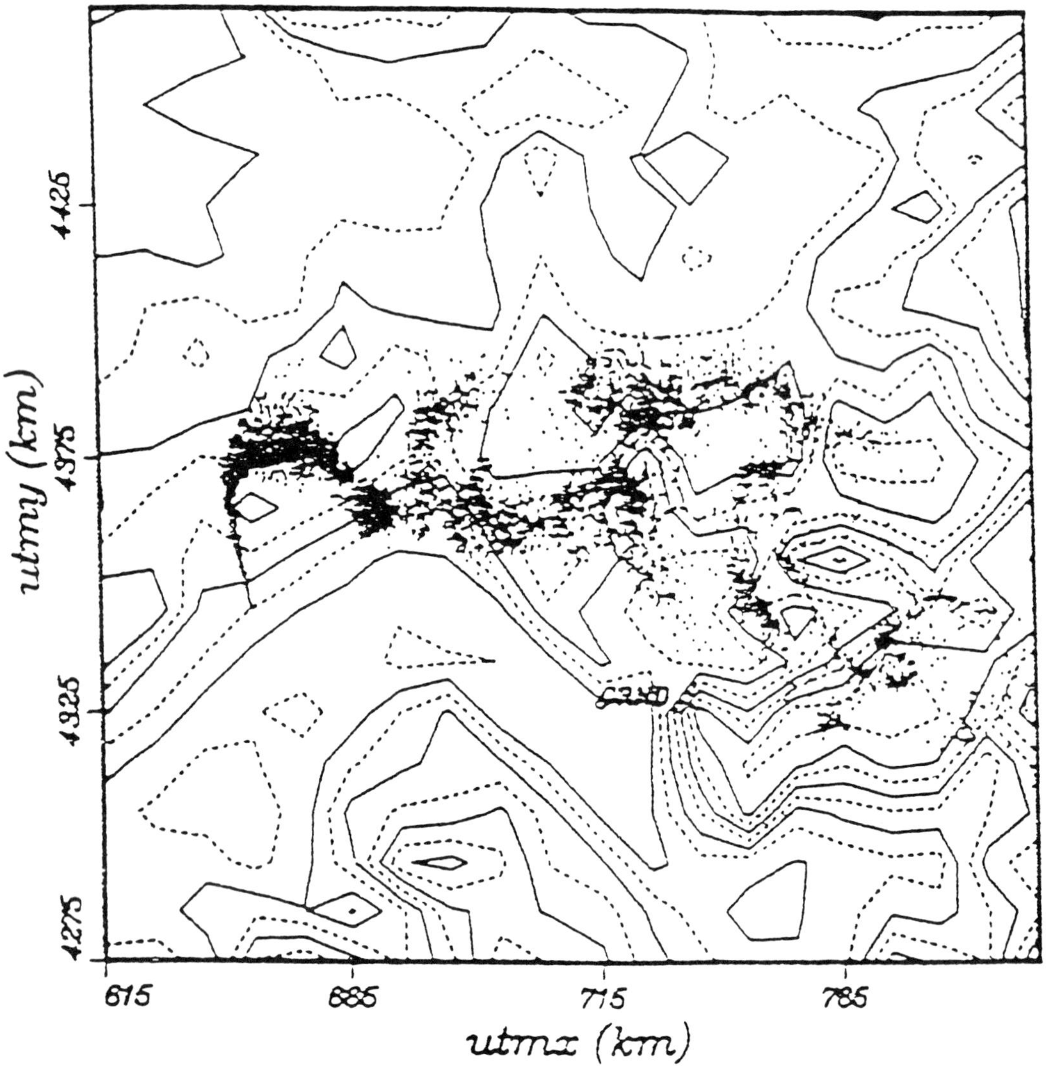

FIG. 6.28. Dispersion pattern calculated by Yamada's (1985) Lagrangian particle model, for a rugged terrain application.

$$u\frac{\partial C}{\partial x}+v\frac{\partial C}{\partial y}+w\frac{\partial C}{\partial z}$$
$$=\frac{\partial}{\partial x}\left(K_x\frac{\partial C}{\partial x}\right)+\frac{\partial}{\partial y}\left(K_y\frac{\partial C}{\partial y}\right)+\frac{\partial}{\partial z}\left(K_z\frac{\partial C}{\partial z}\right). \quad (6.31)$$

This equation can be applied in finite difference form at each grid point, where it is assumed that the velocity components u, v, and w are given by the mesoscale model and the diffusivities K_x, K_y, and K_z are given by boundary-layer parameterizations used in the mesoscale model.

Padro (1987) discusses an application of this approach using the mesoscale meteorological model developed by Walmsley et al. (1986) for flow over rugged terrain. Reynolds et al. (1979) also use this approach in their Airshed model, which is often applied to estimate ozone concentrations in mesoscale regions. It should be pointed out that the Airshed model uses a mass consistent or smoothed version of an observed wind field, rather than solving for the wind field using the equations of motion.

This approach is valid only for pollutant plumes whose dimensions exceed the grid size used in the model. That is, the diffusion terms on the right side of Eq. (6.31) may be approximated by finite differences only if the gradients in concentration, C, are smooth over the grid distance. Because most models use horizontal grid distances of about 1 to 10 km, they may be applied to plumes from point sources only after the plume has diffused over this grid distance, which typically requires a travel distance of 10 to 100 km. Some models, such as Airshed, have provisions for calculating small-scale diffusion by means of the simple Gaussian model until the plume scale exceeds the grid distance, after which the grid model "takes over."

The MATHEW/ADPIC model (Sherman 1978; Lange 1978) is a hybrid model based on the diffusion equation. The MATHEW model generates a three-dimensional mass-consistent gridded mean wind field including the effects of terrain using observed winds and information on topography. The ADPIC model uses these winds fields to simulate dispersion through a particle-in-cell model that combines the principles of Lagrangian particle models with those of the diffusion equation. It rewrites Eq. (6.31) in the form

$$\frac{\partial C}{\partial t} + \nabla\cdot(C\mathbf{U}_p) = 0, \qquad (6.32)$$

where $\mathbf{U}_p$ is a "pseudovelocity" defined as the sum of the mean wind velocity $\mathbf{U}_A$ (from MATHEW) and a diffusive velocity $\mathbf{U}_D$

$$\mathbf{U}_p = \mathbf{U}_A + \mathbf{U}_D$$

$$\mathbf{U}_D = -K\cdot(\nabla C)/C \qquad (6.33)$$

and K is the diffusivity in the x, y, z directions [as in Eq. (31)]. The diffusivities K_x and K_y are functions of the size of the plume, the lateral turbulence energy, and the Lagrangian time scale. The vertical diffusivity K_z is a function of boundary-layer stability, vertical turbulence energy, and height.

6.5.2. *Evaluation of mesoscale grid-based diffusion models in rugged terrain*

The mesoscale grid models have not been evaluated in the same comprehensive fashion as the simple Gaussian-based models for diffusion in rugged terrain (see section 6.2). A few limited evaluations have been completed, such as that reported by Tesche et al. (1987). In that report, data from tracer experiments at the Geysers site in Northern California were used to test the predictions of four grid-based diffusion models.

The following models were evaluated:

- MATHEW/ADPIC (Lange 1978). Described in section 6.5.1.
- Hybrid model. This model is an adaptation of the Airshed model (Reynolds et al. 1979), using the diagnostic wind-flow model developed by Yocke and Liu (1978). Subgrid scale diffusion is approximated by a Gaussian plume model. The model can simulate complex photochemical reactions, although this capability was not tested by the inert tracer study used in the model evaluation.
- SMOG model (Fabrick et al. 1977). This is a revised version of the IMPACT model that was included in some of the evaluations by Wackter and Londergan (1984a,b) reported in section 6.2. It consists of five submodels: 1) wind field generator, 2) pollutant transport, 3) diffusivity field development, 4) plume rise calculation, and 5) chemical transformations.
- 2D FLOW model (Garrett and Smith 1984). This is a dynamic model for drainage flow systems that includes a Lagrangian particle diffusion model.

The data consist of measurements from 16 tracer tests and one historical H_2S episode at the Geysers site. The topography of the region was defined on a 250 m grid over a domain 25 × 40 km. The experiments were conducted on a smaller domain (about 7 × 10 km), which is depicted in Fig. 6.29.

The results of the model evaluation exercise lead to the following general conclusions:

- On the average, the models tend to underpredict by about a factor of 2 for the nocturnal drainage experiments.
- With the exception of ADPIC, the models tend to overpredict the average concentration by a factor of 3 to 4 in the limited mixing experiments.
- The average absolute deviation, $\overline{|C_o - C_p|}$, is approximately equal to the mean observed concentration.
- Correlations can range anywhere from −0.13 to 0.81, with a median of about 0.36.
- Looking only at the maximum concentrations, the models underpredict by about a factor of 3 or 4 for the nocturnal drainage experiments, and overpredict by about a factor of 2 for the limited mixing experiments.

It is difficult to conclude whether any one of these models is a significant improvement over the others. We can conclude that, for this set of experiments, the expected error of any of them is about a factor of 2 to 4, with correlation coefficients in the range of 0.0 to 0.5.

6.6. Summary of findings and recommendations

This chapter has provided an overview of the objectives and results of several field investigations, laboratory investigations, and theoretical analyses of the effects of rugged terrain on diffusion. Since the multiyear and multimillion dollar EPA CTMD and DOE ASCOT programs are nearly completed, it is appropriate to summarize their major findings and assess the needs for future research in this area.

- It is possible to satisfactorily model diffusion using generalized model algorithms only in certain idealized terrain situations such as symmetrical hills and ridges or uniform linear slopes or valleys. Furthermore, only the first impact of the plume on the nearest hill is usually modeled. In these situations, predictions of models such as CTDM are within about ±30% of the mean and maximum observed concentrations; root mean square errors are approximately equal to the mean, and correlations are about 0.50.

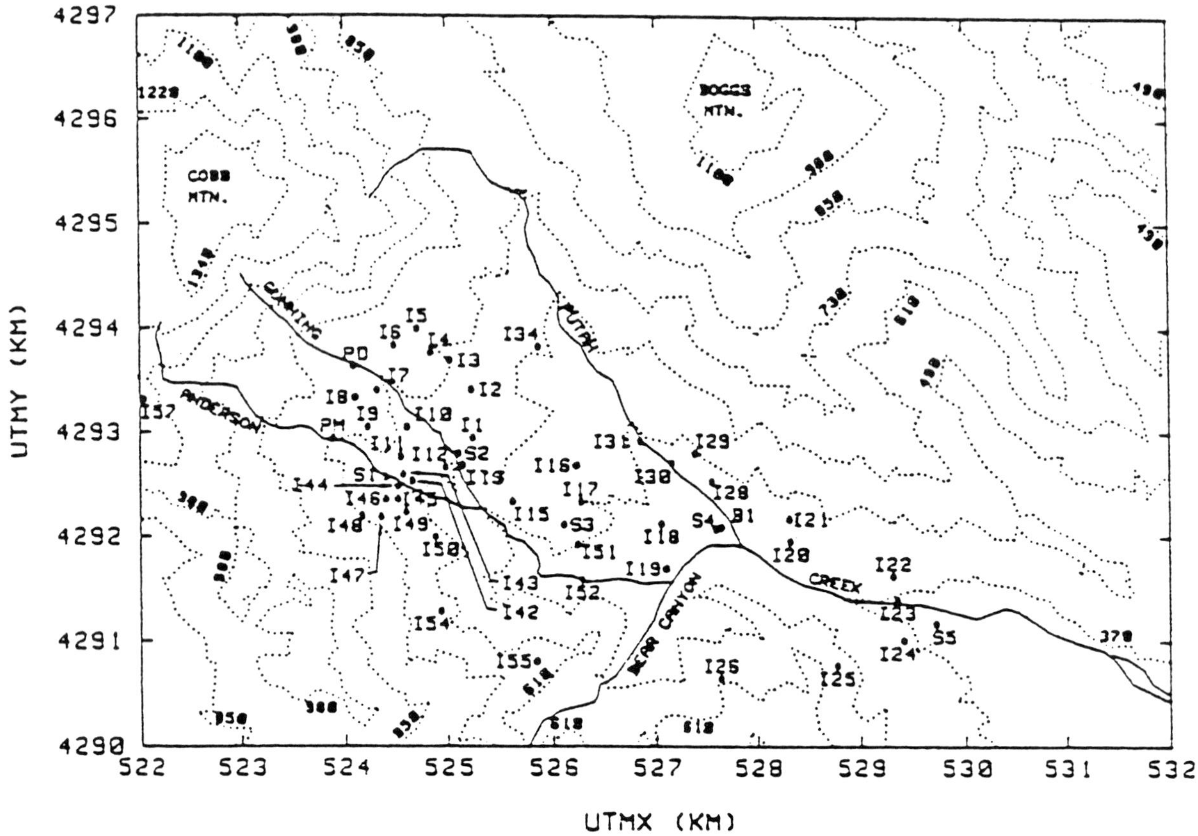

FIG. 6.29. Location of perfluorocarbon sources and samplers during 1980 experiments at the Geysers site, California (from Tesche et al. 1987).

• Model accuracy is closely tied to the availability of accurate, representative input data. Most models can employ only a single wind observation. Wind fields in complex terrain are highly variable both vertically and laterally, and it is important to observe the wind at the elevation and lateral position where the plume is expected to be.

• There is growing recognition of the inherent uncertainty in any kind of atmospheric modeling, due to the effects of stochastic turbulent fluctuations. These fluctuations contribute to model uncertainties even in flat terrain dispersion experiments and model applications with steady-state conditions and simple sources.

• Most of the models discussed in this chapter are intended for application to dispersion near the terrain obstacle that is closest to the source. With the exception of the mesoscale models, the models are not intended for application in the presence of many hills or valleys. Furthermore, the models contain no wake algorithms for simulating the mixing and recirculation found in cavity zones in the lee of a hill. These research areas should receive attention in the future, so that the straight-line Gaussian models can be extended to broader applications.

• Interactions of terrain-generated wind fields with sea breezes or other mesoscale flows have received limited attention in mesoscale models. At many locations, it is important to account for these external features. Field observations at a complex terrain site influenced by other complicating mesoscale effects should be planned and used to improve mesoscale models.

• The CTMD and ASCOT programs emphasized stable conditions, since those conditions are most likely to result in high concentrations. These analyses and models should be extended to convective daytime situations, using the convective scaling models that have recently been successfully applied in flat-terrain situations.

• Not all of the data from field experiments conducted during the past 10 years have been analyzed. Before new expensive field experiments are begun, the existing data from previous field experiments should be thoroughly studied.

Acknowledgments. S. R. Hanna acknowledges the support of the Department of Energy, the National Oceanic and Atmospheric Administration, the Electric Power Research Institute, and the Westvaco Corporation. D. G.

Strimaitis appreciates the support of the EPA throughout the course of the Complex Terrain Model Development Program. Both authors acknowledge the typing skills of N. Kennedy and T. Willard of Sigma Research Corporation.

APPENDIX

List of Symbols

Symbol	Units	Definition
C	$(\mathrm{g\,m^{-3}})$	concentration
C_o	$(\mathrm{g\,m^{-3}})$	observed concentration
C_p	$(\mathrm{g\,m^{-3}})$	predicted concentration
e	$(\mathrm{m^2\,s^{-2}})$	turbulent kinetic energy
$\mathrm{Fr} = u/NH$		Froude number
g	$(9.8\ \mathrm{m\,s^{-2}})$	acceleration of gravity
h	(m)	thickness of drainage flow
h_R	(m)	vertical position of receptor
H	(m)	hill height
H_d	(m)	dividing-streamline height
H_0	(m)	height of plume centerline above flat terrain
H_s	(m)	height of source
K	$(\mathrm{m^2\,s^{-1}})$	eddy diffusivity
K_n	$(\mathrm{m^2\,s^{-1}})$	eddy diffusivity
K_x	$(\mathrm{m^2\,s^{-1}})$	along-wind eddy diffusivity
K_y	$(\mathrm{m^2\,s^{-1}})$	lateral eddy diffusivity
K_z	$(\mathrm{m^2\,s^{-1}})$	vertical eddy diffusivity
ℓ	(m)	turbulent mixing length
L	(m)	Monin–Obukhov length
n	(m)	direction perpendicular to slope
$N = [(g/T)(d\theta/dz)]^{1/2}$	$(\mathrm{s^{-1}})$	Brunt–Väisälä frequency
Q	$(\mathrm{gm\,s^{-1}})$	pollutant emission rate
Q_H	$(\mathrm{m^2\,s^{-3}})$	sensible heat flux
R		autocorrelation function
s	(m)	direction along fall-line of slope
t	(s)	time
T	(K)	temperature
T_h		distortion factor for streamline spacing in vertical
T_ℓ	(s)	distortion factor for streamline spacing in lateral
T_L	(s)	Lagrangian time scale for vertical dispersion
T'_L	(s)	Lagrangian time scale for lateral dispersion
u	$(\mathrm{m\,s^{-1}})$	wind speed
u'_i	$(\mathrm{m\,s^{-1}})$	turbulent velocity component autocorrelation function
u_*	$(\mathrm{m\,s^{-1}})$	friction velocity
v	$(\mathrm{m\,s^{-1}})$	lateral wind speed
w	$(\mathrm{m\,s^{-1}})$	vertical wind speed
w_d	$(\mathrm{m\,s^{-1}})$	drift velocity correction for Lagrangian particle models
x	(m)	along-wind distance
y_R	(m)	lateral position of receptor
y_s	(m)	lateral position of source
z	(m)	height
z_o	(m)	surface roughness
z_s	(m)	vertical position of source
β	(radians)	angle of slope
ρ	$(\mathrm{g\,m^{-3}})$	air density
σ_v	$(\mathrm{m\,s^{-1}})$	lateral turbulence standard deviation
σ_w	$(\mathrm{m\,s^{-1}})$	vertical turbulence standard deviation
σ_y	(m)	lateral dispersion parameter
σ_z	(m)	vertical dispersion parameter
θ	(K)	potential temperature

CHAPTER 7

Fluid Dynamics of Flow over Hills / Mountains—Insights Obtained through Physical Modeling

ROBERT N. MERONEY

Fluid Dynamics and Diffusion Modeling Laboratory, Colorado State University, Fort Collins, Colorado

ABSTRACT

Physical modeling studies of atmospheric flow over hills and mountains span 60 years of research in more than seven countries. Early work in the 1920s led to evaluation of air currents around the Rock of Gibraltar for airfield safety. Water channel work by Long in the 1950s examined lee-wave development over the Sierra Nevada of California. During the 1970s and 1980s laboratory measurements examined speedup over idealized two- and three-dimensional hills and escarpments for wind-power climatology purposes. Since the mid-1960s many fluid-model tests have examined transport and dispersion of pollutants over complex terrain. Considered together there is a wealth of information about the influence of hills and mountains on streamline patterns, speedup, turbulence, and separation under many topographic and atmospheric conditions. This chapter will consider the advantages and limitations of physical modeling, review the principle conclusions drawn through physical modeling, and suggest experiments that might result in a better understanding of the flow processes over complex terrain.

7.1. Introduction

Meteorologists are often faced with problems requiring quantitative estimates of airflow patterns and turbulence characteristics over complex terrain. Use of wind flow information includes air pollution zoning, prediction of smoke movement from forest fires or slash burning, mine tailing dust dispersal, estimation of the movement of vegetative disease vectors or pests, and the siting of fossil fuel burning industrial facilities or power plants. There are many practical difficulties in obtaining useful results, whether by analytical, numerical, or field investigation means. Thus it is natural to explore the possibilities of simulating flow and diffusion over irregular terrain by physical model experiments on the laboratory scale.[1]

Successful modeling of some of the more complex atmospheric surface layer phenomena in wind tunnels and water channels dates back over 50 years. Guidelines for modeling flow over complex terrain are essentially the same as those for modeling around buildings and obstacles; nonetheless, a few unique features are reviewed here briefly. Characteristics of neutral flow over hills, ramps, and escarpments are discussed, and deductions from stratified and drainage flow experiments are noted. Principal conclusions drawn through physical modeling are summarized, and experiments that might result in a better understanding of the flow processes over complex terrain are proposed.

[1] Sometimes also called "fluid" modeling or "hydraulic" modeling.

7.1.1. Advantages and disadvantages of fluid modeling

There are three succinct reasons physical modeling retains its value in meteorological analysis of terrain flows. First, fluid modeling does some things much better than current analytic and numerical alternatives; e.g., it provides realistic subgrid scale information about surface fluxes, separation, and three-dimensional flow interactions. Second, wind tunnels are in effect analog computers that have the advantage of near-infinitesimal resolution and near-infinite memory. A fluid modeling study employs real fluids not models of fluids; hence, the fluid model is implicitly nonhydrostatic, non-Boussinesq, compressible, and it includes variable fluid properties, nonslip boundary conditions, and dissipation. Real fluids permit flow separation and recirculation. All conservation equations are automatically included in their correct form in a laboratory model without truncation or differencing errors, and there are no missing terms or approximations. Third, the fluid model bridges the gap between the fluid mechanician's analytical or numerical models of turbulence and dispersion and their application. Fluid modeling may be used to plan field experiments, to provide conservative estimates of transport, and to validate modules of numerical code.

The most serious limitations of physical modeling are normally related to fluid model facility size. Limitations of facility size can introduce errors associated with wall reflections, sidewall boundary layers, and low Reynolds number viscous distortions. Turbulence scaling also reduces the span of the inertial subrange of eddies. Large weather-size turbulence eddies and the diurnal cycle are absent during physical model experiments, since current facilities do not include nonstationary aspects of atmospheric motion. Only recently have some facilities produced useful elevated inversions or convective boundary layers.

7.1.2. Historical perspectives

Physical modeling studies of atmospheric flow over complex topography span some 60 years of research in more than seven countries. Early work at the National Physical Laboratory, United Kingdom, by Field and Warden (1929–30) led to the evaluation of air currents around the Rock of Gibraltar for airfield safety. The wind field up to 7000 feet high over a 1 to 5000 scale model of the Gibraltar peninsula was mapped using visualization flags and thread streamers. Subsequent measurement of the actual flows around the Rock of Gibraltar with pilot balloons and kites revealed that the model "closely forecast what occurred in nature at Gibraltar, in regard to wind directions and the distribution of vortices and vertical currents."

About the same time Abe (1929) used cold CO_2 sublimated from dry-ice flowing over a 1 to 50 000 scale model of Mount Fuji, Japan, to study mountain-wave clouds. Visualization photographs revealed wavelike motions near the mountain peak that correspond to the presence of laminar wave clouds over the actual volcano.

Theodore von Kármán consulted on wind tunnel studies of flow over several mountainous areas in New England at scales ranging from 1 to 5000 to 1 to 8000 in order to identify good wind power sites (Putnam 1948). Unfortunately, the researchers failed to consider the effects of the surface shear layer; hence, the results did not agree with field measurements. Later, English wind-energy researchers estimated wind speeds over terrain by measuring voltage potential variations over a conductive model in an electrolytic tank (Golding 1955). The method presumed airflow over complex terrain was inviscid potential flow and streamlines obeyed only the Laplace equation.

During the 1950s Robert Long initiated an extensive series of channel flow experiments to investigate stratified flows over mountainous terrain (Long 1959). During his experiments Long towed two-dimensional models of various mountain shapes over stratified brine solutions divided into two or more layers. These experiments successfully reproduced many of the features of mountain lee-wave structure, and the data provided verification for a variety of linear lee-wave models. Later tests focused on improving and varying the stratification, removing the influence of wave reflections, and adding three-dimensional models.

These and other experiments laid the foundations for the new field of wind engineering. Requirements for simulation of atmospheric motions and the development of meteorological laboratory facilities were the subjects of intense study from 1950–70. Modifications to conventional aeronautical wind tunnels to enable simulation of neutral and thermally stratified boundary layers were proposed by Strom (1952), Jensen (1958), Cermak (1958), and Counihan (1969). Cermak et al. (1966) developed similarity criteria for simulation of the atmospheric boundary layer based on the development of thick thermally stratified boundary layers. Orgill (1971a) used barastromatic simulation criteria to predict transport and diffusion of ground-based cloud seeding generators over mountainous terrain. Snyder (1972, 1981) critically evaluated the advantages and limitations of physical modeling when simulating atmospheric transport over irregular terrain. Meroney (1986) extended the simulation discussion to the motion of dense gas clouds over irregular terrain.

7.2. Similarity considerations

The concept of similitude is simple. Two systems at different geometric scales will exhibit similitude if a one to one correspondence exists in space and time between fluid particle kinematics (locations, velocities, accelerations, and rotations) caused by fluid particle dynamics (pressures, gravity, viscous forces, etc.) when properly scaled by characteristic scales of fluid properties, force, length, and time. To achieve this similarity, however, is not trivial. The specification of dimensionless parameters that guarantee similarity has historically been the subject of much discussion and debate.

In the nineteenth century many workers (most notably Lord Rayleigh) solved problems by direct use of the similarity principle with the intuitive identification of relevant force ratios. During the twentieth century, the force ratio methods lost favor and were replaced almost entirely by dimensional analysis, as represented by the Buckingham Pi theorem. Several authors including Cermak (1975), McVehil et al. (1967), Bernstein (1965), and Snyder (1972) derived the governing parameters for atmospheric heat, mass, or momentum transport by dimensional analysis, normalization, and inspectional analysis. Another group justifies similitude by considerations of turbulence theory and recent reviews of full scale wind data that present the characteristics of the prototype atmospheric wind on a parametric basis (Nemoto 1961, 1962; Counihan 1969, 1975; Cook 1977; Melbourne 1977). Although investigators do not agree about details, most would concur that the dominant mechanisms can now be identified and are understandable.

For complete flow similarity in two systems of different length scales, geometrical, kinematical, dynamical, and thermal similarity must be achieved. In addition, boundary conditions upstream in the upper atmosphere and downstream should also be similar.

Geometrical similitude exists between model and prototype if the ratios of all corresponding dimensions in model and prototype are equal. This is realized by using an undistorted scale model of the prototype geometry. Kinematic similitude exists between model and prototype if the paths of homologous moving particles are geometrically similar and if the ratio of the velocities of homologous particles are equal. Dynamic similitude exists between geometrically and kinematically similar systems if the ratios of all homologous forces in model and prototype are the same. Thermal similitude exists if the temperature or density stratification are similar.

7.2.1. *Similitude parameters*

The proper similitude parameters governing the phenomena of interest may be established by dimensional analysis, similarity theory, or inspectional analysis. Good descriptions of these methods are available in various textbooks and publications (e.g., Kline 1965; Meroney et al. 1978b; Meroney 1986). The pertinent parameters for transient, turbulent airflows are:

$\mathrm{Ro} = U(L\Omega)^{-1}$	the Rossby number
$\mathrm{Eu} = \Delta P(\rho U^2)^{-1}$	the Euler number
$\mathrm{Re} = \rho UL\mu^{-1}$	the Reynolds number
$\mathrm{Ri} = gL\Delta T(TU^2)$	the Richardson number
$\mathrm{Pe} = \rho CpULk^{-1}$	the Peclet number
$\mathrm{Pr} = \mu Cpk^{-1}$	the Prandtl number
$\mathrm{Sc} = \mu(\rho D)^{-1}$	the Schmidt number
$\mathrm{Ec} = U^2(Cp\Delta T)^{-1}$	the Eckert number.

"Exact" similitude requires equality of the nondimensional coefficients listed above for the physical model and the prototype situation. If separate length scales are chosen for the different coordinate directions additional parameters are generated; however, current wisdom is that distorted geometric scaling is not normally justified (Robins and Hertig 1986).

Furthermore, boundary conditions governing the flow domain of interest must be similar for the model and prototype. Boundary conditions would require similarity of the following features:

Surface boundary conditions

(i) Topographic relief.

(ii) Surface roughness distribution.

(iii) Surface temperature distribution.

(iv) Reproduction of the shapes and sizes of associated obstacles, buildings, fences, etc.

Approach-flow boundary conditions

(i) Distributions of mean and turbulent velocities.

(ii) Distributions of mean and fluctuating temperatures and humidities.

(iii) Distribution of turbulent scales and energies.

Aloft boundary conditions

(i) The upper streamline should follow a similar trajectory with respect to the ground surface.

(ii) The longitudinal pressure gradient should be nearly zero.

7.2.2. *Partial simulation of complex terrain flows*

The seven governing equations used to specify the parameters above and their associated boundary conditions contain seven unknowns, U_i, T, p, ρ, and X, so that (in principle) their solutions can be determined. If all the foregoing requirements were met simultaneously, all scales of motion ranging from microscale to mesoscale, 10^{-3} to 10^5 m could be simulated within the model flow field. All requirements cannot be satisfied simultaneously by existing laboratory facilities, however, and "partial" or "approximate" simulation must be used. This limitation requires that atmospheric simulation for a particular geophysical application must be designed to simulate most accurately those scales of motion that are of greatest significance for that application. By considering each similarity requirement separately it is possible to determine for which flow features "exact" similarity between the laboratory and the atmospheric boundary layer is lacking.

The effects of equal *Rossby numbers* cannot be obtained in nonrotating wind tunnels or channels. The Rossby number is a measure of the relative magnitudes of the advective or local accelerations resulting from unsteadiness or divergence in the flow field and the Coriolis accelerations associated with the earth's rotation. The Ekman spiral in the earth's wind profile results from such Coriolis accelerations. The laboratory boundary layer is an adequate model for atmospheric flow when time scales or transport distances are not too large. Atmospheric measurements made over complex mountainous terrain rarely suggest the presence of an Ekman spiral. Laboratory simulations of complex terrain fetch distances from 10 to 20 km have compared reasonably well with field observations. For example, Meroney et al. (1978a) observed that the Rakaia Gorge, New Zealand area has a characteristic length of 20 km, northwesterly winds are at least 20 m s^{-1} at gradient wind height, and $\Omega = 9 \times 10^{-5}$ s^{-1} at $\phi = 40$ degree latitude; hence, Ro = 11. That is, the inertial effects are an order of magnitude greater than the Coriolis accelerations. Hoxit (1973) and Scorer (1978) observe that most of the time the classical Ekman spiral does not exist due to thermal wind effects.

The *Euler number* compares the relative magnitude of pressure fluctuations and inertial accelerations. This parameter is usually of order 1 and is automatically simulated.

Equal *Reynolds numbers* are not attainable (except in pressurized or cryogenic gas tunnels). This, however, does not seriously limit capabilities for modeling the atmospheric boundary layer over irregular terrain at high wind speeds, since the significant flow characteristics are weakly dependent on Reynolds number. Townsend (1956) suggested turbulence structure will be similar at all sufficiently high Reynolds numbers. This hypothesis of Reynolds number independence is called "Reynolds number similarity." Exceptions must be (i) very small scale turbulence structure at dissipation, and (ii) flow fields very close to a boundary. Surfaces that are sufficiently rough exhibit Reynolds number similarity. All natural surfaces are aerodynamically rough; hence, flow structure will be similar if the scaled-down roughness is large enough to prevent the formation of a laminar sublayer. A requirement for fully rough flow is $U_* k \nu^{-1} > 100$ or Re $> 10^4$–10^6. The Rakaia Gorge model study satisfied both these criteria. Replication of the main streamline features, regions of turbulent eddies, and the structure of the turbulence thus becomes dependent upon geometric similarity and equivalence of relevant length scale ratios, such as $z_o L^{-1}$, $\Lambda_x L^{-1}$, etc.

A large value of the *Richardson number* implies that buoyancy forces are very large compared to the inertial forces. Thermal effects or density differences become less important as the Richardson number decreases toward zero. Stratification over complex terrain influences leewave production, upstream waves, blocking, streamline separation, and mixing processes. Batchelor (1953) has established that, if the pressure and density everywhere depart by only small fractional amounts from the values for an equivalent atmosphere in adiabatic equilibrium, and if the vertical length scale of the velocity distribution is small compared to the scale height of the atmosphere, then the Richardson number distribution governs dynamic similarity. Flow fields with sufficiently large wind speeds and modest density gradients perceive only modest perturbations because of stratification. Unfortunately, elevated inversions may produce strong surface effects even when surface stratification is modest.

The Richardson number is not necessarily a difficult parameter to duplicate in a fluid model. Wind tunnel temperatures may be controlled by upstream heat exchangers, injection of heated air, or the use of a thermal boundary layer grown over long segments of heated or cooled surfaces (Plate and Cermak 1963; Teunissen 1975; Ogawa et al. 1981; Schon and Mery 1971). Water channels maintain stratification using either heat or, more often, layered salt water (Hunt et al. 1978; Snyder et al. 1979). Unfortunately, to match model and prototype Richardson numbers for typical scale reductions of 1 to 500 to 1 to 5000 using reasonable model temperature or density differences, it is necessary to decrease the mean flow speeds substantially. Yet to match the Reynolds number requires large flow speeds; hence, a conflict arises.

The *Peclet number* is a measure of the ability of the fluid to advect heat or mass compared with its ability to disperse heat or mass by molecular transport. The Peclet number becomes important when Reynolds number independence does not exist. In such cases the relative ability of the fluid to transport heat or mass by molecular collision and the rate of transport provided by turbulent motions become comparable. Such a situation can produce incorrectly simulated plume entrainment and transport rates.

The *Prandtl number* and *Schmidt number* suggest the relative ability of the fluid to transport momentum versus its ability to transport heat or mass by molecular processes. If air is used to simulate the atmosphere, the Prandtl number is automatically satisfied, whereas the Schmidt number will be dependent on the model tracer gas chosen; however, both are usually close to a value of 1. In water the Prandtl number is about 10 times larger than it is in air and varies with temperature. The Schmidt number for typical tracers such as sodium chloride or alcohol dispersing in water is nearly 800; hence, Schmidt number equality is unlikely in water facilities. Since both parameters always appear in the governing equations multiplied by Reynolds number, the heat or mass Peclet number is the critical governing parameter.

The *Eckert number* is the ratio of kinetic to excess internal energy, and it is equivalent to the Mach number squared. It is usually small compared to unity for both atmospheric and laboratory model flows. The Eckert number magnitude's small value in the energy equation suggests that viscous dissipation or compression do not affect atmospheric temperature distributions significantly.

7.2.3. Performance envelopes for fluid modeling

The viability of a given simulation scenario is not only a function of the governing flow physics but also the availability of a suitable simulation facility and the measurement instrumentation to be employed. It would seem appropriate, therefore, to suggest bounds for the range of field situations that can be reasonably treated by physical modeling. This discussion is limited to wind tunnels and a separate evaluation can be made for water channels. Typical meteorological wind tunnel characteristics are listed in Table 7.1. A comparison between field and laboratory parameter ratios is provided in Table 7.2.

TABLE 7.1. Typical boundary-layer wind tunnel characteristics.

Wind tunnel property	Range
Tunnel height H_T (m)	$0.5 \leftrightarrow 2.5$
Tunnel width W_T (m)	$0.5 \leftrightarrow 4.0$
Tunnel length L_T (m)	$1.0 \leftrightarrow 30.0$
Boundary depth δ (m)	$0.10 \leftrightarrow 2.0$
Roughness z_o (mm)	$0.01 \leftrightarrow 15.0$
Velocity U_o (m s^{-1})	$0.0 \leftrightarrow 36.0$
Integral scale L_{ux} (m)	$0.06 \leftrightarrow 0.6$
Temperature gradient $\lvert d\theta/dz \rvert$ (°C m^{-1})	$0.0 \leftrightarrow 200.0$
Hill height h (m)	$0.02 \leftrightarrow 0.5$
Hill half-width L (m)	$0.0 \leftrightarrow 2.0$
Resolution Δz (mm)	0.1
Surface friction $C_f/2$ (dimensionless)	$0.0005 \leftrightarrow 0.0040$
Richardson number $\lvert Ri_B \rvert$ (dimensionless)	$0.0 \leftrightarrow 1.0$
Monin–Obukhov length $\lvert L_{mo} \rvert$ (m)	$0.065 \leftrightarrow \infty$
Turbulence intensity $[u'^2]^{1/2}/U_o$	$0.01 \leftrightarrow 0.30$
Power-law $1/\alpha$ coefficient	$0.05 \leftrightarrow 0.80$

7.2.3.1. NEUTRAL AIRFLOW MODELS

When one combines various operational constraints into a performance envelope, a clear picture appears of the performance region for wind tunnel facilities. Figure 7.1 is such a performance envelope prepared for a large facility (3 m × 4 m × 25 m test section dimensions). The criteria selected to specify operational simulation ranges are:

1) Maximum model height ($h \leqslant 0.5$ m).
2) Minimum convenient model height ($h \geqslant 0.02$ m).
3) Minimum Reynolds number ($\mathrm{Re} = Uh\nu^{-1} \geqslant 10^4$).
4) Maximum model integral scale ($\Lambda_x \leqslant 0.5$ m).
5) Minimum model integral scale ($\Lambda_x \leqslant 0.05$ m).
6) Minimum model measurement resolution ($\Delta z \geqslant 0.1$ mm).
7) Maximum model boundary depth ($\delta \leqslant 2$ m).
8) Minimum model boundary depth ($\delta \geqslant 0.1$ m).

Since field values for some parameters are uncertain the prototype value of δ and Λ_x are assumed to range over complex terrain as follows:

$$300 \text{ m} < \delta < 1000 \text{ m},$$

$$100 \text{ m} < \Lambda_x < 1000 \text{ m}.$$

Not all previous laboratory studies meet such similitude restrictions. Some experiments were performed to meet objectives other than similitude of turbulence or mean velocity profiles; nevertheless, several studies are shown in Fig. 7.1. In most cases values fall within or just outside the proposed operational envelope.

Based on Coriolis force considerations, Snyder (1972) suggests 5 km should be the maximum distance for horizontal length scales when modeling diffusion under neutral or stable conditions over flat terrain. Mery (1969) suggests a 15 km limit, Ukejurchi et al. (1967) suggest 40

TABLE 7.2. Typical wind tunnel and field parameter ranges.

Dimensionless parameter	Wind-tunnel range	Field range
Reynolds number $\mathrm{Re}_h = U_o h/\nu$	$0 \leftrightarrow 7.0 \times 10^5$	$0 \leftrightarrow 10^9$
Richardson number $\mathrm{Ri}_B = \dfrac{g(d\Theta/dz)h}{TU_o^2}$	$-1.0 \leftrightarrow 1.0$	$-1.0 \leftrightarrow 1.0$
Prandtl number $\mathrm{Pr} = \mu Cp/k$	0.72	0.72
Skin friction coefficient or Stanton number $C_f/2 = (u_*/U_o)^2$; $\mathrm{St} = q/(\rho Cp \Delta T)_o$	$0.006 \leftrightarrow 0.004$	$0.001 \leftrightarrow 0.004$
Insolation parameter $\mathrm{Fr}_I = \dfrac{(U_o^2 T_w)}{[g(T_w^o - T_n)L]}$	$0.002 \leftrightarrow \infty$	$2.5 \times 10^{-5} \leftrightarrow \infty$
Scaling lengths		
z_o/h	$2 \times 10^{-5} \leftrightarrow 0.25$	$3.0 \times 10^{-7} \leftrightarrow 0.15$
δ/h	$0.20 \leftrightarrow 50.0$	$0.10 \leftrightarrow 100$
h/L	$0.05 \leftrightarrow \infty$	$2 \times 10^{-4} \leftrightarrow \infty$
Λ_x/L	$0.06 \leftrightarrow \infty$	$0.001 \leftrightarrow \infty$
h/H_T	$0.01 \leftrightarrow 0.15$	—
L_T/L	$0.0 \leftrightarrow 30.0$	—
h/L_{mo}	$0.0 \leftrightarrow 5.0$	$-360 \leftrightarrow 240$
Power-law $1/\alpha$ coefficient	$0.05 \leftrightarrow 0.8$	$0.05 \leftrightarrow 0.7$

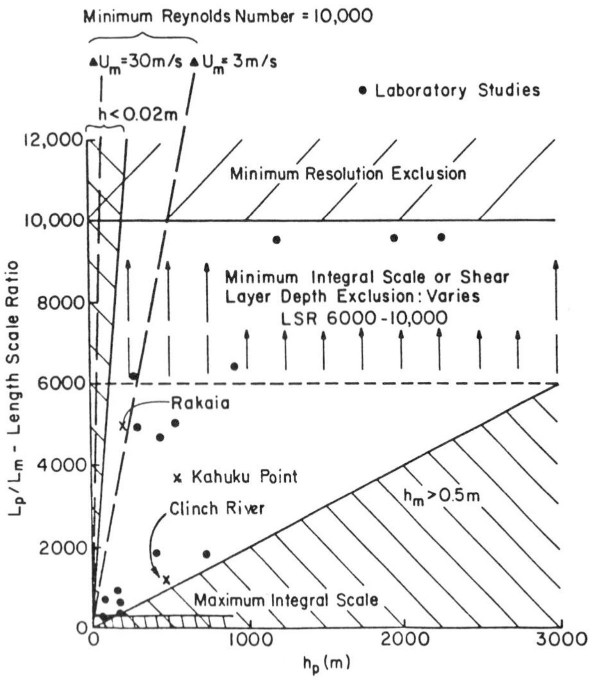

FIG. 7.1. Performance envelope for physical modeling of shear flows over complex terrain in a typical boundary-layer wind tunnel.

to 50 km, and Cermak et al. (1966) and Hidy (1967) recommend 150 km. A middle road value would be that of Orgill et al. (1971a) who suggest for rugged terrain in high winds that a length scale of 50 km is not unreasonable.

Assuming an upper limit for a length scale ratio of 10 000 and a tunnel length of 25 m, then a 50 km fetch in the windward direction is well within the capacity of many existing facilities. Assuming a lateral width restriction of 4 m suggests a 40 km lateral maximum for the field area modeled.

7.2.3.2. VALLEY DRAINAGE FLOWS

Local heating and cooling of hill surfaces are the driving mechanisms for anabatic and katabatic winds that inhibit or enhance separation over hillcrests. Early laboratory work includes simulation of urban heat islands and heated mountains by Yamada and Meroney (1971) and simulation of flow and dispersion at shoreline sites by Meroney et al. (1975). Equality of a Monin–Obukhov type length scale ratio would be appropriate for simulating surface heating situations with a finite flow wind field. The linear numerical analysis of Olfe and Lee (1971) and the experimental studies of Yamada and Meroney (1971) and Meroney et al. (1975) suggest that the intensity of heating or cooling by the land surface may be characterized by a heating ratio, $HR = [\Delta\theta_z L_x][\Delta\theta_x L_z]^{-1}$, where $\Delta\theta_z$ are characteristic temperature changes over a vertical scale L_z, and $\Delta\theta_x$ are similar changes over some horizontal scale L_x. Usually, since the vertical-to-horizontal modeling scale is undistorted, the parameter reduces to a single temperature ratio. In those situations where the gradient wind is small and where the lateral variations in surface temperature are negligible, another parameter called the Rayleigh number, $\mathrm{Ra} = g\beta(\theta_w - \theta_H)h^3(\alpha_h\nu)^{-1}$, is appropriate. Note that the scaled temperature potential increases as the model scale cubed!

Stipulation of strict Rayleigh number similarity would eliminate the possibility of modeling valley drainage flows; however, gravity current movement down steep valley walls will result in high local Reynolds numbers and turbulent mixing. In such circumstances the appropriate viscous parameter is the eddy viscosity, $\epsilon \approx UL$, not ν. Substitution of this variable transforms the Rayleigh number into the Richardson number introduced earlier.

7.2.3.3. VERIFICATION EVIDENCE

Not all the terrain model study cases identified in Fig. 7.1 were successful. Many early studies had model approach flow velocity exponents near zero and were modeled as neutral flows, whereas the field situation observed strong stratification effects, or simulated unrealistic boundary layer depths, integral scales, or turbulence intensities. In 17 of some 40 terrain model studies examined by the author there are counterpart field data available (Meroney et al. 1978b). Few studies claimed unreasonable correlation, and some were strongly self-critical. Nonetheless, most studies accomplished their prestated limited objectives. Meroney (1981) considers verification evidence from three such studies in detail.

7.3. Facilities for fluid modeling of complex terrain meteorology

The characteristics and capabilities of the fluid modeler's hardware determine whether program objectives can be attained in laboratory facilities. No single wind tunnel or water facility is adequate for all possible atmospheric tests. This section considers the size and performance characteristics of wind engineering facilities and instrumentation. The selection of an air versus water medium for modeling flow over hills or mountainous terrain will depend upon the availability of the facility, economics, and the type of problem studied. The kinematic viscosity of water at normal room temperature is a factor of 16 less than that of air; hence, at the same scale and fluid speed the Reynolds number may be 16 times higher for a water experiment. Unfortunately, because water is so much heavier than air, structural and pumping requirements result in water facilities that tend to be much smaller than wind tunnels. Thus the larger Reynolds number potential of water facilities is seldom attained.

7.3.1. Wind tunnels

Low-speed wind tunnel design is described by Rae and Pope (1984), who tabulate characteristics of over 300 low-speed wind tunnels in use worldwide. These facilities include 64 in the United States—of which 12 are nonaeronautical industrial-aerodynamic facilities. Special design attributes of meteorological wind tunnels are considered by Cermak (1981, 1984). Meteorological wind tunnels range in size from cross sections of 0.5 m × 0.5 m × 3 m × 4 m. Several of these facilities are equipped with moveable sidewalls or ceilings to adjust for model blockage. By using a variety of devices such as vortex generators, fences, roughness, grids, screens, or jets a wide range of turbulence integral scales can be introduced into the shear layer (Fig. 7.2). Varying surface roughness permits control of surface turbulence intensity, dimensionless wall shear, and velocity profile shape. Density stratification can be induced by heat exchangers, different molecular weight gases, or heat release during phase change.

Air with its low heat capacity is easy to stratify using heat. A few special meteorological wind tunnels have been designed to reproduce some aspects of the stratified atmospheric boundary layer (Meteorological Wind Tunnel, Colorado State University; Environmental Wind Tunnel, Mitsubishi Industries, Nagasaki, Japan; Meteorological Wind Tunnel, Bundeswahr Hochschule, Munich, West Germany; and CERL Stratified Flow Tunnel, Leatherhead, United Kingdom). Recently Rau and Plate (1986) proposed a new design for a low-speed wind tunnel to study strongly stratified meteorological flows. Using a multiduct return flow design it is anticipated undesirable standing waves and facility blockage may be avoided.

7.3.2. Drainage flow facilities

Aluminum shell models (scales of 1 to 1920 and 1 to 2650) cooled by dry ice have been used by Petersen and Cermak (1980) to study drainage flows in the mountains near Crested Butte, Colorado. Richardson numbers based on the maximum surface jet velocity and jet height were found to vary from 0.5 to 2.0. These produced surface velocities equivalent to 1.0 to 3.5 m s^{-1} between 50 to 150 m above ground. Specific field conditions were not chosen in advance; rather, model conditions were permitted to develop, and the equivalent field conditions were evaluated from the measured Richardson numbers. Hertig and Liska (1983) report drainage flow studies over a vertically distorted scale model of mountainous areas in Switzerland. They used a model with a horizontal scale of 1 to 100 000 and a vertical scale of 1 to 20 000 to adjust for large vertical eddy scales produced in the laboratory shear layers.

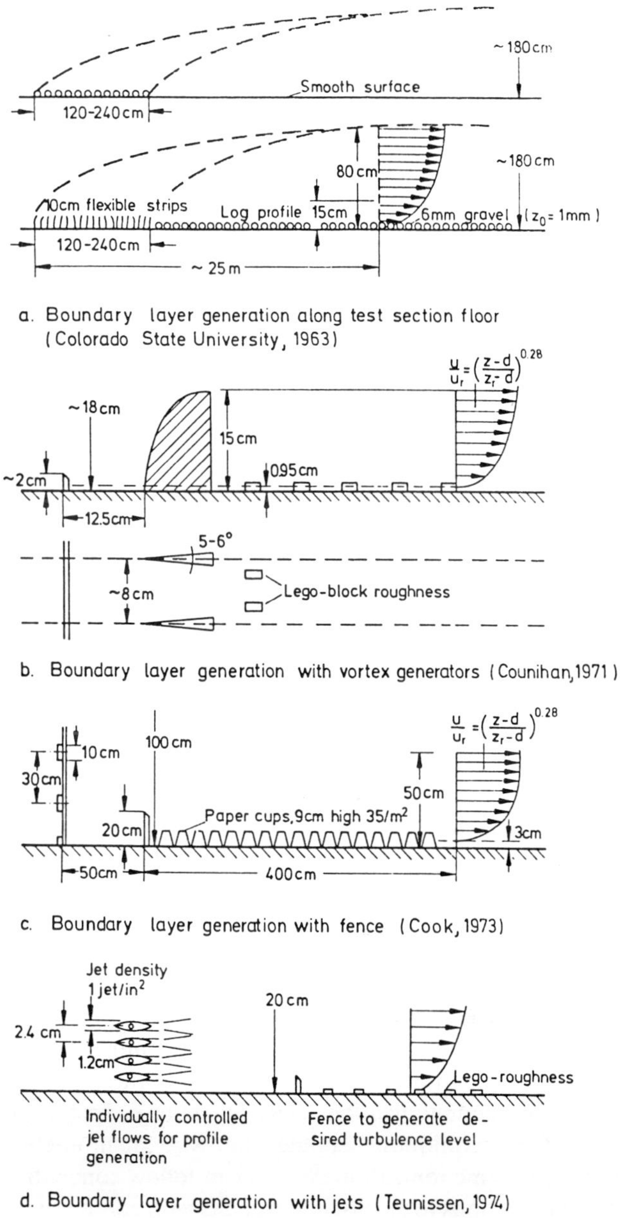

FIG. 7.2. Alternative methods for generating boundary layer flows in a meteorological wind tunnel (Plate 1982).

7.3.3. Water channels and rotating tanks

Several investigators have used water towing tanks to examine flow over hills or other obstacles (Hunt et al. 1978). This method is appropriate when considering atmospheric flows dominated by blocking, lee waves, and inversions. The towing tank method may distort surface layer predictions because the uniform approach profile simulated is not equivalent to the shear flow found near the earth's surface. In water most experiments using salt as a density ingredient were run with specific gravities between 1.0 and 1.1. The highest relative density obtainable in a water soluble solution is about 1.4. Nonetheless, excellent visualization studies have provided understanding of mountain wave and terrain blocking phenomena.

Faller (1966) used a water filled rotating basin to study instabilities in laminar boundary layers. Later, Caldwell and Van Atta (1970) studied Ekman boundary-layer flows in a turbulent boundary layer produced in a rotating air facility. Falvey and Dodge (1977) describe a rotating water tank for simulating the influence of Coriolis accelerations over mountainous terrain. The device was used to predict the dispersion in cloud seeding studies over the Sierra Nevada.

7.3.4. Instrumentation

A variety of visualization and flow measurement techniques have been developed to allow accurate measurement of flow over model complex terrain in the laboratory. Conventional air and water metrology includes pitot-static tubes, hot-film anemometry, and laser-Doppler velocimetry as described by Rae and Pope (1984) or Goldstein (1983). Pulsed wire anemometers are also useful to measure flow in separated regions up or downwind of hill ridges (Bradbury 1976).

Traversing camera and lighting systems are often used with water channel system. Hydrogen bubbles may be used to tag water channel path lines over terrain models. Smoke wire probes can be combined with stereoscopic camera systems to measure wind shear and wind veering over complex terrain models. Huber (1988) and Lee et al. (1988) have shown how video image analysis can produce velocity fields and concentration patterns in complex flow geometries.

Concentrations in plumes moving over complex terrain can be measured in wind tunnels by using thermal-conductivity, flame-ionization, electron-capture, or photoionization techniques. Aspirated hot-wire anemometers of in situ flame-ionization devices can follow concentration fluctuations to at least 60 Hz. Salts with conductivity meters, acids with pH meters, temperature with thermistors, and dyes with colorimeters have been used as tracers for quantitative measurements of concentration in water. Conductivity probes in water respond to frequencies of 10 to 20 Hz.

7.4. Neutral flow over hills, ramps, and escarpments

Mountains may alter atmospheric airflow characteristics in many different ways. These effects can generally be grouped into inertial–viscous interactions associated with a thick neutrally stratified shear layer and thermally induced interactions associated with stratification or surface heating. Near-neutral or adiabatic atmospheric boundary layers will exist over hills and mountains during situations when wind speeds are large. High wind speeds occur when intense synoptic pressure fields exist, when continuous cloud banks impede surface heating, and when sharp terrain features produce separation eddies that mix the flow field vigorously in the vertical (case a, Fig. 7.3). When the static stability is neutral, airflow over mountains creates pressure gradients in the flow direction, which together with surface friction may produce separation, flow reversal, and reattachment. Separation eddies at the windward or leeward side of a mountain can alter the effective shape of the mountain, resulting in a modified wind profile at the crest. Scorer (1978) describes eight different variations of the separation phenomenon. He notes that separation may be changed in character by insolation, blocking, diabatic changes, and three-dimensional effects.

7.4.1. Idealized two-dimensional terrain flow studies

Flow over slight or moderate relief may result in enhanced wind speeds as a result of wind "overshoot" or "speedup." A few field measurement programs over such terrain features were carried out specifically to estimate wind power potential (Putnam 1948; Golding 1955; Frenkiel 1962, 1963; Archibald 1973; Hewson 1973).

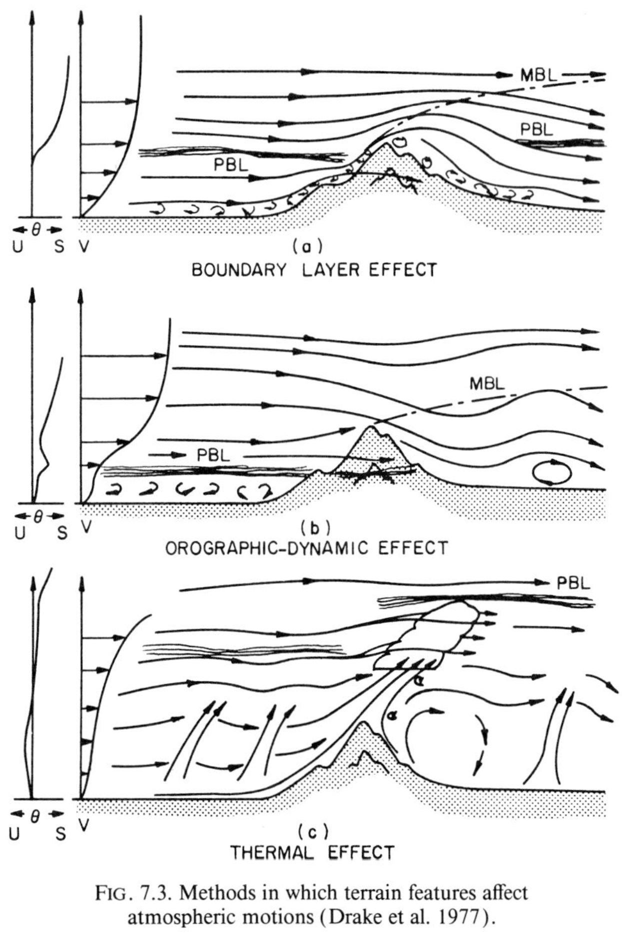

FIG. 7.3. Methods in which terrain features affect atmospheric motions (Drake et al. 1977).

These early results are to a large degree site specific and do not cover a wide enough range of terrain types to allow more than a very limited and qualitative generalization to other situations. The combination of hill features, roughness, upstream topographies, and stabilities studied appear to lead to a set of contradictory conclusions (Davidson et al. 1964). Several laboratory investigations of flow over topography have also been completed. Meroney et al. (1978b) identified 10 studies among 40 laboratory simulations of flow over terrain devoted to flow over idealized hills, ridges, and escarpments. These studies span some 46 years and more than seven countries, and include gentle hills, cones, ridges, escarpments, and mountains.

Flow over wedge, polynomial, and sinusoidal shaped hills have been examined by Plate and Lin (1965), Meroney et al. (1978b), and Arya et al. (1987). Measurements were made in neutral and stably stratified flows with various combinations of surface roughness upwind or over the hills. Bowen and Lindley (1974, 1977), Bowen (1981), and Pearse (1982) examined various escarpment and hill shapes. Bowen also made field measurements at the bank escarpments formed by the wide New Zealand braided rivers. The implications of these measurements are discussed in the following paragraphs.

7.4.1.1. Effects of ridge shape

Wind speed variation over a hill or ridge depends strongly on the upwind and downwind hill slopes.

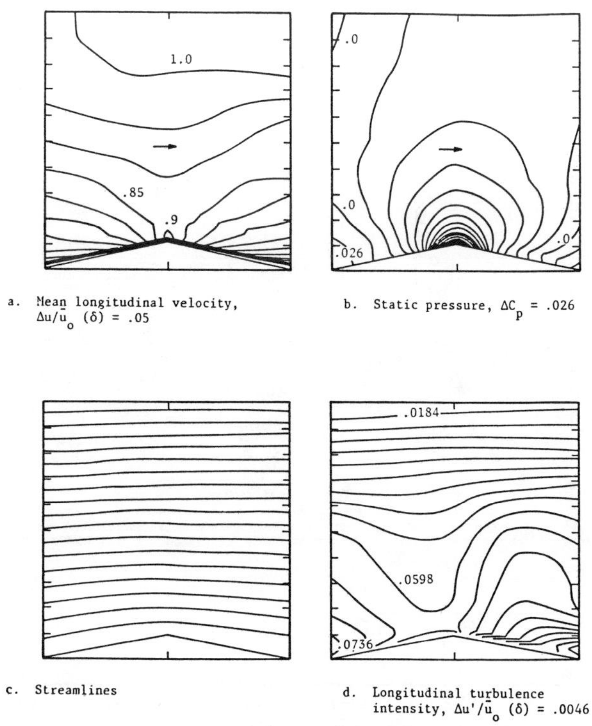

FIG. 7.4. Contours of flow characteristics over a triangular ridge, $h/L = \frac{1}{6}$ (Meroney et al. 1978b).

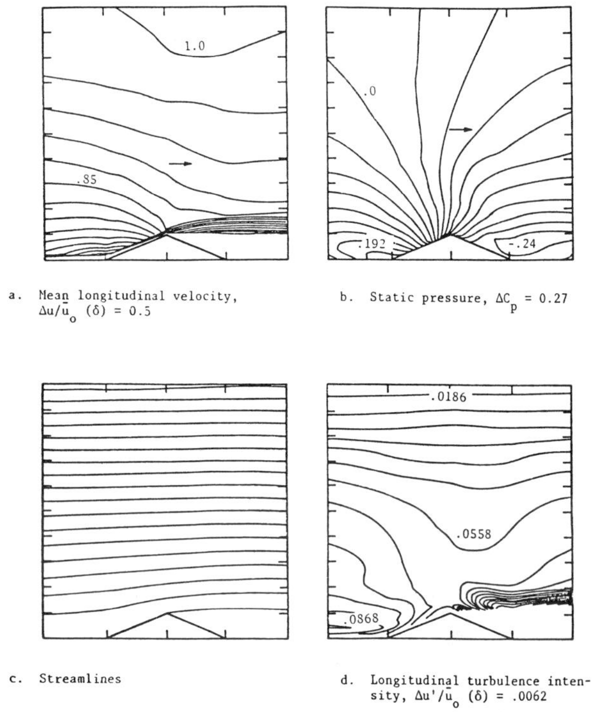

FIG. 7.5. Contours of flow characteristics over a triangular ridge, $h/L = \frac{1}{2}$ (Meroney et al. 1978b).

Speedup of the wind is highest for ridge slopes that just avoid flow separation. Such a speedup causes reduction of the minimum static pressure by a separated wake at either the upwind or downwind side of the crest. In Figs. 7.4 and 7.5 we present results that show the dramatic change in velocities and static pressures over symmetrical triangular hills as a result of flow separation. No flow separation occurs for the hill defined by $h/L = \frac{1}{6}$, whereas flow separation occurs for the hill defined by h/L and $\frac{1}{3}$. [Note L is twice the half height distance proposed by Jackson and Hunt (1975) taken between the crest and the point at the upwind side where the hill height is 0.5 h.]

The character of flow separation occurring at the upwind side of a hill is different from that on the downwind side. The upwind eddy nestles at the foot of the hill. The eddy in the downwind separation region interacts strongly with the main flow producing an extended wake. For steep downwind slopes the separation region may extend to a distance of 10 to 20 times the obstacle height. The interaction between the separation eddy and the main flow at the upwind side is constrained by the presence of the hill. With a slight variation that depends on the parameter h/δ the upstream separation region does not extend farther than two hill heights upwind. The location of the flow separation point depends to some degree on the Reynolds number. A Reynolds number effect is avoided when studying the phenomenon of separation over triangular hill models that fix the separation point at the crest. A round crested hill results in flow separation downwind of the crest. The results obtained for the triangular hill are somewhat conservative if applied to round crested hills.

The extent of the downstream separation region depends strongly on the strength of the eddy just downwind of the separation point. For gentle downwind slopes only weak eddies can develop. Weak eddies permit early reattachment of the separating streamline. The effect of h/L_d = 1, 0.3, 0.25, and 0.167 on the mean velocity field when $h/L_u = 0.5$ and $h/\delta = 0.1$ is illustrated in Fig. 7.6. Huber et al. (1977) showed that the separation phenomenon is not affected significantly by the upstream velocity distribution. Essentially identical downwind separation regions were obtained for a uniform approach velocity profile and for values of $h/\delta = 0.3$. Bouwmeester (1978) identified the parameter space in h/L_d and h/L_u over which separation will occur. Separation is unlikely to occur if $h/L_u \leqslant 0.5$ and $h/L_d \leqslant 0.25$.

Hill perturbations in wind speed may be characterized by a fractional speedup ratio, defined as $\Delta S_{\text{crest}} = [u_{\text{crest}}(z) - u_o(z)]/u_o(z)$, where $u_o(z)$ is the upstream velocity distribution and z is the distance from the surface. By normalizing the increase in wind speed with the upstream profile it was found that the fractional speedup ratio does not strongly depend on the approach velocity distribution. The effect of detailed hill shape on the velocity field has been investigated by comparing velocity fields over symmetrical triangular hill models with those measured over sinusoidal shaped hill models but with the same average slope. Almost identical velocity fields were measured. Rider and Sandborn (1977) considered alternative hill shapes with the same height and distance from crest to the foot of the hill. The models included full sine wave, half sine wave, triangular, trapezoidal, and box-shaped hills (Fig. 7.7). Speedup effects over the crest of the different hill models varied substantially since length scale L varied by a factor of 2. Moreover, separation regions existed upwind and downwind of the box-shaped hill. Nonetheless, hills with similar average slope had similar speedup factor profiles.

A comparison between the laboratory data of Meroney et al. (1978) and an inviscid, rotational solution of the Poisson equation and vorticity transport equations produced by Derickson and Meroney (1977) is displayed in Fig. 7.8. The success of this numerical approach demonstrates the effectively small role played by surface turbulence in the mean wind fields over small hills. Jackson (1979) also compared the Colorado State data with the

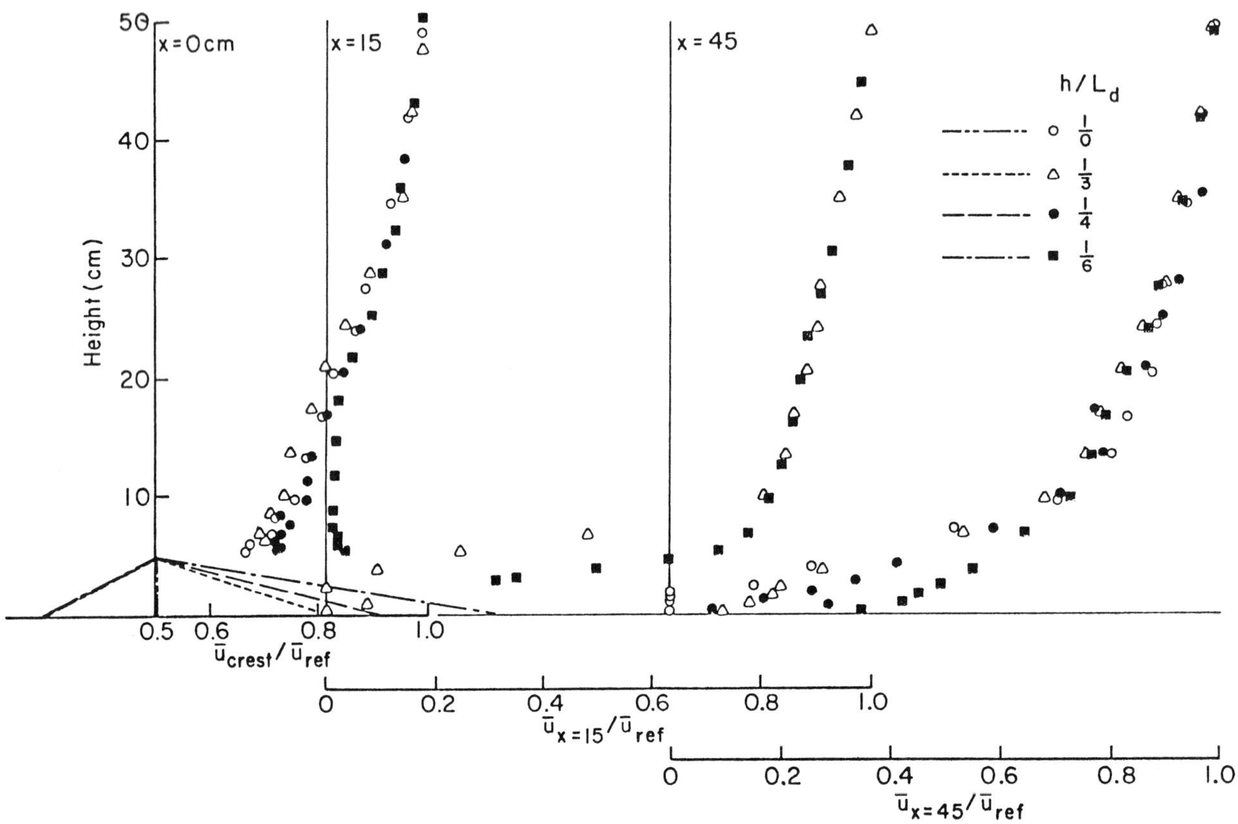

FIG. 7.6. Mean velocity profiles downwind of a triangular hill $h/L_u = 0.5$ and $h/L_d = 1, 0.33, 0.25, 0.17$ (Bouwmeester et al. 1978).

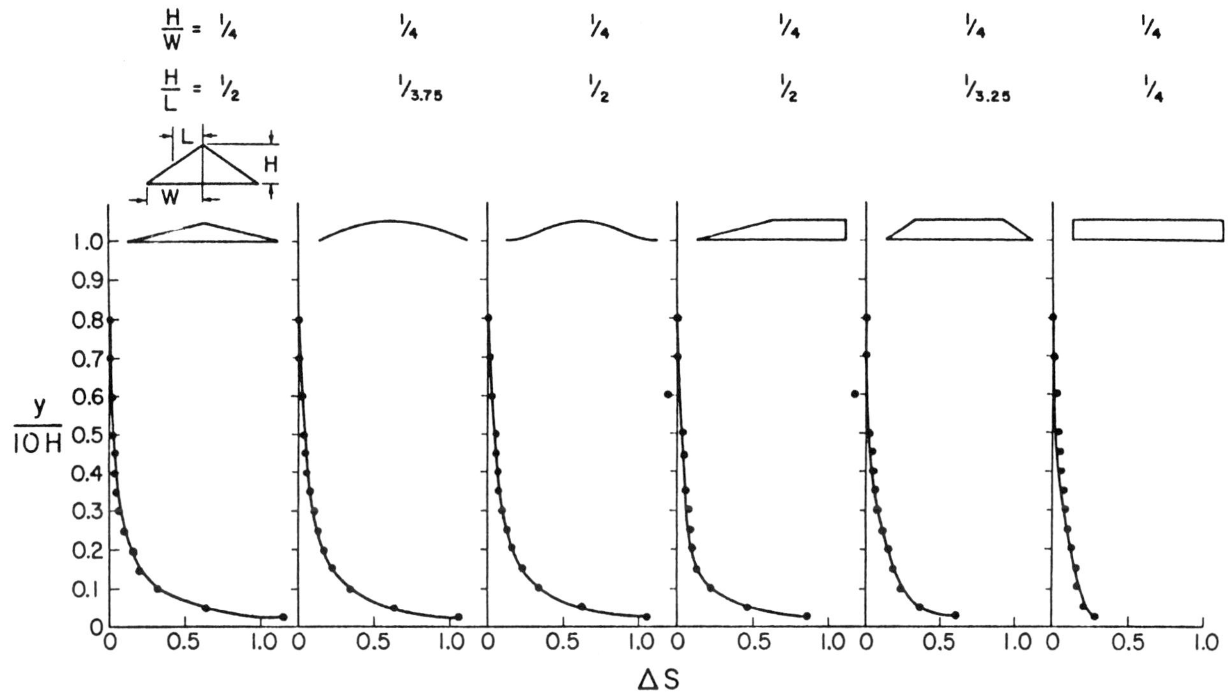

FIG. 7.7. Velocity speedup between approach flow and crest over different hill shapes (Rider and Sandborn, 1977).

Jackson and Hunt (1975) solution method and achieved similar agreement.

7.4.1.2. EFFECTS OF TURBULENCE

When air passes over a hill the turbulence structure is distorted causing Reynolds stress gradients different from those found under equilibrium flow conditions. As a result the momentum transfer from one streamline to another is not at the same rate, and the total head across streamlines does not decrease uniformly over the height of the atmospheric boundary layer. Bouwmeester et al. (1978) provided an order of magnitude analysis to estimate the total head losses caused by the distorted turbulence. Laboratory data show that for the case of short hills the longitudinal turbulence intensity increases toward the base of the hill, then decreases over the crest. The vertical turbulence intensity shows a decrease at the base of the hill and an increase over the crest. Thus, additional total head losses close to the surface are approximately equal to $\Delta P' \approx 2\overline{u'^2_{\max}}$. With maximum local turbulence intensities of 20% the total head close to the crest may then change by 8%. Contour plots of longitudinal intensities for triangular hill models are given in Figs. 7.4 and 7.5. The figures show that the decrease in turbulence intensity between the base and the crest along streamlines close to the surface is 50% for the $\frac{1}{4}$-hill and 25% for the $\frac{1}{6}$-hill of $\overline{u'^2_{\max}}$.

The changes of dynamic head close to the surface are as large as 400%. Supposing that the change of additional total head causes an equal change in dynamic head, it may be concluded that the effect of the turbulence on the mean flow at the crest is not significant, at least for $h/\delta < 2h/L_u$. The effect of turbulence on the dynamic head may be expected to be larger for hills with $h/\delta > 2h/L_u$ (long hills).

Downwind of the crest turbulence production increases and turbulence intensities in this region exceed upstream intensities. The relative effect of turbulence changes on the dynamic head increases, since the dynamic head returns to values approximately equal to upstream values. These considerations are illustrated by examining flow over a symmetrical hill. According to inviscid flow theory the velocity and static pressure field over a symmetrical hill is also symmetric. Therefore the measured degree of flow symmetry or asymmetry gives a direct indication of the effects of turbulence on the mean flow.

7.4.1.3. EFFECTS OF SURFACE ROUGHNESS

Two length scale parameters determine the velocity distribution of a fully developed turbulent boundary layer over a surface perturbation, z_o/h and h/δ. The effect of surface roughness, z_o/h, is most significant for small values of relative hill height, h/δ, for then mean velocity and turbulence structure are significantly different over a thickness of order h. Surface roughness changes the fractional speedup ratio at the crest significantly only if $h/\delta < 1$. Flow separation may occur as a result of increased surface shear stresses over the top of the hill due to the speedup of the wind and adverse pressure gradients on

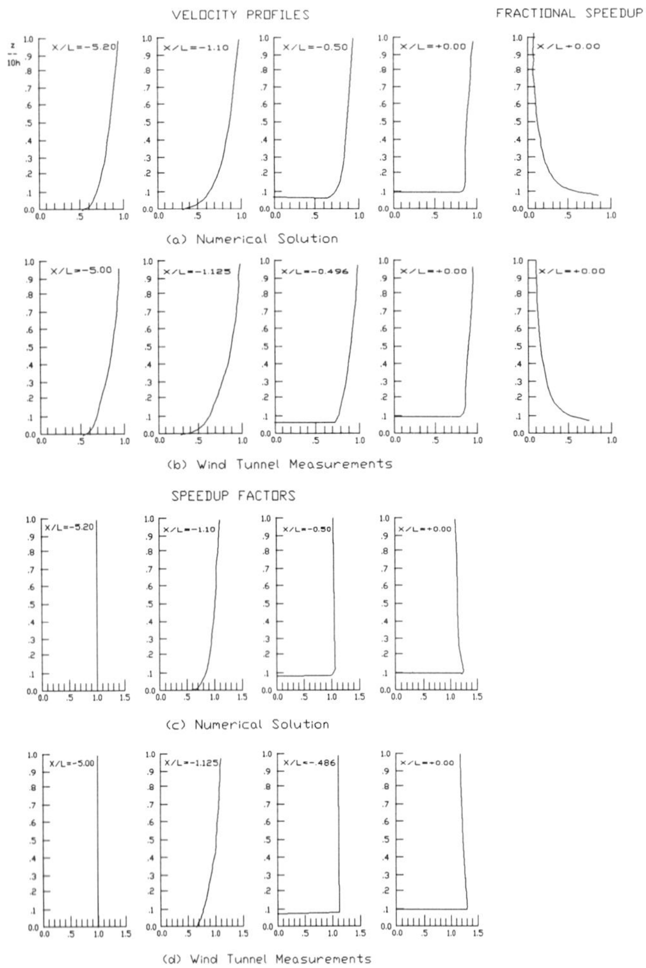

FIG. 7.8. Comparison between wind tunnel (w.t.) and numerical simulation (n.s.) of velocities over a 1 to 4 sinusoidal hill. L is half width of hill (Derickson and Meroney 1977).

the downwind side. Once flow separation takes place the interaction between the eddy in the wake and the main flow will dominate the flow. If the approach surface roughness is not uniform then location of flow separation or even the occurrence of slow separation may be affected. Consider, for instance, an upwind surface roughness that is larger than the surface roughness over the hill. In this case surface shear stress at the hillcrest may be less than for uniform surface roughness. Consequently, flow separation may occur later or may not occur at all. Opposite effects can be expected if the surface roughness over the hill is larger over the hill than upwind.

7.4.2. Idealized three-dimensional terrain flow studies

Irregular three-dimensional surface obstructions are much more prevalent than the idealized two-dimensional ridge. Because of the large number of irregular geometric shapes, the problem has not been amenable to mathematical characterization except for a few simple cases such as flow around spheres or cylinders. Therefore the wind tunnel has often been used to investigate scale models of bluff bodies.

Early work tended to suggest that fields of turbulence and secondary flow around bluff bodies would be similar to those identified downwind of two-dimensional fences or ridges. Hence, a cavity region downwind of separation was hypothesized, which communicated with the extended flow primarily by diffusion and pressure perturbations across free shear layers. Improved visualization techniques have revealed a more complex flow pattern.

Wakes generated by hills or other obstacles are characterized by increased turbulence, mean velocity defect, and (in certain situations) by organized strong vortices with axes parallel to the main flow direction. The characteristics of the wake (the extent of the momentum wake, the strength and extent of the vortex wake, the rate of decay of excesses and defects in the wake, etc.) are highly dependent on the obstacle height, the aspect ratio of the hill, the hill width–height ratio, the shape (projecting corners, step structure, round portions, etc.), the approach wind azimuth, the character of the surrounding terrain (extent of vegetation), and the atmospheric stability. For these reasons, a simple description or generalization of terrain wakes do not appear possible.

Measurements downwind of a hemisphere immersed in a turbulent shear layer are different from the wake of a rectangular body or building in two important ways. First, in the far-wake profiles there is a relative maximum in the velocity (or a relative minimum in the velocity deficit) on the wake centerline (Hansen and Cermak 1975). A significant portion of the wake contains velocity excesses (Fig. 7.9). The near-wake velocity profiles have the shape of a momentum wake. A minimum velocity is observed at the center of the wake, but this profile shape quickly changes into a shape with a local maximum velocity on the wake centerline. Velocity profiles measured in the lower region of the wake show this transition from a momentum-wake profile to what can appropriately be called a vortex-wake profile. The upwind distortion of vortex lines by the presence of a hill in a stream with shear is responsible for the formation of "horseshoe" shaped vortex cells that wrap around the base. A general downflow is induced in the hill wake by these vortices tending to increase velocities where a momentum defect might otherwise exist. The profiles measured above the height of the obstacle show a momentum-wake character far downwind; however, the region that contains the characteristic vortex-wake velocity profiles slowly grows upward giving almost all of the far wake a vortex-wake structure.

The second feature of the hemisphere wake is the extreme persistence of the wake. Mean velocity wakes of simple cubes and blocks in turbulent boundary layers ex-

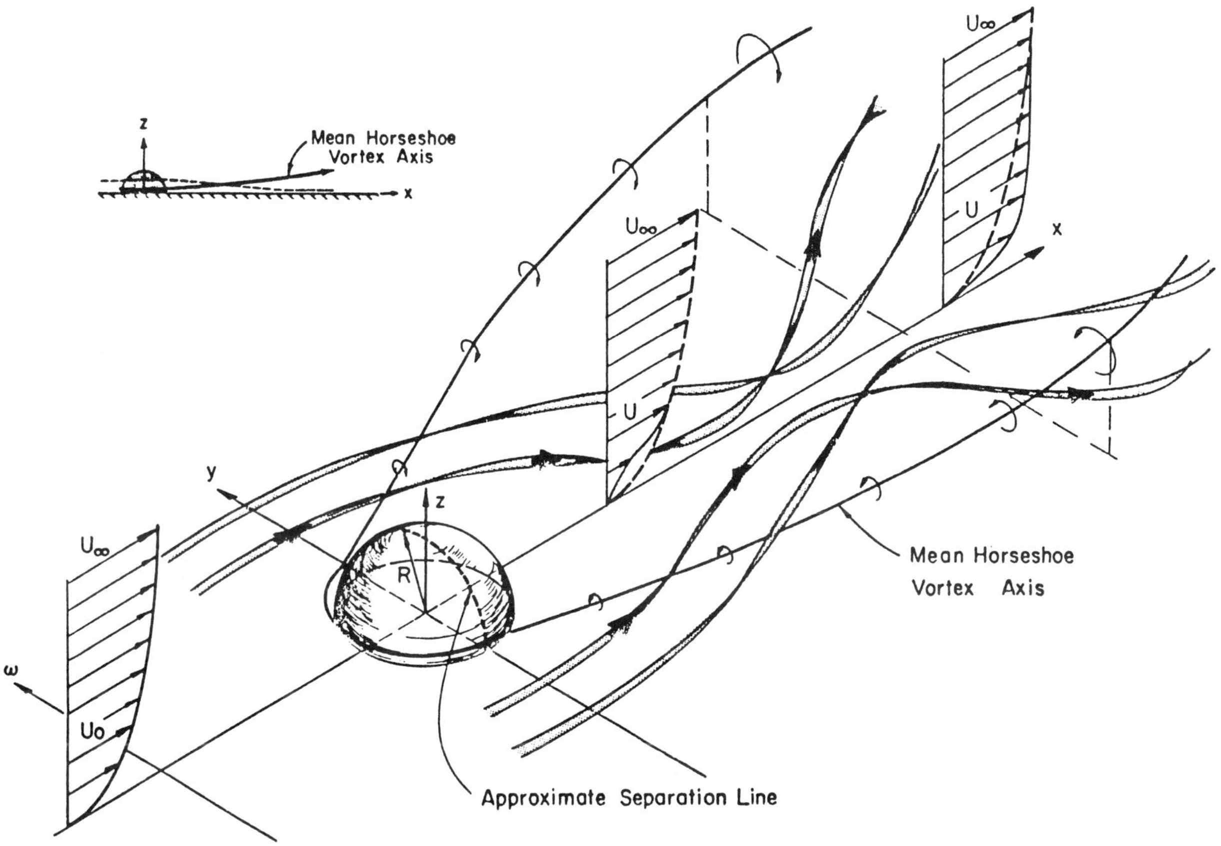

FIG. 7.9. Schematic of the vortex containing wake produced by a hemisphere (Hansen and Cermak 1975).

tend at most to distances of 10 to 15 h, but the wake of a hemisphere may still be evident beyond 60 h.

Separation is generally the dominant feature perturbing flow over surface obstacles; however, thermal stratification or surface heating may further complicate the already confusing flow picture. Hawthorne and Martin (1951) considered flow with vertical velocity and density gradients over a heated hemisphere attached to a horizontal wall. For the stable case, vortex lines, which are horizontal and lie perpendicular to the flow far upstream, are held back by the obstacle. The lines are stretched and rotated so that, when seen from above, the vortex vector points upwind on one side and downwind on the other side of the obstacle. This introduces a secondary component, which produces downward motion behind the obstacle. Over a hot hump, the secondary vorticity generated by gravity is in the opposite direction. Close to the floor where the vertical displacement is negligible the effect is very small, whereas the effect of holding back the vortex lines is a maximum. As a result there is on each side a surface above which the vorticity is dominantly produced by gravity. There is therefore an inflow toward the axis of the flow along these surfaces (see Scorer 1978, section 3.6).

Chien et al. (1978) measured velocities, pressures, and turbulence over Gaussian and cone-shaped hills with crest height to radius ratios of 1 to 6, 1 to 4, and 1 to 3. Pearse (1982) also made similar measurements over a range of smooth and roughened conical shaped hills. Arya and Gadiyaram (1986) examined the wake distributions of velocity, turbulence, and shear out to 20 hill heights and found evidence for well-defined recirculation and trailing vortices. Snyder and Britter (1987) studied flow and diffusion around a bell-shaped hill and moderately steep hills of triangular cross-section and varying crosswind aspect ratio. Surface oil-film visualization revealed that even over a 1 to 6 slope hill an area of separation occurs starting near the crest and spreading downwind laterally at an included angle of 35°. This separation region was fluctuating and nonstationary. The separation regions, increasing in size with increasing aspect ratio, appeared to dominate the entire flow structure. Regions around the crest on the shoulders may produce local high wind velocities. Compared to the two-dimensional ridges, the isolated three-dimensional hills of similar slope do not produce as large magnitudes of fractional speedup.

7.4.3. Field/laboratory comparisons

Flow over complex terrain varies significantly as a function of topographical details and meteorological conditions. For this reason measurement programs were designed to distinguish between the generalizations outlined in the sections above and site-specific types of flow phenomena. Meroney et al. (1978a, 1979) and Neal (1979, 1983) considered flow through mountain gorges and passes in New Zealand, respectively. Chien et al. (1979, 1980) examined flow over the Kahuku peninsula of Oahu, Hawaii. Teunissen et al. (1987) reported the results of a multilaboratory simulation of flow over Askervein Hill, Outer Hebrides, Scotland.

7.4.3.1. Rakaia River Gorge, New Zealand

Both terraced and contoured models of the Rakaia River Gorge region of New Zealand were prepared to an undistorted geometric scale of 1 to 5000 (Meroney et al. 1978a). Flow over the contoured model was examined for three separate surface roughness conditions—smooth, uniform roughness, and detailed shelterbelts. Laboratory measurements included horizontal and vertical profiles of mean wind velocity, longitudinal turbulence, wind direction, turbulence spectra and correlation using hot-wire anemometry, pitot-static pressure probes, and cobra pressure probes.

On two spring days, selected for strong adiabatic down-valley wind flow, three teams of investigators surveyed up to 27 sites on either side and within the Rakaia River Gorge. Measurements consisted of wind speed and direction at a 10 meter height on lightweight portable towers. All measurements were completed during the course of a 5 hour stationary wind event and normalized against continuous records taken from a New Zealand Wind Energy Task Force anemometer located near terrain center. The success of such a procedure depends upon the spatial correlation of wind velocities over the same 100 square kilometer region, quasi-stationarity of the wind event over a 3 to 6 hour period, and the statistical significance of a 15 minute sample at a given point taken once during a 3 to 6 hour recording period. Climatological analysis by Corotis (1977) suggested high correlation (0.76–0.83) would exist over distances less than 22 km and high autocorrelation over periods from 3.5 to 7 hours.

A series of contour diagrams (e.g., Fig. 7.10) were prepared from the laboratory velocity and turbulence intensity measurements into isotach and isoturb charts as shown in Figs. 7.11 and 7.12. Note the wide variation in wind speed near ground level between points within the gorge and the nearby hilltop. Simultaneously large relative gustiness exists within the river gorge when compared with the hillcrest.

The laboratory results were compared with the field data by statistical correlation and scatter diagrams. The sample correlation coefficients ranged from 0.68 to 0.78, and the rank correlation of wind stations by wind speed ranged from 0.78 to 0.95. Correlation between equivalent field measurements at the same sites taken on independent days was 0.68. This suggests that there is an inherent limitation to the paired replication of any single realization of a wind flow pattern by any model whether physical or numerical.

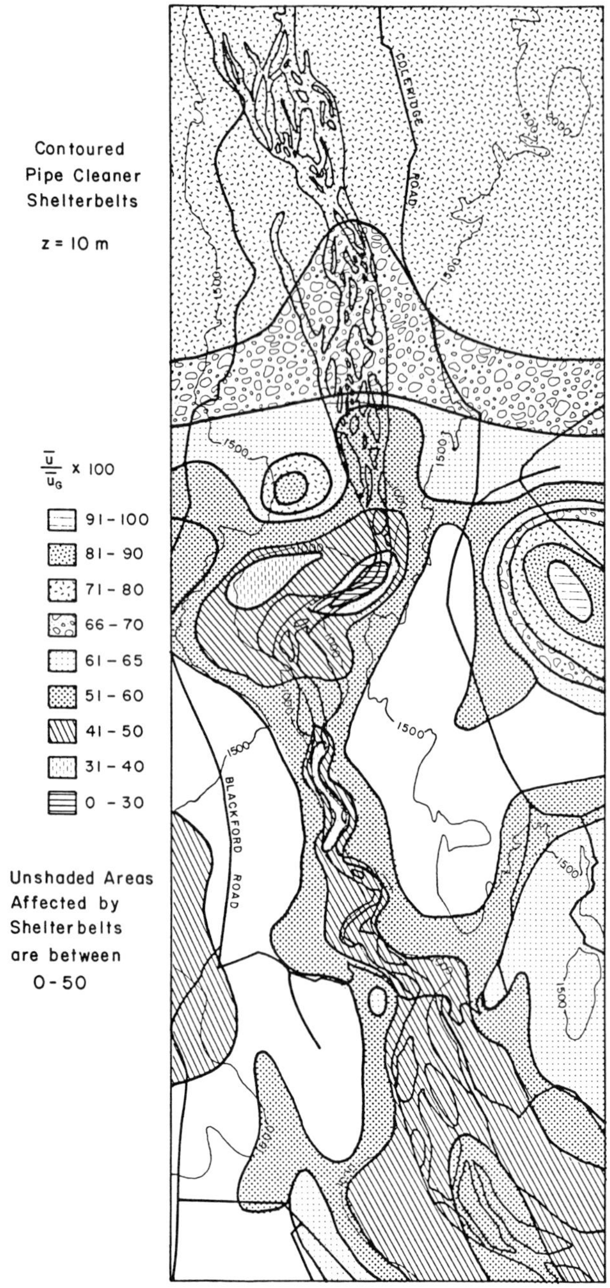

FIG. 7.10. Horizontal isotachs, contoured model with tree shelterbelts at a height of 10 m (Meroney et al. 1978).

7.4.3.2. Gebbies Pass, Banks Peninsula, New Zealand

In a similar research program Neal (1979) prepared a set of 1 to 4000 scale models of the pass (or saddle) that

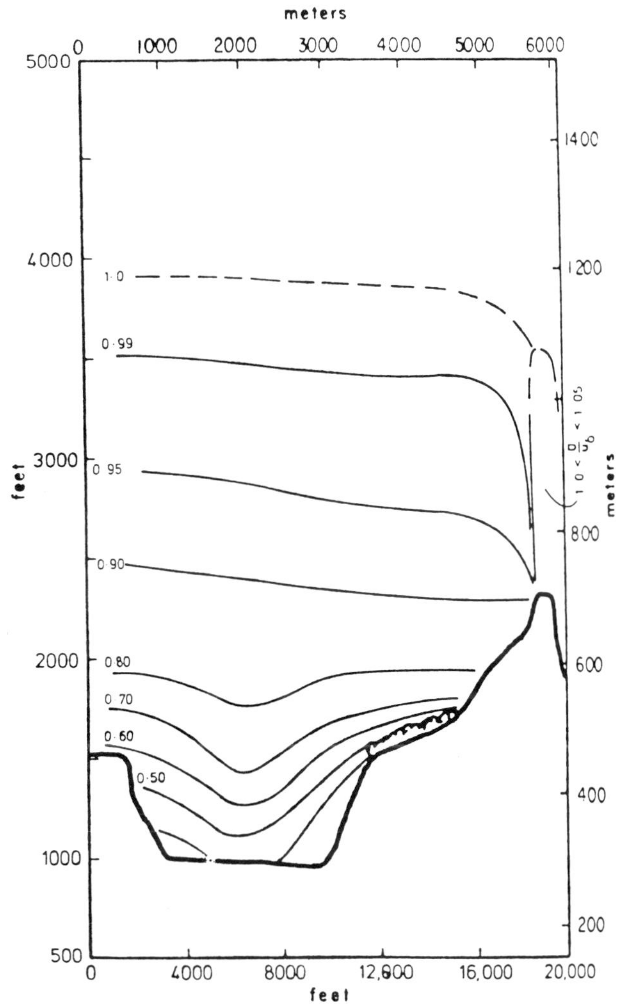

FIG. 7.11. Vertical section of isotachs over Rakaia Gorge contoured model with shelterbelts (Meroney et al. 1978a).

crosses the lip of an ancient volcano today known as Banks Peninsula, New Zealand. Measurements of wind speed and turbulence were compared with field measurements. The sample correlation coefficients ranged from 0.88 to 0.90, and the rank correlation of wind stations by wind speed was 0.98.

7.4.3.3. KAHUKU POINT, OAHU, HAWAII

Chien et al. (1980) measured flow perturbations over an undistorted 1 to 3840 scale model of the Kahuku Point peninsula in a meteorological wind tunnel. Daniels and Oshiro (1980) measured continuous hourly averaged data over the site for a two month period during the fall of 1978 using mobile measurement stations and tethered balloons. Relative wind speeds for each measurement site were compared based on a normalization by an equivalent fixed reference site. Based on field measured wind directions an equivalent relative speed of each site was obtained from the laboratory data using a weighted average approach. The sample correlation coefficient obtained between the field and laboratory data was 0.71, and the relative wind speed rank test correlation was 0.84. Wind speedup over the hilly terrain was also predicted using the methods of Bouwmeester et al. (1978) and Hunt (1980), which were based on experience with two-dimensional idealized hills. Bouwmeester's technique tended to overpredict crest speedup by about +16%, whereas the Hunt method underpredicted winds by −18%. Thus the two methods bracketed typical topography amplification.

7.4.3.4. ASKERVEIN HILL PROJECT, OUTER HEBRIDES, SCOTLAND

The Askervein Hill Project was a collaborative study of boundary-layer flow over low hills (Taylor and Teunissen 1987). Two field experiments were conducted during fall 1982 and 1983 near and around Askervein, a 116 m high hill on the west coast of the island of South Uist in the Outer Hebrides of Scotland. Over 50 towers were deployed and instrumented for wind measurements.

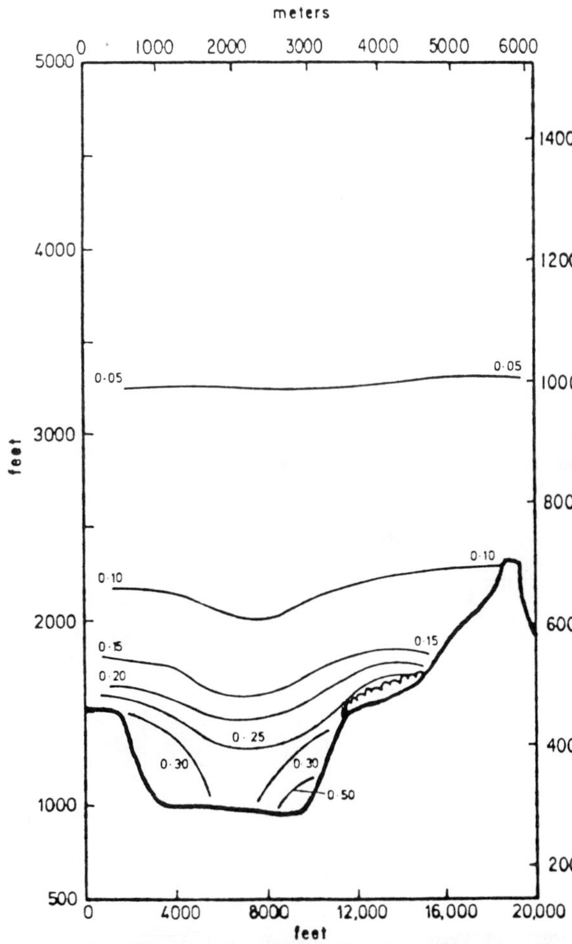

FIG. 7.12. Vertical section of isoturbs over contoured Rakaia Gorge model with shelterbelts (Meroney et al. 1978a).

Most were cup anemometers mounted on simple 10 m posts, but two 50 m, one 30 m, one 16 m, and thirteen 10 m towers were instrumented for three-component turbulence measurements. Subsequently, wind tunnel simulations of neutrally stable atmospheric boundary-layer flow over the hill were carried out at three different length scales (1 to 800, 1 to 1200, and 1 to 2500) and in two wind-tunnel facilities (Teunissen et al. 1987). The objective of the laboratory measurements was to determine the relative impact of varying surface roughness, model scale, and measurement techniques. The wind-tunnel results compared well with each other and with full-scale data (Figs. 7.13 and 7.14).

Changes in mean flow speedup over the physical models were reproduced very well, including those due to small, local terrain features that may be physically small at model scale. Relaxation of the aerodynamic roughness criterion affected the flow only on the lee side of the hill. Turbulence changes induced by the hill did not depend on the nature of the surface roughness. An excessively smooth surface reduced the degree and extent of separated flow and resulted in overestimation of hillcrest wind speeds. Simulations in two facilities using three models at three different scales showed a gratifying degree of consistency. The only effect of model scale was a predictable increase in difficulty in making measurements very close to the surface as the size of the model decreased. The depth of the turbulent inner layer was less than that suggested by Jackson and Hunt's (1975) relationship, but similar to the value predicted by Jensen et al. (1984).

7.4.4. Conclusions from neutral airflow terrain studies

Physical modeling experience from studies performed over idealized two- and three-dimensional hills and scale models of complex actual complex terrain reveals:

- The gradient of velocity with height at ridgecrests is usually small.
- Except for very small slopes (<1 to 10) the amplification factor, A, or fractional speedup, ΔS, varies with height.

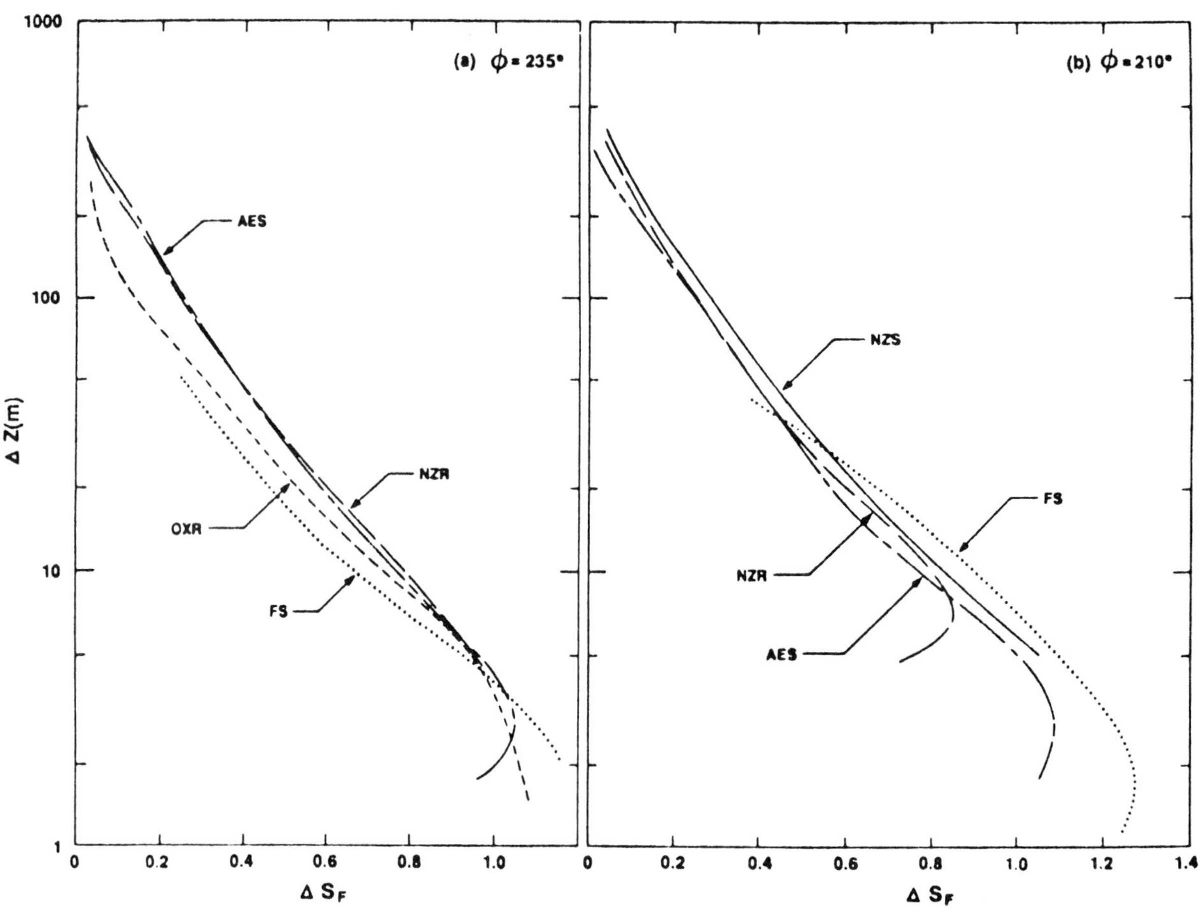

FIG. 7.13. Vertical profiles of mean speed and speedup at the hilltop location for two wind directions in wind-tunnel and full-scale flows (Teunissen et al. 1987).

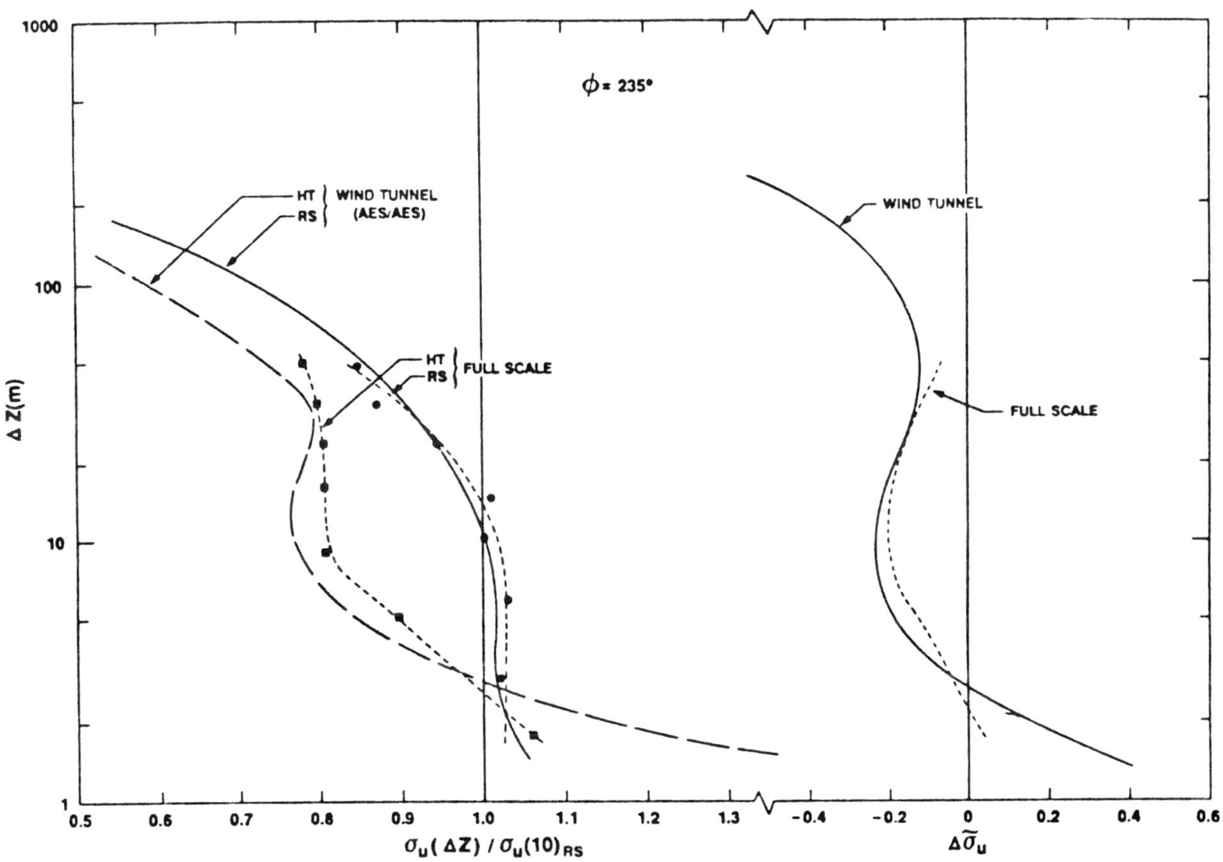

FIG. 7.14. Vertical profiles of σ_u and $\Delta\sigma_u$ at the hilltop (*HT*) and reference location (*RS*) in wind-tunnel and full-scale flows for $\phi = 235°$ (Teunissen et al. 1987).

- Hills of different shape, but equivalent average slope give similar profiles when separation is absent.
- Separation at hillcrest is a function of both upwind and downwind slopes.
- Three-dimensional hills do not produce speedup as large as similar section ridges.
- For small and moderate length hills turbulence intensity decreases at hillcrest since only preexisting turbulent energy is convected over the hill.
- Potential or inviscid-rotational numerical calculations predict unseparated flows very well.
- Undistorted models at scales as large as 1 to 5000 permit accurate resolution of velocity and turbulence details.
- Replication of near-ground wind speeds and separation requires careful attention to local surface features and the presence of vegetation.
- Relaxation of the criteria of roughness Reynolds number to values near $Re_* = Uz_o\nu^{-1} = 0.4$ is permissible.
- Careful measurements over different models in different facilities produce equivalent results.

7.5. Stratified flow over hills and ramps

There is extensive literature that deals with stratified channel flow and tank experiments performed to examine the perturbations generated by obstacles in stratified fluids (Turner 1973; Gebhardt 1988; Simpson 1987; etc.). Similar experiments were performed in wind tunnels (Lin and Binder 1967), but normally it is easier to produce stratification and attractive visualization in water. The purpose of such experiments is to gain understanding about the role stratification, upper boundaries, and terrain form play in airflow patterns, wave generation, blockage, and turbulent mixing. Such experiments provide the foundation for the analytic and numerical concepts about mountain waves and downslope winds discussed in Durran, Chapter 4, and Carruthers and Hunt, Chapter 5.

7.5.1. Idealized two-dimensional flow domains for waves and blocking

Flows induced by combinations of stratification and surface obstructions can be extremely complex and intriguing; however, even flows of a single laboratory liquid layer over a two-dimensional model ridge can induce surface motions very similar to the inversion layer movements observed in the atmosphere. Figure 7.15 displays the five domains of flow behavior observed for different combinations of Froude number ,$U^2(gh)^{-1}$, and hill height ratio, hD^{-1}. When the ridge to inversion height ratio is small and wind speeds are large winds will flow over a ridge in a symmetric rapid-flow pattern (Fig. 7.15a). As the relative ridge height increases or wind speeds decrease then mountain waves (Fig. 7.15b), foehn walls (c), hydraulic jumps and rotors (d), and even upwind blockage (e) may occur. Gradients in vertical stratification and the presence of additional elevated inversions will change both the quantitative and qualitative appearance of the winds.

Baines (1979) performed finite-water-depth tow-tank experiments over small obstacles ($hD^{-1} = 0.076$) for a wide range of stratification conditions ($K = ND(\pi U)^{-1} = 0.5$ to 3.0). He determined that no upstream waves occurred for $K < 1$, but complex waves, columnar upstream disturbances, overturning, and blockage occurred when $K > 1$. There was some concern expressed whether the results were distorted by the presence of a reflective water surface lid during the experiments. Baines and Hoinka (1985) presented additional tow-tank experiments in a water channel designed to remove surface reflections and simulate an infinite depth stratified fluid (Fig. 7.16). They determined that upstream disturbances could be induced in stratified flows even without an upper boundary. Criteria for the presence of columnar upstream motion and blocking in terms of K for various obstacle shapes were proposed, as shown in Fig. 7.17.

7.5.2. *Downslope winds, valley flows induced by crosswinds*

Tampieri and Hunt (1985) used a small water channel to examine the behavior of stratified flow over two successive humps. The two ridges were of small to moderate slope (between 0.25 to 0.5); the distance between them varied between being slightly less than to much greater

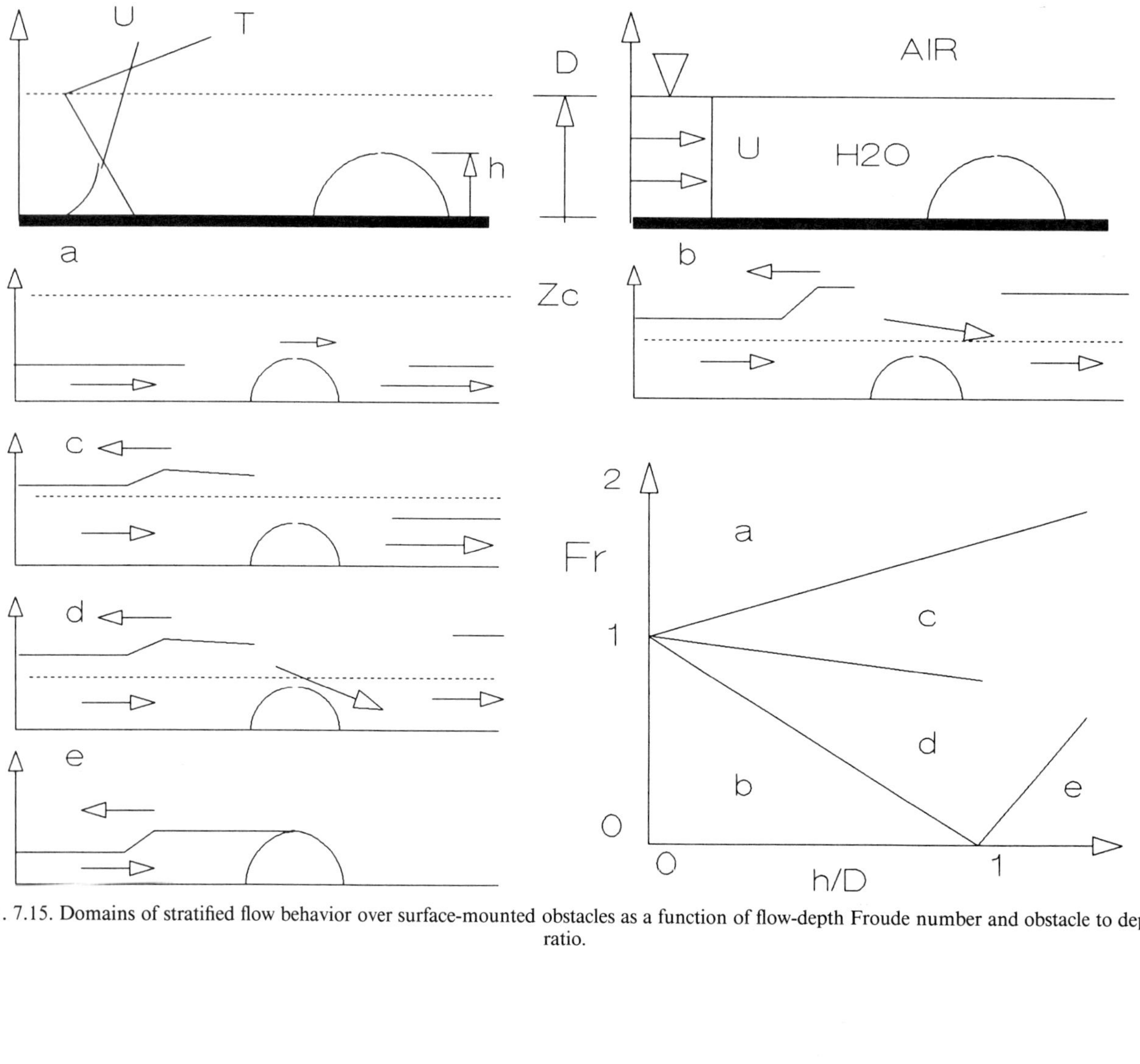

FIG. 7.15. Domains of stratified flow behavior over surface-mounted obstacles as a function of flow-depth Froude number and obstacle to depth ratio.

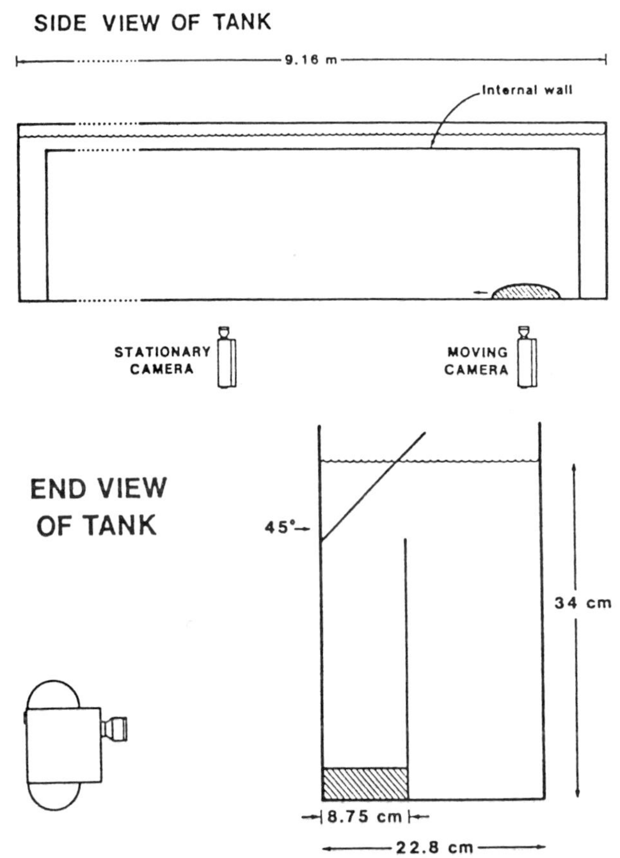

FIG. 7.16. Schematic arrangement of baffles in a water filled towing tank to simulate an infinite depth flow (Baines and Hoinka 1985).

than the wavelength, λ, of internal gravity waves or lee waves (Fig. 7.18). Over a narrow range of Froude number, F_L (based on hump half length L) such that the natural lee wavelength is about equal to the valley width D, and for a particular ratio D/L near 6, the valley is ventilated without significant regions of separation. At other values, there was some recirculation and stagnation of the flow (Fig. 7.19). They suggested that it is not sufficient for the length of the natural lee waves to equal the spacing between the humps to produce ventilation. It is also necessary for the higher speed flow on the lee of the first hump to overcome the slowing down of the flow by the downwind hump. Finally, surface winds are expected to be greatest on the lee slope of the second hump.

7.5.3. Idealized three-dimensional terrain studies

Stratified three-dimensional laboratory studies have been directed to find (i) whether streamlines from upwind impinge, go around, or go over a hill, (ii) the size and location of internal jumps, waves, and separated or recirculating flow, (iii) the effects of heating or cooling the surface, and (iv) the differences between two- and three-dimensional flow fields.

The work of Hunt and Snyder (1980; discussed in Chapter 5) focused on the idea of a separating streamline that divided regions where flow went around a hill versus over the crest (Fig. 7.20). Colored dye visualization of stratified flow about a polynomial-shaped hill revealed that for $F \ll 1$ the flow is approximately horizontal except in narrow regions at the top and bottom of a hill. For $F > 1.5$ flow over the hill was neutral. When $F < 0.8$ separation is suppressed at the top of the symmetrical hill. A criterion was proposed for determining whether the plume will impact on the hill surface and go around the sides or over the top. They proposed the relation $H_s = h(1 - F)$, where H_s is the dividing-streamline height, h is the hill height, and F is the Froude number characterizing the stratification. The analytic justification for this concept is critiqued by Smith in the Appendix to Chapter 5. Snyder et al. (1985) compared stratified flow tank results for flow over a 1 to 690 scale model of Cinder Cone Butte, Idaho, with field data and the separated streamline theory. Laboratory experiments also alerted investigators to the extreme sensitivity of surface concentration patterns around hills to wind direction—a 5° shift in wind direction shifted surface plumes from one side of the hill to the other (Fig. 7.21).

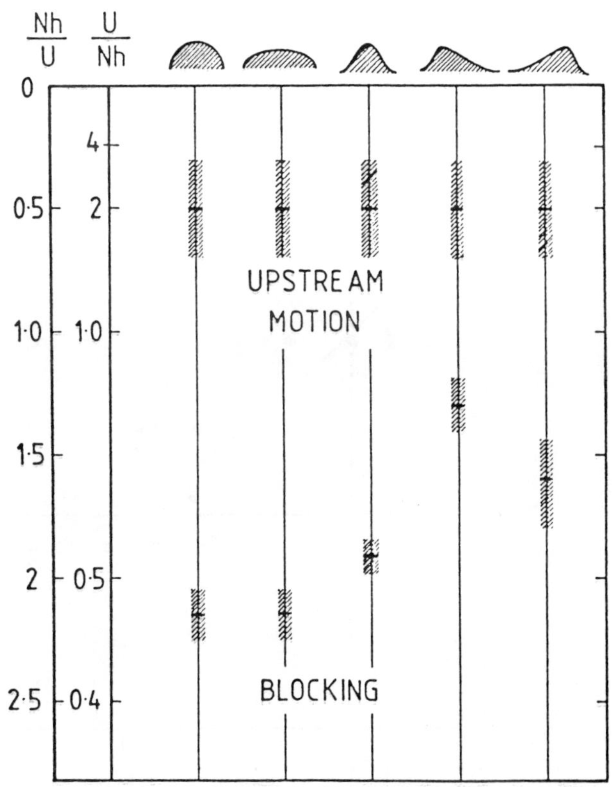

FIG. 7.17. Criteria for the presence of columnar upstream motion and upstream blocking in terms of Nh/U for a range of surface mounted obstacle shapes. Hatched regions denote error bars. (Baines and Hoinka 1985).

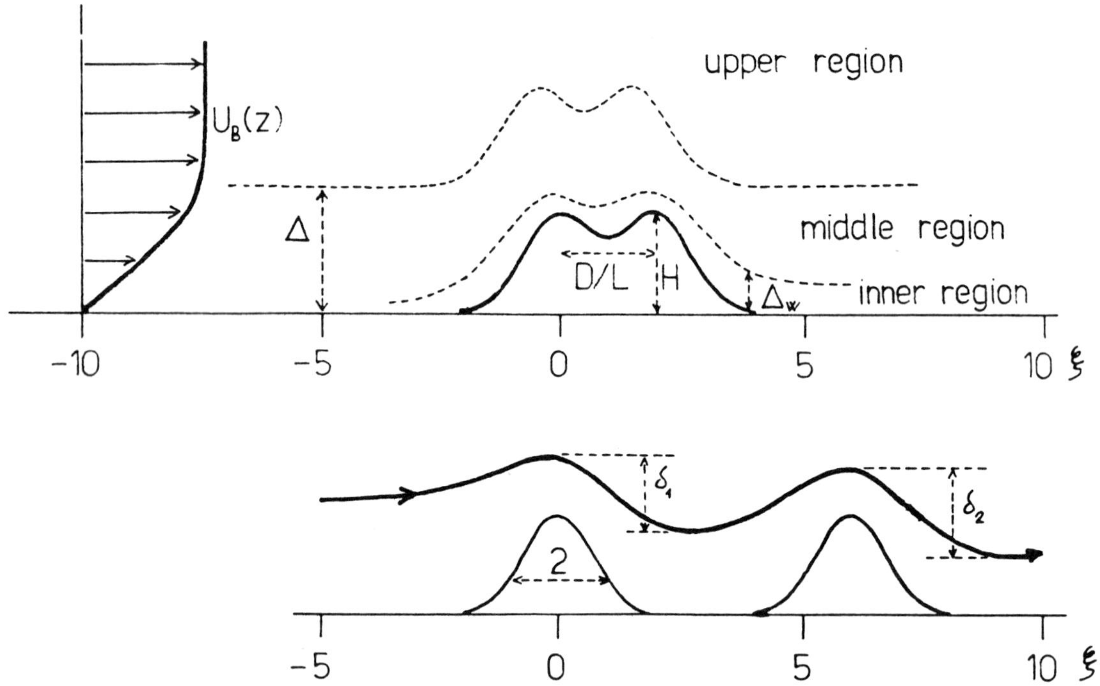

FIG. 7.18. Sketch of stably stratified flow over two humps where the surface height is equivalent to the sum of their separate heights, two spacings (Tampieri and Hunt 1985).

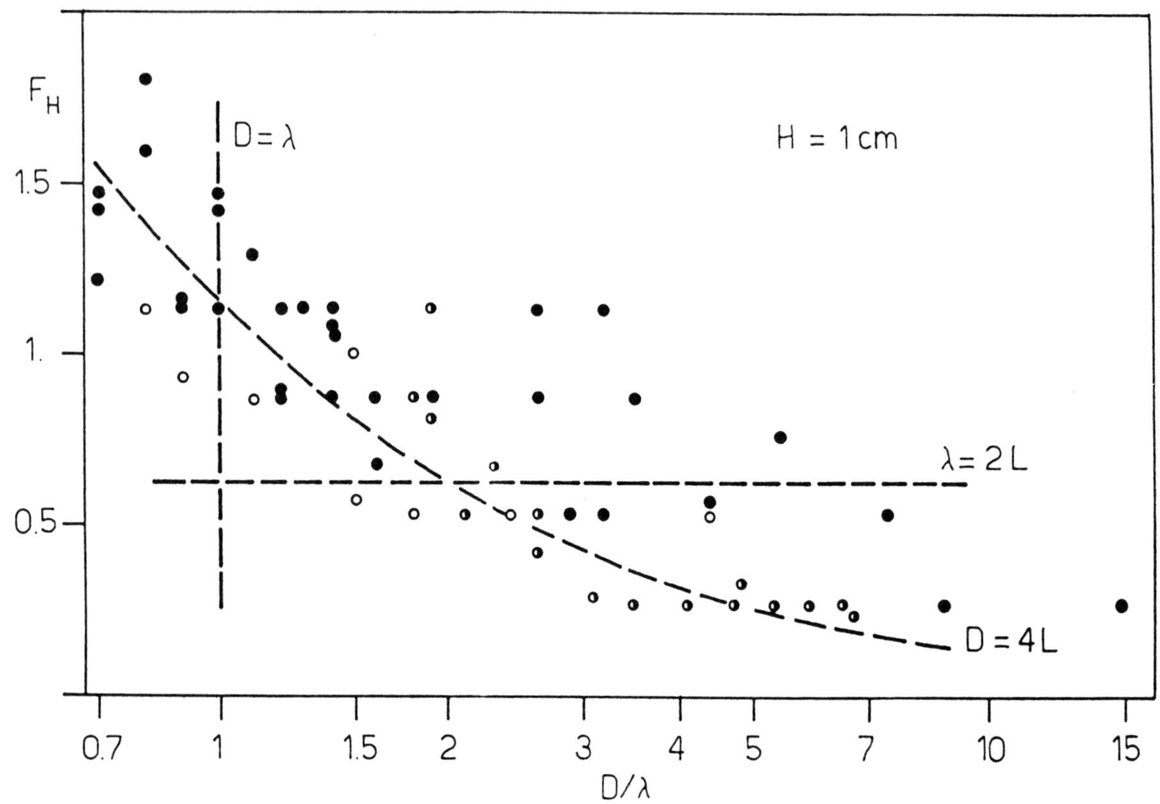

FIG. 7.19. Water channel experiments performed with obstacles 1 cm high. Full symbols correspond to good penetration; open symbols to no penetration; half-filled symbols to intermediate cases (Tampieri and Hunt 1985).

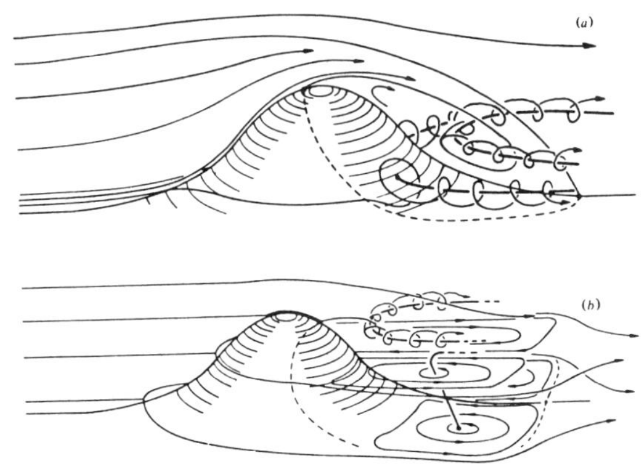

FIG. 7.20. Schematic of dividing-streamline concept for stratified flow over hills (Snyder 1980).

An experimental study of the stratified flow over triangular-shaped ridges of various aspect ratio revealed that wave amplitude can be maximized by "tuning" the body shape or hill aspect ratio, $\alpha = W/h$, to the lee-wave field (Fig. 7.22; Castro et al. 1983). In certain circumstances steady wave breaking can occur or multiple recirculation regions (rotors) can exist downstream, and vortex shedding in horizontal planes is possible even at $F = 0.3$. Significantly, it was found that hill shape and the ratio of the cross-stream width of the body to its height has a negligible effect on the dividing-streamline height. As expected hill span affected upstream separation zones. The authors ended their paper with Scorer's (1978) warning: "There is almost no systematic way of discussing the flow behind a three-dimensional obstacle of arbitrary shape when separation occurs!" Nonetheless, they believed certain generic phenomena could be studied and identified. Snyder et al. (1985) also studied the structure of strongly stratified flow over two- and three-dimensional hills. They concluded aspect ratio had small effect on the dividing-streamline height. They also expressed concern that the upstream influence of stratified flow over long ridges takes so long to develop that it may be unlikely that the stationary results can be obtained in finite length towing tanks. Snyder (1988) also presented a summary of the various contributions made by the EPA Fluid Modeling Facility to a multiyear program known as the Complex Terrain Model Development Program (CTMDP).

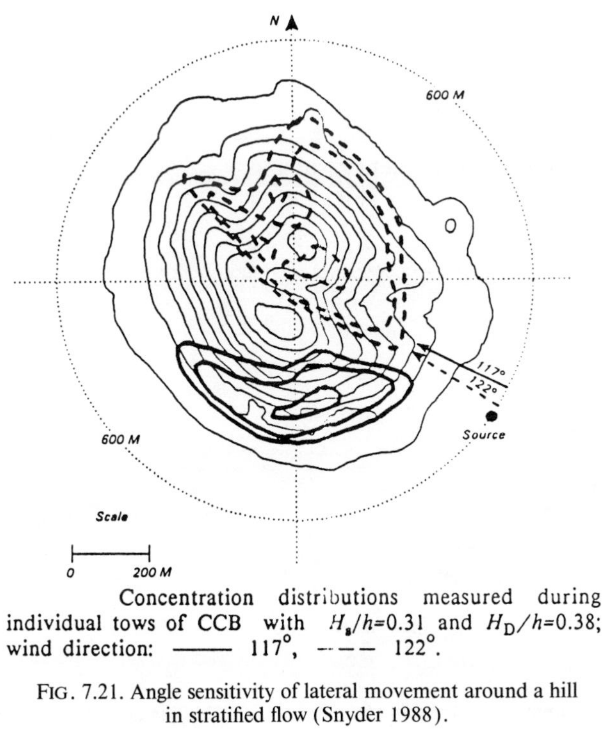

FIG. 7.21. Angle sensitivity of lateral movement around a hill in stratified flow (Snyder 1988).

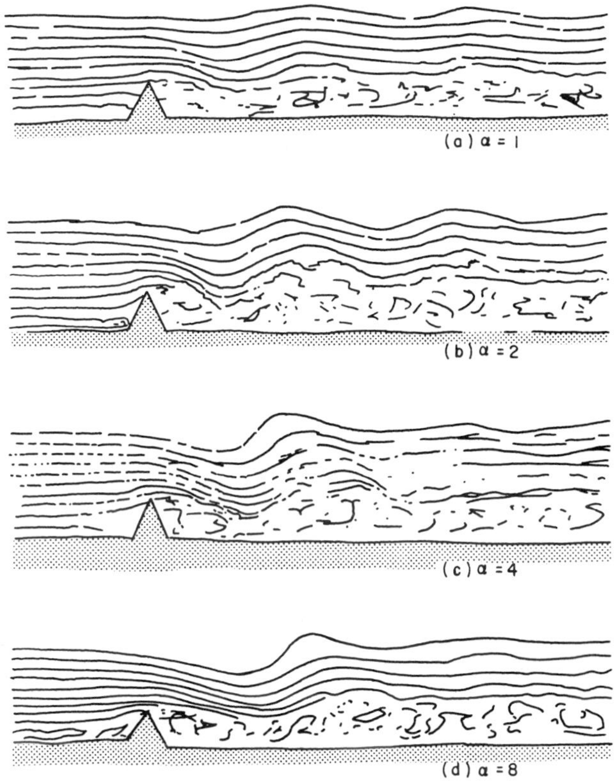

FIG. 7.22. Dye streaks obtained during stratified tow-tank experiments of flow over triangular-shaped ridges of various aspect ratio (Castro et al. 1983).

7.5.4. *Field/laboratory comparisons*

Abe (1929) compared CO_2 streamline flows over a model of Mt. Fuji with cloud patterns in the atmosphere, and Long (1959) simulated mountain lee waves over the Sierra Nevada using a brine filled water channel. Later Cermak and Peterka (1966) and Meroney and Cermak (1967) performed barostromatic[3] wind tunnel experi-

[3] Airflow with stable thermal stability in the upper levels and near-neutral thermal stability in the lower levels.

ments over 1 to 12 000 scale models of Point Agruello, California, and 1 to 6200 scale models of San Nicolas Island, California; however, only plume trajectories and concentrations were available for field/model comparisons. Kitabayashi et al. (1971) compared wind flow measurements over a 1 to 9600 scale model of Elk Mountain, Wyoming, with extensive airplane soundings. Given rather coarse measurement resolution the two studies agreed very well.

During a field and laboratory program to evaluate the efficacy of ground-based cloud seeding generators, Orgill et al. (1971a, 1971b) made extensive measurements of wind flow, turbulence, temperature, and dispersion over a scale model of the Eagle River Valley–Climax and the San Juan Mountain region of Colorado. Both neutral and barostromatic boundary layers were simulated over a 1 to 9600 scale model of the Eagle River/Climax area (Fig. 7.23). Wind speeds and wind direction variations with height were measured using a novel stereo smoke–wire method. Temperature profiles were measured with thermocouples, and carbon dioxide and air density measurements were made using thermal-conductivity cells. Concentrations from model plumes released at nuclei generator sites were determined using a Krypton-85 radioactive detector system.

Field data were taken during the winter snow season using rawinsondes, pilot balloons, constant-volume balloon flights, aircraft, and surface sampling of silver-iodide nuclei. The vertical rise or extent of the plume measured over the model during neutral and barostromatic simulations with field data is compared in Fig. 7.24. The barostromatic model simulated both the vertical and the lateral spread of the plume very well. Model concentrations fell within the range of field surface concentration measurements. There was large data scatter, but this was expected given the limited number of field measurements and the inherent variability of atmospheric turbulence. The laboratory and field data revealed that near-neutral stability conditions within the valley, orographically induced eddies, directional wind-shear, and convection during the orographic storm events dispersed the seeding material in the vertical and cross valley. Laboratory simulation, despite its inherent limitations, was found to be a practical tool for estimating airflow and dispersion characteristics over mountainous areas.

7.5.5. Conclusions from stratified airflow terrain studies

Physical modeling experience from studies performed over idealized two- and three-dimensional hills and scale models of actual complex terrain reveals:

- Dependent upon stratification, hill geometry and spacing various combinations of waves, downslope winds, valley penetration, streamwise division, upwind penetration, and blocking can be reproduced in the laboratory.
- Hilltop profiles of wind speed and turbulence are generally more dependent upon hill height than hill shape or aspect ratio.
- The dividing-streamline concept works over a wide range of hill shapes and hill aspect ratios.
- Two-dimensional ridges may block flow very far upwind.
- Given two parallel ridges the largest downslope winds will occur on the second ridge.
- Two-dimensional ridges will generally produce larger vertical perturbations than three-dimensional hills.
- Undistorted models at scales as large as 1 to 10 000 permit resolution of velocity, turbulence, and plume diffusion details.
- With some effort studies of the effect of raised inversions and convective heating from below are possible.
- Careful tests over different models in different facilities produce equivalent results.

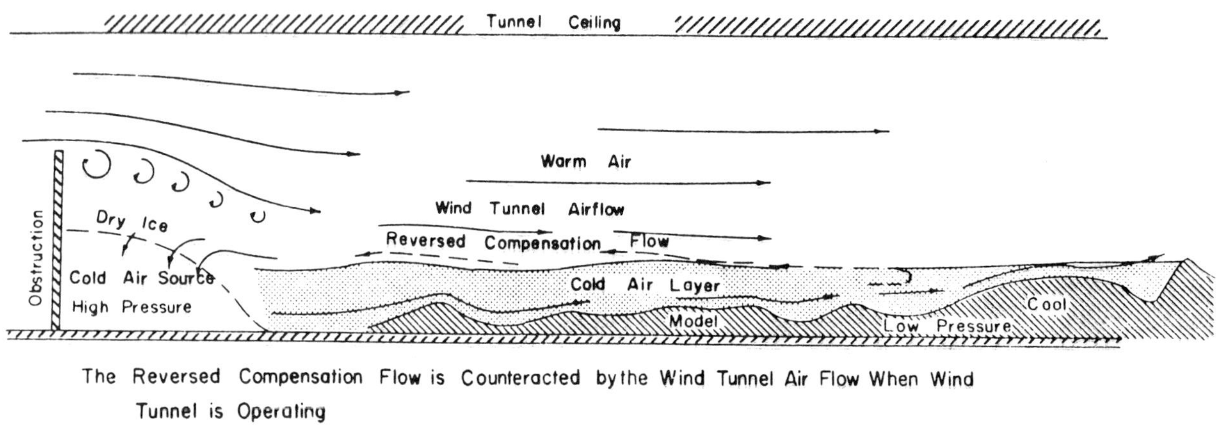

FIG. 7.23. Schematic representation of the wind tunnel airflow from the combination of wind tunnel and dry ice (Orgill et al. 1971a).

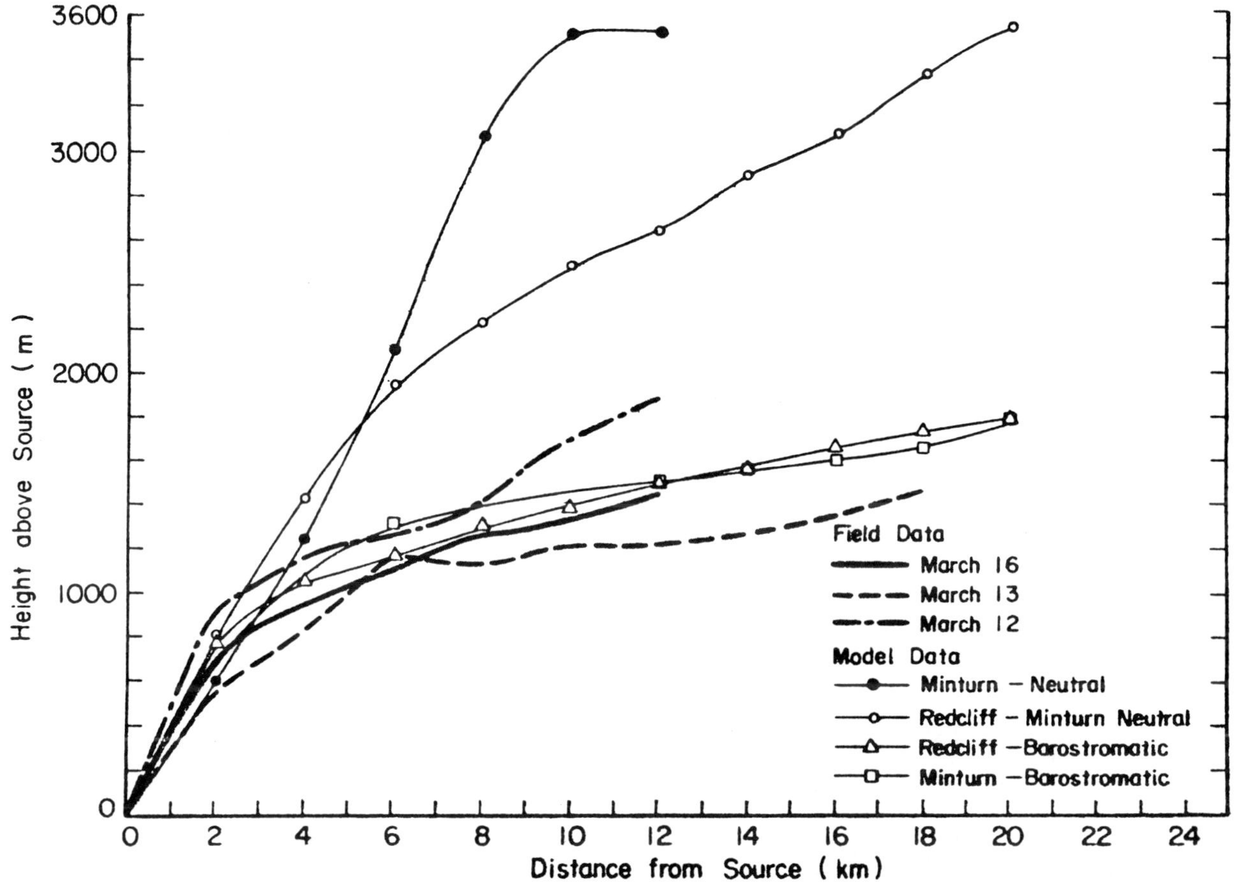

FIG. 7.24. Comparison between the vertical rise of model and field tracer plumes during the Eagle Valley/Climax cloud seeding experiments (Orgill et al. 1971a).

7.6. Drainage flow phenomena

Physical modeling of gravitational convection over complex terrain caused only by buoyancy forces requires that a model be isolated in a chamber free from other air disturbances. Whiteman, Chapter 2, surveys those situations dominated by such thermally forced flows. Cermak (1984) describes a convection chamber used to study mountain drainage flows and their influence on plume dispersion. A schematic of such a drainage flow facility is shown in Fig. 7.25. The model must be constructed of a metal foil conductive surface to permit surface cooling by sublimation of dry ice or convection of cryogenic gases. Hertig (1986) argued that in the special case of drainage flows vertical geometric distortion justifies horizontal scales to 1 to 100 000 and vertical scales of 1 to 20 000. He constructed a scale model of Switzerland to simulate the hydrostatic contributions to the foehn winds. Heating of the upper air temperature in a convection chamber produces a greater temperature potential and larger resultant flow velocities. Wind speeds are usually measured by visualization methods (e.g., smoke–wire).

Cermak and Petersen (1981) report the use of a convection chamber to study the drainage flows at two geothermal power plant sites at the Geysers site in Northern California and at a molybdenum mine site in the Rocky Mountains near Crested Butte, Colorado. An aluminum foil shell model was constructed at the molybdenum site of the surrounding Coal Creek canyon area. The 1 to 1920 scale model was cooled by circulating gas cooled with two tons of dry ice placed in the insulated space below the model. Profiles of temperature and velocity measured over the mine site appear in Fig. 7.26. Notice the up-valley recirculation that occurs at field heights above 500 m, the evidence of a mountain wind maximum at 150 m, and the peak in valley drainage winds at about 60 m. The magnitude of the drainage winds was found to increase as the square root of distance from the canyon origin. Davidson and Rao (1963) observed a similar behavior during field observations. A few prototype measurements were made during drainage events in Coal Creek canyon; peak winds magnitudes found were about 5 m s^{-1} at 100 m. Model measurements at a similar location predicted wind speeds of 5.3 m s^{-1} at 125 to 150 m.

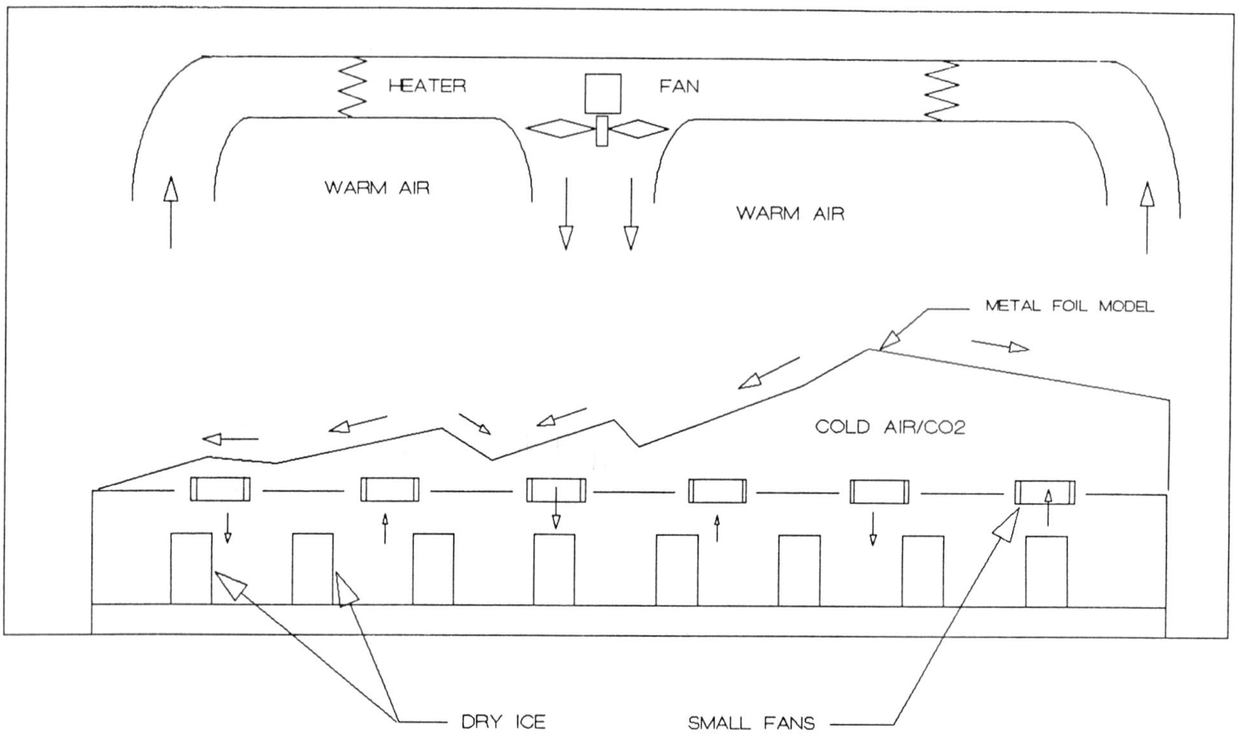

FIG. 7.25. Schematic of convection chamber at Colorado State University used to simulate mountain–valley drainage patterns.

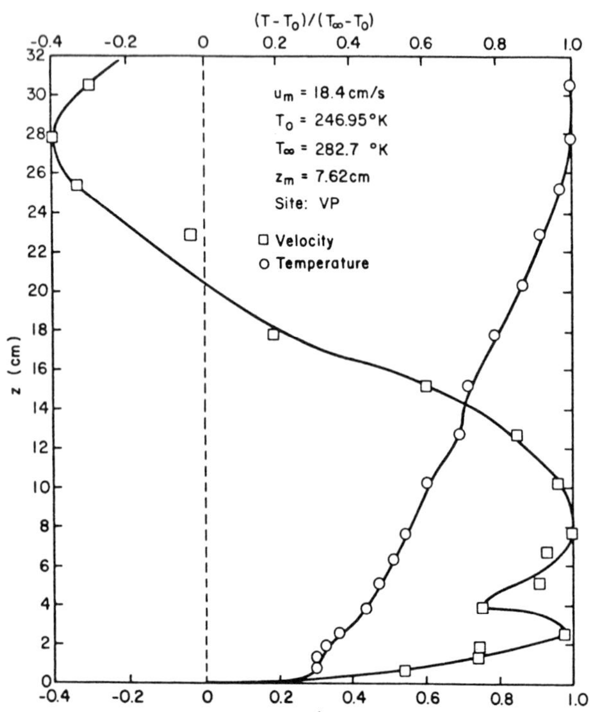

FIG. 7.26. Velocity and temperature profiles taken over the mine site for the Coal Creek drainage flow tests; Ri = 1.69 (Petersen and Cermak 1980).

Limited data suggest that physical modeling of drainage flow can be achieved with sufficient accuracy to provide useful predictions of flow and dispersion; however, further evaluation and verification studies must be completed before the approach can be used with confidence.

7.7. Diffusion phenomena in complex terrain

Given accurate simulation of the dynamics and kinematics of flow over complex terrain there is good reason to expect similar simulation of the transport and dispersion of tracers or pollutants. Physical modeling of terrain effects on diffusion has provided many of the insights discussed in Hanna and Strimaitis, Chapter 6. In the absence of flow data over complex terrain similarity between field measurements of air pollution and model predictions is often used to validate numerical and physical model procedures. Snyder (1981) considered the criteria and evidence for realistic air pollution simulation in his report "Guidelines for Fluid Modeling of Atmospheric Dispersion," which was associated with the EPA good engineering practice stack height regulations. Meroney (1986) extended this discussion to the criteria and evidence for fluid modeling of dense gas dispersion. Two studies have been selected to display the skills of fluid modeling while predicting dispersion over complex terrain.

A combined physical modeling and airborne monitoring study was completed for terrain near the Kingston Steam Plant (KSP), Tennessee, to evaluate the effects of terrain-induced turbulence and vertical velocity upon the plumes emitted from the plant (Graham et al. 1978). The physical modeling tasks, conducted in the EWT at Colorado State University, included simulations both with and without local terrain features and for both short (100 m) and tall (300 m) stack configurations. Modeling techniques employed included flow visualization, gas tracer measurements, and hot-wire anemometry. Airborne monitoring was conducting using a Cessna 207 aircraft equipped to measure wind speed, turbulence, temperature, humidity, and concentrations.

The somewhat limited airborne data provided qualitative verification of the wind tunnel results. Both model and field produced turbulent intensities in the 0.1 range over the first 500 m above ground. Because model experiments were run with and without underlying terrain it was possible to determine the effect of the terrain on ground level concentrations. Terrain effects were shown to increase surface concentrations by a factor of 2 (Fig. 7.27).

Liu and Lin (1975) simulated plume dispersion from a lead smelter in Glover, Missouri, over four terrain models with a 1 to 2500 scale in a stratified towing tank. Figures 7.28 and 7.29 show side and top view visualizations of buoyant stack plumes released into neutral ($\mathrm{Fr}_H = U(NH)^{-1} = \infty$), slightly stable ($\mathrm{Fr}_H = 5.4$), and stable ($\mathrm{Fr}_H = 2.7$) conditions. Stack velocity ratio increases from $K = W_s U^{-1} = 1.4$ to 4.4 as stability increases; however, plume buoyancy remains constant at $\mathrm{Fr}_D \approx 3$, and plume rise stays nearly constant. Note that rapid lateral and vertical spreading decreases as stability increases, but as the stratification increases further elevated waves induce local shear and the plume plunges toward the ground downwind of the mountains. Under neutral conditions the plumes were considerably diluted within the simulated atmospheric boundary layer. Under stable conditions the plumes were aloft upstream of the mountains, but high ground concentrations were measured downstream of the mountains at locations where downslope winds occur.

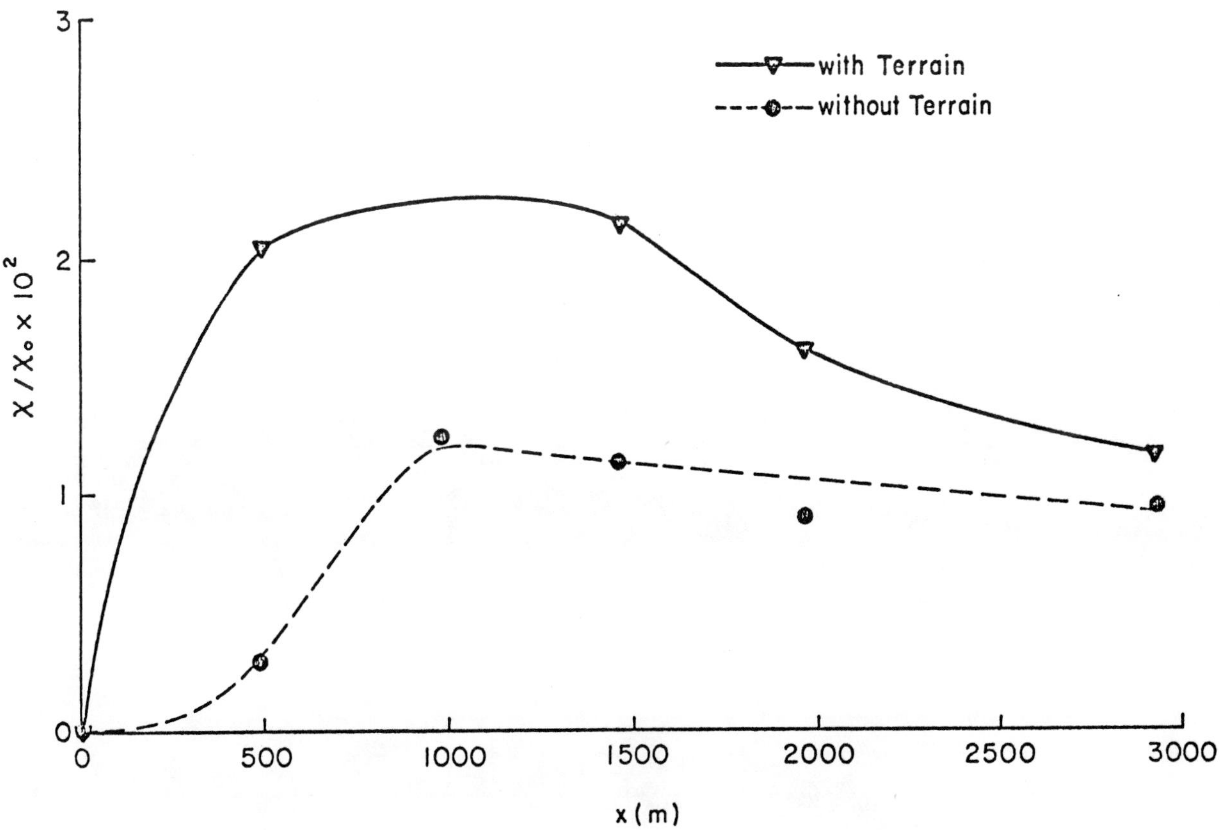

FIG. 7.27. Maximum ground level concentrations versus downwind distance with and without model terrain for a 6.7 m s^{-1} prototype wind speed and Kingston Power Plant Units 1 through 9 operating (Graham et al. 1978).

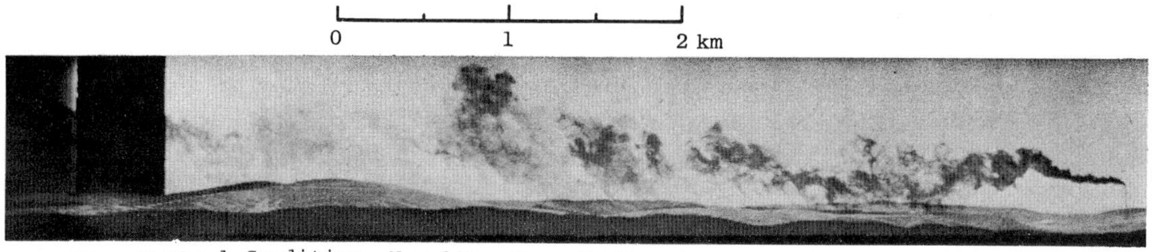

a. Run IVa: Neutral Condition, $N = 0\,\mathrm{Hz}$, $U_\infty = 12.6$ cm/s, $K = 1.4$, $F_H = \infty$, and $F_D = 3.9$

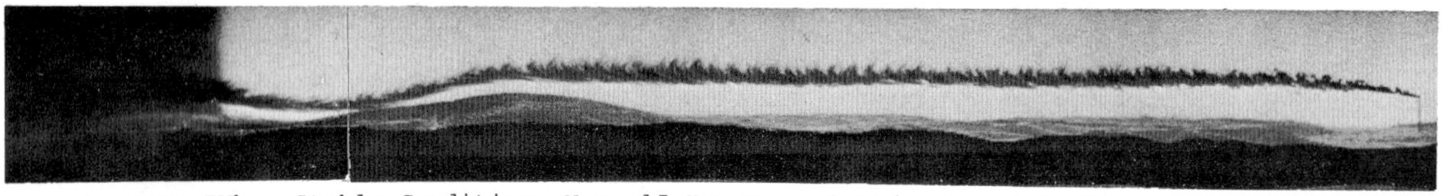

b. Run IVb: Stable Condition, $N = .15$ Hz, $U_\infty = 6$ cm/s, $K = 2.2$, $F_H = 5.4$, and $F_D = 3.1$

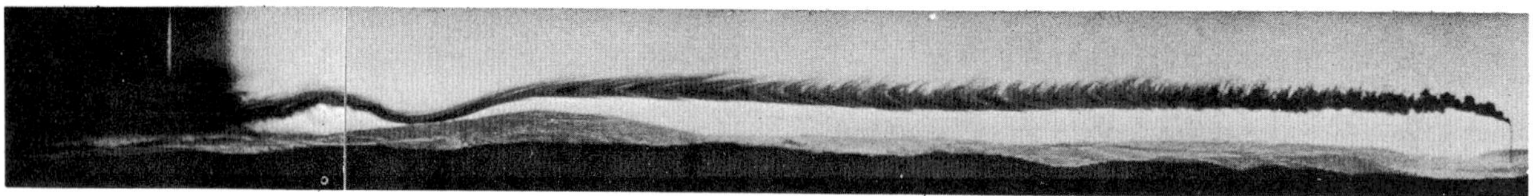

c. Run IVc: Stable Condition, $N = .15$ Hz, $U_\infty = 3$ cm/s, $K = 4.4$, $F_H = 2.7$, and $F_D = 3.1$

FIG. 7.28. Plume visualization of side view of plume dispersion from a lead smelter in Glover, Missouri. Simulation in a stratified towing tank for neutral and stratified flows (Liu and Lin 1975).

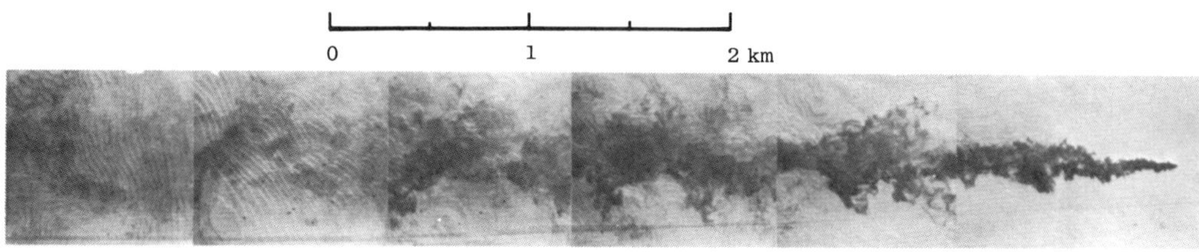

d. Run IVa: Neutral Condition, $N = 0$ Hz, $U_\infty = 12.6$ cm/s, $K = 1.4$, $F_H = \infty$, and $F_D = 3.9$

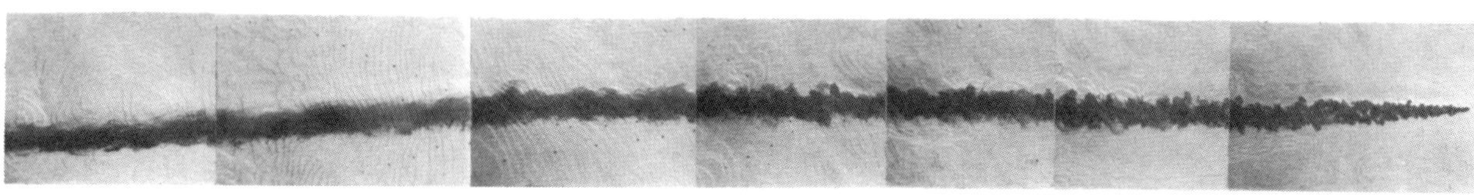

e. Run IVb: Stable Condition, $N = .15$ Hz, $U_\infty = 6$ cm/s, $K = 2.2$, $F_H = 5.4$, and $F_D = 3.1$

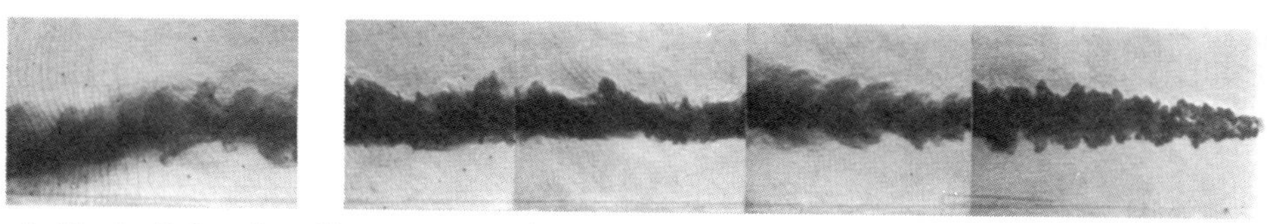

f. Run IVc: Stable Condition, $N = .15$ Hz, $U_\infty = 3$ cm/s, $K = 4.4$, $F_H = 2.7$, and $F_D = 3.1$

FIG. 7.29. Plume visualization of top view of plume dispersion from a lead smelter in Glover, Missouri. Simulation in a stratified towing tank for neutral and stratified flows (Liu and Lin 1975).

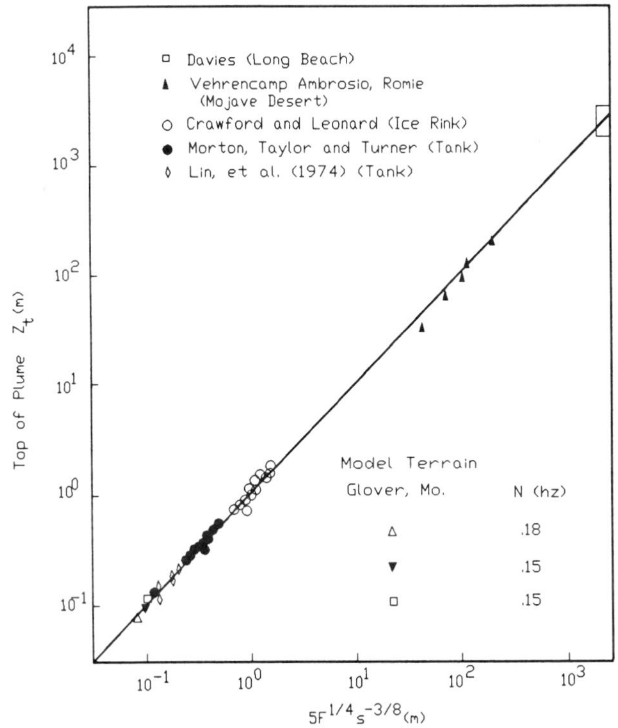

FIG. 7.30. Laboratory and field data from Glover lead smelter plume simulation compared with field data of plume rise in calm and stable flows (Liu and Lin 1975).

Measurements of initial plume rise, Z_T, upwind of the mountain (Fig. 7.30), agree with the prediction of Briggs (1969).

7.8. Summary

During the last 50 years there has been a period of growth in the science of fluid modeling and its contribution to the understanding of atmospheric processes over mountainous terrain. The science has moved from ad hoc and intuitive experiments designed to provide engineering design answers to systematic fluid mechanic studies based on sound principles of fluid dynamics, similitude, and metrology. Broad insights gained through this half-century of study suggest that:

- The turbulent atmospheric boundary layer over complex terrain can be simulated in long test-section wind tunnels and water channels except for effects caused by Coriolis accelerations.
- The characteristics of wind speedup, lee waves, valley penetration, upwind columnar disturbances, and blocking have been simulated successfully in stratified wind tunnels and water channels.
- Transport and diffusion over complex terrain for neutral and stratified flows has been successfully simulated in laboratory facilities.
- Laboratory results have been useful to guide the development of analytic and numerical models.

Nonetheless, much of wind engineering methodology is still more an art than a science; hence:

- Further development of simulation methods for the convective boundary layer and elevated inversions is necessary.
- Further attention should be directed to the lower critical Reynolds numbers required for correct simulation of neutral, stratified, and drainage flow situations over complex terrain.
- The value and validity of using vertically distorted complex terrain models to examine very large surface areas should be examined further.
- Consideration should be given to the value of using inverted topographical models with electrical heating for studies of drainage flows with nonuniform surface temperature distributions.

Finally, new research areas for fluid modeling and complex terrain flows that should be fruitful include:

- Research on the joint effects of surface heating, terrain shape, and raised inversion to evaluate their interactive nature.
- Experiments to examine the sensitivity of drainage flow characteristics to upper air stability and weak synoptic winds.
- Systematic homogeneous and nonhomogeneous heating and roughness studies should be performed to act as a database to guide the surface flux parameterization methods used in mesoscale numerical models.

CHAPTER 8

Remote Sensing of Atmospheric Processes over Complex Terrain

W. D. NEFF

NOAA/ERL/WPL, Boulder, Colorado

ABSTRACT

This chapter describes the application of a variety of remote sensors to complex terrain studies, including sodars, lidars (aerosol mapping and Doppler), clear-air radars and wind profilers, and optical crosswind sensors. It outlines sensor characteristics relevant to operation in complex terrain settings and examines the origin of turbulence microstructure in complex terrain flows, a significant factor in the use of clear-air radars and sodars. It also summarizes the major programs that have made extensive use of remote sensors and promoted their integration with conventional in-situ instruments. Examples and case studies from these programs are then provided including observations of the initiation and destruction of drainage winds. The behavior of quasi-steady drainage winds and their irregularities during marine layer intrusions, seiches, waves, and strong ambient winds are also described. Recent results are presented that extend the use of Doppler lidar in channeled and merging valley flows as well as in the observation of large-scale mountain–plain circulations. The integration of remote and in-situ sensors in a major transport and dispersion study yields further examples of how different measurement methods can complement each other. Future directions in the study of complex terrain processes are suggested that can take advantage of newly developing remote sensors especially in application to larger-scale mountain–valley circulations.

8.1. Introduction

During the past decade remote sensors that measure the wind and map the distribution of aerosol and turbulence have increasingly contributed to our understanding of atmospheric processes over complex terrain, particularly thermally driven circulations. Because remote sensors measure the atmosphere in a fundamentally different and often more complete fashion than in-situ sensors, they have provided new insight into the initiation, evolution, and destruction of nocturnal drainage winds, and into the complexities of transport and dispersion in an environment in which synoptic, mesoscale, and microscale processes interact so strongly. Remote sensing instruments have filled two increasingly important roles. First, they mimic conventional instruments in many respects, but they also can provide much more extensive temporal and spatial coverage. Second, because they depend on the presence of aerosols or turbulence-produced refractive index fluctuations, they see the atmosphere differently than conventional instruments. For this reason, their data often provide a visualization of atmospheric processes that is not easily available from those of in-situ instruments.

The dependence of remote sensing techniques on the spatially and temporally intermittent presence of naturally occurring or flow-dependent tracers, however, often limits their ability to duplicate the height coverage of measurements from conventional instruments. Furthermore, techniques to measure high-resolution profiles of temperature and humidity are only now emerging. For this reason, the application of remote sensing techniques to the study of complex terrain flows has required the deployment of both remote and in-situ sensors. Combined, their data show considerable promise in advancing our study of the physical processes important to airflow over irregular terrain.

Accompanying chapters in this monograph describe the variety of physical phenomena possible in complex terrain flows and their underlying dynamics and modeling. This chapter examines how ground-based remote sensors can be used to study these phenomena quantitatively (usually through the wind field), qualitatively (usually as flow-visualization tools through depiction of the spatial and temporal distribution of turbulent microstructure or aerosol), and how in-situ instruments support such analyses. Because extensive literature on remote sensors exists, we describe only briefly the properties of the atmosphere on which they depend, their operating principles, and those features most relevant to their application to complex terrain flows. We then outline some of the major experimental programs that have exploited the various features of remote and in-situ sensors and describe the combinations of sensors that have proved most useful. Subsequent discussion then draws from several studies in which extensive arrays of both remote and in-situ sensors were deployed. In this discussion we use examples derived from experiments carried out in a variety of mountain–valley environments, ranging from the micro-α (200 m–2 km) through the low end of the meso-β (20–200 km) scales. These examples illustrate many of the physical phenomena described in this monograph and show how remote sensors can be deployed to visualize and study specific aspects of thermally driven complex terrain flows. As part of this discussion we examine how remote sensing data can also help guide the analysis and interpretation of in-situ measurements and, conversely, how in-situ data (towers and balloon-borne sensors) can be used to interpret remote sensing data.

The application of remote sensing to the study of atmospheric processes over complex terrain is an evolving field of research. To the extent that remote sensors provide data that fit conventional analysis methods, they have been well integrated into complex terrain studies. As accurate, high-resolution remote measurements of temperature and humidity become practical, they will be rapidly integrated into the suite of instruments available for these studies. The ability of remote sensors to map the microscale structure of the atmosphere in terms of turbulent refractive index structure or aerosol, however, has been slower to be absorbed into the main course of complex terrain research in other than a qualitative fashion. To help in this process, many of the examples provided in this chapter interpret the patterns of turbulence seen in complex terrain flows using both in-situ data and simple model results. This is done with the expectation that increasingly sophisticated three-dimensional models will eventually be closely integrated with such observational tools and greatly enhance our understanding of the physics underlying complex terrain flows.

8.2. Remote sensing techniques

The use of acoustical, optical, and radio waves to probe the lower atmosphere has received considerable attention over the past two decades. Lenschow (1986) provided the most recent comprehensive reference on remote sensors and their application to studies of the lower troposphere. Current advances in remote sensing technology are described in "Extended Abstracts: Lower Tropospheric Profiling: Needs and Technologies" (AMS 1988).

Sodar, lidar, and radar (sound-, light-, and radio-detection and ranging) are all active remote sensing devices that emit a short pulse of radiation into the atmosphere within a narrow beam. A small fraction of the radiation scatters toward a receiver where it is amplified and isolated (through heterodyning and/or filtering) in a frequency band pass centered on the carrier frequency. The intensity and/or frequency shift is then analyzed as a function of range from the transmitter (using the time of flight to and from the scattering volume). The mean motion of scatterers within the scattering volume produces a mean frequency shift from which one component of the wind can be deduced. Given a sufficiently strong scattered signal from which to deduce the frequency shift, suitable averaging, and radial measurements in several discrete directions (or in a continuous conical scan), the u, v, and w components of the wind can be obtained as a function of height. Detailed tabulations of the operating characteristics as well as a discussion of the general advantages and disadvantages of sodars, lidars, and radars can be found in a succinct form in Schwiesow (1986a).

8.2.1. Scattering mechanisms

Underlying each of the remote sensing techniques is the need for scatterers: aerosol for most lidar techniques, and refractive index variations at small scales (usually less than 1 m) for sodars, clear-air radars, and optical crosswind sensors. In the case of lidars, little has been documented about the natural distribution of aerosol within and over complex terrain. Limited observations with a 10.6-μm Doppler lidar indicate a typical aerosol layer depth of several kilometers but with a strong dependence on atmospheric conditions (e.g., Post and Neff 1986; Nieman et al. 1988). Observations with a downward-looking airborne lidar operating at 0.532 and 1.064 μm over the coastal basins and mountains of California (McElroy and Smith 1986; Wakimoto and McElroy 1986) reveal highly variable aerosol layers extending as much as 2 km above the surface. Of note in many of these observations has been the highly irregular distribution of aerosol relative to terrain features. In their analysis, Wakimoto and McElroy suggested that the layering of aerosol results from the diurnally varying cycle of daytime convection and nighttime stabilization coupled with the eastward flow of marine air toward the coastal mountains. In such cases the distribution of naturally occurring as well as anthropogenic scatterers depends on particular atmospheric circulations and their history.

It is also possible to use lidars to map artificially produced particle plumes that are released into a prescribed region of a flow. For these cases, an oil fog can be produced that allows the lidar to map its subsequent transport and dispersion, particularly within complex terrain (Eberhard et al. 1987). Lidar observation of such artificial fogs, with a radial resolution of about 10 m, can then be used to map, for example, the effect of upward vertical mixing from surface-based sources or the effect of various flow regimes on the downward mixing from elevated sources.

In the case of radars and sodars, tracers, namely turbulence-produced refractive index fluctuations, depend almost instantaneously on wind shear and stratification. In convective conditions, strong refractive index variations develop near the surface because of surface heating and evaporation of moisture from the surface. Subsequent vertical turbulent mixing then results in small-scale refractive index fluctuations detectable by radars and sodars. In statically stable conditions, an increased wind shear is required to produce turbulence and, consequently, refractive index fluctuations. This can occur either near the surface or in free shear layers. As a result, delineation of these flow-dependent tracers by remote sensors can define the surface mixing-layer depth as well as identify regions of dynamical instability and turbulent mixing aloft (Chadwick and Gossard 1986; Neff and Coulter 1986). Thus, properly interpreted, the spatial distribution and temporal evolution of turbulence microstructure should provide direct insight into the dynamics of many complex terrain processes.

8.2.2. *The role of turbulence microstructure in remote sensing*

Remote sensors that depend on the presence of turbulence as a mechanism for the scattering of sound or radio waves suffer, at times, from the absence of echoes. Simultaneously, however, graphical displays of the time–height cross-sections of echo intensity can often reveal previously unexpected phenomena. Unfortunately, because of the ground clutter problem and inadequate minimum range, clear-air radars have not been used in confined complex terrain settings. Sodars, on the other hand, have received extensive application and have some distinct advantages compared with radars, beyond that of cost and availability. For example, radar scattering cross-sections depend on the turbulent refractive index of both temperature and humidity and their cross correlation as described by Chadwick and Gossard (1986). In contrast, sodars, the most commonly used remote sensor in complex terrain studies, depend (in the backscatter or monostatic mode) very strongly on the temperature refractive index microstructure. For this reason their data reflect, albeit indirectly, the temperature structure of complex terrain flows much more clearly than do radars. Furthermore, sodars have a unique characteristic, not yet fully exploited; they can map the distribution of actual velocity turbulence.

The scattering of sound in a turbulent atmosphere is best described using the equation for the differential scattering cross-section:

$$\sigma(\theta_s) = \frac{1}{8} k^4 \kappa^{-11/3} \times \cos^2(\theta_s) \left[\frac{0.33 C_T^2}{T_0^2} + \frac{\cos^2(\theta_s/2)}{\pi C_0^2} 0.76 C_V^2 \right], \quad (8.1)$$

which represents the fraction of the incident acoustic power scattered, by refractive index inhomogeneities, per unit distance into a unit solid angle oriented in the direction θ_s from the direction of transmission and is written in terms of the temperature structure function

$$C_T^2 = \langle [T(\mathbf{x}+\mathbf{r}) - T(\mathbf{x})]^2 \rangle / r^{2/3} \quad (8.2)$$

and the velocity structure function

$$C_V^2 = \langle [U(\mathbf{x}+\mathbf{r}) - U(\mathbf{x})]^2 \rangle / r^{2/3}, \quad (8.3)$$

where $\mathbf{x}$ and $\mathbf{x} + \mathbf{r}$ are two measurement locations and where $\langle \ \rangle$ indicates an ensemble average (but for practical purposes normally derived from a time average). In (8.1), k is the acoustic wavenumber; κ the Bragg wavenumber, $\kappa = 2k \sin(\theta_s/2)$; T_0 the local temperature; and C_0 the local speed of sound.

Equation (8.1) introduces an interesting angular dependence in the acoustic scatter; namely, temperature microstructure alone determines the backscatter cross-section [$\theta_s = 180°$] and neither temperature nor velocity microstructure contributes to scatter at right angles [$\theta_s = 90°$]. In principle, measurements at two angles should allow determination of both C_T^2 and C_V^2. Velocity inhomogeneities contribute to the scattering of sound waves but not electromagnetic radiation because of the relatively slow speed of sound and the fact that typical velocity inhomogeneities in the atmosphere can significantly affect the phase speed of the sound, producing scattering (Batchelor 1957).

Equation (8.1) is written in terms of structure functions C_T^2 and C_V^2 because they provide a conceptually and experimentally useful description of the small-scale refractive index structure that produces acoustic scattering. Underlying the derivation that leads to (8.1), however, are the results that refractive index variations at spatial scales corresponding to one-half an acoustic wavelength (or the Bragg scale for off-axis scattering) contribute most strongly to the acoustic scatter and that C_T^2 can represent the isotropic three-dimensional spectral density of temperature $\phi_T(\kappa)$ through

$$C_T^2 = \phi_T(\kappa)/(0.033\kappa^{-11/3}). \quad (8.4)$$

In addition, C_V^2 can similarly represent the turbulent kinetic energy spectral density $E(\kappa)$ through

$$C_V^2 = E(\kappa)/(0.76\kappa^{-5/3}) = 2\epsilon^{2/3}, \quad (8.5)$$

where ϵ is the dissipation rate of turbulent kinetic energy [m^2/s^3] and κ indicates the spectral wavenumber sensed by the sodar. It is important to note that (8.4) and (8.5) only apply to the portion of the spectra within the inertial subrange (Stull 1988). Because the portion of the spectra at larger scales, outside of the inertial subrange, can be highly variable in magnitude and extent in wavenumber space and because variances are often obtained by integrating arbitrarily over some portion of the spectra, it is not always possible to relate the structure functions C_T^2 and C_V^2 directly to the temperature variance and the total turbulent kinetic energy.

For most sodars, the scale of refractive index variation most important for scattering is a small fraction of a meter. In view of this, the structure functions used in (8.2) and (8.3) have both an experimental and interpretive value. From the experimentalist's point of view, differences of temperature and speed over small separations are often easier to measure and thereby test relationships such as (8.1). As an aid in interpretation, facsimile displays of echo intensity (involving either C_T^2 or C_V^2) reveal regions defined in height and time in which there is a significant cascade of temperature or velocity variance from large to small scales (e.g., Stull 1988). This provides a pictorial representation, albeit qualitative, of turbulence generation/dissipation regions in space and time.

The discussion here has been brief; however, Neff and Coulter (1986) provide a short review of the literature leading to (8.1) through (8.5) for acoustic scattering. General discussion of spectra and structure function relationships can be found in Tennekes and Lumley (1973) and Stull (1988). Tatarskii (1971) treats acoustic and electromagnetic wave propagation and scattering in a turbulent medium in detail.

8.2.2.1. Statically Unstable Conditions

Sodars, under statically unstable conditions, produce distinctive facsimile records that show the passage of a succession of echoes associated with thermal plumes. In the presence of an elevated capping inversion, sodars have documented the erosion of the stable layer aloft by the underlying convection (Kaimal et al. 1982). Similar echo patterns have also been observed in complex terrain over elevated steep slopes (Neff 1988). Within basins, more complicated patterns result, as described in section 8.4.8.2.

Interpretation of sodar convection-echoes over flat surfaces has been aided by some simple surface layer similarity concepts. In particular, the temperature structure parameter C_T^2 in (8.2) depends on the surface heat flux H and the height z above the surface through the relation (Wyngaard et al. 1971)

$$C_T^2 = 2.68(g/T)^{-2/3}(\rho c_p z/H)^{-4/3}, \qquad (8.6)$$

where g is the acceleration of gravity, ρ the density of air, and c_p the heat capacity of air at constant pressure. The decrease with height of C_T^2 arises from the decay of the temperature variance as fluctuations generated near the surface are convected aloft and destroyed through molecular diffusion. The use of sodars to verify (8.6) has shown reasonable agreement (Neff 1975), suggesting that sodar measurements might well provide a measure of the surface sensible heat flux; however, broadening of the sodar beam by turbulence can lead to a faster apparent falloff with height (Neff 1978), the presence of a capping inversion can cause a slower falloff (Frisch and Clifford 1974), and large values of humidity can complicate the interpretation of the scattering (Coulter and Wesely 1980). No detailed studies of the behavior of the structure function under convective conditions in complex terrain have been carried out. For this reason, interpretation of the echo patterns must remain somewhat qualitative.

Clear-air radar observations (were they to be obtained) of the inversion capping convective layers in complex terrain might well prove interesting because, as reported by Chadwick and Gossard (1986), the radar cross-section depends approximately linearly on the temperature and humidity variances and negatively on their covariance. As shown by Wyngaard et al. (1978) the covariance of temperature is such as to subtract from the scattering intensity near the surface and to add to the scattering in the vicinity of the capping inversion. This suggests that clear-air radars, if used in large-scale complex terrain, or in more confined terrain if sidelobe removal methods improve, could be more effective than sodars in observation of the growth of the convective mixed layer over terrain, once the strong surface inversion has broken. This is particularly true because of the rapid attenuation of sound with range in the atmosphere that, for the most part, limits conventional sodar observations to less than 1 km.

8.2.2.2. Statically Stable Conditions

In statically unstable conditions the turbulence microstructure that leads to the acoustic scatter described by (8.1) arises from the temperature gradients created by heating of the surface and/or by the mixing produced by parcels that have been convected vertically and have then encountered a stable layer aloft. In the statically stable atmosphere, however, mechanisms producing scattering are much more diverse and depend on flow conditions that can increase the wind shear sufficiently to produce turbulence. The resulting mixing of fluid parcels of different temperature (and hence density) leads to the microstructure that can produce acoustic scattering. As has been described by Neff (1988) even slight wind shear during light and variable winds can lead to turbulent echo strata. Because of the irregular distribution of temperature in complex terrain, complex thermal forces can develop over slopes and within basins leading to intermittent turbulence and, hence, the complex echo patterns frequently observed by sodars.

As an aid to the interpretation of sodar echo patterns, it is helpful to rewrite (8.1) for the simpler case of backscatter in a stably stratified environment as (Neff and Coulter 1986)

$$\sigma(180°) = (0.013)(k^{1/3}/T_0^2) \times K_H^{2/3}[\mathrm{Rf}/(1-\mathrm{Rf})]^{1/3}\left(\frac{\partial\theta}{\partial z}\right)^{5/3}, \qquad (8.7)$$

where $\partial\theta/\partial z$ is the vertical gradient of potential temperature and Rf is the flux Richardson number defined by

$$\mathrm{Rf} = \left(\frac{K_H}{K_M}\right)\mathrm{Ri} = \left(\frac{K_H}{K_M}\right)\frac{g}{T}\left(\frac{\partial\theta}{\partial z}\right)\left[\left(\frac{\partial U}{\partial z}\right)^2 + \left(\frac{\partial V}{\partial z}\right)^2\right]^{-1}, \qquad (8.8)$$

where K_H and K_M are the eddy diffusion coefficients for heat and momentum, respectively, and Ri is the conventional gradient Richardson number. The expression (8.7) reveals explicitly the dependence of acoustic backscatter on local turbulence (primarily through K_H) but more so on the mean vertical gradient of potential temperature under statically stable conditions. Equation (8.7) was derived by neglecting tendency and triple-correlation terms in the equation for conservation of temperature variance. Thus, (8.7), although limited, does possess a pedagogical value. For this reason, we return in later sections to this formulation as a means of explaining some of the echo patterns observed in complex terrain flows, particularly because of its potential use in higher resolution numerical models that calculate eddy diffusion coefficients in a direct manner from mean vertical gradients.

8.2.3. Sampling geometries

Each of the active remote sensing technologies is different in characteristics such as beamwidth, sidelobe radiation, minimum range, maximum range, scattering volume size, pulse repetition rate, dwell and/or sample averaging time, and possible beam orientations and scanning capability, many of which are summarized by Schwiesow (1986a,b). The different technologies fall into two classes, fixed-beam systems and scanning systems, based on sampling geometries of relevance to complex terrain studies as described below.

8.2.3.1. FIXED-BEAM SYSTEMS

Fixed single- or multiple-beam systems, the first class, probe the same volume(s) repeatedly, producing time series of scattering intensity and of Doppler-shifted frequencies from which various wind components can be deduced. The first and most common of the fixed-beam geometries uses a monostatic mode in which the transmitter and receiver are collocated. For each such monostatic system beam, the scattering volume size (Fig. 8.1) is approximated by the length of the transmitted pulse, the beam halfwidth α, and the range from the transmitter $R = H/\cos(\theta)$ where H is the height above the surface and θ is the angle from the vertical. Typical values for θ range from 20° to 30°. The physical length of the transmitted pulse is given by $L_p = \tau C_0$, where τ is the pulse transmission time and C_0 is the speed of propagation. At any instant, however, the backscattered signal can come only from a volume of length $L_s = \tau C_0/2$. (The factor of 2 comes from the fact that the sound scattering from the leading edge of the pulse must travel a distance $\tau C_0/2$ back toward the receiver in order to add to the sound scattering from the trailing edge of the pulse at the same instant.) This argument, however, assumes that scatterers

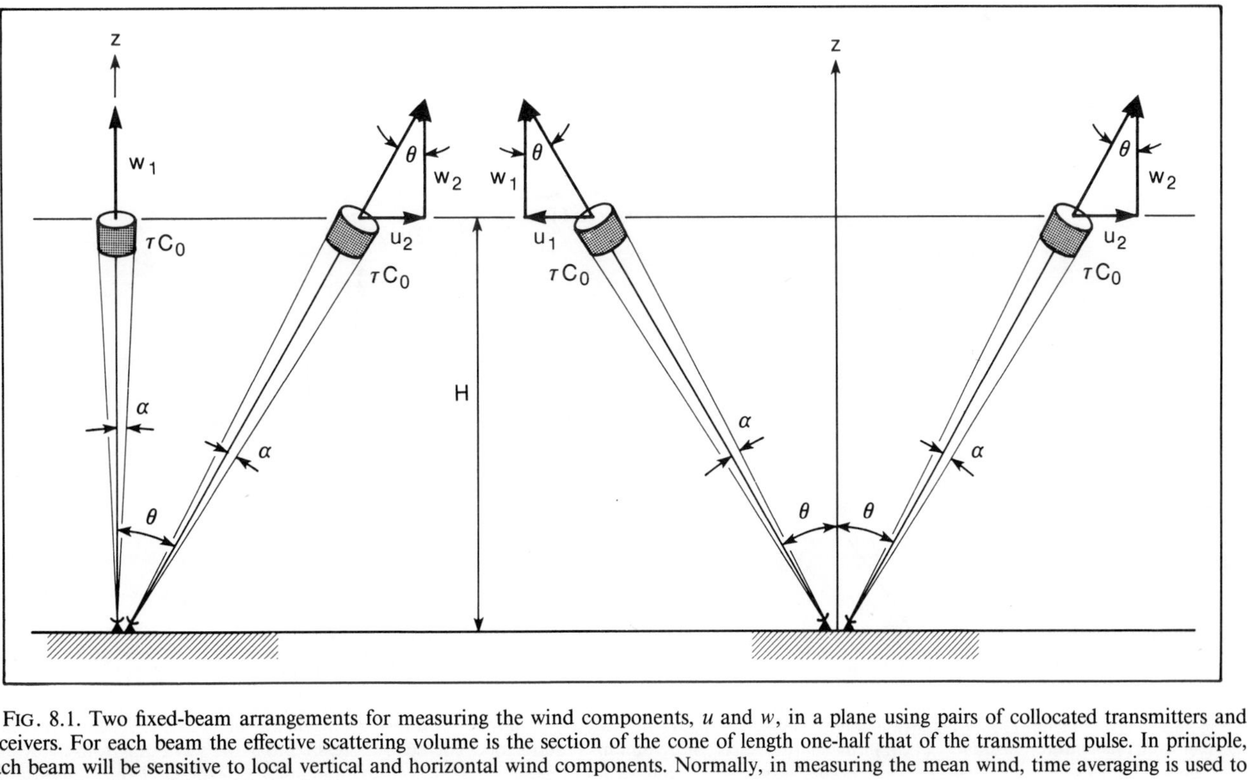

FIG. 8.1. Two fixed-beam arrangements for measuring the wind components, u and w, in a plane using pairs of collocated transmitters and receivers. For each beam the effective scattering volume is the section of the cone of length one-half that of the transmitted pulse. In principle, each beam will be sensitive to local vertical and horizontal wind components. Normally, in measuring the mean wind, time averaging is used to eliminate the effect of the vertical velocity.

are uniformly distributed in space. In reality, particularly under statically stable conditions, lamina of scatterers occur that are thin compared with the pulse volume. In such cases, measured echoes will have the duration of the original transmitted pulse. For sodars and aerosol-mapping lidars, scattering volume lengths range from a few meters to 100 m or more; for clear-air radars and 10.6-μm lidars, scattering volume lengths generally exceed 100 m.

In Fig. 8.1 two monostatic configurations possible for the measurement of the horizontal wind are also shown, each of which is affected by the spatial separation of the scattering volumes. The first, and most common, uses collocated transmitters/receiver antennas, one directed vertically and one directed at some angle θ from the vertical. The Doppler shift is determined along the two radial beams as shown in the figure. Because the Doppler shift along the tilted beam is determined by both the vertical and horizontal components of the wind, either the vertical velocity w_2 must be approximated by w_1 or a long-time average obtained such that $\overline{w_2} = \overline{w_1}$. As the height of interest increases, the horizontal separation of the scattering volumes increases, causing an increased susceptibility to inhomogeneity in the flow, particularly for short averaging times or highly localized wind variations or terrain influence (McAfee et al. 1989). A further assumption is that the scatterers uniformly fill the scattering volume, an assumption subject to uncertainty in regions of intermittent turbulence or irregularly distributed aerosol as described above. In these cases it is difficult to attribute the measured wind components to a well-defined height.

In the second mode shown in Fig. 8.1, the separation of the scattering volumes is greater than in the first; however, some redundancy is provided for the evaluation of long-time averages because if $\overline{w_1} = \overline{w_2}$ then $\overline{u_1} = -\overline{u_2}$. One modification of this configuration can also be obtained by shifting the transmitters on the baseline so that the two beams create a common scattering volume at one height of interest, say just above ridge height in a channeled valley flow, or lower, in the vicinity of low-level jet. Data from this common volume could be used to study short-term variability in the flow such as that due to internal waves or large-eddy structure without the limitation introduced by spatial separation. Such configurations are possible only for systems with separable antennas such as conventional sodars and UHF radars. Other systems such as VHF radars usually use phased arrays because of economic constraints and hence are limited to the use of separated scattering volumes (McAfee et al. 1989).

The second scattering geometry, or bistatic mode, uses separated transmitters and receivers as shown in Fig. 8.2,

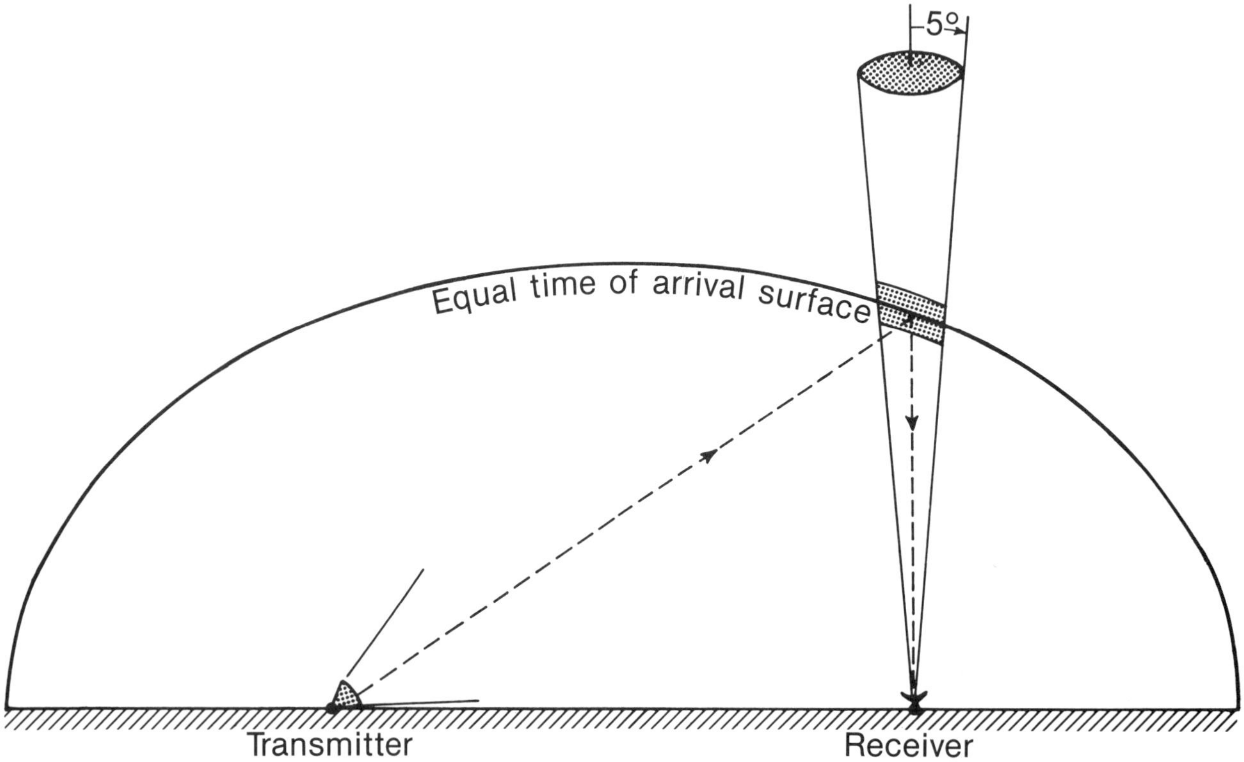

FIG. 8.2. Typical bistatic sounding arrangement using a fanbeam transmitter and narrowbeam receiver at the foci of an ellipse. The narrowbeam receiver limits reception to a small section of the ellipse. Such configurations require good sidelobe suppression to avoid scattering from the region between the transmitter and receiver. The Doppler shift results from air motions along the bisector of the angle subtended at the ellipse by the transmitter and receiver (the local normal to the ellipse).

and is of most interest for sodar applications. At any instant, sound received at the receiver could have come from any point on the ellipsoidal surface indicated. In a pulsed mode, the pulse length can be used to define a family of ellipsoids and the narrow beamwidth of the receiver used to limit the scattering volume to the segment shown. The Doppler shift arises from air motion along the normal to the ellipsoidal surface with the transmitter and receiver at the foci. Equations relating the measured Doppler shifts to wind components in a conventional orthogonal system can be found in Neff and Coulter (1986). Note the derivation of the horizontal wind components requires either that the vertical wind be measured directly (using an independent transmission from the vertical sodar antenna) or that the data be averaged over a long enough period that the vertical velocity becomes almost zero (within the errors of the measurements). A second bistatic mode is also possible; i.e., narrow-beam transmitters and receivers may probe a common volume above the midpoint of their baseline, so that vertical velocity fluctuations as well as temperature and velocity structure parameters can be sensed directly. The latter technique has not yet been applied to complex terrain processes, however, perhaps because of the absence of confirming measurements from in-situ instruments.

One virtue of the bistatic scattering arrangement is that the wind is derived in a single vertical profile rather than from separated scattering volumes. In confined terrain, this arrangement may be necessary to overcome the problem of horizontal inhomogeneity in the flow. One limitation of the bistatic arrangement should also be noted: at greater heights above the surface, this measurement is increasingly sensitive to the vertical component of the wind; greater ranges therefore require a greater separation of transmitter and receiver and/or a longer averaging time to reduce the contribution of random errors. For typical applications (Neff and King 1987) a baseline of 300 m is required for a maximum height of about 600 m. Such an installation is shown in Fig. 8.3 where the extensive terrain required for such an installation is readily evident. A spe-

FIG. 8.3. Photograph showing extensive domain required for a bistatic sounding arrangement (Neff and King 1987).

cial case of the bistatic arrangement is line-of-sight propagation directly from a transmitter to a receiver, such as that used in optical crosswind sensors described in section 8.2.4.5.

8.2.3.2. Scanning systems

Scanning systems, the second class, produce measurements within a hemispherical volume in which the scattering intensity and Doppler-shifted frequency are obtained as a function of azimuth, elevation angle, and range along radial directions from the sensor. Derivation of the wind speed and direction throughout the volume requires multiple systems scanning the same region. One exception can occur when the flow has some natural symmetry that can be exploited with a single system, as in the observation of winds channeled in a narrow valley where a single radial component is sufficient to resolve the along-valley wind (e.g., Post and Neff 1986), or where other natural flow symmetries exist.

For single scanning systems, several modes are possible. For radars or lidars, an important scan is the conical scan in which the azimuth is varied at a fixed elevation angle [referred to as the plan position indicator (PPI) or the velocity–azimuth display (VAD), (Battan 1973)]. The VAD scanning mode showing the geometry of the scan on the left and an example of the variation in radial wind component on the right is illustrated in Fig. 8.4. Because of noise and random atmospheric turbulence the variation in the radial wind component is not smooth. For most analyses, a least-squares fit sine curve that removes these effects is used to deduce the wind speed and direction from its amplitude and phase; the offset of the sine curve from zero reflects any area-integrated vertical velocity or instrumental bias. Variations from the sine curve can also be used to estimate turbulence variances and covariances in the convective boundary layer, as shown for a Doppler radar by Kropfli (1986a) and for a Doppler lidar by Eberhard et al. (1989a).

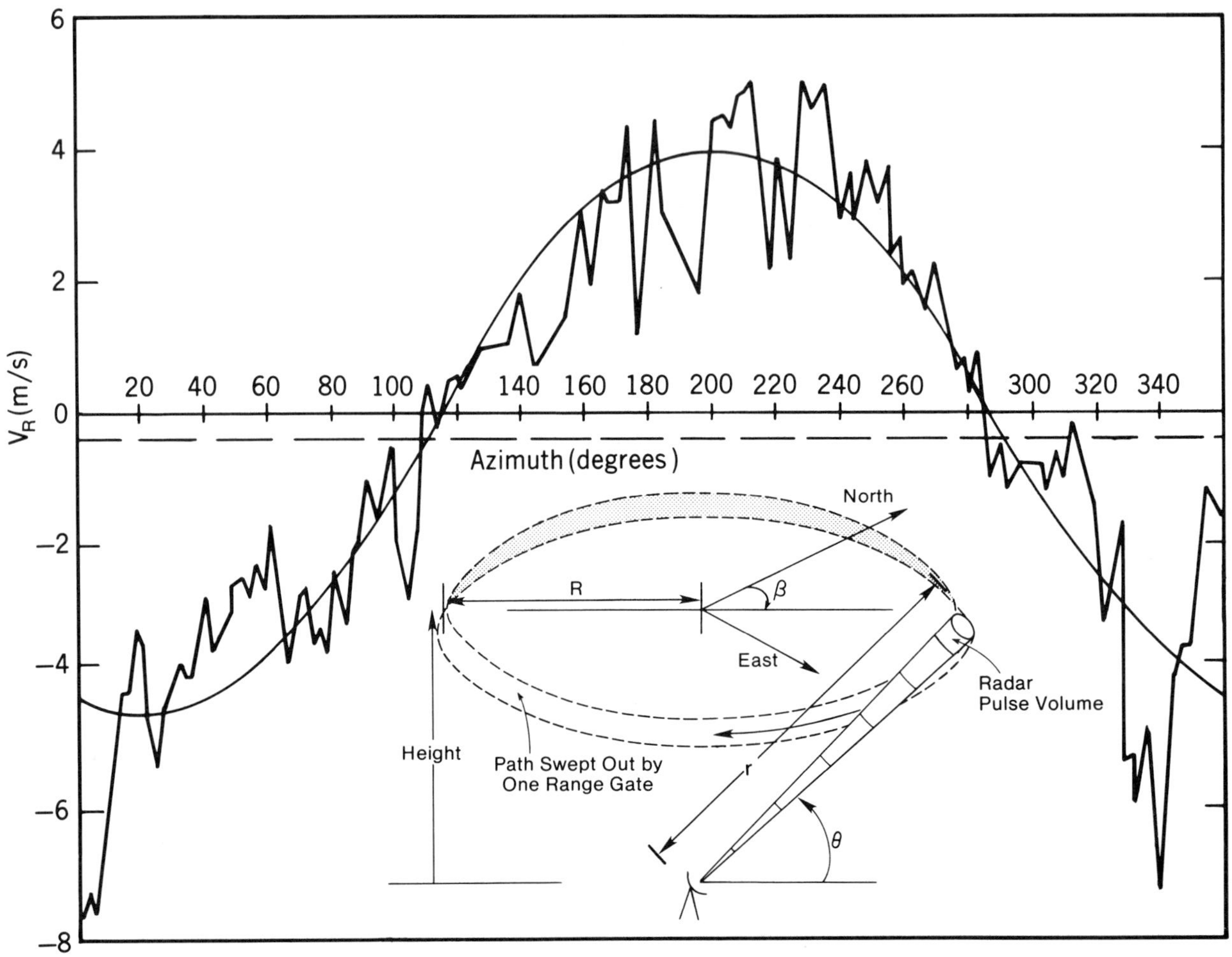

Fig. 8.4. VAD scanning geometry (left) and sample data (right) showing the sinusoidal mean trace produced by a horizontally homogeneous wind field with superimposed turbulent fluctuations. The trace on the right shows the radial velocity as a function of azimuth with a small offset that represents either a small instrumental bias or a nonzero mean vertical velocity averaged over the circle shown. (Kropfli 1986.)

A second scanning mode is one in which the elevation of the beam is swept between two limits while the azimuth is fixed. This is referred to as a range height indicator (RHI) mode. Radars, because of their relatively broad beamwidth (a few degrees, usually) would not be expected to provide good resolution of shallow complex-terrain flows with this type of scan. Lidars, however, with their extremely narrow beamwidth, have used this technique with great success in the study of complex terrain flows. A variation of the technique is called the stepped-RHI (Eberhard et al. 1989b). In this case, scans with a high resolution in the vertical were obtained to some maximum elevation angle, say 10°, at successive increments in azimuth. This mode has value when several flows are being monitored, each of which has a strong directional preference, and when one also desires good resolution in the vertical, as in a basin with drainage flows emerging from several adjacent valleys.

A third scanning mode is the raster scan in which the beam is swept between two azimuth limits at successive small increments in elevation angle. This mode, described by Post and Neff (1986), provides high resolution in both horizontal and vertical directions but is limited in most cases to relatively narrow azimuthal and elevation angle limits because of the time required to scan a prescribed volume. Application of this mode is discussed in more detail in section 8.4.5 because of its value to the study of flows in narrow valleys.

8.2.4. Instruments

8.2.4.1. SODARS

Sodars, because of their low cost, portability, and ability to operate unattended for long periods of time, have been extensively used in complex terrain studies. The U.S. Department of Energy's (DOE) Atmospheric Studies in Complex Terrain (ASCOT) Program first used them extensively in 1979 (Neff et al. 1980). Since then their use has expanded in both ASCOT and the Environmental Protection Agency's (EPA) Complex Terrain Model Development (CTMD) Program as well as in shorter-term studies. Papers that review and trace the development of sodars in general include those by Brown and Hall (1978) and Neff and Coulter (1986). Other references that focus on complex terrain applications include papers by Neff (1986, 1988), Neff and King (1987, 1988), Coulter and Martin (1983), and Coulter et al. (1989).

Sodar beamwidths are, in general, limited by the effects of diffraction; for a given antenna aperture, the beamwidth and sidelobes decrease with increasing frequency (decreasing wavelength) as shown in Fig. 8.5 for a single antenna aperture of 1.25 m. Typical sodar antenna apertures range from 1.0 to 2.0 m for frequencies between 1 and 3 kHz. At the lower frequencies the beam is fairly broad and sidelobes are much stronger than at higher frequencies. In confined terrain, the direct reflection of this sidelobe acoustic energy from local terrain could dominate signals backscattered from the main lobe, obscuring echo patterns representing atmospheric phenomena. In the analysis of the frequency shift, the terrain-scattered component would have to be removed from the power spectrum, if possible, to determine the local wind, and then only if the wind were sufficiently strong to separate the spectral peaks, one corresponding to the Doppler-shifted signal from the atmosphere and one at zero-frequency shift from the terrain. For this reason, high-frequency minisodars, using frequencies in excess of 3 kHz, and hence very narrow beams, have proved quite useful in confined canyon settings (Coulter et al. 1989). Not only do they provide good Doppler-wind height resolution of 10 m and low minimum range, 5 to 10 m; they also provide a high-resolution picture of boundary-layer phenomena using a physical pulse length and minimum range above the surface of only a few meters (Asimakopoulos and Cole 1977; Asimakopoulos et al. 1987). Because of these characteristics they show considerable potential in the study of shallow (<150 m) flow phenomena in complex terrain.

Various fixed-beam geometries are possible for the measurement of the wind as shown in Figs. 8.1 and 8.2. Following (8.1), bistatic configurations can use scattering from both velocity and temperature inhomogeneities; monostatic arrangements, however, can use only scattering from the temperature structure. Neff (1988) therefore analyzed the relative merits of bistatic and monostatic systems for wind measurement under statically stable conditions, using a second-order turbulence closure parameterization, and found that bistatic systems provide stronger signals for small Richardson number regimes, whereas monostatic systems excel under large Richardson number, weak wind-shear conditions. The choice of a bistatic or monostatic Doppler sodar would thus depend on the complex terrain application and on logistical limitations. For example, studies of sea-breeze flows over coastal mountains or flows in strongly draining valleys would benefit from bistatic systems, whereas studies of closed basins at nighttime or with snow cover where stability is strong and winds are light would benefit from the use of monostatic Doppler systems. Sodars have not been used in a scanning mode because of the acoustic noise that would be generated by a motorized drive system, because of the long acoustic propagation times that would limit the scanning rate, and because of the low cost of acoustic antennas that easily allow several independent beams. Sodars do suffer several limitations. First, attenuation of sound with range limits their use except at very low frequencies. Second, acoustic signal interference can be a problem in noisy environments, although most complex

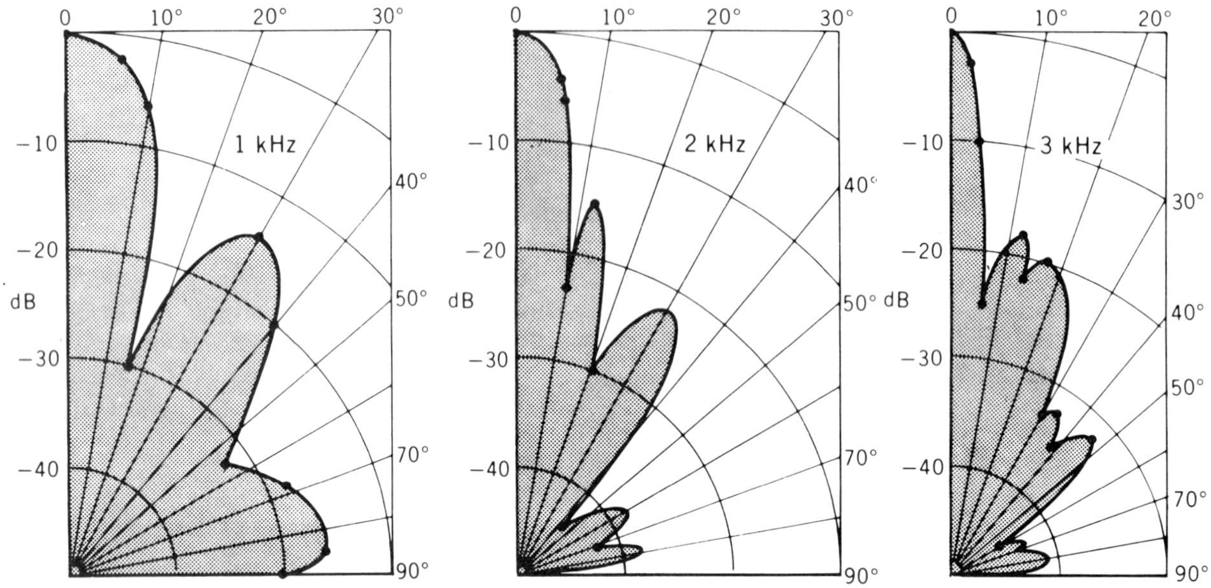

FIG. 8.5. Sodar beamwidths at acoustic frequencies of 1, 2, and 3 kHz, normalized to the maximum intensity on the vertical axis, showing the progressively narrower beams obtained with higher frequencies. (Neff 1975.)

terrain study areas have been relatively quiet acoustically except in the vicinity of power generation sites.

While the use of sodars to measure the wind has had the most immediate impact in complex terrain studies, the development of tools with which to interpret the time–height distribution of refractive index structure has been relatively limited. Until now, the interpretation of sodar facsimile recordings has followed two paths. The first, a phenomenological approach, has involved the use of tethered balloons or tower profiles of wind and temperature to interpret observations of interest (Neff and Coulter 1986; Neff and King 1988). The second approach has used higher order closure models to simulate turbulence profiles (Burk 1980, 1981; Neff 1980, 1986; Neff and Coulter 1986). More recently, tethered-balloon profiles of wind and temperature have been used with second-order closure parameterizations through (8.7) for stably stratified conditions to predict profiles of sodar echo intensity, which have then been qualitatively compared with sodar facsimile recordings (Neff 1988). In the future, as three-dimensional numerical models achieve better vertical grid resolution and as turbulence parameterizations and large-eddy simulation or direct-turbulence calculation methods improve, better comparisons may be possible between observations and model output. In these applications, sodars may well play a key role as the only remote sensor that can measure the dissipation rate of turbulent kinetic energy directly, rather than through inherently noisy and ambiguous Doppler estimation techniques (especially under statically stable conditions).

8.2.4.2. RADARS

Radio waves can be used to probe the atmosphere in the same manner as sound, but without the significant range-limiting attenuation that acoustic waves suffer. Two types of radar systems, scanning and fixed-beam radars, are available but have not been used to study local circulations in confined complex terrain. Scanning radars, for example, have been used to study orographically induced winter storms in combination with measurements of water vapor, using passive radiometers (Uttal et al. 1989) as well as to provide layer-averaged (often hundreds of meters) boundary-layer winds and mesoscale eddy structure (Kropfli 1986b). Such radar methods, however, depend on the presence of either hydrometeors or radar-reflecting chaff that has been artificially introduced into the flow and dispersed by boundary-layer turbulence. Under very stable conditions, the weakness of turbulence will limit the dispersion of the chaff and hence limit radar spatial coverage. Combined with strong reflections from terrain, these factors have limited radar studies to the more turbulent and larger scale terrain-influenced circulations (Kropfli et al. 1986).

The development of the convective layer above complex terrain is, however, one area that might be studied effectively with dual-Doppler chaff-sensitive radars. By scanning common volumes from two locations, assuming that the flow near the surface follows the local terrain, and applying the continuity equation, these Doppler systems be used to derive the three-dimensional wind field (Kropfli

1986b) in regions where radar-reflecting chaff has been dispersed. Such methods could well reveal many facets of airflow over complex terrain not yet observed, assuming suitable observing sites could be located.

Clear-air radar wind profilers, a more recent development, provide a potential means to monitor winds within and above complex terrain, using beam combinations like those used for Doppler sodar and depending on similar turbulence-backscatter mechanisms (except that clear-air radar backscatter cross-sections depend on both temperature and humidity variations at small scales). These systems exist in a long-range (2 to 10 km) version (Strauch et al. 1984), a medium-range (0.5 to 5 km) version (Moran et al. 1989), and a short-range (0.15 to 2.5 km) version (Ecklund et al. 1988). The last, a UHF radar system, is the only one with a short enough minimum range and range resolution (less than 150 m) to observe local circulations in mountain–valley systems. It would be particularly effective if used in combination with Doppler minisodars having high-resolution range gates from 10 to 200 m. Furthermore, recently developed Radio-Acoustic Sounding Systems (RASS) allow the radar to measure the speed of acoustic waves (generated by a collocated acoustic source) propagating along the radar beam and hence deduce the profile of virtual potential temperature on which the speed of sound depends (May et al. 1988, 1989a; Currier et al. 1988).

In the RASS technique, the variations in temperature produced by the compression/expansion phases of the acoustic waves provide the refractive index structure from which electromagnetic waves can scatter. The local propagation speed of the sound produces a Doppler-shifted signal from which the virtual temperature can be calculated. The major limitation in the system comes from the rapid attenuation of sound with range. For many complex terrain applications, however, measuring the virtual temperature profile continuously in time in the first kilometer of the atmosphere (a region easily available to RASS) would be a significant advance.

For the most part, radar systems are limited by their minimum range, by range resolution, by receiver saturation produced by ground clutter, and by reflections from slightly moving objects such as tree leaves that prevent removal of ground clutter. As UHF radar technology advances and these problems are solved, however, such systems will be cost competitive with sodar technology and have the potential for vastly improving our ability to observe local circulations in complex terrain.

8.2.4.3. Aerosol-Mapping Lidars

Aerosol-mapping, pulsed, monostatic lidars have been available since the early 1970s (Hinkley 1976), but only in the last decade have they been coupled with good graphic displays, more flexible scanning capability, and both airborne and ground-based platforms. These lidars, operating at wavelengths below 1 μm, have proved useful in mapping both naturally occurring aerosol distributions and artificially produced oil fogs (Eberhard and McNice 1986). With a range resolution as small as a few meters and the very narrow beams characteristic of lidar systems, they supply a high-resolution picture of aerosol distributions. Newer and more sophisticated techniques involving differential absorption lidars measure various species of pollution particles and gases (Browell 1988). A variety of artificial tracers may also become available such as fluorescent particles (Uthe et al. 1985). Extension of this method to multiple fluorescent tracers (Eberhard and Chen 1989) may provide the means to distinguish transport and dispersion from multiple sources in complex terrain, thereby overcoming the limited spatial sampling characteristic of conventional tracer technology (Gudiksen and Shearer 1989).

Lidars operating in the submicrometer range have also been used to measure water vapor profiles using Raman scattering under nighttime conditions (Melfi et al. 1989). In the results reported by Melfi et al., water vapor profiles were obtained from the surface to almost 7 km at 100-m intervals and revealed the passage of warm and cold fronts through the observation site. Comparison with rawinsonde profiles revealed good agreement except at humidities less than 20%, the level below which the rawinsonde data were suspect.

A further wind-measuring method available with pulsed, monostatic lidars involves measuring the wind and turbulence in convective or neutral conditions by observing the correlation of aerosol inhomogeneities as they move along radial beam directions or from beam to beam (Eloranta et al. 1975; Kunkel et al. 1980; Sroga et al. 1980). Because of the narrow lidar beams and short pulses such techniques could prove useful for the study of large eddies in complex terrain flows.

Ground-based lidar backscatter systems have proved useful in monitoring the local atmosphere with good spatial and temporal coverage. They have helped delineate, for example, the downwind variability of elevated plumes in three dimensions with time. Some challenges in analysis include the need to correct for the attenuation of the signal with range and the need for good definition of the background aerosol level in order to provide a correct three-dimensional distribution of scatterers (Eberhard et al. 1987). Because they have some eye-safety restrictions, they must be used with caution in complex terrain settings [within a few years, aerosol mapping lidars may operate in the eye-safe region near 2 μm]. Despite the current limitations they have been used successfully to monitor the behavior of both ground-based and elevated plumes in complex terrain (Eberhard and McNice 1986).

Airborne, downward-looking lidars fulfill a need for coverage of a wider domain. The high data rates and the ability of the aircraft to cover large geographical areas make possible a detailed, nearly instantaneous picture of the variation of naturally occurring aerosol layers, the downwind evolution of power-plant plumes, and the mesoscale variation of pollution. One advantage of the airborne system is that aerosol or oil fog distributions can be compared over extensive horizontal dimensions with direct graphical displays. These lidars also have good vertical resolution (usually less than 10 m), and the lidar minimum range is not of concern because the region of interest is somewhat below the airplane (McElroy and Smith 1986). The limitations of the airborne systems include the high cost of the aircraft operation as well as the inherent limitation of the two-dimensional vertical slice-type sampling in the analysis of rapidly changing phenomena.

8.2.4.4. Doppler lidars

Pulsed Doppler lidars have been used in airborne (Blumen and Hart 1988) and ground-based (Post and Neff 1986) modes to study processes above and within complex terrain, respectively. In each case, the lidar has been able to measure only the wind component along the beam and hence special techniques have been developed to overcome the limitation of a single radial wind component.

In an example of a complex terrain application of an airborne Doppler lidar, Blumen and Hart studied stratified airflow around an isolated mountain. In their case, the airborne lidar measured wind components along two radial directions fore and aft of a line transverse to the aircraft line of motion. This results in a diamond-shaped grid pattern (Fig. 8.6) as the aircraft advances. At each point in the grid, two components of the wind are measured. The accuracy of the measurements then depends on the lidar performance, the extraction of the airplane motion relative to the ground, and the direction of the wind relative to the measurements, as well as the separation in time of the measurements relative to the time scale of the motions (Carroll 1986). Despite uncertainty in relative wind field vectors, Blumen and Hart found that observations of the divergence fields in the vicinity of the mountain agreed in many respects with linearized internal wave theory.

A ground-based infrared Doppler lidar has also contributed to complex terrain studies. A 10.6-μm lidar (Post

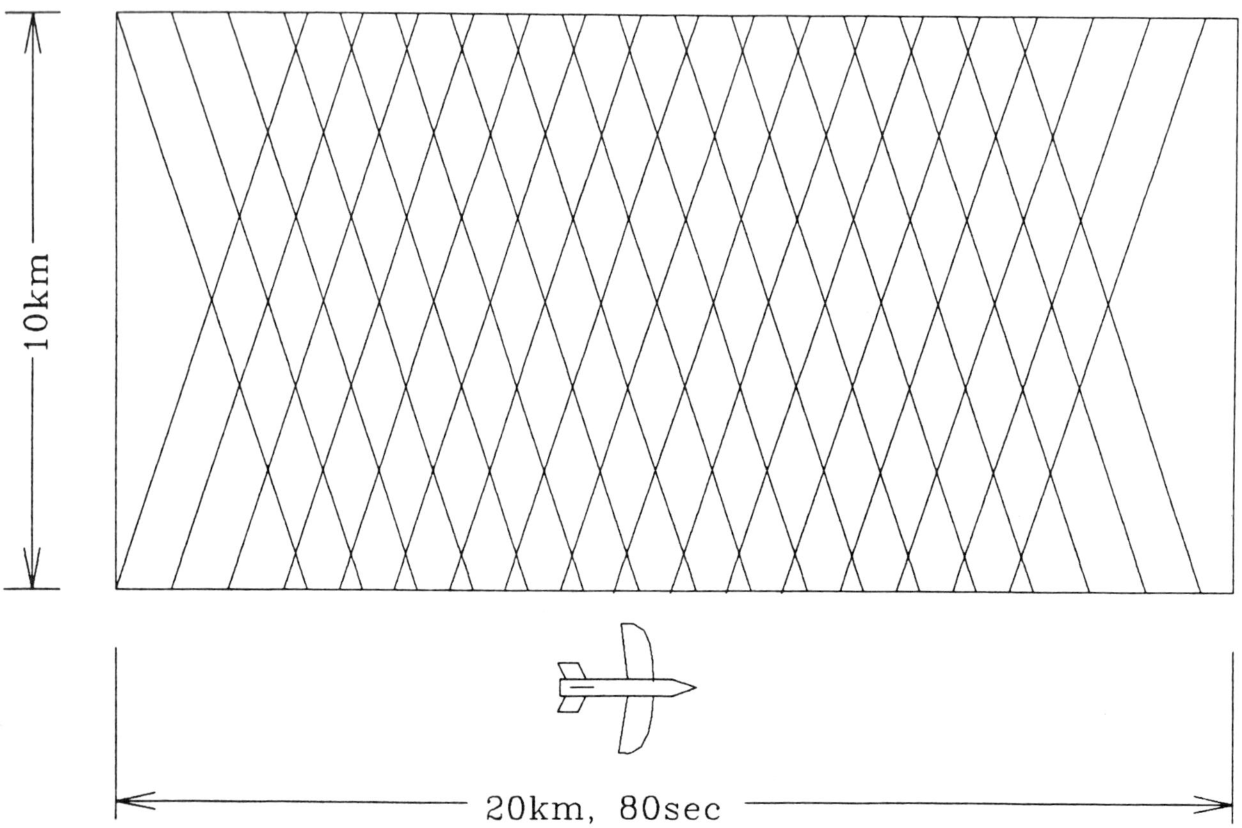

Fig. 8.6. Horizontal Doppler lidar sampling pattern obtained using fore and aft lidar observations from a moving aircraft. In principle, two wind components can be obtained at each intersection point in the grid. (Carroll 1986.)

et al. 1982) was used in one ASCOT experiment (Post and Neff 1986) to study drainage flows in narrow valleys and also in a larger-scale mountain–valley study (Eberhard et al. 1989b) to observe flows confined in a wide shallow valley in the lee of the Rocky Mountains. Data available from the lidar include the radial component of the wind and backscatter intensity measured with a typical along-beam resolution of 300 m and a transverse dimension of about 1 m. Computer control of the lidar allows for rapid sequences of VAD-derived wind profiles, raster scans, and stepped RHI scans.

The pulsed Doppler lidar has two major limitations in measuring the vertical profile of the wind in complex terrain. The first lies in the requirement for a horizontally homogeneous wind field, a necessary condition for using the VAD technique. The other is a substantial minimum range of 1.5 km. A typical VAD elevation angle of 60 degrees leads to the winds measured across a 1.5-km-diameter circle at a minimum range of 1.5 km. Requiring that the sample diameter represent less than one-tenth of the transverse dimension of the flow precludes the use of this technique in complex terrain on scales less than about 15 km. Furthermore, the minimum range of 1.5 km limits the first range gate to a height >1 km. Lower scanning angles have even less utility because of the systematic irregularity of many terrain-dominated flows. Improvements in technology may permit shorter minimum ranges, such as the 0.6 km achieved by Eberhard et al. (1989a) at some cost in signal intensity.

It is still possible to use the lidar effectively in complex terrain when the flow has a directional preference along which the lidar scans can be directed. For example, in nighttime drainage flows in narrow canyons, the lidar can be situated within the canyon and measure the along-valley component of the wind with high resolution over the section of a spherical surface bounded by the valley sidewalls, the valley floor, and the top of the drainage winds (where the normal component of the wind is assumed to be almost zero). Because the lidar measures the normal component of the wind over this surface, one can calculate the volume flux of air through it (multiplying by the density of the air provides the mass flux). Because this measurement can be made at range gates spaced at 300-m intervals along the valley axis, the change of the flux with range can be measured (Post and Neff 1986). Observations like these take advantage of the most valuable feature of a lidar; i.e., the narrow beam and absence of sidelobe radiation allow the lidar to scan very close to the surface, if positioned properly. Furthermore, such special scans can be combined with VAD wind profiles obtained well above the surface to relate changes in local flows to changes in external winds.

In cases where the lidar cannot be positioned to take advantage of a particular flow symmetry, supplemental wind speed and direction profiles must be measured (e.g., using Doppler sodars, radar wind profilers, or balloon-borne instruments) to determine how to apply the lidar data. As Doppler lidars become more prevalent and more automated, dual- or triple-Doppler wind measurements will become available, thus advancing the study of complex terrain wind fields. It can also be hoped that technological advances will reduce the lidar's minimum range, pulse length, and cost.

8.2.4.5. Optical crosswind sensors

Optical crosswind sensors measure the wind component normal to the path between a transmitter and receiver. The instrument uses naturally occurring optical refractive index inhomogeneities, related to C_T^2, as a tracer for the wind. [Because of this feature its spatially averaged scintillation measurements have been compared with measurements obtained from a sodar situated below an optical path that traverses a narrow canyon (Porch et al. 1988) in order to determine the fluctuating properties of the flow.] It does this by observing a distant light source through two optical receivers; by correlating the time delay in the scintillation patterns between the two receivers, it is possible to deduce the wind component normal to the beam and weighted somewhat more heavily toward the center of the path (Lawrence et al. 1972). These instruments normally operate with pathlengths between 100 m and several kilometers. Their advantages include the ability to provide a path-averaged wind with no mechanically limited threshold for low wind speeds. Their disadvantages lie primarily in logistical constraints; sometimes appropriate transmitter and receiver sites are not easily accessible in complex terrain settings. Also, because they measure only a single component of the wind, they can be used only in unidirectional flows or in settings where a second orthogonal path can be provided. Furthermore, vertical profiles require many transmitter/receiver pairs at successively higher elevations, an arrangement that is often difficult in complex terrain. For this reason the sensor's major use has been to measure drainage flows channeled in narrow valleys or in flows with a well-defined directional preference and height profile (Porch 1982; Porch et al. 1989b; Fritz and Neff 1989).

8.3. Major complex terrain field studies using remote sensors

8.3.1. Atmospheric Studies in Complex Terrain (ASCOT) Program

The U.S. Department of Energy began its ASCOT program in 1979 with a preliminary field experiment in the Geysers geothermal area of California in order to study the development of nighttime drainage winds in complex terrain. These thermally driven winds (described by Whiteman, Chapter 2) were expected to grow in magnitude as they accumulated and progressed down the slopes of the study area. In order to study this evolution, a variety of conventional instruments were deployed. In addition, the first extensive use was made of sodars and optical crosswind sensors. In particular, sodars were deployed along a 10-km section of Anderson Creek (Fig. 8.7). The four monostatic sodars in the upper valley, equipped with facsimile recorders, monitored the growth in depth of the drainage wind. The combined monostatic/Doppler sodar at the east end of the valley provided a continuous quantitative measurement of the drainage winds exiting the valley. At a number of sites, tethered balloon systems provided intermittent profiles of wind and temperature.

Early in the planning process, simply structured drainage winds were expected in the valley; however, sodar records revealed a complicated structure with three levels: a drainage layer next to the surface, a transition layer, and a mesoscale inversion layer. The data also revealed many transient events including short-period internal waves and longer period seichlike oscillations (Neff et al. 1980).

A second major experiment was conducted during 1980 with the addition of multiple inert tracers, additional Doppler sodars, crosswind sensors, and a plume-mapping lidar. A third and final experiment in the geothermal area was carried out in 1981 in a seaward-facing valley, Big Sulphur Creek, just west of the Anderson Creek valley shown in Fig. 8.7, with a smaller suite of instruments but with the assistance of a downward-looking airborne lidar. Early results from these experiments, the instrumentation, analysis of data, and resulting numerical model development have been described by Dickerson and Gudiksen (1984).

In summary, these ASCOT experiments provided for the systematic deployment and testing of new remote sensing devices for the study of drainage winds in a coastal environment where sea breeze and marine inversion effects predominated. The data from these new instruments revealed a more complicated drainage wind structure than originally expected.

As a result of the complicated flows found in the Geysers geothermal area of California, a more ideal valley topography was sought as the ASCOT program moved to the oil shale area of western Colorado in 1982. Following pre-

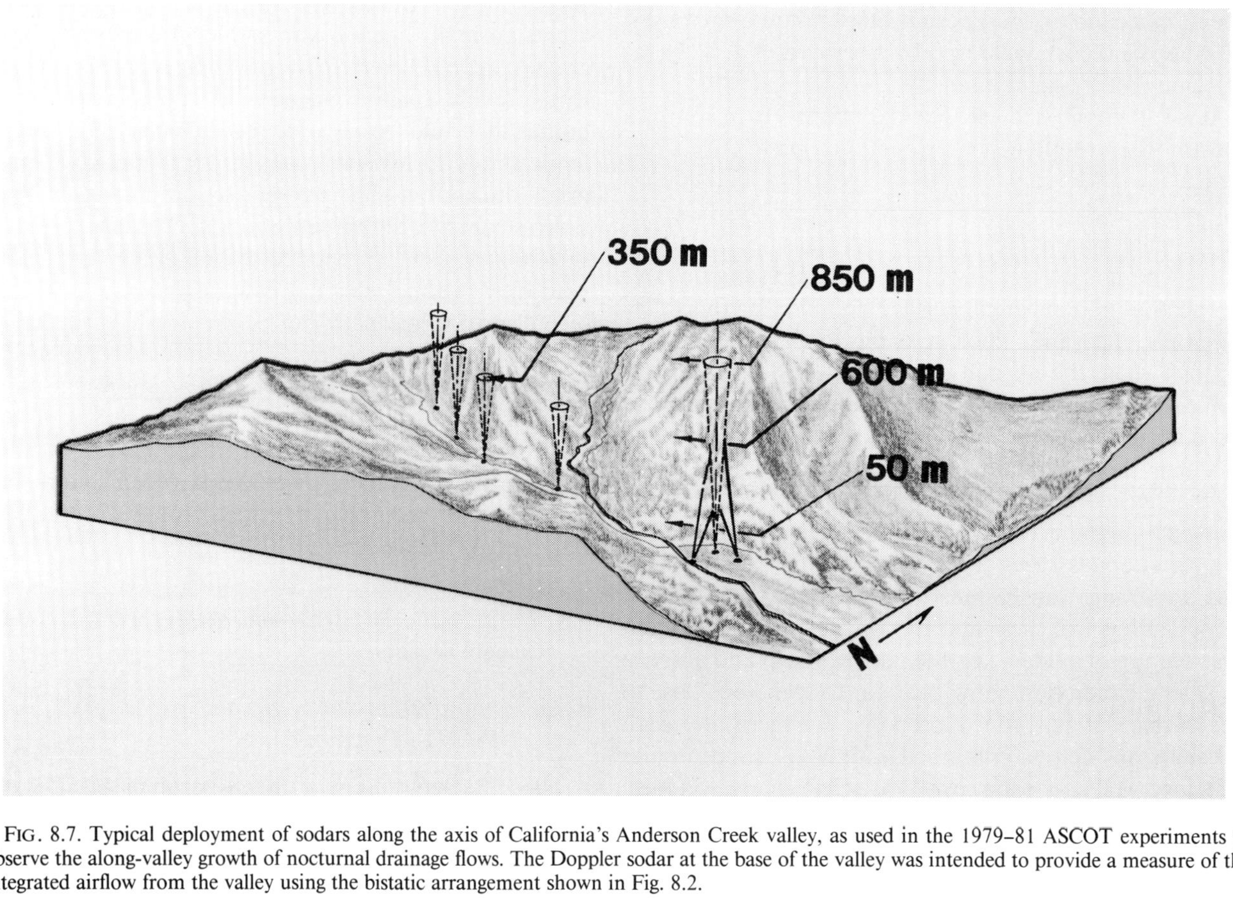

FIG. 8.7. Typical deployment of sodars along the axis of California's Anderson Creek valley, as used in the 1979–81 ASCOT experiments to observe the along-valley growth of nocturnal drainage flows. The Doppler sodar at the base of the valley was intended to provide a measure of the integrated airflow from the valley using the bistatic arrangement shown in Fig. 8.2.

liminary studies in the Brush Creek valley in 1982 that revealed nearly steady drainage flows, a major experiment was carried out in 1984 introducing for the first time the use of an infrared Doppler lidar together with an array of sonic anemometers along the valley floor and minisodars in side canyons to supplement now conventional remote and in-situ sensors. Of particular interest was the deployment of the Doppler lidar. Although this instrument normally measures only a radial wind component, the Brush Creek valley, because of its straight uniform structure, furnished a rare opportunity to the lidar's high-resolution raster-scan mode to measure spatial and temporal details of the along-valley evolution of a drainage flow and thus overcome the earlier uncertainties inherent in the use of single balloon profiles. Furthermore, it provided an opportunity to examine the representativeness of such profiles through mapping of the along-valley component of the wind in the same volume probed by conventional sensors such as Doppler sodars, crosswind sensors, and tethered balloons. Subsequent smaller experiments have been carried out in the same area using combinations of surface tower networks, tethered balloons, sodars, optical crosswind sensors, and smoke releases as a visualization technique. Because of the continuity in the experimental program and its research orientation, considerable progress has been made in understanding the physical processes affecting complex terrain flows. The results of the program's research orientation are described in the June and July 1989 issues of the *Journal of Applied Meteorology*.

8.3.2. EPA Complex Terrain Model Development (CTMD) Program

Whereas the ASCOT program sought to study in detail the diverse facets of nighttime drainage winds, the EPA complex terrain program developed and tested a potential regulatory transport and dispersion model intended to predict surface concentrations of pollutants in complex terrain. The CTMD experiments were carried out in valleys where drainage winds could transport elevated plumes onto higher surrounding terrain. The sequence of experiment sites progressed from an isolated 100-m-high hill in Idaho, to a hogback ridge near Farmington, New Mexico, to a more complex valley/basin system east of the Sierra Nevada and Reno (see Hanna and Strimaitis, Chapter 6).

The CTMD program promoted improvements in ground-based plume-mapping technology, including the development of better processing methods to accommodate the vast amount of data required to map elevated plumes in three dimensions and time. Significant technical improvements in the lidar were made for the hogback experiment in 1982 as well as the first integration of lidar observations with tower-mounted sonic anemometers and sodars. The final phase of this program resulted in the Full-Scale Plume Study in 1984 using a combination of a plume-mapping lidar, sodars, tower-mounted sonic anemometers, tethered-balloons, Doppler and monostatic sodars, and surface meteorological stations to document the effect of atmospheric processes on plume behavior in a complex terrain setting and to evaluate the efficacy of new plume models developed for EPA applications (Eberhard et al. 1987). Because the program focused on the development of a regulatory dispersion model, however, the full research value of the remote sensing dataset collected has not yet been realized.

8.3.3. Urban and regional air quality studies

A majority of the world's large urban areas are embedded in complex terrain settings, either in coastal-facing basins or in river valleys, or at the base of mountain ranges. Therefore, further understanding of atmospheric processes over complex terrain can have immediate application to the practical problem of urban air quality. Current and potential areas of interest include the transport, interchange, and mixing of biogenic and anthropogenic gases through the thermally and dynamically driven diurnal mountain–valley circulations; venting of pollutants into the mid- and upper-troposphere through sidewall heating; and the trapping, recirculation, and chemical transformation of pollutants in cold-air pools in large valleys and basins. Examples include the South Platte River Basin in Colorado (Haagenson 1979; Watson et al. 1988; Neff 1989a,b,c), the San Joaquin valley (Blumenthal et al. 1985), the Sierra Nevada (Carroll and Baskett 1979), the Los Angeles basin (Wakimoto and McElroy 1986), the Kanto Plain in Japan (Kimura 1986; Sasaki et al. 1988), the Athens basin (Katsoulis 1988), Calgary (Nkemdirim and Leggat 1978; Mathews et al. 1984), and Munich (Hoinka and Roesler 1987). In many of these cases, the local complex terrain circulations have been discovered to be more complex than a simple case of air stagnation. In fact, in Calgary, Munich, and Denver, higher pollution potential has been found for conditions normally favorable for strong downslope wind storms.

It is only recently, however, that remote sensing instruments have been applied in any substantial degree to these problems. A recent study of the Denver Brown Cloud (Watson et al. 1988; Neff 1989a,b) made use of a broad range of sensors including radar wind profilers, Doppler lidar, monostatic and Doppler sodars, and optical crosswind sensors, as well as conventional in-situ and balloon-borne instruments. Because many of the regional and urban air quality issues involve both complex terrain and mesoscale circulations at scales from tens to hundreds of kilometers, they provide significant challenges for the application of remote sensing.

8.4. Application of remote and in-situ instrumentation to complex terrain studies—case studies

Individual remote sensors can rarely provide all the information needed for studying flows in complex terrain. Various combinations of remote and in-situ sensors have proved useful, however, depending on the objectives of a particular study. Thus, the choice of instruments is best determined by the scientific issues that must be addressed. For example, in preliminary or exploratory experiments, where the phenomena of interest have not been defined, a single monostatic sodar can be useful because of its ready availability, its low cost, and the pattern recognition potential that it provides. At a second level, tethered-balloon- or tower-derived wind and temperature profiles can be provided and/or individual surface weather stations can be used to help quantify the sodar observations.

Often, however, much is already known about a particular flow regime and more specific and quantitative information is required. For example, in the 1984 ASCOT experiment, preliminary measurements had examined the evolution of the volume flux in a drainage flow in the long, narrow Brush Creek valley (Whiteman and Barr 1986). Questions remained, however, about the representativeness of the tethered-balloon profiles used in the analyses, both in space and in time. This is a particularly relevant issue because in-situ profiling techniques are often limited by logistical, not scientific, constraints. For this reason, among others, the Doppler lidar appeared to be a useful tool with which to overcome the limitations of conventional profiling methods. By using the lidar in the raster-scan mode to remotely probe the horizontal and vertical structure of the drainage jet in Brush Creek, a number of questions could be answered about the spatial structure of the jet and, as we show below, much was revealed about the growth of the drainage volume flux along the valley axis, its source, and the merging of multiple valley flows.

In the sections that follow we describe a number of complex terrain flow phenomena to which remote sensing methods have been applied and the analysis methods used. In sections 8.4.1 and 8.4.2 we describe simple nocturnal drainage flows and the complexities that one is likely to encounter. These sections draw heavily from sodar observations and use a simple turbulence/mean-flow parameterization to explain some aspects of the observed echo structure. Section 8.4.3 focuses on sodar observation of long-period waves, their relation to thermally driven circulations, analysis of supporting surface observations, and finally, remote sensor observation of waves generated by flow over terrain. Sections 8.4.4 and 8.4.5 address the use of a Doppler lidar to study channeled drainage flows during the 1984 ASCOT experiment in Brush Creek. Following the pattern from simple to complex, section 8.4.4 addresses drainage flow growth in the almost linear valley of Brush Creek. In section 8.4.5 we focus on the merging of drainage flows from valleys with somewhat different physical characteristics. These sections include some recently completed analyses that have sought to overcome earlier limitations in the lidar data processing and to examine additional processes to which lidar data could be applied. Section 8.4.6 outlines results from an experiment on dispersion in complex terrain in which lidar plume mapping, sodar pattern interpretation, and in-situ measurements were integrated. The first six sections deal almost exclusively with well-developed drainage flows and their complications in confined terrain; little is described about the transition to and from drainage flow conditions and the information available from remote sensors during these periods. For this reason, we examine in section 8.4.7 some of the insights that remote sensing can provide into the initiation and destruction of nighttime drainage flows in narrow valleys and basins, and into the structure of drainage flows as they emerge from mountains onto plains.

8.4.1. Sodar observations of simple drainage flows

Simple drainage flows can be identified in both sodar and tethered-balloon data through inspection of wind profiles and echo patterns. For example, Neff and King (1987) contrasted the behavior of the sodar-derived drainage wind speeds and their standard deviation in both draining and pooling valleys. In the case of draining valleys the standard deviation of the wind speed, as observed by a Doppler sodar averaging over 20-min periods, was much less than the mean wind speed averaged over a 6-h period. Furthermore, the wind profile was characterized by a single wind speed maximum. In poorly draining valleys, the mean and deviation were comparable and multiple speed maxima were present.

The evolution of the wind speed profile as observed by Doppler sodars in two draining valleys can be seen in Fig. 8.8, reflecting slow changes in speed and the presence of a single wind-speed maximum throughout the night. Differences in airshed size and time of sunset account for the timing and scale of the flows. By way of contrast, Doppler-derived wind-speed isopleths in a poorly draining valley show multiple drainage wind speed maxima (as can be seen in Fig. 8.9 shortly after 0100 PST) and even subsequent wind reversals (after 0130 PST). In each of these cases, the higher time resolution (20 min) and continuity of the remote sensing data are of value; e.g., tethered-balloon systems typically provide snapshot-type profiles at intervals of 1 to 1.5 h, intervals often greater than that characterizing major variations in the wind, and a factor that can lead to aliasing of the observations. Another feature in Fig. 8.9, to which we return below, is the presence

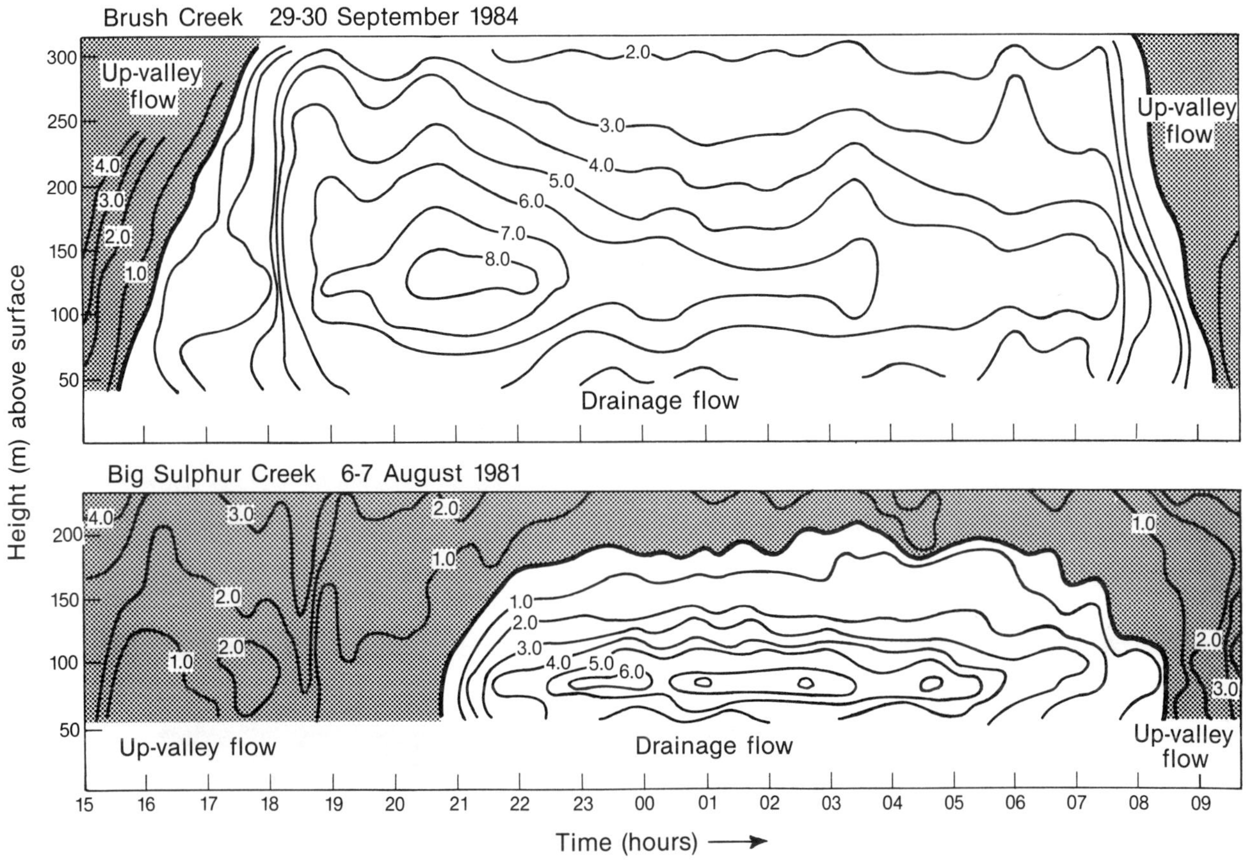

FIG. 8.8. Doppler wind observations showing the development of drainage flows in two valleys in the presence of opposing winds using 20-min averaged sodar-derived wind profiles. Note the gradual decrease in drainage winds in Colorado's Brush Creek valley compared with the nearly steady drainage winds in the California's Big Sulphur Creek valley from 2300 to 0500 PST. (Neff and King 1987.)

of layering and intermittency in the echo patterns, a feature often associated with strong stability and intermittent drainage.

Sodar echo patterns also provide a straightforward tool for identifying simple drainages. A sodar facsimile record obtained in a simple channeled drainage flow, tethered-balloon-derived wind and temperature profiles, and a model-calculated scattering profile [in terms of C_T^2 as in (8.2)] is shown in Fig. 8.10. In this case, the sodar minimum range was not sufficient to resolve the low-lying echo layer that the turbulence model predicts. With the advent of high-frequency minisodars, however, this region will be better resolved. Of interest for pattern recognition in the figure are the strong echo layer in the region of decreasing winds above the wind maximum and the minimum in echo intensity in the region of the wind maximum. This latter effect reflects the weak shear and hence small K_H at the nose of the jet. Figure 8.10 also suggests a limitation in models that calculate turbulence properties directly in terms of mean gradients, namely that the turbulence is normally set to zero when Ri is greater than some critical value, usually 0.25. This modeling limitation avoids the observational evidence that once turbulence occurs, laminar flow does not become reestablished until Ri exceeds 1 (Woods 1969; Blumen 1984). This may explain the discrepancy between the heights of the observed and predicted echo layers. In particular, the sodar reveals the strongest echo layer in the strong-stability/strong-shear region at the top of the surface inversion, whereas the model predicts the strongest echoes in the weak-stability/strong-shear region just above the top of the inversion. This suggests the value of high-resolution modeling efforts that use higher levels of turbulence closure and that account for the "hysteresis" effect of Woods as well as the turbulent transfer of temperature variance. Such efforts would advance the current state of sodar pattern interpretation considerably.

Sodar observations, such as those in Fig. 8.10, provide perhaps the severest test of simple models. At times, however, the atmosphere provides simpler circumstances. For example, a turning of the wind through the nose of a low-level drainage jet will create a larger vector shear and hence maintain a low Ri throughout the region of the jet and result in a somewhat different echo pattern as shown in

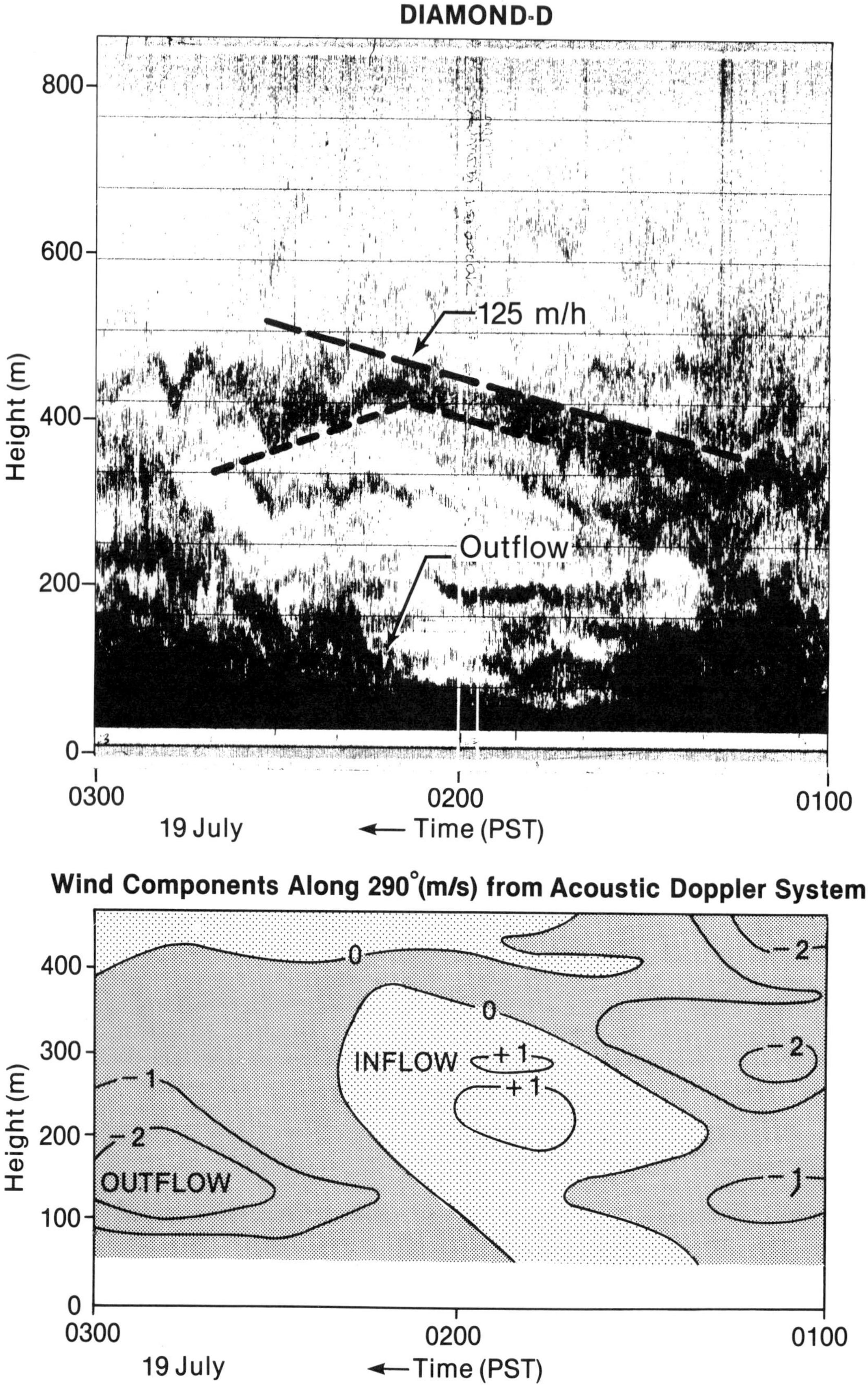

FIG. 8.9. Sodar observations obtained at the valley exit site of California's Anderson Creek valley shown in Figs. 8.2 and 8.7. (Lower) Doppler wind profiles show multiple wind maxima (0100 PST), inflow into the valley (0130–0200 PST), and a drainage-like outflow (after 0200 PST). (Upper) The period of inflow is coincident with a rising inversion over the site as noted by the dashed lines. (Neff and King 1988.)

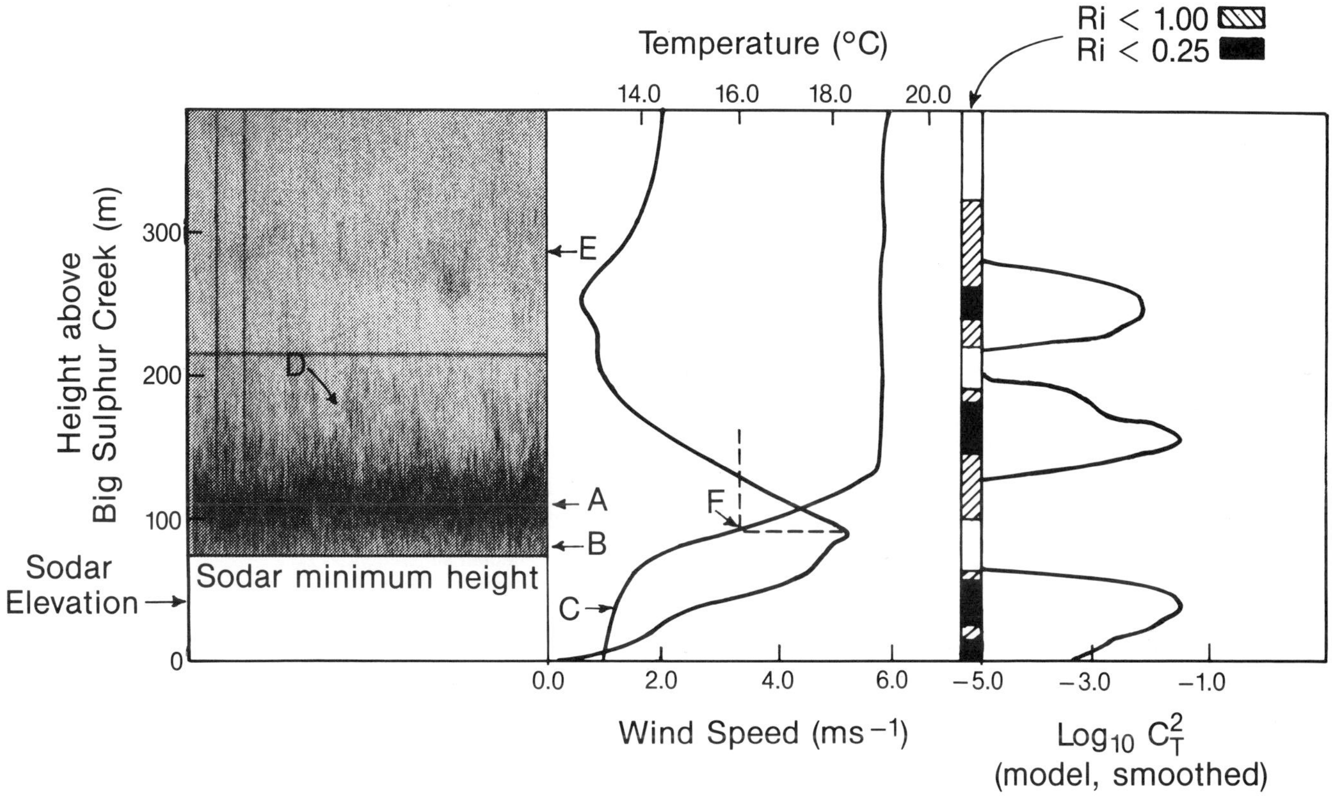

FIG. 8.10. Sodar facsimile recording obtained during nearly steady drainage flow in California's Big Sulphur Creek valley with corresponding tethersonde-derived wind and temperature profiles and numerical model calculated scattering profiles and regions of reduced Richardson number. Features in the figure are keyed as follows: (A) Echo layers form typically in the wind shear region above the height of the wind maximum. (B) Weaker echoes form in the region of large Ri near the nose of the jet. (C) The nearly isothermal lapse rate near the surface may represent an almost balanced surface energy budget. (D) Intermittent turbulence occurs above the jet. (E) A weak elevated inversion marks the top of the drainage flow. (F) The wind speed maximum occurs at the midpoint of the temperature profile. (Neff 1988.)

Fig. 8.11. This observation from Neff (1986) reveals the echoes and predictions associated with a large-scale downslope flow in Antarctica where Ekman turning from the geostrophic wind direction maintains a strong shear through the wind maximum; here the strongest echo layer occurs at the height of maximum wind speed as observed with both the model and the Doppler sodar. The interpretation of echo patterns associated with low-level jets channeled in complex terrain will thus depend on the scale of the terrain and the orientation of the jet with respect to the ambient winds. A further effect in very narrow valleys that has not yet been addressed is that of lateral mixing from the valley sidewalls. Study of this effect will require further observations as well as more detailed three-dimensional models.

8.4.2. Sodar observation of complex drainages and their interaction with ambient flows

In the preceding section we described a simple drainage flow over a slope. In such cases the layer of cold air overlying the slope moves downhill in response to gravity and its isothermal surfaces are usually assumed to be parallel to the underlying slope. The turbulence structure associated with such simple flows was also simple: a layer of turbulence due to shear with the lower boundary and an elevated turbulent layer due to wind shear in the region of decreasing wind speed above the drainage jet. Often, however, drainage flows develop in the presence of preexisting inversion layers and/or in the presence of time-varying ambient winds. Furthermore, slopes and valleys that drain initially may, in fact, lie within larger-scale terrain, including larger valleys and basins. Eventually, these areas may fill with cold air, thereby modifying the pressure gradients causing the drainage flow. In such cases, complex flow patterns develop that result in the complex echo patterns (Neff 1988). As we show below, these complex patterns may simply identify turbulent layers at the interface between air masses moving relative to each other.

8.4.2.1. THE ROLE OF DIFFERENTIAL ACCELERATION OF AIR MASSES IN ECHO CREATION

The situation depicted in Fig. 8.12 is one of the simplest complications occurring within a two-layer flow over a

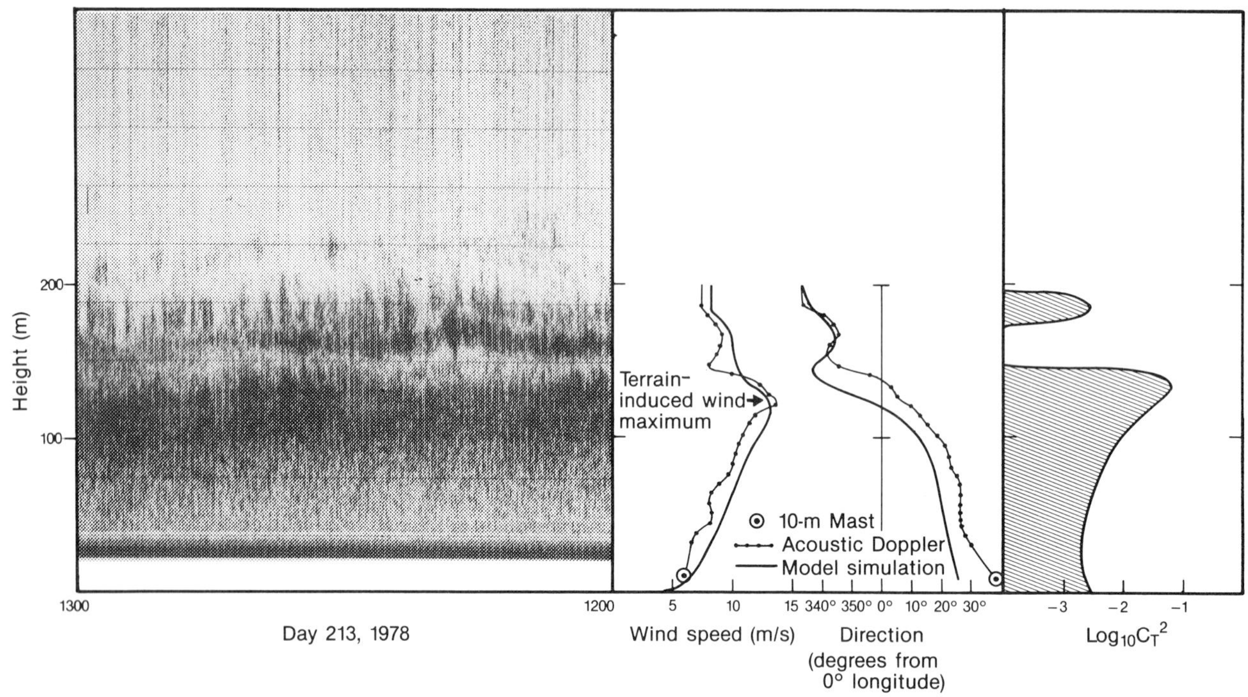

FIG. 8.11. Sodar observation and model-simulated profiles in a quasi-steady large-scale drainage system at South Pole Station on the East Antarctic ice sheet with no diurnal insolation cycle. Doppler-sodar-measured drainage jet and Ekman turning are well reproduced by the model. (Neff 1986.)

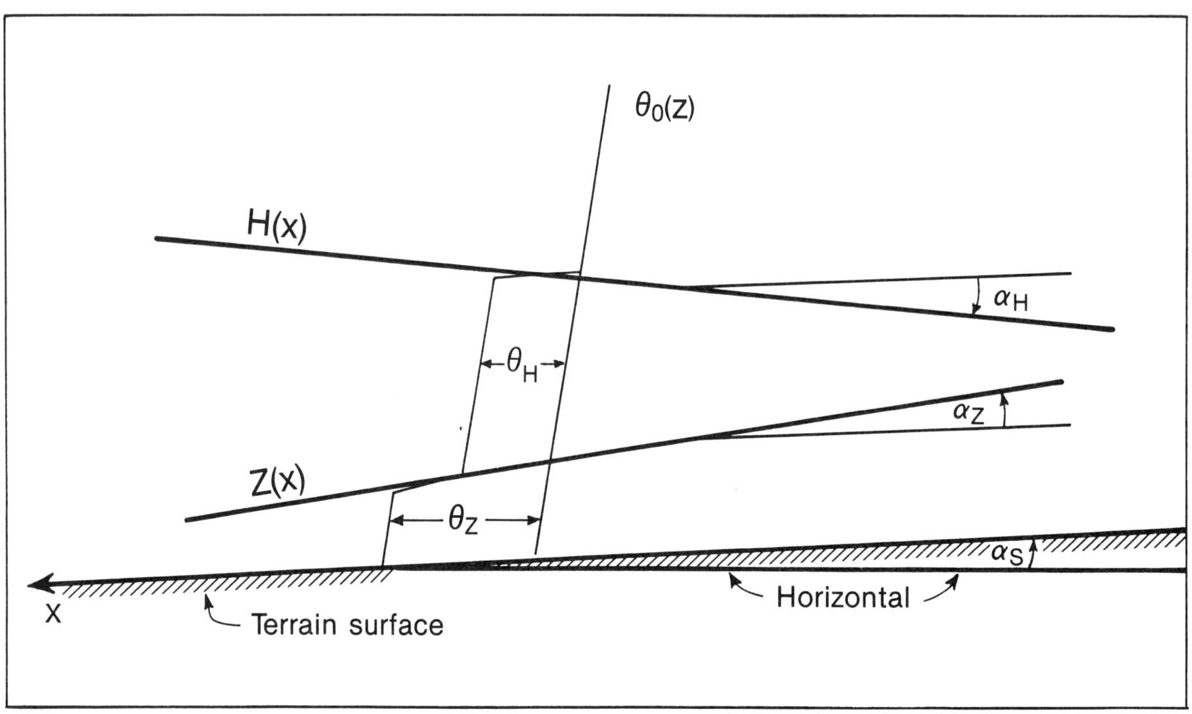

FIG. 8.12. Schematic view of two-layer fluid in which the surface pressure gradient depends on the relative slopes of the elevated inversion interfaces.

slope as described in more detail by Neff (1986); e.g., the lowest layer could represent a local slope flow whereas the upper inversion layer might represent a mesoscale inversion whose slope has been perturbed, perhaps through an internal seiche. In this situation the upper layer will accelerate to the right because of the horizontal pressure gradient caused by the slope of $H(x)$. The lower layer below $Z(x)$ may move in either direction depending on the relative values of the slopes α_H and α_Z and the respective temperature deficits Θ_H and Θ_Z from the initial potential temperature profile $\Theta_0(z)$ (in the classical slope flow problem, $\alpha_Z = \alpha_S$ and $\alpha_H = 0$).

When such individual flows accelerate differentially, the interfaces $H(x)$ and $Z(x)$ will eventually become unstable and fully developed turbulence will most likely follow a series of dynamical instabilities. Once turbulence develops, its further evolution will depend on changes in Ri described by Neff (1988):

$$\frac{\partial \mathrm{Ri}}{\partial t} = \left[-\frac{g}{T_0 S^2} \frac{\partial^2 \overline{w\theta}}{\partial z^2} + 2 \frac{\mathrm{Ri}}{S} \frac{\partial^2 \overline{uw}}{\partial z^2} \right] + \left[2 \frac{\mathrm{Ri}}{S} \frac{\partial}{\partial z}\left(W \frac{\partial U}{\partial z} \right) - \frac{g}{T_0 S^2} \frac{\partial}{\partial z}\left(W \frac{\partial \Theta}{\partial z} \right) \right] - \frac{g}{T_0 S^2} \frac{\partial^2 F_R}{\partial z^2}, \quad (8.9)$$

where $S = \partial U / \partial z$ for unidirectional flow, T_0 is the local temperature, and horizontal thermal gradients and nonlinear terms in the horizontal velocity have been neglected. The first term in brackets on the right describes the effect of curvature in the stress ($\overline{uw}$) and temperature flux ($\overline{w\theta}$) profiles; the second term includes the effects of subsidence W on wind shear and potential temperature profiles; and the third term accounts for changes in the temperature profile due to curvature of the longwave radiative flux (F_R) profile. From a heuristic point of view, a turbulent layer, once formed, should entrain adjacent fluid and momentum and spread, resulting in progressively smaller vertical gradients of temperature and wind. Because Ri, and hence turbulence intensity, depend inversely on the square of the wind shear but only linearly on the potential temperature gradient, Ri should gradually increase until turbulent mixing is quenched. From (8.9) we see that several effects can inhibit or reverse this process, including weakening of local temperature gradients through radiative transfer as well as differential acceleration of the two layers, a process that can maintain the shear across the layer, countering the effect of turbulent diffusion.

In Fig. 8.13 the result of a model calculation of this process is displayed at an interface at which two layers maintained a weak relative acceleration of about 2 m s^{-1}/h, first in the absence of subsidence or radiative transfer effects (upper) and then in the presence of a strong radiative diffusivity (lower). In the latter calculation, the effect of the radiative flux divergence was simulated with a diffusivity (1000 cm^2 s^{-1}) multiplying the curvature of the temperature profile, following the method of Deardorff (1972). In the upper figure, the reduction in Ri due to curvature of the stress profile dominates at the boundaries of the turbulent layer whereas the curvature of the temperature flux profile dominates in the interior, causing Ri to increase there. The same result holds for the case with radiative diffusivity except that the reduction in Ri at the top boundary is less, and at the lower boundary the terms corresponding to the temperature and momentum flux nearly cancel. Of interest for the interpretation of sodar facsimile records is the formation, in Fig. 8.13, of two strong echo layers some 20 meters apart as a result of interfacial mixing. In situations such as these, short pulse lengths are required to delineate the separate layers. (Note that a bistatic sodar would see uniform echos lying between the two thin backscatter echo layers.) Thus, four distinct echo layers can form in the two-layer idealized flow portrayed in Fig. 8.12. Such echo layers are most often seen in light wind situations typical of basins or of drainage flows in the presence of mesoscale inversions such as that associated with the Pacific coast marine layer (Neff and King 1988).

8.4.2.2. Entrainment by the external wind—turbulence and instability

These results suggest that the differential acceleration of one stably stratified layer overlying another in complex terrain may create the multiple echo strata often observed in sodar facsimile records. A by-product of the calculation is the suggestion that a continuing differential acceleration of the two layers is required to maintain the process by which laminar fluid is entrained into a turbulent layer. An example of this is shown in Fig. 8.14 where 20-min averaged winds derived from a Doppler sodar reveal the erosion of an external, accelerating wind into a cold pool (Neff 1988). An artifact of the shear and stability within this layer is the presence of echoes suggestive of Kelvin–Helmholtz billows (Neff and Coulter 1986), particularly from 0520 to 0540 PST. Although additional mechanisms may be at work here, these observations indicate the potential importance of accelerating ambient winds for the erosion of cold-air pools and suggest that they may also be important in the erosion of already established slope flows on exposed, elevated terrain.

Even though the model applications described by Neff (1988) are highly simplified, they do show that models can be used to provide insight into the physical processes important to complex flows over topography when they

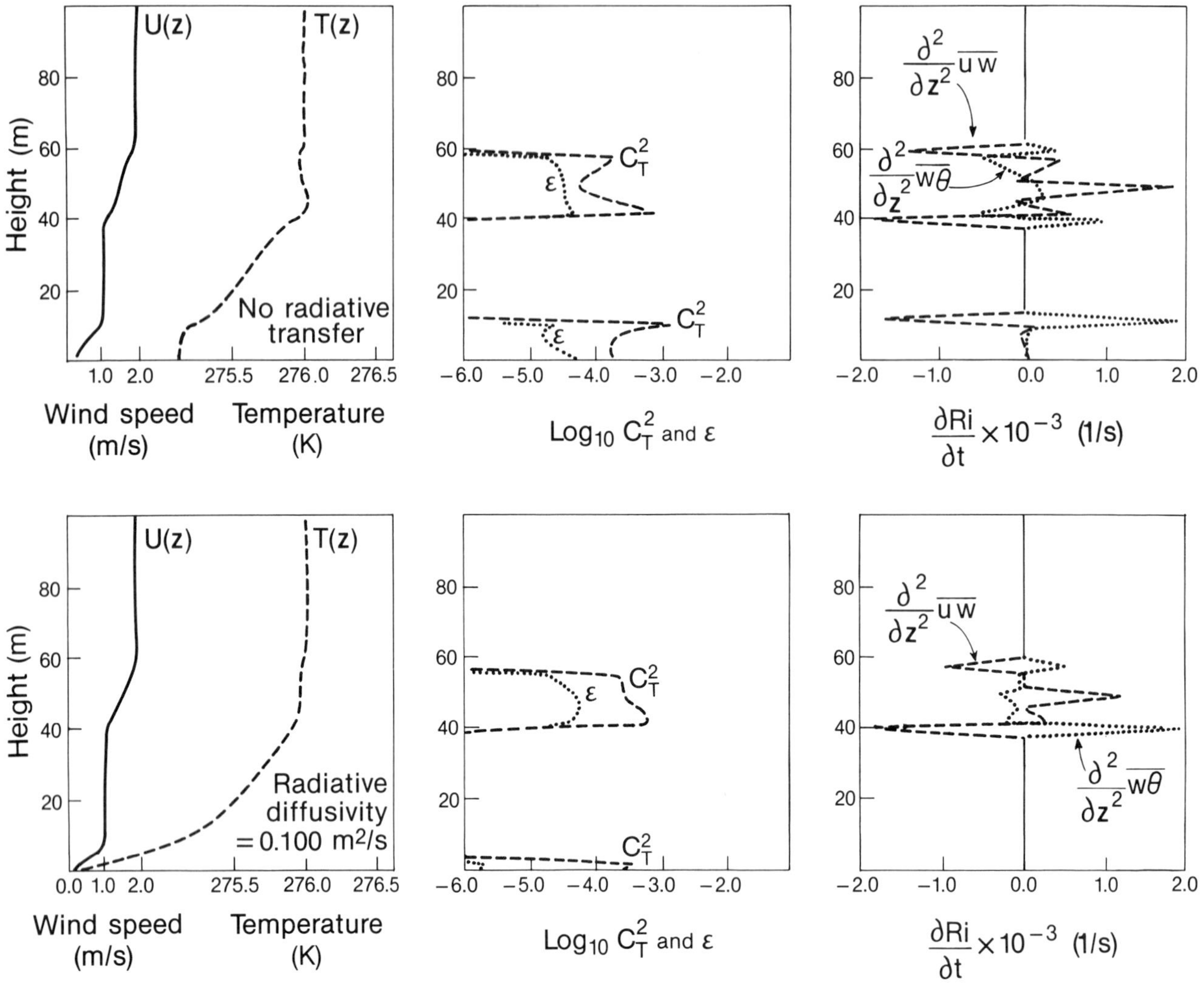

FIG. 8.13. Model-simulated interfacial turbulence including double echo layer often observed in stratified flows. Richardson number tendency terms are also shown from the model output that reveal the continuing erosion at the layer boundaries and a tendency to equilibrium in the interior. (Neff 1986.)

are more closely integrated with in-situ and remote sensing data. A further example can be seen in Fig. 8.15 for a model calculation of a downslope flow developing in the presence of steady cross-slope flow (to maintain Ri below 0.25 throughout the jet), a moderately accelerating upslope flow component, and then a strongly accelerating upslope flow. In each case, the effect of larger upslope accelerations is to reduce the magnitude of the drainage flow and to create a thicker region of strong shear above the drainage in which instabilities can grow as shown in Fig. 8.14. Similar patterns in local tributary flows have been described by Neff and King (1989) in which echo patterns appear to trace out large eddies that extend well into the region of strong shear created by accelerating opposing ambient winds.

8.4.3. Remote sensor observation of waves in complex terrain

In Chapter 4, Durran describes the basic physics of internal waves relevant to flow over irregular terrain, particularly those aspects underlying the generation of mountain lee waves. Because internal waves produce a periodic shear (in space and time) in stably stratified flows and can transfer momentum from one part of the flow to another (e.g., to a wave critical level), they can produce localized regions of turbulence detectable with sodars and clear-air radars. Furthermore, the layered regions of interfacial turbulence described in the preceding section can also be perturbed by propagating internal waves resulting in oscillatory echo strata on sodar or radar time–height

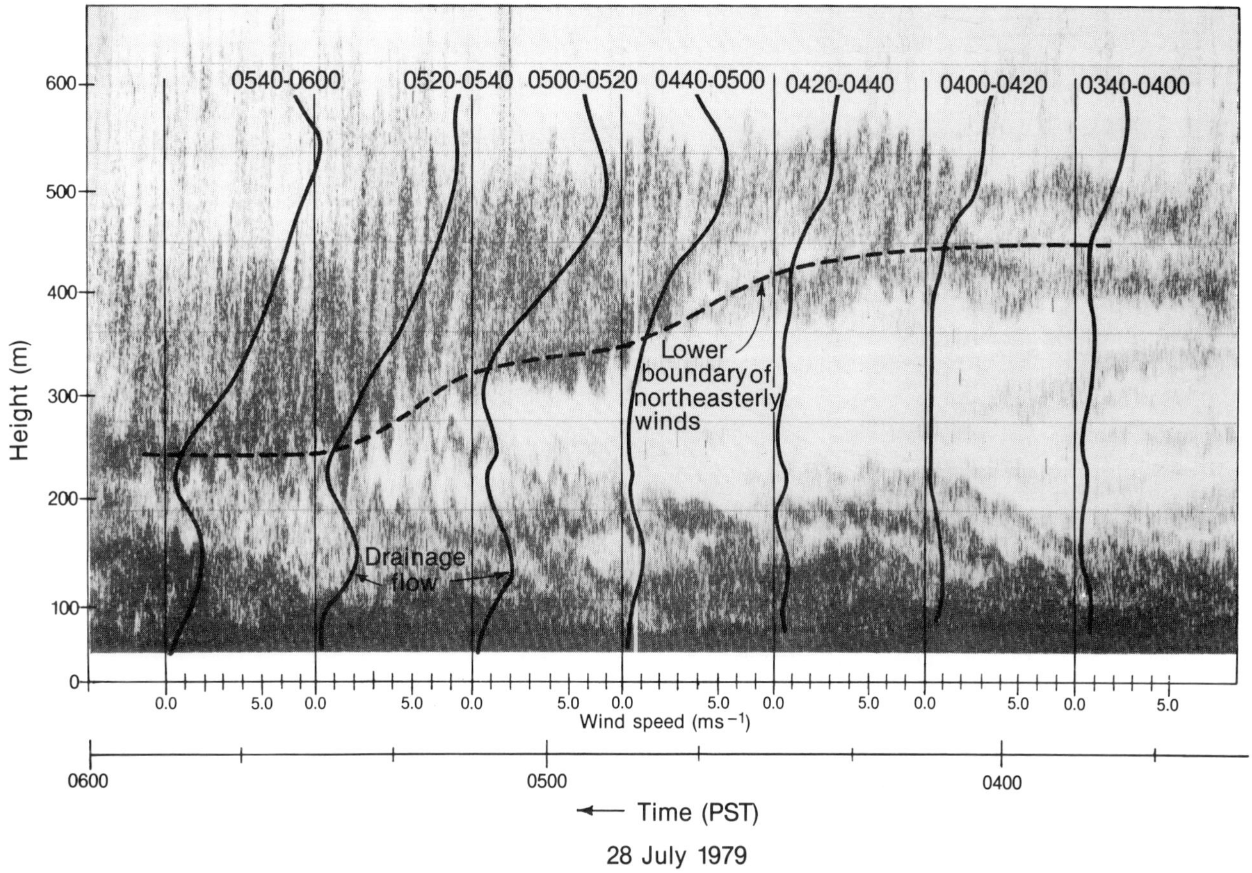

FIG. 8.14. Sodar echoes showing dynamical instability and erosion at the top of a cool-air pool that formed in the lowland areas below the Anderson Creek valley (Fig 8.7). Drainage flows entering the basin are confined below 200 m. The increasing wind shear produced by accelerating northeasterly winds (opposing the drainage flow direction) was documented by a bistatic Doppler sodar. (Neff 1988.)

facsimile recordings. In many complex terrain studies, observations have been obtained of waves in basinlike topography where cold air can pool. In such topography long-period seiches can occur; furthermore, the very stable interface at the top of the cold-air pool can support a broad spectrum of internal waves.

8.4.3.1. LONG-PERIOD OSCILLATIONS OBSERVED IN COMPLEX TERRAIN FLOWS

Neff and King (1985) described long-period (1 to 2 h) motions as seen in sodar records in basinlike terrain. In particular, they provided an analysis of Doppler-sodar derived wind profiles and sodar measurements of inversion height at several points along an axis across a basin and onto the Anderson Creek valley sidewall as shown in part in Fig. 8.7. Of interest in their observations were long-period waves that propagated across the basin with a phase speed approximating that of a shallow-layer internal wave mode and showing a doubling of the vertical displacement of the wave as it impinged on the side slope of the basin. In the immediate vicinity of the sidewall, Neff and King used the Doppler-sodar derived wind profile to measure the volume inflow and outflow underneath the inversion. An example of the variation of the inversion height in response to the influx and efflux of air under the inversion can be seen in Fig. 8.16 where we have reproduced the Doppler wind isopleths from the case in Fig. 8.9 (at the mouth of the valley) together with the sodar facsimile record from a site higher within the valley. The period of inflow corresponds closely with the rise in the inversion (marked by echo strata) 5 km to the west. Furthermore, the more rapid rate of rise of the inversion at the Thorne-7 site indicates possible wave steepening over the side slopes of the basin as the wave phase speed slows over the slope in response to a shallower inversion layer. As a further example of the quantitative data available for this configuration of sodars, Fig. 8.17 shows a scatter plot, resulting from data obtained during many such events, that reveals a relation between the change in the volume of air under the inversion (detected by the monostatic sodar within the valley) and the volume of

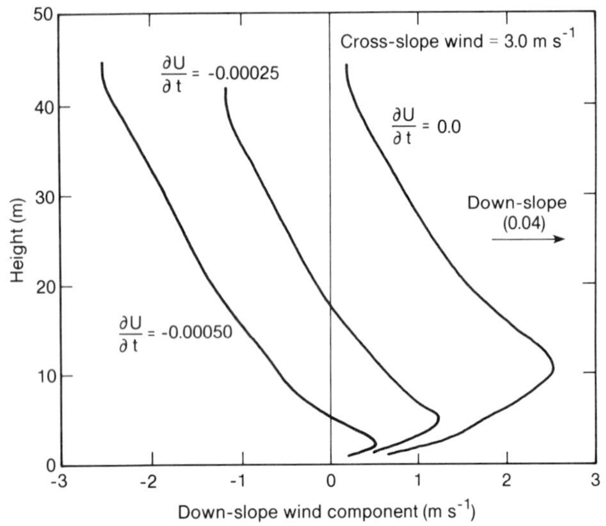

FIG. 8.15. Model calculation of slope flows (terrain slope = 0.04) in the presence of a 3 m s^{-1} cross-slope ambient wind. Three calculations are shown with progressively larger upslope accelerations of the ambient wind [0.0, −0.00025, and −0.00050 (m/s) s^{-1}] opposing the developing drainage flow for a period of 3 h. Stronger accelerations show greater erosion of the developing drainage flow.

air emerging from the valley (as estimated from the Doppler wind profile and integrated over the period of the event). Again, it should be noted that events such as these could not have been observed with the typical 1-h sampling interval available with tethered-balloon systems.

Long-period oscillations usually appeared to be related to the presence of an elevated inversion overlying the slope flows. Sodar observations above the upper slopes of Anderson Creek on 19–20 and 24–25 September 1980 are presented in Fig. 8.18. The locations of the sodar and of the various Portable Automated Mesonet (PAM) stations present for the experiment are shown in Fig. 8.19. On the night of 19–20 September oscillations occurred on an elevated echo layer, with periods ranging from a few minutes to more than an hour. On the night of 24–25 September a shallow drainage flow, almost steady in depth, persisted through the same period of the night. The major difference between the two nights was the deeper marine layer on the first night (Neff and King 1988), which provided an environment supportive of internal waves.

8.4.3.2. Scale analysis

The importance of wave motions relative to drainage winds can be seen from a scale analysis of the relative horizontal accelerations due to the pressure gradients associated with geostrophic winds, katabatic forces, and long-period internal waves:

- Geostrophic:

$$fU_g = 1 \times 10^{-3}\ \mathrm{m\ s^{-2}}$$

for

$$U_g = 10\ \mathrm{m\ s^{-1}}$$

and

$$f = 10^{-4}\ \mathrm{s^{-1}}. \tag{8.10}$$

- Katabatic:

$$\frac{\Delta\Theta}{\Theta} g \sin\alpha = 1.6 \times 10^{-3}\ \mathrm{m\ s^{-2}}$$

for

$$\Delta\Theta = 5^\circ\mathrm{C}$$

and

$$\alpha = 0.01\ \mathrm{rad}. \tag{8.11}$$

- Internal wave pressure perturbations (P_W):

$$\left[\frac{1}{\rho_0}\frac{\partial P_W}{\partial x}\right]_{\mathrm{surface}} \approx i\left[\frac{\omega N_B}{\sin(kN_BH/\omega)}\right]\eta_H = 1.7 \times 10^{-3}\ \mathrm{m\ s^{-2}}. \tag{8.12}$$

Here $\omega = 1.2 \times 10^{-3}$ rad s^{-1}; $k = 3.14 \times 10^{-4}$ m^{-1}; H = 300 m; and the amplitude of the vertical displacement of the wave is given by $\eta_H = 50$ m at the top of a layer of height H above the ground and characterized by a Brunt–Väisälä frequency $N_B = 2 \times 10^{-2}$ rad s^{-1}. The factor of i in (8.12) indicates a 90° phase shift between the displacement η at height H and the surface pressure gradient fluctuations. The factors used in evaluating (8.12) were representative of the long-period waves described earlier. [Equation (8.12) was derived from Gossard and Hooke (1975) using their Eqs. (68.1)–(68.5). It is assumed that $\omega^2 \ll N^2$ and that the horizontal direction of propagation is along the x-direction.] This scale analysis thus reveals the complicated force balance that can occur over complex terrain with slopes of a few parts in a hundred. In particular, it is clear that the role of long internal waves in the strong modulation of the drainage flow is consistent with this scale analysis.

8.4.3.3. Surface wind analysis

Unfortunately, because of limited resources, the major field experiments of the ASCOT and CTMD programs were not designed to investigate explicitly the character of internal waves in nocturnal flows. Because of this, events such as those on 19–20 September (Fig. 8.18) presented a challenge for analysis of wave motion using data collected for study of the mean state of nocturnal drainage flows. Fortunately, during the 1980 ASCOT experiment the 26 PAM stations (Dickerson and Gudiksen 1984) recorded data at 1-min intervals (Fig. 8.19). The possibility therefore arose that the combination of surface in-situ

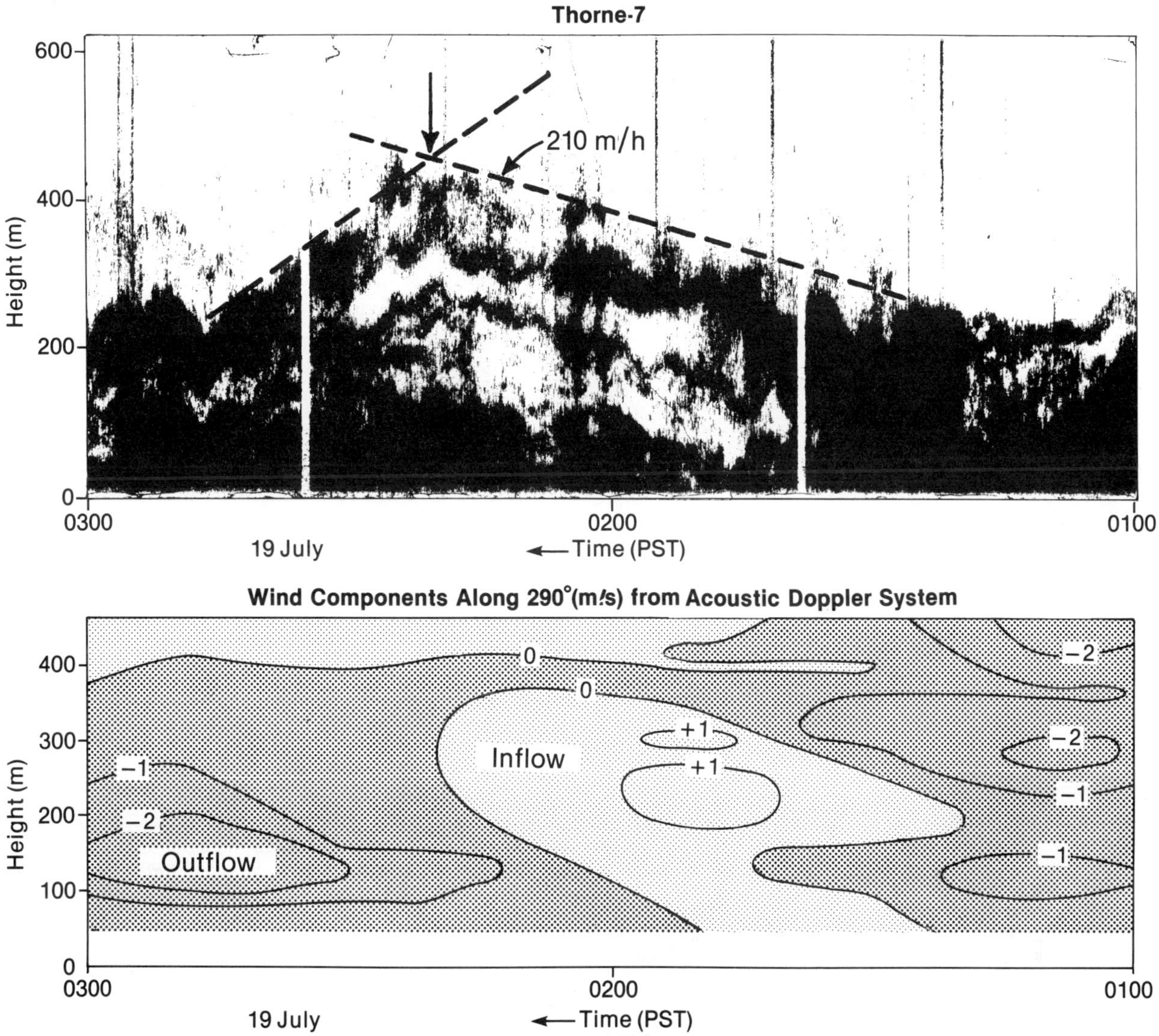

FIG. 8.16. Doppler sodar wind profiles (from Fig. 8.9) at mouth of valley compared with sodar inversion height measurements within the valley (obtained using the second sodar from the top of the valley shown in Fig. 8.7). Note the phase difference between the maximum inversion height and the time of maximum outflow.

instrumentation and sodar observing systems might prove a simple approach to observing at least a limited class of internal waves in complex terrain.

The rationale for using the 1-min PAM data was that long-period wave motion might well be revealed by periodic fluctuations in wind speeds. In particular, Neff (1987), following a suggestion of Gossard and Munk (1954), showed that wave motion (because of its two-dimensional character) could be identified in tower-derived wind direction and speed when the data were band-passed at wave periods identified in sodar facsimile records (see section 8.4.6.3.). Evidence of such wind speed oscillations for three PAM stations (with the mean removed): one in the bottom of the valley, one on an intermediate slope, and one on an exposed ridge is shown in Fig. 8.20. All show long-period oscillations in wind speed. Wind direction fluctuations (not shown), however, were not as well behaved, presumably because of deflection and channeling by very local terrain.

Spectral and cospectral analyses of wind speed data from pairs of surface stations appeared to be the next logical step in the analysis of the fluctuating data; however, Kaimal and Gaynor (1983) described a standard analysis package available for the analysis of phase speed and direction from an array of surface microbarograph stations sensitive to wave-associated pressure perturbations. [The theory underlying this so-called beamsteering analysis can be found in Gossard and Hooke (1975, section 65)]. We

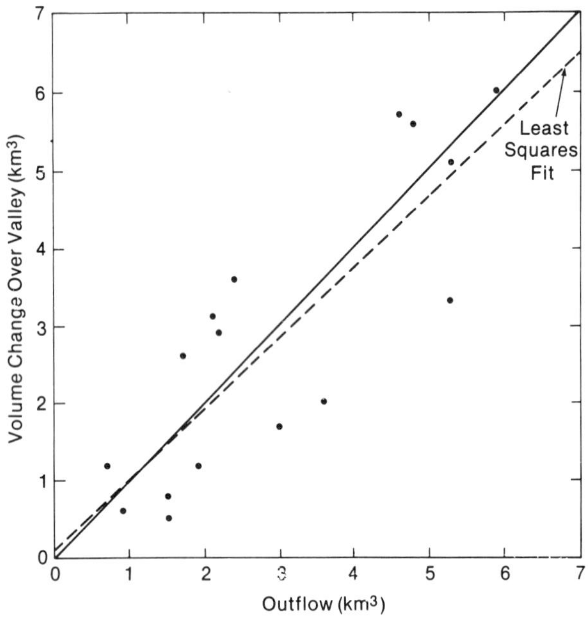

FIG. 8.17. Decrease in volume of air underneath the inversion and the volume of outflow measured at the valley exit for a number of events such as those shown in Fig. 8.16. (Neff and King 1985.)

therefore adapted this program to analyze phase speed and direction using wind speed from triads of PAM stations. In particular, we used PAM Stations 1, 6, and 9 on the southeastern flank of Cobb Mountain (Fig. 8.19), an area of relatively uniform slope. A composite spectrum from these stations indicates two strong low-frequency spectral peaks, one with a period slightly less than one hour and one slightly more than 2 hours. The narrow band-passed time series containing these two spectral peaks can be seen in Fig. 8.21 together with time series isolating two individual frequency components. The 3-hour extent of the event corresponds to the duration of the long-period oscillations shown in the sodar facsimile recording that was obtained just west of PAM Station 6 (Fig. 8.18). Comparison of the two figures also shows that the times during which the elevated inversion is the deepest correspond to the maxima in wind speed fluctuations (shortly after 0000 PST, about 0100 PST, and again at 0140 PST).

Finally, we addressed the direction of propagation of the wave by using the same analysis package (Kaimal and Gaynor 1983; Gossard and Hooke 1975) to produce a

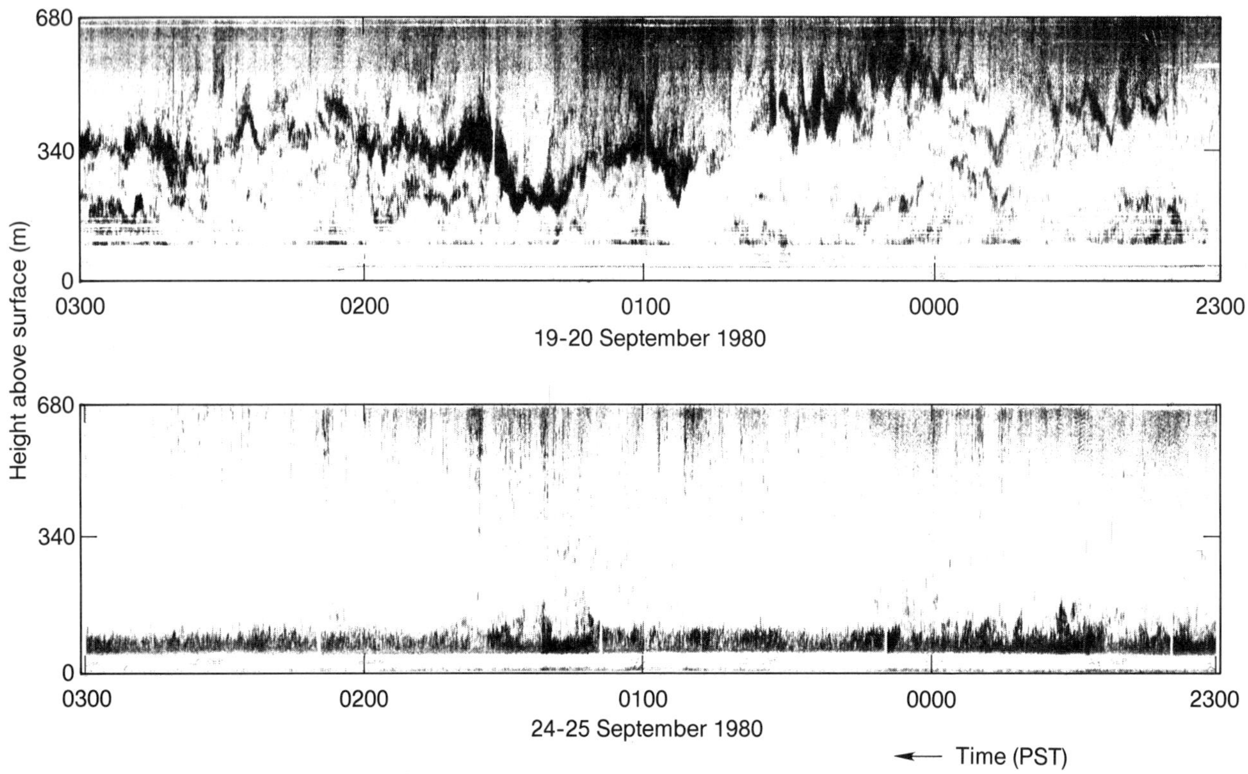

FIG. 8.18. Sodar records on nights with a marine inversion (19–20 September) and without (24–25 September). In the first case, long- and short-period internal waves are both evident in the inversion height oscillations. In the second case, a nearly steady drainage flow is observed as in Fig. 8.10 where the sodar sees only the top echo layer.

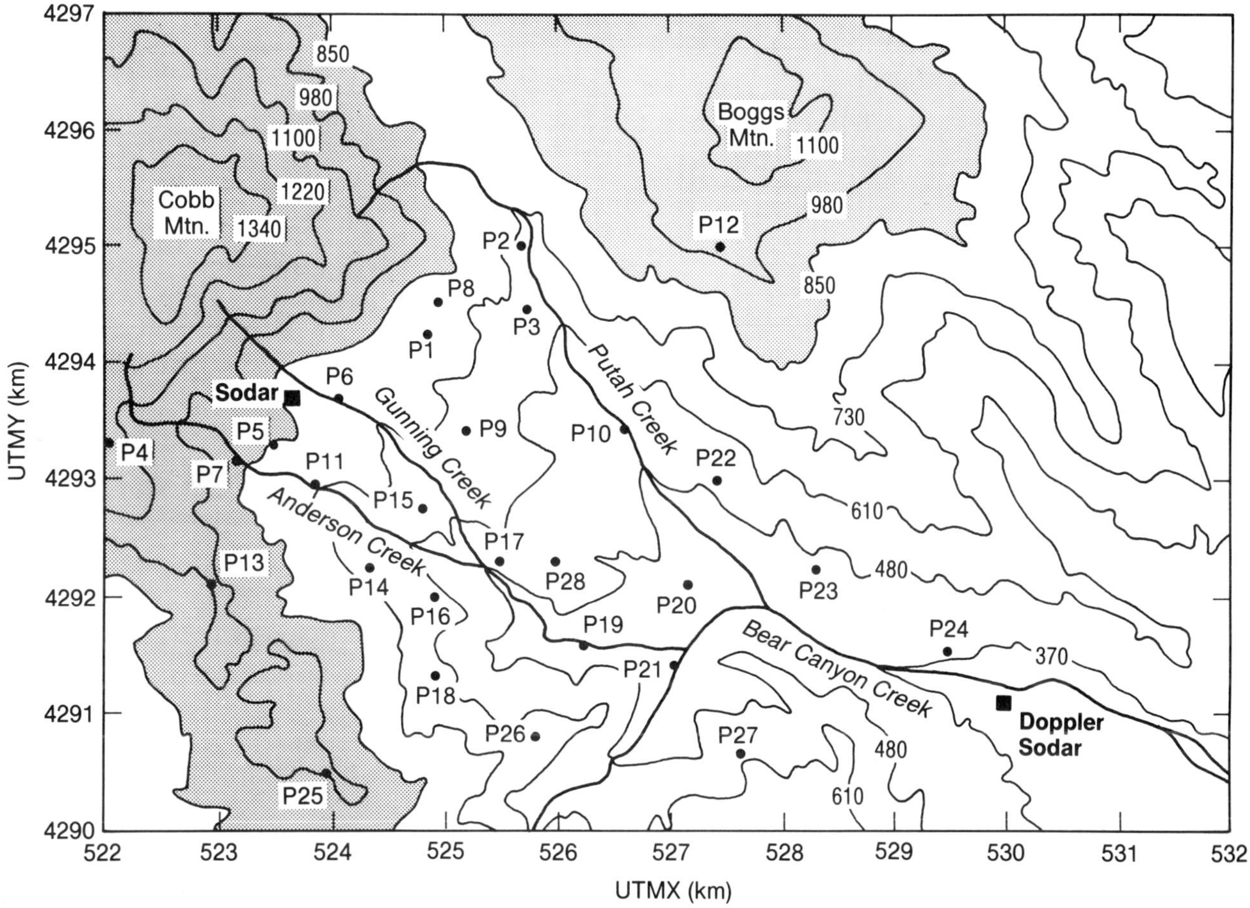

FIG. 8.19. Topography of California's Anderson Creek valley and location of PAM stations (*P*__) and monostatic sodar (near P6).

correlation matrix as shown in Fig. 8.22. In this case, the matrix reveals a wave propagating at a speed of about 2 m s^{-1} from the northeast, from the vicinity of Boggs Mountain. A separate calculation for the longer period wave shows a direction of propagation more from the east, the direction of the Putah Creek basin. Of interest is the possibility that Barr and Orgill (1989) suggested—that Boggs Mountain may be a source of more vigorous lee waves on the two nights subsequent to the one described here. It may thus be possible that two sources are present, one mountain-related and one associated with an internal seiche. Resolution of issues such as these awaits more detailed and focused experiments.

Of importance to complex terrain studies, however, is the evidence that sodar facsimile recordings together with short-period averaging of surface parameters may provide an effective means of studying wave motions in complex terrain flows. Among the variety of instruments used in complex terrain studies, such combinations may prove to be among the more cost effective. For example, sodar records could be used to identify periods in which wave effects occur; the data could then be related to the variance of horizontal wind fluctuations, a quantity important to the lateral diffusion of pollutants; triads of surface wind stations (recording data using short averaging times) could be used to determine wave phase speeds and directions. Eventually, with the advent of high-resolution wind and temperature profiles from UHF/RASS profilers, detailed stability analyses should become possible, allowing more effective tests of wave theory with observations. Until that time, tethered-balloon profiles can provide effective snapshots of the wind and thermal structure associated with many wave events in complex terrain.

8.4.3.4. WAVES ASSOCIATED WITH FLOW OVER RIDGES AND MOUNTAINS

Remote sensor observations of distinct wave motion due to orographic flow are limited to a few simple cases. The development of oscillations in an elevated jet passing over a two-dimensional ridge as observed by two sodars, one upwind and one downwind is shown in Fig. 8.23. The sodars were deployed in this fashion in order to characterize the drainage winds on two sides of a ridge. In the

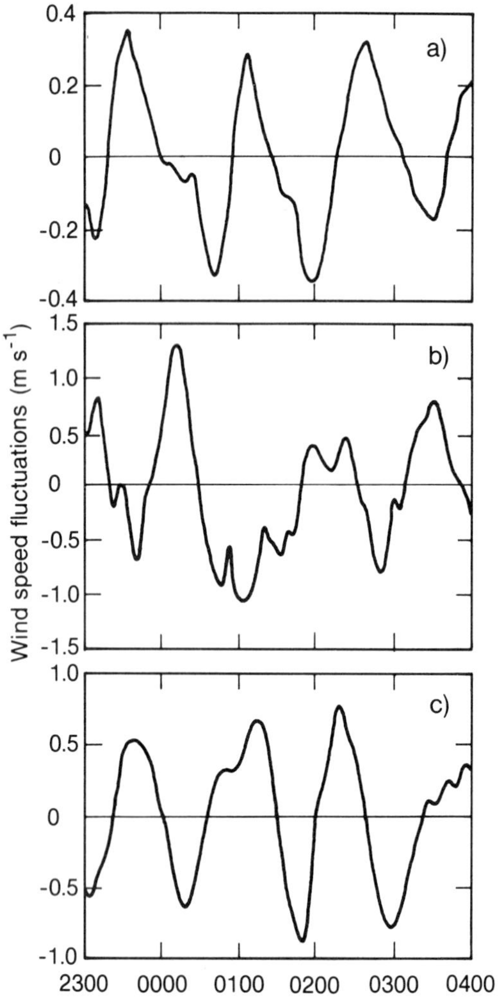

FIG. 8.20. Wind speed traces (mean removed) from PAM stations (a) 13 (ridgetop), (b) 9 (midvalley), and (c) 20 (lower-valley) showing long-period oscillations at all sites.

situation shown, an easterly wind over the ridge had developed and then increased after 0200 PST as shown in tethered-balloon profiles. In the lee of the ridge the elevated jet is marked by 1-h oscillations and distinct echo layers above and below the wind maximum; the elevated wind maximum is indicated in this case by oppositely sloping echo patterns above and below, a pattern described by Neff and Coulter (1986). The extent to which these oscillations modulate the underlying drainage flow, however, remains unknown.

8.4.4. Volume flux in simple drainage flows

A recurring issue in the study of nocturnal drainage flows channeled by valleys has been the mass balance of the flow and the mechanisms by which the air draining

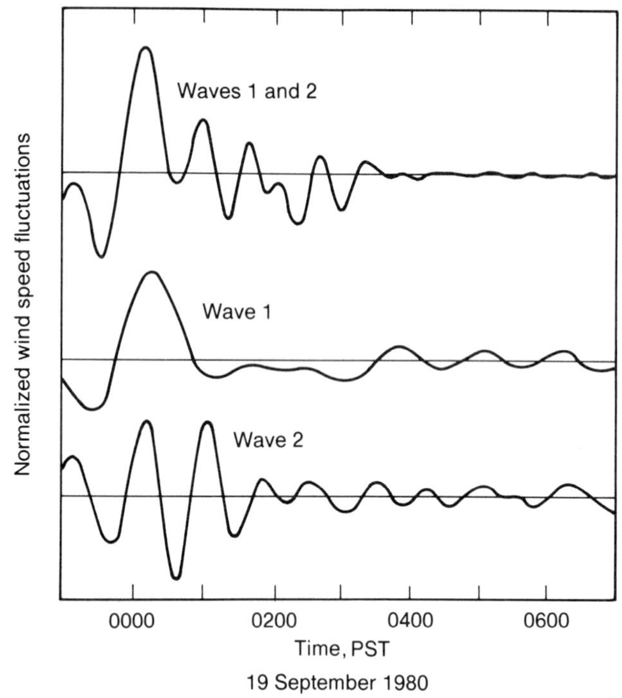

FIG. 8.21. Low-pass filtered waves observed in surface wind speeds decomposed, by use of a beam-steering program, into two waves, one about half the period of the other. The analysis used data from PAM sites 1, 6, and 9.

from the valley is replaced. Resolution of this issue, however, has been limited by the lack of suitable measurement techniques. For example, Whiteman and Barr (1986) and Neff and King (1987) examined the growth of drainage flows along the axis of simple valleys using tethered-balloon-derived and Doppler-sodar-derived wind-profile data. Similarly, McKee and O'Neal (1989) examined data

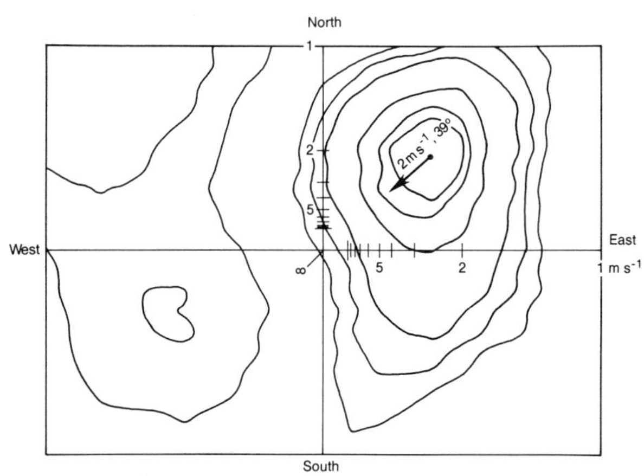

FIG. 8.22. Direction of propagation of higher frequency wave (see Fig. 8.21) as shown in correlation-matrix output of beam-steering program. Coordinates are in reciprocal speed, and quadrants indicate direction from which the wave is moving.

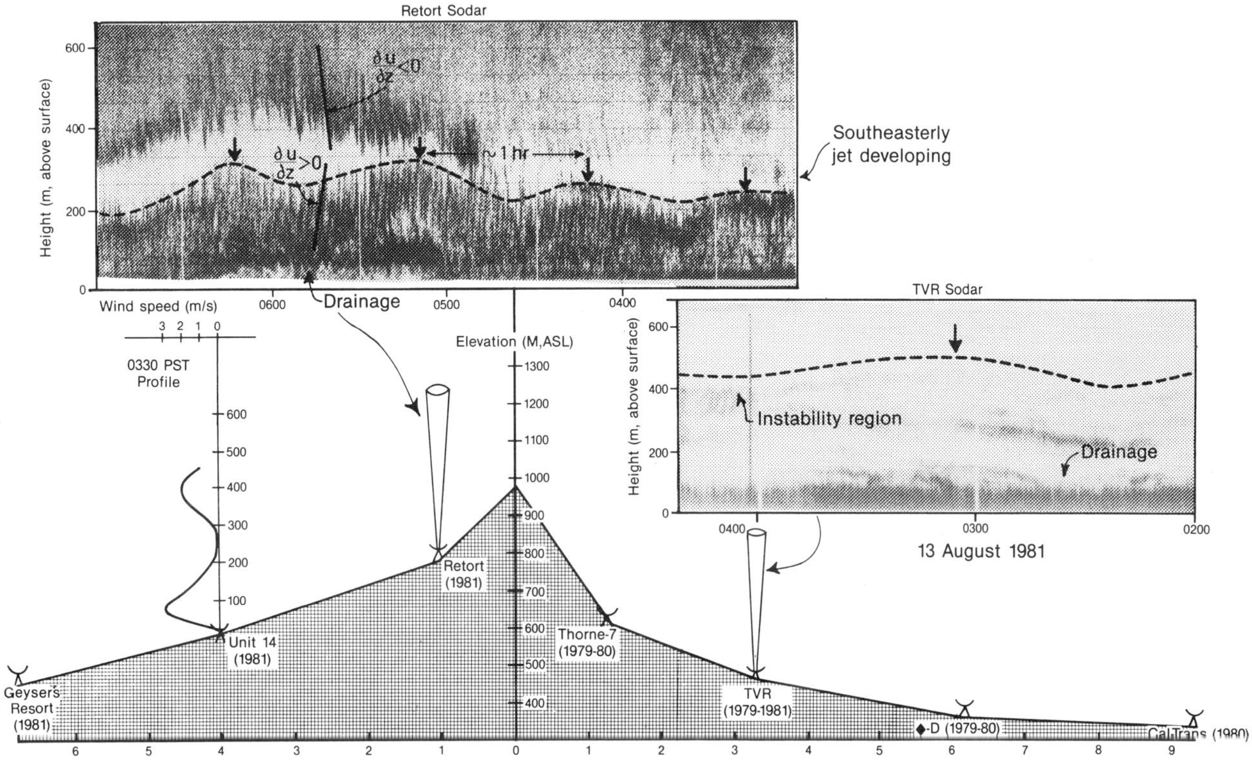

FIG. 8.23. Sodar facsimile recording (top) showing 1-h period waves in elevated jet passing over a ridge (bottom). The wind accelerated from the northeast after 0400 PST, as shown for a different case in Fig. 8.14. In this case only minor echo strata are seen to the east of the ridge (middle). (Neff 1986.)

from several valleys in Colorado, using topographic factors of valley width and cross section to determine the valley's draining potential. Neff and King analyzed the growth of the drainage flow along the axis of Big Sulphur Creek (a valley without significant tributaries used in the 1981 ASCOT experiment), and found a nearly linear growth in volume flux with airshed area.

Similarly, Whiteman and Barr (1986) examined the growth of the drainage flux along the axis of Brush Creek in western Colorado. The Brush Creek valley, unlike those used in earlier ASCOT experiments, is composed of several distinct topographic segments: a gently sloping valley floor, steep sidewalls, and sharply cut tributary canyons that reach into an extensive, partially eroded plateau some 600 to 800 m above the valley floor. In their analysis Whiteman and Barr concluded that the mass feeding the drainage flow originated primarily from air subsiding over the central portion of the valley, with lesser contributions from the side canyons. Neff and King (1985), however, from an inspection of topographic factors, suggested the tributary canyons as a channeling region for drainage flows developing on the plateau.

The more extensive instrumentation deployed during the 1984 ASCOT experiment thus provided an opportunity to study the issue in more detail, particularly with a larger suite of remote and in-situ sensors. Furthermore, tributary flows were of considerable interest, particularly because they represented a subgrid-scale process in most numerical models being applied to drainage flows (Porch et al. 1989b). To help resolve the issue, extensive instrumentation was deployed in Pack Canyon, a tributary canyon to Brush Creek and the only one with ready access. The instrumentation included meteorological towers on the sidewalls, tethered balloons on the plateau above and within the canyon itself, Doppler sodars within the canyon, and optical crosswind sensors across and within the canyon (Clements et al. 1989a). Furthermore, inert tracers were released in the shallow drainages feeding the canyon from the plateau. Observations were obtained for a short period in an adjacent side canyon, Willy's Gulch, for comparison. Preliminary analysis suggested that Pack Canyon contributed a much larger percentage of the mass flux (15%) than its airshed area (2%) would have suggested. Estimates from Willy's Gulch suggested a mass flux more consistent with its airshed area. The uncertainty in these results led to further observations in 1986 that confirmed the earlier measurements. Subsequently, a number of additional papers have addressed the analysis

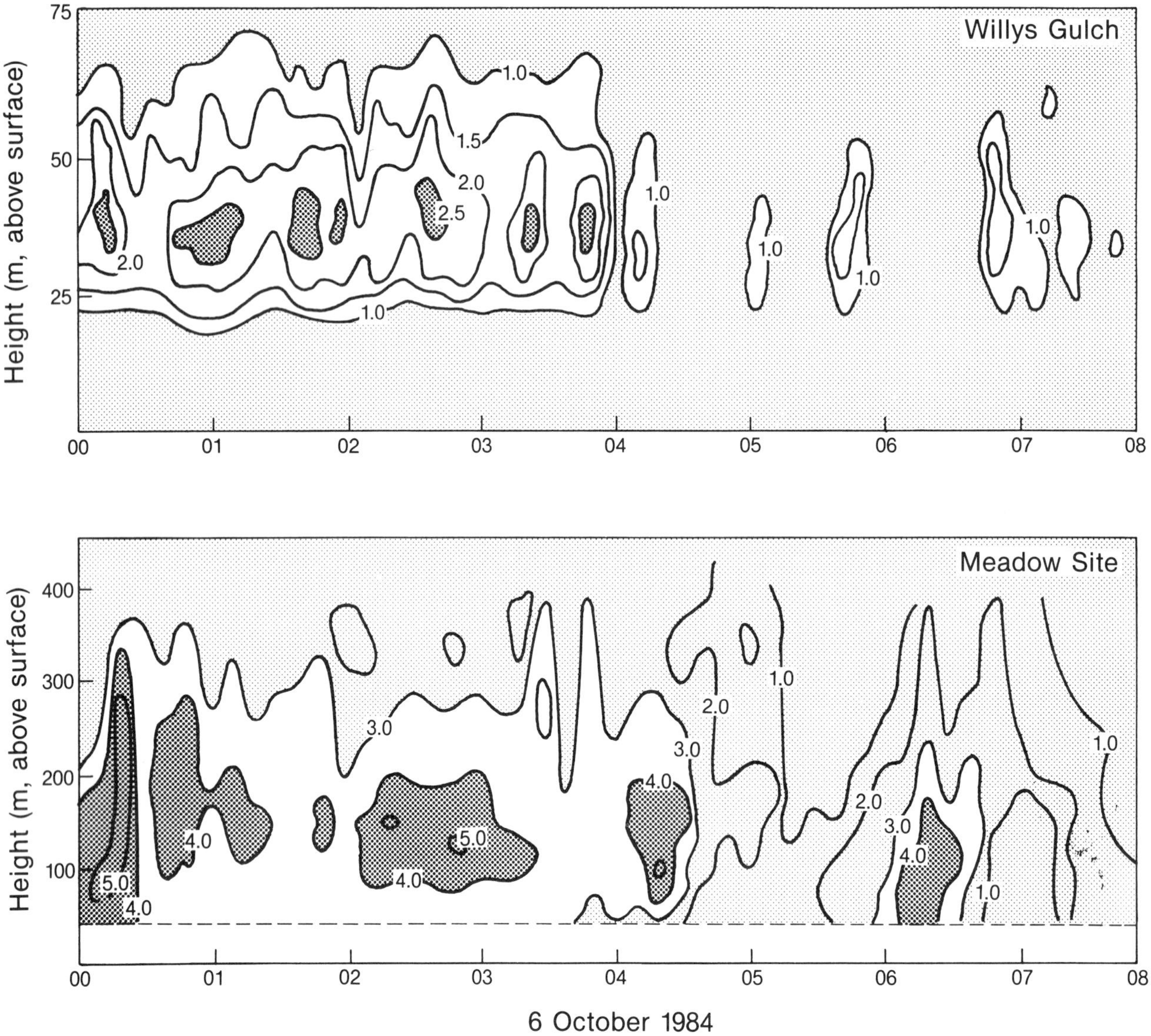

FIG. 8.24. Doppler sodar wind speed isopleths from main canyon and small tributary showing a slow down in speed coincident with accelerating winds at ridgetop (above 6 m s^{-1}). (Neff and King 1987.)

and interpretation of the tributary flows and fluxes (Kyle 1987; Coulter et al. 1989; Gudiksen and Shearer 1989; Shinn et al. 1989; Neff and King 1989).

Although the tributary issue has not been fully resolved, Coulter et al. (1989) suggested the existence of a large topographically related eddy that introduces extra mass into the tributary from the main canyon flow and/or the ambient circulation similar to the large eddies observed by Start et al. (1975). In addition, Kyle (1987) argued from an analysis of the energetics of the valley flow, including the total radiative loss, that the entire valley watershed including plateau tops must be included in calculating the energy and mass budget for the valley. A further issue raised from the analysis of Neff and King was that acceleration of the wind at ridge level might result in a rapid decrease in the magnitude of the valley flows, even though the winds do not intrude into the valley.

The effect of such increasing ambient winds is shown in Fig. 8.24 where a Doppler minisodar monitored the flow in a side canyon while a standard Doppler observed the flow in the main valley. Shortly after 0400 MST the drainage in Willy's Gulch ceases, followed by a rapid decrease in wind speed in the main valley. At this particular time, tethered balloons reaching heights greater than those available with the standard sodars in the valley showed winds just above the ridgetops accelerating rapidly from about 3 m s^{-1} to almost 7 m s^{-1}. This result suggested the significant influence of ambient winds eroding the shallow drainages that originate on the plateau. (This is a case in which high time resolution but limited vertical range of the Doppler sodars could be complemented by the greater height coverage but limited temporal sampling capability of tethered-balloon systems.) The complexity of the results in the study of tributary flows suggests that new remote sensing devices, such as mini-Doppler lidars, will be required to provide complete three-dimensional mappings of such subgrid-scale circulations within the tributary canyons. Furthermore, because of the relative inaccessibility of the plateau top surrounding Brush Creek, detailed measurements of the plateau top flows have not been possible; this lack offers additional possibilities for research in the future, using, for example, tower-mounted sonic anemometers, minisodars, and tethered balloons to study the accumulation process and its susceptibility to accelerating ambient winds.

A future research opportunity in the design of experiments that deploy Doppler sodars with high time resolution, perhaps with a 1-min averaging time, is also suggested in Fig. 8.24. Specifically, amplitude and phase velocity relationships could be determined through cospectral analysis or multistation spectral analysis between spatially separated stations, similar to the analysis done earlier (Fig. 8.22) for surface wind sites but farther removed from local terrain influences. Such analyses have not yet been attempted because of the long time-averages usually required for the horizontal wind components. That need could be avoided, however, with sodar or radar wind profiler systems that probe at least one common volume above the surface in a height range of interest.

8.4.5. *Main canyon mass fluxes and merging*

8.4.5.1. LIDAR DATA ANALYSIS AND CORRECTION

Whatever the details of the tributary contributions, the deployment of the Doppler lidar in Brush Creek in 1984 provided a new opportunity to study the growth and structure of nocturnal drainage flows. Deployed in a central position in the valley, the lidar could scan both up-valley and down-valley. In particular, Post and Neff (1986) described the lidar operation and data in some detail and reported early results, including observations of the horizontal and vertical structure of the drainage flow, its monotonic increase along the valley axis, and the nature of the thermally driven wind reversal during the morning transition. In this early analysis, the lidar scans obtained at each range gate were processed using terrain elevations and locations derived from topographic maps and the velocities were estimated to be within 0.5 m s^{-1}, a relatively good value compared with conventional in-situ instruments. In the analysis of Post and Neff, terrain masking had become an important element in the accurate calculation of the drainage flow volume flux. The difficulties inherent in such masking can be seen in Fig. 8.25 where the cross- and along-valley sample volumes are indicated for one hypothetical elevation. Each of the 300-m sample gates corresponds to sampling of the scattered signal for 2 μs. Note that small errors in ranging or in horizontal positioning can lead to significant errors in individual velocity estimates, particularly in the vicinity of tributary entrances into the main canyon flow.

Further evaluation of these concerns revealed several complications that had to be overcome to pursue the analysis further. First, it was discovered that the positioning of the lidar beam varied systematically from its expected value by as much as 0.5°, small for most applications but not for use in confined terrain. Second, the 2-μs pulse contained a tail with a frequency content different from that of the primary pulse. When this trailing part of the transmitted pulse scattered from a solid object on the valley sidewalls, even at grazing incidence, an apparent frequency shift and scattered power could be measured in time gates corresponding to ranges farther from the lidar, that were similar to returns from the actual atmosphere. When the blockage of the pulse occurred within the minimum range of the lidar, it became even more difficult to distinguish these events. These first two difficulties combined to create another problem. In particular, reliance on inaccurate terrain masking from topographic data allowed data to appear at certain positions where none should have been because of terrain blockage; because signals and frequency shifts appeared normal, any outlier-removal method would thus fail. In Fig. 8.25 it is shown how data from a blocked region, for instance within the entrance to a tributary, could result in spurious conclusions about the circulations near the tributary. Finally, it was discovered that the pulse monitor, used to calculate the power-weighted frequency of the transmitted signal, averaged over several samples, had saturated on occasion, leading to slight errors in the calculated frequency difference (usually less than 0.5 m s^{-1} in radial velocity). Unfortunately, in the down-valley direction, where the drainage flow volume flux had been integrated over a large surface area and average drainage speeds were smaller than in the up-valley direction (because of the diverging valley cross-section), the error quickly manifested itself as a large error in the volume flux. Because the error added to the calculated flux in the up-valley direction and subtracted in the down-valley direction, a drastic and physically implausible reduction in the volume flux was observed.

For these reasons, terrain and velocity correction methods were needed that could overcome these difficulties in the Brush Creek data. In particular, detection algorithms were developed that could identify terrain features and thereby map the actual terrain as observed by the lidar. This eliminated the need for extremely precise positioning of the lidar and highly accurate topographic data. In Fig. 8.25 it is shown how even a slight error in

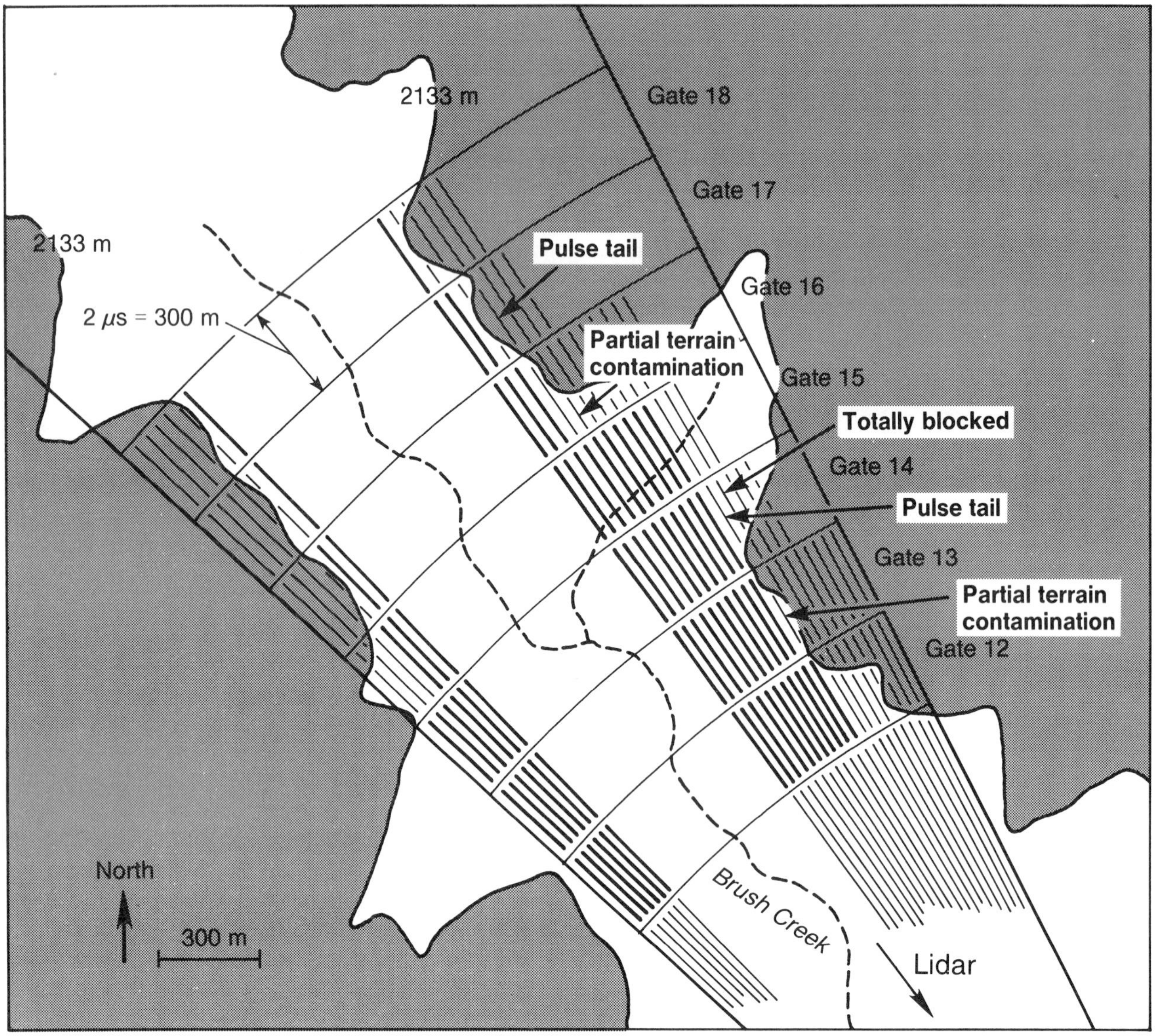

FIG. 8.25. Lidar sampling geometry in Brush Creek, showing contaminated, noncontaminated, and blocked data regions.

positioning could lead to some small fraction of terrain contamination; because the terrain scattering cross section is so much larger, it would dominate the aerosol contribution. Second, by using the vertical velocity derived from VAD scans well above the valley and averaged through the night, one set of data could be corrected directly for the pulse monitor error; in this case an offset of 0.74 m s^{-1} was found. Data for other nights were more difficult to correct directly because the absence of aerosol aloft limited the vertical extent of the VAD scans above the terrain. In these cases, because vertical velocities were measured close to the terrain, terrain-induced motions may have contaminated the results. Inspection of the data from the one directly corrected case, however, revealed that the volume flux observed down-valley lay on the extension of the least squares fit line to the data from up-valley. By using this as guidance, correction factors for subsequent nights could be obtained by requiring that the calculated down-valley flux lay on the least squares extension of the up-valley flux. The small relative errors in the fluxes in the up-valley direction and the linear growth in airshed size down-valley resulted in some confidence in this method.

Despite these minor corrections and the newness of the technique, the Doppler lidar data provided fresh insights into the structure of the nocturnal drainage flows in Brush Creek and have answered a number of questions about the growth and structure of the drainage flow in narrow canyons such as Brush Creek. This success supports risking the use of new remote sensing methods, particularly in the face of fiscal limitations. New observing tools do, in fact, enhance our understanding of the atmosphere.

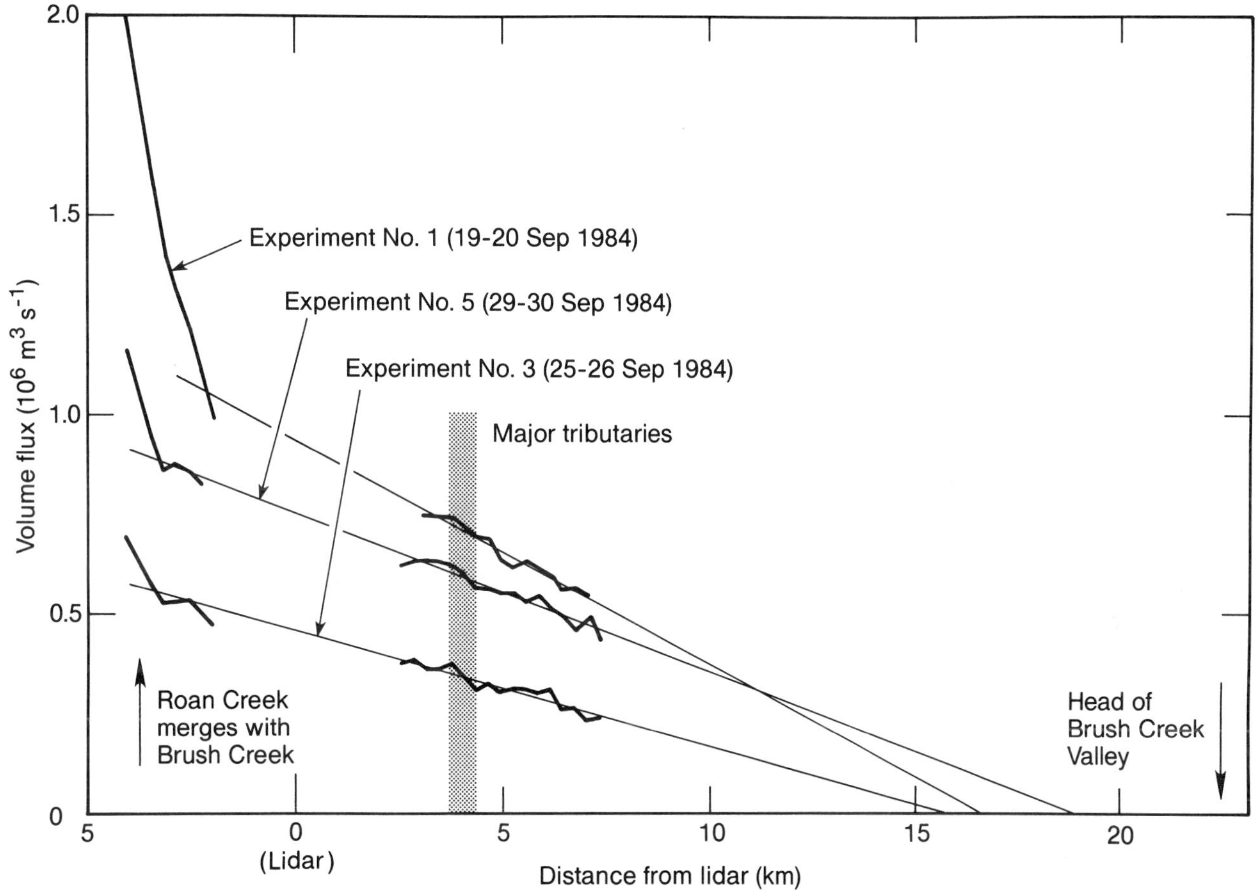

FIG. 8.26. (a) Lidar-derived volume flux in Brush Creek on 19–20 September 1984, (b) 25–26 September 1984, and (c) 29–30 September 1984.

8.4.5.2. VOLUME FLUX GROWTH AND THE ROLE OF TRIBUTARIES

In Fig. 8.26 data are presented documenting the along-valley growth in the nocturnal drainage flow volume flux in Brush Creek for three experiment nights: 19–20, 25–26, and 29–30 September 1984. These are identified as expts 1, 3, and 5, respectively. On each of the nights, about eight up- and down-valley lidar scans were averaged where pairs of scans in each direction occurred within about 10 min of each other. In each case, data are presented only for those range gates in which the drainage flows were totally contained within the lidar field of view. In the down-valley direction the volume flux rapidly increased because of the merging of Brush Creek with Roan Creek. Gates 11 and 12 provide the clearest view of the exiting Brush Creek drainage. At this point the total Brush Creek airshed is about 110 km^2.

The along-valley flux data in Fig. 8.26 provide some insight into the role of tributaries in Brush Creek. In particular, during expts 3 and 5, when the drainages were weaker than in expt 1, the fluxes at gates 12 and 17 were very close to the least squares fit line, with a slower rate of increase up-valley and a greater rate of increase down-valley. These range gates, each 300 m wide, lie across sections of Brush Creek with uniform, steep sidewalls that in each case are viewed well by the lidar. Between this set of gates are two major side canyons on opposite sides of the valley, comparable in size with Pack Canyon (which lies immediately down-valley of gate 10). In both experiments the change in volume flux between gates 16 and 14 shows an increase of almost 5×10^4 m^3 s^{-1}, much larger than the average growth rate. For several range gates, just before and just after the tributaries, the change in flux is almost zero. Because these areas are free of large tributaries these lidar observations confirm the hypothesis that the tributaries of Brush Creek concentrate the flow of mass into Brush Creek. It also implicates further the role of the plateau as a significant source for drainage mass.

8.4.5.3. THE MERGING OF DRAINAGE FLOWS FROM VALLEYS WITH DIFFERENT PHYSICAL CHARACTERISTICS

Further analysis suggested that the Roan Creek drainage into which Brush Creek merges is characterized by a considerably smaller fraction of plateau top airshed area than

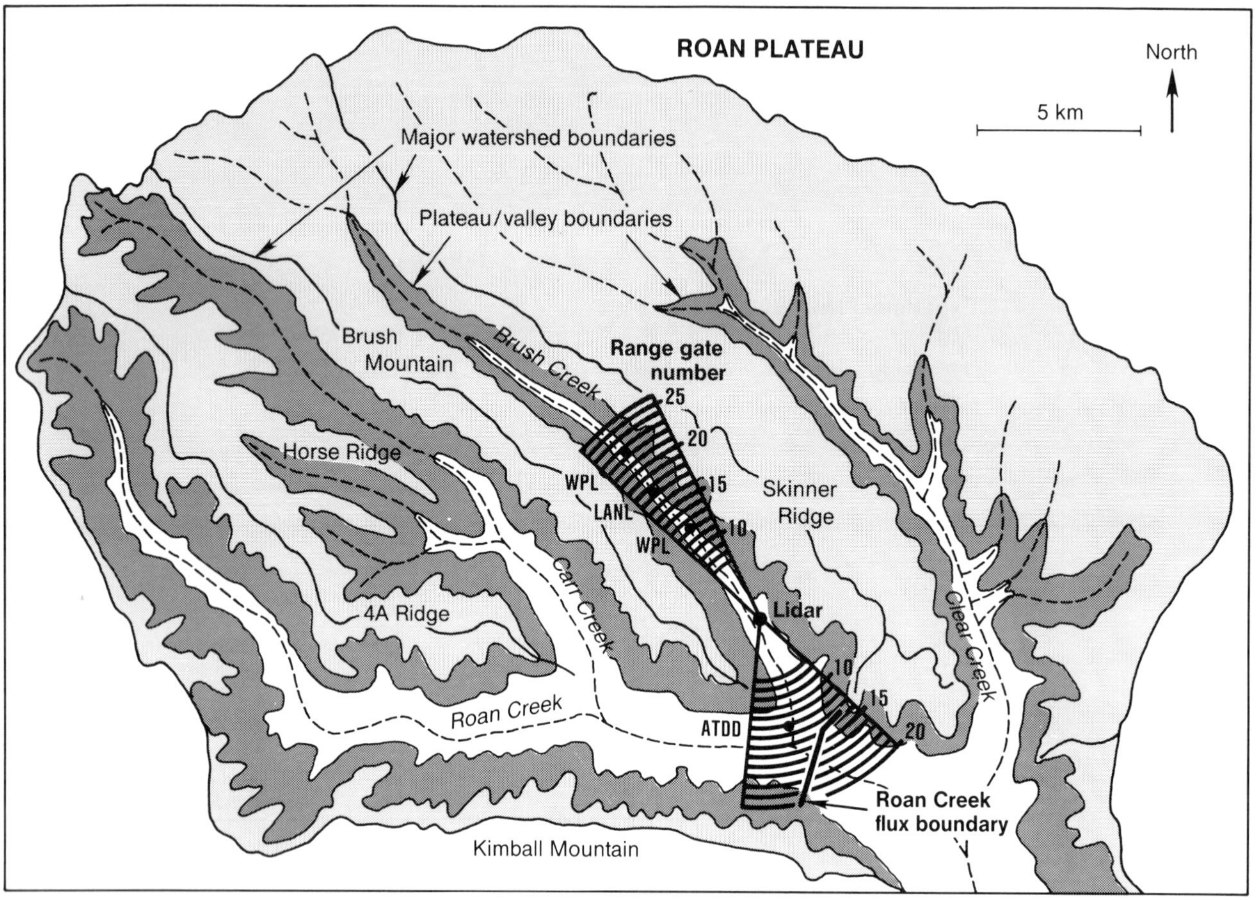

FIG. 8.27. Lidar flux-measuring geometry relative to the major airsheds, showing the ability of the lidar, properly positioned, to observe several drainage flows simultaneously. The sampling geometry shown was used to compare the drainage flow volume flux from the much larger Roan Creek airshed with that of Brush Creek alone.

is Brush Creek. This suggested the use of the Doppler lidar to measure the volume flux in Roan Creek for comparison with the volume flux from Brush Creek (Fig. 8.27). Were the elevated plateau areas to contribute significantly to the valley drainage flows, then the volume flux from Brush Creek should represent a greater fraction of the total Roan Creek volume flux than total airshed values would indicate.

These calculations, however, required further assumptions. In Fig. 8.28, an expanded view from Fig. 8.27, it is shown that no complete surface, over which the normal component of the wind could be integrated, was visible to the lidar from its position in Brush Creek. From its vantage point, however, the lidar viewed a major portion of the Roan Creek drainage, albeit at an angle. An artificial, segmented surface was then defined, as shown, across the combined Brush Creek and Roan Creek drainages. Next it was assumed that the wind direction in the flow along Roan Creek was tangent to the local terrain; this provided the orientation of the segments. By applying a cosine correction based on the angle between the lidar radial direction and the orientation of each segment, the combined volume flux from both the Roan and Brush creeks could be estimated. A selected cross section from Roan Creek derived in this manner is shown in Fig. 8.29. The location of the main Roan Creek drainage jet axis lies on the south side of the valley, well away from access roads in the area. Eight individual scans provided an average total volume flux with which the volume flux from Brush Creek could be compared (Table 8.1) for the three lidar experiment nights. Of note is the remarkably consistent ratio of the volume fluxes on the three nights.

These results suggest that the volume flux from Brush Creek is larger than expected from considerations of total airshed area. This occurs even though Roan Creek has a much larger total airshed area as can be seen in Fig. 8.27. This figure also indicates the lidar scanning region, the Roan Creek volume flux boundary, and the contrast between plateau top, steep sidewall, and valley flow regions. Comparing total airsheds one finds that Brush Creek represents only 0.20 of the total airshed. If one compares total plateau top areas, however, Brush Creek possesses about 0.40 of the total, a ratio more consistent with the ratio of drainage flow volume fluxes.

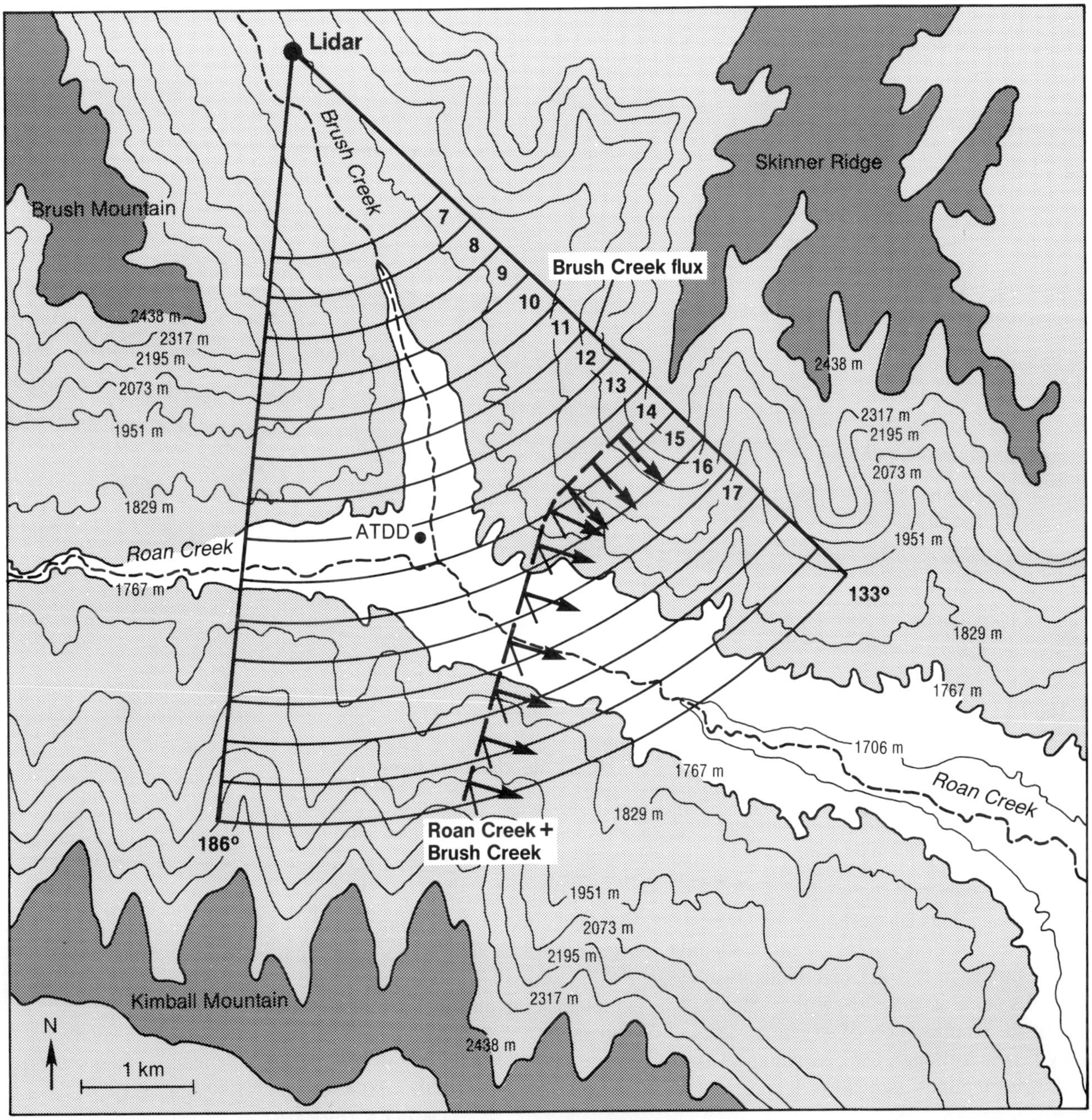

FIG. 8.28. Geometry used to calculate the volume flux in Roan Creek from data obtained with the lidar in Brush Creek. The flow is assumed to be approximately parallel to the local sidewalls. A segmented flux surface is shown with the assumed mean flow direction and the direction of the radial component measured by the Doppler lidar. Range gate 12 provides the best view of the total flux from Brush Creek, just downstream of a major tributary.

The volume flux estimates described above used model calculations of Yamada and Bunker (1988, 1989) for guidance in making the assumption that the drainage flow followed the local tangent to the sidewalls; however, the model results also suggested that strong winds at ridge level from the southwest or south could cause the wind direction in the upper levels of Roan Creek to shift more to the west, aligning more with the upper-level winds rather than the valley. Recomputation of the combined flux, assuming that the Roan Creek flow followed, not the local tangent to the terrain, but rather the direction of the valley upwind of the flux boundary, thus leads to an upper bound on the Roan Creek volume flux error because it provides a maximum correction for the lidar data. Table 8.1 includes these results, where the ratio is now reduced but remains larger than the ratio of the total

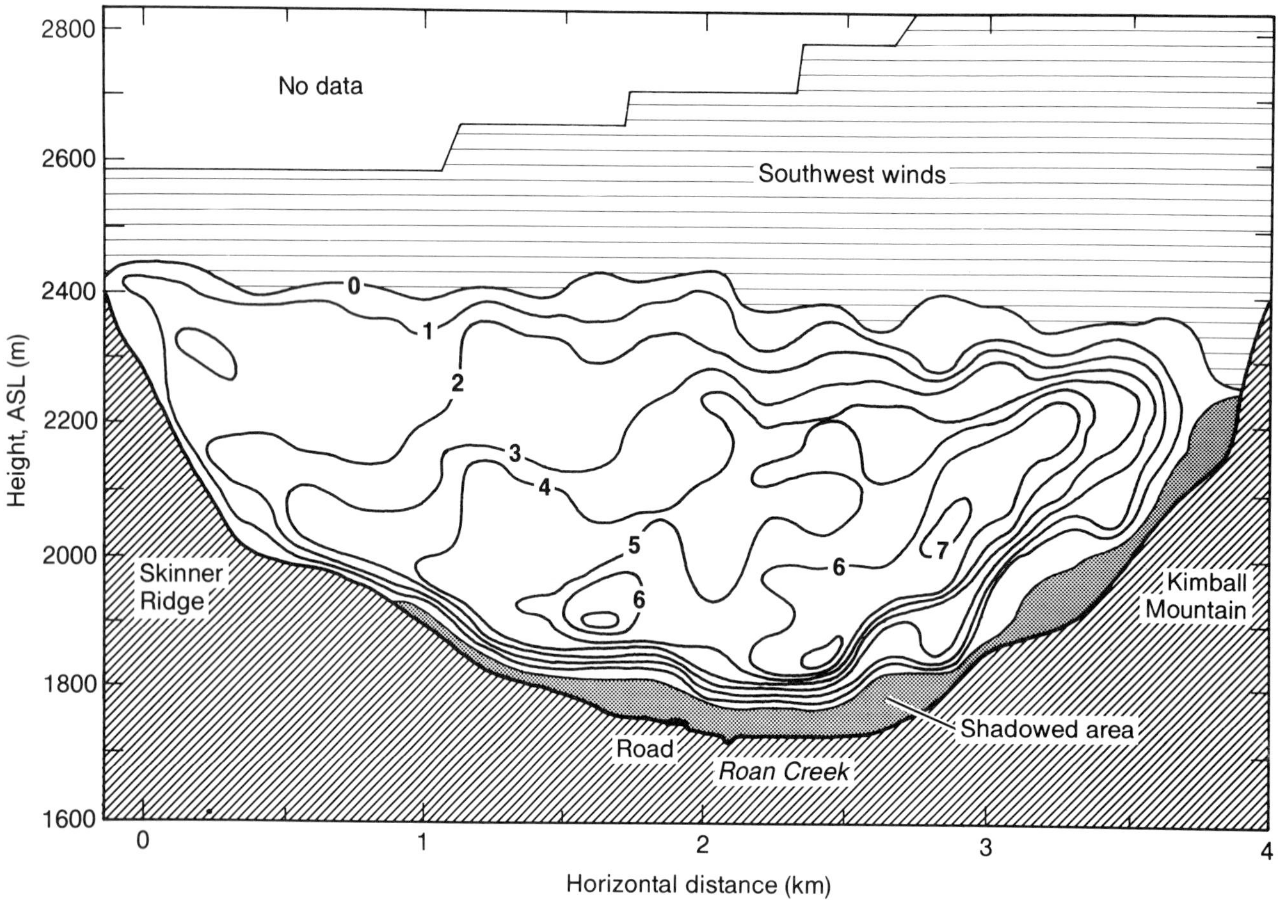

FIG. 8.29. Sample view of along-valley wind speed estimate from oblique lidar view into Roan Creek, showing the ability of the lidar to see the majority of the Roan Creek drainage. Only the lowest 50 meters or so are not visible. Of particular note is the location of the main drainage jet along the far southwest side of the valley.

airshed areas. This supports the hypothesis that the elevated plateau drainage areas are more effective sources of volume flux than the interiors of steep valleys.

Results such as these show the value of the Doppler lidar in the examination of specific aspects of nocturnal valley flows. Although the experiment was not designed to address merging valley flows, it is clear from these limited observations that more effective and conclusive sampling geometries could be used to study the merging process in order to avoid the logistical limitations of conventional profile methods.

TABLE 8.1. Volume flux, m^3/s.

	Brush Creek	Roan Creek (total)	Fraction (Brush Creek/total)
expt 1	1.4×10^6	3.4×10^6	0.41*
		(4.7×10^6)*	0.30
expt 2	0.53×10^6	1.2×10^6	0.46*
		(1.6×10^6)*	0.33
expt 5	0.86×10^6	2.1×10^6	0.42*
		(2.8×10^6)*	0.31

* Upper bound on Roan Creek flux.

8.4.5.4. COMPARISON OF LIDAR OBSERVATIONS WITH OTHER MEASUREMENT METHODS IN COMPLEX TERRAIN

The use of the Doppler lidar in Brush Creek also provided an opportunity to evaluate other measurement techniques including a determination of the representativeness of single profile measurements such as those from tethered balloons (Clements et al. 1989b; King 1989). These studies found simple mathematical descriptions of cross valley and vertical speed distributions in the narrow up-valley regions of Brush Creek where the authors concluded that simple correction factors could be applied to single profile data in order to estimate volume fluxcs. In the down-valley direction, King found that single profiles obtained during the 1984 experiment would have overestimated the flux by a factor of 2 in comparison with the more complete lidar data. Overestimation would result primarily from the offset of the drainage jet to one side of the valley.

The use of single profile data to determine the mass and momentum flux and flux divergence has also been evaluated by Dobosy et al. (1989) using Doppler lidar data as a standard against which to judge the individual profile techniques. They concluded that single-profile techniques were inadequate for evaluating the flux divergence of mass and momentum along Brush Creek. In one case, they found that even the deployment of two tethered-balloon systems across Brush Creek did not make it possible to observe the location of the drainage jet wind maximum, a measurement critical to the calculation of the momentum flux.

8.4.6. *The role of remote and in-situ sensors in the study of elevated plumes during the 1984 CTMD experiment*

8.4.6.1. Background

During this study, which was carried out in August 1984, a plume-mapping lidar was deployed to map the transport and dispersion of a power plant plume embedded in a large-scale drainage flow moving east from the Sierra Nevada (Eberhard et al. 1987) along the Truckee River (Fig. 8.30). The lidar, positioned to the south of the expected direction of motion of the plume, mapped the plume in vertical slices oriented along the azimuths indicated in the expanded view shown in Fig. 8.31. Of interest in the experiment were the various complex terrain processes that could affect the transport and dispersion of an elevated plume. For this reason, a number of supporting meteorological sensors were used that included both in-situ and remote sensors (Fig. 8.32). [Of note in the terrain profile is the constricted exit from the valley along the Truckee River. Neff and King (1989) suggested that this type of topography produces a pooling tendency; the more slowly moving air lying below the height of the constriction.] The Doppler sodar measured winds west of the Tracy Power Plant stack; the tower site, downwind of the stack, was instrumented with sonic anemometers at three levels. A monostatic sodar, situated near the tower, provided a pattern recognition capability, and a tethered balloon provided intermittent wind and temperature profiles above the 150-m height of the tower. A third monostatic sodar was deployed at the far east end of the valley. This suite of instruments provided the means to examine transport and dispersion in complex valley/basin flow regimes.

8.4.6.2. Lidar aerosol plume mapping

Eberhard et al. (1987) described the extensive data processing required for the extraction of accurate two-dimensional plume maps including corrections for extinction, background aerosol distribution, and terrain identification. In Fig. 8.33 an example is shown of the hour-averaged lidar returns from each of the five azimuths identified in Fig. 8.31. This sequence shows the ability of the lidar to trace the evolution of the plume from its inception to its impact on the elevated terrain of Beacon Hill (azimuth = 21°) and finally its exit from the study area where some blockage of the lidar's view may have occurred. In this case oil fog was injected into the power plant stack together with the gaseous tracer SF_6 to provide for an intercomparison of the aerosol and gas tracer techniques near ground level. Eberhard et al. found generally good agreement between the two techniques in the vicinity of Beacon Hill; they found disagreement only where the lidar field of view was blocked by terrain or where there may have been residual SF_6 from releases prior to that of the oil fog. They argued that such disagreements between independent tracer methods may reveal new facets of plume behavior that would otherwise not be noticed.

8.4.6.3. Complex terrain processes affecting plume transport and dispersion

Plume behavior can be affected by motions at a variety of time and space scales, particularly in complex terrain. Oftentimes, these motions are composed of slow meanders, wave motions, and turbulence. In one study using CTMD data, Neff (1987) made use of tethered-balloon data, sodar facsimile records, and tower-based sonic anemometer data to study these motions. In this analysis, sonic anemometers proved useful because of their ability to measure wind fluctuations under light-wind, stable conditions.

In Fig. 8.34 the case of interest is shown in which the sodar facsimile record revealed a region of potential dynamical instability 300–500 m above the surface together with a fortuitous tethered-balloon profile obtained at the beginning of the event. At the time of the tethered-balloon profile, Ri had decreased to a value of 0.17 from an earlier one of about 0.40. The onset of instability followed an increase in easterly winds above the Tracy basin. Except for a brief reversal period, winds within the basin were from the west, the nominal drainage flow direction. The combination of sodar and tethered-balloon data thus showed the potential for wave instability aloft. Of further interest was whether the short-period waves created by the instability would be felt through the fluid below and hence affect plume dispersion.

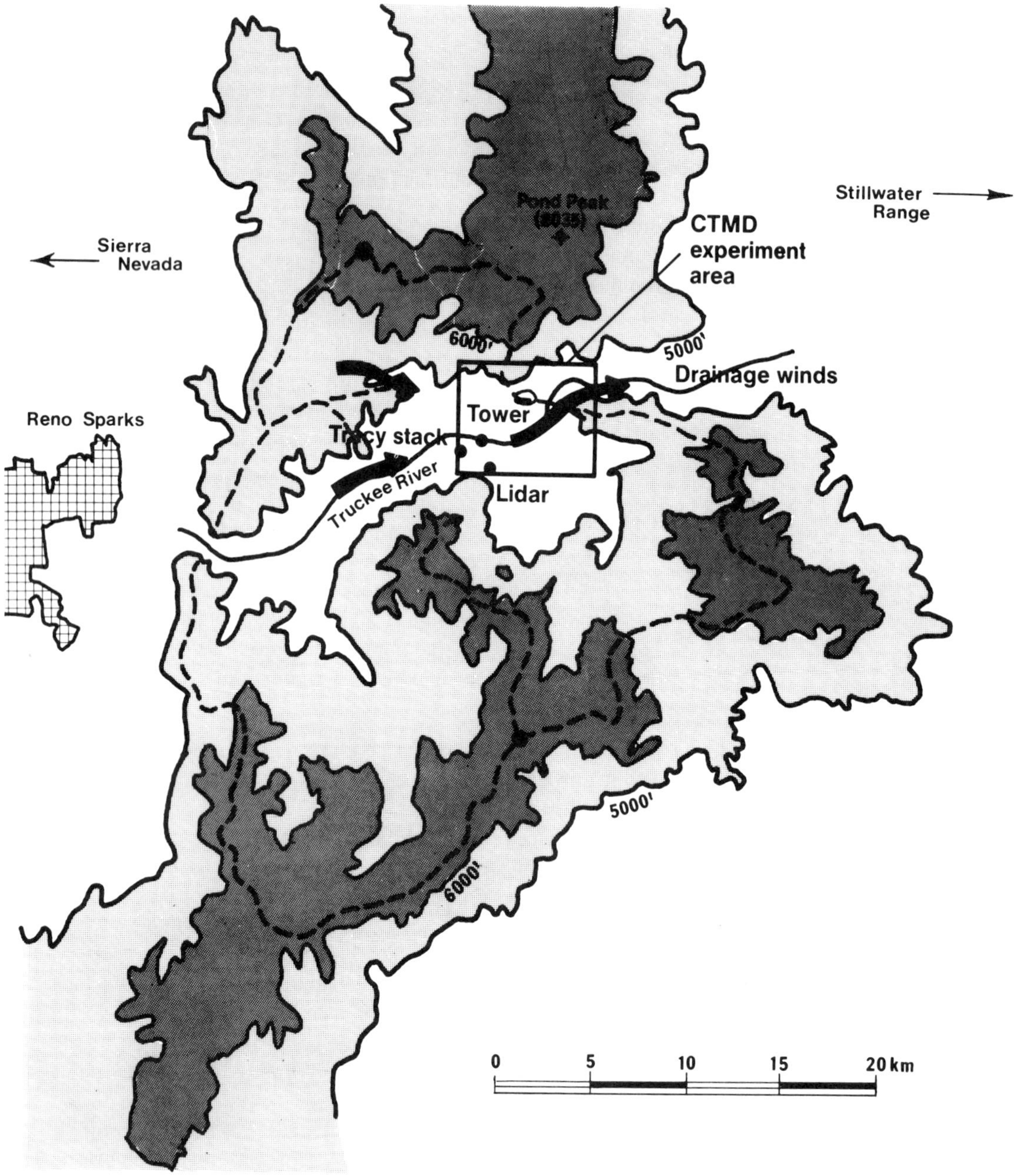

FIG. 8.30. Valley system in which the 1984 EPA Tracy Experiment took place (boxed area). Large-scale drainage flows move to the east from the Sierra Nevada and are channeled along the Truckee River. The terrain below 1524 m (5000 feet) forms an almost closed basin in which cold air could pool at night. The small exit to the east allows some drainage out of the basin. In this case, the Tracy Power Plant was within the basin, so its plume at night would drift to the east.

Sonic anemometers at the 100-m level of the tower provided the data with which to address this question. In particular, use was made of the property of internal waves that the fluctuations in the wind should lie along the horizontal direction of propagation of the wave (Gossard and Munk 1954). For this reason, *x–y* plots were obtained of the vector wind fluctuations in four frequency ranges (Fig. 8.35). Figure 8.35a shows the raw data; Fig. 8.35b shows the slow meanders (>30 min); Fig. 8.35c shows the high-frequency fluctuations (<5 min); and Fig. 8.35d shows the fluctuations in a narrow band pass centered on the apparent period of the waves at 400 m in the sodar record. This last vector-fluctuation plot shows a distinctly two-dimensional character to the fluctuations. At this point

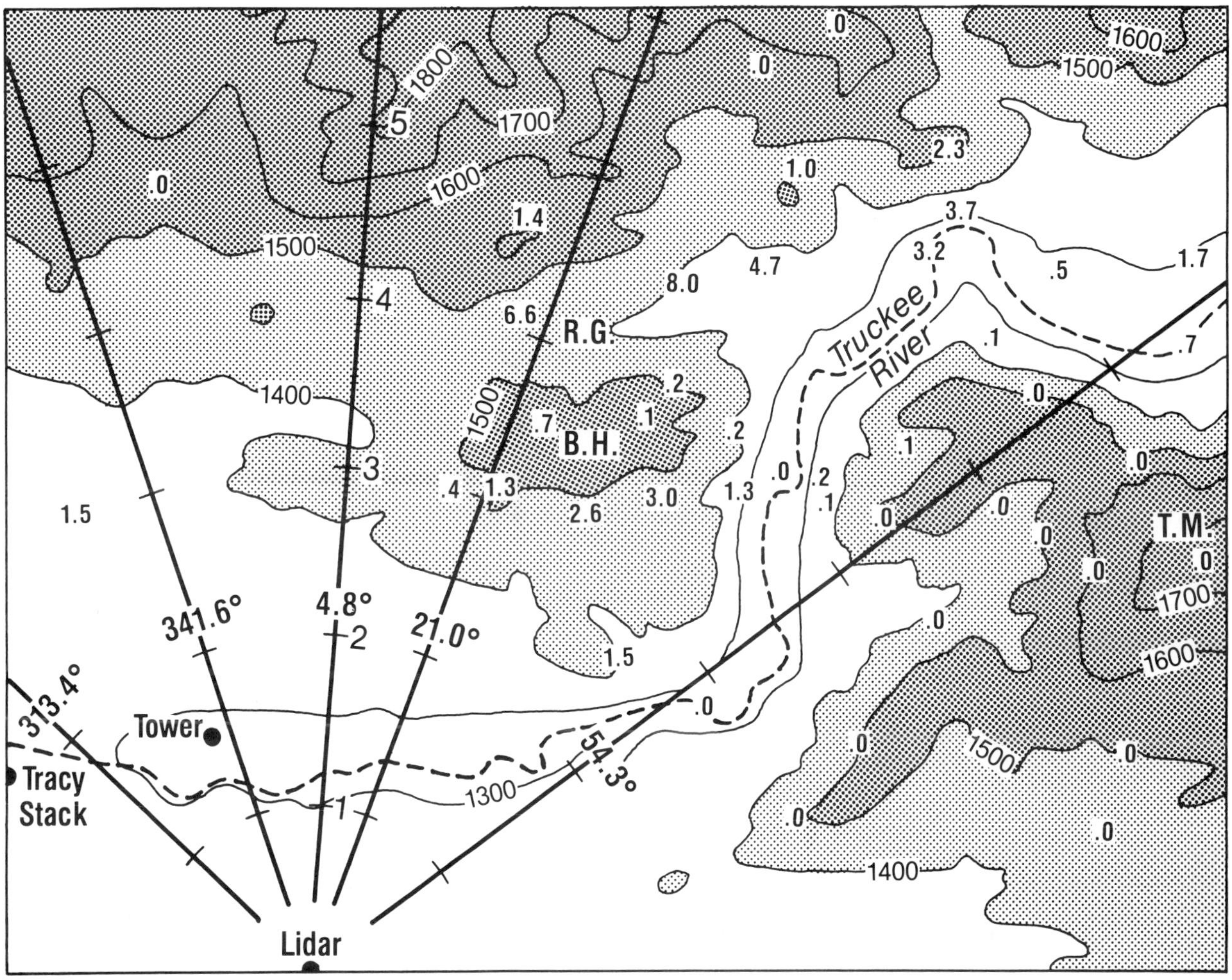

FIG. 8.31. Expanded view of 1984 EPA Tracy Experiment area identified in Fig. 8.30, showing the deployment of the lidar to the south of the expected plume path such that it could obtain cross sections across the plume in a plane also containing a 150-m tower, sodar, and tethered balloon system (341.6°). Tick marks along the radial directions from the lidar are at 1-km intervals. (Eberhard et al. 1987.)

the vector wind shear across the layer of instability was obtained from the tethered-balloon data: the rationale was that the instabilities should produce wind fluctuations along this direction. These results are summarized in Fig. 8.36 showing excellent agreement between observations and expectations. The direction of the wind at plume height is also shown in this figure. In this case, wave-induced wind fluctuations lie along the direction of plume motion rather than transverse to it. For this reason, lidar cross-sections did not reveal any significant lateral dispersion; however, the data shown do reveal how suites of sensors can be deployed to better understand the processes affecting plumes in complex terrain. In particular, remote sensors may well provide an effective means with which to identify influences originating outside the region probed by conventional in-situ sensors.

8.4.6.4. INTERPRETATION OF ELEVATED PLUMES

Because both lidar plume maps and sodar data were available, we can compare plume behavior (as measured with the lidar) with the atmospheric behavior (as indicated on sodar records). A variety of data were available along the 348° azimuth. In particular, the tower-mounted turbulence instrumentation, the sodar, and the tethered-balloon system were well situated to help interpret the behavior of the plume as observed with lidar.

As an example of how sodars may help characterize the vertical dispersion potential within the inversion layer, Figs. 8.37a,b show two cases in which both sodars and lidar indicated considerable difference in vertical distribution of echo structure. In Fig. 8.37a, the stratified, isolated echo layers suggest light and variable winds with

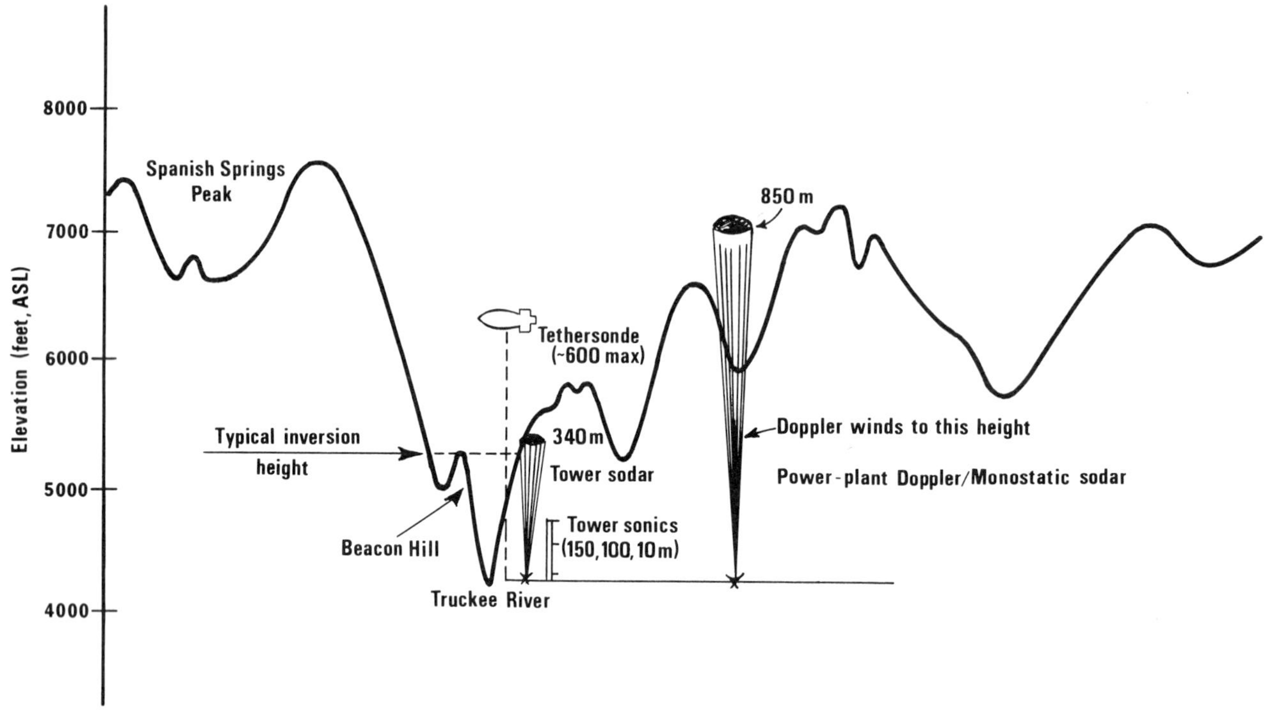

FIG. 8.32. Heights probed by remote and in-situ sensors in the 1984 EPA Tracy Experiment relative to the terrain in which the instruments were embedded. Strong nighttime inversions in the basin typically reached heights between 300 and 400 m above the surface, well above the height of the 150-m meteorological tower used in the experiment.

limited vertical mixing, as is verified by the lateral spread of the plume and its limited vertical extent. In Fig. 8.37b, organized echo patterns extend throughout the region, suggestive of large vertically organized motions. In this case, the lidar shows vigorous vertical mixing through the inversion layer. Cases such as these demonstrate the usefulness of sodar facsimile records as a means to classify the stability of the atmosphere for dispersion calculations, a technique used in a recent large air-quality study (Neff 1989a,b).

8.4.7. *Interpretation of ground-based plumes during the 1980 ASCOT experiment*

Ideally, sodar facsimile records should give a qualitative indication of mixing layer depth and the processes governing plume dispersion. During the 1980 ASCOT experiment, an aerosol-mapping lidar was deployed at the base of the Anderson Creek valley in California in a position to view the down-valley spread of an oil fog released near ground level (Gudiksen et al. 1984). At the same time the ASCOT program had deployed an acoustic sounder at a midpoint on the slope (Coulter and Martin 1983) through which the evolving ground-based plume should have been transported by the nocturnal drainage flow. Neff and King (1985) used the data obtained to examine the relation between echo layer depth as determined by the sodar and the depth of the layer through which the oil fog had mixed as measured by the lidar. In Fig. 8.38 lidar backscatter profiles show the same vertical variation as the sodar echo intensity in the region corresponding to the boundary layer. In a sense this should be expected because the backscatter in each case arises as a result of mixing from surface sources, a ground-based oil fog for the lidar and cold air due to surface cooling for the sodar. In some of the profiles, elevated echo layers appear that are decoupled from the surface mixing layer and do not correspond to lidar echoes. Neff and King (1985) showed a similar comparison with SF_6 profiles in which the tracer mixes to the top of the echo layer revealed by the sodar.

8.4.8. *Remote-sensor observation of day/night transitions in complex terrain*

8.4.8.1. BACKGROUND

In Chapter 2, Whiteman describes observations of the inversion destruction that occurs with the onset of solar insolation on valley sidewalls in the morning. In particular, he describes a process by which mass is removed from the lower levels of a valley by means of thermally

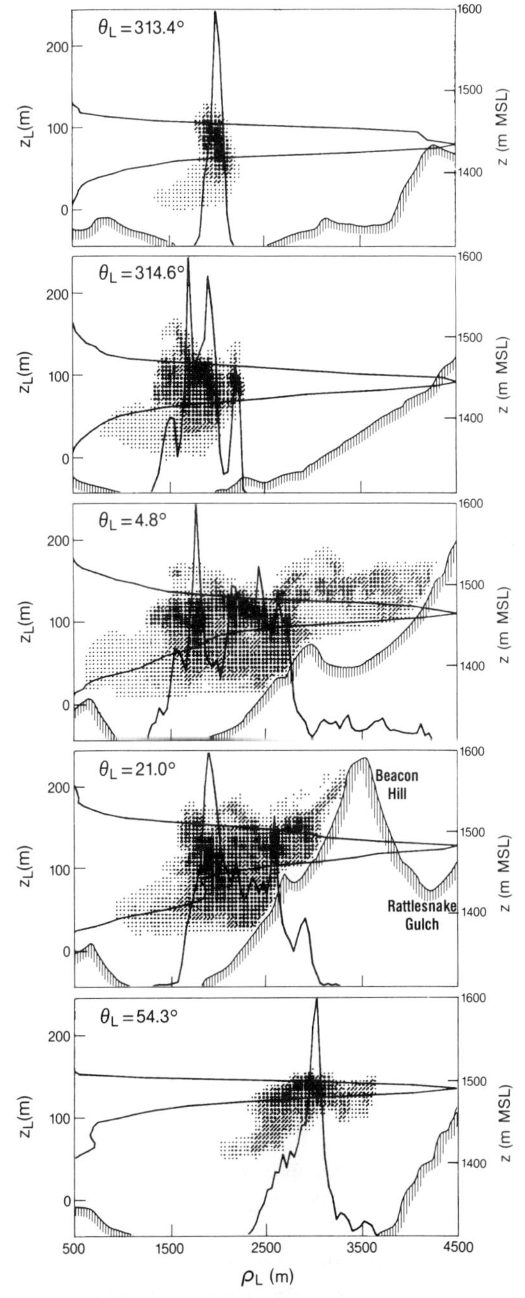

FIG. 8.33. Sequence of lidar plume mappings at the five azimuths shown in Fig. 8.31 and averaged over an hour. The horizontal and vertical profiles of integrated backscatter are shown on the edges of each figure, normalized to their peak values. Note the initial plume height of about 150 m and narrow cross-section near the power plant and subsequent dispersion. This night was characterized by strong stability and limited vertical dispersion. (Eberhard et al. 1987.)

driven convective sidewall slope flows. This causes a compensatory sinking of air over the valley and a subsequent downward translation of temperature profiles. Heating of the surface together with adiabatic heating of the sinking air should result in an overall increase in the temperature of the entire column. Within the remaining "stable core" the flow often continues down-valley, a final remnant of the nocturnal drainage. Because of the sensitivity of sodars to temperature inversions and convection, they are an ideal instrument with which to measure the depth of the convection and the sinking of the elevated capping inversion layers.

Whiteman also noted the paucity of observations of the evening transition in complex terrain. In such cases, sodars should see a continuous evolution of echo patterns from those of convection, to an absence of echoes in the adiabatic transition, to a monotonic growth of the statically stable boundary layer with the onset of surface cooling. Furthermore, to the extent that the main drainage source is distant from the point of observation, such as an exit onto a plain from a valley, the sodar should reveal a gravity current type head at the onset of drainage, such as that described by Blumen (1984).

8.4.8.2. SODAR OBSERVATIONS OF NOCTURNAL INVERSION DESTRUCTION

We have already described the evolution of the nocturnal inversion in partially draining valleys and the complex echo structures seen by sodars in the early ASCOT California experiments. Dominant in the observations were echo strata that evolved in concert with temperature profiles during the night in the Anderson Creek valley where strong low-level inversion occurred to heights of 300 m. With the onset of morning insolation, the sodar often revealed the erosion of the nocturnal inversion from below as shown in Fig. 8.39. At the same time, the elevated echo strata, decoupled from the convection below, began to sink, suggesting subsidence above the convective layer. The fact that the sinking motion begins at 0630 PDT when the sun first illuminates the valley sidewalls suggests that the sodar is, in fact, observing the subsidence of an elevated stable region ("stable core") described by Whiteman and McKee (1982). By 0900 PDT, the stable core appears to be largely eroded, and only a convective region and capping inversion remain. After 0900 PDT, the individual convective elements maintain an almost constant depth. Presumably, this reflects periods during which the subsidence-induced sinking of the inversion is balanced by the upward erosion of the convective layer into the overlying stable atmosphere. The continuing effect of subsidence aloft is seen in the figure as the temperature warms above the inversion as well as below. Finally, convection echoes become obscured from 1000 to 1030 PDT, and as the inversion weakens, vertical mixing increases; finally the inversion is destroyed.

Whiteman and McKee (1982) in their thermodynamic model of valley inversion destruction showed that the rate of destruction of the inversion depends on the inversion strength and depth. Sodar observations in the coastal ma-

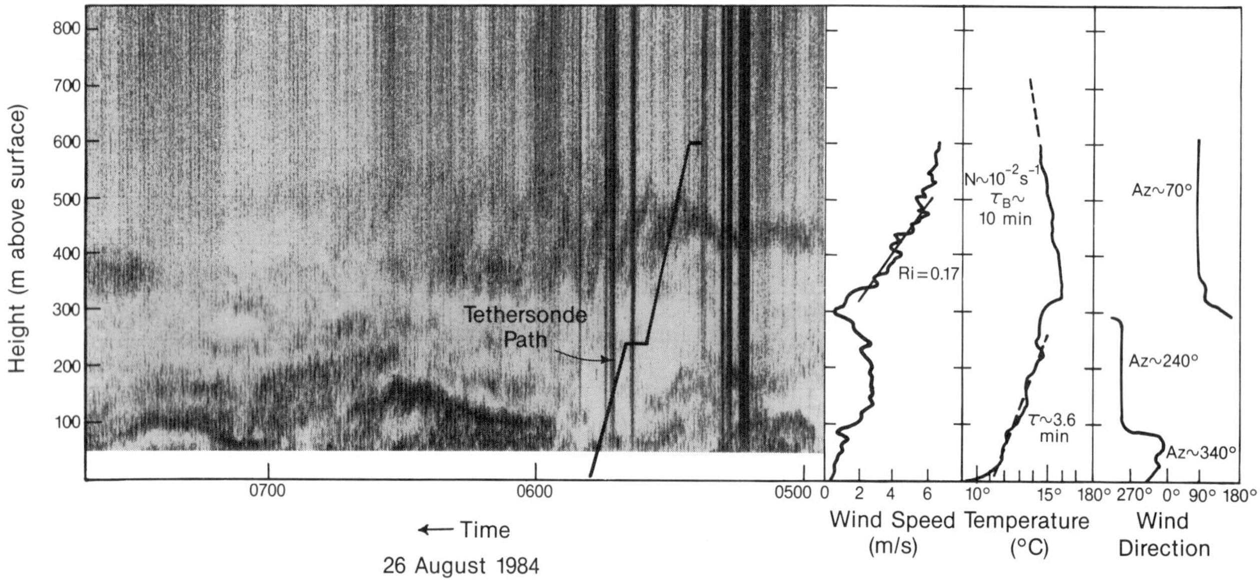

FIG. 8.34. Sodar record obtained during the EPA Tracy Experiment, showing stratified echo layers below 300 m (within the basin) and an elevated region of shear instability aloft. Three flow regimes are evident: a low-level drainage from the slopes to the north (340°), a down-valley drainage flow of 2 m s^{-1} along the Truckee River (240°), and an east-northeasterly flow aloft that accelerates after 0500 PST to speeds exceeding 5 m s^{-1} (compare with the topography in Fig. 8.30). Buoyancy frequency (N), buoyancy periods (τ), and wind azimuths (Az) are shown. (Neff 1987.)

rine environment of the Anderson Creek valley, where the marine inversion determines the initial inversion strength and depth and where sidewall heating is essentially one-sided, reveal that the time of destruction of the inversion, relative to local sunrise, depends also on the depth of the initial marine inversion. In Fig. 8.40 the time of inversion destruction was determined from sodar records such as those in Fig. 8.39 and under conditions of well-defined sea breeze/marine intrusions into the coastal valleys of California. In this analysis, the depth of the marine inversion was determined from early-morning rawinsonde soundings obtained at Oakland, some 120 km to the south.

Such results suggest the potential value of multiple sodar observations of inversion morphology in confined valleys. In particular, it would be of interest to study the spatial distribution of inversion sinking rates relative to the local flux of mass moving up the valley sidewalls as well as the local turbulent entrainment at the interface between the convective layer and its capping inversion.

8.4.8.3. Lidar observations of morning flow reversals

In Chapter 2, Whiteman describes tethered-balloon wind profiles in large valleys in which winds continue down-valley while the convective slope flows remove mass from the valley and surface heating continues to erode the stable core from below. In Brush Creek this phenomenon had not been observed in preliminary experiments, presumably because the narrow confines of the valley led to a more rapid destruction of the stable core with the onset of sidewall heating. During the 1984 ASCOT experiment, however, the Doppler lidar provided an opportunity to search for such an effect with a greater spatial resolution than that provided by tethered balloon profiles. In fact, during one morning of the experiment (30 September 1984) a special scan sequence was carried out, looking in the up-valley direction at about 5-min intervals as the lidar performed repeated raster scans (Post and Neff 1986). These wind measurements show the reversal from down-valley to up-valley flow but without an intermediate stage in which down-valley flow persists in the elevated stable core; however, they do reveal a local minimum in wind speed near the former top of the inversion. The observations also show remnants of the drainage flows emerging along the main valley sidewalls from tributaries that are still shadowed.

The lidar description of the flow-reversal process on several mornings was somewhat complicated by the presence of prevailing winds blowing up-valley, in the same direction as the convective slope flows. On the last night of the 1984 ASCOT experiment, however, a morning flow reversal occurred in the presence of down-valley prevailing winds (Post and Neff 1986). In this case (Fig. 8.41), the thermally driven upslope flow was confined near the valley floor and defined clearly by its interface with the countervailing ambient winds.

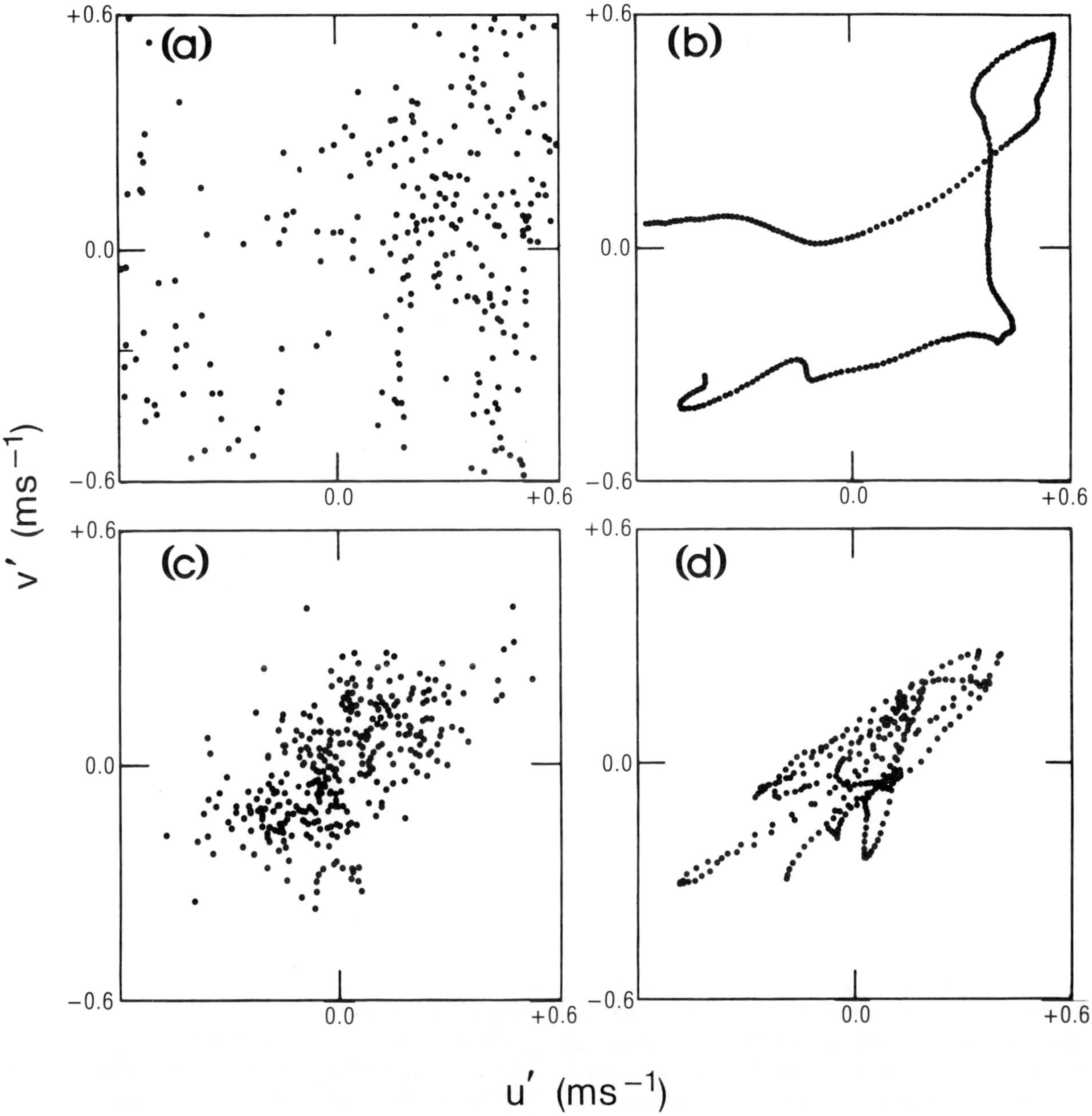

FIG. 8.35. Vector wind fluctuation (u', v') plots obtained at the 100-m tower level, using different frequency band passes. (a) Raw data showing the northeasterly drift, (b) slow meander (1-h period), (c) high-frequency vector fluctuations with periods shorter than 5 min, (d) fluctuations in band pass centered on 7-min, showing two-dimensional structure associated with wave motion. (Neff 1987.)

8.4.8.4. THE EVENING TRANSITION TO NOCTURNAL DRAINAGE IN A CONFINED VALLEY

Monostatic and Doppler sodars have, to a limited extent, documented some of the characteristics of the evening transition from unstable to stable temperature structure in complex terrain, particularly near the surface where the strongest variations in the energy budget occur. For example, Neff and King (1987, 1988) described the evening transition in the presence of both aiding and opposing ambient winds (in their case, the evening sea breeze). In the case of aiding winds, they found some ambiguity in drainage wind signatures until, for example, the sea breeze had diminished sufficiently to reveal a low-level wind maximum. In contrast, opposing winds provided for the clearest drainage-wind signature (Fig. 8.42); tethered-balloon profiles provided a means to interpret the sodar records. In the top part of Fig. 8.42 the tethered-balloon

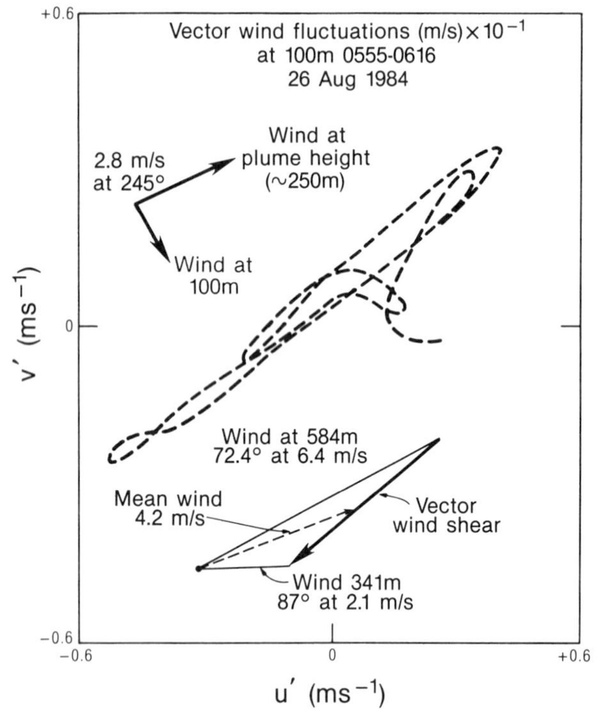

FIG. 8.36. Vector wind fluctuations observed at 100 m. Orientation of vector shear between 400 and 500 m in region of instability, revealed by sodar record, shows agreement with the predictions for the vector wind fluctuations of Gossard and Munk (1954).

profiles define regions where Ri < 1.0 as well as the location of the wind maximum, the midpoint and top of the surface temperature inversion, the top of the drainage (indicated by a 180° wind reversal), and the top of the stable region where cooler air from the drainage flow mixes into the opposing ambient flow, creating a weak elevated inversion. Evident in the sodar record are the initial statically stable layer, the onset of drainage and mixing into the ambient flow (just after 1900 PST), the strong echo layer (temperature variance at small scales) just above the wind maximum, and the minimum in scattering just below the wind maximum (after 2030 PST).

8.4.8.5. THE EVENING TRANSITION AND THE EMERGENCE OF DRAINAGE FLOWS ONTO PLAINS

The onset of drainage within the Big Sulphur Creek valley (Fig. 8.42) provides an example similar to many other locations; however, drainage flows have also been observed to progress from valleys onto plains with somewhat different initial structure. In such situations the cool air must first accumulate within the valley; as the flow exits the valley onto the plain, a distinctive gravity-current head develops as described, for example, by Blumen (1984) using data from the Boulder Atmospheric Observatory (BAO) situated some 20 km from the base of the Rocky Mountains. The subsequent airflow can then become more complicated as seen in Blumen's work where waves and instabilities develop and provide a complicated sodar echo structure. In a similar case (Neff and Juan 1987), the subsequent drainage flow became quasi-steady for several hours and then rose slowly through turbulence instrumentation on the BAO as shown in Fig. 8.43. As described by Neff and Juan, five regions of interest in the progression of the drainage current could be identified. Features *A* and *B* indicate a precursor wave and gravity current head, respectively. Feature *C* is a quasi-steady low-level boundary phenomenon. Feature *D* marks a period during which the feature *C* lifts slowly through the 100-m tower level. Feature *E* indicates a region of shear between the low-level flow and the ambient atmosphere into which the drainage has intruded. This case provides an example of how sodar facsimile records can be used to select in-situ data, such as those from tall towers, for analysis. It can also be hoped that analyses from events such as these will eventually help in the interpretation of remote sensor data from a variety of complex terrain settings and not just this simple mountain–plains circulation.

Feature *C* in Fig. 8.43 can be seen as a low-level jet in Fig. 8.44, where a spline-under-tension fit to the tower fixed-level data reveals a wind maximum just below the height of the maximum echo layer (for the period averaged from 2320 to 2340 MST). In this case the strong echo pattern occurs in the same relative location shown earlier in Fig. 8.42 for a confined valley flow, just above the wind speed maximum. Another feature of interest is the weak lapse rate and strong shear between 100 and 150 m, a region of the facsimile record bounded by stronger echo layers and a feature previously found in model simulations such as those shown in Fig. 8.13.

The quasi-steady nature of the drainage jet during the period *C* suggested an analysis of the momentum flux above and below the jet, using the sonic anemometer turbulence data from the 50- and 100-m tower levels. As an aid in the analysis and to anticipate examination of the slowly varying period in feature *D*, we examined the behavior of the vertical flux of westerly momentum given by

$$\overline{uw} = \frac{1}{N} \sum_{i=1}^{N} u_i w_i \tag{8.13}$$

where i indicates the succession in a time series of data sampled at 0.1-s intervals but recorded every 10 s. [This data-recording technique preserves the high-frequency contributions to the covariance as described by Kaimal and Gaynor (1983). The technique avoids the necessity

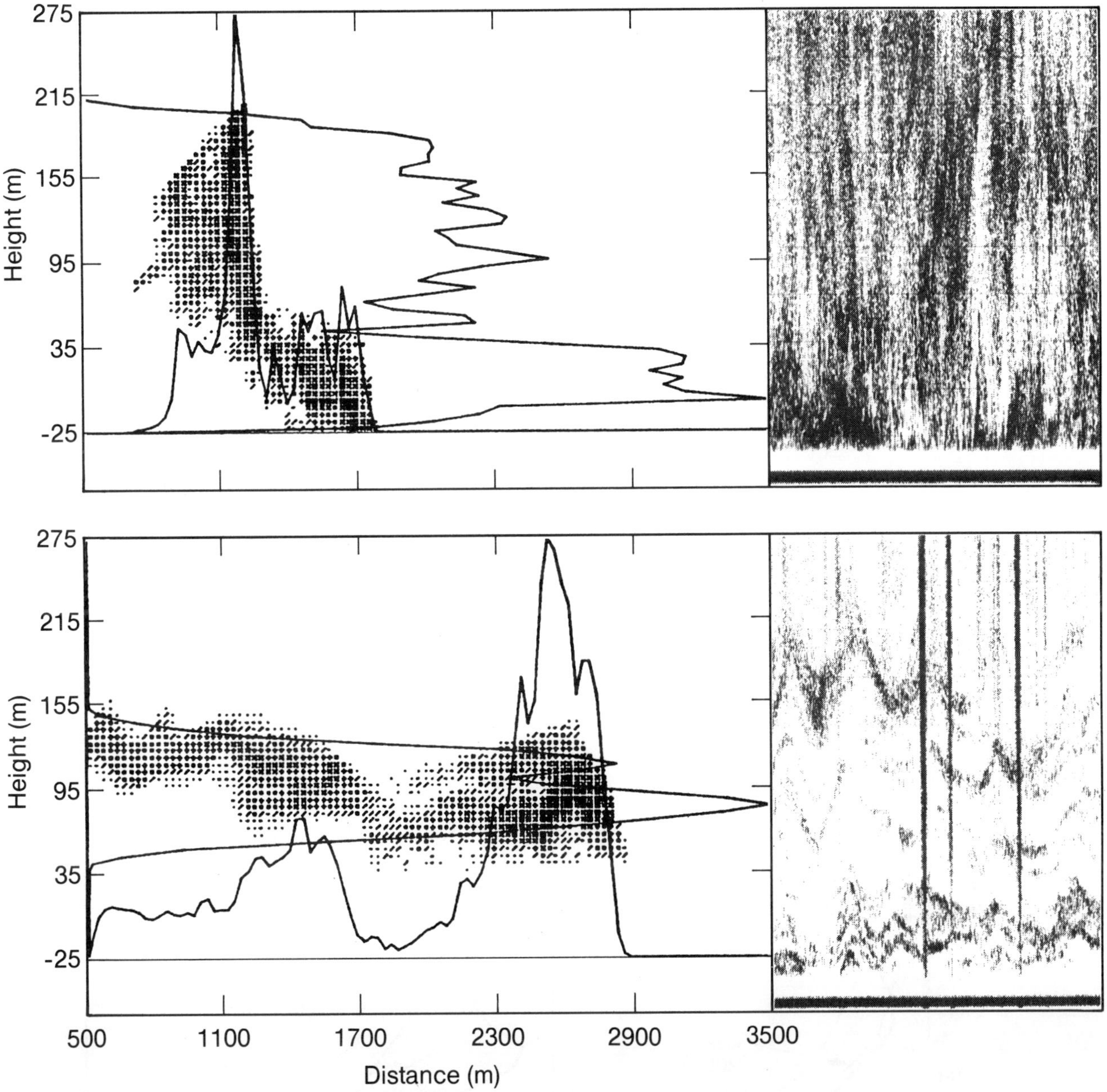

FIG. 8.37. (a) Plume cross-section compared with sodar record during period of strong stability and light winds. Note the horizontal extent of the plume under these conditions when the sodar shows a highly layered structure. (b) Similar to (a) but under conditions of stronger winds and more vertical mixing as indicated by the vertically oriented echoes on the sodar record. In this case the lidar indicates that the plume mixes almost to the surface.

for recording massive amounts of data but prevents subsequent spectral analysis of the high frequencies.] If we now define a discrete cumulative sum defined by

$$F(N) = \sum_{i=1}^{N} u_i w_i \tag{8.14}$$

and then calculate the slope of this function between N and $N + \Delta N$, we obtain

$$\frac{\Delta F}{\Delta N} = \left[\frac{\sum_{i=1}^{N+\Delta N} u_i w_i - \sum_{i=1}^{N} u_i w_i}{\Delta N} \right] = \frac{1}{\Delta N} \sum_{i=N}^{N+\Delta N} u_i w_i . \tag{8.15}$$

Therefore the slope of $F(N)$ for any N can be interpreted as the flux in the vicinity of the time corresponding to the

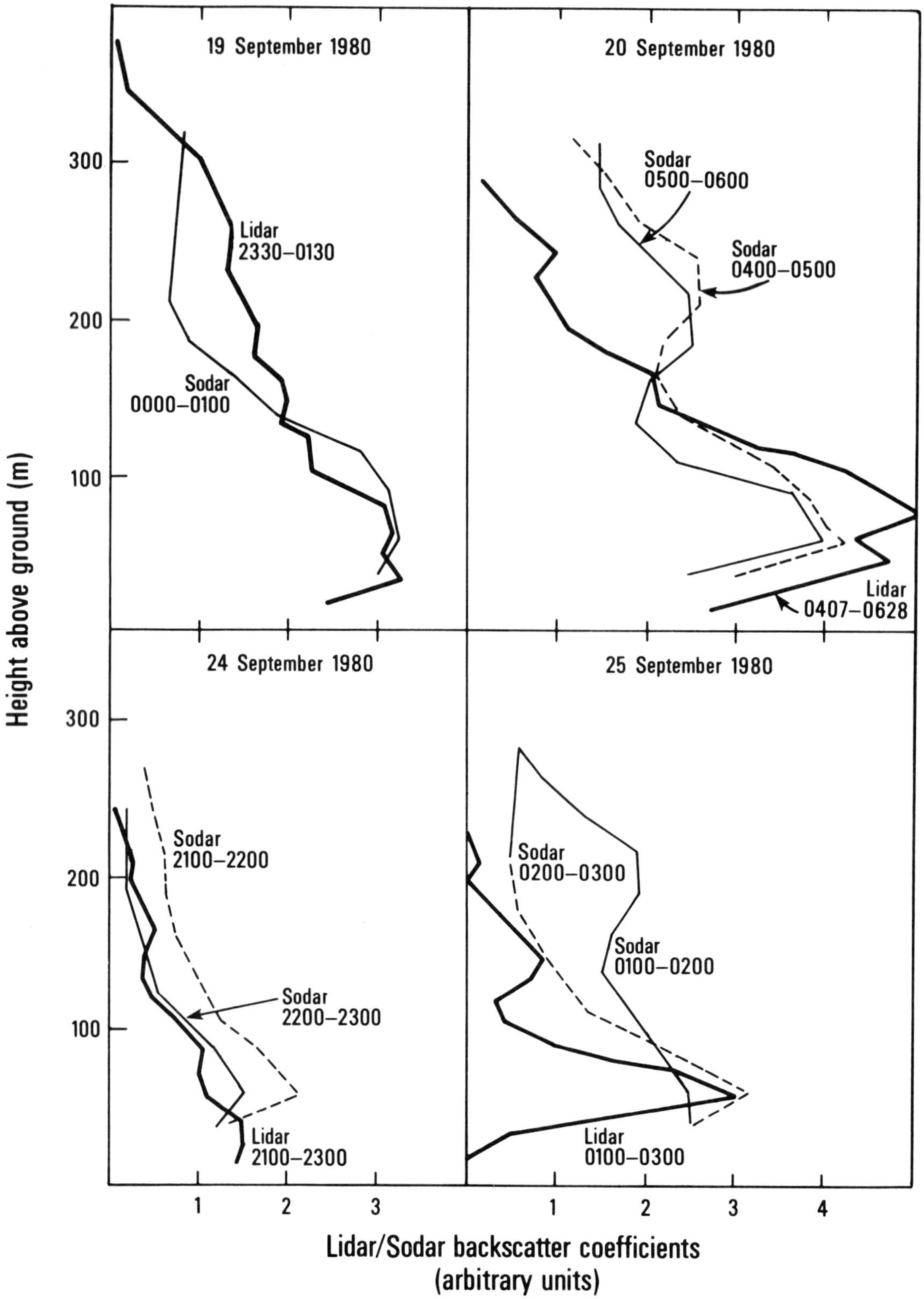

FIG. 8.38. Sodar echo intensity profiles compared with lidar backscatter profiles in the same area along a developing drainage flow (in the vicinity of the fourth sodar from the top of the valley shown in Fig. 8.7). In this case the oil fog was released in the creek bed nearest the upper sodar shown in Fig. 8.7. In general the oil fog is confined in the strong ground-based echo layer. These data correspond to the two nights represented in Fig. 8.18. On the first night, in the presence of significant wave activity, there is more vertical dispersion than on the second night when the drainage flow is shallow and unperturbed.

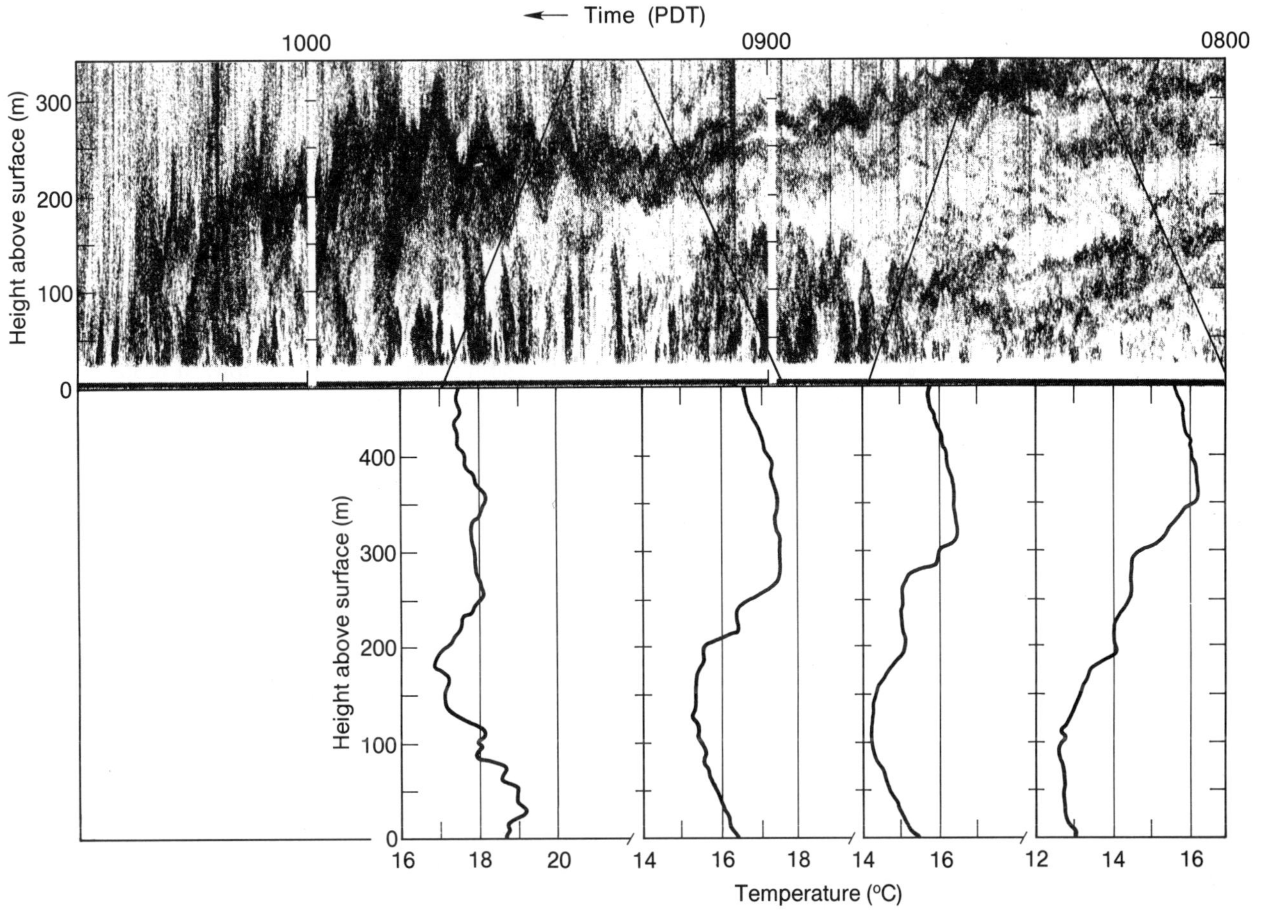

FIG. 8.39. Tethersonde profiles obtained on the morning of 26 August during a period of inversion descent in the basin setting of the 1984 EPA Tracy Experiment (Fig. 8.30) where sidewall heating may be removing mass from under the inversion, causing the air above to sink and thus warm adiabatically. Temperature profiles show the growth of the convective layer together with the sinking of the inversion aloft, corresponding closely with the sodar facsimile recording. Note also the warming in the profiles above the inversion and the final destruction of the inversion shortly before 1030 PDT.

sample N. Plotting $F(N)$ versus N (time) can thus reveal subtle changes in the flux due to events revealed on the simultaneous sodar facsimile recording, and thus determine appropriate flux averaging times. It can also reveal whether the covariance is well behaved or whether major contributions to the covariance arise from discrete burst-like events (e.g., see Lu et al. 1983). In Fig. 8.45 the functions $F(N)$ are shown for the 50- and 100-m levels on the tower for the period C where a high-pass (10-min) filter has been used on the data. Two periods are identified during which the slope of $F(N)$ is well behaved, one before and one after 2355 MST. Here $F(N)_{100\text{m}}$ has a positive slope throughout the period, indicating an upward momentum flux from the jet. Of interest is the rapid increase in the flux after 2355 MST, a period on the sodar record when instability type echoes rise through the 100-m level. Below the jet the slope of $F(N)_{50\text{m}}$ is almost constant except for occasional small bursts every 10 or 20 minutes and a large burst at 0015 MST. The momentum balance of the jet can also be compared with the katabatic acceleration over the local slope for a temperature deficit of 2°C at the nose of the jet:

- Momentum flux divergence:

$$-\frac{\partial \overline{uw}}{\partial z} \approx -\left[\frac{\overline{uw}_{100\text{m}} - \overline{uw}_{50\text{m}}}{50\ \text{m}}\right] = -1.2\ (\text{m/s})\ \text{h}^{-1} \qquad (8.16)$$

- Katabatic acceleration:

$$\frac{g}{T}(\Delta\Theta)\sin(\alpha) = +1.0\ (\text{m/s})\ \text{h}^{-1} \qquad (8.17)$$

where the average large-scale slope in the vicinity of the BAO is about 0.004. The similar values in (8.16) and (8.17) suggest that the momentum flux divergence almost balances the local katabatic force for the quasi-steady drainage jet revealed by the sodar.

A short time later, just before 0100 MST, the sodar

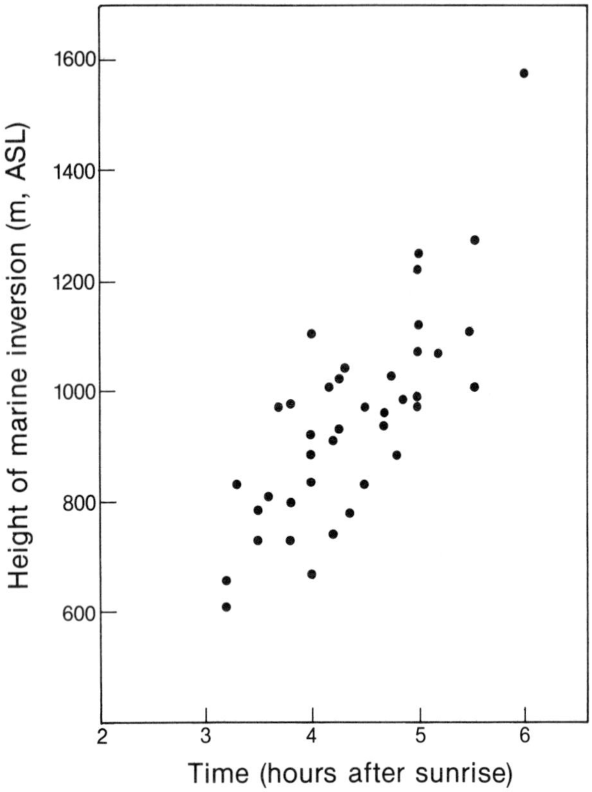

FIG. 8.40. Time of destruction of the inversion observed below the Anderson Creek valley versus the depth of the marine inversion in California. The marine inversion depth was determined from the Oakland morning rawinsonde data obtained 120 km south of the sodar site. (Neff and King 1988.)

reveals the low-level jet rising through the 100-m level of the tower. This can also be seen in the time series of wind speed and direction shown in Fig. 8.46. With the insight provided by the sodar, the wind speed increase to 7 m s^{-1} can be interpreted as being due to the rise in the height of the low-level jet. The relatively slow and smooth rise of the wind speed maximum suggested that the previous cumulative flux functions might reveal the character of the fluxes above and below the jet. In Fig. 8.47 similar plots are shown of momentum flux and temperature flux functions during this period. In the case of $\overline{uw}$ the cumulative sum function describes upward momentum flux above the jet, zero flux at the nose of the jet, and downward flux below the jet. (The movement of the jet through the level was determined by the time of maximum wind speed.) The temperature flux $\overline{w\theta}$ is everywhere negative except at the nose of the jet where it appears to be near zero, consistent with our earlier discussion. The temperature profile in Fig. 8.44 shows a steepening of the lapse rate at the nose of the jet consistent with a convergence of heat above and divergence below, as suggested by Neff and King (1988) in their analysis of channeled valley flows where they found the height of the drainage wind maximum at the midpoint of the temperature inversion.

8.5. The use of remote sensors in large-scale complex terrain flows

The previous examples have all dealt with local aspects of complex terrain flows, for the most part at scales less

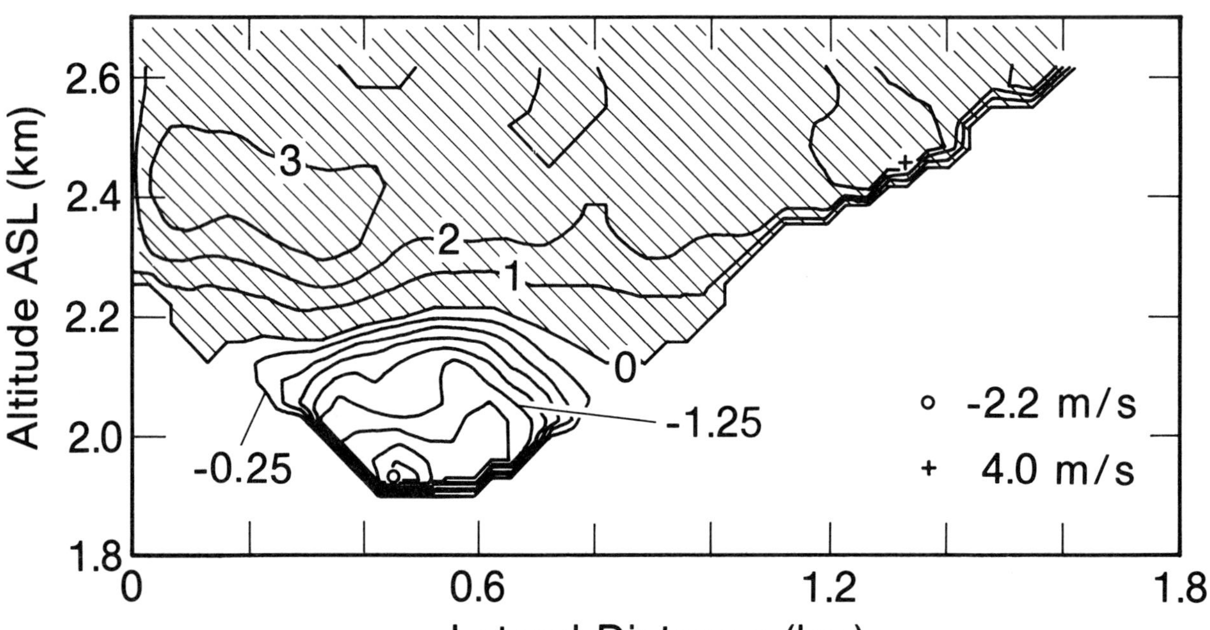

FIG. 8.41. Lidar view of convectively driven up-valley flow in bottom of the Brush Creek valley during the morning of 6 October 1984 in the presence of prevailing down-valley winds aloft (shaded). Up-valley winds are contoured at 0.25 m s^{-1} intervals; down-valley isopleths are contoured at 1 m s^{-1} intervals. (Post and Neff 1986.)

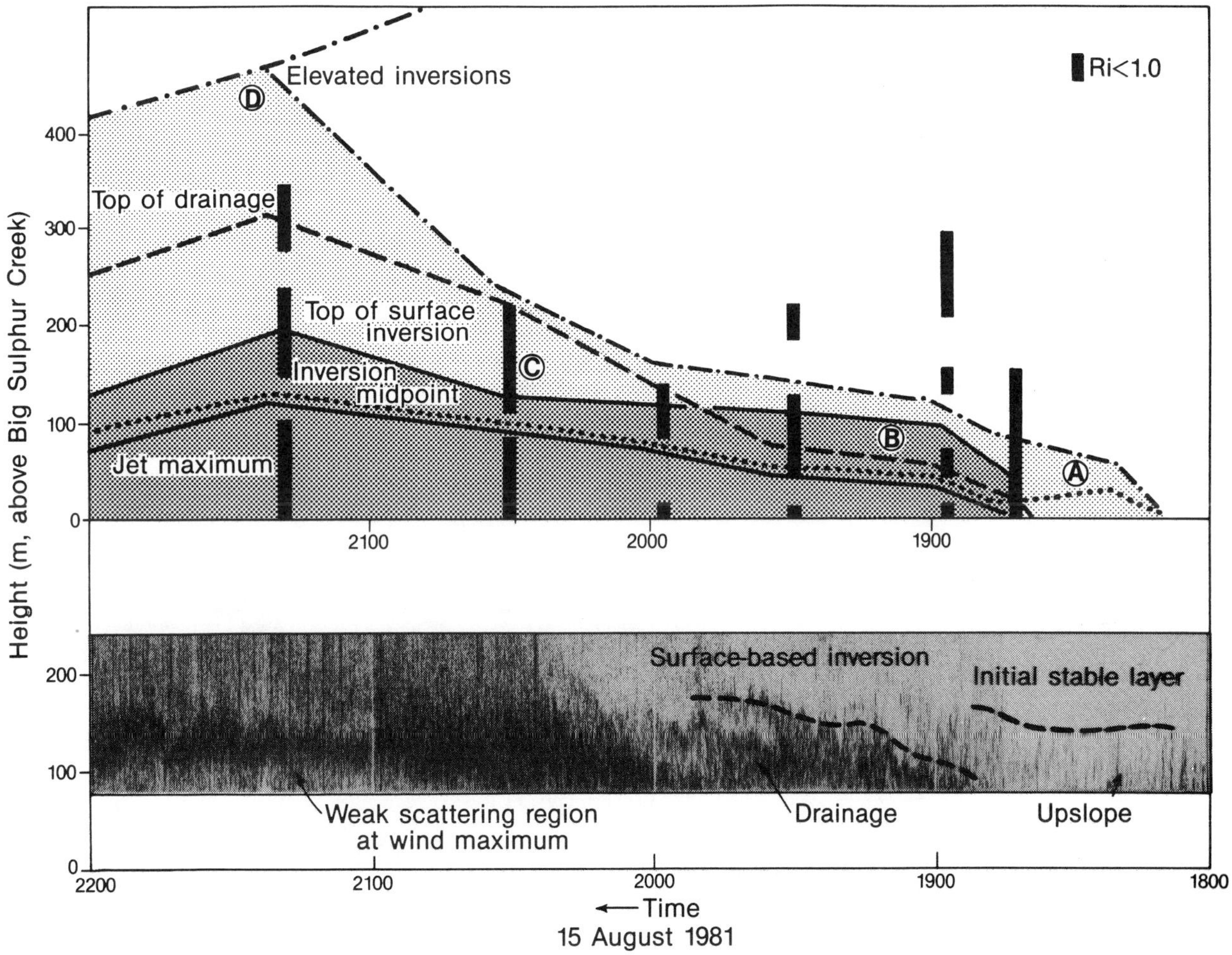

FIG. 8.42. Evening initiation of drainage flow in the presence of an opposing sea breeze flow. Initial cooling of the surface begins to form a statically stable layer between 1800 and 1845 PST in which the flow is still upslope. The drainage flow actually begins at about 1845; the sodar shows a strong echo layer at the interface between the cool downslope flow and the warmer westerly winds. Solid vertical bars indicate regions in the occasional tethersonde profiles for which Ri < 1.0. The weaker echoes below 100 m after 2100 PST correspond to the height of the drainage wind maximum; stronger echoes are just above in a stronger inversion, stronger (reverse) shear region. Extensive vertical mixing occurs above the jet. Serious noise contamination from a venting steampipe occurred in the sodar record after 2030 PST.

than about 20 km. There have also been efforts to define flow regimes at larger scales using both satellite- and ground-based remote sensors. For example, Bromwich (1989) used high-resolution infrared imagery from satellites to observe katabatic winds emerging from the slopes of Antarctica. Schaaf et al. (1988) described the use of satellite imagery to define preferred locations of convective development over mountainous terrain. Uttal et al. (1989) studied orographic effects on winter storm development using radars and passive microwave radiometers. McElroy and Smith (1986) used airborne lidar to study regional air quality in coastal basins as well as in the South Platte River basin in Colorado (McRae and McElroy, personal communication). Combined modeling efforts and dual-Doppler radar observations (Kropfli et al. 1986) have documented topographic influences on the creation of the Santa Barbara Eddy. A diverse suite of remote sensors has been used to study the effect of mountain–valley circulations on regional air quality (Neff 1989a,b).

8.5.1. *Large-scale drainage flows*

The studies described thus far have all dealt with drainage flows on the scales of a few kilometers or at the most a few tens of kilometers. The scale of the mountain–valley systems, particularly in the western United States, suggests that such circulations should exist over scales of hundreds of kilometers and involve extensive transport of mass. A variety of observations have measured the scale and extent of these circulations (Toth and Johnson 1985; Neff and King 1989; Bossert et al. 1989); however, the magnitude and scale of such drainage wind systems may not always

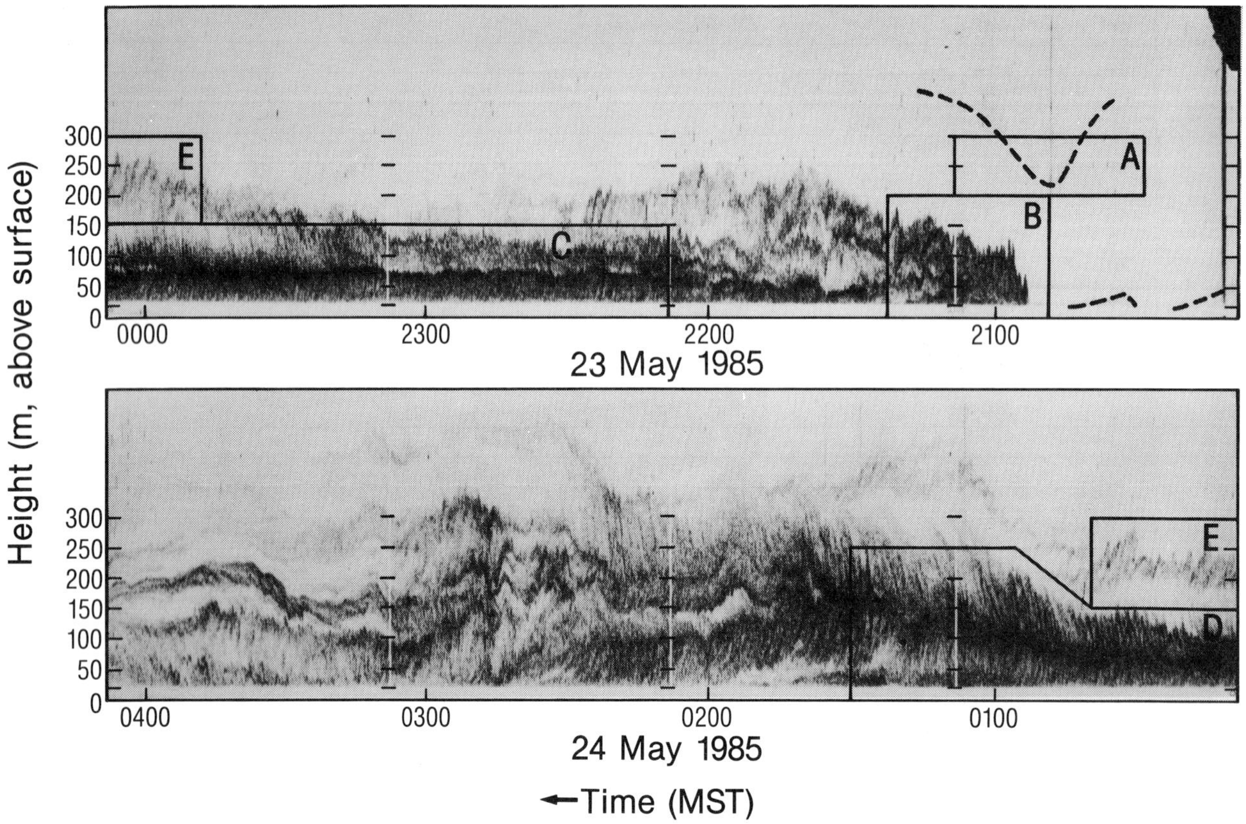

FIG. 8.43. Progression of a drainage flow onto a plain (see text). Of particular interest are the quasi-steady period *C*, and the period *D* during which the drainage jet lifts slowly through the 100-m level of the BAO tower. Heights are instrumentation levels.

be evident because of the absence of adequate vertical profiles. For example, Toth and Johnson studied surface airflow in the South Platte River basin using an array of surface weather stations. These observations, however, give no hint of the deep vertical structure of the drainage wind systems. In particular, observations from the BAO tower near the foothills indicate drainage flows with a depth of 100 to 200 m and speeds greater than 6 m s^{-1} (Blumen 1984). More recent measurements in Denver show a drainage jet some 300 to 400 m deep and reaching speeds of 8 m s^{-1} at heights of 150 to 200 m.

8.5.2. Denver Brown Cloud Study

Analysis of wintertime observations from a Doppler sodar situated beside the South Platte River reveals an essentially bimodal distribution of wind directions along the river axis; down-valley flows are observed at night, up-valley flows during the day. This occurs despite the visual impression of almost flat terrain; however, following Haagenson (1980), computer exaggeration of the terrain along the South Platte reveals a well-defined valley (Fig. 8.48). Because major pollution sources are located along this river valley, the diurnally oscillating winds can often recirculate pollutants for long periods of time (Neff 1989b). For this reason a better understanding of the dynamics of the drainage flows and more gentle upslope circulations was sought in a major pollution study (Denver Brown Cloud Study; DBCS) carried out in the winter of 1987–88 (Watson et al. 1988; Neff 1989a).

To accomplish a more detailed study of these flows, the DBCS made use of a number of remote and in-situ sensors already in place along the foothills of the Rocky Mountains; in particular, the BAO 300-m tower, NOAA's network of 26 surface weather stations, and the Stapleton Airport wind and thermodynamic profilers (Strauch et al. 1984). In addition, three Doppler sodars, several monostatic sodars, a portable wind profiler, and a Doppler lidar were used (Fig. 8.49). Two additional sites were instrumented with tethered balloon and rawinsonde observing systems. For a short period the portable wind profiler, on a site north of the BAO, operated in a RASS mode to provide temperature profiles during a particularly poor visibility period (May et al. 1989b). It should be noted that efforts like these represent a slowly evolving research area, particularly in the integration of new ob-

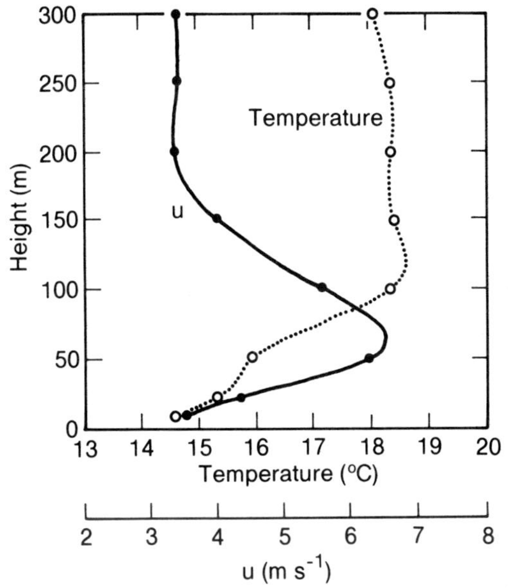

FIG. 8.44. Spline fit profiles of mean-*U*-component of the wind and the temperature at 2320 MST during the steady period shown in Fig. 8.43. Note that the wind speed maximum occurs near the midpoint of the temperature profile.

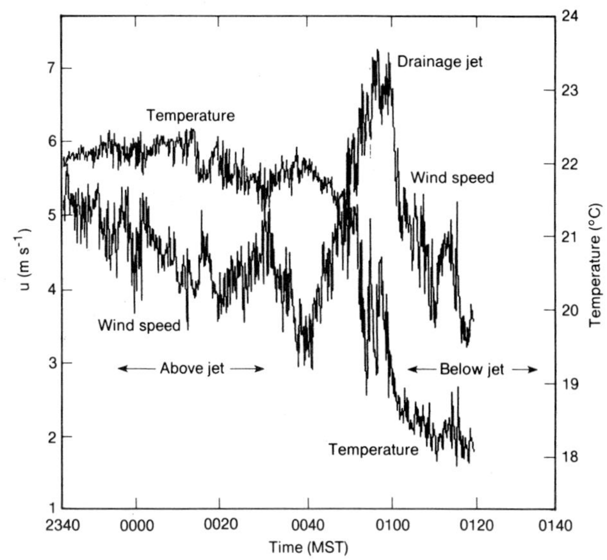

FIG. 8.46. Time series of wind speed and temperature at 100 m, indicating the vertical movement of the low-level jet through the tower sonic anemometer. Viewing the figure sideways provides a quasi-vertical profile of the jet that can be compared with the profile shown in Fig. 8.44.

servations of the meteorology of local circulations with air quality related measurements and analyses.

8.5.2.1. DOPPLER LIDAR OBSERVATIONS

Observation of the along-valley circulation during high-pollution episodes provided one of the goals for the

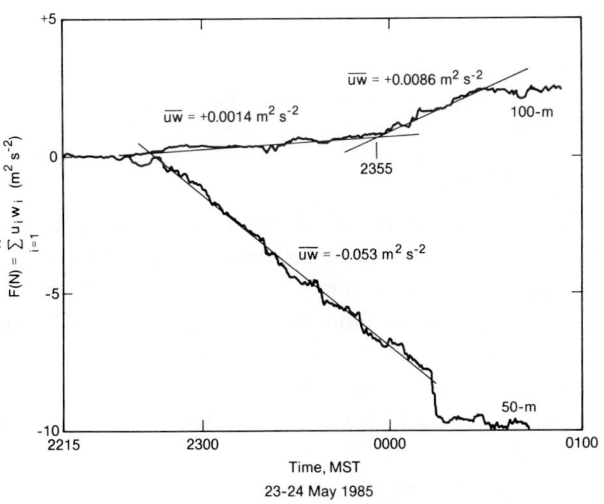

FIG. 8.45. Cumulative sums of covariance terms described in the text, obtained below (lower trace) and above (upper trace) the low-level jet revealed by the sodar record in Fig. 8.43. The slope of the trend lines indicates the magnitude and reveals steadiness of the momentum flux for prolonged periods. The period of increased vertical flux after 2355 MST corresponds to the increased vertical mixing indicated on the sodar record at the same time.

Doppler lidar during the experiment. In particular, BAO tower measurements and sodar records had often revealed up-valley circulations with a shallow frontal structure during periods when surface pressure had been dropping along the foothills of the Rocky Mountains. Such recirculations most often occurred with warm southwesterly or westerly winds aloft. In order to monitor such episodes, the stepped-RHI mode of the Doppler lidar was used to obtain vertical sections every 7.5° in azimuth (Eberhard et al. 1989b). Three sections of particular interest are indicated, together with the location of several of the supporting profiling systems. Two lidar RHI scans were oriented along the valley, and a third looked toward the mountains. On the afternoon of 9 December 1987 a series of lidar scans was carried out in anticipation of the onset of strong downslope winds. One set of observations is shown in Fig. 8.50. In the top plate, scanning toward Denver, the lidar observed flow toward Denver from the northeast along the river valley some 800-m deep as seen in both the aerosol backscatter and in the wind field. In the down-valley scan to the northeast, shown in the second plate, a low-level upslope flow is observed while winds aloft have a strong component away from the lidar corresponding to winds from the west or southwest. In this view, strong lidar backscatter is not observed, suggesting that the aerosol cloud observed in Denver is locally generated. In the third view, toward the mountains, strong downslope winds are observed with gusts to 30 m s^{-1} and a periodic structure of about 4 km horizontal scale, suggestive of breaking waves. Closer to the lidar at about

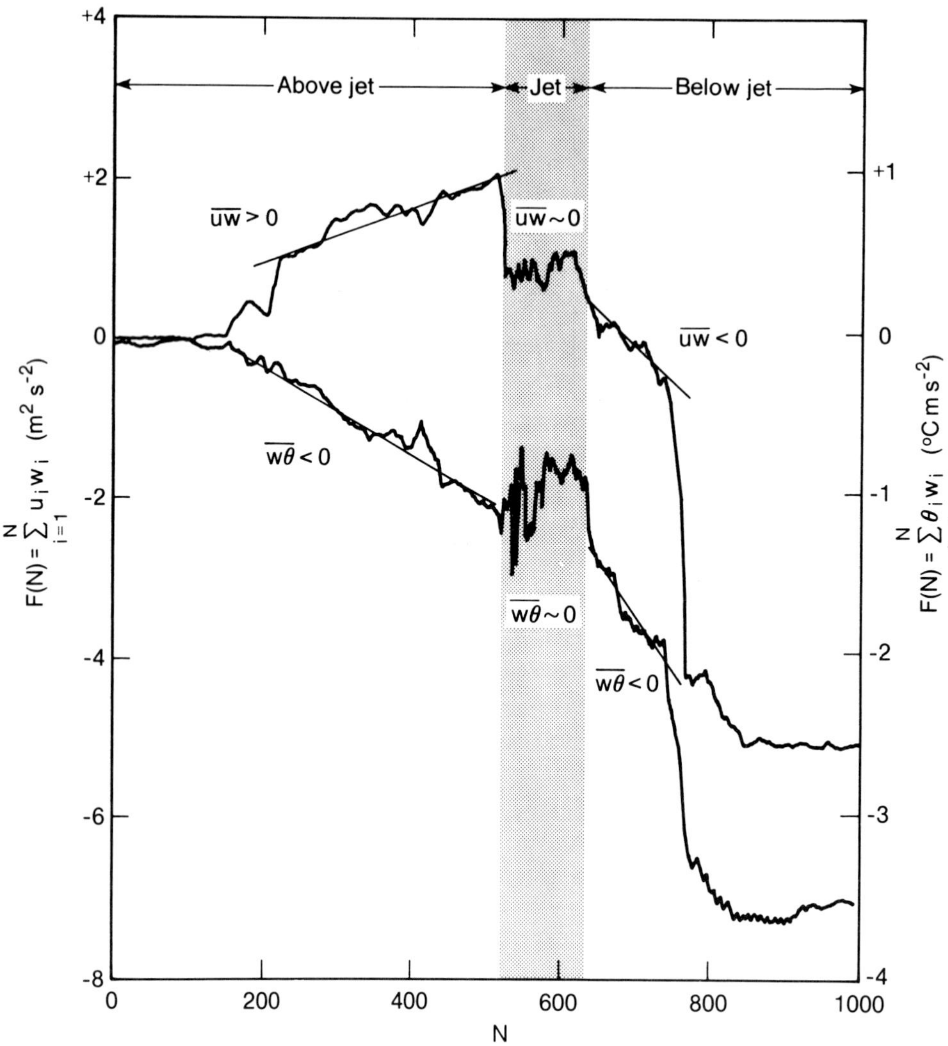

FIG. 8.47. Cumulative covariance sums derived from the data shown in Fig. 8.46, after removal of frequencies less than 5 min. The upper trace shows upward momentum flux above the jet, and downward momentum flux below the jet. Although the mean flux is near zero near the nose of the jet, the slope of the cumulative sum is somewhat erratic, probably because the jet does not pass smoothly through the 100-m level. Similarly, for the temperature flux (lower trace) heat transfer is everywhere downward except near the nose of the jet where some erratic behavior is seen even though the mean is near zero. Data such as these make clear the interpretative value of sodar records in the analysis of direct turbulence data.

10 km, strong aerosol scatter can be seen in the bottom of the river valley underneath the strong westerly winds.

The example in Fig. 8.50 made use of the lidar's stepped RHI mode. The VAD mode can also be used to examine the vertical structure of the wind fields using a small elevation angle as shown schematically in Fig. 8.49 (essentially tracing out a VAD cone across the top of the stepped-RHI scans shown in the figure). Such lidar scans were obtained during high pollution episodes with the inflow from the northeast appearing clearly at low levels in the VAD display and countervailing southerly flow above. The use of a low scanning angle increases the lidar's susceptibility to inhomogeneities in the flow due to local terrain and natural variability. For this reason such scans must be interpreted carefully.

These examples demonstrate how the Doppler lidar, properly positioned, can observe several aspects of a complex flow situation simultaneously. To remove ambiguity in the radial wind component, Doppler sodars and balloon sounding systems were positioned within the scanning radius of the lidar. As these data are analyzed further, better guidance for the use of Doppler lidars will emerge.

8.5.2.2. RASS OBSERVATIONS

From 8 to 9 January 1988 a portable wind profiler operated in an experimental RASS mode to provide real-time remotely sensed temperature profiles as described by May et al. (1989b). The minimum height for the RASS temperature measurements was 300 m in this early test

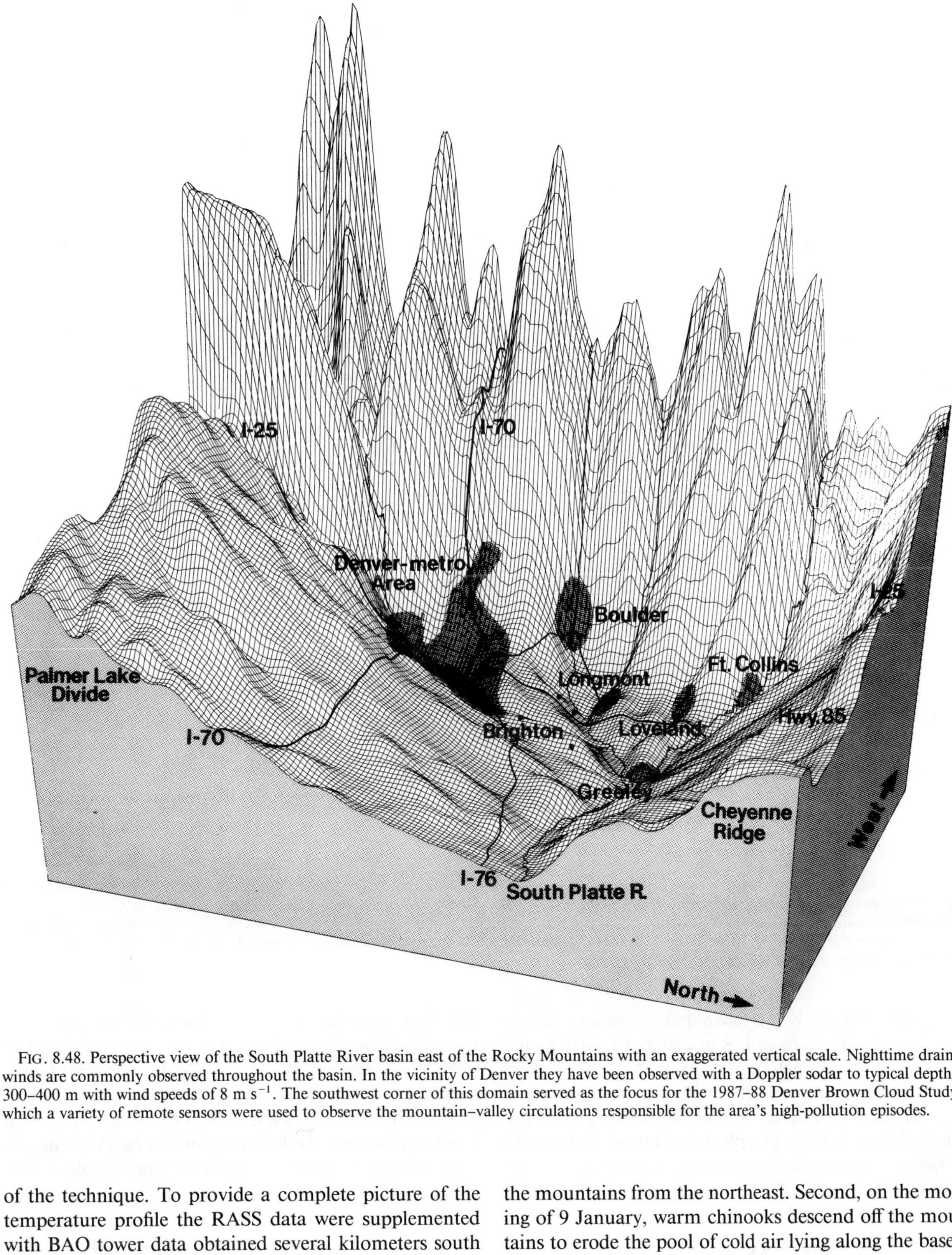

FIG. 8.48. Perspective view of the South Platte River basin east of the Rocky Mountains with an exaggerated vertical scale. Nighttime drainage winds are commonly observed throughout the basin. In the vicinity of Denver they have been observed with a Doppler sodar to typical depths of 300–400 m with wind speeds of 8 m s^{-1}. The southwest corner of this domain served as the focus for the 1987–88 Denver Brown Cloud Study in which a variety of remote sensors were used to observe the mountain–valley circulations responsible for the area's high-pollution episodes.

of the technique. To provide a complete picture of the temperature profile the RASS data were supplemented with BAO tower data obtained several kilometers south of the RASS site. Using the combined dataset, two significant events are revealed in Fig. 8.51. First, in the early evening of 8 January, a cold mass of air moves toward the mountains from the northeast. Second, on the morning of 9 January, warm chinooks descend off the mountains to erode the pool of cold air lying along the base of the mountains. RASS thus provides the ability to continuously monitor the temperature structure of the lower atmosphere. As new systems become available, the min-

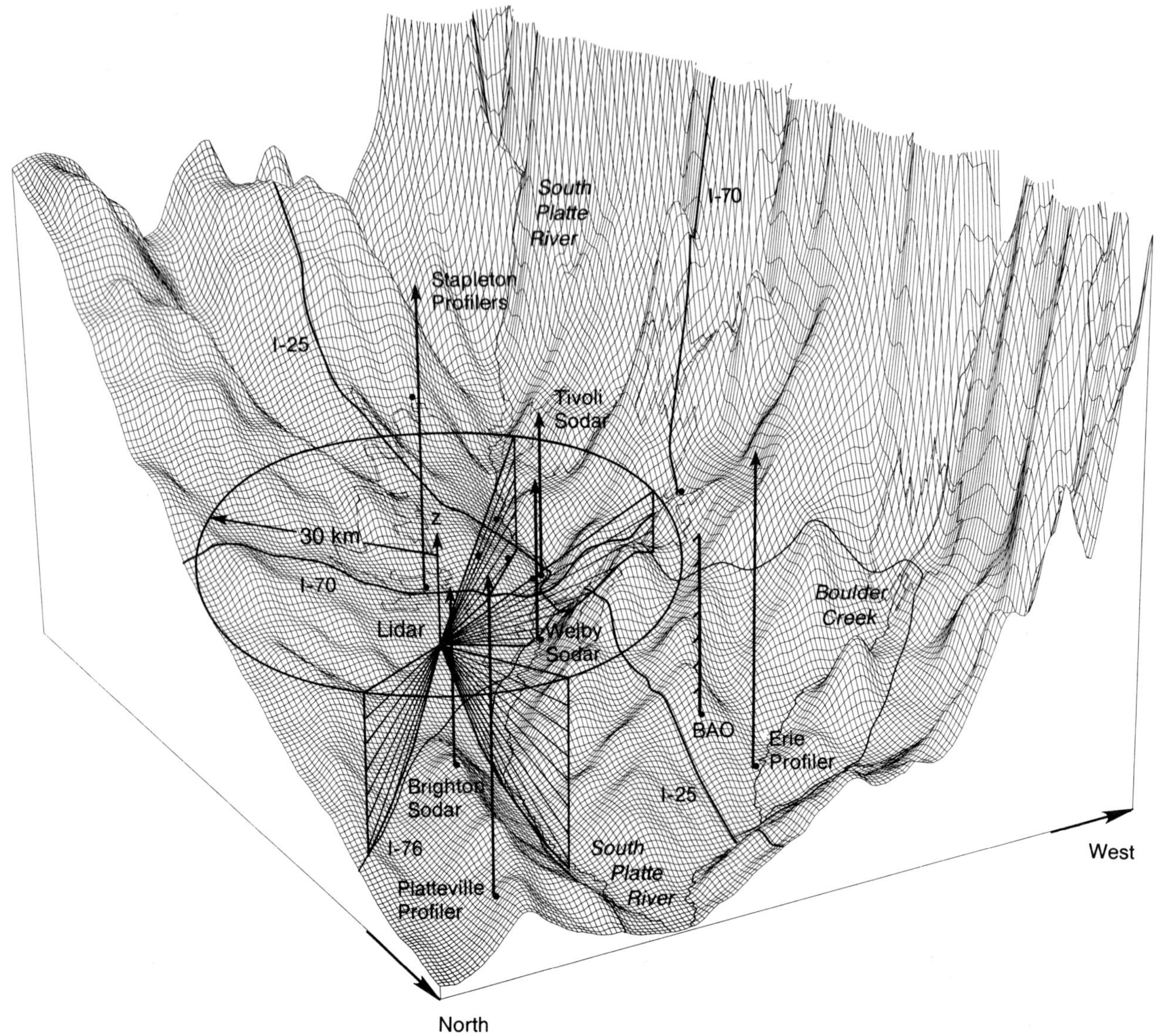

FIG. 8.49. Detailed view of study area and deployment of remote sensors within the basin. Heights are not to scale but serve to show relative ranges and locations. In addition to already existing systems, several sodars (monostatic and Doppler) were located along the axis of the river valley (Tivoli, Welby, and Brighton) as well as a Doppler lidar (on high terrain to the southeast of the river). Wintertime climatologies of wind profiles obtained at Welby show a predominantly bimodal distribution along the axis of the river to heights of 300 m and a weaker third mode corresponding to strong winds from the west.

imum range for the RASS technique can be expected to become lower, perhaps to 100 m above the surface, and the range resolution better, perhaps to 75 m. When these RASS measurements are combined with a wind-profiling capability and embedded within the scanning radius of instruments such as the Doppler lidar, a much wider range of complex terrain phenomena will be observable with a level of detail never before possible.

8.6. Summary and future possibilities

Progress in applying remote sensors to the study of complex terrain flows has been slow in some respects. The data from these new technologies that are similar to those from conventional instruments have been the most freely used; however, new types of data from remote sensors that reflect the complex distribution of turbulence, particularly in complex terrain flows, have been more difficult to assimilate. To help overcome these problems, we have presented a variety of analyses using simple models and in-situ data to aid in the interpretation of remote sensor observations.

Complex terrain processes that have been explored with remote sensors include the initiation, growth, and merging of drainage flows, the decay of nocturnal flows in basins and valleys, the erosion of cold-air pools, the perturbations

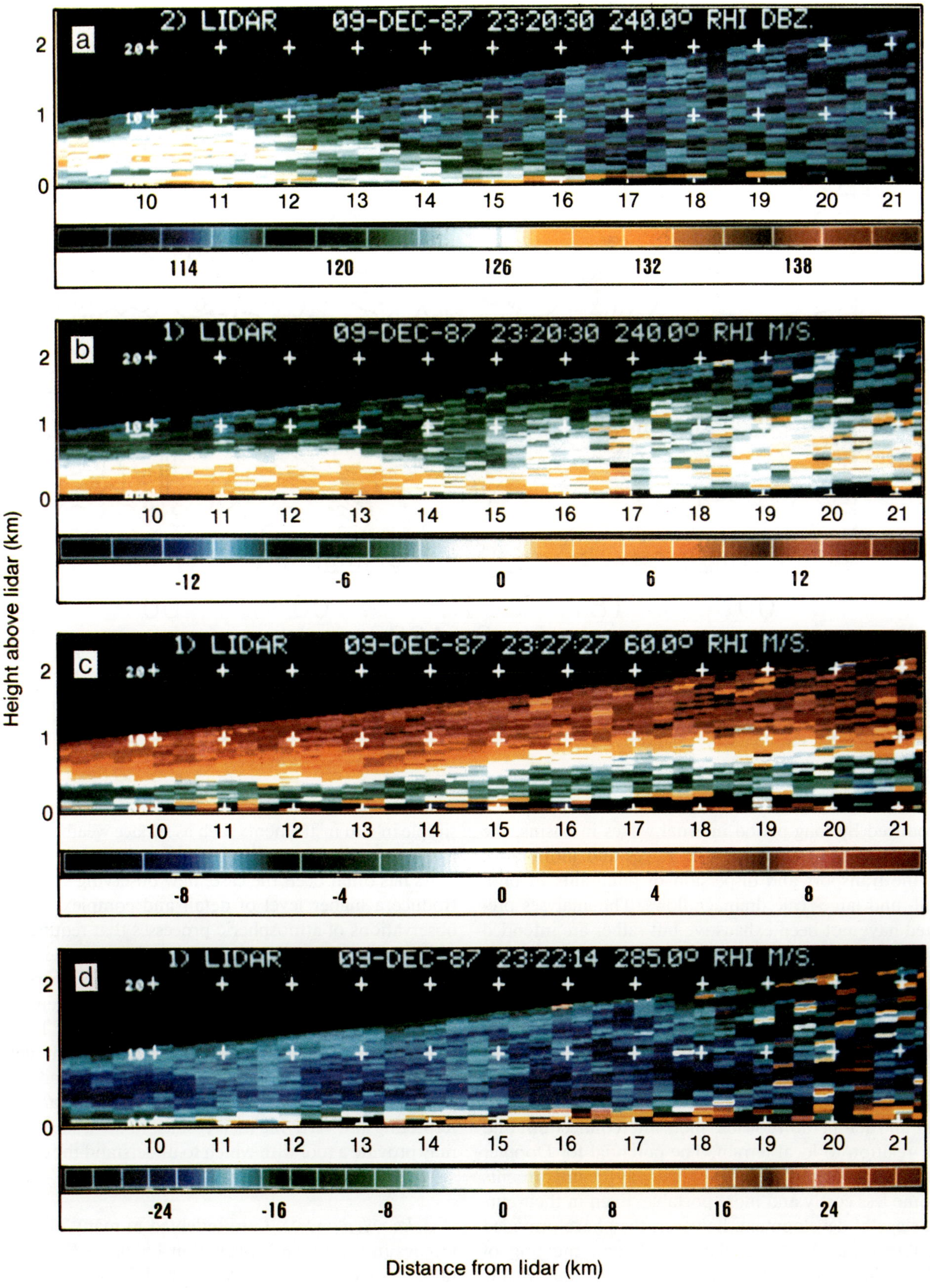

FIG. 8.50. Four stepped-RHI lidar views obtained from the scan sequence shown in Fig. 8.49. (a),(b) Radial winds and backscatter observed over Denver (16 km) displayed on a 1-km grid. Yellow to red color coding represents flow away from the lidar and green and blue represent flow toward the lidar. Backscatter coding uses colors representing larger decibel values to indicate stronger aerosol return. (c) Doppler radial winds observed in the northeast, showing a strong return (up-valley) flow toward the lidar at low levels and a strong component away at higher levels. The interface appears to have a long wavelength oscillation, perhaps due to local erosion by winds aloft or displacement by internal wave motions. (d) View into a high-wind region closer to the foothills, showing wind components reaching 30 m s^{-1} from the west with a spatial periodicity of about 4 km. (Eberhard, personal communication.)

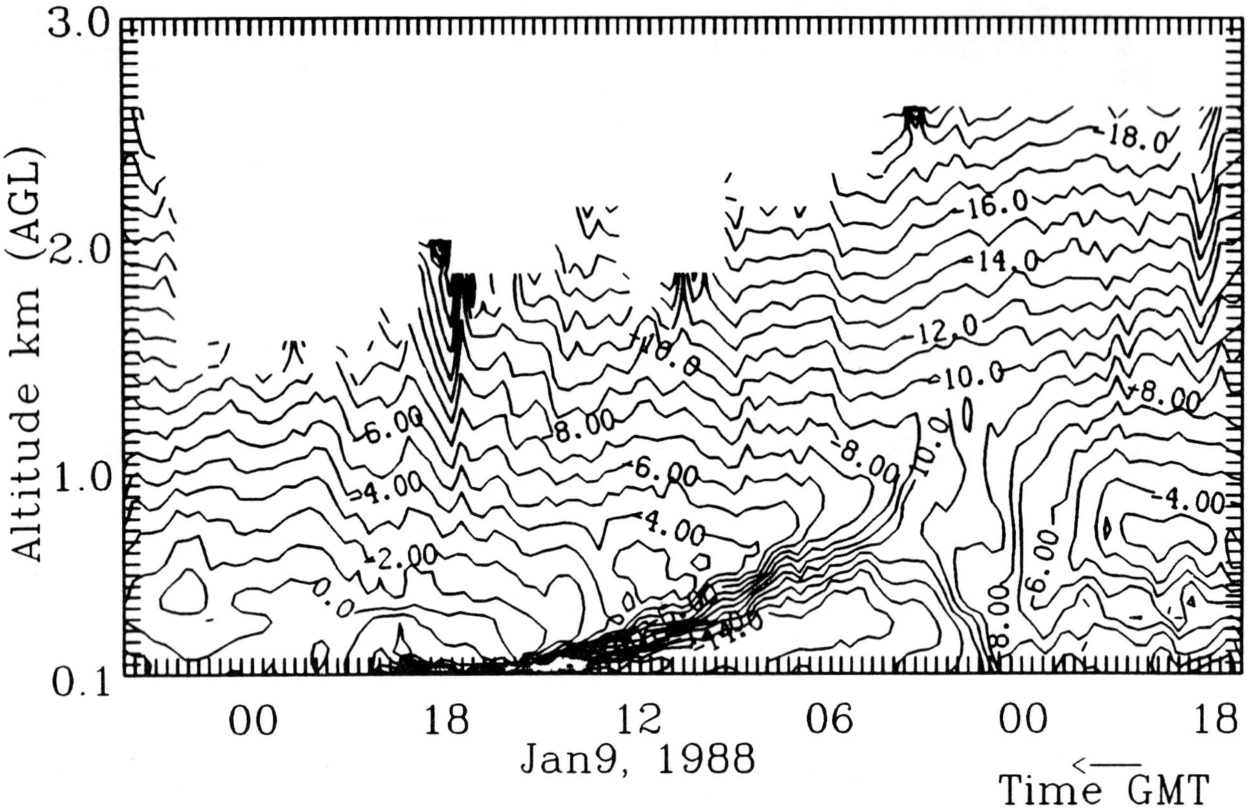

FIG. 8.51. Combined RASS/tower temperature profiles obtained during the Denver Brown Cloud Study. Strong stability is present prior to 0000 UTC; a microfront (less than 1 km in depth) passes shortly after 0000 UTC, followed by erosion of warmer air into the cold-air pool with the onset of higher winds after 0300 UTC. (May et al. 1989b.)

introduced by long-period internal waves in basins, the waves produced by airflow over ridges and mountains, and the transport and dispersion of pollutants in both small- and large-scale drainage flows. The analyses presented have not been exhaustive but rather are intended to stimulate further research by demonstrating the broad range of phenomena observable with remote sensors.

As scales of interest expand into the meso-β the application of remote sensors to their study will become more and more important. In particular, with the advent of high-resolution wind and temperature profiling, much more will be learned about the mesoscale variability of wind and temperature fields, their interconnection, and their relation to local terrain. The potential for Doppler lidar applications will improve greatly as the instruments become less costly and more portable. Even in their current stage of development, however, they have provided valuable information on the growth and merging of drainage flows with a quantitative precision unavailable from conventional measurements. Many studies can also benefit from the deployment of relatively inexpensive remote sensors such as sodars, optical crosswind sensors, and UHF radar/RASS profilers in combination with simple in-situ instruments such as surface weather stations and tethered balloon profiling systems.

As has often been the case, new observing systems introduce a higher level of detail and complexity in our observations of atmospheric processes that require more sophisticated analysis methods to assimilate and interpret their data. Remote sensors are no exception. In particular, interpretation of the microstructure of turbulence and aerosol distributions found in the atmosphere has presented a significant challenge. As numerical models of small-scale processes advance over the next decade, however, remote sensors may provide the only means of interpreting and verifying their results. In turn, such models may provide a tool with which to understand the complex structure observed using remote sensors.

Acknowledgments. I am indebted to many of my colleagues in the Wave Propagation Laboratory, including in particular W.L. Eberhard and M.J. Post, for their willingness to share their data, ideas, and suggestions throughout the preparation of this chapter. The many comments and detailed reading of the manuscript by C.D. Whiteman are also greatly appreciated.

CHAPTER 9

The Role of Mountain Flows in Making Clouds

ROBERT M. BANTA

Wave Propagation Laboratory, NOAA/ERL, Boulder, Colorado

ABSTRACT

Mountains disrupt basic airflows to force ascending air currents and initiate clouds. Clouds are produced when air becomes saturated and water condenses. Air can be brought to saturation by evaporation, by cooling, or by expansion from adiabatic lifting—adiabatic lifting being the most important process in the atmosphere. Mountains produce rising motion and adiabatic ascent in two ways, forced ascent and convergence caused by the heating of elevated topography.

Two factors must be considered in assessing the ability of a mountain range to produce clouds under a given flow regime: 1) The properties of the atmosphere, and 2) the characteristics of the flow disturbance; i.e., the lifting created by the mountain. The properties of the atmosphere include two important vertical levels; the lifting condensation level (LCL), and the level of free convection (LFC). The LCL is where air first reaches saturation by lifting, and the LFC is where saturated air becomes positively buoyant. The properties of the atmosphere thus determine the amount of lifting needed to produce clouds, and the flow characteristics determine the amount of lifting provided. If air is lifted to its LCL but not to its LFC, *stable* or stratiform clouds will be produced. If air is lifted to its LFC, *unstable* or cumuliform clouds will be produced. It is important to remember, however, that the two factors are not independent because characteristics of the atmosphere (especially its stability) affect how much lifting the flow will be able to produce—and lifting destabilizes the atmosphere.

Stable clouds are formed by forced lifting as flow ascends a mountain barrier. A conceptually simple case is considered that seems to apply when moisture is confined to the lowest 1 or 2 km above the terrain and the flow is nearly parallel to the topography. The main issues in this case are physical rather than dynamic: the ability of these dynamically simple clouds to produce precipitation on the ground is governed by microphysical factors and may be influenced by radiation. In general, the dynamics of stable orographic flow are more complex, however, and mountain wave and blocking effects provide regions of enhanced rising motions that determine the occurrence and distribution of precipitation.

The major focus of this chapter is on unstable mountain clouds (especially mountain thunderstorms) that result from a mixture of large- and small-scale effects. Whereas larger-scale processes control whether or not thunderstorms will be able to form, the role of the mountain (i.e., smaller-scale) circulations is in the initiation or triggering of thunderstorms. Thus mountains have an important role in the temporal and spatial distribution of storms. Three major kinds of lifting that lead to thunderstorm initiation are direct orographic forced lifting to the LFC, convergence of thermally forced circulations, and aerodynamic or obstacle effects.

Warm season convection in relatively arid mountainous regions such as the western United States is generally triggered by convergence and updrafts produced by thermally forced, daytime upslope circulations. Temporally, such clouds and thunderstorms form first over the mountains, later moving downwind onto or reforming on the plains. It is conceptually useful to define two types of thunderstorm, primary and secondary. During its initiation, secondary convection contains a feature generated by a preexisting thunderstorm, such as a gust front or a gravity wave, whereas primary convection does not. Thus primary convection may be forced, for example, by orographically or thermally forced mountain flows. Spatially, thunderstorms develop in preferred areas of the topography, depending on external factors such as ridgetop winds and moisture distribution.

Forecasters traditionally recognize three ingredients necessary for thunderstorms. Applied to mountain storms of the arid western United States these are 1) a conditionally unstable temperature lapse rate, which is generally the case in the warm season, 2) an initiation or triggering mechanism, which is provided daily by mountain circulations, and 3) adequate moisture, which is generally advected in by the larger-scale flow, although local sources can be important.

Two other considerations that are important in mountain thunderstorm initiation are the roles of larger-scale processes and surface moisture. Larger-scale processes can be prohibitive, permissive, or actively forcing in their effect on thunderstorms. Soil moisture reduces surface heat flux by partitioning increased amounts of incoming energy during the day into latent heat. The reduction in heat flux reduces the strength of thermally forced circulations and retards the rate of growth of the daytime boundary layer.

9.1. Introduction

Mountains have a strong effect on atmospheric flow, especially on the vertical component of the motion of air. On the average (and at large scales) the atmosphere is stable, meaning that vertical displacements of air increase the potential energy of the atmosphere, and the atmosphere resists vertical air movement. Smith (1979) points out from this larger-scale perspective that the atmosphere is stable, implying laminar and horizontal flow, and the sources of moisture are at the surface. Mountains disrupt this stratification and force vertical displacement of air trajectories.

We expect such regions where flow processes disturb the overall stratification of the atmosphere to be regions in which moist air ascends and clouds can form. In general, determining where clouds will form in the mountains is based on where the mountains will force upward motion, given sufficient moisture in the lifted air.

The purpose of this chapter is to discuss the role of mountains in generating clouds. There are actually two major roles that mountains have in forming clouds, first as obstacles to the flow and second as "high-level heat sources" (Braham and Draginis 1960). In the first role, the approach flow (usually stable) is forced to ascend the obstacle or barrier. In the second role, daytime heating of the surface causes convergence of flow toward the mountains and rising motion over the mountains. In both cases clouds form in the rising air, again provided enough moisture is present.

As in the other chapters in this volume, the emphasis is on processes that occur on the scale of the mountain or mountain range itself; i.e., smaller mesoscale processes with length scales up to, but usually less than, a few hundred kilometers.

Detailed reviews concerning stable orographic precipitation and unstable wintertime orographic precipitation have been published (see Smith 1979; Cotton and Anthes 1989). Therefore these topics are not strongly emphasized. In section 9.2 some basic background principles of condensation and stability are described. Section 9.3 contains a discussion of several issues in those reviews relating to the role of mountain flows as opposed to other physical processes (such as atmospheric radiation or cloud microphysics) in producing clouds. Section 9.4 is the focus of this chapter; it is a review of the role of mountains in generating summertime convective clouds, including thunderstorms, and of the relative roles of mountain circulations and the large-scale environment. Section 9.5 contains a summary and conclusions.

9.2. Condensation and stability effects

9.2.1. Condensation

Condensation occurs after air becomes saturated. The thermodynamic relationship between the vapor pressure e and temperature T of air at saturation is expressed by the Clausius–Clapeyron equation:

$$\frac{de_s}{dT} = \frac{Le_s}{R_v T_s^2}, \tag{9.1}$$

where e_s is the saturation vapor pressure, T_s the temperature at saturation, L the latent heat of condensation, and R_v the ideal gas constant for water vapor. This equation can be integrated using a mean value of L/R_v to give

$$\ln\left[\frac{e_s}{0.611}\right] \approx \frac{L}{R_v}\left[\frac{1}{273} - \frac{1}{T_s}\right]. \tag{9.2}$$

This leads to the solid curve on Fig. 9.1, which is the familiar temperature, vapor pressure (T, e) phase diagram for water. Basic atmospheric science texts such as *Introduction to Theoretical Meteorology,* Hess (1959); *Atmospheric Thermodynamics,* Iribarne and Godson (1973); and *The Ceaseless Wind,* Dutton (1976) contain further details on the interpretation of these diagrams.

Since T and e are not conserved in general for atmospheric processes, it is convenient to use potential temperature θ and specific humidity q, which is nearly interchangeable with mixing ratio. These variables are conserved during adiabatic dry ascent and descent, and thus can be used to trace the properties of an air parcel until it becomes saturated. In terms of θ and q, Eq. (9.1) can be written:

$$\left[\frac{dq_s}{d\theta}\right]_p = \left[\frac{Lq_s}{R_v\theta_s^2}\right]\left(\frac{1000}{p}\right)^{R/c_p}, \tag{9.3}$$

where q_s is the value of q at saturation, the subscript p denotes a constant pressure process, R is the ideal gas

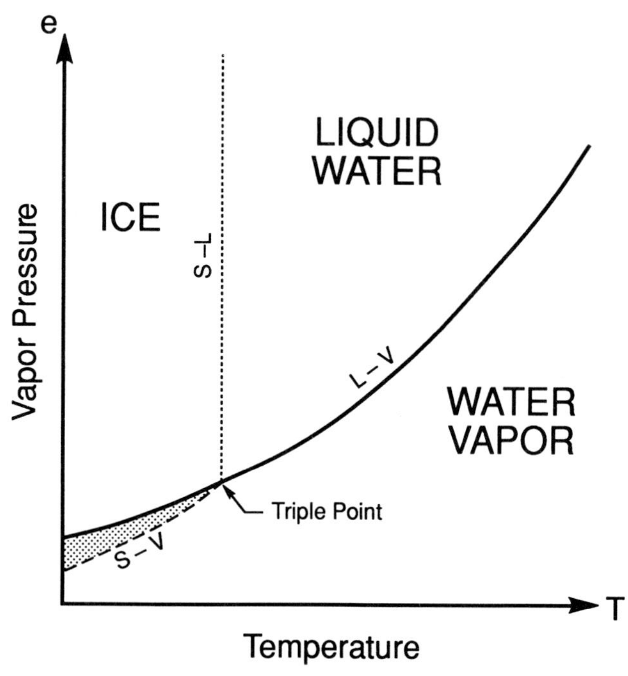

FIG. 9.1. Temperature, vapor pressure phase diagram for water. The solid curve passing through the triple point is the liquid-vapor saturation curve for water represented by Eq. (2). The dashed line is a similar curve for ice-vapor sublimation/deposition, and the dotted line represents ice-liquid melting/freezing. The shaded area represents an important area where supercooled water can exist; i.e., water can exist (at nonequilibrium) in all three phases simultaneously. In this region ice is supersaturated with respect to the vapor, and liquid water is subsaturated.

constant for dry air, and c_p is the specific heat at constant pressure for air. The integrated expression can be obtained by substituting expressions for q_s and θ into (9.2) and taking the exponential of both sides:

$$q_s \approx .611\left(\frac{\epsilon}{p}\right)\exp\left\{\left(\frac{L}{R_v}\right)\left[\frac{1}{273}-\left(\frac{1000}{\theta p}\right)^{R/c_p}\right]\right\}, \quad (9.4)$$

where $\epsilon = R_v/R$. Thus q_s is a function of θ and p, and one can graph the behavior of q_s as a function of θ for a given value of p (Fig. 9.2a). One can also draw saturation "isobars" on a (θ, q) plot (Fig. 9.2b).

Both figures can be used to illustrate atmospheric processes. For example, in Fig. 9.2a process A shows an increase in q at constant θ, representing evaporation (at constant pressure) of enough water to bring the air to saturation. Process B represents cooling by nonadiabatic processes (such as radiation or conduction from the ground) until the air is saturated (after saturation the process moves along the saturation line). Figure 2b illustrates cooling by adiabatic expansion. Initially, at a pressure of 100 kPa, air with values of θ and q indicated on the chart

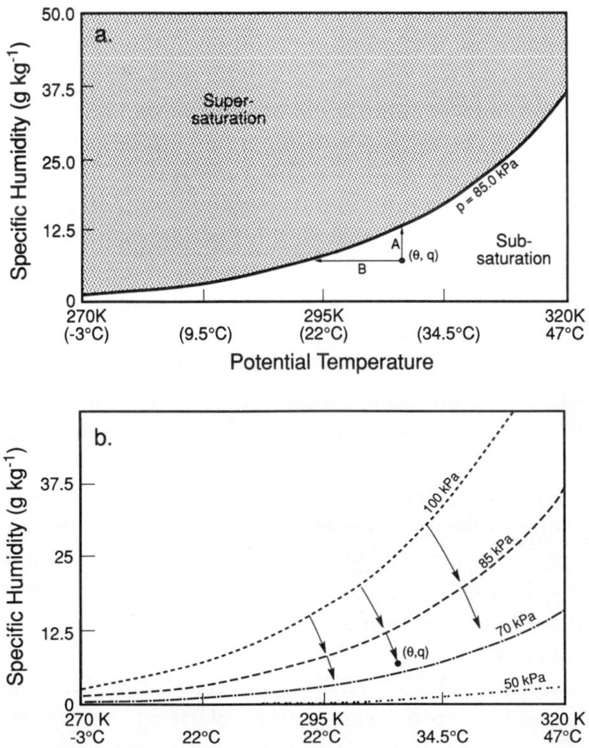

FIG. 9.2. (a) Potential-temperature, specific-humidity phase diagram for water vapor and liquid at 85 kPa pressure. Here (θ, q) represents properties of an arbitrary "parcel" of unsaturated air; A and B represent processes discussed in the text. (b) Saturation isobars on a (θ, q) diagram. Arrows represent adiabatic lifting of a "parcel" starting out at 100 kPa and becoming saturated at about 72 kPa. Parcel properties are conserved for this process, but as lifting proceeds the pertinent saturation curve moves toward the point representing the parcel.

is unsaturated. As the pressure of the air decreases and the air cools by expansion, its saturation isobar moves downward on the (θ, q) chart, eventually intersecting the parcel values of θ and q, which are conserved for this process, at about 72 kPa. At this point the air is saturated. Further expansion (cooling) causes water to condense, releasing latent heat so that θ and q are no longer conserved. Betts (1973, 1982) discussed variables that are conserved during the vapor–liquid phase changes.

Thus, three processes that can produce saturation in the atmosphere are 1) the addition of moisture, which increases q, 2) nonadiabatic cooling, which decreases θ (and q_s), and 3) adiabatic expansion and cooling, which decreases q_s. The last process, adiabatic expansion, includes the important process of adiabatic lifting (ascent). Each of these processes can be applied to mountain clouds. For example, the addition of moisture causes post-precipitation fractostratus (scud) or stratocumulus clouds (not discussed in this review). Radiational (nonadiabatic) cooling produces valley fog and stratus, which are discussed briefly in section 9.3. Adiabatic ascent is the most important mechanism for producing clouds in the atmosphere and thus will be the main process discussed in this chapter.

Can clouds be dissipated by the reverse processes? The first process is not reversible. There is no normal atmospheric mechanism for removing enough water mass from a saturated region of the atmosphere to leave the air unsaturated. The atmosphere can, however, be heated diabatically to dissipate clouds, as often happens when morning fog or stratus burns off, so the second process is reversible. The third process, adiabatic ascent, is also reversible, since adiabatic descent (subsidence) dissipates clouds.

9.2.2. *Lifting profiles*

In examining the role of mountains in producing clouds, one needs to consider 1) the properties of the atmosphere flowing over the mountain, and 2) the characteristics of the flow disturbances created by the mountain. It is important to realize at the outset that these two considerations are not independent. The stability of the inflow profile affects the maximum lifting experienced by air trajectories in passing over a barrier: greater stability tends to produce less vertical displacement. On the other hand, the lifting of a stable layer destabilizes the layer. As a first step in understanding these processes, however, it is convenient to discuss the two considerations as though they operated independently.

The first consideration, properties of the atmosphere, can be represented by an inflow sounding; i.e., profiles of temperature, moisture, and winds approaching the mountains. According to parcel theory, two important

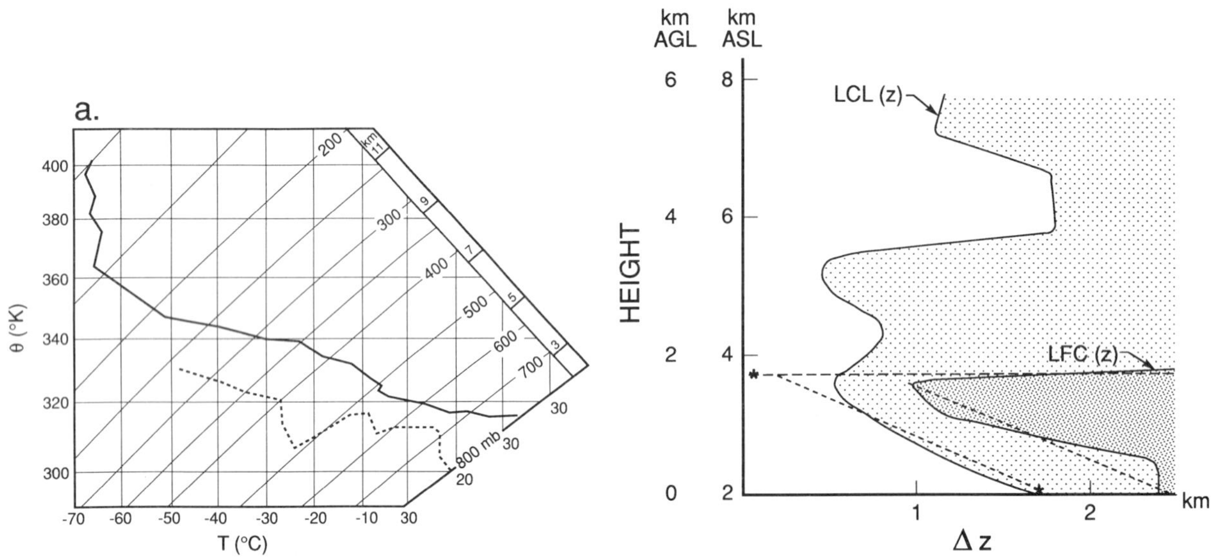

FIG. 9.3. (a) Mountain thunderstorm proximity sounding from Brown (1966), taken at 1201 MST, 1 August 1961, near Flagstaff, Arizona. The solid line is the temperature profile, and the dashed line, the dewpoint. (b) LCL(z) and LFC(z) functions calculated from the sounding in (a), as discussed in the text. Asterisks represent the mountaintop elevation on both the height and the Δz axes.

threshold values are determined by the thermodynamic properties of this inflow sounding: the lifting condensation level (LCL) and the level of free convection (LFC). The LCL is the level to which air at a given height must be lifted to become saturated. The LFC is the level to which air at a given height must be lifted, once saturated, to become positively buoyant. This implies that the air has entered a conditionally unstable layer of the atmosphere; i.e., a layer in which $d\theta / dz < \gamma_m$, where γ_m is the moist adiabatic lapse rate (e.g., see Hess 1959; Iribarne and Godson 1973; Dutton 1976).

The LCL and LFC each can be regarded as a function of height z by taking the inflow sounding properties at each level and calculating the LCL and LFC for that level. For example, the temperature and moisture properties for a sounding at the surface can be used to determine the LCL and LFC for a surface parcel, and similarly the sounding properties at 85, 70, and 50 kPa yield LCL and LFC values for parcels displaced upward from those levels. The same procedure produces LCL and LFC values for the other levels of a sounding. One then has profiles of these quantities, LCL(z) and LFC(z), where z represents the starting level of a lifted parcel. If clouds form, cloud base is generally the lowest value of LCL(z). In Fig. 9.3 an example of the functions LCL(z) and LFC(z) is shown, expressed as the difference Δz between the LCL or LFC and the starting level of the parcel obtained from a sounding.

In cases of forced lifting, where surface heating is not a significant factor, the appropriate cloud-base indicator is the LCL (as just defined). Where surface heating is important and a convective, well-mixed boundary layer forms, however, it is often more appropriate to use the convective condensation level (CCL) as the indicator of cloud base. Assuming that a well-mixed boundary layer will become the subcloud layer, one finds the CCL by estimating the mean water-vapor mixing ratio that the well-mixed layer will have. Often this is done by finding the mean mixing ratio in the lowest perhaps 500 or 1000 m (5 or 10 kPa) of the morning radiosonde moisture profile. That mean mixing ratio is equal to the saturation mixing ratio at some level on the temperature profile, and this level is the CCL. In practice the CCL is normally determined using a thermodynamic diagram: the CCL is where the mean mixing ratio line intersects the temperature curve (Saucier 1955; Iribarne and Godson 1973). The assumption of a well-mixed subcloud layer means that the "LCL" (i.e., CCL) and the LFC functions are constant in the subcloud layer. The short-dashed lines in Fig. 9.3 indicate how the well-mixed assumption would affect the LCL(z) and LFC(z) difference curves. Note that cumulus clouds appear only if there is sufficient surface heating for the well-mixed boundary layer to grow upward to the CCL. This implies that the surface temperature reaches a certain value, referred to as the convective temperature, during the course of the day.

The second consideration involves the degree to which inflow air is lifted by the flow disturbance produced by the mountain. This orographic lifting function $L(z)$ expresses the amount of lifting that inflow air at each level experiences as a result of the flow disturbance. For example, Fig. 9.4 shows how $L(z)$ is obtained for potential

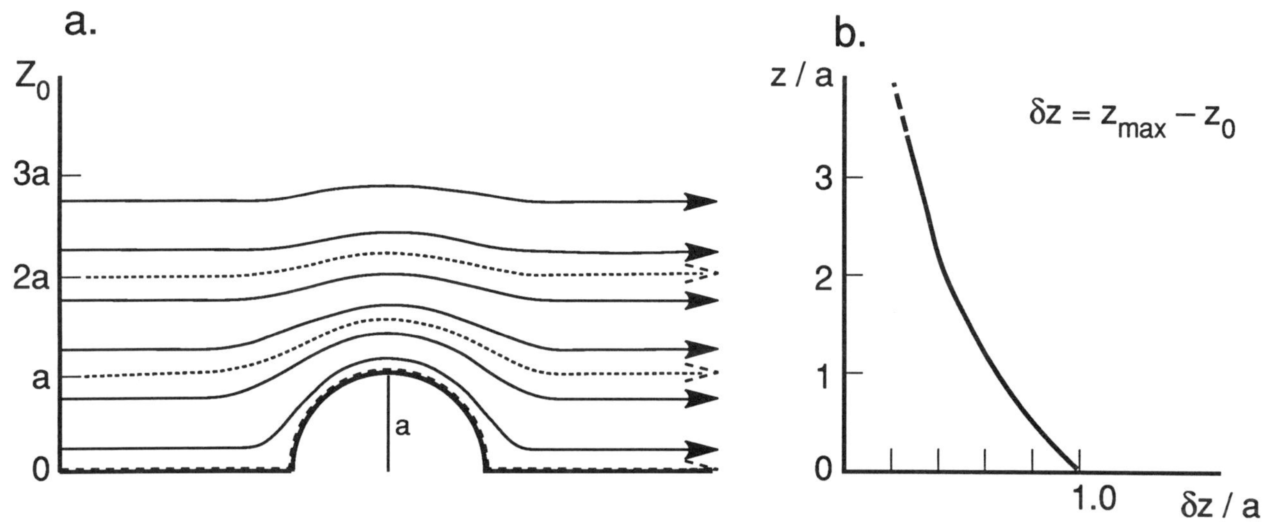

FIG. 9.4. Lifting function for potential flow over a semicylinder. (a) Streamlines and (b) normalized lifting curve, where δz represents the maximum lifting experienced by a parcel starting at a level z_o; i.e., $\delta z = z_{\max} - z_o$, where $z_{\max}$ is the maximum height attained by the parcel. Variables are normalized by the radius a of the semicylinder, which also represents maximum height of the obstacle.

flow over a semicylinder. Potential flow implies neutral stability, so in this case the flow is unaffected by stability.

From this discussion the functions LCL and LFC based on the inflow profiles thus represent the amount of lifting *needed* to produce clouds or moist instability, whereas the lifting function L describes the amount of lifting *realized* by the flow over the mountain. The unstable portion of this approach follows a procedure developed by Foote (1984), and used by Wilson and Mueller (1987), to predict whether a gust front will provide enough lift to trigger a thunderstorm.

9.2.3. Stability effects on cloud forms

A crude test helps determine what kind of cloud is apt to be produced by a given combination of the inflow profile and flow configuration: 1) Can the flow pattern lift air to condensation, or can air at any level be lifted to its LCL? If so a cloud will form; if not there will be no cloud. 2) Can the flow pattern lift air at any level to its LFC? If so unstable cumuliform clouds can form; if not the clouds will be stable or stratiform.

These two questions can be answered by superimposing the $L(z)$ curves derived from the flow model over the LCL(z) and LFC(z) curves derived from the inflow sounding, as shown in Fig. 9.5. Air starting at a level (as indicated on the height axis) where the lifting curve $L(z)$ stays in the white will form no clouds. Air at levels where $L(z)$ is in the light-shaded area will produce clouds, since the flow produces enough lift to bring a parcel starting at that level to saturation. The clouds will be stratiform but not convective because air is not lifted to its LFC. Convective (cumulus) clouds will be produced, however, by air starting at levels where $L(z)$ is in the dark-shaded area; the flow will produce enough lift to carry a parcel up to its LFC.

Figure 9.5 shows the situation for the given sounding and a lifting curve representing potential flow over a

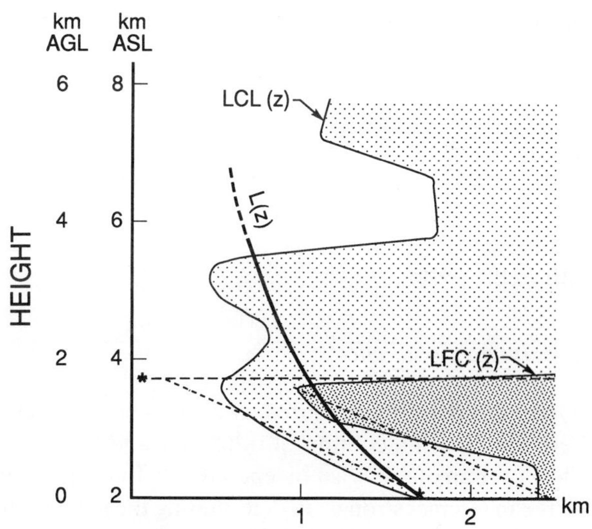

FIG. 9.5. $L(z)$ curve for potential flow over a semicylinder, superimposed over the LCL(z) and LFC(z) curves from Fig. 9.3. Mountain height was chosen from Brown's (1966) case study. The figure shows that if this lifting mechanism were operating, air from near the surface at the base of the mountain up to over 5 km would be lifted to saturation, and air in a layer just below mountaintop would reach its LFC. Dotted lines represent sounding values of LCL(z) and LFC(z) below cloud base if the sounding were well mixed in this layer and cloud base were taken to be the cloud condensation level (CCL), instead of being determined by the LCL at each level. The figure is for illustration only, since potential flow over a cylinder is probably not a good representation of the actual flow in Brown's (1966) case.

semicylinder. The example is contrived, since the sounding was taken on a thunderstorm day (unstable), whereas the flow is for neutral stratification, but it illustrates the principle. A parcel starting out upstream at the surface would be lifted to saturation at the top of the mountain (the asterisk along the Δz axis), forming a stratiform cap cloud. Air starting just below mountaintop level (the asterisk on the height axis and the horizontal dashed line at that level) will be lifted to its LFC and therefore will produce convective clouds.

Since the sounding was for a sunny day during which a well-developed mixed layer formed, the short-dashed lines represent the situation where the CCL replaces the LCL below cloud base. LCL(z) (equals CCL) and LFC(z) are constant with height in the subcloud layer, meaning that Δz for each decreases linearly with height in that layer, the LCL(z) curve going to zero at cloud base. As before, a parcel starting at the surface upstream does become saturated at mountaintop, indicating that a capping stratus will form; air starting just below mountaintop level will be lifted to its LFC, indicating that it will produce convective clouds. The major difference between this example and the real case is, of course, that the lifting curve would be different for a mountain heated by the sun.

Some general rules exist on what kind of cloud to expect for certain configurations of the inflow profile. If $d\bar{\theta}/dz > \gamma_m$ everywhere on the inflow sounding, the sounding is absolutely stable and only stable clouds can occur, neglecting effects of lifting on stability. (The overbar represents an undisturbed inflow value well upstream of the barrier.) In this case LFC(z) is (infinitely) large, and therefore $L(z) <$ LFC(z) for each z on the sounding. If $d\bar{\theta}/dz < \gamma_m$ in some interval of the inflow sounding, unstable clouds can occur in that interval, if $L(z) >$ LFC(z) for some z. The depth of cloud is often related to the depth of the unstable layer.

9.2.4. *Interactional effects*

Up to this point the idealized case has been considered where stability of a column of inflow air and lifting due to flow over an obstacle act independently. In reality these effects can interact strongly in determining the final nature of the flow over the obstacle. It is beyond the scope of this review to describe systematically all the possible ways that flow and stability can interact. Studies presented later in this chapter illustrate only a few. In the following section some general principles are considered that guide or give insight into how the interactions occur.

One can appreciate the possible effects of stability on the flow property $L(z)$ by considering a trajectory of a surface parcel under strongly stable conditions or under unstable conditions. Strongly stable flow inhibits the ability of air to ascend a barrier, to the extent that a surface parcel may not make it over the top (see Chapter 5 and its Appendix). When such blocking occurs, the maximum height reached by a surface parcel is less than the height of the barrier a, or $L(0) < a$. On the other hand, under unstable conditions buoyant updrafts often carry air originating near the surface high above the ridgetop, or $L(0) > a$.

Unstable cumulus clouds grow when lifting "releases" moist instability that is latent in a column of the atmosphere, as determined by its vertical profiles of temperature and moisture. In discussing how lifting by flow over an obstacle can affect the stability of the lifted column of atmosphere, it is important to clarify the ways the column can be unstable.

There are two types of inherent moist instability. Terminology has been ambiguous, since both types have frequently been called potential instability (in addition to other terms). The two types of moist instablility are referred to in this chapter as potential instability and latent instability, after Iribarne and Godson (1973):

1) *Potential instability.* When equivalent potential temperature θ_e decreases with height (i.e., $d\theta_e/dz < 0$) in a layer, forced lifting to saturation makes the layer unstable and produces cumulus convection (in this definition, the wet-bulb potential temperature θ_w may be substituted for θ_e). Also referred to as convective instability, it often occurs when stratus or orographic stratus is forced up the slope of a mountain range, producing cumuliform clouds embedded in the stratus deck. Because θ_e variations are generally dominated by moisture variations, a moist layer near the surface with dry air aloft is generally sufficient to make θ_e decrease with height, even when the surface is somewhat cooler (in terms of θ) than air 2–3 km above the ground. Lifting does not change the sign of $d\theta_e/dz$, so the potential for this type of instability does not change as the layer ascends adiabatically.

2) *Latent instability.* According to parcel lifting theory, a sounding is from an unstable column of the atmosphere if it has an LFC; i.e., if a parcel can be lifted to free convection. Processes such as surface heating or surface convergence "release" this instability if they lift a parcel to its LFC, as often happens with daytime cumulus clouds produced by surface heating. Latent instability is affected by forced lifting of a layer of atmosphere. For unsaturated air that remains unsaturated, lifting a layer makes the lapse rate closer to dry adiabatic; i.e., for a stable layer ($d\theta/dz > 0$), *lifting destabilizes the layer* with respect to vertical parcel displacements. A layer that starts out absolutely stable can be lifted (e.g., orographically) to become conditionally unstable. Ascent occurring over a broad area can destabilize a stable air mass sufficiently to produce a potential for cumulus convection. Another way of viewing the lifting–destabilization process is to consider the effect on the LFC. For a column of air that has a finite

LFC, this destabilization lowers the LFC and thus reduces the amount of lifting (Δz) required to bring a parcel to its LFC. For a column that is too stable to have an LFC (all LFCs are infinite), destabilization can produce layers with finite LFCs.

Thus a major difference between potential and latent instability is that lifting does not affect whether a layer is potentially stable or unstable with respect to vertical displacements of the layer, but lifting can destabilize a column of the atmosphere with respect to vertical parcel displacements in such a way that a stable column becomes unstable.

The two types of moist instability are not mutually exclusive; before saturation occurs a sounding can exhibit both types. In fact soundings with latent instability generally have $d\theta_e/dz < 0$ at least in the lowest kilometer or so; therefore they also have potential instability. The kind of instability realized by the lifting thus depends on how the process unfolds.

For example, if a layer with $d\theta_e/dz < 0$ is sufficiently moist and exposed to sufficient forced lifting that a portion of the layer at least a few hundred meters deep becomes saturated, then potential instability can be released. The dimensions of eddies in a turbulent layer are generally proportional to the depth of the layer; thus, it is necessary to develop a deep enough layer to produce eddies of sufficient horizontal dimension to grow into cumulus clouds. Although the minimum depth required to produce such eddies is unknown, something in excess of 100 m seems appropriate. According to this line of reasoning, therefore, a precursor to the release of potential instability is a deck of stratiform clouds produced by forced ascent.

On the other hand, if a deep enough saturated layer is not formed by forced lifting, then other mountain flow mechanisms may be able to lift a parcel at some level to its LFC, releasing latent instability.

To put the lifting diagram of the previous section into perspective, it is necessary to consider the effects of lifting on LCL(z) and LFC(z). LCL(z) is unaffected by lifting, since lifting does not change θ_e of a parcel; however, because lifting cools, levels above the LCL are cooled, and the effect is to lower the LFC for that layer. The largest values of LFC(z) for each z are in the upstream profile, before any lifting occurs. The upstream values of LFC(z) at each level, which are the ones on the lifting diagrams in the previous section, thus represent an upper bound to values actually encountered over the obstacle.

In addition to the interactions between flow and instability, the simple picture of cloud production by mountains presented in section 9.4.1 has a number of other complications: 1) Latent heating and turbulence enhance lifting above theoretical values for the airflow being considered, as often happens in cumulus clouds. 2) Radiation at cloud top destabilizes the cloud; the longer-lived the cloud, the greater the effect. 3) Horizontal inhomogeneities in the flow field or differential advection due to shear in the inflow winds can change the stability. 4) The lifting aloft caused by the upstream tilting with height of a mountain wave can stretch the column of air upwind of the mountain. This destabilizes the column by cooling it aloft (Smith 1979). 5) Moisture evaporation at the surface and other nonadiabatic effects also change the stability of the inflow.

One can construct $L(z)$ curves that include some of these effects as well as the effects of stability and lifting using numerical model results. For example, in Fig. 9.6 two θ cross sections are presented, representing steady-state solutions to two kinds of mountain wave simulated by Durran (1981); also Durran and Klemp (1982). At steady state, isentropes, streamlines, and trajectories all coincide. The $L(z)$ curves thus were determined from the maximum vertical displacement of the isentropes pictured in Fig. 9.6, since this also represents the maximum vertical displacement of the trajectories. When the appropriate physics are included in the model, numerical modeling

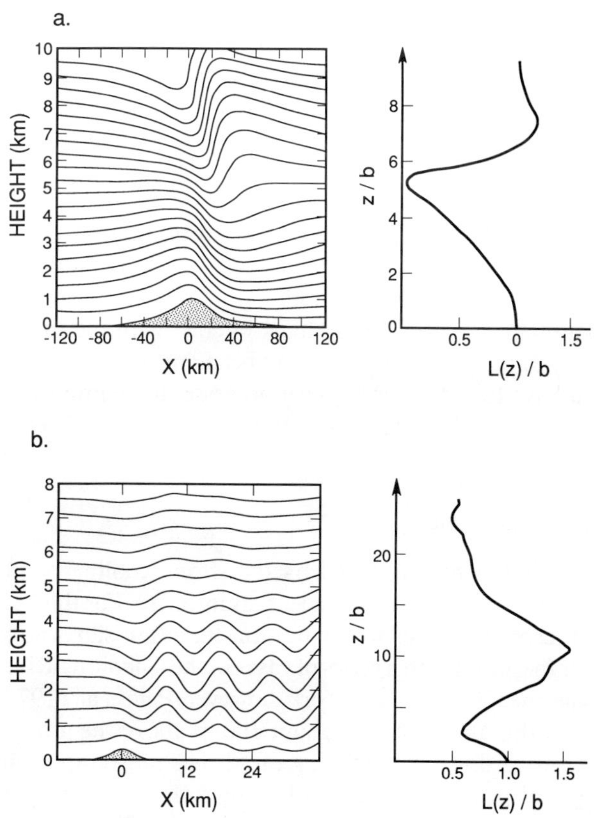

FIG. 9.6. Examples of lifting curves from numerical mountain-wave solutions from Durran (1981); b represents the height of the mountain. (a) $L(z)$ from a vertically propagating mountain wave over a 1-km mountain. (b) $L(z)$ from lee waves over a 300-m mountain.

can be used in this way to determine trajectories that take into account the complicating effects listed above.

Setting these complications aside for the moment, let us review the more simplified view of mountain-induced clouds discussed in this section by dividing these clouds into two categories. *Stable* clouds occur when air at some level is lifted to its LCL (saturation), but not to its LFC at any level. *Unstable* clouds occur when air at some level is lifted not only to saturation but also to its LFC. In the following discussion each classification is further divided into warm-season and cold-season clouds, which are referred to as *rain* or *snow* clouds, depending on the predominant type of precipitation expected at the ground from that kind of cloud. Thus, for example, "rain" clouds can produce some frozen precipitation (graupel, hail) at the ground, but the predominant type of precipitation expected is liquid. "Snow" clouds can produce ice pellets (sleet), snow pellets, or graupel, for example, in addition to many different crystal forms of snow. These divisions yield four general types of mountain-induced clouds, which are discussed in the following order: stable rain, stable snow, unstable snow, and unstable rain.

9.3. Fog, stable clouds, and unstable snow clouds

Diabatic cooling and adiabatic ascent are the two condensation processes considered in this section. Radiational fog in mountain valleys is briefly discussed as an example of diabatic cooling. The remainder of the section is devoted to three types of clouds produced by adiabatic ascent: stable rain, stable snow, and unstable snow.

Comprehensive reviews of these kinds of clouds were published by Smith (1979) and Cotton and Anthes (1989). Since the reviews described these clouds in detail, this section examines in more general terms how the mountain-induced flow systems interact with the topography to produce observed cloud behavior. Recent studies that have found other physical processes to be important (such as cloud microphysics and radiation) are also discussed.

9.3.1. Valley fog

Radiation fog is caused by the strong cooling of the ground surface and the adjacent atmosphere. For this reason, especially over flatter terrain, fog often starts at or near the ground. In a series of cases in the Chemung River Valley near Elmira, New York, however, Piliè et al. (1975) found that the fog forms about 100 m above the ground and grows downward to the ground. The most important physical process that causes the fog to form first aloft is the deposition of dew at the cold surface. This removal of vapor from the lower few tens of meters of the atmosphere produces a dewpoint inversion, in which the dewpoint increases with height through this layer. Such a dewpoint profile, along with the fact that just before fog formation the atmosphere becomes isothermal in the same layer, ensures that fog will form first aloft.

In the prefog stage, radiational cooling produces a temperature inversion to higher than 200 m above the valley floor. By midnight the dewpoint inversion forms as just described. The maximum rate of prefog cooling occurs at ~100 m above the valley, which is a little below the eventual fog top (about one-third below).

Within 2 h before sunrise the temperature profile over the valley becomes isothermal and fog forms aloft; 15 min later the temperature lapse rate in the saturated layer turns moist adiabatic and the fog begins to propagate downward to the surface. Piliè et al. (1975) hypothesize that destabilization by cloud-top radiation causes the moist adiabatic temperature profile and the downward propagation of the fog.

After sunrise the dew again plays a role in fog evolution. Surface heating occurs, but evaporation of the dew keeps the fog layer saturated, and often maintains the fog layer for 2–3 h. In some cases in the study by Piliè et al., fog did not even form until after sunrise, and then it formed because of dew evaporation. These cases exemplify the mechanism of saturation by the addition of moisture.

Thermally forced mountain circulations play a role in the development of valley fogs (Piliè et al. 1975). Ridgetop winds are always light when fog forms. These conditions are favorable for the occurrence of well-developed downslope and down-valley wind systems, as described in Whiteman, Chapter 2. The downslope wind produces upward motion over the center of the valley, which prevents strong cooling near the surface by distributing the cooling in the vertical. About 3 h before fog forms, the mountain breeze takes over, wiping out the upward motion over the valley center. This confines the cooling to the lower levels of the valley, where the fog eventually forms.

Occasionally stratus forms in a valley at night instead of fog. As with fog, stratus is dissipated by solar heating, often a few hours after sunrise. An interesting example of the interaction of the morning upslope circulation with a dissipating valley stratus deck was presented by Hindman (1973). Aerial photographs from a height of 3 km (Fig. 9.7) show a clear slot forming over the center of the valley, where the descending branch of the upslope circulation is producing subsidence. The surprising narrowness of the slot indicates that the downward branch is rather narrow, at least in this case where clouds are present.

9.3.2. Stable rain clouds

Observations show that hills—even small hills—receive more rainfall than adjacent low areas, both climatologically and for individual events in general. In this section we discuss reasons for this in warm, stable orographic

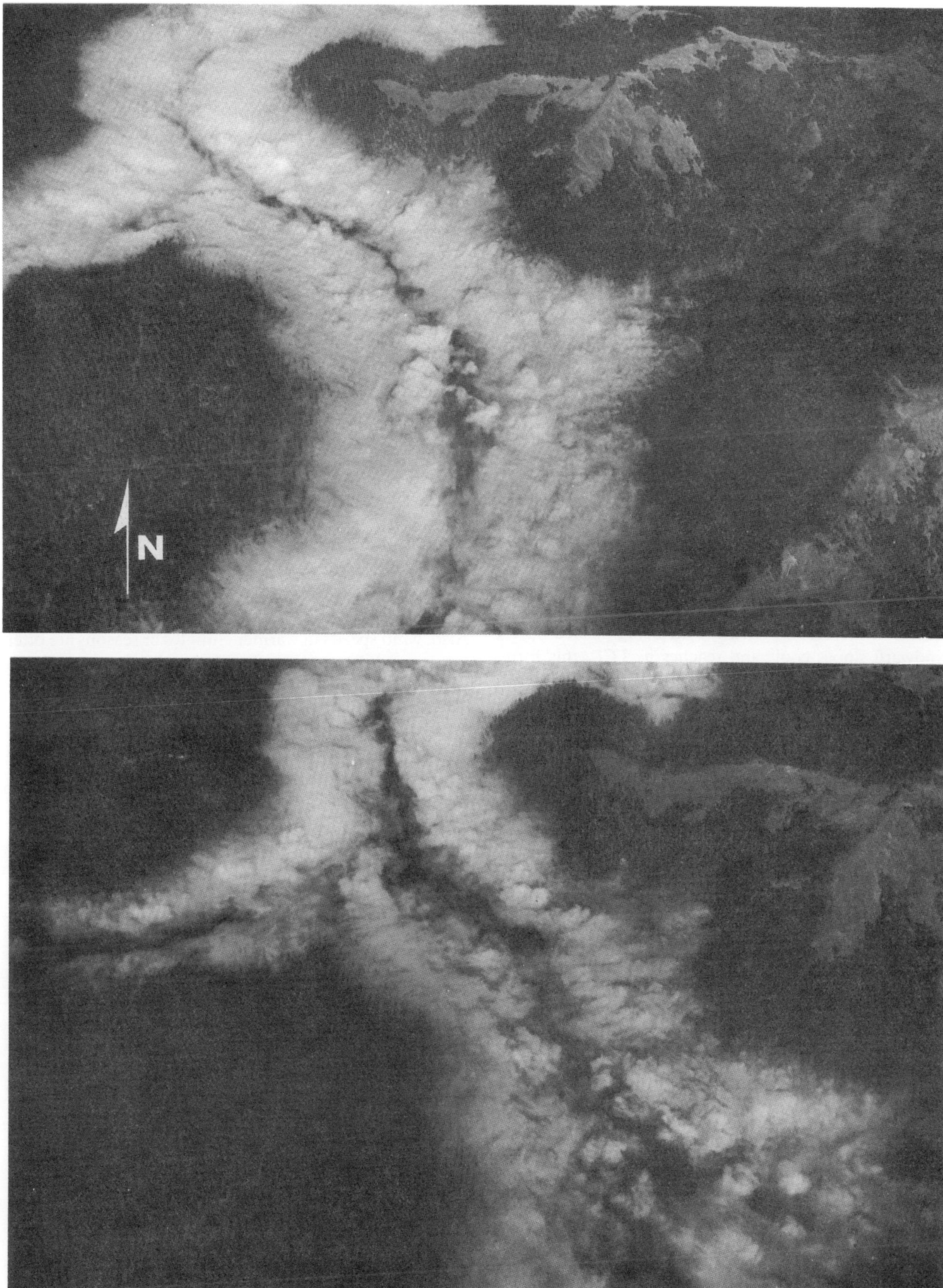

FIG. 9.7. Aerial photograph of breakup of a stratus deck during late morning hours in a valley in Northern California. Subsidence compensating for the morning upslope flow up the valley sidewalls appears over the center of the valley. Thus breakup occurs there first. Photographs were taken at an altitude of 3 km MSL, and 9 min apart: (a) at 0959 PST and (b) at 1008 PST (Hindman 1973).

clouds, first considering only the effects of moist flow over a mountain barrier, and then considering interactions with larger-scale weather systems. One type of stable cloud that belongs in this category but will not be discussed in detail here is the mountain-wave cloud, which is produced by the flows discussed in Durran, Chapter 4.

9.3.2.1. OROGRAPHIC STRATUS

Stable orographic clouds form when moist, stable air flows over an obstacle and reaches its LCL. Stable flow over an obstacle generally involves complex processes including mountain waves (discussed in Durran, Chapter 4); blocking (discussed in Carruthers and Hunt, Chapter 5 and in the next section); and other hydrodynamic effects (such as those described in Carruthers and Hunt, Chapter 5, and Hanna and Strimaitis, Chapter 6). Thus orographic stratus clouds usually form in complicated airflows—flows in which the streamlines are not parallel to the topography. It is convenient, however, to consider first the dynamically simple case and then the more complex cases.

9.3.2.1.1. Simple orographic flow. Reduced to its most basic conceptual level, warm orographic stratus results from simple airflow over a mountain ridge—up one side and down the other. When the air being pushed up the windward slopes cools and reaches its LCL, this forms the leading edge of the cloud. As the ambient flow continues to push the cloudy air up the slope, more water condenses. Subsiding flow down the lee slopes causes this water to evaporate, so that orographic stratus generally caps the ridgetops.

The maximum rate of condensation occurs where the maximum updraft occurs; in the simple flow model with uniform horizontal velocity U, this occurs over the steepest terrain slope. If precipitation is negligible, peak liquid water contents are found near the region of maximum vertical displacement of the air stream; i.e., near the top of the ridge. The location of maximum rainfall depends on the microphysical processes and how far condensate is advected downstream by the flow. If one adopts the unrealistic assumption that all condensate falls out immediately as it forms, then the maximum precipitation coincides with the maximum condensation rate at the location of maximum terrain slope. On the other hand, if larger droplets or particles are carried downwind significant distances and do not evaporate completely before striking the ground, the maximum rainfall could occur downwind of the peak. In the simple case the actual rainfall maximum probably would occur somewhere in between, but in reality the situation is complicated by more complex airflows, by the presence of embedded convective cells, and by ice processes.

9.3.2.1.2. Microphysics. The major question in this simple cloud system is microphysical: How do we get precipitation out of these clouds? If the droplets stay small, they will have negligible fall speeds with respect to the air. Thus they follow the air trajectories over the barrier and evaporate on the lee side, and no precipitation hits the mountain (Fig. 9.8a).

The warm-rain, collision-coalescence process that produces larger droplets requires the growth of a few large droplets to get the process started. The problem is timing. Grant and Kahan (1974), for example, observe that most of the mountain barriers in the western United States and other places have a "lifting section" of 30–65 km. Because mountain-level wind speeds are generally 10–20 m s^{-1} or greater for the wintertime cases Grant and Kahan describe, they conclude that "it is apparent that the microphysical properties are primarily established in a short time interval of consistently less than 1 h, and frequently less than $\frac{1}{2}$ h. This restricted time for cloud particle growth strongly . . . limits the size to which liquid droplets can grow." Of course there is considerable variability in these numbers. Many major mountain ridges have upwind lifting sections of less than 15 km, whereas others, such as areas of the Cascade and Sierra Nevada ranges of the West Coast of the United States, are much broader and can provide upwind lifting for several hours (Grant and Kahan 1974). Also, wind speeds in warmer climates or seasons are often lighter, and therefore produce longer lift times but slower rates of condensation. These estimates indicate that the

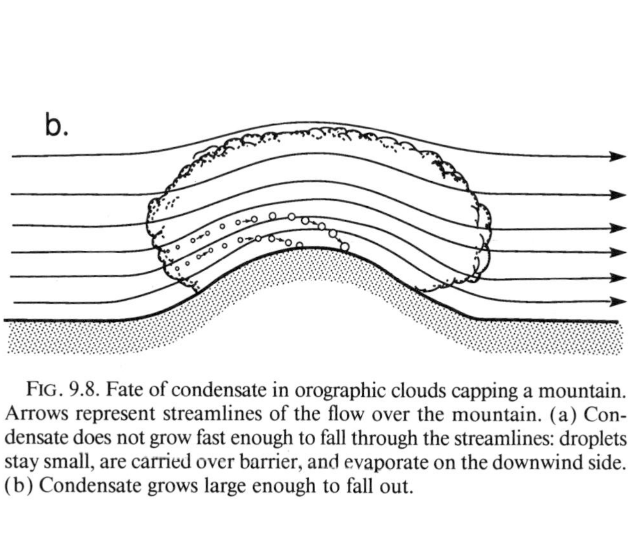

FIG. 9.8. Fate of condensate in orographic clouds capping a mountain. Arrows represent streamlines of the flow over the mountain. (a) Condensate does not grow fast enough to fall through the streamlines: droplets stay small, are carried over barrier, and evaporate on the downwind side. (b) Condensate grows large enough to fall out.

time available for microphysical processes to produce rain is often between $\frac{1}{2}$ and 2 h.

In a cumulus cloud with active updrafts and strong turbulence, calculations indicate that precipitation-sized droplets may take 20–30 min to form, if the cloud forms in clean, maritime air, or 1–2$\frac{1}{2}$ h in continental air with higher aerosol concentrations (Fig. 9.9; Braham 1968). The difference between the maritime and continental clouds is that the higher aerosol concentrations in the latter produce high concentrations of small droplets; the population of small droplets takes a long time to produce the larger droplets needed for collision and coalescence. In the gentle flow and weaker vertical wind components of stratiform clouds, Mason (1952) found that growth to precipitation-sized droplets may take 3–4 h.

The fact that warm orographic rain is observed indicates that droplets large enough to fall through the airstream can be produced (Fig. 9.8b); the problem is how. One such dilemma is the steady orographic rains that fall in the Koolau Mountain Range of Oahu, Hawaii, which Woodcock (1975) dubbed "anomalous orographic rains." Extensive mountain locations average over 50 mm of rainfall per day during these events. The events occur when moist, onshore flow advects nonprecipitating stratocumulus clouds over the coast and up the mountains. It is tempting to speculate that the precipitation process begins offshore. Woodcock (1975) argues, though, that if this were the case, significant rainfall would also be expected at times in the lowlands along the coast—but this is not observed. Therefore he concludes that the growth of larger droplets begins only when the air begins to ascend the mountain barrier. The problem is that the Koolau Range is very narrow, and according to Woodcock's calculations, droplet growth to precipitation size must occur in 5 min to explain his observations!

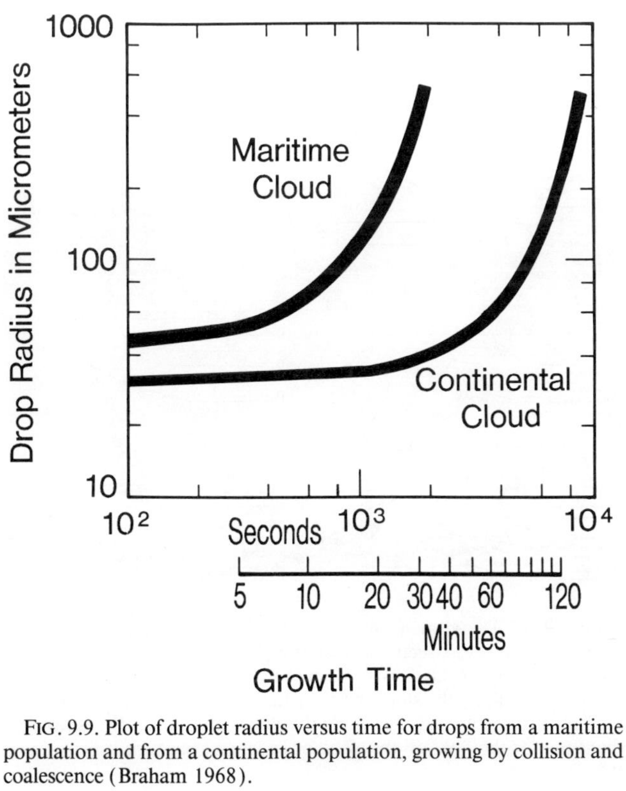

FIG. 9.9. Plot of droplet radius versus time for drops from a maritime population and from a continental population, growing by collision and coalescence (Braham 1968).

Although upstream dynamic effects of the Koolau barrier were not conclusively ruled out as a factor in initiating the droplet growth process, it seems that in this and many other orographic cases the formation of precipitation-sized droplets occurs in less time than that predicted by the traditional microphysics approach (e.g., Fig. 9.9). A number of other mechanisms recently have been proposed to explain the short times often observed for precipitation formation by warm-rain processes:

1) Giant aerosol particles can immediately produce large droplets. In maritime cases, giant sea-salt particles that act as condensation nuclei make large droplets. In continental cases, a cloud ingests large dirt particles that become wetted by collisions with small cloud droplets and then act as large droplets (Johnson 1982).

2) Turbulence in the airstream over the barrier, and turbulent entrainment of clean, dry air from above the cloud (Cooper et al. 1986), can contribute to faster growth rates for cloud droplets.

3) Clouds with warm bases (cloud-base temperatures greater than $\sim +15°C$) immediately produce clouds with high liquid-water contents. Such clouds seem to produce large droplets quickly, regardless of other factors such as aerosol concentrations (D. B. Johnson 1989, personal communication).

These processes help to explain how a mountain barrier, which can be relatively narrow, can provide enough lift to produce rain by collision and coalescence. Other important processes (which are not so important in tropical clouds) are ice processes. In his account of rain cloud processes, Ludlam (1956) generalizes that in simple layer clouds, ice mechanisms produce rain in half the time of coalescence mechanisms ($\frac{1}{2}$ h versus 1 h). This is mostly a result of the difference between the saturation vapor pressure for liquid water and that for ice, as indicated in the shaded region in Fig. 9.1.

The physical explanation is as follows. Because liquid water can be supercooled (i.e., can exist as a liquid below freezing), water can exist simultaneously in all three of its phases, vapor, liquid, and solid, at temperatures below freezing. Evaporating liquid droplets keep the vapor pressure close to its liquid-water-saturation value, but this environment represents a considerable supersaturation with respect to ice. At $-10°C$, for example, the ice supersaturation exceeds 10%, and this value increases at colder temperatures. This can be contrasted with values of supersaturation generally encountered by liquid droplets growing in an active cumulus cloud, typically less than 1%. Thus ice crystals, which are also generally fewer in number (concentration) than water droplets, are highly favored for growth in a three-phase, subfreezing environment, at the expense of the evaporating liquid droplets. Large ice particles that grow in this environment can then sweep out and capture smaller ice particles and smaller water droplets, by virtue of the greater fall speeds of these larger particles, and this mechanism of "accretion" considerably accelerates their growth. In orographic cloud systems whose tops penetrate the freezing level, ice processes, because of their greater efficiency, would produce much of the precipitation in most cases.

In the simple airflow case, the most interesting issues are microphysical. There are many orographic cases in which 1) most of the water vapor in the air flowing over the barrier is in the lowest kilometer or so above the ground, and 2) the flow in this region is more or less parallel to the surface of the mountain. In these cases, simple models may predict rainfall well. This situation occurs often in wintertime snow clouds, and as discussed in section 9.3.3.1, simple models seem to be useful in predicting precipitation amounts. In the majority of cases the airflow is more complex.

An extreme example of more complex flow leading to cloud formation is the banner cloud, which is often observed to the lee of the Matterhorn and other peaks in the Alps (Douglas 1928) and has also been documented to the lee of Mount Everest (Hindman and Wick 1989). According to Douglas (1928), the banner cloud "streams away on the lee side of a sharp, prominent peak in a strong wind, and presents the appearance of a great mass of vapor pouring out of the mountain itself." The cloud forms when lee-side flow separation and the accompanying upslope pressure gradient force produce upslope flow in the wake of the peak. When the upslope flow contains sufficient moisture, the banner cloud appears.

Complex airflows produce cloud and precipitation patterns that are very different from those predicted by simple airflow models, especially when significant moisture is distributed through a deep layer of the troposphere. The next two sections, therefore, consider two more-complex effects of stable airflow over a barrier: blocking and waves.

9.3.2.1.3. Blocking. Smith (1979) states that "the blocking of low-level air is one of the most important ways in which mountains affect the air flow. . . . It is probably fair to say that this windward-side slowing is due to the difficulty that the heavy [cold] surface air has in running upslope." In a local sense, blocking can be described as follows: as the stable, low-level air is pushed up the windward slope, it becomes potentially colder than the "ambient" air at that level (Fig. 9.10). This creates a relative high pressure next to the slope, decelerating the flow coming up the slope and often causing it to flow back down the hill. Smith (1979) cautions, however, that "blocking phenomena cannot be considered strictly local phenomena," since mathematically, the solutions to the equations of motion that give rise to blocking simultaneously produce mountain waves that propagate away to infinity.

The importance of blocking for cloud formation is that it produces a region of near-surface flow upstream of the mountain barrier that is decoupled from the synoptic flow aloft, as pointed out by Peterson (1989) in his study of wintertime stable clouds. Flow approaching the barrier encounters the decoupled region and decelerates, producing convergence and rising motion ahead of the barrier. Some of the orographic lifting occurs well upstream of the mountain range, and clouds and precipitation can thus also occur upstream. This process is important not only in the case of stable rain clouds, but also for the other kinds. The importance of blocking to cold orographic clouds (already referred to) is further mentioned in section 9.3.3.1, and its importance in providing lift to trigger unstable rain clouds is discussed in section 9.4.1.3.1.

If this blocking configuration and the high pressure next to the slope persist long enough for geostrophic adjustment, Coriolis turning causes a low-level jet to blow parallel to the barrier— referred to as a *barrier jet*. Parish (1982) studied numerically the wintertime, southerly barrier jet that forms over the western slopes of the Sierra Nevada in California.

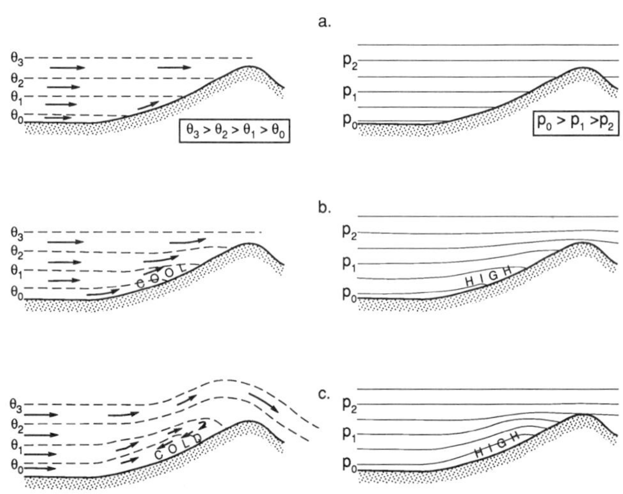

FIG. 9.10. Two-dimensional, x–z potential temperature (left) and pressure (right) cross-sections illustrating blocking. (a) Cold air at lower levels in the approach flow is pushed up the slope by the winds. (b) Since the air along the slope is cooler than "ambient" air at each level, the pressure becomes higher along the slope. This adverse pressure gradient retards the flow up the slope or (c) may even turn it back down the slope.

An interesting example of an effect of microphysics on the dynamics of the barrier jet was described by Marwitz (1983). This effect was observed in conditions where 1) rain forms by the melting of frozen precipitation that falls from higher levels, and 2) the freezing level is along the slope of the barrier. Melting occurs in a layer several hundred meters thick, and the release of latent heat makes this layer isothermal (at 0°C). Because an isothermal layer is very stable, the region above the layer is dynamically decoupled from the layer below. The layer above continues to blow over the barrier and produce snow, which falls into and maintains the melting layer. The layer below, however, is blocked; the blocking is intensified by the strong stability of the isothermal layer, which persists long enough to also intensify the barrier jet. Marwitz (1983) found speeds of 18 m s^{-1} in the southerly barrier jet, and he estimated that this was twice as strong as the jet would have been in the absence of the melting layer.

9.3.2.1.4. Mountain-wave effects. Important features often present in stable flow over obstacles are mountain waves of many kinds (discussed in Durran, Chapter 4). When clouds associated with mountain waves are considered, an important issue discussed in Durran, Chapter 4 is that condensation reduces the stability of the atmosphere, thus decreasing the gravity-wave frequency; i.e. increasing the wavelength of mountain waves. For example, an unsaturated layer with a moist adiabatic lapse rate, which is stable, becomes neutrally stable when saturated.

Bradley (1985) performed a series of numerical sensitivity experiments to demonstrate the effects that waves have on orographic precipitation. A major effect of waves is to provide regions of enhanced upward motion that produce higher liquid water contents than the simple orographic lifting models. Microphysically, the regions of high liquid water contents are effective regions for rain production.

Bradley's (1985) numerical study examines the precipitation along the west coast of Wales that results from moist southwesterly flow in the warm sector of cyclonic storm systems. Such regimes have attracted much attention, since they produce considerable rainfall over their lifetimes; summaries of this research appear in Smith (1979), Browning (1980), Bradley (1985), and Cotton and Anthes (1989). A major controversy that resulted from these studies, which is nicely reviewed by Smith (1979) and Bradley (1985), is whether the heavy rains were produced mostly by stratiform clouds of high efficiency or mostly by embedded convective clouds, which would have more effective precipitation-forming mechanisms.

Bradley (1985) cited the observations of Hill et al. (1981) that the atmosphere was stable, rather than neutral (moist adiabatic) or conditionally unstable, in seven out of eight orographic, warm-sector rain cases observed in Wales. These observations indicate that mountain waves must have been present and that the flow did not follow the terrain. Bradley's simulations show that for the stable case where mountain-wave structure was present, the flow produced higher vertical velocities, higher condensation rates, and slower horizontal flow over the upwind slope than for either the moist-adiabatic (neutral) case, which did not support wave activity, or the potentially unstable case.

Bradley's (1985) conclusion regarding the high-efficiency stable versus convective rainfall controversy is that one can reproduce the observed orographic enhancement and sensitivities of the rainfall by properly accounting for the complex wave structure of the flow. In related experiments, Bradley also showed that the details of the topography are important in reproducing local maxima in rainfall, but that smoothed topography adequately describes the gross behavior and sensitivity to external influences of the system.

9.3.2.1.5. Radiation. In cases when the airstream takes a long time (several hours) to traverse the barrier, cloud-top radiation becomes a factor in the development of precipitation. Diabatic cooling at cloud top destabilizes the cloud layer, causing overturning in the cloud layer, local enhancement of updrafts, and entrainment of dry air from above the cloud. These processes contribute to the formation of droplets of different sizes that are needed for collision and coalescence. In addition to these effects, Roach (1976) showed that cloud-top radiation also produces a favorable environment for condensation that enhances the growth rate of droplets. By adding a radiative flux divergence term to the heat transfer equation for a droplet and using this relationship to derive the droplet growth equation, he shows that droplets will grow in a subsaturated environment or will grow faster in a saturated environment.

Thus, in the puzzle over how orographic stratiform clouds can produce rain, radiation at cloud top increases the likelihood of forming precipitation-sized droplets. In cases where precipitation would form anyway, radiation can increase the amount.

Destabilization at cloud top by radiation also leads to changes in wave propagation characteristics of the atmosphere there. Weissbluth and Cotton (1989) performed cloud-model simulations of an orographic mountain-wave cloud that included atmospheric longwave radiation. In a series of experiments with moisture but without radiation, waves form that extend vertically out of the 2-km-deep cloud layer and through the troposphere. When radiation is added to the simulation, cloud-top radiation produces an unstable layer near cloud top that traps the waves, confining them to the region below cloud top; the trapped waves then produce vertical motions that trigger convective instabilities in the cloud layer. The trapping of the waves inhibits vertical momentum flux by the waves out of the cloud layer. Thus Weissbluth and Cotton hypothesize that the presence of clouds in large-amplitude mountain waves with strong cloud-top cooling (i.e., with a dry atmosphere above) significantly reduces the wave drag produced by the wave.

9.3.2.1.6. Other upslope flows. A deep layer of moist air flowing over a mountain barrier is one large-scale mechanism by which stable orographic cloud can be generated. Another is a cold front, which may be relatively shallow, pushing equatorward along the eastern slopes of a north–south mountain range such as the Rocky Mountains. The resulting flow brings moisture up the slope from lower areas farther east and produces upslope fog and stratus. These systems can produce light precipitation (drizzle and occasionally light rain) but larger amounts or intensities require augmentation by larger-scale weather systems or unstable clouds.

Upslope flow can also be produced by more local circulations. An important source of precipitation on mountainous tropical islands is the orographic stratus clouds produced by thermally forced upslope winds. On the island of Hawaii (Fig. 9.11), for example, orographic cloud encircles the major peaks (Fig. 9.12) as morning upslope carries moist air up the mountain from all sides. The cloud rarely extends up to the tops of the peaks,

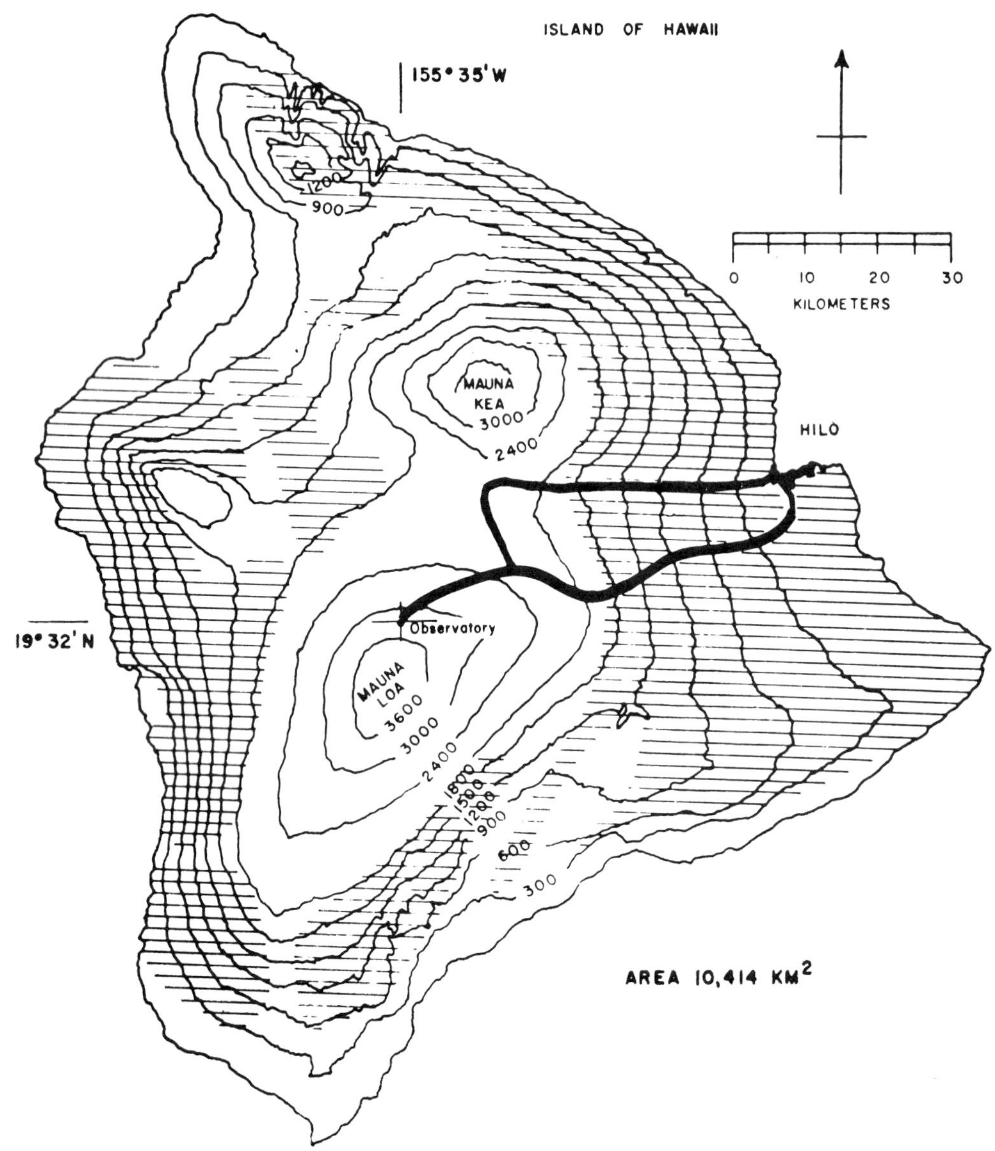

FIG. 9.11. Contour map of the island of Hawaii (Mendonca 1969). Areas of vegetation are hatched. Contour interval is 300 m up to 1800 m, 600 m above.

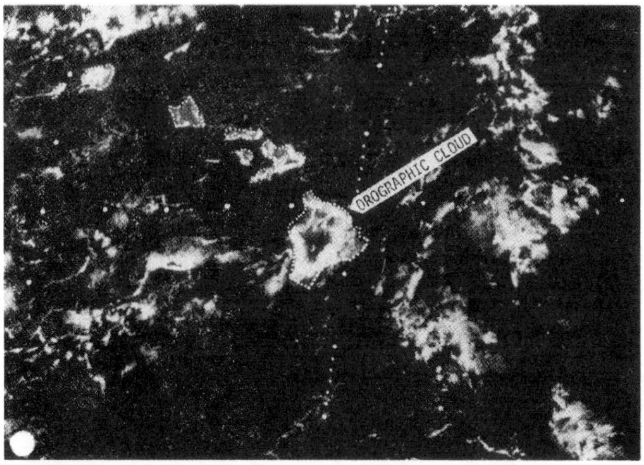

FIG. 9.12. High-resolution GOES satellite image showing the orographic cloud forming a ring around the high mountains of Hawaii. Image taken on 2 June 1978 at 1115 LST (Garrett 1980).

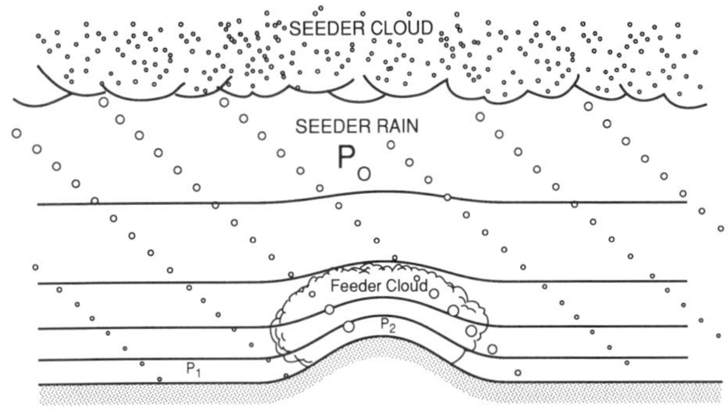

FIG. 9.14. Conceptual model illustrating orographic enhancement of rain by the seeder–feeder mechanism, as depicted by Browning (1980). The size of a droplet falling from the cloud indicates the precipitation intensity at that point. Here P_0 indicates the intensity of seeder precipitation entering the top of the feeder cloud, P_1 is the intensity of the rain at the surface upwind of the hills, and P_2 is the intensity of the rain at the surface in the orographic maximum near the top of the hill.

though, because of the capping trade-wind inversion, as shown by Garrett (1980; Fig. 9.13) and Mendonca (1969). Two factors favor precipitation formation. First, the clean maritime air mass has low aerosol (cloud condensation nucleus—CCN) concentrations and, at least within a kilometer or two above sea level, large sea-salt nuclei, which reduce the time necessary to form larger, precipitation-sized droplets. Second, as indicated in Fig. 9.13, the orographic stratus often contains embedded cumulus convection, which increases the effectiveness of the precipitation process.

9.3.2.2. Interaction with Larger-Scale Processes

As discussed in the previous section, orography can produce cloud and precipitation in situations where none would have formed without the terrain effects. Mountains also redistribute and enhance precipitation that falls from

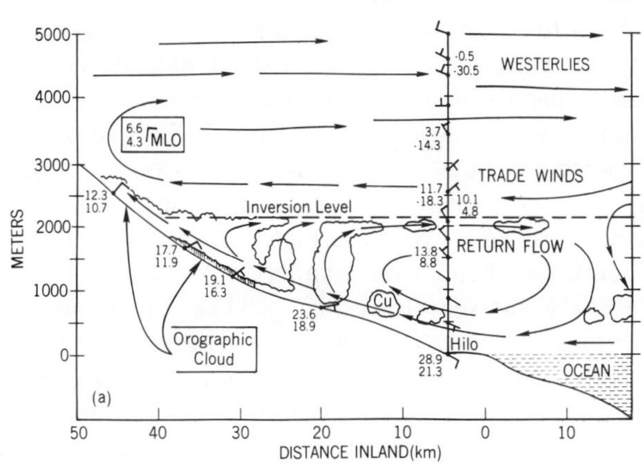

FIG. 9.13. Cross section of the slope of Mauna Loa, Hawaii, showing the sea breeze/upslope flow producing an orographic cloud, with embedded convection (Garrett 1980).

larger-scale systems. Two examples of this interaction are the "seeder–feeder" mechanism and the effect of mountains on midlatitude cyclonic-storm and frontal systems.

The seeder–feeder mechanism has two components: an orographic stratus cloud (as discussed earlier), and at some level above that an upper-level precipitating cloud, often from a larger-scale storm system. The hill on which the orographic cloud sits can be very small—elevation differences of as little as 50 m above the surrounding terrain. Orographic cloud droplets being carried over the hill are not in the cloud long enough to grow big enough to fall out, so the orographic stratus by itself would produce no precipitation on the ground.

The larger-scale system provides a means for getting the water from the orographic cloud to the ground before it evaporates on the lee side of the hill: raindrops from this upper-level system fall through the orographic cloud, grow by collecting the smaller cloud droplets in their paths, and carry this excess water to the ground. Even over small hills the enhancement can be 25% to 50%. In the warm-sector orographic precipitation in the hills along the western coast of Wales, the studies of Browning et al. (1974) and Hill et al. (1981) show that seeder–feeder precipitation is a major component of the heavy rainfall events there. The process is illustrated in Fig. 9.14 (from Browning 1980). Numerical simulations of Bradley (1985) indicate that enhanced lifting by mountain-wave processes increases the liquid water available to seeder droplets and thereby increases the amount of rainfall on the ground. These simulations show that proper representation of the flow structure also reproduces the observed sensitivity of the rainfall to wind speed, a result that simpler models failed to achieve.

The seeder–feeder mechanism has been studied extensively in recent years both by observations and by numerical modeling, and Cotton and Anthes (1989) give

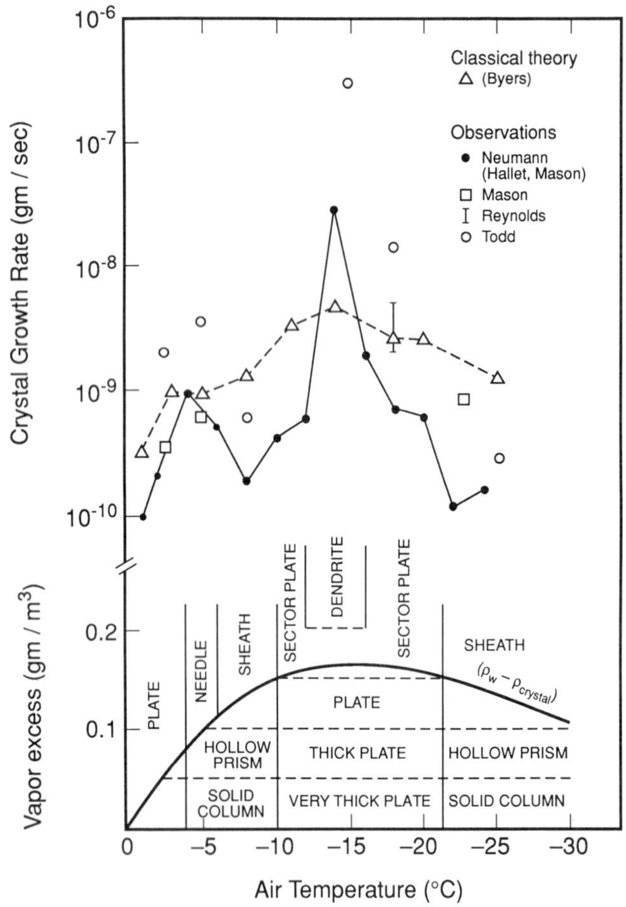

FIG. 9.15. Top: ice crystal growth rates as a function of temperature. Bottom: dependence of ice crystal habit on the temperature and humidity environment in which it grows (Braham 1968). Solid curve in the bottom part of the drawing (marked $\rho_w - \rho_{crystal}$) represents a crystal growing in a liquid-water-saturated environment.

further details of these studies. Overall, the studies indicate that seeder-feeder precipitation is the major reason that hills, especially small hills, have greater precipitation than nearby low areas.

9.3.3. *Stable snow clouds*

Most of what was said about stable rain clouds applies to stable snow clouds. The division between stable rain and snow clouds is often blurred in the case of orographic stratus clouds: the distinction between those with bases a few degrees warmer than freezing that form rain via ice processes and those with bases a few degrees cooler than freezing that form snow is sometimes quite arbitrary. Orographic stratus clouds often rain at lower elevations and snow at higher elevations of the same mountain (e.g., Reynolds and Dennis 1986).

The main difference between stable warm-rain and stable snow clouds is that ice microphysics has a much richer variety of forms and interactions than water microphysics. The form or *habit* of ice crystals has a strong effect on how they interact with other hydrometeors, and crystal habit in turn depends on the conditions in which the crystal grew; i.e., temperature and relative humidity (Fig. 9.15). We saw in the previous section that precipitation from warm orographic clouds was strongly influenced by microphysical effects. This is even more true of cold orographic clouds, because of the diversity of crystal types and because of the coexistence of two condensed-water phases, ice and supercooled water.

9.3.3.1. OROGRAPHIC STRATUS

It is surprising in light of this diversity of microphysical possibilities that simple models have done a very good job of predicting orographic snowfall. Rhea (1978) devised such a model. The input is the topography of Colorado and the six rawinsonde soundings in the vicinity of western Colorado. These soundings are interpolated to profiles located along the upwind border of the state. The profiles form the upwind initial conditions for a set of parallel, two-dimensional (vertical and horizontal), side-by-side Lagrangian models, with the horizontal dimension aligned along the mean 70-kPa wind (to the nearest 10°). Previously stored in computer memory are fine-resolution terrain data (Rhea used 5- and 10-km resolution) for each slab of mountain topography, starting at the location of the initial profile and extending downwind across the domain; a different set of topography slabs is used for each 10° of wind direction.

Precipitation is squeezed out of the parcels as they ascend mountain barriers, according to moist adiabatic condensation (and evaporation) theory, to continuity of condensate (that did not fall out previously) being advected over the terrain, and to an "efficiency factor" that represents the fraction of condensate allowed to fall out. The efficiency factor, which parameterizes the microphysical processes, is constant over the domain for a run and is a function of a calculated cloud-top temperature. A typical value was 25% for cloud-top temperatures below −25°C, decreasing linearly to 0% at 0°C.

Rhea (1978) developed the model for hydrological purposes, so he verified it against seasonal snowfall, streamflow (runoff), and climatology. The model performance was impressive, with high correlation coefficients (often 0.8 or better) for a variety of comparisons. The model also showed surprisingly good skill in quantitative precipitation forecasts (QPF) for individual snow events.

The model performs very well for pure orographic snowfall in Colorado. It nominally considers large-scale vertical motion, and model performance is degraded when this process is left out; however, for reasons discussed later in this section, the model best predicts orographic precipitation. Tests in other parts of the country showed that

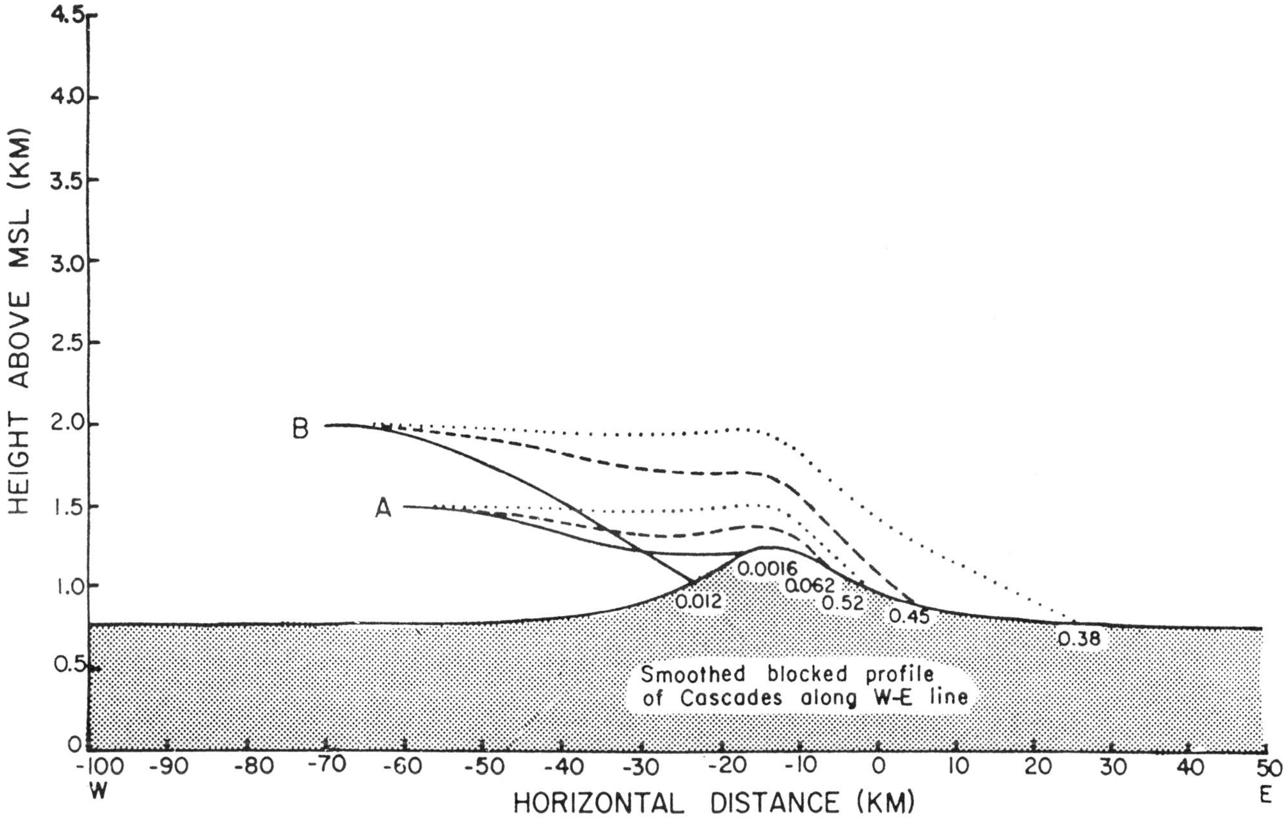

FIG. 9.16. Precipitation-particle trajectories computed by Hobbs et al. (1973) for ice crystals growing by riming for a prescribed ambient flow from left to right. Model follows an initial volume of 1 L from points A and B containing the precipitating ice, and the numbers at the end of each trajectory represent the mass (mg) of precipitation that reaches the ground from the original volume. Line types represent different concentrations of crystals as follows: solid, 1 L^{-1}; dashed 25 L^{-1}; and dotted, 100 L^{-1}.

in instances where precipitation involves a larger-scale storm system, model predictions do not verify as well as in Colorado. Thus, a major reason for the success of the model in Colorado is apparently that 80% of Colorado's snowpack results from orographic snowfall (Grant and Kahan 1974).

A tempting conclusion from the success of this model is that orographic precipitation amount is a poor choice of quantities to verify more sophisticated models against. It is hard to get excited about a model that runs for an hour on a Cray computer and properly predicts *only* orographic snowfall amounts, when the same result can be obtained from a model that contains very simple physics and easily fits onto a personal computer. More complex models should be verified against more exacting dynamic and thermodynamic constraints. A good reason to use a more sophisticated numerical model is to identify important physical processes, as in the simulations of Abbs and Pielke (1987) or Cotton et al. (1986), which are mentioned again later in this section, or when it is important to consider more-complex flows, as in the simulations of Bradley (1985; discussed in section 9.3.2.1.4). Abbs and Pielke showed the importance of the interaction of topography and flow in determining surface snowfall, and Cotton et al. showed the importance of microphysics for regulating surface snowfall.

The microphysical variability of the orographic clouds is represented in Rhea's model by the "efficiency factor." Other modeling studies have investigated how variations in microphysics affect the amount of precipitation reaching the ground. Hobbs et al. (1973) prescribed the airflow over a two-dimensional ridge and followed the trajectories of ice particles falling through the airstream to the surface. The ice crystals increase in size by riming; i.e., collecting supercooled droplets as they fall. The amount of snow that reaches the ground is sensitive to the concentration of ice crystals and to the height of the initial upstream volume (Fig. 9.16). At lower concentrations of ice crystal, there is less competition for the supercooled water, the ice crystals grow and fall out faster, and more snow ends up on the ground. If the ice particles blow too far downstream, however, they enter the region of subsidence to the lee of the barrier, and some of the condensate sublimates. Thus the trajectories that extend too far downstream show a reduction of snowfall on the ground.

The microphysical processes that are active in a given

region of a cloud are a complicated function of the temperature and humidity of the air and the structure, size, and mixture of hydrometeors present. The structure of hydrometeors is determined by their history. With all this complexity it is important to sort out which processes are critical in the production of particles that are large enough to fall out. Cotton et al. (1986), hypothesizing that the *aggregation* of ice crystals that are too small to produce significant precipitation is an important process, ran experiments with a dynamic numerical model to test the sensitivity of snowfall to aggregation. They found that the amount of snow on the ground varies by a factor of 50%, depending on whether aggregation is active or inactive. Thus the details of which microphysical processes dominate can introduce considerable uncertainty into the amount of snowfall predicted by a model.

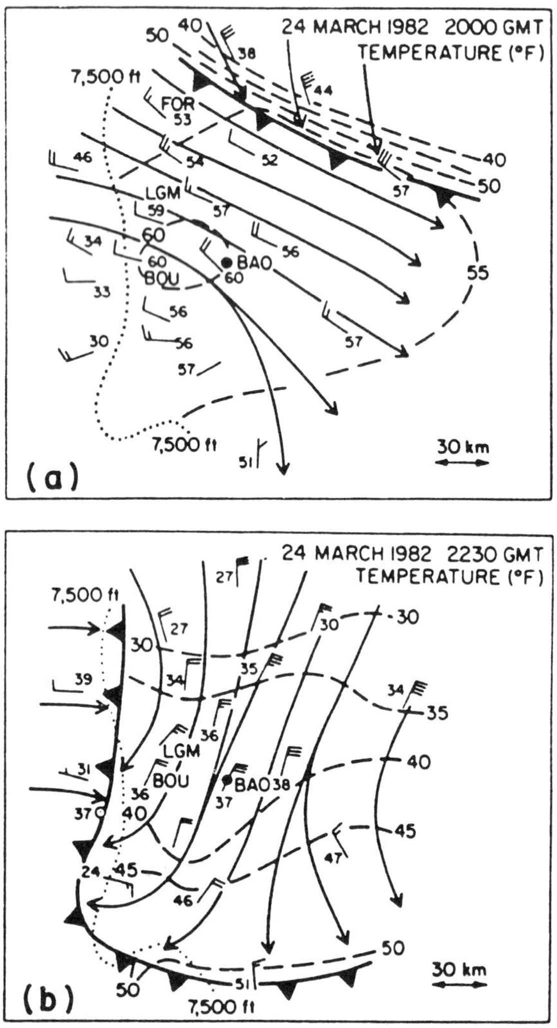

FIG. 9.17. Movement in $2\frac{1}{2}$ h of an upslope cold front down the eastern slopes of the Rocky Mountains in Colorado; mountains are to the left (Shapiro 1984).

In light of this uncertainty it is intriguing that Rhea's model has performed so well. Apparently the "efficiency factors" that parameterize all this microphysical diversity have been chosen judiciously. Cotton and Anthes (1989) ascribe the success of the simple models to two causes. First, the heaviest precipitation results from lifting by the largest orographic barriers—an unsubtle effect. Second, the highest moisture contents are in the lowest levels, so that most of the liquid-water production is there. Thus, "A model forced simply by continuity of air flowing over a barrier can capture the dominant lifting of the moisture-rich air" that produces the precipitation.

As a result of the effectively simple dynamics of wintertime orographic clouds and their sensitivity to microphysical details, they are a natural "laboratory" for studying the complexities of ice microphysics. This fact has been the motivation for a number of field projects, including a large number of weather modification projects (discussed by Grant and Kahan 1974) and more recent experiments in the San Juan Mountains of southwestern Colorado (Marwitz 1980; Cooper and Saunders 1980), in the Park Range of north-central Colorado (Rauber et al. 1986), and in the Sierra Nevada of California (Reynolds and Dennis 1986; Marwitz et al. 1985).

As with stable rain clouds, the stable flows are often complex, with blocking and mountain-wave effects evident. Peterson (1989) showed that upstream decoupled flow that can be caused by blocking and other processes produces lifting upstream of a barrier. The upstream lifting distributes cloudiness and snowfall upstream, and this can result in more total snow on the ground because of the longer distances and times that the condensate has to grow.

As in the case of warm stable clouds, cold stable clouds can also be produced by upslope flow from anticyclones on the east slopes of mountain ranges, normally as a result of the passage of a cold front from polar regions. In addition to Cotton and Anthes' (1989) review, Lilly and Durran (1984) and Reinking and Boatman (1986) discuss these events. Upslope flow from anticyclone systems occurs, for example, in the United States along the east slopes of the Rocky, Cascade, Sierra Nevada, and Appalachian ranges, along the east coast of southern Mexico, and in Europe along the north slopes of the Alps. Cold fronts at the leading edge of the upslope anticyclonic flow can move very fast, as shown in Fig. 9.17.

East of the Rocky Mountains the basic anticyclonic upslope occurs when an Arctic or polar cold front pushes south along the mountains. These fronts are shallow (1–2 km deep), the air mass is capped by a strong inversion, and the air above the inversion is often dry. The vertical component of the upslope velocities over the sloping plains is a gentle 1–3 cm s^{-1}. When the flow reaches the foothills, however, Lilly and Durran reported an increase in the ascent rate to 5–10 cm s^{-1} because of the increase in terrain slope. Only a little over half of these systems produce precipitation, and those that do rarely produce more than 2–3 cm of snow.

The ability of these simple systems to precipitate is strongly controlled by the microphysics (Reinking and Boatman 1986). At temperatures warmer than −10°C, droplets are too small and too few for the warm-rain process, and ice nucleation is very inefficient. Colder than −10°C, ice can nucleate effectively, and precipitation formation depends primarily on the rates of nucleation and growth of ice particles. At temperatures between −6° and −10°C there is an interesting growth regime in which one type of ice crystal, an isometric crystal consisting of a short column or a thick plate, can grow efficiently by riming. As a result, in this regime stratus clouds only 1 km deep can produce small graupel particles ("snow pellets" or "snow grains") that fall to the ground (Weickmann 1981; Reinking and Boatman 1986).

Variations in the simple upslope anticyclone pattern can lead to situations that increase the amount of precipitation the system produces. For example, in the Rocky Mountain systems described by Reinking and Boatman (1986), the air above the cold air mass may still contain Pacific Ocean moisture, or even more dramatically, a southerly flow ahead of the cold front may contain significant moisture from the Gulf of Mexico. In these cases the cold front lifts the moist air and can produce copious precipitation. On smaller scales more germane to the topics of this monograph, moisture aloft being carried over the barrier can condense and fall, providing a cold-cloud seeder–feeder. The upslope flow can also induce its own seeder cloud, as moisture originating in the upslope layer is carried aloft at the edge of the front into the westerly flow aloft. There it condenses and precipitates into the cold, orographic feeder cloud.

9.3.3.2. Interaction with larger-scale processes

Major precipitation in mountain regions from stable clouds occurs when orographic effects occur in cooperation with large-scale cyclonic storm systems. Effects include seeder–feeder enhancement and orographic upslope enhancement of the cyclonic precipitation, and modification of the basic cyclonic structure of the storm. These topics are covered extensively in Smith (1979), Reinking and Boatman (1986), and Cotton and Anthes (1989). Because they involve generally larger scales than those examined in this chapter, only two topics are discussed here. The first is the role of topography in regulating the spatial distribution of upslope–cyclonic snowfall, and the second is the role of microphysics in very large snowfall events.

Heavy snowfall from major winter storms occurs as a result of ascending motion produced by synoptic-scale dynamics. Even with large-scale rising motion, topography can play a major role in regulating how much snow a given area receives. For example, on Christmas Eve 1982 a major winter storm dropped over 20 cm of snow over much of northeastern Colorado. In areas south and west of Denver actual recorded snowfall reached 0.5–1 m, but less than 100 km to the north there was a broad area of very light snowfall (0–3 cm). This variability, which is actually characteristic of snowstorms in this region, was reported by Schlatter et al. (1983).

The causes of the variability were investigated by Abbs and Pielke (1987), using a dynamic numerical model. Their model, a dry, mesoscale model of northeastern Colorado at 15-km resolution, was initialized with Denver's morning temperature and wind sounding and run for 8 h of simulated time. The resulting surface winds were similar to observations, and interestingly, the areas of strongest rising motion at the 250-m level AGL corresponded to the areas of heaviest snowfall. The vertical-velocity analysis at this level also indicated several areas of terrain-induced subsidence, corresponding to the regions of minimum snowfall. These results indicate that terrain-forced rising or sinking can strongly amplify or almost negate the organized rising motion of large-scale cyclones and that such terrain effects control the spatial distribution of snowfall from these systems.

Another Front Range snowstorm on Thanksgiving 1979 in southern Wyoming and northern Colorado provided a case study for Auer and White (1982), illustrating the importance of cloud microphysics in major snowfall events. It is implicit in Auer and White's argument that in cyclonic snowstorm systems (where precipitation is formed in gentle updrafts usually of 10 cm s^{-1} or less) that the most efficient microphysical pathways to precipitation involve ice processes, and that if significant amounts of (supercooled) liquid water are formed, then the process is less efficient. The most important ice process is the growth of ice crystals by direct deposition from vapor to solid ice. As shown in Fig. 9.15 (Braham 1968), this process is most efficient at around −15°C (between about −13° and −16°C), temperatures at which dendritic ice crystals grow. Since temperature decreases monotonically with height in a moist adiabatic layer, this temperature corresponds to a vertical level that one can call the level of peak ice-growth rate. For their study, Auer and White used radiosonde temperature profiles in the vicinity of the storms to determine this level.

The other important level of the atmosphere in Auer and White's study is the level where condensation (or deposition) occurs at the fastest rate. The authors reasoned that the condensation rate should be largest where the vertical velocity w is largest; i.e., at the "level of nondivergence" (LND), or the level where convergence below turns to divergence above. Climatologically, this level occurs near 60 kPa, and it occurred at about 55 kPa in the Front Range snowstorm. The actual magnitude of the peak in condensation rate depends on the temperature (and pressure) there: for a given value of peak vertical velocity at the LND, lower temperatures yield lower condensation rates.

Auer and White hypothesized and illustrated with the Front Range snowstorm that when these two levels coincide, major snowfalls occur. In practice this means that when a significant storm is developing in a region where the 60 kPa temperature is near −15°C, a forecaster should anticipate heavy snowfall. The authors observe that heavy accumulations of snow generally fall as aggregates of dendritic crystals, as one would expect if the major source region for crystals were near −15°C. They surveyed 75 snowstorms over the United States that produced more than 13 cm of snow and found that the 60-kPa temperature was between −13° and −15°C in 70% of the cases. They also looked at 42 mountain snowstorms. In these cases the maximum w (LND) was assumed to be near mountaintops, which is around 65 kPa in the western United States. The 65-kPa temperatures were −12° to −14°C in 70% of the cases. Thus if the actual LND were a little above mean mountaintop height, it would fall comfortably within the −13° to −15°C temperature range.

9.3.4. *Unstable snow clouds*

Unstable wintertime clouds are caused by destabilization of a layer either through processes discussed in section 9.2 or through destabilization of an otherwise stable air mass by differential temperature advection; e.g., cold advection aloft over warm or neutral advection. Processes in section 9.2 included the forced lifting to saturation of a potentially unstable layer; i.e., a layer in which θ_e decreases with height, and the forced lifting of a nonsaturated layer, which can destabilize the layer with respect to parcel displacements in the vertical.

Destabilization by differential advection can occur because of larger-scale processes, when air with trajectories from two different source regions is advected over the same point (with cooler air over warmer air). An important mechanism over the mountains is blocking, as illustrated in Fig. 9.10, when a cold front approaches a mountain range. Figure 9.18 from Smith (1982) illustrates this mechanism. A stable cold air mass approaches the mountain range, and the cold air in the lowest layers is pushed up the slope. It is cooler than environmental air at its new level, as in the blocking mechanism discussed in section 9.3.2.1.3. The resulting low-level high pressure against the mountains retards the progress of the cold front near the ground, but cold air continues to flow over the mountains

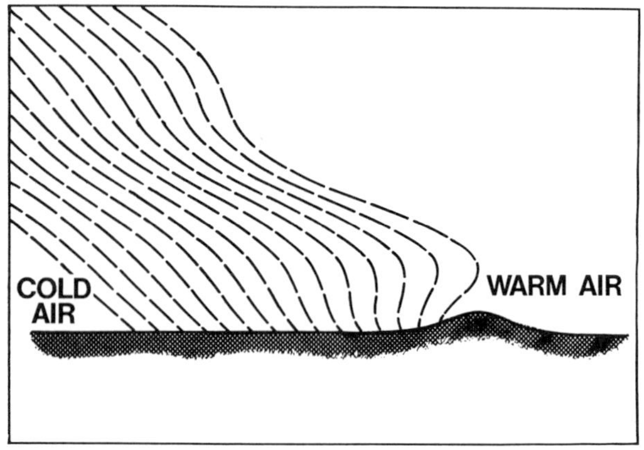

FIG. 9.18. Distortion of a cold front (dashed lines) approaching a mountain range by retarding of the low-level stable airflow as it is pushed up the slope. The cold front starts at the left of figure and proceeds toward the mountain range; positions are at hourly intervals (Smith 1982).

aloft, where the air is still warm at low levels. This advection of cold air over the warm air produces a destabilized region over the mountains, often producing stratiform upslope clouds with embedded convection.

Embedded convection is important because updrafts are much stronger than those generated during forced orographic ascent. Thus these cells are more effective than stable orographic clouds in producing precipitation-sized particles in the cloud and precipitation on the ground. They often contain supercooled liquid water, at least early in their lifetimes. The presence of supercooled water in clouds, reviewed by Cotton and Anthes (1989), is very important in a number of applications, including those involving weather modification and propagation or attenuation of electromagnetic energy.

Browning (1980) observed this configuration in the cold-front portion of cyclonic storm systems approaching the British Isles. He referred to this structure as a *split-front* configuration, since the cold front aloft arrives ahead of the cold front at the surface. Heggli and Reynolds (1985) identified this structure in cold fronts of storms hitting the Sierra Nevada of California.

9.4. Unstable rain clouds

Unstable rain clouds, or moist warm-season convection, constitute the fourth division of orographically forced clouds. These clouds grow when lifting "releases" moist instability that is latent in a column of the atmosphere, as determined by its vertical profiles of temperature and moisture. Since this unstable layer often produces overturning that extends upward through the depth of the troposphere to the tropopause, the results are often dramatic; the resulting deep moist convection (cumulonimbus or thunderstorm clouds) can produce severe weather, including hail, damaging winds, heavy rains, and tornadoes.

Because unstable rain clouds range from shallow cumuli to thunderstorms and severe weather, they are in many ways the most interesting clouds. There is considerable practical interest in mountain thunderstorms. They can cause flash floods, and produce lightning that can severely injure or kill exposed individuals, ignite forest fires, and affect refueling operations (e.g., for airplanes) and the handling of explosives during mining operations. Moreover, the eastern Rocky Mountains often serve as a breeding ground for thunderstorms that propagate onto the Great Plains and then organize into mesoscale convective complexes (MCCs); these can bring summertime rain to areas hundreds or sometimes even thousands of kilometers to the east (Cotton et al. 1983; Wetzel et al. 1983; Maddox 1980).

Because the release of moist instability is vital to the appearance and growth of these clouds, it is important to reemphasize here the point made in section 9.2.3—that there are two types of moist instability, referred to as latent and potential instability. A major difference of interest here is the effect of lifting on each type of instability: lifting can create latent instability from a layer that is stable with respect to parcel displacements, but lifting cannot make potential instability from a layer that is (potentially) stable for layer displacements.

Forecasters have traditionally recognized three ingredients as being necessary for thunderstorm formation: a conditionally unstable temperature lapse rate (dT/dz), sufficient moisture, and a "trigger" or updraft pulse. Although mountain ranges do play a role in directing or denying moisture to regions, the major direct role of mountains in thunderstorm occurrence is in the triggering or initiation process. Flow in the vicinity of mountains provides initiation mechanisms, or flow configurations that produce updrafts and initiate thunderstorms.

This section on unstable rain clouds is divided into six parts. The first three parts address the three ingredients necessary for thunderstorms. The important role of mountains in thunderstorm initiation is discussed at some length in section 9.4.1, and the significance of lapse rate and moisture sources are the subjects of sections 9.4.2 and 9.4.3. A number of studies have described the structure of mountain-forced updrafts and the clouds into which they grow, and section 9.4.4 describes the evolution of the structure of mountain thunderstorms according to these studies. The interaction of local-scale initiation processes with larger-scale atmospheric systems in producing mountain thunderstorms, and recent studies on upscale feedbacks in which mountain convection organizes into larger-scale systems downstream, are discussed in section 9.4.5. Section 9.4.6 includes a discussion and summary of the material on the effects of mountains on unstable rain clouds.

9.4.1. Initiation mechanisms

The strongest link between mountain meteorology and the occurrence of moist convection is through those local processes or mechanisms that initiate the clouds. Such local processes govern the locations of storm development by encouraging cloud growth in some areas (e.g., by providing moist upcurrents) while suppressing growth in others (e.g., through the subsidence branches of thermally forced mountain circulations). Local circulations also regulate the timing of mountain storm evolution. Figure 9.19 shows three general mechanisms (discussed next) by which mountains initiate (trigger) thunderstorms: direct lifting of air to its LFC, thermally generated circulations, and obstacle (aerodynamic) effects.

FIG. 9.19. Three types of thunderstorm initiation mechanism: (a) direct orographic lifting to LFC, (b) thermally forced (solenoidal) circulations, and (c) obstacle or aerodynamic effects.

9.4.1.1. Direct lifting to the LFC

Forced ascent of moist air up the slopes of a mountain barrier can lift the air directly to its LFC, when the LFC at some inflow level is near or below the level of the ridgetop. Two ways in which this instability occurs are by embedded stratocumulus or cumuliform convection that begins when a potentially unstable layer is forced to ascend, and by orographic thunderstorms. In the latter case, the layer above the LFC through which the instability acts extends through the troposphere, whereas in the former the unstable layer is generally shallower (often 2 km or so deep).

9.4.1.1.1. Potential instability release. Rainfall episodes along the western coasts of Europe and the United States illustrate the role of hills and mountain ranges in producing unstable rain by lifting of a shallow surface-based layer. In two studies, potentially unstable layers were found between the surface and 2 km in the warm sectors of midlatitude cyclone systems (i.e., the warm air between the surface warm front and the cold front) as these systems approached the shores of south Wales, Great Britain, (Browning et al. 1974), and of Washington state on the West Coast of the United States (Parsons and Hobbs 1983). In both cases orographic lifting by hills or mountains just inland of the coast produced unstable rain (showers). Parsons and Hobbs (1983) also found low-level potential instability in the post cold-frontal air, which produced showers as it ascended the inland mountains.

In Southern California, Elliot and Shaffer (1962) and Elliot and Hovind (1964) found a strong effect of stability on rainfall distribution over two mountain ranges. When the approach flow coming off the Pacific Ocean was stable, coastal areas received more rain than mountain areas, probably because of blocking effects. When the inflow was unstable, however, ascent of the air released potential instability over the mountains, produced showers, and resulted in more rain over the mountains than along the coast. In Browning et al.'s (1974) case the overall rainfall augmentation from the orographic showers was small compared with the action of these low-level orographic clouds as feeder clouds, and compared with showers produced by the release of middle-tropospheric potential instability, which is discussed in section 9.4.1.3.

9.4.1.1.2. Flash flooding. Thunderstorms that grow when moist, near-surface air blowing up a slope is forced to its LFC have produced spectacular and tragic results, in the form of flash floods. Floods in the Big Thompson River canyon, west of Loveland, Colorado, in 1976, in the Black Hills and Rapid City, South Dakota, in 1972, and in Oahu, Hawaii, in 1974 (Schroeder 1977) were all initiated by this mechanism. The Big Thompson storm in particular has been well documented in a series of observational papers (Caracena et al. 1979; Maddox et al. 1977, 1978), and has recently been the subject of a numerical sensitivity study (Yoshizaki and Ogura 1988). Details of that storm provide an interesting example of a storm initiated by the forced lifting mechanism.

Conditions that can lead to flooding in a mountain valley occur when 1) a cloud (system) produces copious rainfall, and 2) rainclouds either continually reform over a certain area of the valley or remain stationary with respect to the valley topography. Deep cumulonimbus clouds that form in an environment with little shear in the cloud layer can be prodigious rain producers (Marwitz 1972a). Light cloud-layer winds blowing toward a barrier can keep thunderstorms that form along the barrier nearly stationary, provided low-level (boundary-layer) moisture is supplied to the storm. These conditions prevailed during the Big Thompson storm, as well as the other storms mentioned.

The Big Thompson River canyon opens to the east near the city of Loveland, Colorado, and its western boundary is the Continental Divide in the mountains of Rocky Mountain National Park (Fig. 9.20). Loveland is at an elevation of ~1500 m MSL, and the Continental Divide rises to over 3700 m, and has many peaks over 4000 m. Easterly winds bring moisture from the plains into the canyon, where the flow then ascends toward the divide. Table 9.1 shows the elevations of several locations in the Big Thompson canyon; note that the major rain-producing clouds originated between about 2300 and 2750 m MSL. The following account was taken from Caracena et al. (1979) and Maddox et al. (1978) except where noted.

On 31 July 1976 synoptic-, mesoα-, and large mesoβ-scale processes conspired to produce a situation ripe for a stationary, heavily precipitating storm system in the mountains. The midtropospheric (50 kPa) synoptic pattern showed Colorado between a negatively tilted (oriented northwest–southeast) ridge receding to the east and a short-wave trough advancing from the southwest, which left central Colorado in weak south-southeasterly flow with weak shear in the cloud layer.

At the surface a cold front had passed through the area the day before, but a somewhat stronger secondary cold front had moved into northeastern Colorado from the northeast and was moving toward the southwest (i.e., toward the mountains) that day. The secondary front had three important characteristics: 1) the flow behind the front was strong and easterly below ~$1\frac{1}{2}$ km AGL, 2) the band of air behind the front had a very high moisture content (over 14 g kg^{-1}) for the region and also had very strong latent and potential instability, and 3) the front had a capping inversion that prevented deep moist convection behind the front; i.e., prevented release of the moist instabilities. Although the air was cooler behind the front, its θ_e was higher; witnesses observed that cloud bases

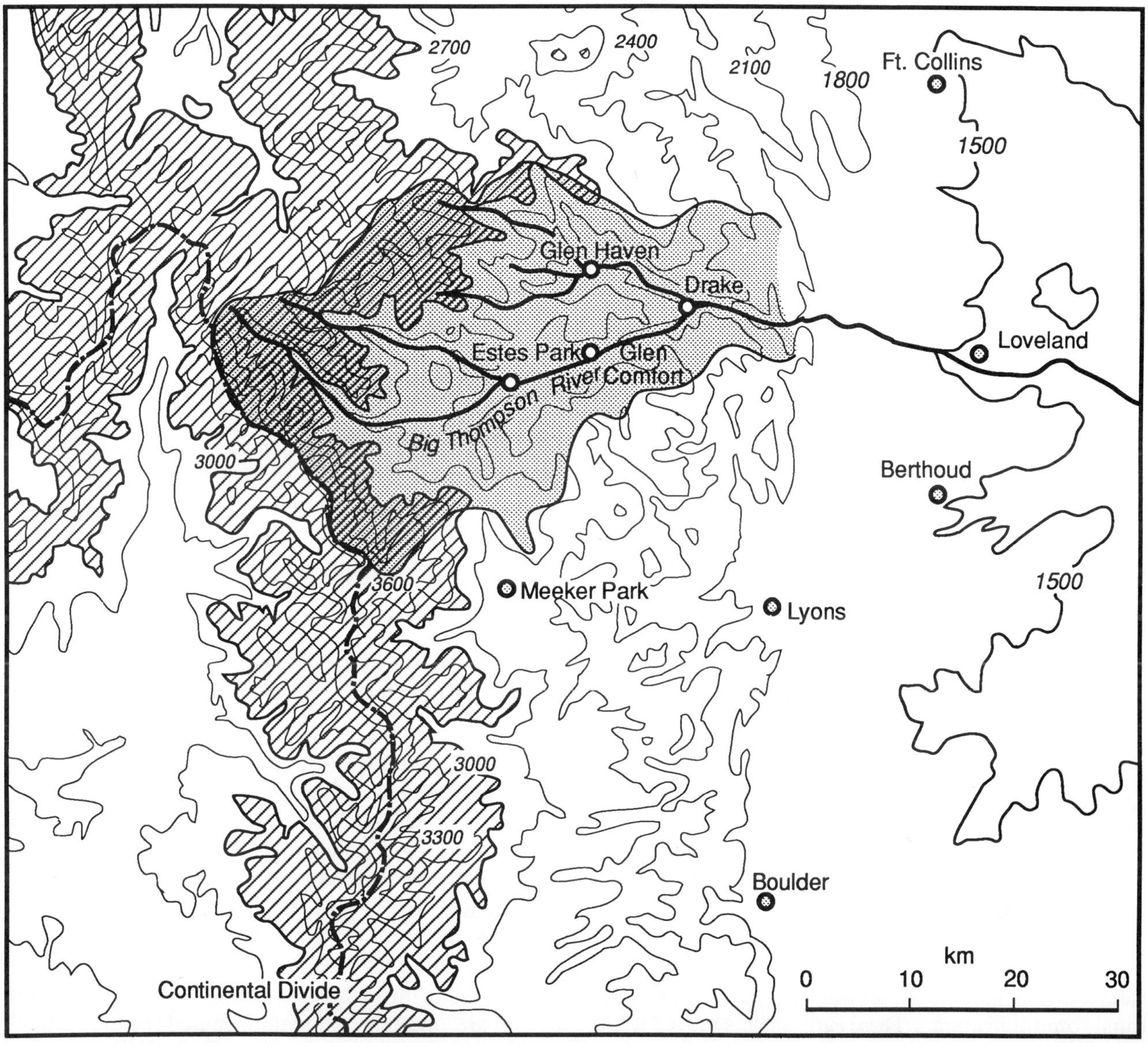

FIG. 9.20. Topographic map of the Big Thompson Canyon area. The shaded region indicates the Big Thompson drainage. Contours are at 300-m intervals, the cross-hatched region represents elevations of over 3000 m MSL, and the dash–dotted line indicates the Continental Divide. (After Yoshizaki and Ogura 1988.)

were lower behind the front than in the warmer air ahead of it.

Also at the surface, a low-pressure center intensified in western Colorado (probably associated with the advancing short-wave trough in the midtroposphere) and high pressure continued to build in behind the cold front in northeastern Colorado. This intensification of the east–west pressure gradient accelerated the easterly flow in the low-level jet behind the cold front to over 20 m s^{-1}. This jet would be the feature that continuously fed high-moisture air upslope to the storms as they grew in the mountains.

Mesoscale effects also contributed to the environment that produced the storms. A line of thunderstorms ahead of (to the south of) the cold front produced a mesoscale high to the south of Loveland near Boulder (see Fig. 9.20). This high retarded the advance of the cold front and caused it to stall, confining the moisture-rich air and the easterly jet behind the front to the region north of Boulder. The jet thus funneled moisture into the mountain area to the north and west of Boulder.

The resulting postfrontal atmospheric sounding, reconstructed for Loveland at about the time the storms were starting in the foothills on the evening of 31 July, is given in Fig. 9.21a; the lifting diagram for this sounding is given in Fig. 9.21b. The strong, low-level easterly flow, weak south-southeasterly flow aloft, capping frontal in-

TABLE 9.1. Elevations above mean sea level of locations along the Big Thompson River, Colorado (McCain et al. 1979).

Location	Elevation (m, MSL)
Long's Peak	4346
Continental Divide	3660
Big Thompson headwaters (near Continental Divide)	3350
Area from which flood derived	2290–2750
Western limit of flooding	2290
Drake, Colorado	1870
Canyon mouth	1600
Loveland, Colorado	1500
Confluence with South Platte River (La Salle Colorado)	1425

version, and extreme instability in the air above the inversion are all evident. The stage has been set by larger-scale processes for heavy precipitation in the mountains. It only remains for the extreme latent instability to be released, by lifting moist air from below the frontal inversion to its LFC above the stable layer.

These studies of the storm are a postanalysis based on available data. The data network was too coarse to determine the precise manner in which the thunderstorms were initiated, but there are several possibilities that illustrate processes discussed in this section:

1) Existing thunderstorms in the prefrontal air had been initiated by cold-frontal lifting. As these storms moved west into the foothills northwest of Boulder, they were overrun (undercut) by the front. Instead of dissipating, they were able to tap into the moisture-rich postfrontal air, invigorating the storms (Caracena 1989, personal communication). These early storms moved to the west and northwest more in response to steering winds and the advance of the front than to the orientation of the terrain (Caracena et al. 1979). Gust fronts from these moving storms could lead to new cell development, illustrating secondary storm development, as discussed in the next section.

2) θ_e decreased with height in the inflow layer below the frontal inversion. Thus forced lifting could have released potential instability and initiated convective clouds. Orographic stratus clouds (indicating forced lifting and identified in section 9.4.2 as a frequent precursor to the release of potential instability) were observed over the foothills and the plains just east of the foothills (Caracena et al. 1978).

3) In the nonsaturated layers of the ascending atmospheric column, lifting destabilizes the layers with respect to parcel displacements. Specifically, lifting cools the frontal inversion layer; e.g., lifting of only 3 kPa (about 300 m) cools the inversion enough that a surface parcel is at its LFC as soon as it becomes saturated, as shown in Fig. 9.22. (Note that even if the strength of the inversion layer were underestimated in the constructed Loveland sounding, it would only take a little more lifting to overcome the effects of stronger stability.) This process, which illustrates destabilization by nonsaturated lifting, could help trigger deep convection in combination with either process 1 or 2.

The mechanism that actually initiated the Big Thompson storm combined with a highly favorable larger-scale environment to produce a considerable amount of rain in a small area over the Big Thompson drainage (Fig. 9.23). Caracena et al. (1978) estimate that 250 mm of rain fell over the town of Glen Comfort in 4 h. This tremendous amount of water, funneled into a small canyon, produced rises of up to 6 m in the water level along some stretches of the river (McCain et al. 1979). The flood took approximately 140 lives and left over $35 million in property damage.

The structure and microphysics of the storm in its mature stages (when it was multicellular, with successive cells forming, growing, and decaying over the same region) also contributed to its heavy rains. The troposphere was very warm, with the freezing level above 5.6 km; so much of the precipitation was probably formed by warm-rain (collision and coalescence) processes, similar to those in many tropical clouds. In a low-shear environment this normally leads to a cloudburst in which showers are strong but generally last for less than one hour. The short lifetimes in this case occur because the cumulonimbus towers are nearly vertical, and the rain falls into the updraft, killing the storm. In the Big Thompson storm, however, the cloud ingested air with high easterly momentum from below cloud base, and this momentum caused the towers to tilt to the west just enough for the rain to fall out of the updraft. This allowed the storm to remain at near steady state for several hours. Caracena et al. (1979) estimated the precipitation efficiency of the storm to be nearly 85%, a very high value for a Colorado thunderstorm.

The detailed discussion of the well-studied Big Thompson flood in this section is justified by the striking similarities it has with the Rapid City and Oahu floods. The synoptic conditions in all three cases featured light winds aloft in the troposphere, strong, moist flow in the subcloud layer blowing up the slope of a barrier, and a warm, tropical air mass with a high freezing level. The only major difference among the three cases is that in the Big Thompson and Rapid City storms the LFC was well below the height of the ridgeline, so that the heavy rains fell midway up the windward slope, with respect to the subcloud layer winds. In the Oahu case, the LFC was just above the major ridgeline, and the heavy precipitation fell just to the lee

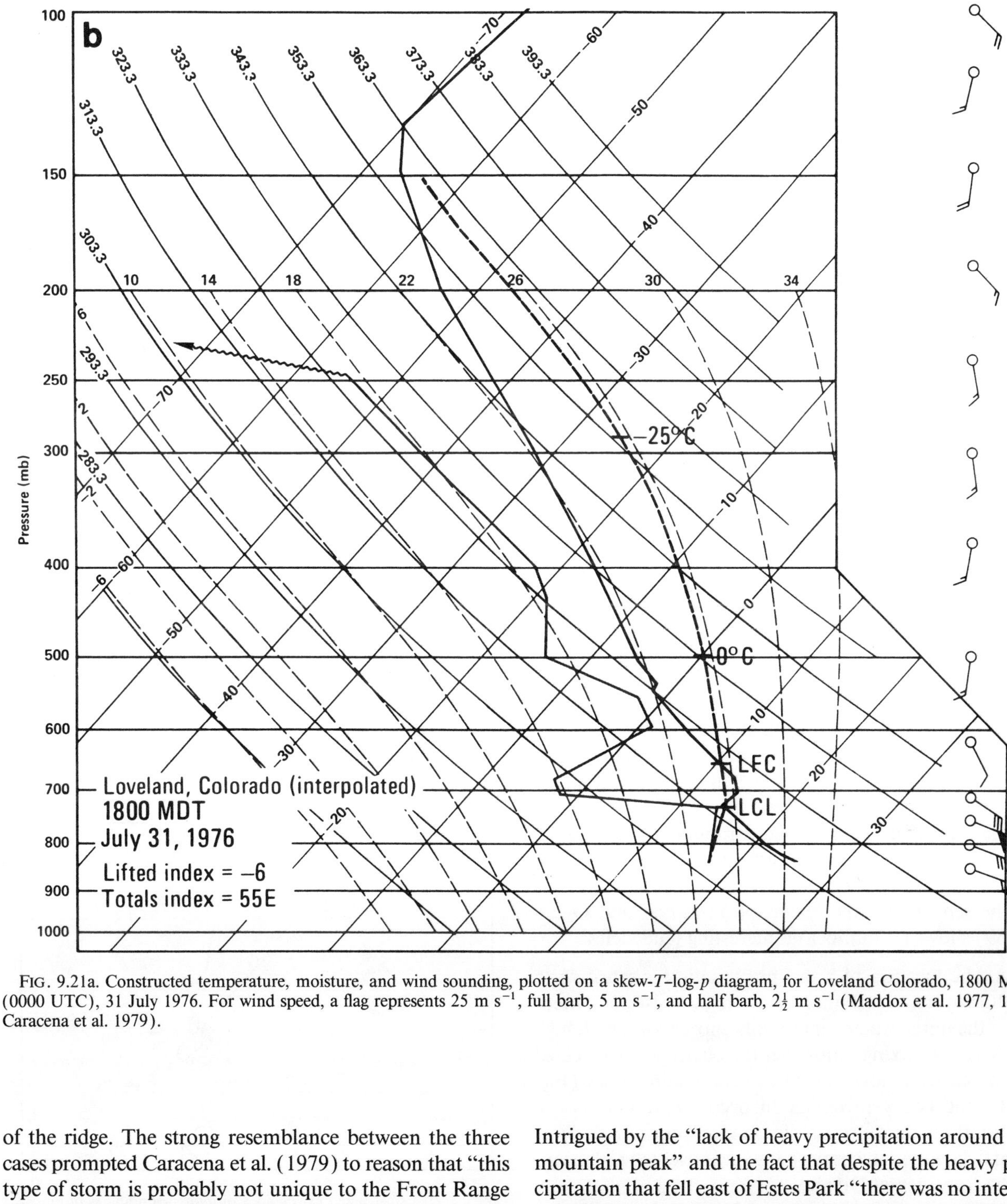

FIG. 9.21a. Constructed temperature, moisture, and wind sounding, plotted on a skew-T–log-p diagram, for Loveland Colorado, 1800 MDT (0000 UTC), 31 July 1976. For wind speed, a flag represents 25 m s^{-1}, full barb, 5 m s^{-1}, and half barb, $2\frac{1}{2}$ m s^{-1} (Maddox et al. 1977, 1978; Caracena et al. 1979).

of the ridge. The strong resemblance between the three cases prompted Caracena et al. (1979) to reason that "this type of storm is probably not unique to the Front Range area, but it can occur anywhere . . ." that the environmental conditions listed above are met.

Another indication of the generic nature of the storms is the numerical study by Yoshizaki and Ogura (1989). They used a nonhydrostatic cloud model in two-dimensions and three-dimensions with a warm-rain microphysical parameterization and highly smoothed and simplified terrain shapes to simulate the Big Thompson storm. Intrigued by the "lack of heavy precipitation around the mountain peak" and the fact that despite the heavy precipitation that fell east of Estes Park "there was no intense precipitation on higher mountains from Estes Park to the Continental Divide," they sought to determine factors that are important in the distribution of precipitation. They used a smoothed version of the constructed Loveland sounding (see Fig. 9.21), but considering the generic topography of the simulations, one should regard their results as general rather than specific to the Big Thompson area.

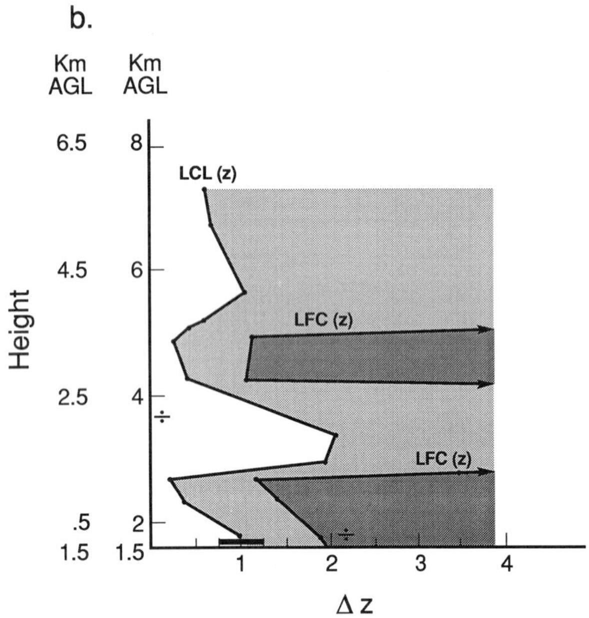

FIG. 9.21b. Lifting diagram from sounding in (a). Height of Continental Divide is indicated along each axis by a ÷ sign, and the bold line segment along the Δz axis indicates the heights of the terrain over which the flood-producing storms formed. It is interesting that simple lifting to this height does not raise a surface parcel to its LFC. If destabilization by forced lifting of the air column lowers the LFC by 1 km or perhaps slightly less, however, then a surface parcel will reach its LFC in the elevation interval where convection was observed to begin.

Yoshizaki and Ogura's simulated results for their control experiment were similar in many respects to the observations reported by Caracena et al. (1979). The storm formed midway up the windward slope and rained heavily there, it rained little on the higher mountains to the west, and the storm itself tilted westward with height, all in agreement with observation.

The sensitivity experiments of Yoshizaki and Ogura (1989) showed a number of interesting tendencies:

1) The location of heavy precipitation is sensitive to the horizontal distribution of the high-moisture-content air. If the initial water vapor is inhomogeneous, with high water-vapor mixing ratios over the plains being advected by the easterly flow toward dry mountain slopes (Fig. 9.24), the heavy precipitation occurs midslope as observed. If the moisture distribution is uniform in the horizontal, however, the heavy rains occur over the high peaks of the Continental Divide.

2) The location of heavy precipitation is sensitive to the location of the first occurrence of deep convection. In the uniform-moisture simulation, heavy precipitation falls at midslope rather than at the Continental Divide if cumulus convection is first started artificially by a warm bubble at the base of the slope.

3) The numerical solution is sensitive to the presence of the frontal inversion. When the inversion is smoothed out, as in the experiments described above, the cumulonimbi resemble those observed. When the inversion is added, however, orographic stratus clouds form in agreement with observation, but the simulated cumulonimbi are unrealistic.

4) The location of heavy rainfall is sensitive to some changes in terrain shape but not to others. For example, the addition of a bump to the slope does not change the midslope occurrence of the heavy rain, but the combination of a bump with a longer slope up to a higher Continental Divide confines the rains to locations between the bump and the divide; i.e., to higher elevations than observed.

These experiments indicate that either an inhomogeneous moisture distribution or preexisting convection (or both) may have been important in determining the location of the heavy precipitation. These factors were probably more important than the details of the terrain shape, indicating that topographies other than the Big Thompson canyon can produce midslope heavy rains under similar atmospheric temperature, moisture, and wind conditions. The role of the inversion was not clarified by the simulations, since it was present in the observations, but produced unrealistic cumulonimbus clouds in the simulations.

9.4.1.2. THERMALLY GENERATED MOUNTAIN CIRCULATIONS

Wind systems generated by surface heating in mountainous terrain, such as those discussed in Whiteman,

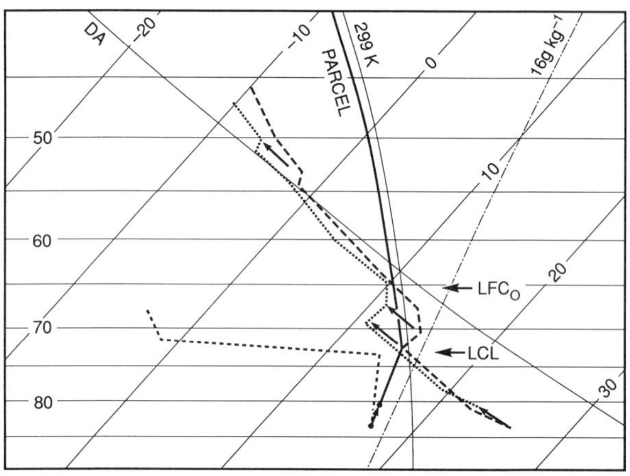

FIG. 9.22. Constructed Loveland, Colorado, sounding as in Fig. 9.21 (dashed lines: long dashed is temperature and short dashed is moisture), and the same temperature profile lifted 3 kPa, or ~300 m (dotted line). The solid line below the LCL indicates parcel mixing ratio, and above the LCL, parcel mixing path conserving θ_e. Horizontal lines are pressure in kilopascals, skewed lines are temperature in degrees Celsius; 16 g kg^{-1} mixing ratio line, 299 K moist adiabat, and a dry adiabat (DA) are indicated; and LFC_0 represents the LFC of the original, unlifted sounding. Note that the inversion caps convection before lifting, but after lifting the inversion cools sufficiently to allow free convection as soon as the parcel is saturated.

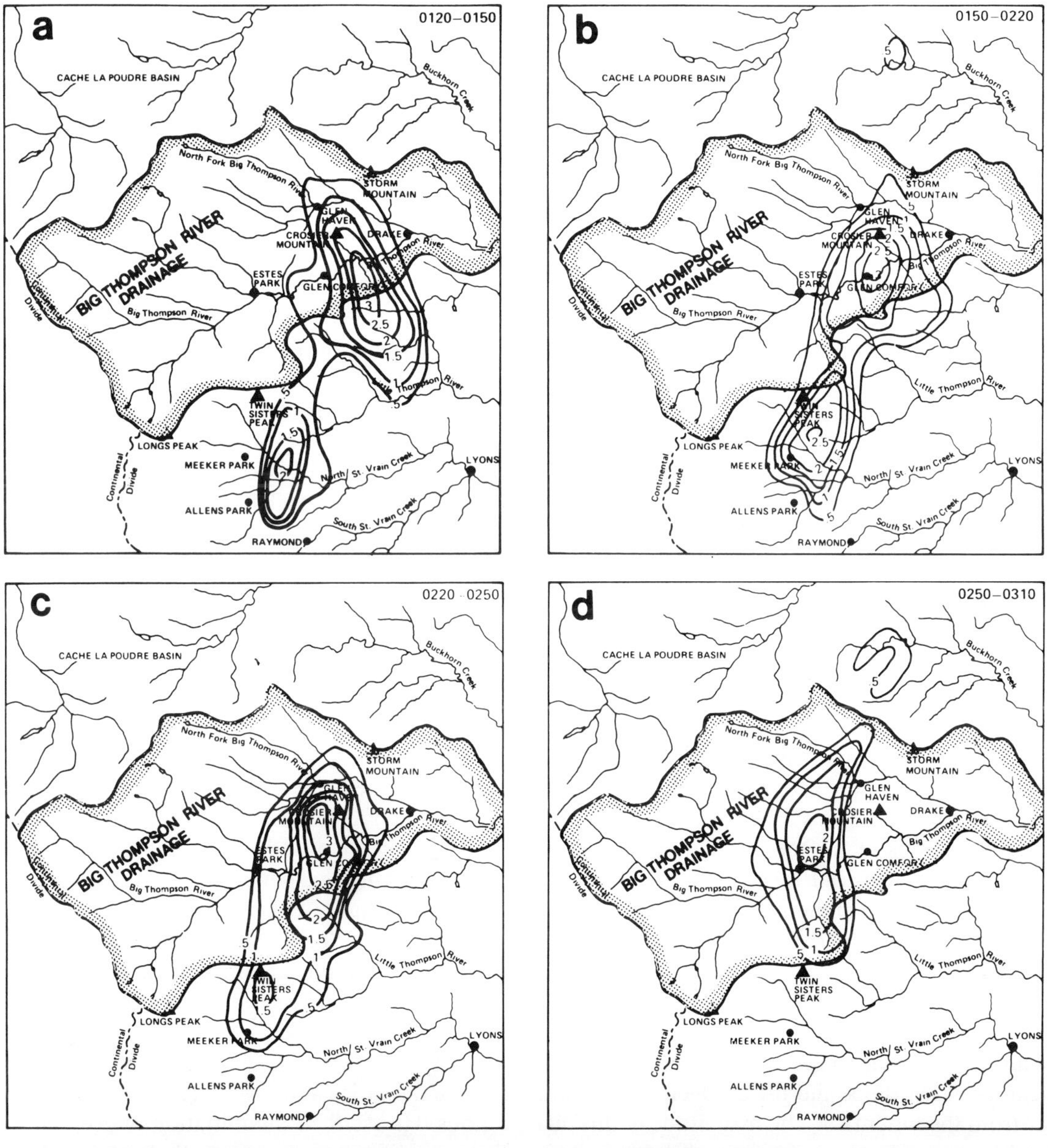

FIG. 9.23. Isohyets of heavy rain component, derived from radar echoes of VIP level 3 or higher, for 30-min intervals from 0120 to 0310 UTC (1920 to 2110 MDT) on 1 August 1976 (Caracena et al. 1979).

Chapter 2 and Egger, Chapter 3, also produce updrafts that initiate cumulus cloud growth. Generically, a blanket of warm air often several hundreds of meters thick covers a mountain heated by the sun. The mountain protrudes into the atmosphere, and the blanket of atmosphere next to the mountain is warmer than the atmosphere at the same level away from the mountain over lower land. This configuration produces relative low pressure in the atmosphere next to the mountain, convergence of flow toward the mountain, and updrafts somewhere over the mountain. When viewed in this way, mountains are referred to as elevated or high-level heat sources.

More specifically, solar heating of a mountain's surface generates upslope and up-valley wind systems (again, as discussed in Whiteman, Chapter 2 and Egger, Chapter 3). These winds blow toward the peaks, producing convergence and updrafts near the peaks. Factors that affect where updrafts form, and thus eventually where clouds form, are discussed in this section.

A significant topic of this section, therefore, is dry mountain flows and circulations that lead to cumulus initiation. Since it is the unstable atmospheric boundary layer (ABL) that becomes the subcloud layer once clouds form, it is important to understand the structure and evolution of the entire ABL, not just the region within 100–200 m of the surface. It is also important to understand the phys-

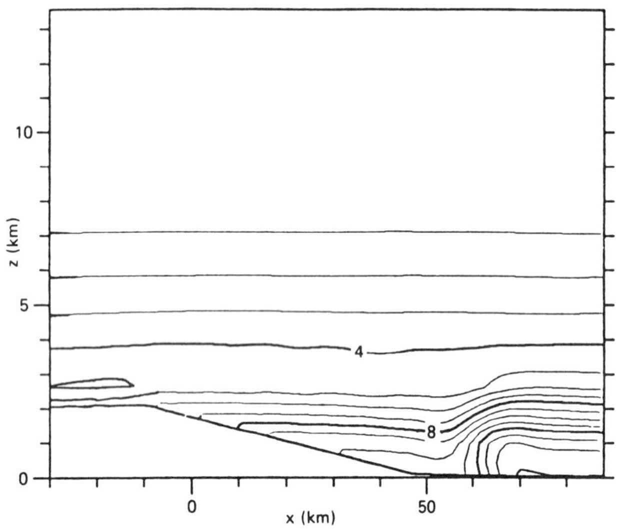

FIG. 9.24. Vertical cross-section of water-vapor mixing ratio distribution for Yoshizaki and Ogura's (1988) horizontally inhomogeneous case near the beginning of the simulation (at 30 min of simulated time). Contours are at intervals of 1.0 g kg^{-1}.

ical processes that affect the strength of the mountain flow systems that lead to cloud initiation and growth. Therefore, the following subsections give examples and case studies illustrating how flow patterns that lead to thunderstorm initiation arise from surface heating. They also discuss how variations in surface heat flux and surface moisture availability can lead to regional and seasonal differences in the strength of thermally forced, solenoidal circulations and thus in their importance as cloud-producing mechanisms.

9.4.1.2.1. Heat flux and soil moisture effects. Thermally forced wind systems that lead to cumulus cloud initiation are driven by the temperature difference between the air near the warm slope and the air over the lowland area away from the mountain. The magnitude of this difference is determined by the magnitude of the "sensible-heat" or enthalpy flux to the atmosphere at the surface, referred to as the surface heat flux. The surface heat flux is controlled by the surface energy budget, as discussed in Whiteman, Chapter 2. The major influence on the partitioning of energy during the daytime is generally the amount of moisture available for evaporation: the more energy that is partitioned into latent heat of evaporation, the less goes into the surface heat flux. Higher latent heat fluxes, and consequently lower sensible heat fluxes, occur over surfaces with high soil moisture. Following this sequence backward, one would expect thermally forced wind systems to be weaker over mountains where the soil moisture content is higher, and stronger over drier mountains. Numerical simulations of upslope flows over a 1-km-high triangular mountain support this expectation. Ookouchi et al. (1984) found a reduction from peak upslope speeds of 6 m s^{-1} for a mountain with dry soil to peak speeds of 3 m s^{-1} for one with moist soil.

More recently Segal et al. (1988b) found a similar reduction for a mountain with moist, vegetated slopes as opposed to one with dry, bare soil. At night Banta et al. (1987) also found reductions in the thermally forced downslope circulation with increased soil moisture. In this case higher soil moisture produces higher thermal conductivity of the soil. Reductions in the magnitude of the sensible heat flux and surface cooling result from increases in the downward flux of heat into the soil, rather than the latent heat flux of the daytime case.

The implication of this argument for cumulus initiation is that cumulus initiation mechanisms that result from thermally forced flows should be more prevalent in arid regions than in moist regions. In other words, since increased moisture availability reduces the strength of solenoidal circulations, one expects the effects of mountains on daytime cumulus cloud initiation to be strong in drier regions and more subtle in moister areas.

9.4.1.2.2. Regional differences. The strength of thermally forced mountain flows is strongly controlled by the amount of moisture available at the surface for evaporation (or transpiration from plants); therefore there are regional differences in the strength of these circulations and in their effectiveness in initiating cumuli. Moister regions, such as eastern North America and much of Europe, are often covered by deciduous forest or other vegetation with high evapotranspiration characteristics. These areas should have weaker solenoidal wind systems, which would have a strong role in convective cloud initiation only under special circumstances. For example, Hagemeyer (1984, 1985) studied thunderstorm initiation in the Upper Current River valley of the Ozark Mountains in Missouri. He found that storms form regularly at the head of the valley under very light tropospheric wind conditions. He ascribed the initiation of deep cumuli under these conditions to the thermally forced upslope and upvalley wind systems. Local thermally forced circulations thus are able to have a role in thunderstorm initiation in this case only when the winds and shear in the troposphere are weak; i.e., when the synoptic- and mesoα-scale forcing is weak.

In moist midlatitude areas including the eastern United States, summertime thunderstorms are often associated with organized larger-scale systems, such as cold fronts, squall lines, or other mesoscale convective systems. In arid mountainous regions, however, such as the Rocky Mountains, cumulus and thunderstorm formation seem to be more tied to the mountains and mountain circulations. Short waves, jet stream maxima, and other synoptic and larger mesoscale influences can determine how severe the thunderstorms will be, but initiation is generally tied to terrain features.

This may be due in part to how difficult it is to get a thunderstorm started in each environment, as well as to the fact that larger (mesoα-) scale systems often tend to be stronger in the eastern United States. We know from numerical cloud modeling experience for that area, for example, that it is often relatively easy to produce deep moist convection using input from a thunderstorm proximity sounding. Often simply initializing a model with a few saturated grid points in such an environment will cause a cloud to grow the depth of the troposphere. In simulating clouds from the western United States, though, it is generally necessary also to provide a substantial initial vertical velocity perturbation in the saturated region (5 m s^{-1} or more), to overcome the stability of the dry environment and generate a cloud that can grow to the tropopause (e.g., Banta and Hanson 1987b). In other words, because of the reduced boundary-layer moisture in the drier climates, stronger forcing is required to produce deep moist convection.

Thus, in drier regions deep moist convection needs a greater vertical velocity pulse to get started. In these regions the stronger solenoidal circulations around mountains do provide a stronger boost. Because these circulations can produce the strong vertical velocities necessary for cloud initiation, they are an important factor in the generation of mountain thunderstorms, and they influence where storms form.

In addition to the differences in forcing, the moisture disparities between the two regions of the United States lead to differences in the nature of the storms. Eastern thunderstorms tend to have low cloud bases (a few hundred meters) and rain copiously on the ground, whereas western storms have high bases (1–2 km or so) and often have virga or sparse rainfall because of evaporation below the high bases. Doswell (1980) pointed out that severe weather in the arid High Plains just east of the Rocky Mountains arises from synoptic or large-mesoscale environments that are relatively benign by central and eastern United States standards. The same could very likely be said of the mountain thunderstorm environment farther west.

The discussion in the rest of this section focuses on cumulus initiation in drier regions where topographical influences are less subtle. Most of the studies cited here are from the dry western United States, where this phenomenon has been studied since the late 1950s. In this section, "western United States" refers to the arid, mountainous regions, excluding the moist Pacific coast.

9.4.1.2.3. Spatial and temporal distribution. For reasons just discussed, mountain cumuli are strongly tied to the peaks in regions where thermally forced initiation mechanisms prevail, especially in the clouds' formative stages. Thunderstorms tend to form in preferred areas of the topography (sometimes referred to as "hot spots") and at characteristic times of the day. This has long been known to inhabitants, travelers, and vacationers in the mountains, and has led to many peaks with names like Storm Mountain and Storm King. In an effort to be more quantitative about the relationship between thunderstorm occurrence and topography, researchers have used a variety of instrumentation systems. Early studies using raingages found that summertime precipitation tended to favor high ground, the areas around mountain peaks. Studies using weather radar (Karr and Wooten 1976; Henz 1974; Wetzel 1973) showed not only that radar echoes in the morning and early afternoon favored the mountains, but also that certain regions of the mountains were favored over others (Fig. 9.25).

A recent addition to our ability to monitor thunderstorm activity over data-sparse regions is the ground-strike lightning detector. Ground-strike data were used to build a short "climatology" of thunderstorm occurrence over Colorado for a summer (Lopez and Holle 1986) and over the western United States for two summers (Reap 1986). Again, cloud-to-ground lightning strikes, especially early in the day (up to 1300 MST), preferred certain mountain locations (Fig. 9.26).

Another useful tool for studying thunderstorm behavior has been the geostationary satellite. One approach is to use data from many sample days to composite additively all the images for a given time of the day. White areas in the composite images represent areas of repeated cloudiness at that hour, black areas are usually clear, and gray areas are sometimes cloudy. Klitch et al. (1985) did this for Colorado and found differences in early-morning composites between days that have severe weather and days that do not.

Another approach for using geosynchronous satellite data is to animate the images in reverse, to trace thunderstorms back to their area of initiation. Banta and Schaaf (1987) and Schaaf et al. (1986, 1988) reported on results over Colorado and northern New Mexico using this technique. Thunderstorm anvils were identified on late morning or early afternoon satellite images. Then, employing the "looping" capability of the McIdas computer system being used, the images were looped backward, tracing each thunderstorm back to where it originated. Thunderstorm initiation data gathered for three summers were plotted on a terrain map. As in the radar and lightning detector studies, the thunderstorm genesis areas grouped together into preferred regions of the mountains. The study also found that the ridgetop wind direction, determined from the 70 or 50 kPa flow, had an influence on where thunderstorms started; i.e., on which genesis areas were activated on a given day. This is discussed again in section 9.4.1.2.5.

In those studies two terms were defined that pertain to the temporal (diurnal) behavior of thunderstorms. *Pri-*

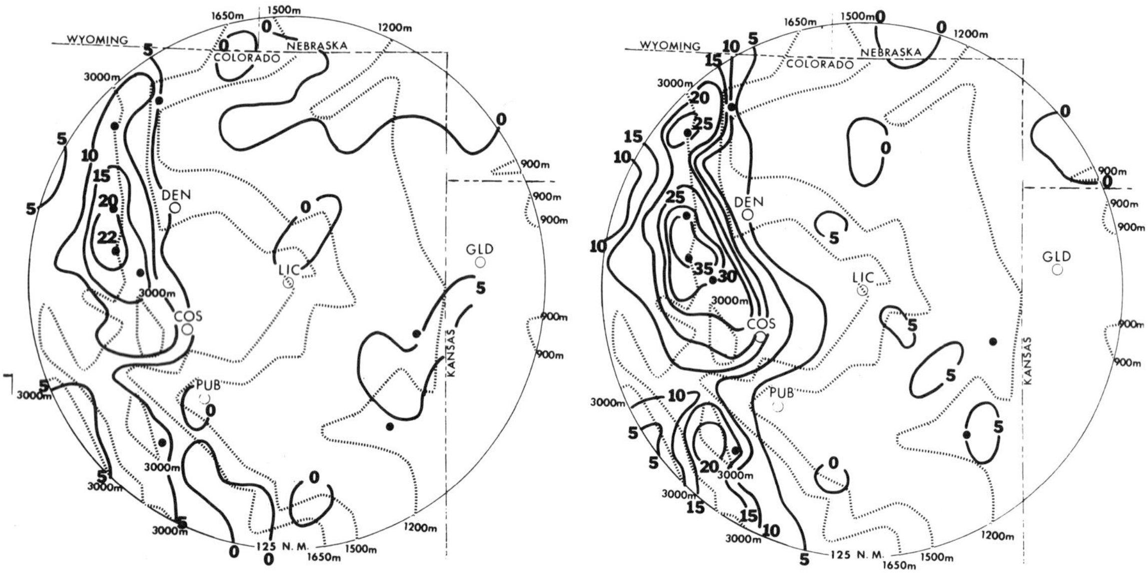

FIG. 9.25. Spatial distribution of contours of frequency (percentage of all days when echoes are present in the area) of summer radar echoes over eastern Colorado for the late morning hours of (a) 1030 MST and (b) 1130 MST (Karr and Wooten 1976).

mary convection refers to those cumulus clouds that originate solely from upward air motion driven by terrain or thermally induced flows. These storms form over the mountains, the clouds form in the morning or early afternoon, and they produce afternoon precipitation maxima. *Secondary convection* has some type of forcing from previously existing cumulus convection as an ingredient in its formation. This forcing could consist of a gust front or a gravity wave, for example. Secondary thunderstorms tend to occur more over the plains, and they produce late-afternoon or evening precipitation maxima, as in the case of the well-known nocturnal maxima over the Great Plains. Hales (1975), for example, gives an example of a severe thunderstorm that formed at the confluence of the gust fronts from two thunderstorm systems (that incidentally originated in the mountains), an excellent representation of secondary convection. Many of the storms observed to form along intersecting convergence lines by Wilson and Schreiber (1986) were secondary storms, since one of the convergence lines was often a gust front.

Studies in Colorado, Utah, and Arizona verify that storms tend to form earlier in the day over higher ground. A radar study in Colorado showed that for areas near the mountains the peak in echo frequency was at about 1430 MST, whereas regions farther out in the plains exhibited maxima at about 1700 MST or later (Karr and Wooten 1976). At hourly reporting stations in Utah, those stations above 2100 m elevation had their peak in measurable precipitation at 1430 MST, stations between 1800 and 2100 m had the peak at 1500 MST, with the larger amounts occurring somewhat earlier, and stations below 1800 m had peak frequencies at 1700 MST (Astling 1984). In Arizona the frequency maxima in the mountains, especially in the northern half of the state, occur in the afternoon, but the maxima occur progressively later in the evening in the valley of south central Arizona; as shown in Fig 9.27, Phoenix has a maximum at midnight (Balling and Brazel 1987; Hales 1977). The point that moist convection occurs earlier in the day at higher elevations than at lower elevations is nicely summarized for the western United States by Reap's (1986) analysis of ground-strike lightning-detector data (Fig. 9.28).

Two important questions about mountain thunderstorms are: 1) Where do they form? 2) Where do they go once they have formed? This section on cloud initiation covers factors that affect where and when storms are likely to form, but there is little information on where thunderstorms in the mountains move once formed. It is observed that sometimes thunderstorms move with the midtropospheric or "steering" winds, that sometimes mountain storms move along valleys, and that sometimes a thunderstorm can even propagate into the wind and up the side of a mountain following its gust front (Fujita et al. 1962). Obviously more work is needed in this area.

9.4.1.2.4. Isolated peaks and small ranges. What are the processes that determine where clouds will be initiated over a mountain? In general which processes become important depends on the winds near and below mountaintop. In order to be more specific, it is convenient to consider isolated peaks or small ranges of mountains (with diameters of less than several tens of kilometers) and larger mountain ranges separately. This section examines isolated peaks and small ranges. Material presented by

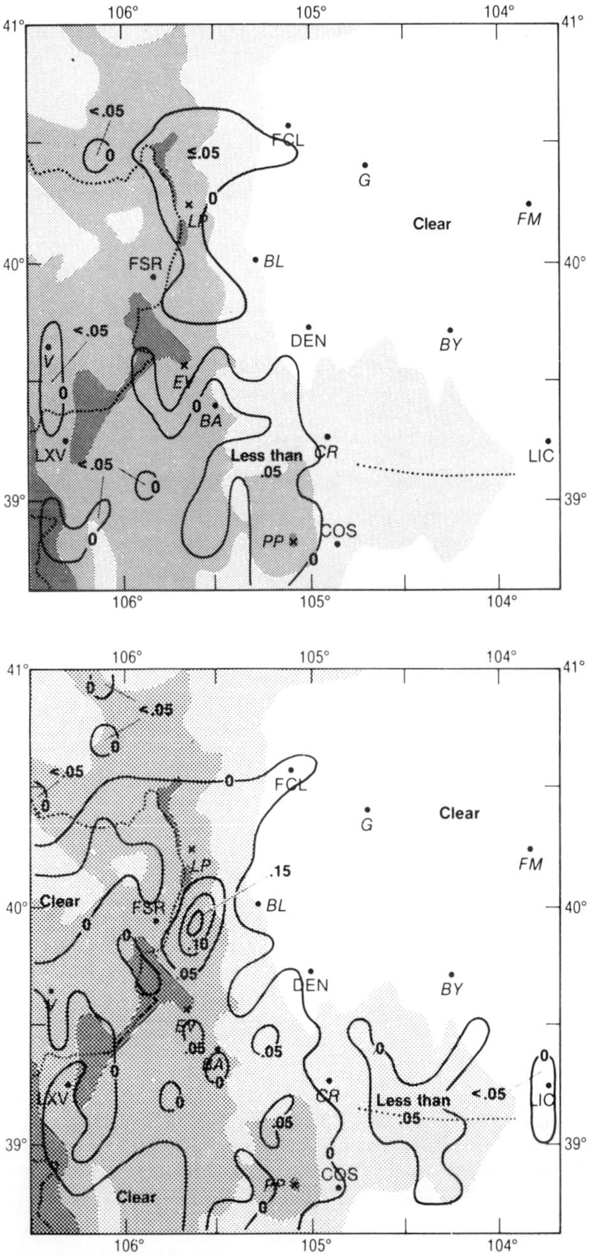

FIG. 9.26. Spatial distribution of lightning flashes, as determined from lightning detector data. Contours give percentage of total echo population (104 019 echoes) during July and August 1983 in each $\frac{1}{8}$ by $\frac{1}{8}$° lat/long box (Lopez and Holle 1986).

MacCready (1955), based on his experiences as a glider pilot, is a suitable background to this discussion. A glider pilot, of course, needs to find updrafts to stay aloft, and updrafts are where cumulus clouds form.

MacCready noted that early in the morning the only upward currents are over hills; later in the day some updrafts appear over valleys or lowlands, but the strongest are still over the hills. He observed that surfaces sloping toward the sun heat up faster and that on a hot day the difference in surface temperature between the valley and high on the mountain was usually less than dry adiabatic ascent would suggest. Thus potential temperature θ near the surface must be greater at higher elevations. Since the air is also usually drier on the mountain, he reasoned that cloud base over the mountain must be higher than cloud base over the valley. Orville (1965) later found photogrammetric evidence for this over the mountains near Tuscon, Arizona.

MacCready attributed the characteristic clear weather (at least during morning hours) that we see over valleys and plains adjacent to mountain ranges to compensatory subsidence. This sinking motion represents the downward branch of the circulation driven by the intense rising motions over the mountains. He also found that ambient horizontal winds strongly affect the structure of the thermals (updrafts), and that once cumulus clouds form, they can exert a strong upward pull on the atmosphere below.

With these qualitative observations in mind, one can appreciate a number of important field projects in Arizona in the late 1950s and early 1960s, which studied cloud growth over isolated peaks and relatively small mountain ranges. Several investigators studied cumulus development in low-wind environments.

Near Flagstaff, Arizona, Fujita et al. (1962) and Brown (1966) documented the life histories of two thunderstorms that formed in weak flow over the San Francisco Peaks, a small range of mountains north of the city (the lifting diagram for Brown's case was presented in Figs. 9.3 and 9.5). They found that the most important factor in initiating the first cumuli is terrain aspect: clouds first form over the east-facing slopes of the mountain, which face the early-morning sun. Fujita et al. defined a deviation surface pressure as the difference between the observed station pressure and the mean for each station for that day. Using this pressure to analyze the surface pressure field for stations at different elevations, they show a low-pressure center forming over these heated slopes. Their surface analysis also shows convergence of the mesonet surface wind field toward the low-pressure center as the clouds form there (Fig. 9.29).

Todd (1964) noted the strong influence of terrain aspect and favorable sun angle on where morning cumuli appear during a clear-sky and light-wind day. Valleys on the south and southeast side of the peaks north of Flagstaff have an east-facing sidewall that is an especially effective source region for cumuli. Todd found, too, that higher peaks produce earlier clouds: the highest peaks in the vicinity (over 3800 m MSL) produce clouds about 50 min before lower peaks nearby (2900 to less than 3200 m).

The location of cloud initiation is affected by the presence of valleys running up the mountain and also by the fact that moisture advection toward the peak may be greater from one direction than from another. For ex-

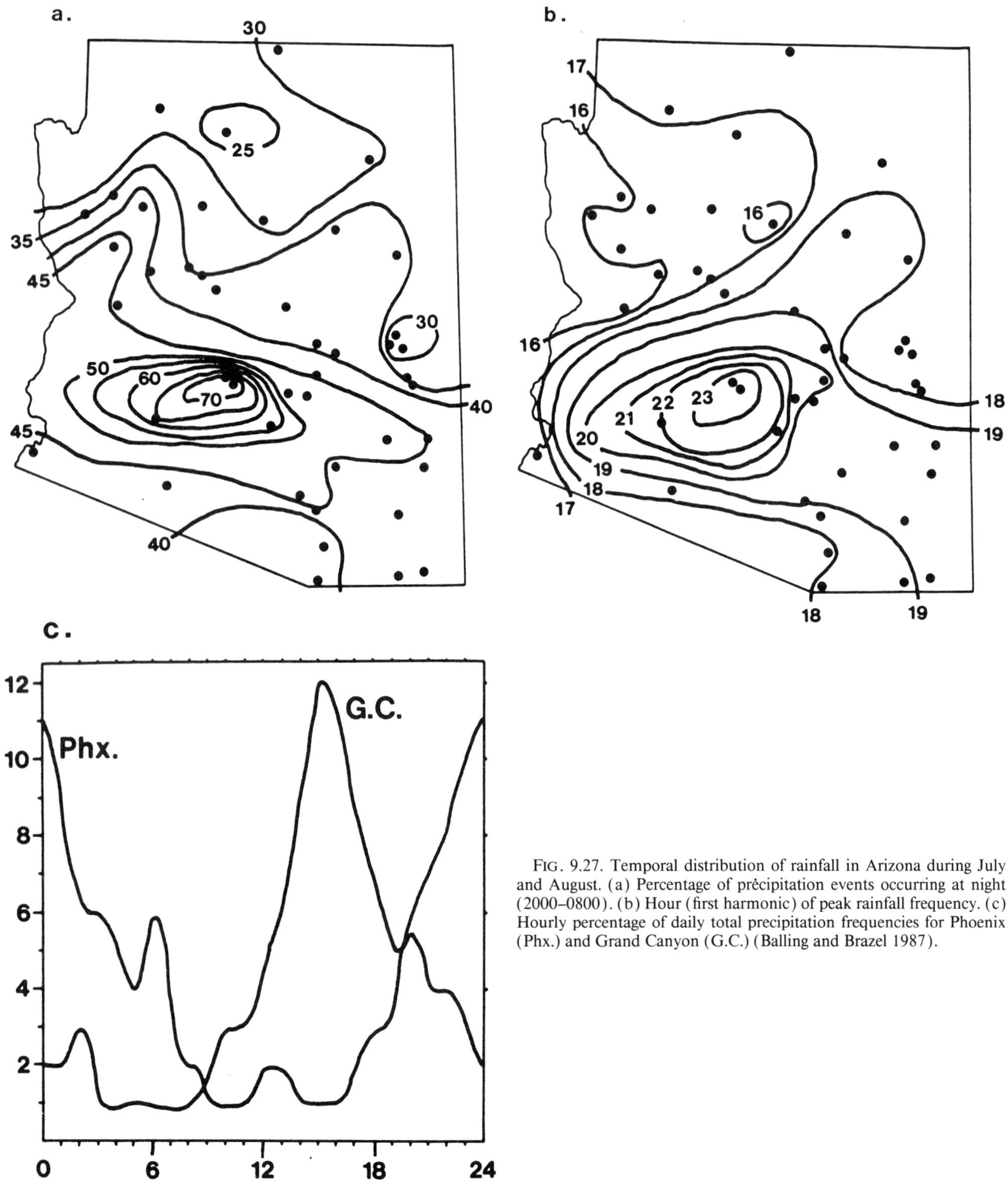

FIG. 9.27. Temporal distribution of rainfall in Arizona during July and August. (a) Percentage of prĕcipitation events occurring at night (2000–0800). (b) Hour (first harmonic) of peak rainfall frequency. (c) Hourly percentage of daily total precipitation frequencies for Phoenix (Phx.) and Grand Canyon (G.C.) (Balling and Brazel 1987).

ample, Raymond and Wilkening (1980) used aircraft to sample the inflow around a mountain in New Mexico under low wind conditions and found that southeasterly flow had mixing ratios of 3.5 g kg^{-1}, whereas drier air approaching from the northwest had mixing ratios of 2.0 g kg^{-1}. Although this was a dry day, clearly the southeastern slopes would be favored for cloud growth. Another feature of importance is the stable nocturnal inversion layer over the plains that inhibits updrafts and vertical mixing over the lower terrain. The near-surface flow to-

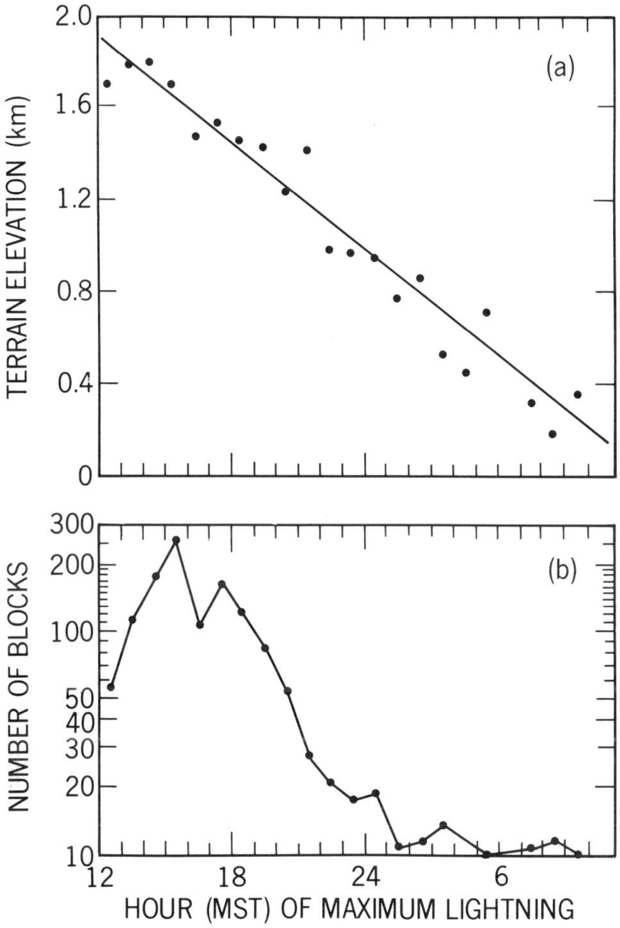

FIG. 9.28. (a) Hour (MST) of maximum frequency of lightning versus terrain elevation, and (b) logarithmic plot of frequency (total number of occurrences) in each hourly category (Reap 1986).

ward the peak cannot escape vertically until it reaches a place where the inversion has broken, and at least early in the day this occurs only over the mountain, as noted earlier by MacCready (1955). The convergence is thus focused so that it is greatest close to the peaks, and the resulting updraft speeds are also greatest over the mountain.

The previously mentioned case studies of Fujita et al. (1962) and Brown (1966) also show two interesting aspects of deep convection in the mountains: the significance of the rainfall pattern and the control of the boundary-layer winds. The total precipitation pattern (as determined by a raingage network) for the storm does not necessarily give an accurate picture of the storm's life history. In the storm studied by Fujita et al. the precipitation pattern resulted from the storm propagating back into the wind and up the slope of the mountain, following its gust front. By itself the total daily rainfall pattern might suggest that the storm started near the peak and advected downslope with the winds. Thus it is important to have the detailed time sequence of the precipitation record and other supporting data in order to reconstruct the history of a storm.

These studies also clearly show the cumulus cloud circulations taking control of the boundary-layer wind field from the local terrain-forced circulations. This happens during the middle part of the day (1300–1500 MST) while the thunderstorm cells are strong and are shading the slopes of the mountains. Late in the day (after 1600 MST), however, the clouds break up over the peaks, and the western slopes are heated by the sun. As had happened earlier in the day on the eastern slopes, a relative low pressure forms over the western slope, resulting in a local convergence region. By sunset highly complex wind and pressure fields (resulting both from the moist convection and from the renewed surface heating) precede the advent of the nocturnal regime.

An important issue is the relationship between cumulus clouds that form over high ground as opposed to those that form over adjacent lower ground. Braham (1958) conducted a thorough study of radar echoes in southern Arizona near Tucson, in which he contrasted the characteristics of echoes that form over the mountains with those that form over the valley. He found that there are more first echoes over the mountains than over the valley, and mountain clouds form earlier in the day. Valley echoes, though, are more vigorous; more valley echoes show growth after first being seen. Valley clouds individually produce more rain, but surprisingly, mountain and valley clouds are equally likely to rain. The observed maximum in summer rainfall over the mountains is therefore due to the larger number of smaller showers that occur there. Finally, because of the low-wind environment, the echoes move little after formation. Although individual cells move very little, Braham did note an apparent movement or propagation of storms because of the "progressive development of new cells at successively greater distances from the mountains."

The effects of ridgetop winds on the location of cumulus initiation is, according to Brown (1966), mostly a matter of blowing over or tilting of the updraft core. One of the consequences of this view is that if cloud base is below mountaintop, clouds will form upwind of the peak, but if cloud base is higher than the mountain, clouds will form downwind. Orville (1965) noted the effects of ambient winds in causing high-based clouds to form downwind of the peaks, and he concluded that winds (when present) are more important than solar angle in determining the location of cumulus initiation.

Thus the processes that decide where clouds will form over an isolated peak or small range of mountains depend on the strength of the winds near and below the peak. With light or calm winds solar angle is most important: east-facing slopes are favored for early-morning cloud initiation. This tendency can be modified, however, by

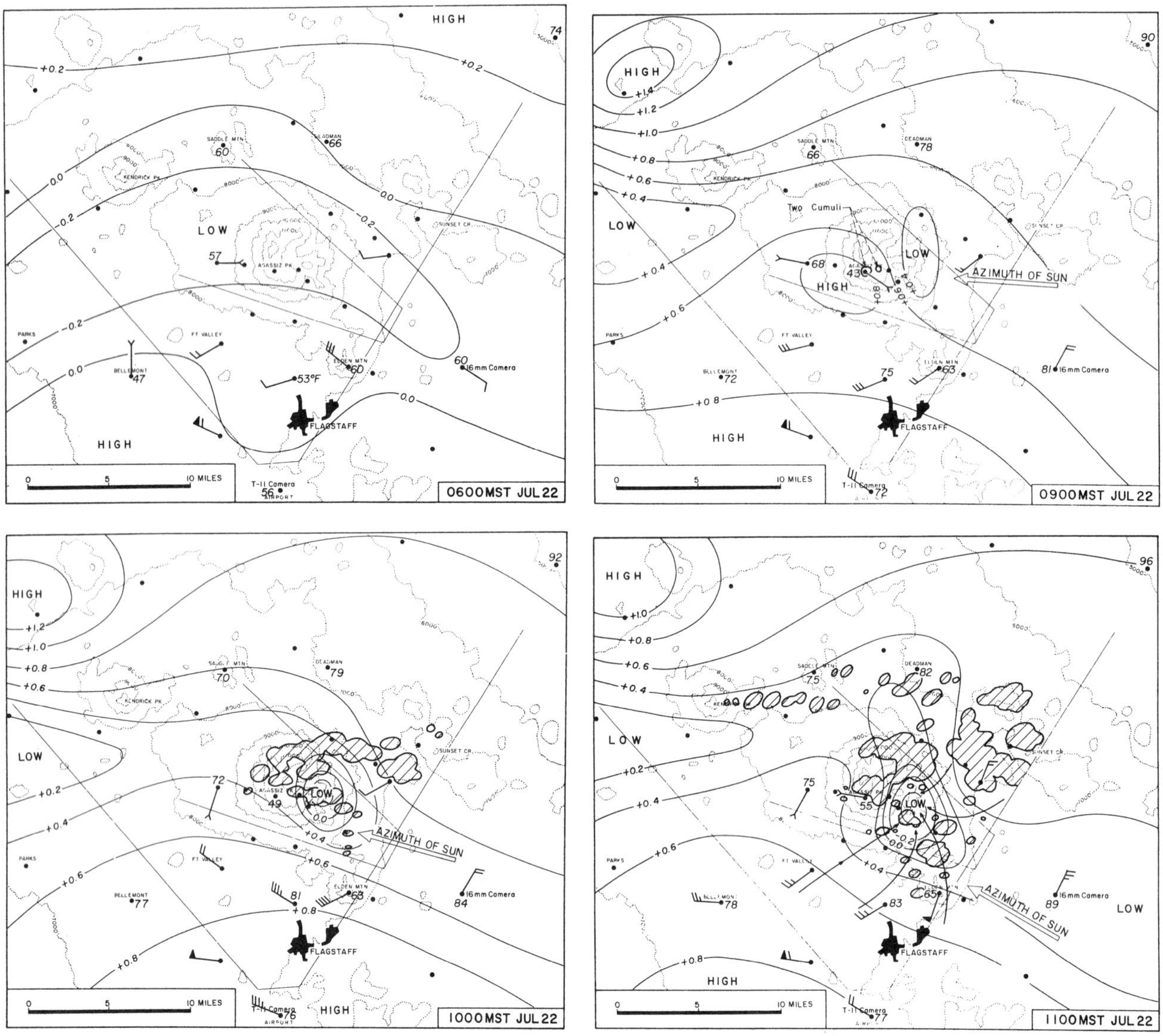

FIG. 9.29. Development of pressure field (deviation from daily mean at each station), winds (wind barbs at each station), and clouds (cross-hatched areas) over the San Francisco Peaks, Arizona. Numbers plotted next to stations are surface temperature in degrees Fahrenheit (Fujita et al. 1962).

asymmetries around the mountain in topography and moisture. Stronger winds (probably ~3 m s^{-1} or greater) when present have a bigger influence than solar angle on cumulus initiation, and clouds tend to form downwind of the peak unless the LCL is below the mountaintop (in which case clouds will tend to form upwind of the peak).

9.4.1.2.5. Larger ranges of mountains. Mountain ranges that are more than 100 km long and several tens of kilometers or more wide present massive thermal sources to the atmosphere and are generally long enough so air does not easily flow around. Ranges oriented north–south present an extensive east-facing slope that favors morning convection when ridgetop winds are light. Ridgetop winds are defined as the winds at or just above the tops of the mountains, as discussed by Banta (1984). When ridgetop winds are not light, the complexity of the terrain undoubtedly dictates that a variety of mechanisms can be responsible for cumulus initiation. These mechanisms probably are composed of a mixture of thermally forced wind systems and obstacle (including wave) effects, since the flow interacts with the topography.

A mechanism that results from the interaction of surface heating effects and ridgetop winds (but not obstacle effects) is lee-side convergence. Lee-side convergence, which is described in greater detail in section 9.4.4.1, arises because high on the lee slopes of a mountain range at sunrise the nocturnal inversion layer is shallower than it is lower

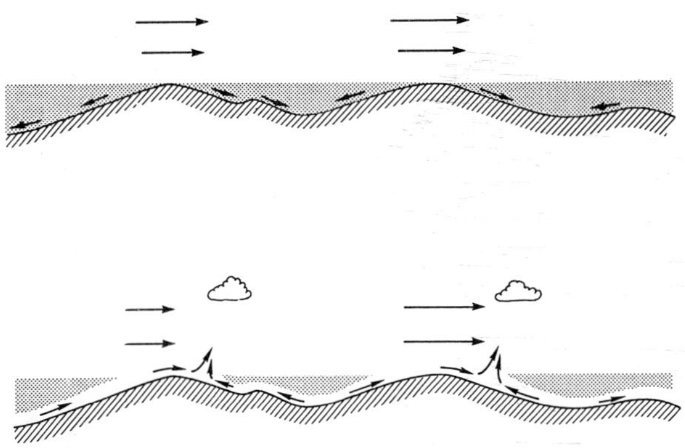

FIG. 9.30. Schematic cross sections of (a) the nocturnal inversion layer, or "cold pool," build-up at night, and (b) the appearance of the shallow mixed layer the next morning. The inversion layer is shaded. The convergence zone forms at the uphill edge of cold pools and leads to cloud initiation (Banta 1984).

down the slopes (Fig. 9.30a). As solar heating erodes the inversion layer from below, three things happen: 1) upslope flow forms beneath the inversion layer, 2) the inversion breaks first where it was shallowest; i.e., high on the slope, and 3) turbulence from the heated surface mixes momentum downward to the surface; on the lee slope that momentum is downslope. This results in a flow configuration similar to that shown in Fig. 9.30b, in which convectively mixed downslope flow from higher elevations on the slope meets upslope flow from lower elevations in a convergence zone (Banta 1984). This convergence zone produces updrafts to the lee of the mountain range that can be fed moisture either by the upslope flow or by the flow blowing over the ridge. With the moisture, cumulus clouds can form in the updrafts.

Studies in South Park, Colorado, a wide valley to the east of the north–south Mosquito Range, documented the initiation of a number of thunderstorms to the lee of a long mountain range (Cotton et al. 1983; Motallebi 1982; Hahn 1980). These studies and the one discussed in the next paragraph indicate the importance of the lee-side processes in thunderstorm initiation.

It is the direction of the ridgetop winds that determines the lee side of a mountain range. The importance of wind direction in determining where thunderstorms will form was demonstrated in results from the satellite traceback study described in section 9.4.1.3.3. Figures 9.31 and 9.32 from Schaaf et al. (1988) show contours of thunderstorm initiation frequency focused on one mountain range, the Sangre de Cristo Range in southern Colorado and northern New Mexico. In Fig. 9.31 the data are stratified into three basic ridgetop wind direction patterns in the first three panels, and all initiations are shown in the last panel. The first three panels are for southwesterly flow (SW), northwesterly flow (NW), and a pattern called "curving southerly" flow (CS), in which the winds were southerly or southeasterly in about the lower two-thirds of the panel and southwesterly in the upper one-third.

In the lower central portion of Fig. 9.31, the initiations shift from the east side of the range under westerly (SW and NW) flow to the west side for easterly (or southerly) CS flow. Thus the initiations stay to the lee side.

This figure also shows the important effect of the ridgetop-level wind regime on several isolated peaks and small mountain ranges. Mount Wheeler, a major source of thunderstorms just northeast of Taos, New Mexico (T), is active only under NW flow. The Wet Mountains northwest of Rye, Colorado (Y), and especially Pike's Peak, just west of Colorado Springs (COS), are mostly active under CS flow, even though the flow in their vicinity is southwesterly, as in the SW panel. Obviously the southeasterly component in the low-level flow to the south brings in moisture from the Gulf of Mexico, and this must be an important source for these thunderstorm-genesis areas.

9.4.1.2.6. Ambient wind effects on thermally forced mechanisms. In severe-storm research the effects of the environmental wind profile on the structure of thunderstorms in general has been studied extensively in recent years. Whereas the terrain-relative winds are an important ingredient in determining the movement of thunderstorms, the structure is most affected by the nature of the storm-relative shear, in addition to the amount of parcel buoyancy represented on the sounding (Weisman and Klemp 1986). Storm environments with little shear (less than 10 m s^{-1}) between the subcloud layer and the mid-troposphere (4–6 km AGL) produce single cell storms that generally last less than one hour. Increasing amounts of shear produce multicell and then supercell storms, and the picture is complicated by directional and curving shear profiles. Shears that are very strong compared with the buoyant energy available are capable of preventing storm development by "tearing apart" clouds as they attempt to grow. The magnitude of shear that shuts storms down is not well known, since storms have been observed to develop in strong shears (Marwitz 1972b). A recent review of this topic is presented by Weisman and Klemp (1986). These findings should apply to mountain thunderstorms as well.

An additional factor in mountain storm occurrence is that strong ridgetop-level winds interfere with the normal thermally forced wind systems that lead to storm initiation. For example, using numerical modeling studies, Banta (1986) showed that thermally forced upslope flow becomes shorter in duration and more disorganized as ridgetop wind speeds increase from 4 to 12 m s^{-1}. Simulations of Orville (1968) show that an initial environment with ambient winds causes earlier cumulus initiation than a calm environment does, but the clouds also dis-

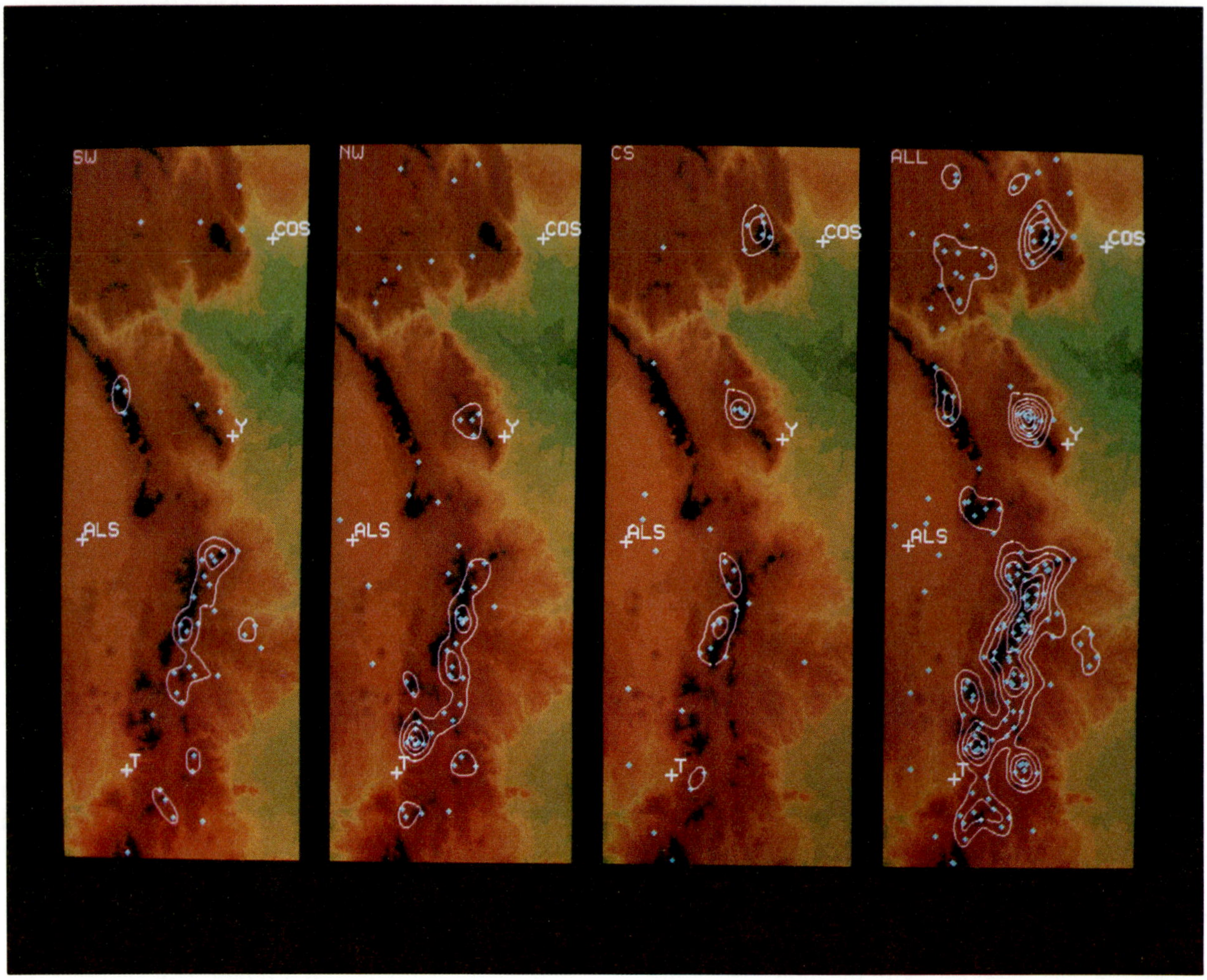

FIG. 9.31. Thunderstorm initiation sites for different ridgetop-level wind regimes, plotted on a terrain map of the Sangre de Cristo Mountains of southern Colorado and northern New Mexico. The four panels show thunderstorm initiation sites that occurred on days with (left to right) predominantly southwesterly (SW) ridgetop winds, northwesterly (NW) ridgetop winds, southeasterly or southerly ridgetop winds in the southern two-thirds of the domain and southwesterly in the northern one-third of the domain (CS, or "curving southerly" winds), and the total number of initiation sites (ALL) identified over the three-summer study. Small crosses indicate individual initiation sites, and contours were normalized to number of initiations per 100 km^2. (Schaaf et al. 1988.)

sipate sooner. Strong winds over the mountains ventilate the surface as it tries to heat up, and they produce mechanical turbulence that also mixes the heat away from the surface. These processes prevent the buildup of a strong slope–atmosphere temperature difference that would drive thermally forced circulations. Another way of viewing this effect is to regard the mountain thermal source region as a stationary heat source. From plume-rise experience, we would expect strong winds to inhibit vertical cloud development.

From the photogrammetric study of Orville (1965) we know that stronger ridgetop flow does discourage the growth of cumuli into thunderstorms. The same study also shows, however, that it does not totally prevent thunderstorm development. Under the right conditions, clouds can grow deep even with strong flow at ridgetops. As in the case of severe weather over more uniform terrain, whether thunderstorms grow undoubtedly depends on several factors, including the amount of buoyancy available in moist lifting.

An interesting issue in several studies, including Orville (1965, 1968), Braham and Draginis (1960), and Booker (1963), is that mountain lee waves seem occasionally to play a role in the initiation and spacing of cumuli downstream of the mountain. Numerical simulations of Orville (1968) show that waves can interact with thermally forced upslope flow to initiate cumuli. This issue is further discussed in section 9.4.1.3.

9.4.1.2.7. Discussion. The significance of thermally forced initiation mechanisms is that in the mountains of the western United States and other drier regions, such

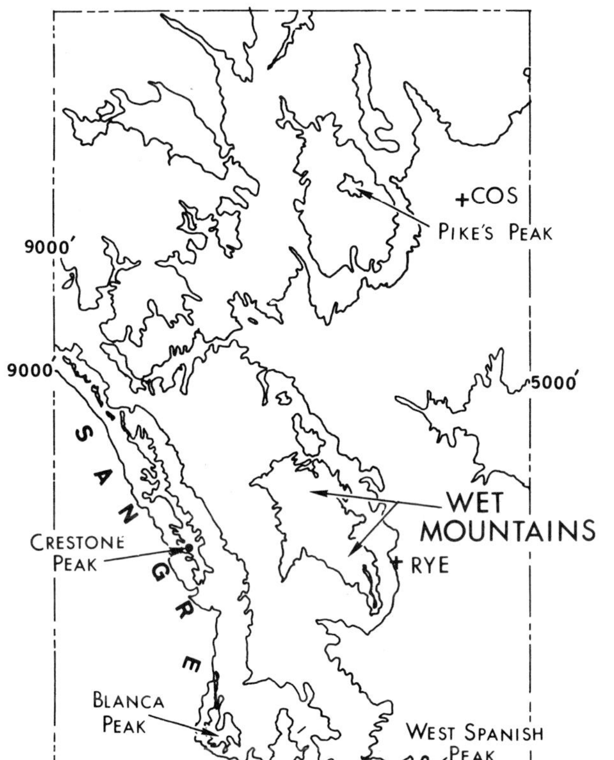

FIG. 9.32. Topographical map of the Sangre de Cristo Mountain region shown in Fig. 9.31.

mechanisms are present nearly every day. Familiarity with a region can provide information on which genesis areas ("hot spots") are apt to be active on a given day.

Over flat land, initiation (triggering) often comes from large-scale (fronts) or larger-mesoscale (gust fronts, gravity waves, etc.) systems. In the absence of these, initiation mechanisms often seem more or less random in location (even though at small scales they are probably not). In the mountains of the western United States, however, we have seen that local circulations routinely provide updrafts that can initiate convection. Thus, depending on external factors such as ridgetop wind direction, cloudiness, and soil moisture, certain areas repeatedly and perhaps predictably act as triggers.

9.4.1.3. Obstacle (Aerodynamic) Effects

The third mechanism for cumulus initiation by mountains is obstacle effects, which include aerodynamic effects such as upstream blocking, gravity waves, upward motion caused by the turbulent wake, and convergent flow downstream of the mountain range. Aerodynamic effects are discussed in Carruthers and Hunt, Chapter 5 and Meroney, Chapter 7, and mountain waves are discussed in Durran, Chapter 4.

The numerical simulations of Smolarkiewicz and Clark (1985) indicate the importance of terrain features as obstacles to the flow in determining the location of cloud development. They modeled the evolution of a field of shallow cumuli over rugged but not mountainous terrain in southeastern Montana. Their model was forced at the lower boundary by the observed surface wind field and vertical gradients of temperature and water vapor, determined from a surface mesonet. Surface heat and moisture fluxes were calculated from the gradient data. In their model, terrain could be either activated or "turned off"; when off, the model topography was flat. The results apply to terrain with elevation differences of less than 200 m in a 50-km square domain, and with initially specified surface wind speeds of 3–4 m s^{-1} or less.

Smolarkiewicz and Clark found that when the topography in the model was flat, the cloud fields did not form as observed, but rather clouds tended to form along the upstream boundary of the model. This occurred even though the boundary-layer convergence and divergence regions arose from the observed surface wind field and even though the surface heat and moisture fluxes were similarly calculated from observed data. When the gridded topography was activated in the model, however, the clouds were distributed more evenly, as observed, and the cloud size distribution was narrower, also as observed. Smolarkiewicz and Clark (1985) conclude that "dynamical inhomogeneities imposed by the flow over terrain play an important role in the development of the convective cloud field. . . . Cloud field characteristics, i.e., the spatial distribution of clouds, the cloud-size distribution, and the cloud sky coverage are determined more by the surface orography than by the thermodynamics of the surface layer."

9.4.1.3.1. Blocking. As stable flow approaches an obstacle or barrier, it slows down because of blocking effects, which are discussed in Carruthers and Hunt, Chapter 5 and section 9.3.2.1.3. The flow retardation results in horizontal mass convergence ahead of the obstacle, and upward air motion. This vertical motion can produce unstable clouds upstream of the barrier if it lifts air to its LFC.

A good example of this process for an isolated obstacle is the action of the island of Hawaii in producing its upstream convective-band clouds, which were investigated observationally and numerically by Smolarkiewicz et al. (1988) and Rasmussen et al. (1989). These clouds form in the upward-moving flow of a wave that resembles the wave made by the bow of a ship, and therefore the clouds are often V- or U-shaped in the horizontal. Smolarkiewicz et al. (1988) point out that the band cloud is strong and often precipitates when the forced updrafts reach the LFC, but it is weak and disorganized when the updrafts reach only the LCL.

The deep, moist, monsoonal convection produced in southwesterly flow upstream of the Western Ghats Mountains, as well as similar monsoon rainfall regimes along the Burma–Thailand coast and in the Phillipines, are examples of convection initiated by blocking by a long mountain barrier, according to Grossman and Durran (1984) and Durran and Grossman (1985). Satellite studies of the Western Ghats monsoon summarized by Grossman and Durran (1984) and by Ogura and Yoshizaki (1988) indicate that the major rainfall from deep convection occurs over the Arabian Sea 100 km or so upstream of the mountain range, and that the clouds directly over the mountains are low—probably orographic stratiform clouds. Grossman and Durran (1984) conclude that "the Western Ghats are capable of contributing to the production of deep convection well offshore by gentle lifting of potentially unstable air as it approaches the coast. This lifting is due to deceleration of the low-level air as it approaches a zone of high pressure caused by an upstream blocking effect of the mountains."

This conclusion is not without controversy. Smith (1985b) pointed out that the effects of latent heating by the offshore deep convection, which Smith and Lin (1983) had found to be important, were not included in Grossman and Durran's (1984) study. He also cited vertical wind shear across the troposphere, surface fluxes from the ocean, and the Coriolis force as effects that were likely to be important but were not included in either study. In an effort to address some of these uncertainties, Ogura and Yoshizaki (1988) ran a series of numerical simulations. They tested the sensitivity of rainfall and cloud-height distribution to the presence of surface fluxes of heat and moisture over the ocean, and to the presence of a reversal from westerly flow at low levels to easterly flow in the troposphere above 5 km. They obtained realistic results only when both effects were included.

In their most realistic simulation, cross sections of the component of the wind perpendicular to the mountain range showed deceleration of the flow as it approached the range. Thus, despite the importance of other factors in the overall distribution of cloudiness and rain, it is the lifting, caused by blocking, of a stable but potentially unstable layer of air that initiates the convection.

A similar situation occurs over the hills along the west coast of Britain, when flow with a southwesterly component in the warm sector of an extratropical cyclone encounters the topography (Browning 1980; Browning et al. 1974). Upstream lifting produces unstable rain, as in the Western Ghats, but the major contributions to rainfall in these cold-season cases are from the release of instability at middle levels, between 3 and 5 km MSL. Significant rainfall at the surface over the hills occurs when the moist flow from the Atlantic Ocean establishes a seeder-feeder situation. The midlevel convective clouds act as seeders, and orographic stratiform clouds produced by forced lifting of the low-level onshore flow act as feeders.

9.4.1.3.2. Flow deflection. One of the most obvious effects of topography as an obstacle to flow is the direct deflection of the flow. Deflection of low-level flow can produce regions of convergence and divergence that lead to updrafts and subsidence, respectively. Examples of flow deflection, depicted in Fig. 9.33, include the splitting of the flow by a linear range of mountains that is nearly (but not quite) parallel to the flow, resulting in convergence on one side of the range and divergence on the other, and convergent or divergent flow in the wake of a more circular mountain or range of mountains.

Orographic splitting of the flow contributed to heavy rains in a case study over Japan, according to Watanabe and Ogura (1987). Deep convective cells formed about 50 km off the coast of southwestern Honshu in a moist environment of large (mesoα-) scale lifting. Cells intensified as they approached the coast, formed a heavily precipitating band along the coast, and dissipated as they crossed over the mountain range. The west-southwesterly surface flow was split by a range of mountains oriented southwest–northeast that reached nearly 1 km in elevation. This produced convergence and ascending air currents north(west) of the range and subsidence to the south(east). The convergence reached a maximum along the coast, as indicated in Fig. 9.33a, and this is where maximum values of rainfall (300 mm in 16 h) were recorded over land. The heaviest precipitation thus did not occur over the peaks.

How active confluent deflections in the wake of an obstacle are in producing vertical motion depends on a number of factors, including stability of the approach flow.

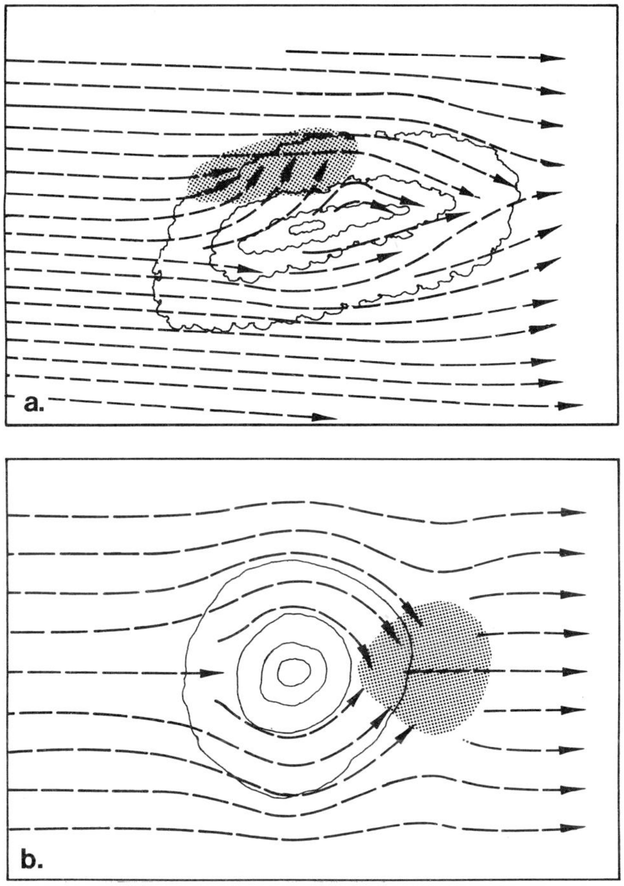

FIG. 9.33. Horizontal flow patterns showing obstacle effects: (a) flow deflection by a mountain range nearly parallel to the flow, and (b) confluence in the wake of an obstacle. In (a), the region of maximum convergence over northern shore is shaded.

For unstable flow around and over the Black Hills in South Dakota, Kuo and Orville (1973) found an enhancement of deep convective activity to the lee using radar echo data. On the other hand, for stable flow over an obstacle resembling the island of Hawaii, Smolarkiewicz and Rotunno (1989) found a tendency for subsidence in the wake.

9.4.1.3.3. Gravity-wave effects. The effects of mountain waves in producing stable clouds have been studied observationally, theoretically, and numerically, as discussed in Durran, Chapter 4. Less is known about the role of waves in unstable cloud development. Studies cited in section 9.4.1.2.6 indicate that wave activity in the stable atmosphere above the mixed layer plays a role in the initiation and spacing of cumulus clouds over peaks and small mountain ranges. Recent numerical studies over a flat surface by Balaji and Clark (1988) indicate that gravity-wave interactions with mixed-layer processes can lead to deep moist convection. Similar effects should be found over mountainous terrain.

As with smaller mountain ranges, there is some indication that gravity-wave or mountain-wave activity can affect the spatial distribution of moist convection over larger ranges. Hahn (1980) noted the presence of waves in his aircraft cross-sections of flow over South Park, Colorado, on days when cumulus convection occurred. Booker (1963) found case-study evidence (using radar data) and climatological evidence that mountain-wave activity affects the distribution of summertime convective rainfall in the mountains of Pennsylvania. Thunderstorm echoes were observed to dissipate in the subsiding air immediately downwind of a mountain barrier only to reform in (presumed) rising air several kilometers farther downwind.

Recent two-dimensional simulations of mountain convection on a day in which strong deep convection developed over the mountains show that gravity waves (mountain waves) can act together with organized solenoidal, slope-flow circulations and contribute significantly to storm development (Tripoli and Cotton, 1989a). The large-scale experiments poorly represented the details and timing of the first stages of cumulus initiation by terrain features because the resolution (1 km horizontal by 250 m vertical near the ground) was too coarse for such processes; however, the role of mountain waves and other gravity waves in modifying the basic mountain-range-scale upslope flow and establishing updrafts and favorable regions for thunderstorm development is well established by the model results.

A gravity wave forms in the stable free atmosphere at night over and just downwind of the Continental Divide, according to Tripoli and Cotton's (1989a) results. An upward-moving branch of the wave intersects the mountain surface about 60 km east (downwind) of the divide. After sunrise, surface heating produces an unstable mixed layer over the peaks, but the gravity wave persists in the stable atmosphere above. The mountain-wave-produced updraft thus remains in place during the morning hours in spite of the establishment of unstable mixed-layer structure over the peak. As morning heating continues, two regions produce deep moist convection—an area just over the divide and an area of mesoscale ascent at the location of the gravity-wave upcurrent (Fig. 9.34). The w impulse that leads to cloud development in the second region is initiated by the superposition of a traveling gravity wave and the stationary mountain wave.

The convection over the Continental Divide is fed by moisture advected up the windward (west) slope by the combined action of the upslope flow provided by larger-scale processes and of the western-slope, thermally forced upslope flow. Convection in the gravity-wave updraft to the east is less intense from the start; it moves farther to the east, abandoning its source region, and dies. The storms that form over the divide become the major prop-

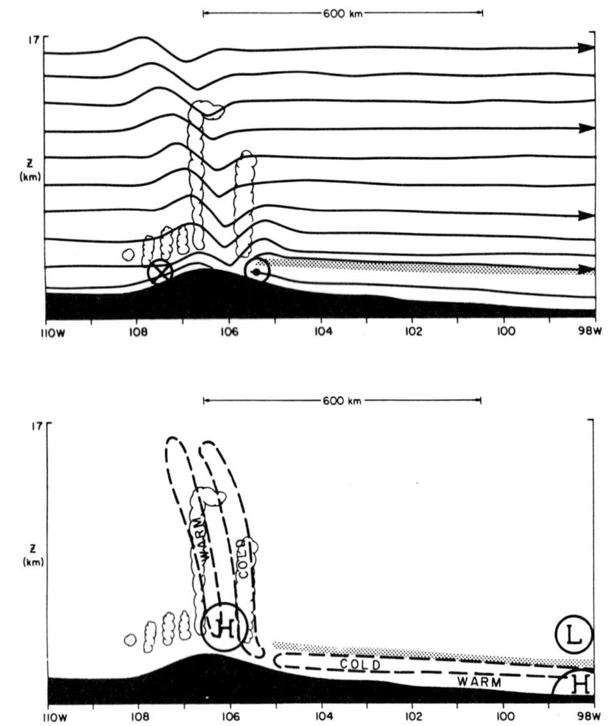

FIG. 9.34. Conceptual model showing the flow field and position of convective elements at the time when deep moist convection forms in the numerical model simulations of Tripoli and Cotton (1989a). The hill represents the Rocky Mountains with the plains to east (right), and the shaded layer over the plains represents the plains inversion. (a) Solid lines are streamlines for flow relative to the ground. Circles denote meridional (north–south) flow developing into and out of the page: southerly and northerly, respectively. (b) Pressure and temperature fields: circles represent pressure centers, and dashed contours enclose temperature regions. The 600-km length scale represents twice the Rossby radius of deformation.

agating storms, as they move eastward into the stationary mountain-wave updraft region downwind of the peak. There they grow explosively and transform a shallow solenoidal circulation (occurring within 2 km of the ground) into a deep circulation that extends through the depth of the troposphere. The deeper circulation begins to feed on plains moisture to the east and to move eastward.

The role of the mountain wave is thus to establish an updraft region that intensifies storms as they move into it from source regions upwind. Details of the initiation of those storms, which were not well represented in the model, are probably not important, except for the timing of the initiation processes: observations showed that the first moist convection occurs earlier in the day than predicted by the model.

The simulations of Tripoli and Cotton (1989a), which investigated the orographic contribution to MCC formation downwind of the mountains, are discussed again in section 9.4.5 on interactions between mountain processes and larger-scale systems. It is worthwhile to note here that in addition to the role of mountain waves and organized slope flows in the initiation of storms, the interactions of various scales of internal gravity waves with the moist convection were found to be extremely important in the development of the larger, mesoα-scale system.

The presence or absence of thunderstorms over a mountain or over a region in a mountain range depends critically on whether the local flows in the area can provide an updraft strong enough to initiate convection (i.e., release the instability); however, in the case of thermally forced mechanisms in dry regions, for example, this forcing is present nearly every day, whereas thunderstorms occur only on some days. Thus, the other conditions for thunderstorms, potential instability and moisture, are equally as important, and these aspects of the environment are generally provided by synoptic- and mesoα-scale processes.

9.4.2. Conditional instability

The second condition for thunderstorms is a conditionally unstable temperature lapse rate in the troposphere. This does not include any effects of moisture. This criterion merely states that the tropospheric lapse rate, though stable, must be less than moist adiabatic. The degree of stability affects how severe the convection can be, and an indication of the lapse rate between the lower and middle troposphere is an ingredient (explicitly or implicitly) in many of the severe weather indices.

Because of the low moisture and high heat fluxes in arid regions such as the western United States, this condition is met on most days during the warm season. Exceptions occur generally with strong frontal or subsidence inversions.

9.4.3. Moisture sources

The third ingredient for thunderstorms is adequate moisture, usually at least in the lower levels. Moisture can be imported from great distances by large-scale wind systems, or it can be injected into the atmosphere by evaporation from local sources. The sensitivity of deep convection to moisture is very evident in drier regions. This section presents examples of moisture availability and its effects on deep convection in a dry region, the western United States.

The major large-scale moisture sources in the western United States are the Gulf of Mexico, the Pacific Ocean, and the Gulf of California. We saw in Fig. 9.31 (third panel) how important access to the Gulf of Mexico moisture is to thunderstorms produced by Pike's Peak. It is not uncommon for meteorologists to track midlevel moisture; e.g., from tropical storms or hurricanes in the Pacific. Wherever the area of moisture goes, thunderstorms follow over the region.

The Gulf of California is also an interesting situation, as shown in Fig. 9.35. The surface waters of this land-surrounded gulf get very warm as summer wears on, and because of the warm water and air temperatures there, a considerable amount of water vapor evaporates into the atmosphere. This water vapor is piped northward between two mountain ranges, and then spreads across the western United States, where it forms the primary moisture source for the "summer monsoon" afternoon convective-rain-shower periods during late July and August (Hales 1972 1974; Brenner 1974).

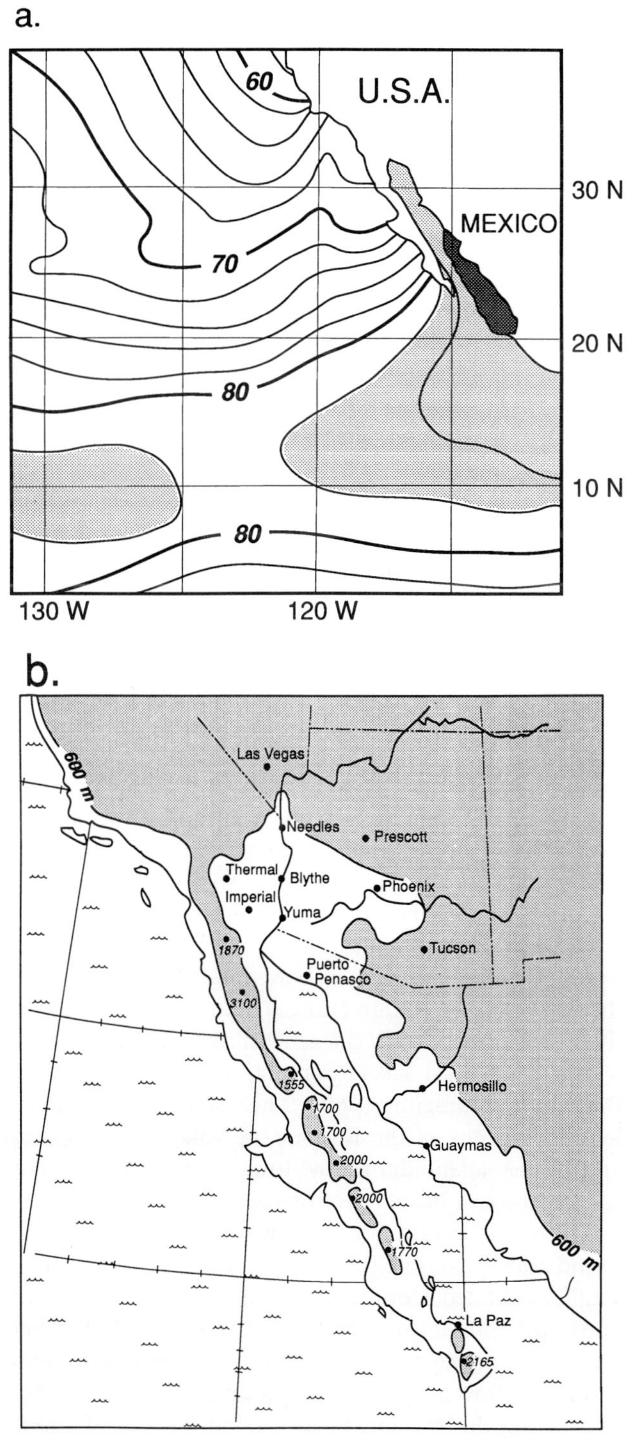

FIG. 9.35. (a) Mean eastern Pacific Ocean sea surface temperatures (degrees Fahrenheit) for August, showing temperatures of 84° to 86°F or more in the Gulf of California. Light shading represents temperatures of greater than 82°F, and dark shading greater than 86°F. (b) Topography map of western Mexico, showing regions of greater than 600 m elevation (shaded). Numbers represent elevations in meters (Hales 1974).

The study of Fujita et al. (1962) near Flagstaff, Arizona, provides a striking example of the importance of wind direction and rain shadowing to convective rainfall. In Fig. 9.36 a time series is shown of the wind profile and the precipitation measured by Fujita et al.'s mesonet. On days when the winds aloft were northerly, precipitation was zero or insignificant. Days when the winds aloft had a southerly component, however, were the rainy days. Figure 9.37 is a topographical map of the southwestern United States centered on Flagstaff. The radial lines emanating from that area indicate data for each day of the study: the direction represents the direction of the 60-kPa wind, and the length, the amount of rain that fell on that day. On the days with southerly wind components, moisture advects into the area from one of the major moisture sources. On days with northerly winds, however, air flows over major barriers (black) before it reaches the region; thus it is most likely that any significant moisture would fall out before reaching Flagstaff.

Even in cases where moisture from large-scale sources is inadequate, evaporation from local sources, such as lakes, marshes, or even regions where it has recently rained, can provide enough moisture for thunderstorm development. Cotton et al. (1982) concluded that increases in surface-layer moisture that led to thunderstorm initiation on a case-study day in South Park, Colorado, had to be from local evaporation sources, since advection was too slow to explain the observed increases. Results of aircraft soundings from New Mexico (Fig. 9.38) show an increase in moisture producing an increase in equivalent potential temperature θ_e of the inflow, from a value that would probably only barely support shallow moist convection to a value that supports deep moist convection. Thunderstorms were observed. Since "significant rain fell on the two previous days, causing the ground to be moist," Raymond and Wilkening (1982) conclude that this increase was most likely due to surface evaporation.

Thus, although large-scale sources of moisture are of prime importance in assessing whether thunderstorms will occur, it is also important to consider whether local sources could be a factor in supplying moisture. In cases where large-scale moisture is marginal, local moisture sources can supply enough moisture for deep moist convection to occur.

9.4.4. Structures and transitions

The previous sections have considered features of mountain flow that lead to the production of updrafts and cloud initiation, and features of the large-scale flow

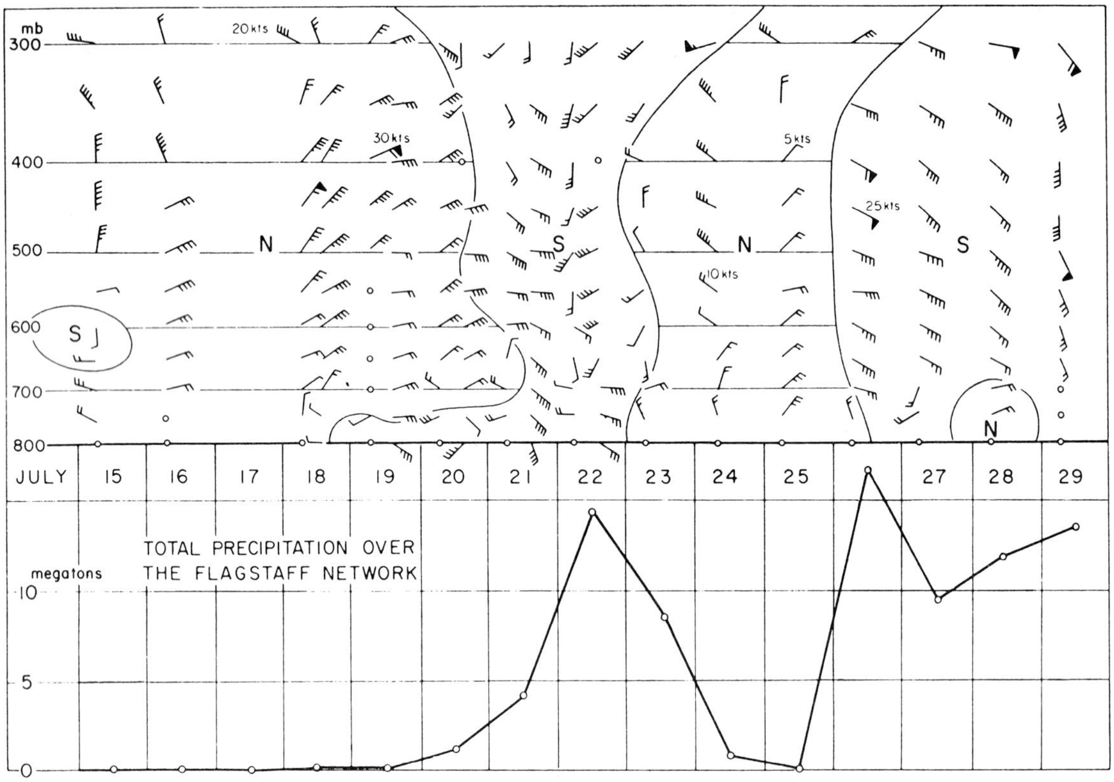

FIG. 9.36. Top: rawinsonde winds at the Flagstaff, Arizona, Airport during the field project in the San Francisco Peaks area during July 1960. Long wind barb is 2.5 m s^{-1}; flag is 12.5 m s^{-1}. Bottom: daily precipitation total in megatons, measured by the experimental raingage network (Fujita et al. 1962).

that affect importation of moisture. It is also interesting to consider the structure of these updrafts and the nature of the transitions from dry to moist convection, from shallow to deep moist convection, and from simple "air mass" cumulonimbus structure, tied to terrain-forced convergence, to propagating storm structure with dynamics independent of terrain-forced updrafts. A number of studies, especially in connection with other studies in the western United States already cited, have addressed these issues, as discussed next.

9.4.4.1. Dry circulations

Aircraft studies over isolated mountain ranges in Arizona and New Mexico investigated the structure of the dry circulations leading to shallow cumuli under light wind conditions. Braham and Draginis (1960) and Silverman (1960) flew vertical cross-sections over the Santa Catalina Mountains northeast of Tucson, Arizona, the former with along-wind legs on a day when shallow cumuli formed, and the latter with cross-wind legs on a clear day with strong large-scale subsidence. Raymond and Wilkening (1980) flew a variety of sampling patterns over and around the San Mateo Mountains in New Mexico. These studies showed that there are two distinct scales of updraft motion, and that a "wake" occurs in the flow over and downwind of the mountain, even in the light winds that were present during these studies. These processes efficiently transport surface properties, such as moisture, into the upper part of the evolving mixed layer.

The two scales of updraft are 1) a general, more-or-less toroidal circulation with a horizontal dimension about the size of the mountain (range), and 2) smaller convective cores embedded in the general rising motion, having widths of about 3 km (Raymond and Wilkening 1980; Braham and Draginis 1960). The rising motions are fed by convergence of the upslope/up-valley wind systems and by the solenoidal inflow toward the high-level heat source, the sun-heated mountain.

The overall mountain-scale rising was studied by Raymond and Wilkening (1980) by flying traverses over the small San Mateo Mountain range and by flying horizontal "box" patterns at various vertical levels around the range. Velocity divergence calculations from the box patterns revealed inflow at levels below mountaintops and outflow above (Fig. 9.39), and mean updraft speeds of several tens of centimeters per second. The resulting circulation is about 10–20 km in horizontal extent, but only about 3 km deep. The updraft is confined horizontally, by the remnant of the nocturnal inversion layer, to the vicinity

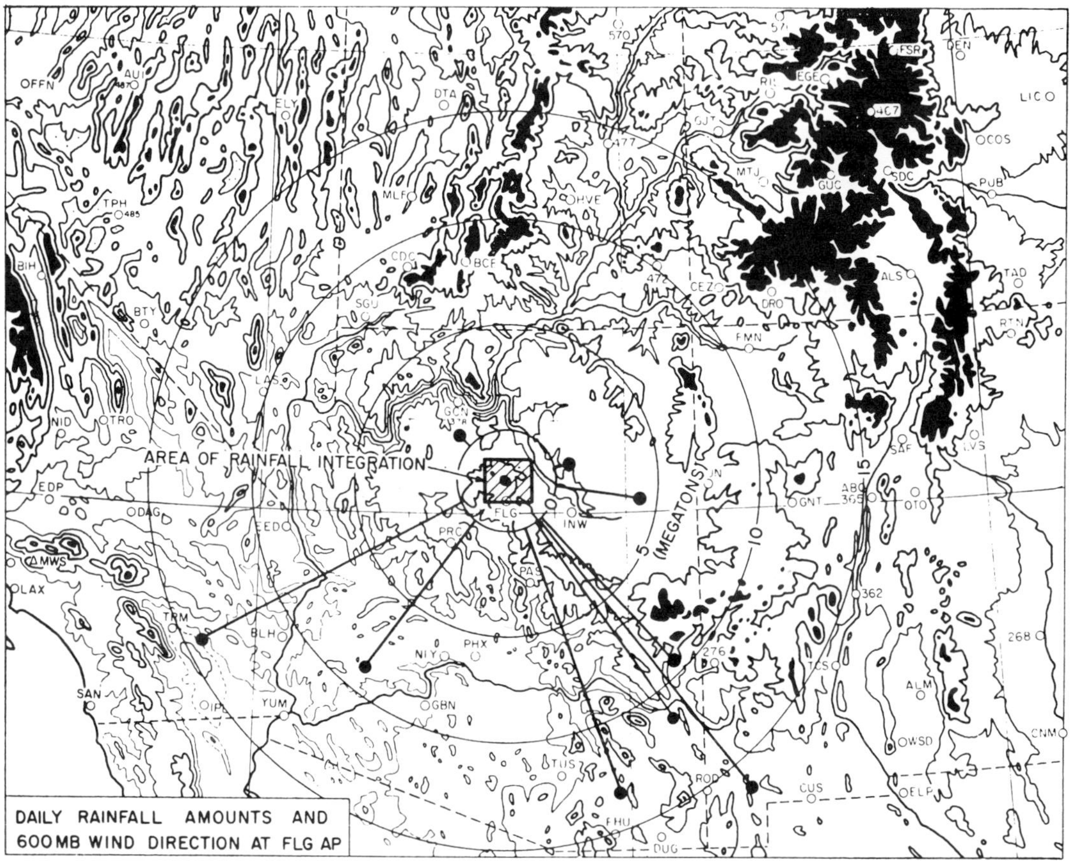

FIG. 9.37. Topographical map of the southwestern United States centered on Flagstaff, Arizona, with elevations above 2744 m (9000 ft) MSL blackened. Rays emanating from Flagstaff indicate the direction from which 60-kPa winds were blowing on each experimental day, with lengths indicating precipitation amounts from Fig. 9.36. Rain shadowing effects of the high terrain to the north are evident (Fujita et al. 1962).

of the mountain; the inversion layer restricts convection vertically to the lowest few hundred meters above the plains until the inversion erodes away. Figure 9.40a shows a schematic representation of the mean circulation, found by Raymond and Wilkening (1980), for a case in which no clouds or only shallow clouds form.

Embedded in the mountain-scale circulation are the 3-km-wide convective cores, called "adiabatic cores" by Braham and Draginis (1960), which carry surface properties aloft. As with thermal plumes in the convective boundary layer (CBL) studied over flatter terrain, however, they are superadiabatic at lower levels (Raymond and Wilkening 1980), as they ascend from the mountain-induced circulations (Fig. 9.41), and their upwind edges are sharp whereas their downwind sides are diffuse. Also as in CBL thermals, these cores rise through a neutral layer into the stable free atmosphere, where they become cooler than their environment. At these upper levels the cores are identifiable in the aircraft data using moisture information. The downward entrainment of warm air from the free atmosphere produces a strong downward eddy heat flux aloft at the mixed-layer inversion. Raymond and Wilkening (1980) observed that the magnitude of this downward heat flux at the interface at the top of the mixed layer is about equal to the magnitude of the surface heat flux. This result is in contrast to the case for the horizontally homogeneous mixed layer, where the magnitude of the downward heat flux at the top is about 20% of the surface value. Raymond and Wilkening (1980) explain this by noting that the heat fluxes in the vicinity of and near the level of the mountaintop often reach 400 W m^{-2}, and as high as 900 W m^{-2}—values that are very high compared with the surface value. They conclude that "such large values . . . indicate that the mountain range is collecting solar heat from a region . . . larger than its actual area."

In Braham and Draginis' (1960) along-wind flight legs over the mountains near Tucson, the flow was smooth upwind of the barrier but disturbed over and downwind of the peaks. Raymond and Wilkening (1980) noted that at lower levels where the ambient flow was weak, inflow came from all directions. At upper levels, though, the ambient flow was stronger, and the outflow was a "clearly delineated wake carried along by the ambient flow." The

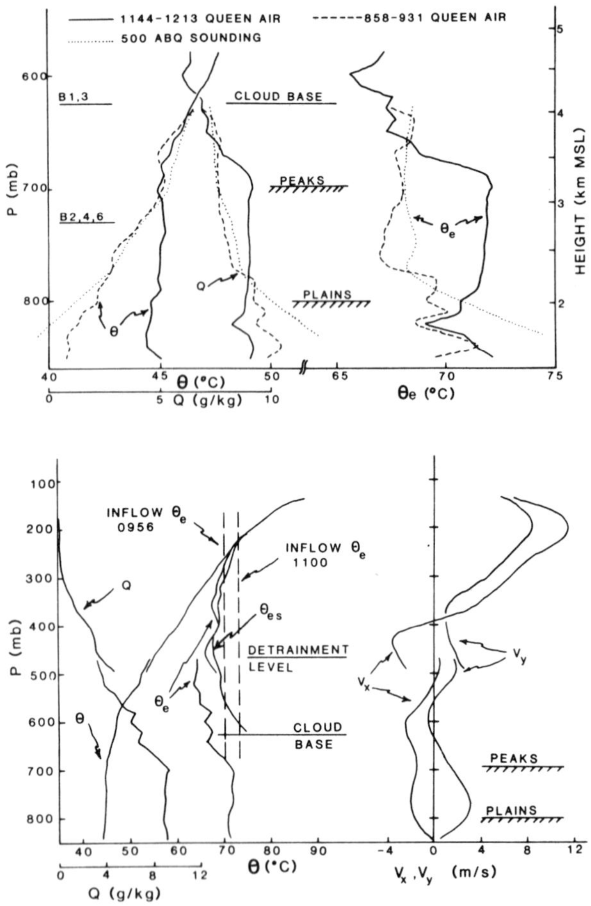

FIG. 9.38. (a) Subcloud layer aircraft soundings in the vicinity of the Magdalena Mountains in New Mexico at about 0900–0930 MST (dashed) and noon (solid) compared with the Albuquerque radiosonde profiles at 0500 MST (dotted). The low-level warming and the moistening in the layer around 70 kPa shows up in the θ, Q, and θ_e profiles. (b) Composite soundings of θ, Q, θ_e, θ_{es}, and the horizontal wind components (V_x, V_y). The aircraft soundings were used below 46 kPa, and the Albuquerque 0500 MST sounding above. Here Q is mixing ratio, and θ_{es} is saturation equivalent potential temperature (Raymond and Wilkening 1982).

boundary layer downwind is a region of higher moisture content than upwind, indicating that the updrafts and the downwind turbulent wake are efficient mechanisms for the upward mixing of surface layer moisture and other properties. Glass and Carlson (1963), for example, found that horizontal ridgetop winds (as determined by photogrammetric tracking of dissipating clouds) are weaker near the peaks and stronger farther away, indicating the effectiveness of these mechanisms in mixing momentum vertically, in this case low-momentum air upward from the surface.

It is the convective cores that form the "roots" of terrain-induced cumuli. Because the western United States is so dry, cloud base is generally well into the stable free atmosphere above the mountain, so the clouds are negatively buoyant at cloud base (e.g., Hahn 1980). In other words, it takes a strong updraft penetrating into the region of forced ascent to make it to the condensation level. As increasingly strong updrafts go higher and higher into the stable layer, however, the upward mixing of moisture also brings the air at this level closer to saturation and condensation (Braham and Draginis 1960).

These results are for light wind conditions (2–3 m s^{-1} or less at ridgetop) and small, isolated mountain ranges. Raymond and Wilkening (1980) note that in a case when the ridgetop winds increased to 5 m s^{-1}, "The resulting mesoscale flow was much weaker and shallower . . . Thus, moderate low-level winds apparently suppress the type of flow we have discussed."

For larger mountain ranges and "moderate" ridgetop-level winds, other mechanisms become important. Lee-side convergence processes and mountain lee waves are two mechanisms that have a role in storm initiation under these conditions.

The mechanism of lee-side convergence arises from the dry circulations, which occupy the mountain boundary layer. Because the boundary layer becomes the subcloud layer when cumuli form, a knowledge of the structure and evolution of the daytime boundary layer throughout its depth is important in understanding how these processes lead to cloud initiation and affect later growth. The next few paragraphs focus on discussion of observational and numerical studies of the structure of the daytime mountain boundary layer, as related to cloud development.

Observations in the Rocky Mountains of Colorado have shown that the diurnal cycle of the potential temperature profile develops very much as it does over flat terrain. Figure 9.42 shows the evolution of the typical θ and wind

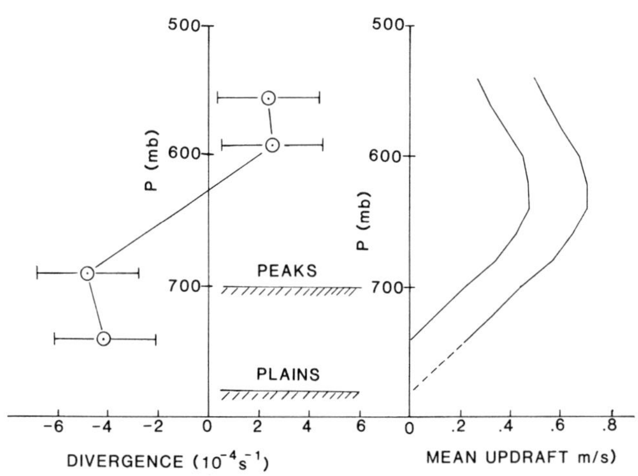

FIG. 9.39. Left: divergence values derived from aircraft-measured winds obtained for "box" patterns flown around mountain at four levels, showing convergence below and divergence above mountain peak height. Right: mean updraft speeds calculated for divergence values in the diagram at left. The two curves are for two different levels where the lower boundary condition $w = 0$ is applied. (Raymond and Wilkening 1980.)

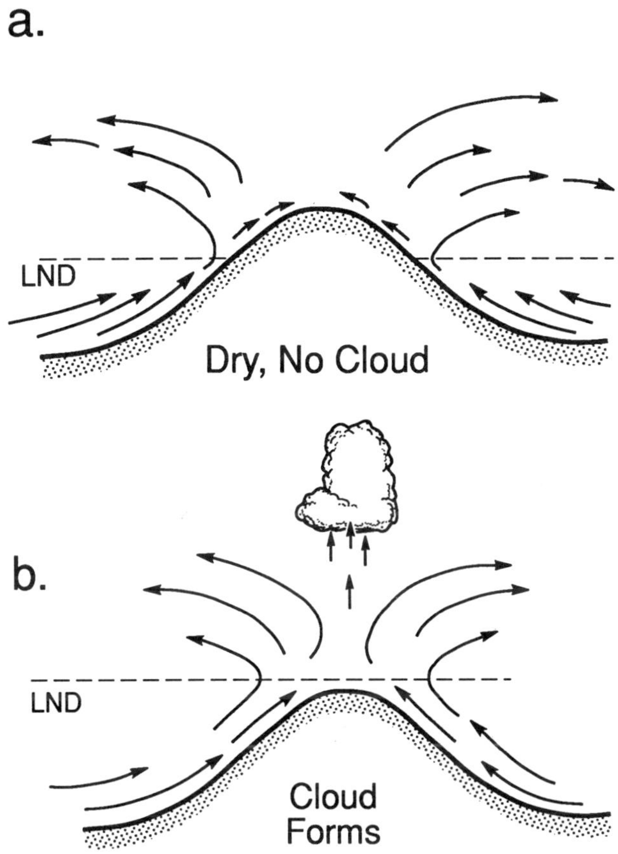

FIG. 9.40. Schematic drawing of mean circulations (ignoring mean wind and convective cores) determined by Raymond and Wilkening (1982) (a) for the case where no cloud or only shallow cloud forms, and (b) for the case where a cumulus forms that grows to congestus. In the former case, the level of nondivergence (LND) often is below the peak. In the latter case, the LND is at or above mountaintop, but the divergence aloft almost completely compensates for the convergence below, as in the dry case, with little *mean* vertical motion extending aloft to force the cloud. Apparently smaller-scale downdrafts compensate for the rising cells of the cloud during this phase of cloud evolution. Later in the life of the storm, when it reaches thunderstorm proportions, mean rising motion extends upward into the troposphere.

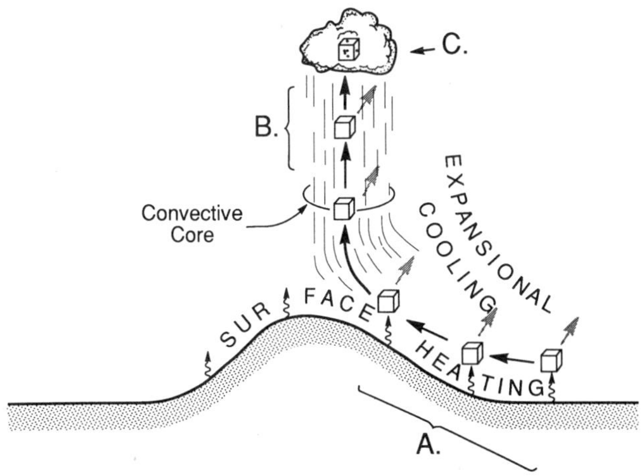

FIG. 9.41. The scenario presented by Braham and Draginis for cumulus roots. Along A, air rises with the upslope flow along the mountain, with surface heating keeping the air warmer than the environment in the presence of cold advection and expansional cooling. The air leaves the slope (B) in a convective core, and cools by adiabatic expansion, having left the warming of the surface. A rising parcel encounters a stable region of the atmosphere (C), overshoots its level of neutral buoyancy, and becomes cooler than its environment. In cases where the parcel reaches its condensation level, a cloud forms.

profiles observed on undisturbed days during a 6-week summer field project in South Park, a wide valley in the mountains of Colorado (Banta and Cotton 1981). The observations were from the eastern slopes of the Mosquito Range, which was the lee side under the usual westerly flow observed in the region.

Figure 9.42 also shows a relationship between these θ profiles and the evolution of the low-level wind field. The stable nocturnal inversion layer acts to isolate or protect the winds at the surface from the winds aloft, by suppressing turbulent mixing between the layers. At night, under downslope flow, this makes little difference along the lee slopes, because the winds near the surface and the winds aloft are from a similar direction (Fig. 9.42a). After sunrise, however, when upslope winds start to form next to the slope, the protective role of the stable layer becomes very important. This layer of suppressed mixing prevents the downward mixing of the ridgetop winds and allows upslope winds, blowing in a direction opposite to the winds aloft, to form. Figure 9.42b illustrates that these upslope winds form in a shallow unstable boundary layer, or mixed layer, that forms at the surface just at the un-

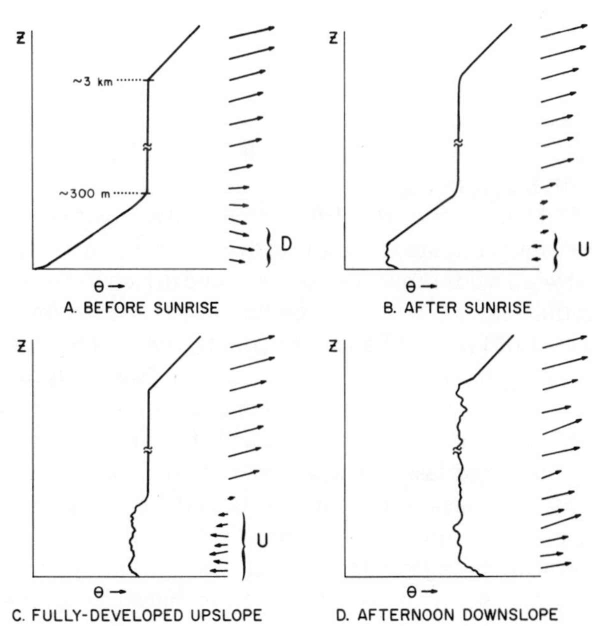

FIG. 9.42. Evolution of vertical profiles of potential temperature and horizontal winds on the lee side of a mountain barrier. At lower levels, D indicates a layer of downslope winds and U, a layer of upslope winds (Banta and Cotton 1981).

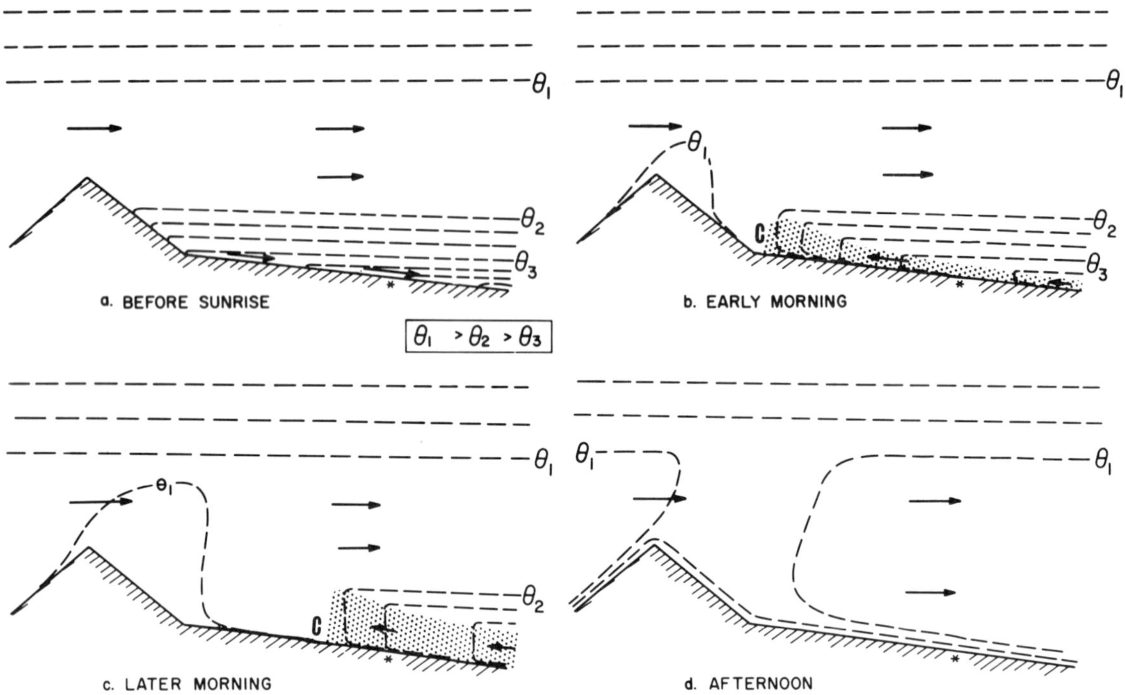

FIG. 9.43. Cross sections of θ and winds corresponding to profiles in Fig. 9.42. The shaded region indicates a shallow mixed layer that contains the upslope flow, and *C* at the upwind edge indicates the convergence zone.

derside of the nocturnal inversion layer. Figure 9.42c shows that the shallow mixed layer continues to grow as daytime heating continues, and the upslope wind layer deepens with it. Finally the inversion layer is eroded totally (Fig. 9.42d). The upslope flow is protected no longer from the strong winds aloft, and these winds aloft rapidly mix downward to the surface. Potential temperature cross-sections corresponding to each of the panels in Fig. 9.42 are given in Fig. 9.43. These cross sections, inferred originally from vertical-sounding data, mesonet data, and theory, were verified by research aircraft flights (Hahn 1980; Banta 1984).

A dynamic numerical modeling study, in which the model was initiated with the early-morning temperature and wind fields shown in Fig. 9.43a and driven by surface heating, also showed the evolution of the θ structure indicated in Figs. 9.42 and 9.43 (Banta 1986). The study indicated that the total time that upslope flow exists on a given day depends on the strength of the ridgetop winds. With strong ridgetop winds, thermally forced upslope flow is weaker and lasts for a shorter time than with lighter winds. This means, in other words, that the convectively mixed afternoon winds appear at the surface earlier on days with stronger ridgetop-level winds. Evidence from surface mesonet observations given in that study support this conclusion.

These results apply to dry days, when deep moist convection does not form over the mountains from thermally forced circulations. On days when deep cumulus convection does form, the afternoon mixed-wind regime shown in Figs. 9.42d and 9.43d does not form. Instead the flow configuration shown in Fig. 9.43b,c leads directly to cloud initiation. At the leading edge of the upslope layer (shaded) near the surface is a region where the upslope flow meets downslope flow in a convergence zone. Figure 9.30 suggests how this convergence zone can trigger cloud development. The convergence zone is favorable for the initiation and continued growth of clouds both because of the persistent upward motion there and because it is a region in which moisture is imported from lower elevations, where mixing ratios are generally higher (Banta 1984). The importance of the zone in producing clouds was simulated numerically, and results of these simulations are shown in the next section (see Fig. 9.44; Banta and Hanson 1987a).

The larger-scale simulations of Tripoli and Cotton (1989a), discussed in section 9.4.1.3.3, address the role of the convergence zone in forming deep moist convection on a day when strong convection forms over the mountains. As mentioned, these studies show that gravity waves (mountain waves) acting together with the organized slope flows under such conditions contribute significantly to storm development. The numerical experiments do not represent very well the details and timing of morning thunderstorm initiation by terrain features, because the resolution was too coarse for such processes. The simulations show that on such days the lee-side convergence zone similar to that described in previous paragraphs can

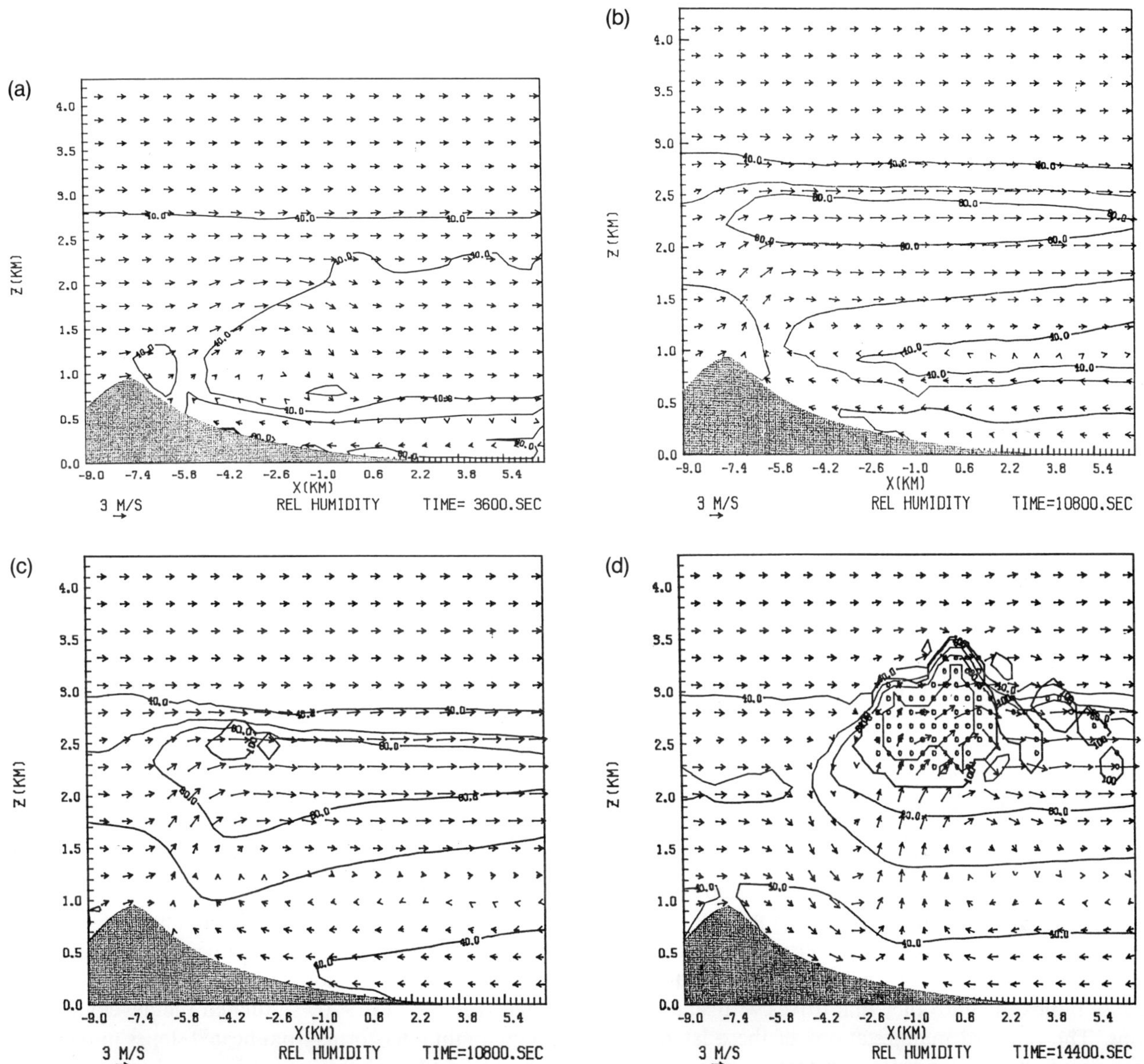

FIG. 9.44. Relative humidity (*RH*) cross-sections showing the evolution of the moisture field in numerical simulations of upslope flow. (a) and (b): case with low surface heat flux (17 K cm s^{-1}); (c) and (d): case with high surface heat flux (25 K cm s^{-1}). (a) After 1 h of simulated time, most of the region between 1 and 2 km height has an *RH* of less than 40%, and surface *RH*s are near 80%. (b) After 3 h *RH*s in the upper mixed layer between 2 and 2.5 km reach 80%, but no condensation occurs. The high case (not shown) looks very similar to the low case (a) after 1 h, but (c) by 3 h a cloud forms over the convergence zone. (d) After 4 h the cloud has continued to grow (Banta and Hanson 1987a).

lead to the initiation of deep moist convection, but that this convection is short lived. The major propagating storms arise from strong cells that form over the peak and move into the mountain-wave updraft that forms just downwind of the peak.

9.4.4.2. Cumulus and cumulus congestus

The production of cumulus by mountain circulations thus depends on several processes, including 1) the strength of the solenoidal circulation, since a stronger circulation is more effective in moistening the upper part of the boundary layer, 2) the vigor of the updrafts in the convective cores, since more vigorous updrafts reach higher into the atmosphere, 3) the descent of the condensation level through the morning, as surface evaporation and the mixing activity of the mountain circulations increase, 4) ventilation by the ambient winds, which tends to keep the upper boundary layer dry, and 5) the growth of the deep convective boundary layer by the action of dry convection off the peaks. On the day studied by Braham and Draginis (1960), for example, "The uppermost height reached by the forced convection was barely sufficient for condensation," so the clouds were small and shallow.

The importance of the upward transport of surface-layer moisture was demonstrated by the numerical simulations of Banta and Hanson (1987a). These simulations of 3 m s^{-1} flow over a heated ridge included a case with low surface heat flux and a case with high heat flux (Fig. 9.44). The lower-flux case generated a weaker circulation and no cloud for over 4.5 h of simulated time, whereas the higher-flux case generated a stronger circulation and did produce a cloud after 3 h. These simulations show that two of the processes are important in determining whether clouds will form: moistening by the solenoidal circulation and ventilation by the ambient winds. The relative strengths of these processes determine the fate of clouds: if the circulations are strong and can supply moisture fast enough to overcome the removal of the moisture by the ventilating winds, then clouds will form. A third process, growth of the deep mixed-layer inversion, also has an effect on the outcome, but since condensation first occurs below the initial level of the inversion, the role of this process is less important in the present case than the other processes mentioned.

The nature of the moist convection; for example, whether it is shallow or deep, is determined by the large-scale environment, as described in section 9.4.5 and by Braham (1958). Shallow cumuli are passive clouds that are pushed to condensation by convective-core updrafts, only to subside and dissipate. Updrafts reach their LCL but not their LFC (and thus ironically would be considered stable by the definition in section 9.2.4). When the environment favors continued vertical growth of the cloud; i.e., when cloudy air can reach its LFC, latent heat release builds the buoyancy that drives vertical accelerations. The outward manifestations of these latent-heat-driven updrafts are cloud elements that have been referred to as thermals (Glass and Carlson 1963), towers (Todd 1964), or sometimes cells, but that are referred to as *turrets* in this section to avoid ambiguities that the other terms suffer from. The diameters of these turrets are about a kilometer or less for shallow cumuli or for congesti, different from the 3-km diameters of the convective cores that initiate the clouds. The scale of these updrafts thus seems to be determined more by the moist instability mechanism than by the 3-km scale of the subcloud updraft.

A discussion of details of the in-cloud structure of cumulus congestus is beyond the intended scope of this chapter. Still, to complete the picture of evolving cumulus in low-wind environments over isolated peaks, as studied in Arizona, it is useful to briefly summarize some of the findings. Two studies near Flagstaff, Arizona, described the development of cloud turrets as the clouds grew from shallow cumulus to congestus. Glass and Carlson (1963) used photogrammetry and radiosonde data, and Todd (1964) coordinated aircraft penetrations of cumuli with all-sky camera, photogrammetry, and radar data, to study the evolution of clouds and turrets. The following discussion has been distilled from those reports.

Through the congestus stage, mountain cumuli consist of one or a few easily definable turrets. Larger clouds have bigger turrets, rather than more. Turrets have a distinct life cycle, consisting of active growth, deceleration, and dissipation (evaporation of cloud water) stages. The turrets of shallow cumuli undergo one growth cycle and dissipate, whereas those of congesti go through a series of growth stage cycles or pulsations. The cloud system as a whole goes through a series of surges in turret growth, followed by periods of leveling off or even retreat.

The width, duration, and height of turrets all increase with time through the morning. Early shallow clouds have turrets with lifetimes of 2 min, whereas later, turrets of the deeper clouds last for 20 min, and some last up to 30 min. Aircraft penetrations showed that updrafts deep within the clouds occur on many successive penetrations, indicating that updraft lifetimes are similar to the lifetimes of the turrets, and that updrafts continue through several pulsations in rise rate of the turrets. The widths of the updrafts increase gradually to about a kilometer until precipitation develops (i.e., just before the first radar echo), at which time there was observed a sudden broadening to widths of a few kilometers.

9.4.4.3. CUMULONIMBUS

In a sufficiently unstable environment, mountain cumuli grow all the way to the tropopause and become cumulonimbi or thunderstorms. Liu and Orville (1969) and Orville and Sloan (1970) simulated the development of mountain thunderstorms from heated slopes numerically. Raymond and Wilkening (1982) found in one of their observed case studies that this transition to deep moist convection is accompanied by a change in the low-level inflow pattern. Prior to the transition, inflow comes from all around the mountain, but afterward it prefers the southern and eastern sectors of the mountain, with outflow toward the west (Fig. 9.45). Interestingly, however, the magnitude of the convergence and divergence associated with the large mountain-scale circulation is very similar to that which is observed for the dry case (Fig. 9.40a), and the residual updraft going through cloud base is surprisingly small (Fig. 9.40b). This pattern is maintained throughout the growing stages of the cumulonimbus, and it gives way to a deeper more characteristic thunderstorm inflow/outflow pattern only after the change in the low-level inflow pattern takes place.

The evolution of a mountain thunderstorm in South Park, Colorado, was investigated using triple-Doppler ra-

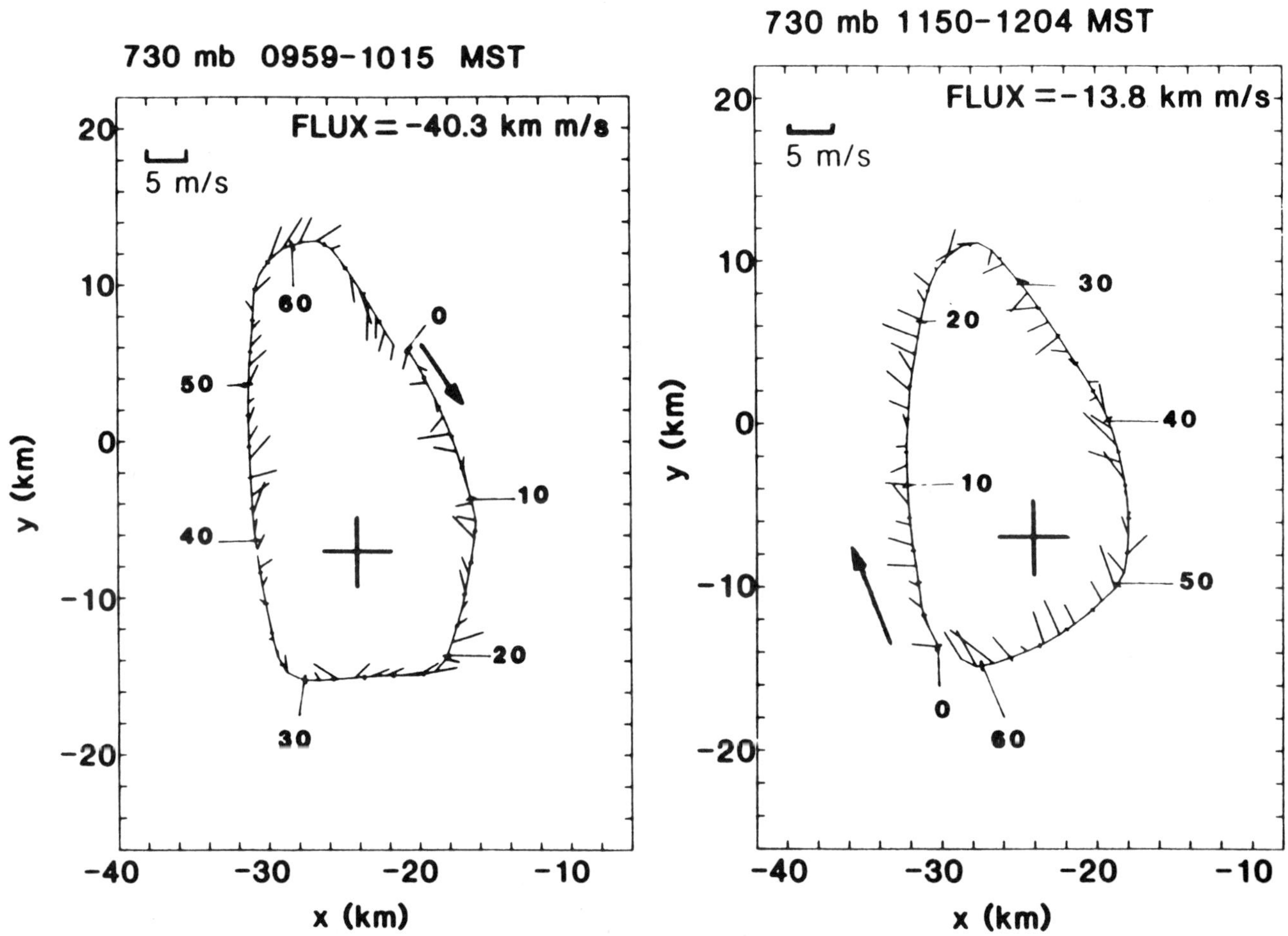

FIG. 9.45. Low-level, aircraft "box" patterns flown around a mountain during the development of a cumulonimbus cloud. At around 1000 MST inflow (barbs pointing inward) comes from all directions, but by 1200 inflow prefers south and east, and some outflow is indicated to the west (after Raymond and Wilkening 1982).

dar by Motallebi (1982). She observed that the storm initially has a simple updraft structure related to the inflow pattern set up by the lee-side convergence zone (Fig. 9.46a). As the storm matures, the main downdraft is pushed ahead of the storm by the westerly shear, and the gust front cuts off the inflow from the convergence zone (Fig. 9.46b,c). The storm begins to propagate eastward and organizes its own relative inflow ahead of the gust front, thus attaining the structure more typical of propagating midlatitude thunderstorms (Fig 9.46d). The moist downdraft structure found by Motallebi (1982) was also found in other mountain thunderstorms by Knupp (1987).

Severe weather is less common in mountain thunderstorms than in storms occurring over lower ground. This undoubtedly is related to the conclusion of Braham (1958), that mountain cumuli are less vigorous than valley cumuli in Arizona. This is not to say that severe weather does not occur over mountainous terrain, however; mountain thunderstorms do produce hail and damaging winds. Recently, Fujita (1988, 1989) documented a "devastating" (F4) tornado and strong microburst activity produced by a thunderstorm that swept through the mountains of northwest Wyoming, crossing the Continental Divide.

9.4.5. *Roles for large-scale forcing*

A working hypothesis for the relationship between the small scales and the large scales of forcing might be stated as follows: the large scales regulate whether thunderstorms form or not, whereas the small scales govern the details of precisely where and when convection will occur.

Previous examples in this chapter have shown how the large-scale flow can be an agent in importing or denying moisture to a region. Besides this activity one can define three roles that large-scale forcing can have in regulating thunderstorm occurrence: suppression, permission, or direct and active forcing. Suppression occurs, for example, through strong subsidence fields, which one might see in

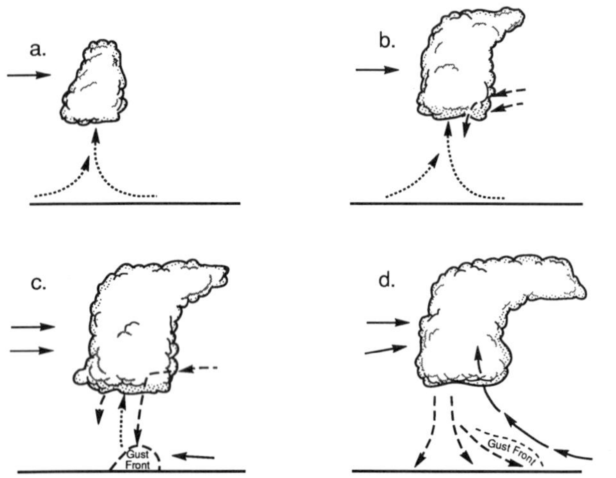

FIG. 9.46. Stages in the life of a mountain thunderstorm, after Motallebi (1982). (a) Convergence-zone forcing initiates the cloud. (b) Downdraft starts on the downshear side of the cloud. (c) Downdraft produces beginnings of the gust front at the surface, which cuts off inflow from thermally forced upslope. (d) Old updraft dies and is replaced by a fully developed downdraft and gust front. The new updraft, driven by storm dynamics (buoyancy), rides up over the gust front (Cotton, personal communication 1983).

a high-pressure system. In its permissive role, the large-scale atmosphere is passive, "allowing" deep convection to occur if the other necessary factors are supplied by smaller scales. Direct forcing occurs, for example, through the rising motion of a short wave, the lifting action of a front, or other larger-scale activity.

Examples of one kind of large-scale direct forcing are the flash flood events discussed in section 9.4.1.1. A layer of moist air, forced to ascend a mountain barrier by large-scale easterly flow, produced large, persistent, nearly stationary thunderstorm systems over the mountains. The heavily precipitating storms and their lack of movement dumped considerable rainfall into small watersheds, leading to flooding.

Two other examples of direct forcing by the larger scale are provided from studies from South Park, Colorado: a cold frontal passage and the passage of a large mesoscale ridge. Figure 9.47 shows the cold front passing through South Park and radar data of thunderstorms accompanying the front (Cotton et al. 1982). This represents initiation of thunderstorms by a synoptic-scale feature.

Forcing by features larger than mountain circulations but smaller than synoptic also occur. On the morning of 4 August 1977 a mesoscale ridge in atmospheric pressure passed through South Park, leaving a deep layer of easterly flow to bring moisture up from the lower elevations to the east. In Fig. 9.48 tethered balloon soundings are shown from that morning, indicating normal boundary-layer development until after 0730 MDT. The wind profile at 0836 MDT shows the deep layer of easterly upslope flow into South Park that resulted from the passage of the mesoscale ridge at around 0800 MDT. This easterly flow, extending to over 400 m above the surface, is more than twice as deep as the upslope layer at this hour on a normal, undisturbed day (Banta 1982). In Fig. 9.49 the location of the ridge at 0600 and 1200 MDT is shown (George 1979). This inflow persisted all morning and forced a very strong thunderstorm system that moved into the plains and produced a tornado near Colorado Springs, and later evolved into an MCC that moved into Kansas (Cotton et al. 1983; McAnelly and Cotton 1986; Tripoli and Cotton 1989a,b).

Although the low-level mesoscale ridge was a factor in the initiation of early deep convection in the morning, the overall synoptic environment on 4 August 1977 in the troposphere over the mountains of central Colorado was quite undisturbed, according to Tripoli and Cotton (1989a,b). They chose this day as a case study to demonstrate, using their two-dimensional numerical model, that topographic heating on the scale of a large mountain range (such as the Rocky Mountains) is capable of producing a large- (mesoα-) scale precipitation system, without benefit of synoptic-scale support; i.e., in a "permissive" synoptic-scale environment. As discussed in sections 9.4.1.3.3 and 9.4.5, the details of the initiation of the first moist convection in this case are unimportant according to the simulations, because it is sustained mountain-range-scale slope flows and a stationary mountain-wave vertical velocity pattern that determine where strong intensification will take place.

The discussion in section 9.4.1.3.3 (on gravity waves) ended with the explosive growth of a thunderstorm in the stationary updraft that had been set up by the mountain wave. The thunderstorm tapped into moisture being advected up from the plains by the upslope flow, and it drove the upward branch of a deep, direct circulation cell. This deep cell extended through the depth of the troposphere.

Deep moist convection persists in the region at the western edge of the plains inversion for several hours (Fig. 9.50). The strong upward velocities in this convection produce compensating subsidence in the troposphere downwind, and the subsidence intensifies the inversion that caps the mixed layer over the plains. The strong capping inversion 1) prevents competing deep moist convection from developing over the plains, and 2) channels high-moisture air in the shallow mixed layer over the plains up the slope to feed the deep convection over the mountains. As the thunderstorm system persists and drifts eastward with the mean tropospheric wind, latent heating warms the upper-tropospheric layer between 8 and 10 km MSL, producing low pressure hydrostatically in the upper and middle troposphere below the upper-level warming (Fig. 9.50). The atmosphere begins to adjust geostrophically to the persistent, upper-tropospheric heat source,

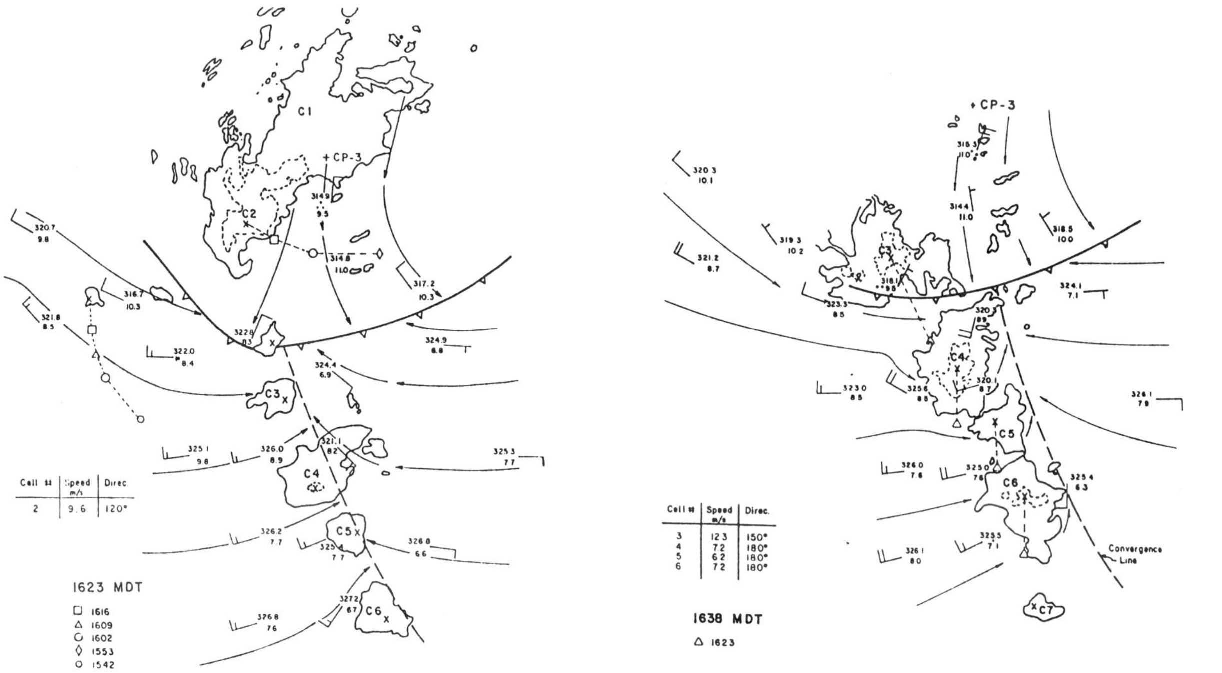

FIG. 9.47. Thunderstorms initiated by a cold frontal passage in South Park, Colorado, 19 July 1977 at (a) 1623 MDT and (b) 1638 MDT. Convective cells, labeled C1, C2, . . . , are radar echoes, and surface temperatures, mixing ratios, and winds (knots) are from mesonet (PAM) data. Mean velocities of selected cells are tabulated in the lower left of each analysis. Westerlies ahead of the cold front are gust front winds of thunderstorms to the west (Cotton et al. 1983; George 1979).

and a larger-scale, longer-lived component of the convective system is born.

Geostrophic adjustment thus creates a portion of the convective disturbance that is in geostrophic balance. Conceptually the disturbance can be divided into two parts, a geostrophically balanced part and a geostrophically unbalanced part. The balanced part or "system" is characterized by convergence and cyclonic shear at lower lev-

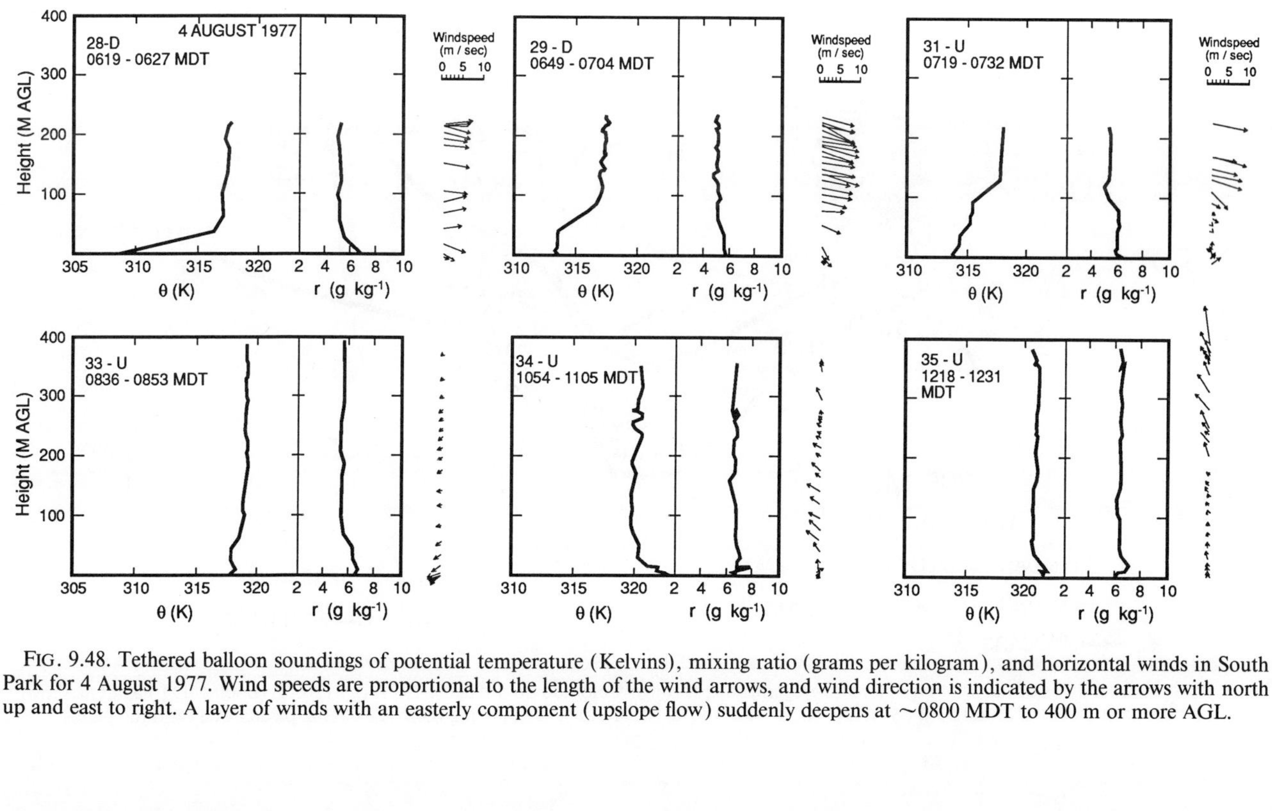

FIG. 9.48. Tethered balloon soundings of potential temperature (Kelvins), mixing ratio (grams per kilogram), and horizontal winds in South Park for 4 August 1977. Wind speeds are proportional to the length of the wind arrows, and wind direction is indicated by the arrows with north up and east to right. A layer of winds with an easterly component (upslope flow) suddenly deepens at ~0800 MDT to 400 m or more AGL.

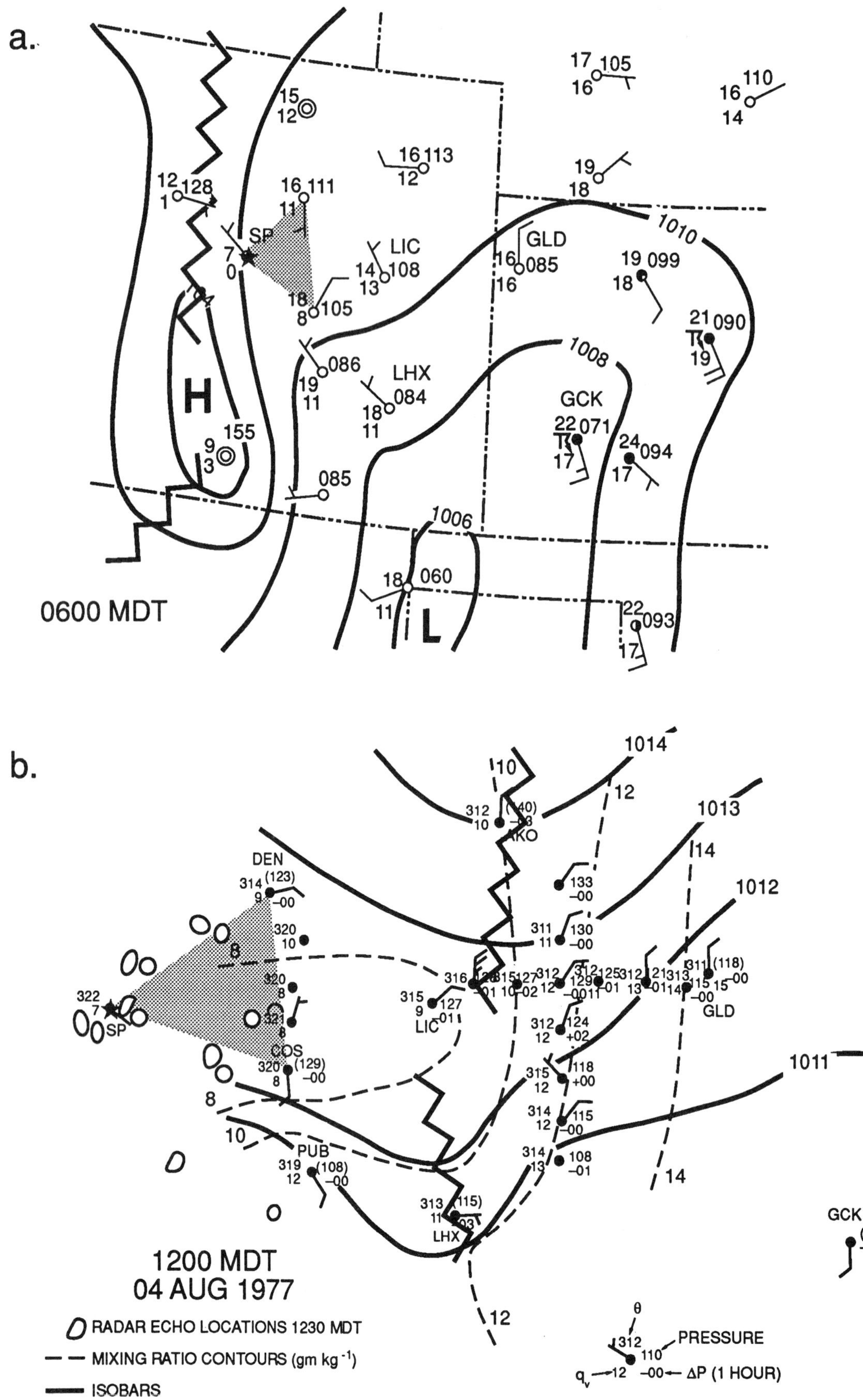

FIG. 9.49. Surface analyses of the location of a meso-ridge passing through Colorado on 4 August 1977. (a) 1200 Z (0600 MDT) analysis showing the ridge to the west of South Park (SP), and (b) 1800 Z (1200 MDT) analysis showing the ridge having passed well onto the plains.

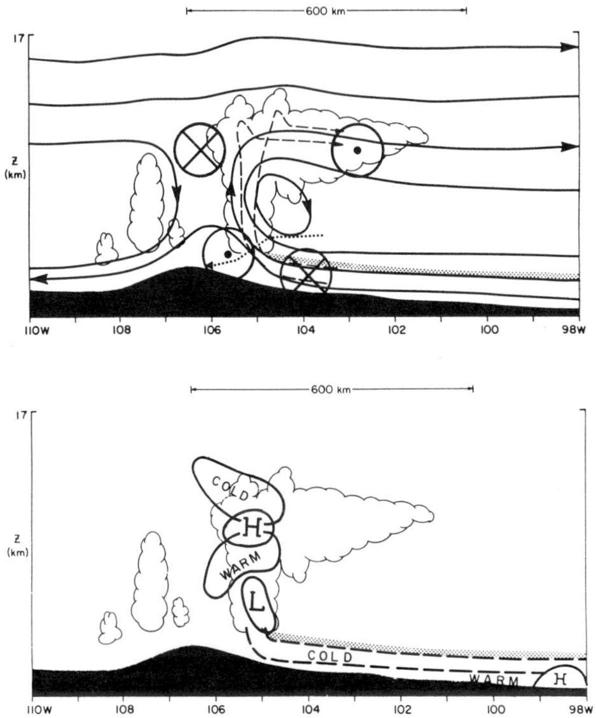

FIG. 9.50. Conceptual model of flow field and position of convective elements at the time when one major cell exists in the numerical simulation of Tripoli and Cotton (1989a) and convective heating aloft begins to induce meridional flow by geostrophic adjustment. (a) Streamlines and meridional flow as in Fig. 9.34, except streamlines are storm relative. (b) Pressure and temperature regions as in Fig. 9.34.

els, divergence and anticyclonic shear in the upper troposphere, and upward motion in midtroposphere.

The balanced system spins up slowly over many hours, but also spins down slowly. This inertia allows the balanced system to survive as the convective disturbance moves eastward through regions unfavorable to convection in the immediate lee of the mountain range. The balanced system survives even though most of the geostrophically unbalanced part (convection and shorter-wavelength gravity waves) dies out in this region. When the balanced system emerges from the eastern extremity of the unfavorable region after a time period of perhaps 2–4 h, it enters a region of high low-level moisture that is favorable to convection. At this stage the balanced system rapidly regenerates convection.

Tripoli and Cotton (1989a,b) show that this larger-scale system continues to develop in spite of undergoing several cycles of deintensification and regeneration. By the evening of the simulation day, the system achieves a structure similar to the MCC composite of Maddox (1981). The system does not continue to develop into a mature MCC, because the two-dimensional simulation does not represent meridional features, such as the moisture-rich, southerly low-level jet over the "Kansas" region of the simulation. Gulf of Mexico moisture advected into the Great Plains by southerly winds is probably an important ingredient in the intensification and maintenance of such nocturnal precipitating systems in that region.

The significance of the results is that "it has been shown that the action of the mountain valley [i.e., solenoidal] circulation in a basic horizontally homogeneous zonal flow will, in itself, deliver a suitable trigger to the region of the nocturnal southerly jet. . . . The Rocky Mountain slope flow system, in the absence of synoptic effects, has delivered to central Kansas by 0200 MST a warm-core, partially geostrophically balanced wind system . . ." (Tripoli and Cotton 1989a).

This system is somewhat larger than other systems discussed in this chapter and in the monograph. It is interesting, though, because it shows a different role for mountain solenoidal forcing, that of organizing the convection on scales larger than individual thunderstorm cells. Other studies have demonstrated how mountain-related flow systems have the ability to initiate individual convective clouds. Tripoli and Cotton's study shows that under the right circumstances mountain flow systems can have a role in determining where convection *persists*. In their case, interactions between thermally forced mountain flows, warming (and cooling) effects of deep moist convection, and atmospheric gravity–inertia wave dynamics hold the deep, moist convection in place for a significant portion of a pendulum day; i.e., long enough for geostrophic adjustment processes to build a persistent, large-(mesoα-) scale circulation system.

Conditions that Tripoli and Cotton found favorable to the establishment of long-lived systems are moderate vertical wind shear in the troposphere, a supply of moisture to the western slopes of the mountain range, and adequate soil moisture over the mountains and sloping plains to the east. Moderate wind shear (midtropospheric winds of 10 m s^{-1} or so) favors propagating systems, whereas weaker shear would produce more stationary systems that could cause flash flooding. The supply of moisture to the western slopes often occurs in the Rocky Mountains as a result of the summer monsoon flows discussed in section 9.4.3. The soil moisture provides moisture to the boundary layer by evaporation and replenishes moisture lost when the deep moist convection transports moisture upward out of the boundary layer. If this moisture were not replenished, excessive drying of the subcloud layer could destroy the moist convection before a long-lived system could get started.

9.4.6. Discussion

It is important to remember that convective clouds are by their nature highly turbulent and therefore stochastic;

thus there is a random component to their occurrence. The studies cited in the previous section have merely indicated tendencies in their behavior. For example, in considering which thunderstorm genesis areas will produce thunderstorms on a given day, there is undoubtedly a competition among nearby sites as to where the first deep, moist convection will form. Once one site establishes its dominance, deep convection at other nearby sites is suppressed. The processes that select which of several neighboring sites will dominate are controlled by variations in ridgetop wind direction, surface moisture, and other factors, and of course these variations have a random component. As another example, on the evening of the Big Thompson flood, there were several quasi-stationary thunderstorms over the mountains in other neighboring watersheds. Flooding did occur in these rivers, but only the Big Thompson storm had the necessary combination of storm and topography characteristics to produce the devastating results that followed.

This section has covered the relationship between the large- and small-scale forcing in determining primary thunderstorm generation. The large scales, generally synoptic and mesoα-scale, set the stage for and determine the likelihood of deep moist convection on a given day, and the smaller-scales govern the details of timing and location. Large-scale processes generally provide the stability and moisture environment, and occasionally provide triggering (e.g., by frontal activity) or large-scale vertical motion. Mountain-scale processes, probably the most susceptible to randomness, provide localized pulses of vertical motion that can initiate moist convective cells.

The mountain-scale initiation processes fall into three categories. 1) Direct, forced orographic lifting can release potential instability or lift a layer of the atmosphere to its LFC. 2) Aerodynamic or obstacle effects produce airflows, such as blocking or waves, that have upward branches that can lift air to its LFC. 3) Heating of the mountain surfaces produces convergence of upslope and up-valley flows toward the peaks and updrafts over the peaks that are capable of lifting air to its LFC.

This third mechanism, initiation of convection by thermally forced solenoidal circulations, is the important mechanism for daytime cumulus convection over mountains, especially in more arid regions where the circulations are well developed. One of the most striking aspects of these mountain cumuli in their formative stages in the morning is their strong tie to the peaks and ridges. An advantage we have in working in the mountains in this case, as we saw in the discussion of initiation, or triggering, mechanisms, is that initiation sites for thunderstorms tend to occur in the same locations when meteorological conditions are similar.

Mountain thunderstorms initiated by thermally forced mechanisms form earlier in the day and are in general less intense than valley or plains thunderstorms. Because they form earlier, mountain thunderstorms produce gust fronts and gravity waves that propagate onto the plains and participate in the initiation of secondary convection there. Thus it is likely that mountain thunderstorms have an effect on the spatial distribution and timing of thunderstorms and severe weather later in the day over the plains.

The numerical simulations of Tripoli and Cotton (1989a,b) established a more direct link between strong thunderstorm activity over the mountains and the organization of larger mesoscale systems over the plains. These simulations showed that under proper tropospheric wind and moisture conditions, the large, mesoα-scale, mountain–plains, solenoidal circulation is capable of organizing a long-lasting mesoscale convective system that is in partial geostrophic balance, and all this can happen without benefit of a preexisting synoptic- or mesoα-scale disturbance.

9.5 Conclusions

What is the role of mountain flows in making clouds? It is obvious that local flows in the vicinity of mountains and mountain ranges provide the lift that is necessary to achieve saturation and condensation. The kinds of lift that a mountain can provide are direct, such as forced orographic ascent when flow encounters an obstacle, or more indirect, such as when stable flow produces blocking or waves or when mountain slopes are heated and the mountain acts as a high-level heat source.

Mountain flows do not make clouds by themselves. They require inputs of moisture, proper stability, and at least a nonsuppressive synoptic- or large-mesoscale environment. Cloud characteristics can be significantly altered by radiation and cloud microphysics. The purpose of this chapter has been to put the role of the mountain flows into perspective with the roles of the other physical and dynamic processes.

In all cases the larger scale exerts a controlling influence on whether mountain flow systems will be able to make clouds or not. Clouds were divided into two main classes, stable and unstable. Under special circumstances, when moisture is confined to the lowest kilometer or two above the terrain and when the flow in that layer is reasonably parallel to the topography (i.e., does not exhibit strong wave or blocking effects), microphysics and radiation have strong effects on the ability of stable clouds to produce precipitation-sized droplets or particles. Often, however, wave or blocking effects are significant, and under these conditions the dynamics control the location and strength of the updrafts, and thus control the location and often even the intensity of the precipitation.

For unstable clouds, which include thunderstorms, mountain flows play a significant role in determining where convective cells will initiate, and often how precipitation from the showers will be distributed spatially. When mountains act as high-level heat sources and initiate storms by thermally forced, solenoidal circulations, they regulate not only the location of storm initiations, but the timing, too. Primary thunderstorms, those that form only from terrain forcing, occur during late morning or early afternoon. They help trigger later, more intense secondary storms and convective storm systems that can move through the mountains and onto the plains or low areas nearby.

Acknowledgments. Several reviewers of this chapter made excellent suggestions and gave constructive criticism that led to significant improvements. For these helpful reviews, I thank Bill Cotton, Ray McAnelly, Jim Wilczak, Al Bedard, and an anonymous reviewer. I also thank Fernando Caracena for illuminating discussions on the Big Thompson flood analysis, Christine Sweet for her careful technical editing of the manuscript, and Lisa Olivier for her help with manuscript and figure preparation.

CHAPTER 10

Predictability of Flows over Complex Terrain

J. PAEGLE

Department of Meteorology, University of Utah, Salt Lake City, Utah

ROGER A. PIELKE, G. A. DALU,* W. MILLER**

Department of Atmospheric Science, Colorado State University, Fort Collins, Colorado

J. R. GARRATT

CSIRO, Division of Atmospheric Research, Mordialloc, Victoria, Australia

AND

T. VUKICEVIC, G. BERRI,[†] AND M. NICOLINI[†]

Department of Meteorology, University of Utah, Salt Lake City, Utah

ABSTRACT

We review the relevance of predictability theory to flows in local domains that are modified by boundary conditions produced by topography, surface physical forcing, or imposition of numerical lateral boundary constraints. There is an apparent inconsistency between the results of classical predictability theory and limited domain experiments. We attempt to reconcile the discrepancy by analysis of error growth and chaos. The analysis suggests that boundary conditions provide the dominant flow contribution for many limited domain predictions, and this motivates review of error spread from uncertain boundary specification. Case summaries include free flows of the troposphere, moist convection, topographically bounded local flows, and flows above other complex surfaces. A tentative conclusion is that boundary uncertainty does not provoke chaotic responses so easily as do small imperfections of initial data, even for dynamical systems that are chaotic in global or periodic domains. One exception to this appears to be the case that the boundary uncertainty possesses a spatial scale that is not much shorter than the perimeter bounding the simulation domain.

10.1. Overview of concepts of chaos and relation to predictability

The investigation of dynamical systems in the 1960s and 1970s resulted in the development of the concept of chaos. Lorenz (1963) was one of the original researchers on this topic. A recent qualitative discussion of the topic has been given in Gleick (1987), while an excellent mathematical survey was presented in Bergé et al. (1984).

As summarized in the latter book, for a dissipative nonlinear dynamical system to become chaotic, the phase space of the system must have a dimension greater than or equal to 3. In a chaotic regime, one of the characteristics is a sensitivity to initial conditions; i.e., infinitesimal differences in initial conditions will result in solutions at a later time that are uncorrelated with one another. This concept has been used to define the limits of predictability of weather forecasts.

In this chapter we present limited relevant work related to the predictability of flows over complex terrain. Among the important questions are:

- Under what circumstances are flows in complex terrain dominated by the surface boundary conditions, as opposed to the initial state of the atmosphere?
- To what extent are the flows dominated by linear forcing and response such that a chaotic mesoscale response will not occur?

From Lorenz (1963) and from the rich literature that followed, we know that a convective system becomes chaotic when the diabatic forcing is strong enough. Furthermore, from Lorenz (1980) and from subsequent literature, we know that the large-scale atmospheric quasi-geostrophic flow can have a chaotic behavior. Actually Lorenz (1980) shows that the quasi-geostrophic model, with a suitable change of variables, can be reduced to the same

---

* Permanent affiliation: Institute for Atmospheric Physics, (IFA-CNR), Rome, Italy.

** Permanent affiliation: U.S. Air Force.

[†] Permanent affiliation: Department of Meteorology, University of Buenos Aires, Argentina.

model that describes the nonperiodic deterministic flow of a convective system (Lorenz 1963). Again from Lorenz (1980) the flow described by primitive equations exhibits chaos, in certain regions of the parameters. Therefore it is reasonable to presume that mesoscale atmospheric flow may exhibit chaos (i.e., finite predictability); however, research in this direction is needed. In turbulent flow and in synoptic forecasting, the sensitivity to initial conditions is well established. This is the reason that turbulence is characterized statistically. For thermally forced flow over rough terrain and forced airflow over rough terrain, however, when the flow is dominated by the pressure gradient force (which is essentially a linear term in the equation of motion), and/or by the terrain acting as either an elevated heat source or sink, or as an obstacle that also acts through the pressure gradient force, it is not clear that chaotic solutions (i.e., sensitive to initial conditions) will result. In addition, for the case of forced airflow over rough terrain, a chaotic response could be initiated as the air crosses the topographic barrier, but if the air is swept downward rapidly enough, the significant chaotic effect would occur at a large distance from its initiation. The upwind flow far enough upstream will then not be affected by the chaotic response, but by the pressure gradient force caused by the terrain barrier.

This chapter only initiates a discussion of this topic. The predictability of mesoscale flows in complex terrain, however, must address under what circumstances these types of flows are chaotic.

The results are unexpectedly different from those observed on much larger and smaller scales. This is emphasized by examples of chaotic evolution on global scales in section 10.2 and for convective scales in section 10.3, in contrast to the apparently deterministic solutions above complex surfaces that are emphasized in sections 10.4 and 10.5. The examples lead to a study of the influence of perturbations in boundary conditions. Recent preliminary results of such analyses are presented in sections 10.4 and 10.5. The conclusions are summarized in section 10.6.

10.2. Predictability studies on the synoptic and global scales

The term "predictability" refers to the potential to produce forecasts in which the signal (defined as the actual or forecast evolution) exceeds the noise due to the uncertainty of the forecast system. In meteorology, predictability connotes a somewhat narrower meaning, related solely to the time evolution of uncertainties associated with the specification of the initial state. This emphasis originated from the use of a vorticity conserving model for the first deterministic computerized forecasts of the atmosphere. Forecast states produced by this model are simply due to the fluid inertia and transport of conserved properties (vorticity or potential vorticity) from one place to another. Boundary conditions, physical forcings, and dissipation do not affect the simplest global implementations of this model and the only uncertainty of the resulting forecast is associated with the uncertain specification of the auxiliary initial data.

Many studies of the sensitivity of more complete forecast systems to other auxiliary conditions have been performed. These studies also qualify as predictability studies in a broader context; however, the present review will focus on aspects related most directly to the narrower questions regarding evolution of forecast errors associated with errors of the initial state. This focus allows exploitation of the best developed tools for study of error evolution. As we shall see in section 10.4, the conclusions have broad implication for a variety of predictability questions in local domains, including mountainous regions where boundary constraints are especially relevant. It is also useful to review the large-scale theory because of its strong implications for small scales, and because a major potential inconsistency currently prevails between the theory and small-scale experimental results.

10.2.1. Predictability theory for large scales

Thompson (1957) and Lorenz (1969) performed analytical studies of the time evolution of errors in the barotropic vorticity equation model, in which one of the important dynamical mechanisms is the advection of vorticity. Thompson (1957) points out that this term also includes the mechanism for barotropic instability, which is important for the growth of small initial errors. These analyses are based on simple barotropic models, and make further simplifying assumptions regarding the homogeneity, isotropy, and slope of the spectra. Some of the latter assumptions are weakened by Novikov (1959), and the impacts of these generalizations are outlined later in this review.

Thompson's (1957) analysis applies to bounded regions and assumes no error on lateral boundaries, and Lorenz's (1969) analysis treats a periodic domain. Both specify the initial streamfunction as a superposition of an uncertainty R and the correct value ψ and demonstrate that for synoptic scales, typical errors double on a time scale of a few days. Two conclusions are particularly relevant to small-scale applications on bounded domains.

One of these is that the growth of uncertainty depends upon the relative spatial scale of the uncertainty with respect to the scale of the correct flow field. Both analyses conclude that uncertainty growth occurs only for those interactions involving spatial scales of R that are larger than those of ψ. Other interactions do not produce error growth. One interpretation of this result is that barotropic instability only occurs in the former case (Thompson

1957; Lorenz 1953). Another interpretation is that the former case corresponds to advection of a small-scale wave by an erroneous larger scale that produces loss of phase information in the small-scale wave.

The phase errors increase more rapidly for smaller spatial scales for a given advection error, and this provides partial explanation for the second major conclusion; i.e., that small scales lose predictability more rapidly than large scales. The quantitative details of the loss depend upon the energy spectrum of the model atmosphere. Sample calculations presented by Lorenz (1969) suggest that wavelengths on the order of 8000, 1000, 100, and 10 km lose predictability after approximately 5 days, 1 day, 5 hours, and 1 hour, respectively. The results at smaller scale have obvious practical significance for local flows, reviewed in sections 10.3 and 10.4. We next summarize the extent to which the theoretical predictions are verified in global and synoptic-scale experiments.

10.2.2. Global experiments

The implications of the theory have been most exhaustively tested in a number of global integrations of nonlinear forecast models in which error evolution is represented by the square root of the area-integrated mean of the squared differences between two model forecasts initialized with two slightly different initial states. The earliest experiments using primitive equation global models are summarized by Charney et al. (1966) and by Smagorinsky (1969). In general these experiments support the theory, yet exhibit somewhat slower error growth rates, partly because of relatively strong diffusion and the consequence of generally weak flows, which results in weaker nonlinearity.

Figure 10.1 exhibits one example of a global barotropic model (taken from Vukicevic and Paegle 1989) in which the control experiment is first run for 14 days. Then the initial conditions are perturbed by small, random differences that have an area-averaged rms value of 2 m s^{-1}, and the model is rerun. Three of the experiments use Rossby–Haurwitz waves 1, 2, and 4 to initialize the model, and a fourth experiment is performed on a real-data case using the observed 200 mb wind for 0000 UTC 8 August 1978.

The experiments clearly demonstrate the scale dependence of error growth, with the higher wavenumber Rossby–Haurwitz waves losing predictability more rapidly than those of lower wavenumber. The fastest error growth occurs for the real-data case, which starts with the most energy in the shortest resolved scales. This case loses deterministic predictability by two weeks, when the error variance equals the wave kinetic energy. Similar results are obtained with more complete primitive equation numerical models that retain a fuller spectrum and baroclinicity. Global model studies of the spectra of the error evolution (Baumhefner 1984; Dalcher and Kalnay 1987) also show that the rate of the error growth becomes higher with smaller scales, as anticipated by the theory.

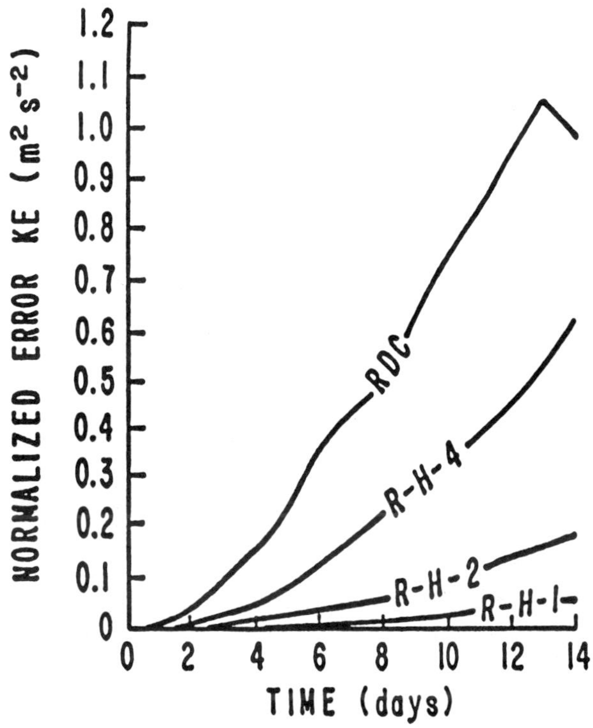

FIG. 10.1. Global rms error kinetic energy normalized by kinetic energy of control flow field (wave part only) for: R–H (Rossby Haurwitz) wave 1, R–H wave 2, R–H wave 3, R–H wave 4, and real-data case (RDC) (from Vukicevic and Paegle 1989).

It is noteworthy that these experiments generally cover only one or two orders of magnitude in wavelength. For example, the Vukicevic and Paegle (1989) study explicitly resolved only waves 1 through 16. Consequently, the results of Fig. 10.1 correspond to situations in which the range of retained scales varies by only one order of magnitude. The most highly resolved real-data global integrations of which we are aware use spectral wavenumber truncations on the order 100, and thus span only two orders of magnitude in the spatial scale.

By contrast, Lorenz's (1969) idealized analysis of the predictability of a flow possessing many scales of motion explicitly retains wavelengths ranging from 38 m to 40 000 km, spanning six orders of magnitude. The similarity of the conclusions of the highly resolved but idealized theory and the more poorly resolved but otherwise more complete model studies appears to validate both approaches for global-scale applications.

It might consequently be anticipated that forecasts made over regional domains, which resolve spatial scales ranging over one or two orders of magnitude, would display relatively fast rates of error growth; however, almost all such integrations either display no error growth or reduced error growth relative to the theoretical predictions. These apparently contradictory results are summarized next.

10.2.3. Predictability experiments in limited area domains

The only currently reported experiments on predictability within local domains, using real-data, are for situations of meso- to synoptic horizontal scales. Many of these models are run with grid dimensions containing about 50 points on a side. The ratio of the wavelength of the fundamental mode to the shortest resolved waves is on the order of 20 to 1; consequently such models span a range of decades of spatial scales similar to that contained in the global integrations of Fig. 10.1. Extensive experimentation with such grids is presented by Anthes et al. (1985), using grid sizes of 80 and 160 km. Thus, the total domain dimensions are on the order of 4000 to 8000 km on a side, and these simulations resolve much of the synoptic scale spectrum as well as the longer mesoscales.

A sample rms error evolution over one such domain is depicted in Fig. 10.2. This figure shows virtually no temporal growth of initial errors through two days of the forecast, even though most of the resolved scales would display a total loss of deterministic predictability within two days or less according to Lorenz's (1969) estimates.

Such results present apparent contradictions with the theory, and suggest that the details of the initial data are not particularly relevant to local prediction. The conclusion has important practical implications for resource allocation for mesoscale forecast development and observation networks, and has prompted a number of recent studies. Errico and Baumhefner (1987) performed integrations similar to those presented by Anthes et al. (1985) for a cyclogenetic situation over North America. They find that the evolution of domain integrated forecast differences associated with small initial data differences damps with time, but in certain locations uncertainty growth occurs, particularly in a region near the developing cyclone. Two of the reasons given for general error decay are that diffusion and geostrophic adjustment limit error growth. Another interpretation of the error suppression near the boundaries is that perfect information comes into the region, principally from the western inflow boundary, while errors are swept out through the outflow boundaries. This "sweeping out" of errors occurs more rapidly than growth of the errors over most regions, especially near the boundaries, but not fast enough to suppress the local instability around the cyclone developing in the middle of the domain. This is only one of several possible explanations for their result.

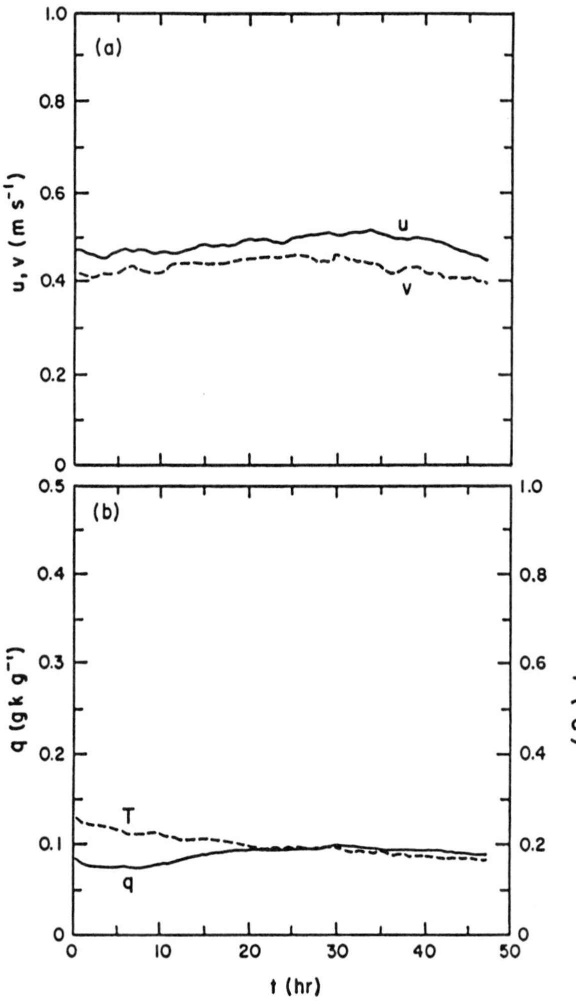

FIG. 10.2. Temporal variation of rms differences between control and initially perturbed experiments: (a) wind components u and v, and (b) specific humidity q and temperature T (from Anthes et al. 1985).

There exist three main possibilities related to the boundary constraint:

1) The imposition of lateral boundary conditions inhibits the range of spatial scales that can react to initial uncertainty. In particular, the largest scales are imposed through the prespecified boundary conditions, and cannot react to the internal data uncertainty in most limited area models. Consequently, the theoretical requirement for error growth, that the uncertainty scale be larger than the scale of the correct solution, is more difficult to satisfy in local area predictions than in global predictions.

2) The transport of exact information into the forecast domain through perfectly specified inflow boundaries, and the export of errors through outflow boundaries may occur at a rate that is as fast or faster than the unstable growth of uncertainties. This mechanism could be effective only if the cross-domain advective time scale is at least as fast

as the error growth rate associated with the physical instability.

3) A third possibility is that many local flows are strongly forced by lower boundary conditions, such as topography, and highly dissipated by friction. In such cases, the forecast system cannot have a long memory of its initial state (or its initial errors) and the forecast may proceed relatively independently of initial data uncertainties.

We will explore these possibilities because they appear to have the most potential impact for the case of smaller scales that are of greatest practical concern to local prediction around complex terrain. We also note that certain small-scale physical systems maintain coherent structures for much longer periods than suggested by the predictability theory, and these may have still different explanations than those given above. Such cases include highly intermittent small-scale events that violate various statistical assumptions of the theory. For example, a hurricane may persist as a coherent entity for many days, and may consequently be highly predictable, at least in principle, far longer than suggested by the theoretical arguments of Lorenz (1969), which impose assumptions of homogeneity, isotropy, and a monotonic energy spectrum. These assumptions may not be appropriate to solitary or intermittent flow structures such as hurricanes. Useful reviews of these incompatibilities of the theory, as well as other turbulence considerations are given by Lorenz (1969), Lilly (1984), and Anthes et al. (1985).

10.3. Predictability studies of turbulence and clouds

On the turbulence and cloud scales, there has been considerable study of the onset of chaos since Lorenz's (1963) ground-breaking work. The chaotic or near chaotic behavior of a convective system makes cloud prediction a difficult task for forecasters.

From DeGregorio and Dalu (1984) we see that when the value of the forcing is close to the chaos threshold the quiescent rate is attractive in the (y_2, y_3) plane and the system from an initial state of convective activity goes to a state of almost no convection. Then the system rapidly switches to the other convective state (Fig. 10.3). Transition to the previous convective state cannot occur; however, this transition can be acquired for this level of forcing with the addition of a very small amount of noise (Sutera 1980), or with a slight increase of the forcing (Fig. 10.4). Therefore prediction of the convective activity is likely to be difficult when the atmosphere is close to this critical state.

Model convection can be increased by destabilizing the atmosphere with low-level heating or through rising motion. Both forms of destabilization commonly arise as a

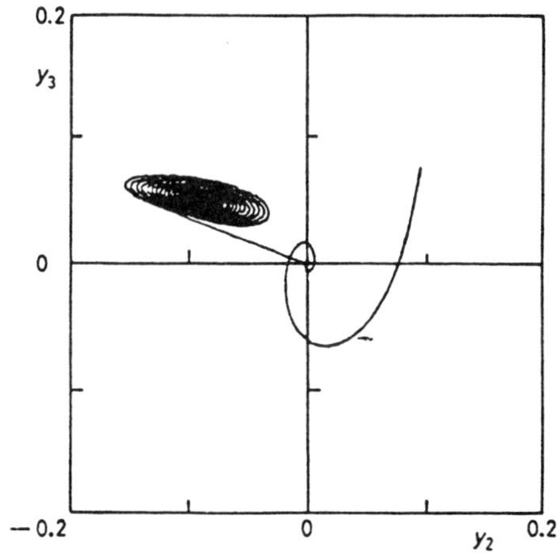

FIG. 10.3. There are two attractive convective states in the second and in the fourth quadrant, and a quiescent state that is attractive in the (y_2, y_3) plane and repulsive in the y_1 direction. If the initial state is in the (y_2, y_3) plane, the system falls toward a state of almost no convection, and prediction on this stage would give decaying convective activity. Only small perturbations (even the roundoff error) would kick out the system toward one of the two convective attractive states.

consequence of boundary modifications, such as heating of the lower boundary, or by forced ascent produced by larger scale influences that are not resolved by convective scale simulations and must be applied as modifications of boundary flow. Nicolini et al. (1987) investigated the latter form of destabilization in a model of shallow moist convection. Unlike earlier studies that imposed the rising motion with a direct modification of the vertical motion or heating term of the thermodynamic equations, Nicolini et al. (1987) imposed the vertical motion through boundary modification of the wind. This is the most natural procedure in nested sequences of convective models, and also relates more directly to the analysis of boundary error spread discussed in the next sections.

The spatial scales of these boundary modifications are the same as the domain perimeter, so that a substantial internal response may occur (section 10.4.4). Their am-

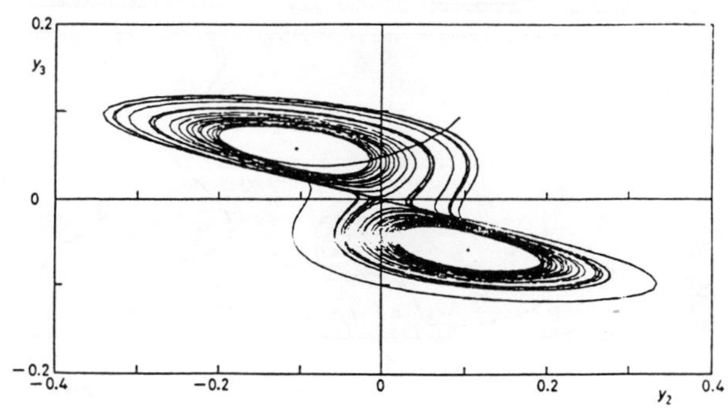

FIG. 10.4. If the forcing is further increased over that shown in Fig. 10.3 the behavior of the convective system becomes chaotic.

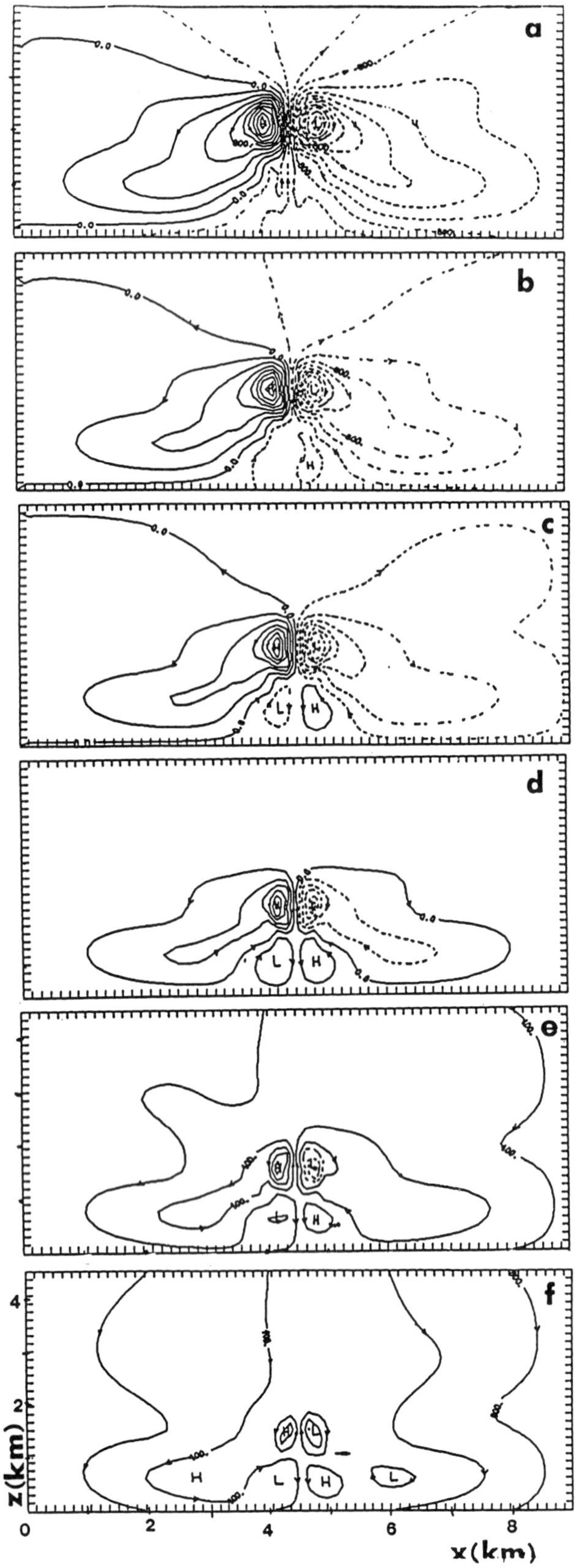

FIG. 10.5. Streamfunction of forecast convective scale motion. All forecasts start with the same unstable stratification and are run for about one hour. The only difference is in the imposition of weak ambient vertical velocity through the boundaries, which has the following values in the different panels: (a) 0.1 m s^{-1}, (b) 0.05 m s^{-1}, (c) 0.02 m s^{-1}, (d) −0.02 m s^{-1}, (e) −0.05 m s^{-1}, (f) −0.1 m s^{-1} (from Nicolini et al. 1987).

TABLE 10.1. Absolute extremes of cloud parameters (from Nicolini et al. 1987). Here W is vertical velocity maximum imposed through boundaries (m s^{-1}), W^* is maximum convective scale vertical velocity (m s^{-1}), Q_L is maximum cloud liquid water concentration (gm kg^{-1}), and Q_{LP} is maximum precipitating liquid water concentration (gm kg^{-1}).

W m s^{-1}	W^* m s^{-1}	Q_L 10^{-3}	Q_{LP} 10^{-3}
0.10	8.50	3.75	10.70
0.05	7.05	3.60	4.20
0.02	6.10	3.75	3.00
0	5.40	3.60	2.35
−0.02	4.90	3.60	2.15
−0.05	4.70	3.10	2.00
−0.10	3.40	2.25	0.85

plitude, however, is only on the order of 1 to 10 cm s^{-1}, which is much smaller than the 10 m s^{-1} motion of the convection. Nevertheless, there is a significant sensitivity of the convective scale circulation to weak boundary modification (Fig. 10.5). Various cloud parameters also respond sensitively to the boundary modifications (Table 10.1).

In section 10.2 we presented evidence that boundary prespecification may limit the chaotic behavior associated with initial uncertainty. This section has described substantial solution sensitivity to boundary modifications. The purpose of the next section is to summarize the available theory regarding initial and boundary uncertainty in flow simulations within limited domains.

10.4. Analysis of limited domain predictability

10.4.1. Theoretical considerations

The theoretical analysis most relevant to limited domain predictability experiments was performed by Thompson (1957). He defines the mean square error by

$$E = \frac{1}{A}\int_A \nabla R \cdot \nabla R dA, \tag{10.1}$$

where A is the area of the forecast domain, dA is an element of A, and R is the error in the streamfunction. Differentiation of (10.1) with respect to time gives a forecast equation for E involving area integrals of R, its gradients and time derivatives. Some of these integrals can be transformed to line integrals along the perimeter of the forecast domain, and these vanish because R, which appears as a factor in the integrands, vanishes on the boundary. Further use of the perturbed and unperturbed barotropic vorticity equation gives:

$$\frac{\partial E}{\partial t} = \frac{2}{A} \int R\mathbf{J}(\psi, \nabla^2 R) dA, \qquad (10.2)$$

where $\mathbf{J}$ is the Jacobian. This equation expresses the time rate of change of domain integrated error energy as a simple area integral of the product of the streamfunction error, R, with the advection of the error vorticity by the correct flow field. We note that the concise form of (10.2) results from the assumption of zero error on the boundaries, which is the condition assumed in local area integrations of Anthes et al. (1985), Errico and Baumhefner (1987), and other studies of regional predictability.

Novikov repeated Thompson's (1957) analysis under less restrictive assumptions regarding spatial homogeneity of the flow statistics. The dimensionless scale factor that determines the dependence of error growth upon the spatial scale of the error (λ) and of the correct flow (L) is then:

$$\frac{\lambda^4}{(L^2 + \lambda^2)^2}.$$

The factor assymptotes to 1 for $\lambda \gg L$ and decreases very rapidly with λ for $\lambda < L$. While this modifies the qualitative nature of Thompson's (1957) conclusions, the predicted growth rates produced by interactions among errors of relatively small scale with correct flows of larger scales is relatively insignificant for all cases where λ / L is on the order of 0.5 or less.

It would be useful to evaluate (10.2), or related expressions from Novikov (1959) over local domains within local regions in experiments such as those performed by Errico and Baumhefner (1987), which have slow error growth in local area integrations and fast error growth in global integrations. For initially uncorrelated errors $\partial E / \partial t = 0$, but as the integrations proceed, this becomes nonzero.

We are unaware of any explicit calculation of (10.2) to compare limited area predictability with global domain predictability; however, we anticipate from Thompson's (1957) and Lorenz's (1969) analyses that the right-hand side of (10.2) should be smaller (perhaps negative) in the local area forecasts run without boundary error. The conclusion follows from the observation that the streamfunction field of the limited area model is given by the solution of:

$$\nabla^2 \psi = \zeta \qquad (10.3)$$

subject to boundary conditions $\psi = \psi_b$ at the perimeter of the domain.

The solution of (10.3) can be decomposed in the sum:

$$\psi = \psi_I + \psi_B,$$

where ψ_I solves

$$\nabla^2 \psi_I = \zeta, \quad \psi_I = 0 \text{ on boundaries}$$

and ψ_B solves

$$\nabla^2 \psi_B = 0, \quad \psi_B = \psi_b \text{ on boundaries.}$$

Only ψ_I can react to the initial interior uncertainty because ψ_B depends entirely upon the lateral boundary condition, which cannot respond to uncertainties in the initial data for cases where boundary conditions do not react to the interior solution. Paegle and Vukicevic (1987a) and Vukicevic and Paegle (1989) show that ψ_B makes the dominant contribution in many regional area integrations. Moreover, this contains all of the information on scales larger than the regional domain, and is error free in local domain predictability experiments invoking externally imposed lateral boundary conditions. In such cases, most of the R field is in scales smaller than most of the ψ field, a situation that does not promote uncertainty growth (Thompson 1957; Lorenz 1969). In a global or periodic domain there is no contribution from boundaries because ψ_B becomes part of the interior solution. Thus, it can react to uncertainty spread, and contribute to efficient error growth.

The degree of boundary control is illustrated by Vukicevic and Paegle (1989) for a circular domain centered in a midlatitude region at 200 mb in the winter hemisphere. In Fig. 10.6 ψ_I, ψ_B, and ψ are shown in the case where the domain diameter is approximately 4400 km, corresponding to a 40° diameter polar cap rotated onto the midlatitude region. The boundary contribution completely dominates the internal contribution. Vukicevic and Paegle (1989) integrated the barotropic vorticity equation in circular limited area domains centered in the midlatitudes. The boundary conditions at the perimeter of these domains were imposed from the forecasts of the global real-data case for which the predictability curve is displayed in Fig. 10.1. This was done at each time step, and the initial state within the limited domain area was perturbed as in the global case.

The rms error evolution over the limited region is displayed in Fig. 10.7 for various domain sizes. All cases with domain diameter less than approximately 65° of latitude or 7000 km lack error growth. These integrations represent an inertially driven flow constrained only by lateral boundary conditions and damped only by weak diffusion. Unlike the limited area integrations presented by Anthes et al. (1985) and Errico and Baumhefner (1987), the present runs do not have a sponge layer of enhanced diffusion near the lateral boundaries. The ex-

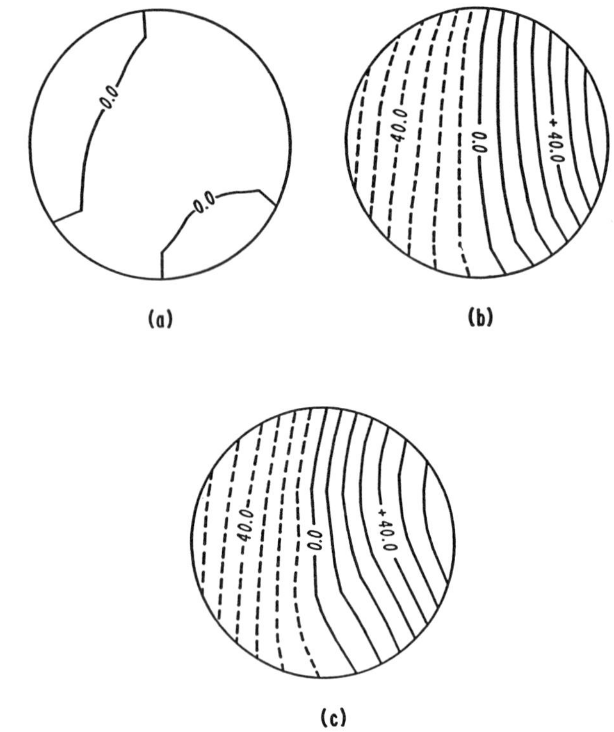

FIG. 10.6. Streamfunction solutions: (a) ψ_I; (b) ψ_B; and (c) ψ for real-data case centered in midlatitudes of the Southern Hemisphere at 0000 UTC 8 August 1979. The outer circle of each panel is located 20° from the central point of each subdomain (from Vukicevic and Paegle 1989). The unit is 10^6 m^2 s^{-1}.

periments support the possibility that error damping in local area forecasts may be entirely produced by boundary prespecification, and furthermore suggest that the critical domain size distinguishing growing and nongrowing errors is approximately 7000 km on a side, at least for the present model.

10.4.2. *Predictability of highly forced and dissipated flows*

Another possibility for control of error growth is the effect of strong forcing and dissipation as represented, for example, by topographically bound low-level jets. Such circulations were frequently in evidence in the Alpine Experiment (ALPEX), which was designed to study the dynamics of topographically modulated flows (Kuettner 1982). Data from this experiment have been used to initialize and verify simulations of flow blocking and splitting, as well as barrier jets produced by the Alps (e.g., Paegle et al. 1984).

The predicted location of low-level jets and circulation maxima are relatively insensitive to both random and systematic analysis differences of initial data, although the strength of the forecast phenomena does have some sensitivity to these data. This is demonstrated by Paegle and Vukicevic (1987b) using a detailed boundary-layer forecast model initialized from smooth operational analyses of the ECMWF (European Centre for Medium Range Weather Forecasts). The model predicts jet maxima on the flanks of the Alps even though these are missing in the smoothed initial and boundary conditions provided by the ECMWF analyses.

The model response is enhanced in experiments based upon the more complete (level IIB) datasets produced by the ALPEX project, because portions of the jets are resolved in these more detailed data, but the overall features are similar to forecasts initialized with the smoother operational ECMWF analyses. Other examples of finescale forecasts emerging from smooth initial and boundary data are given by Paegle et al. (1984) and Astling et al. (1985). These results suggest that the boundary effects of topography enhance the predictability of low-level flows in ways that are unrelated to the boundary condition prespecification discussed above.

This possibility has been further probed by Paegle and Vukicevic (1987a) in a series of predictability experiments based upon the ALPEX datasets. Their model is an extension of the boundary-layer model described by Paegle and McLawhorn (1983). The integrations begin at 0000 UTC 15 April 1982, and include episodes of strong nocturnal low-level jets north of the Alps (Paegle et al. 1984)

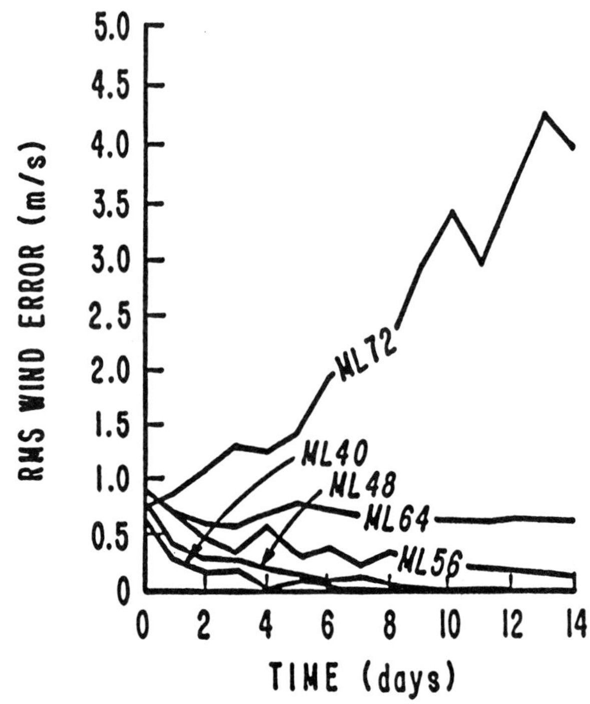

FIG. 10.7. Root-mean-square wind error evolution in the experiments with real initial data for midlatitude domains of diameter 40° (ML40), 48° (ML48), 56° (ML56), 64° (ML64), and 72° (ML72) (from Vukicevic and Paegle 1989).

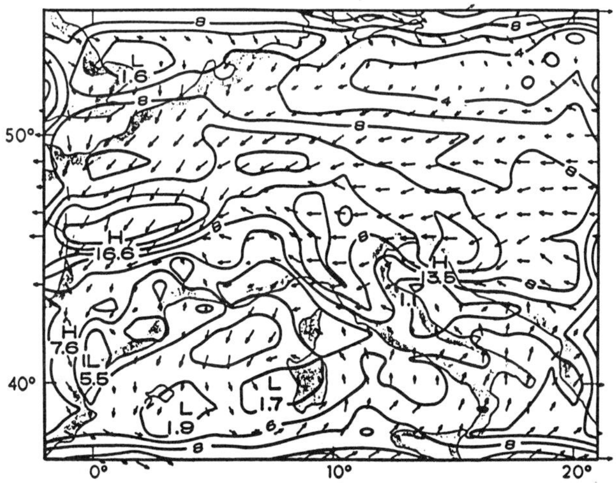

FIG. 10.8. The 12 h forecast of the wind field 119 m above the surface for the control case described in the text. The contours are isotachs spaced at 2 m s^{-1}, and a vector equaling the grid point separation represents a speed of 17 m s^{-1} (from Paegle and Vukicevic 1987a).

and northeast bora wind on the north coast of Yugoslavia (Smith 1982). The forecast domain is centered on the Alps, and a 25 × 19 spherical grid covers a region extending from Western to Eastern Europe and from the Mediterranean to Northern Europe. The model top is located approximately 2 km above the surface; here the observed 850 mb height field gradients are used to specify the pressure variations, and the initial flow field is geostrophically balanced with this state.

The model is allowed to come into approximate balance of radiative forcing, dissipation, and topographic deflection in a 12 h integration, which is displayed at a sample forecast level in Fig. 10.8. A random noise with an rms magnitude of 2.5 m s^{-1} is then superimposed on the 12 h forecast. This perturbation has the same value at all levels, including the upper boundary, where it is maintained.

A sample evolution of the area-averaged uncertainty is displayed by the thick curves in Fig. 10.9a. The thin curves of this diagram show the rms difference of the forecast experiment from the solution at the time of the noise introduction. These curves represent the evolving forecast signal. After only 6 h, the rms of the noise is on the order of only 0.5 m s^{-1}, while the rms magnitude of the forecast signal exceeds 6 m s^{-1}. The ratio of forecast noise to signal continues to decline, illustrating that the predictability is very high in these experiments. The rate of error damping is more rapid than in the prior limited area domain experiments presented in Figs. 10.2 and 10.7. One reason for this may be that the present case emphasizes frictionally dissipated flow above complex terrain. The strong dissipation and low-level forcing constrain the evolution

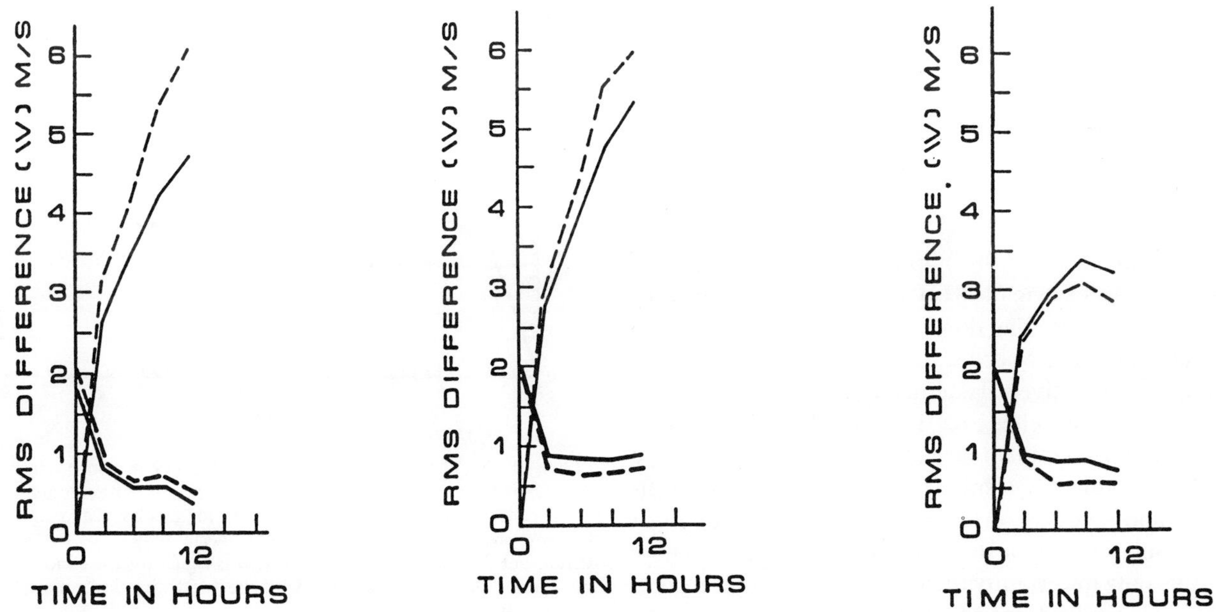

(a) (b) (c)

FIG. 10.9. (a) The thick decaying curves represent the rms difference between the control and the experiment with perturbed initial conditions. The thin growing curves represent the evolution of the rms difference of the initial condition and forecast at the displayed time in the experiment. The solid curves are for the full 25 × 19 grid and the dashed curves are for the 16 × 13 subdomain centered on the Alps (from Paegle and Vukicevic, 1987a). (b) As in (a) for cases where the boundary conditions assume values forecast on the adjacent interior grid point at each time step. (c) As in (a) without topography.

more strongly than is the case of flows of the "free troposphere." Additionally, there remains the possibility that imposed lateral boundary conditions limit the error growth.

The next experiments are designed to investigate the relative roles of lateral boundary versus physical effects in the suppression of error growth by allowing uncertainty growth on the lateral boundaries. The boundary value is now set equal to the value forecast one grid point within the boundary. In this case the boundary is free to adjust to the regional forecast and its error growth is not constrained as in the prior experiments.

The model is run as before for 12 h to allow adjustment of any initial imbalance. A random noise with an rms magnitude of 2.5 m s^{-1} is again imposed on the 12 h forecast. The evolution of the area-averaged rms differences of the perturbed and unperturbed control is illustrated in Fig. 10.9b. After 6 h the noise is less than 1 m s^{-1}, while the rms magnitude of the signal exceeds 6 m s^{-1}, so that the ratio of forecast noise is less than 20% of the signal after 6 h of prediction. It appears that the major factors limiting error growth in this case are physical constraints of forcing and dissipation, and lateral boundary constraints are not required to inhibit growth of errors due to initial data uncertainty.

In order to ascertain how much of the high predictability is due to topographic forcing and how much is due to dissipation, Paegle and Vukicevic (1987a) repeated the experiments without topography. The evolution of the rms differences is displayed in Fig. 10.9c. The forecast signal is only about half as strong as in the case with mountains, while the noise appears to be damped mainly by friction, and topography amplifies the signal.

10.4.3. *Predictability on small scales*

Berri and Paegle (1989) applied the model used by Paegle and Vukicevic (1987a) to a smaller domain, with a smaller grid size. The domain size is of order 300 km, and the grid size is of order 20 km. One experiment was done over the relatively flat terrain of Uruguay and eastern Argentina, and another over the Andes Mountains. Both specify uniform ambient flow of 6 m s^{-1}. Berri and Paegle (1989) introduced systematic deviations of the wind direction and speed as perturbations at every grid point of the domain. Systematic deviations are more representative of initial data uncertainty than random deviations in very small-scale applications characterized by few observation points. One example of predictability error growth for these experiments is displayed in Fig. 10.10a. In one of the perturbed experiments the wind is rotated clockwise by 15° with respect to the control, and in another, it is rotated counterclockwise by 15°. These rotations correspond to the uncertainty in the radiosonde measurement

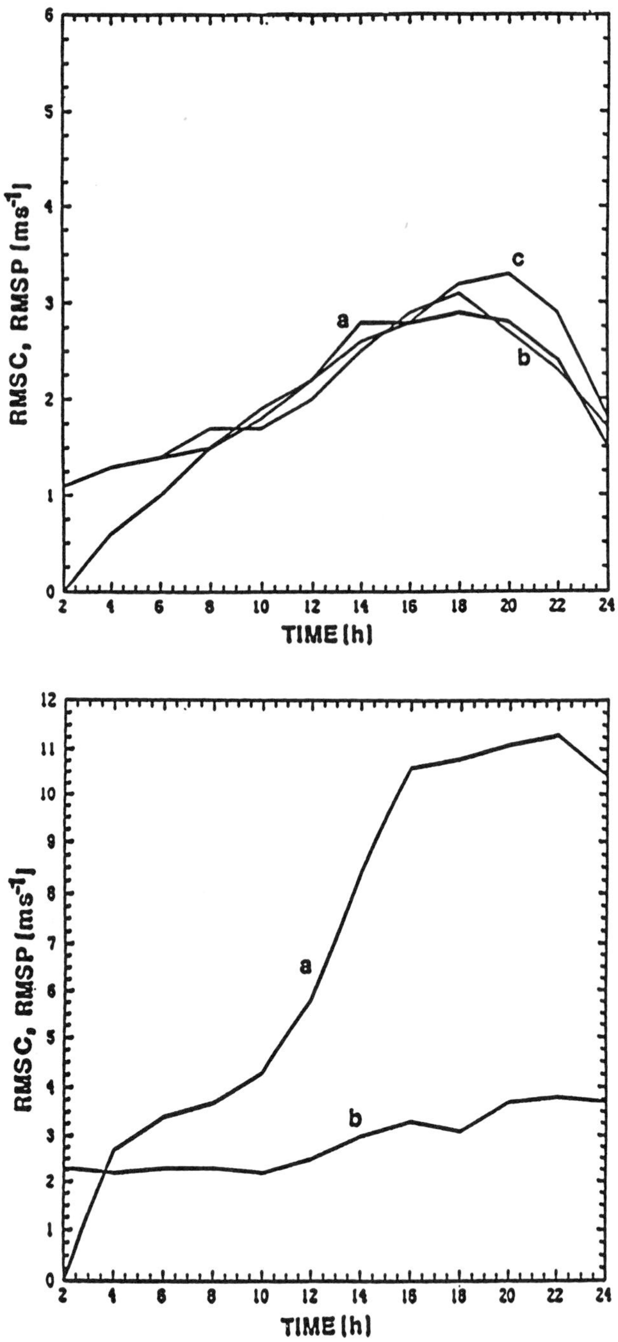

FIG. 10.10a. The rms difference (in m s^{-1}) 119 m above relatively flat surface. Curve *a* is signal for the control run; curves *b* and *c* are noise for the runs with the wind direction rotated 15° clockwise and counterclockwise, respectively (from Berri and Paegle 1989). (b) As in Fig. 10.10a, but for complex terrain. Curve *a* is signal for the control run, and curve *b* is noise for the run with the wind direction rotated 15° clockwise from the north.

of wind direction for a wind speed of 6 m s^{-1} (Hoehne 1980). The curves labeled *b* and *c* in Fig. 10.10a display the rms differences of the two perturbed experiments with respect to the control. The noise associated with the wind direction uncertainty equals or exceeds the signal at all times.

In order to investigate how these conclusions depend upon the geographical location, Berri and Paegle (1989) repeated the calculations above the highly mountainous terrain of the Andes. The elevation ranges from sea level to more than 4000 m. The experiment design is as before, except now the control forecast assumes an ambient wind from the north with speed 6 m s^{-1}. The perturbed run is initialized with a wind direction rotated uniformly 15° clockwise from the wind direction of the control.

Curve *a* of Fig. 10.10b displays the signal for this case, and curve *b* displays the noise. This signal/noise ratio in this case is on the order of 4 to 1. The steep topography amplifies the low-level signal sufficiently to make a deterministic forecast feasible.

10.4.4. Spread of lateral boundary errors

Virtually all of the published real-data experiments on limited domain predictability have appeared within the past half decade. There have, however, been many other attempts to simulate flow over complex terrain. Most of the three-dimensional simulations above real terrain have emphasized air quality applications. The approach in these applications is to diagnose the wind field using available observations, while constraining the analysis to satisfy conditions of nondivergence or weak divergence together with boundary conditions specified by the topography and the ambient flow. Numerous implementations have been described (e.g., Dickerson 1978; Sherman 1978; Lange 1978; Davis et al. 1984; Fosberg 1984; Kitada et al. 1986; Ross et al. 1988). The techniques use variational calculus to minimize both the flow divergence and the deviation of the analysis from the observations. The result is an elliptic boundary value equation in three space dimensions for the velocity potential, whose gradients give the resultant wind field.

Mathematically, the resultant equations are analogous to the Poisson equation (10.3) for the streamfunction, whose gradients also give the wind. The use of these diagnostic models has been largely motivated by considerations of computational economy for real-time applications. For example, the recent study by Ross et al. (1988) states that "predictive models are, in general, time consuming and impractical for real-time applications." We do not share the viewpoint that predictive models are impractical for real-time applications on small scales. However, the large number of predictive studies that point out the dominating role of boundary conditions for the flow evolution in limited domains do partly vindicate the diagnostic approach. Comparative studies are required for a definitive conclusion about the relative merits of prognostic and diagnostic methods in real-data situations.

The principal advantages of the predictive method are that it is less awkward to include physical effects and diurnal oscillations in this approach. There is also some evidence that the forecast signal may exceed the noise in cases of strong boundary forcing by steep topography. Partial justification for the latter viewpoint is given in comparing Fig. 10.9a with Fig. 10.9b, Fig. 10.10a with Fig. 10.10b, and in the sensitivity tests presented by Paegle and Vukicevic (1987a), as well as in the next section.

Recent sensitivity studies of regional models (Anthes and Xiuji 1989; Paegle and Vukicevic 1987a; Vukicevic and Paegle 1989) suggest that accurate lateral boundary conditions may be the most important single ingredient for an accurate local forecast. Consequently, both the predictive and diagnostic methods are sensitive to boundary error, and we close this section with a brief theoretical overview of the spread of lateral boundary uncertainty into the model domain. The analysis will be performed at the level of the vorticity equation, used previously to study the spread of initial data errors, and it follows the simple treatment of Paegle and Vukicevic (1987a). The analysis and its conclusions are easily extended to a complete forecast system in which the vorticity and the divergence are the forecast variables carrying flow information. We now consider only the vorticity equation explicitly, and suppose that it produces a forecast of the vorticity from which the streamfunction must be obtained in order to compute the flow field, using (10.3).

The general problem of error spread can be studied by considering the solution to two problems:

$$\nabla^2\psi = F_A \tag{10.4}$$

with $\psi = \psi_A$ on boundaries and

$$\nabla^2\psi = F_B \tag{10.5}$$

with $\psi = \psi_B$ on boundaries. Here F_A and F_B are the vorticity field forecast from wind observations, and the initial difference of these fields represents the initial vorticity uncertainty. The wind or vorticity within the boundaries can be perturbed by the typical observational uncertainty, while leaving the observations fixed on the boundaries (i.e., $F_A \neq F_B$, but $\psi_A = \psi_B$), as in section 10.4.1.

We now set $F_A = F_B$ and investigate the effect of changing boundary data ($\psi_A = \psi_B$). Subtracting (10.4) from (10.5) gives:

$$\nabla^2\psi = 0 \tag{10.6}$$

as the equation governing the interior perturbation with

$$\psi = \psi_A - \psi_B$$

on the boundary.

Equation (10.6) is most easily solved in a cylindrical geometry (r, λ), where r is the radial distance from the origin, and λ is the azimuth angle. We assume that the azimuthal dependence of the boundary values at a bounding radius (normalized to 1 for simplicity) is given in a Fourier series:

$$(\psi_A - \psi_B)(r = 1, \lambda) = \sum_{m=0}^{M} A_m e^{im\lambda}.$$

For each mode m, a solution of the form

$$\psi_m = A_m r^m e^{im\lambda}$$

satisfies (10.6), so that the total solution is

$$\psi(r, \lambda) = \sum_{m=0}^{M} A_m r^m e^{im\lambda}.$$

From this solution it is clear that the variance of mode m, associated with the erroneous boundary values, decreases as (r^m). The extension of these arguments from the streamfunction uncertainty to the wind field uncertainty is obvious. The only important modification is that the mode m contribution to the wind field has a radial dependence of r^{m-1}. These conclusions are supported in regional boundary-layer integrations with primitive equation models (Paegle and Vukicevic 1987a) as well as in regional barotropic integrations (Vukicevic and Paegle 1989).

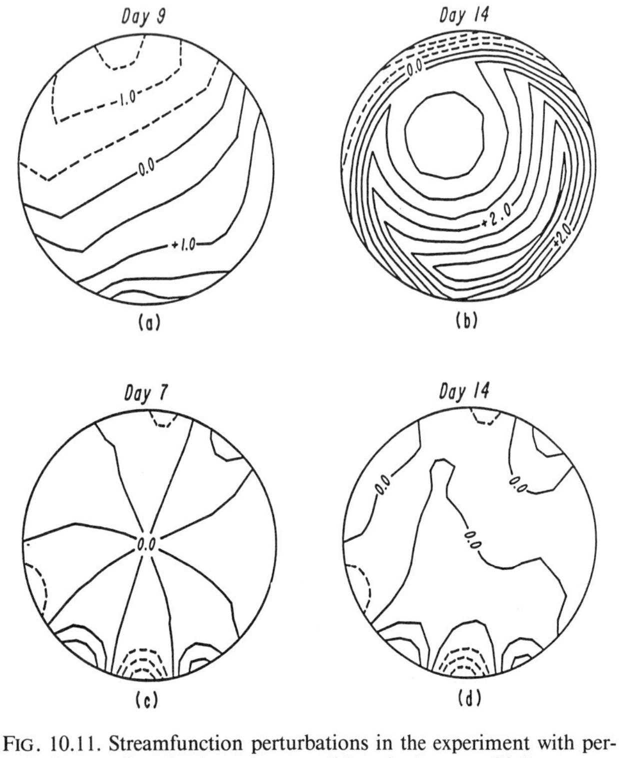

FIG. 10.11. Streamfunction perturbations in the experiment with perturbed streamfunction boundary conditions in the case of full spectrum perturbations, panels (a) and (b), and small scale perturbations only, panels (c) and (d). Contour interval is 5×10^5 m^2 s^{-1}.

The latter study is based on the barotropic model. In one set of limited area experiments, the streamfunction on the boundary is perturbed by using the errors that are produced at a relatively early stage of the perturbed global version of the model. The perturbations are imposed at the perimeter of the local area model, and maintained through the duration of the local area forecast. In the first case, the full spectrum of boundary uncertainty is included at day 9, when the total boundary error is particularly large. The results for day 9 and day 14 are displayed in Figs. 10.11a and 10.11b, respectively. The initial differences between the control and the perturbed forecast of the streamfunction are displayed in Fig. 10.11a. Fourier analysis of the boundary perturbation at this time shows most amplitude in wavenumber 1; i.e., a relatively large-scale pattern that should penetrate the interior domain, according to the above analysis of spread of boundary errors. The large spatial scale of the error field in Fig. 10.11a should favor subsequent error growth according Thompson's (1957) and Lorenz's (1969) analyses. These expectations are borne out in the 14 day predictions displayed in Fig. 10.11b.

The next experiment with disturbed boundary conditions retains only small-scale boundary disturbances. These are selected from day 7, rather than day 9 because day 7 contains more energy in the smaller scales. In this case, zonal wavenumbers less than 4 are filtered from the boundary errors. In Figs. 10.11c and 10.11d the initial (day 7) and 14 day differences between the control and perturbed streamfunction field are displayed. The perturbations remain trapped by the boundary, as anticipated from the analysis of boundary error spread. Consistent with the analysis of Thompson (1957) and Lorenz (1969) their small scale does not permit subsequent error growth.

These results offer hope for accurate local prediction in a nested sequence of models provided the coarse domain forecasts contain accurate boundary information on scales that are comparable to the larger scale structures resolved by the more finely resolved nested subdomain. They furthermore suggest that boundary information on scales much smaller than the perimeter of the boundary may not be especially relevant near the center of the forecast subdomain. Our analysis shows this rather strong conclusion to be entirely consistent for a purely diagnostic model and our experiments support its applicability for a prognostic barotropic model. The degree to which this result applies to more complete forecast models remains to be studied.

The major limitations of this analysis is its dependence on the infinite signal speed associated with the boundary forced ("potential flow") portion of the solution. A similar limitation was cited by Ertel (1944) as a barrier to exact prediction within limited domain models. In practice, it appears that the actual rate of propagation is determined by the group velocity of the wave system, which is always finite, and determines finite domains of influence at finite times. Indeed, the first computerized forecasts made by Charney et al. (1950) using the barotropic vorticity equation model were relatively successful over rather restricted domains. Charney had made earlier estimates of the smallest region for which a one day forecast could be performed without being contaminated by outside influences using the group velocity.[1]

The next section presents examples from a more complete mesoscale model for which the lower boundary condition on heat flux is randomly perturbed.

10.5. Contrast between average and realization parameterization in mesoscale models

Atmospheric modelers currently utilize parameterizations of subgrid scale turbulence, cumulus cloud effects, radiative flux divergence, etc., based on an "average" parameterization. In the case of subgrid scale turbulence, for example, observations from field experiments and theoretical models are coupled to provide a mean value of the turbulence correlations as a function of larger scale averaged conditions (e.g., the 20 minute average wind speed or a 10 km flight-leg average wind).

In contrast to this approach, an investigator could use the probability density distribution from these observations, from which a realization is randomly selected for use in the parameterization in the mesoscale model. By performing a number of mesoscale model runs in which the realizations are used for the parameterization, a sample population of mesoscale simulations will result. By averaging the mesoscale responses, an ensemble mesoscale simulation result will be obtained.

If the equations that comprise the mesoscale system produce chaotic results and the concept of sensitivity to initial conditions can be generalized to sensitivity to boundary conditions we would expect the mesoscale responses to diverge from one another for different sets of realizations.

To illustrate this approach, we will use the model analyses reported in Garratt et al. (1989) in which the sensitivity of sea breeze simulations to stochastic variations in the surface heat flux are evaluated. The model used to perform this analysis is described in Mahrer and Pielke (1977) and McNider and Pielke (1981). The primitive equations used for mesoscale flow are of course analogous to those used in synoptic-scale prediction models that have shown sensitivity to initial conditions, and thus we might suspect chaotic solutions will result here as well.

The ensemble-averaged parameterization scheme for the surface turbulent heat flux used in this mesoscale model is

$$H = \rho C_p \bar{u}_* \bar{\theta}_*, \qquad (10.7)$$

where an overbar has been used to indicate that these are the average values based on the time or space averaged field data.

For this example, extracted from Garratt et al. (1989), the mesoscale model is integrated for a two-dimensional sea breeze simulation with an 84×26 grid array for a location at 30°N, an offshore geostrophic flow of 5 m s^{-1}, and a sea surface temperature of 290 K. The simulated vertical velocity field after 6 hours is shown in Fig. 10.12, representing the control run and based on the standard procedure for integration in mesoscale models.

To obtain realization runs, Eq. (10.7) was replaced with

$$H = \rho C_p u_* \bar{\theta}_* \qquad (10.8)$$

where

$$u_* = \bar{u}_* + u'_* \qquad (10.9)$$

with u'_* obtained from

$$u'_* = f\sigma_{u_*}. \qquad (10.10a)$$

The variable f is a random number from a normal distribution with mean zero and variance of unity. The standard deviation of u_*, defined as

$$\sigma_{u_*} = \bar{u}_*/3.4 \qquad (10.10b)$$

was found to produce, through trial and error, values of surface temperature over land with a standard deviation corresponding to observations; e.g., over South Florida as discussed in Pielke and Cotton (1977).

Equation (10.8) was used to generate ten realizations and the result after 6 hours that is most different from the control is shown in Fig. 10.13. As evident from a comparison of Figs. 10.12 and 10.13 the results are very

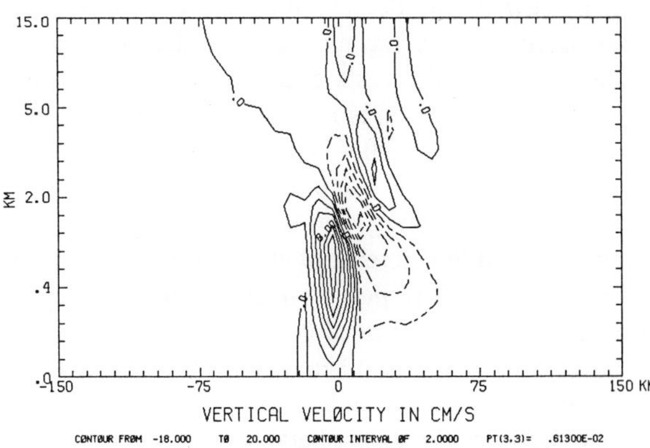

FIG. 10.12. Vertical velocity (W) after 6 hours [control-used equation (10.7)].

[1] This was learned from Office Note 331 of the National Meteorological Center, authored by N. A. Phillips in May 1987.

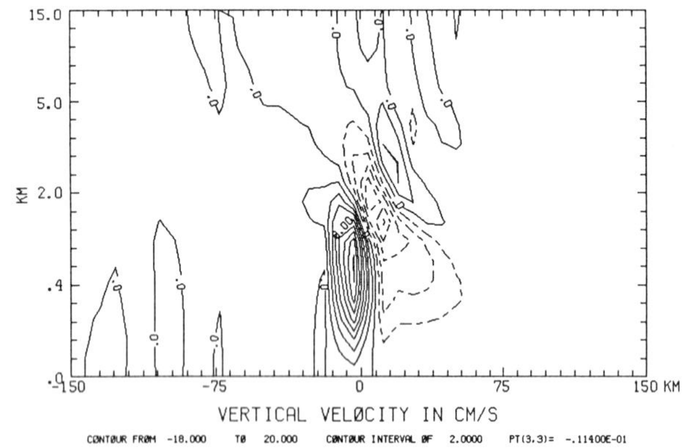

FIG. 10.13. W after 6 hours [most extreme realization using Eq. (10.8)].

similar. In Fig. 10.14 the results are summarized for all ten of the simulations where the domain average surface temperature, $\langle \theta_0 \rangle$, the domain average value of u_*, $\langle u_* \rangle$, and the maximum values of vertical motion, W_m, at 6 hours after sunrise in the model are given. Note that while the model runs using (10.8) yield generally higher values of domain-averaged surface temperature than the control (by about 1.2°C in the most extreme realization), the difference in temperature gradient between land surface and water (290°) is changed by less than 10%. Similarly, the vertical motion intensity is 14 cm s^{-1} in the control experiment in contrast to 16.6 cm s^{-1} in the extreme case; i.e., a small difference.

The conclusion is that for the specific situation modeled in this experiment either the response is not chaotic, or within the diurnal period of heating (which controls the surface boundary condition) the sensitivity to boundary conditions evolves slowly enough such that from a practical perspective chaos is unimportant.

10.6. Conclusions

The extent to which atmospheric modeling skill can be improved in complex terrains has renewed practical interest in view of the development of several advanced mesoscale models during the past two decades, and the need for accurate forecasts and descriptions of the atmosphere in mountainous regions. The modeling of topographically forced airflow has obvious practical difficulties, including the parameterization of turbulent mixing, radiative heating, and soil transfer (physical uncertainties), and the increase of numerical errors for steep terrain slopes. With respect to initial data requirements, however, the strongly forced and highly dissipated local slope circulations may be inherently more predictable than flows over flat terrain that are more sensitive to initial data errors. The potential predictability might be especially good for local flows that are contained within high mountain valleys and for which the forcing is well determined by the local solar cycle and ambient synoptic-scale flow field.

Whether these advantages outweigh uncertainties related to physical parameterizations and numerical resolution has never been unequivocally proven or refuted, and a large portion of air quality studies applied to flows over complex topography specifically avoid the prognostic approach. It is, however, reasonable to suppose that topography may enhance the predictability of certain types of flows.

This chapter has framed this question in terms of concepts of chaos and predictability theory. In the process, we have touched upon an apparently artificial enhancement of predictability due to the imposition of one-way interacting boundary conditions in limited domain predictions. This may make an inherently unpredictable flow appear to be highly predictable, and obscure the role of local topography. Conversely, topography represents a boundary to the flow field, and the same mechanism that controls chaos for bounded domain simulations of dynamical systems that are chaotic in unbounded circumstances may also augment predictability through real physical processes around complex terrain. Because all simulations around small-scale topography use limited area models, the predictability enhancement by topography has been difficult to quantify.

We have shown examples where the forecast signal to

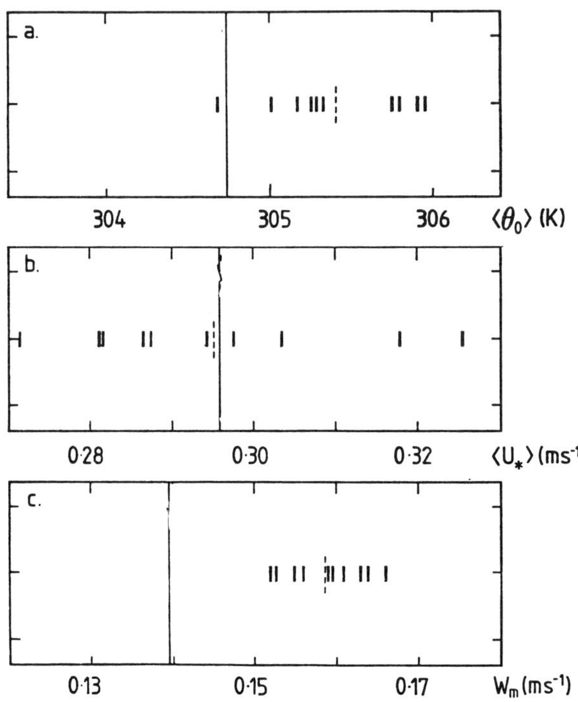

FIG. 10.14. Domain average values of surface potential temperature, θ_0, and friction velocity, u_*, and maximum value of vertical velocity, for the model realizations using Eq. (10.8).

noise ratios are significantly enhanced as a consequence of topography and other strong surface variations. These examples suggest the need to study error spread from uncertain boundary specifications. Our analysis of the spread of boundary errors suggests that this is unimportant for scales that are small with respect to the perimeter of the forecast domain. Although this analysis is strictly correct only for the purely diagnostic models such as those commonly used for air pollution simulations, it may also explain certain more complete forecast systems, and perhaps apply to random spatial uncertainties of surface physical forcings such as surface heat fluxes.

Acknowledgments. This research has been partially funded by NSF Grant ATM-8611951 and ATM-8611952 to the University of Utah and NSF Grant ATM-8616662 to Colorado State University (CSU). We would also like to acknowledge the financial support provided by CONICET (a federally supported Argentine research foundation) for Matilde Nicolini and Guillermo Berri during sabbatical visits to the Department of Meteorology at the University of Utah, and the CSU Cooperative Institute for Research in the Atmosphere (CIRA) who provided partial support to G. A. Dalu and J. R. Garratt during sabbatical visits to CSU. The editorial preparation of this contribution was completed by Dallas McDonald and Tony Smith. Some of the computer calculations were completed at NCAR. NCAR is supported by the NSF. R. Errico provided helpful comments in the final preparation of the manuscript.

REFERENCES

Abbs, D. J., and R. A. Pielke, 1987: Numerical simulation of orographic effects on NE Colorado snowstorms. *Meteorol. Atmos. Phys.,* **37,** 1–10.

Abe, M., 1929: Mountain clouds: Their forms and connected air currents. Part II. *Bull. Centr. Met. Obs., Japan,* **7,** 93–145.

Amer. Meteor. Soc., 1988: *Extended Abstracts, Lower Tropospheric Profiling: Needs and Technologies,* Boulder, Amer. Meteor. Soc., 260 pp.

Annals of the Astronomical Observatory of Harvard College, 1889: Vol. XXII: Meteorological observations made on the summit of Pike's Peak, Colorado: January 1874 to June 1888. E. C. Pickering, Director, John Wilson & Son, University Press, 459–475.

Anthes, R. A., Y. H. Kuo, D. P. Baumhefner, R. M. Errico and T. W. Bettge, 1985: Predictability of mesoscale motions. *Advances in Geophysics, Issues in Atmospheric and Oceanic Modeling. Part B: Weather Dynamics,* Vol. 28, Academic Press, 159–202.

——, and Z. Xiuji, 1989: Joint PRC-US workshop on mesoscale meteorology. *Bull. Amer. Meteor. Soc.,* submitted.

Arya, S. P. S., and P. S. Gadiyaram, 1986: An experimental study of flow and dispersion in the wakes of three-dimensional low hills, *Atmos. Environ.,* **20,** 729–740.

Archibald, P. B., 1973: An analysis of the winds of site 300 as a source of power. Lawrence Livermore Laboratories Report UCLR-51469, University of California.

Arya, S. P. S., M. E. Capuano and L. C. Fagen, 1987: Some fluid modeling studies of flow and dispersion over two-dimensional low hills. *Atmos. Environ.,* **21,** 753–764.

Asimakopoulos, D. N., and R. S. Cole, 1977: An atmospheric acoustic sounder for the remote probing of the lower atmosphere. *J. Inst. Phys. Sci. Instrum.,* **10,** 47–50.

——, C. G. Helmis and G. J. Stephanou, 1987: Atmospheric acoustic minisounder. *J. Atmos. Oceanic Technol.,* **4,** 345–347.

Astling, E. G., 1984: On the relationship between diurnal mesoscale circulations and precipitation in a mountain valley. *J. Climate Appl. Meteor.,* **23,** 1635–1644.

——, J. Paegle, E. Miller and C. J. O'Brien, 1985: Boundary layer control of nocturnal convection associated with a synoptic scale system. *Mon. Wea. Rev.,* **113,** 540–552.

Atkinson, B. W., 1981: *Meso-scale Atmospheric Circulations.* Academic Press, 495 pp.

Auer, Jr., A. H., and J. M. White, 1982: The combined role of kinematics, thermodynamics and cloud physics associated with heavy snowfall episodes. *J. Meteor. Soc. Japan,* **60,** 500–507.

Ayer, H. S., 1961: On the dissipation of drainage wind systems in valleys in the morning hours. *J. Meteor.,* **18,** 560–563.

Bacmeister, J. T., and R. T. Pierrehumbert, 1988: On high-drag states of nonlinear flow over an obstacle. *J. Atmos. Sci.,* **45,** 63–80.

Bader, D. C., and T. B. McKee, 1983: Dynamical model simulation of the morning boundary layer development in deep mountain valleys. *J. Climate Appl. Meteor.,* **22,** 341–351.

——, and ——, 1985: Effects of shear, stability and valley characteristics on the destruction of temperature inversions. *J. Climate Appl. Meteor.,* **24,** 822–832.

——, and C. D. Whiteman, 1989: Numerical simulation of cross-valley plume dispersion during the morning transition period. *J. Appl. Meteor,* **28,** 652–664.

——, T. B. McKee and G. J. Tripoli, 1987: Mesoscale boundary layer evolution over complex terrain. Part I: Numerical simulation of the diurnal cycle. *J. Atmos. Sci.,* **44,** 2823–2838.

Baines, P. G., 1979: Observation of stratified flow over two-dimensional obstacles of finite depth. *Tellus,* **31,** 351–371.

——, and K. P. Hoinka, 1985: Stratified flow over two-dimensional topography in fluid of infinite depth: A laboratory simulation. *J. Atmos. Sci.,* **42,** 1614–1630.

Balaji, V., and T. L. Clark, 1988: Scale selection in locally forced convective fields and the initiation of deep cumulus. *J. Atmos. Sci.,* **45,** 3188–3211.

Balling, R. C., and S. W. Brazel, 1987: Diurnal variations in Arizona monsoon precipitation frequencies. *Mon. Wea. Rev.,* **115,** 342–346.

Banta, R. M., 1982: An observational and numerical study of mountain boundary-layer flow. Atmos. Sci. Paper 350, Ph.D. dissertation, Colorado State University, 203 pp.

——, 1984: Daytime boundary-layer evolution over mountainous terrain. Part I: Observations of the dry circulations. *Mon. Wea. Rev,* **112,** 340–356.

——, 1985: Late-morning jump in TKE in the mixed layer over a mountainous basin. *J. Atmos. Sci.,* **42,** 407–411.

——, 1986: Daytime boundary-layer evolution over mountainous terrain. Part II: Numerical studies of upslope flow duration. *Mon. Wea. Rev.,* **114,** 1112–1130.

——, and W. R. Cotton, 1981: An analysis of the structure of local wind systems in a broad mountain basin. *J. Climate Appl. Meteor.,* **20,** 1255–1266.

——, and C. L. B. Schaaf, 1987: Thunderstorm genesis zones in the Colorado Rocky Mountains as determined by traceback of geosynchronous satellite images. *Mon. Wea. Rev.,* **115,** 463–476.

——, and K. R. Hanson, 1987a: Numerical simulations of the development of mountain cumulus clouds. *Short- and Medium-Range Numerical Weather Prediction,* T. Matsuno, Ed., Meteor. Soc. Japan, Tokyo. 767–775.

——, and ——, 1987b: Sensitivity studies on the continentality of a numerically simulated cumulonimbus. *J. Climate Appl. Meteor.,* **26,** 275–286.

——, K. C. Pantley and P. T. Gannon, Sr., 1987: Soil moisture and ridge height effects on katabatic flow simulations. Preprints, *Fourth Conference on Mountain Meteorology,* Seattle, Amer. Meteor. Soc., 139–145.

Barcilon, A., J. C. Jusem and P. G. Drazin, 1979: On the two-dimensional hydrostatic flow of a stream of moist air over a mountain ridge. *Geophys. Astrophys. Fluid Dyn.,* **13,** 125–140.

Barr, S., and M. M. Orgill, 1989: Influence of external meteorology on nocturnal valley drainage winds. *J. Appl. Meteor.,* **28,** 497–517.

——, R. E. Luna, W. E. Clements and H. W. Church, Eds., 1977: *Workshop on Research Needs for Atmospheric Transport and Diffusion in Complex Terrain* Albuquerque, N.M. Energy Research and Development Administration, 17 pp. [CONF-7609160.]

Barry, R. G., 1981: *Mountain Weather and Climate.* Methuen and Co., Ltd., 313 pp.

Batchelor, G. K., 1953: The conditions for dynamic similarity of motions of a frictionless perfect-gas atmosphere. *Quart. J. Roy. Meteor. Soc.,* **79,** 891–905.

——, 1957: Wave scattering due to turbulence. *Symposium on Naval Hydrodynamics,* F. S. Sherman, Ed., National Research Council, Washington D.C., NAS-NRC, 515, 409–430.

Battan, L. J., 1973: *Radar Observation of the Atmosphere.* University of Chicago Press, 324 pp.

Baumhefner, D. P., 1984: The relationship between present large-scale forecast skill and new estimates of predictability error growth. *AIP Conference Proc. No. 106,* H. C. Wolfe, Ed., 169–180.

Beljaars, A. C. M., J. L. Walmsley and P. A. Taylor, 1987: A mixed spectral finite-difference model for neutrally stratified boundary-layer flow over roughness changes and topography. *Bound. Layer Meteor.,* **38,** 273–303.

Bergé, P., Y. Pomeau and C. Vidal, 1984: *Order Within Chaos: Towards a Determinate Approach to Turbulence.* Wiley and Sons, 329 pp.

Bernstein, A. B., 1965: Dimensional analysis applied to the wind distribution in the planetary boundary layer. *Mon. Wea. Rev.,* **93,** 579–585.

Berri, G., and J. Paegle, 1989: Sensitivity of local predictions to initial conditions. *J. Appl. Meteor.,* submitted.

Betts, A. K., 1973: Non-precipitating cumulus convection and its parameterization. *Quart. J. Roy. Meteor. Soc.,* **99,** 178–196.

——, 1982: Saturation point analysis of moist convective overturning. *J. Atmos. Sci.,* **39,** 1484–1505.

Bilbro, J., G. Fichtl, D. Fitzjarrald, M. Krausse and R. Lee, 1984: Airborne Doppler lidar wind field measurements. *Bull. Amer. Meteor. Soc.,* **65,** 348–359.

Bjorkland, J. R., and J. F. Bowers, 1982: User's instructions for the SHORTZ and LONGZ computer programs. EPA Manual 903/9-82-004, U.S. Environmental Protection Agency.

Blumen, W., 1965: A random model of momentum flux by mountain waves. *Geofys. Publ.,* **26,** 1–33.

——, 1984: An observational study of instability and turbulence in nighttime drainage winds. *Bound.-Layer Meteor.,* **28,** 245–269.

——, and J. E. Hart, 1988: Airborne Doppler lidar wind field measurements of waves in the lee of Mount Shasta. *J. Atmos. Sci.,* **45,** 1571–1583.

Blumenthal, D. L., T. B. Smith, D. E. Lehrman, R. A. Rasmussen, G. Z. Whitten and R. A. Baxter, 1985: Southern San Joaquin Valley ozone study. STI Final Report 94100-510-FR, Prepared for Western Oil and Gas Association (WOGA), Los Angeles, Contract No. 84-8.0.05(2)-07-01 by Sonoma Tech., Inc., and Systems Applications, Inc., 143 pp.

Böhm, H., 1985: *Erweiterung des thermodynamischen Modells für Inversionsauflösung von C. D. Whiteman und seine praktische Anwendung.* [*Further development of the Thermodynamic Model for Inversion Breakup by C. D. Whiteman and its practical application.*] Diplomarbeit, ETH Zürich, Switzerland, 90 pp.

Booker, D. R., 1963: Modification of convective storms by lee waves. *Meteor. Monogr.,* **5,** 129–140.

Bossert, J. E., J. D. Sheaffer and E. R. Reiter, 1989: Aspects of regional-scale flows in mountainous terrain. *J. Appl. Meteor.,* **28,** 590–601.

Bouwmeester, R. J. B., R. N. Meroney and V. A. Sandborn, 1978: Sites for wind-power installations: Wind characteristics over ridges. Part I. U.S. Dept. of Energy Report RLO/2438-78/1 (UC-60), 206 pp.

Bowen, A. J. 1981: Some effects of escarpments on the atmospheric boundary layer. Ph.D. dissertation, University of Canterbury, 322 pp.

——, and D. Lindley, 1974: Measurements of the mean wind flow over various escarpment shapes. *Proc. Fifth Australian Conference on Hydraulics and Fluid Mechanics,* Christchurch, New Zealand, 9 pp.

——, and ——, 1977: A wind tunnel investigation of the wind speed and turbulence characteristics close to the ground over various escarpment shapes. *Bound.-Layer Meteor.,* **12,** 259–271.

Bower, J. B., and D. R. Durran, 1986: A study of wind profiler data collected upstream during windstorms in Boulder, Colorado. *Mon. Wea. Rev.,* **102,** 592–602.

Bradbury, L. J. S., 1976: Measurements with a pulsed-wire and a hot-wire anemometer in the highly turbulent wake of a normal flat plate. *J. Fluid Mech.,* **77,** 473–497.

Bradley, E. F., 1968: A micrometeorological study of velocity profile and surface drag in the region modified by a change in surface roughness. *Quart. J. Roy. Meteor. Soc.,* **94,** 361–379.

——, 1980: An experimental study of the profiles of wind speed, shearing stress and turbulence at crest of a large hill. *Quart. J. Roy. Meteor. Soc.,* **106,** 101–123.

Bradley, M. M., 1985: The numerical simulation of orographic storms. Ph.D. dissertation, Univ. of Illinois, Urbana-Champaign, 263 pp.

Braham, R. R., 1958: Cumulus cloud precipitation as revealed by radar—Arizona 1955. *J. Meteor.,* **15,** 75–83.

——, 1968: Meteorological bases for precipitation development. *Bull. Amer. Meteor. Soc.,* **49,** 343–353.

——, and M. Draginis, 1960: Roots of orographic cumuli. *J. Meteor.,* **17,** 214–226.

Brehm, M., 1986: Experimentelle und numerische untersuchungen der hangwindschicht und ihrer rolle bei der erwärmung von tälern. [Experimental and numerical investigations of the slope wind layer and its role in the warming of valleys.] Wiss. Mitt. Nr. 54, Ph.D. dissertation, Universität München, Meteorologisches Institut, München, FRG, 150 pp.

——, and C. Freytag, 1982: Erosion of the night-time thermal circulation in an alpine valley. *Arch. Meteor. Geophys. Bioklim.,* **B31,** 331–352.

Brenner, I. S., 1974: A surge of maritime tropical air—Gulf of California to the southwestern United States. *Mon. Wea. Rev.,* **102,** 375–389.

Briggs, G. A., 1969: *Plume rise.* AEC Critical Review Series. U.S. Atomic Energy Commission, 81 pp.

——, 1979: Analytic modeling of drainage flows. ATDL Report 79/22. [ATDL/NOAA, Oak Ridge, TN 37830.]

Brighton, P. M. W., 1978: Strongly stratified flow past three-dimensional obstacles. *Quart. J. Roy. Meteor. Soc.,* **104,** 289–307.

Brinkmann, W. A. R., 1974: Strong downslope winds at Boulder, Colorado. *Mon. Wea. Rev.,* **102,** 592–602.

Britter, R. E., J. C. R. Hunt and K. J. Richards, 1981: Airflow over a two-dimensional hill: Studies of velocity speed-up, roughness effects and turbulence. *Quart. J. Roy. Meteor. Soc.,* **107,** 91–110.

Bromwich, D. H., 1989: Satellite analyses of Antarctic katabatic wind behavior. *Bull. Amer. Meteor. Soc.,* **70,** 738–749.

Brooks, C. E. P., 1951: Selected annotated bibliography on mountain meteorology. *Meteorological Abstracts and Bibliography,* **2,** 916–959.

Browell, E. V., 1988: Lidar measurement of tropospheric trace gases. *Extended Abstracts, Lower Tropospheric Profiling: Needs and Technologies,* Boulder, Amer. Meteor. Soc., 99–102.

Brown, E. H., and F. F. Hall, 1978: Advances in atmospheric acoustics. *Rev. Geophys. Space Phys.,* **16,** 788–794.

Brown, R. A., 1966: Three-dimensional growth characteristics of an orographic thunderstorm system. Satellite & Mesometeorology Research Project Paper 61, University of Chicago.

Browning, K. A., 1980: Structure, mechanism and prediction of orographically enhanced rain in Britain. *Global Atmospheric Research Programme,* Ser. 23, WMO, 88–114.

——, F. F. Hill and C. W. Pardoe, 1974: Structure and mechanism of precipitation and the effect of orography in a wintertime warm sector. *Quart. J. Roy. Meteor. Soc.,* **100,** 309-330.

Buettner, K. J. K., and N. Thyer, 1966: Valley winds in the Mount Rainier area. *Arch. Meteor. Geophys. Bioklim.,* **B14,** 125–147.

Burk, S. D., 1980: Refractive index structure parameters: Time dependent calculations using a numerical boundary layer model. *J. Appl. Meteor.,* **19,** 562–576.

——, 1981: Comparison of structure parameter scaling expressions with turbulence closure model predictions. *J. Atmos. Sci.,* **38,** 751–761.

Burt, E. W., and H. H. Slater, 1977: Evalution of the valley model. *Proc. AMS-APCA Joint Conference on Applications of Air Pollution Meteorology,* Salt Lake City.

Businger, J. A., 1973: Turbulent transfer in the atmospheric surface layer. *Workshop on Micrometeorology,* Chapter 2, D. A. Haugen, Ed., Amer. Meteor. Soc.

Caldwell, D. R., and C. W. Van Atta, 1970: Ekman boundary layer instabilities. *J. Fluid Mech.,* **44,** 79–95.

Caracena, F., R. A. Maddox, L. R. Hoxit and C. F. Chappell, 1979: Mesoanalysis of the Big Thompson storm. *Mon. Wea. Rev.,* **107,** 1–17.

Carr, D. D., and J. R. V. Lopik, 1962: Terrain quantification. Phase I: Surface geometry measurements. Technical Report AFCRL-63-208 to the Geophysics Research Directorate, Texas Instruments, Inc., Dallas.

Carroll, J. J., 1986: Accuracy of wind measurements using an airborne Doppler lidar . *J. Atmos. Oceanic Technol.,* **3,** 3–11.

——, and R. L. Baskett, 1979: Dependence of air quality in a remote location on local and mesoscale transports. *J. Appl. Meteor.,* **18,** 474–486.

Carruthers, D. J., and T. W. Choularton, 1982: Air flow

over hills of moderate slope. *Quart. J. Roy. Meteor. Soc.,* **108,** 603–624.

——, and ——, 1984: Acid deposition in rain over hills. *Atmos. Environ.,* **18,** 1901–1908.

——, J. C. R. Hunt and W. Weng, 1988: Computational model of airflow over hills—FLOWSTAR I. Proc. of Envirosoft. *Computer Techniques in Environmental Studies,* P. Zanetti, Ed., Springer–Verlag.

Castro, I. P., and W. H. Snyder, 1982: A wind-tunnel study of dispersion from sources downwind of three-dimensional hills. *Atmos. Environ.,* **16,** 1869–1887.

——, —— and G. L. Marsh, 1983: Stratified flow over three-dimensional ridges. *J. Fluid Mech.,* **135,** 261–282.

Cermak, J. E., 1958: Wind tunnel for the study of turbulence in the atmospheric surface layer. Colorado State University Report CER58-JEC-42, 40 pp.

——, 1975: Applications of fluid mechanics to wind engineering: A Freeman Scholar lecture. *J. Fluid Eng.,* **97,** 9–38.

——, 1981: Wind tunnel design for physical modeling of atmospheric boundary layer. *J. Eng. Mech. Div., ASCE,* **107(EM3),** 623–642.

——, 1984: Physical modelling of flow and dispersion over complex terrain. *Bound.-Layer Meteor.,* **30,** 261–292.

——, and J. A. Peterka, 1966: Simulation of wind fields over Point Arguello, California, by wind-tunnel flow over a topographic model. Colorado State University Report CER65JEC-JAP64, 83 pp.

——, and R. A. Petersen, 1981: Physical modeling of downslope mountain wind and atmospheric dispersion. *Proc. Fourth U.S. National Conference Wind Engineering Research,* Seattle, 11 pp.

——, V. A. Sandborn, E. J. Plate, G. H. Binder, H. Chuang, R. N. Meroney and S. Ito, 1966: Simulation of atmospheric motion by wind-tunnel flows. Colorado State University Report CER66JEC-VAS-EJP-GJB-HC-RNM-SI17, 101 pp.

Chadwick, R. B., and E. E. Gossard, 1986: Radar probing and measurement of the planetary boundary layer. Part I: Scattering from refractive index irregularities. *Probing the Atmospheric Boundary Layer,* D. Lenschow, Ed., Amer. Meteor. Soc., 163–182.

Chamberlain, A. C., 1966: Transport of lycopodium spores and other small particles to rough surfaces. *Proc. Roy. Soc. London,* **A296,** 48–70.

Charney, J. G., R.Fjørtoft and J. von Neumann, 1950: Numerical integration of the barotropic vorticity equation. *Tellus,* **2,** 237–254.

——, R. G. Fleagle, H. Riehl, V. E. Lally and D. Q. Wark, 1966: The feasibility of global observations and analysis experiment. *Bull. Amer. Meteor. Soc.,* **47,** 200–220.

Chien, H. C., R. N. Meroney and V. A. Sandborn, 1979: Sites for wind-power installations: Physical modeling of the wind field over Kahuku Point, Oahu, Hawaii. U.S. Dept. of Energy Report DOE/ET/20292-79/1 (DE 82005971), 162 pp. [In H. C. Chien et al. 1980: *Proc. Third International Symposium on Wind Energy Systems,* Copenhagen, 75–90.]

——, —— and ——, 1980: Sites for wind-power installations: Physical modeling of the wind field over Kahuku Point, Oahu, Hawaii. *Proceedings Third International Symposium on Wind Energy Systems,* Copenhagen, 75–90.

——, ——, —— and R. J. B. Bouwmeester, 1978: Preliminary measurements of flow over model three-dimensional hills. Colorado State University Memo. CEM77-78HCC-VAS-RNM-RJBB29, 34 pp.

Choularton, T. W., I. E. Consterdine, B. A. Gardiner, J. Gaymor, M. F. Hill, J. Latham and I. M. Stromberg, 1986: Field studies of the physical and microphysical characteristics of clouds enveloping Great Dun Fell. *Quart. J. Roy. Meteor. Soc.,* **112,** 131–148.

Clark, T. L., and R. D. Farley, 1984: Severe downslope windstorm calculations in two and three spatial dimensions using anelastic interactive grid nesting: A possible mechanism for gustiness. *J. Atmos. Sci.,* **41,** 329–350.

——, and ——, 1977: On the evolution and stability of finite-amplitude mountain waves. *J. Atmos. Sci.,* **34,** 1715–1730.

——, and ——, 1984: Critical level reflection and the resonant growth of nonlinear mountain waves. *J. Atmos. Sci.,* **41,** 3122–3134.

Clements, W. E., and C. J. Nappo, 1983: Observations of a drainage flow event on a high-altitude simple slope. *J. Climate Appl. Meteor.,* **22,** 331–335.

——, J. A. Archuleta and P. H. Gudiksen, 1989a: Experimental design of the 1984 ASCOT field experiment. *J. Appl. Meteor.,* **28,** 405–413.

——, —— and D. E. Hoard, 1989b: Mean structure of the nocturnal drainage flow in a deep valley. *J. Appl. Meteor.,* **28,** 457–462.

Colson, D., 1954: Meteorological problems in forecasting mountain waves. *Bull. Amer. Meteor. Soc.,* **35,** 363–371.

Cook, N. J., 1977: Determination of the model scale factor in wind-tunnel simulations of the adiabatic atmospheric boundary layer. *J. Ind. Aero.,* **2,** 311–321.

Cooper, W. A., and C. P. R. Saunders, 1980: Winter storms over the San Juan Mountains. Part II: Microphysical processes. *J. Appl. Meteor.,* **19,** 927–941.

——, D. Baumgartner and J. E. Dye, 1986: Evolution of the droplet spectra in Hawaiian orographic clouds. Preprints, *Conference on Cloud Physics,* Vol. 2, Snowmass, Amer. Meteor. Soc., 52–55.

Cotton, W. R., and R. A. Anthes, 1989: *Storm and Cloud Dynamics.* Academic Press, 828 pp.

——, R. L. George and K. R. Knupp, 1982: An intense, quasi-steady thunderstorm over mountainous terrain. Part I: Evolution of the storm-initiating mesoscale circulation. *J. Atmos. Sci.,* **39,** 328–342.

——, ——, P. J. Wetzel and R. L. McAnelly, 1983: A long-lived mesoscale convective complex. Part I: The mountain generated component. *Mon. Wea. Rev.,* **111,** 1893–1918.

——, G. J. Tripoli, R. M. Rauber and E. A. Mulvihill, 1986: Numerical simulation of the effects of varying ice crystal nucleation rates and aggregation processes on orographic snowfall. *J. Climate Appl. Meteor.,* **25,** 1658–1680.

Coulter, R. L., and M. L. Wesely, 1980: Estimates of surface heat flux from sodar and laser scintillation measurements in the unstable boundary layer. *J. Appl. Meteor.,* **19,** 1209–1222.

——, and T. J. Martin, 1983: Remote sensing of winds and turbulence above complex terrain during ASCOT-1980. Argonne National Laboratory Report ANL/ER-83-1, ASCOT 83-4, U.S. Dept. of Energy. [Available from NTIS, U.S. Dept. of Commerce, Springfield, VA.]

——, and ——, 1986: Results from a high-power, high frequency sodar. *Atmos. Res.,* **20,** 257–270.

——, M. Orgill and W. Porch, 1989: Tributary fluxes into Brush Creek valley. *J. Appl. Meteor.,* **28,** 555–568.

Counihan, J., 1969: An improved method of simulating an atmospheric boundary layer in a wind tunnel. *Atmos. Environ.,* **3,** 197–214.

——, 1975: Adiabatic atmospheric boundary layers: A review and analysis of data from the period 1880-1972. *Atmos. Environ.,* **9,** 871–905.

Crapper, G. D., 1959: A three-dimensional solution for waves in the lee of mountains. *J. Fluid Mech.,* **6,** 51–76.

——, 1962: Waves in the lee of mountains with elliptical contours. *Phil. Trans. Roy. Soc. London,* **A254,** 601–623.

Currier, P. E., W. L. Ecklund, D. A. Carter, J. M. Warnock and B. B. Balsley, 1988: Temperature profiling using a UHF wind profiler and an acoustic source. *Extended Abstracts, Lower Tropospheric Profiling: Needs and Technologies,* Boulder, Amer. Meteor. Soc., 121–122.

Dalcher, A., and E. Kalnay, 1987: Error growth and error predictability in operational ECMWF forecasts. *Tellus,* **A39,** 474–491.

Daniels, P. A., and N. E. Oshiro, 1980: Detailed wind survey of Kahuku, Oahu. U.S. Dept. of Energy Report, U. of Hawaii, Dept. of Meteorology Report, UHMET 79-06, 142 pp.

Davidson, B., and P. K. Rao, 1963: Experimental studies of the valley–plain wind. *Int. J. Air and Water Pollut.,* **7,** 907–923.

——, N. Gerbier, S. O. Papagionakis and P. G. Rijkvort, 1964: Sites for wind-power installations. Tech. Note 63, WMO-NO156, TP76, WMO, 38 pp.

Davis, C. G., S. S. Bunker and J. P. Mutschlecner, 1984: Atmospheric transport models for complex terrain. *J. Climate Appl. Meteor.,* **23,** 235–238.

Dawkins, R. A., and D. R. Davis, 1981: The effect of surface topography on momentum and mass transfer in a turbulent boundary layer. *J. Fluid Mech.,* **108,** 423–442.

Defant, F., 1949: Zur theorie der hangwinde, nebst bemerkungen zur theorie der berg- und talwinde. [A theory of slope winds, along with remarks on the theory of mountain winds and valley winds.] *Arch. Meteor. Geophys. Bioklim.,* **A1,** 421–450. (English translation: C. D. Whiteman, and E. Dreiseitl, 1984: Alpine meteorology: Translations of classic contributions by A. Wagner, E. Ekhart and F. Defant. Pacific Northwest Laboratory, Richland, Washington, 121 pp. [PNL-5141/ASCOT-84-3]).

——, 1951: Local winds. *Compendium of Meteorology,* T. M. Malone, Ed., Amer. Meteor. Soc., 655–672.

DeGregorio, S., and G. A. Dalu, 1984: Barotropic (low-order) spectral-model flow in β-plane. *El Nuovo Cimento,* **7,** 179–209.

Derickson, R. G., and R. N. Meroney, 1977: A simplified physics airflow model for evaluating wind power sites in complex terrain. *Proc. Summer Computer Sim. Conference,* Chicago, 14 pp.

Dickerson, M. H., 1978: A mass consistent atmospheric flux model for regions with complex terrain. *J. Appl. Meteor.,* **17,** 241–253.

——, and P. H. Gudiksen, Eds. 1984: Atmospheric studies in complex terrain. Tech. Progress Report FY-1979 through FY-1983, ASCOT 84-1/UCID-19851, U.S. Dept. of Energy, Lawrence Livermore National Laboratory, Livermore, CA.

DiCristofaro, D. C., D. G. Strimaitis, B. R. Greene, R. J. Yamartino, A. Venkatram, D. A. Godden, P. Karamchandani, T. F. Lavery and B. A. Egan, 1986: EPA Complex Terrain Model Development Program. Fifth Milestone Report—1985. EPA 600/3-85/069, U.S. Environmental Protection Agency.

Dobosy, R. J., K. S. Rao, J. W. Przybylowicz, R. M. Eckman and R. P. Hosker, 1989: Mass and momentum balance in the Brush Creek drainage flow determined from single profile data. *J. Appl. Meteor.,* **28,** 467–476.

Dollard, G. J., M. H. Unsworth and M. J. Harve, 1983: Pollutant transfer in upland regions by occult precipitation. *Nature,* London, **302,** 241–243.

Doran, J. C., and T. W. Horst, 1983: Observations and models of simple nocturnal slope flows. *J. Atmos. Sci.,* **40,** 708–717.

——, ——, and C. D. Whiteman, 1989: The development

and structure of nocturnal slope winds in a simple valley. *Bound.-Layer Meteor.*, in preparation.

Doswell, C. A., 1980: Synoptic-scale environments associated with high plains severe thunderstorms. *Bull. Amer. Meteor. Soc.*, **61**, 1388–1400.

Douglas, C. K. M., 1928: Some alpine cloud forms. *Quart. J. Roy. Meteor. Soc.*, **54**, 175–178.

Dozier, J., 1980: A clear sky spectral solar radiation model for snow covered mountainous terrain. *Water Resour. Res.*, **16**, 709–718.

Drake, R. L., 1977: Site selection techniques and methodologies for WECS. *Third Wind Energy Workshop*, U.S. Dept. of Energy, 635–645. [CONF770921/2.]

Drazin, P. G., 1961: On the steady flow of a fluid of variable density past an obstacle. *Tellus*, **13**, 239–251.

Dreiseitl, E., H. Feichter, H. Pichler, R. Steinacker and I. Vergeiner, 1980: Windregimes an der gabelung zweier alpentäler. [Wind regimes at the intersection of two alpine valleys.] *Arch. Meteor. Geophys. Bioklim.*, **B28**, 257–275.

Durran, D. R., 1981: The effects of moisture on mountain lee waves. Ph.D. dissertation, National Center for Atmospheric Research Cooperative Thesis No. 65, 142 pp. [NTIS PB82156621.]

——, 1986a: Another look at downslope windstorms. Part I: On the development of analogs to supercritical flow in an infinitely deep, continuously stratified fluid. *J. Atmos. Sci.*, **43**, 2527–2543.

——, 1986b: Mountain waves. *Mesoscale Meteorology and Forecasting*, P. Ray, Ed., Amer. Meteor. Soc., 472–492.

——, and J. B. Klemp, 1982: The effects of moisture on trapped mountain lee waves. *J. Atmos. Sci.*, **39**, 2490–2506.

——, and ——, 1983: A compressible model for the simulation of moist mountain waves. *Mon. Wea. Rev.*, **111**, 2341–2361.

——, and R. Grossman, 1985: Reply. *Mon. Wea. Rev.*, **113**, 2178–2181.

——, and J. B. Klemp, 1987: Another look at downslope winds. Part II: Nonlinear amplification beneath wave-overturning layers. *J. Atmos. Sci.*, **44**, 3402–3412.

Dutton, J. A., 1976: *The Ceaseless Wind.* McGraw Hill, 579 pp.

Eberhard, E. W., and Z. Chen, 1989: Lidar discrimination of multiple fluorescent tracers of atmospheric motions. *Appl. Opt.*, **28**, 2996–3007.

——, R. E. Cupp and K. R. Healy, 1989a: Doppler lidar measurement of profiles of turbulence and momentum flux. *J. Atmos. Oceanic Technol.*, **6**, 809–819.

——, —— and W. D. Neff, 1989b: Wind fields and aerosol distributions in the Denver Brown Cloud observed by CO_2 Doppler lidar. *Extended Abstracts, Sixth Joint Conference on Applications of Air Pollution Meteorology with APCA*, Anaheim, Amer. Meteor. Soc., 117–119.

Eberhard, W. L., and G. T. McNice, 1986: Versatile lidar for atmospheric studies, including plume dispersion, clouds, and stratospheric aerosol. *J. Atmos. Oceanic Technol.*, **3**, 614–622.

——, —— and S. W. Troxel, 1987: Lidar sensing of plume dispersion: Analysis methods and product quality for light-scattering tracer particles. *J. Atmos. Oceanic Technol.*, **4**, 674–689.

Ecklund, W. L., D. A. Carter and B. B. Balsley, 1988: A UHF wind profiler for the boundary layer: Brief discussion and initial results. *J. Atmos. Oceanic Technol.*, **5**, 432–441.

Egan, B. A., 1975: Turbulent diffusion in complex terrain. *Lectures on Air Pollution and Environmental Impact Analyses; Workshop Proceedings.* Amer. Meteor. Soc.

——, 1984: Transport and diffusion in complex terrain. *Bound.-Layer Meteor.*, **30**, 3–28. [Also in H. Kaplan and N. Dinar, Eds: *Proc. 29th Oholo Biol. Conf. on Boundary Layer Structure*, Zichron Ya'Acov, Israel.]

——, R. J. Paine, P. E. Flaherty and J. E. Pleim, 1985: Evaluation of COMPLEX I and RTDM using 1979–1980 data from the TVA Widows Creek monitoring network. ERT Doc. PD 523-400, Hunton & Williams (UARG), Washington D.C.

Egger, J., 1981a: On the linear two-dimensional theory of thermally induced slope winds. *Beitr. Phys. Atmos.*, **54**, 465–481.

——, 1981b: Stochastically driven large-scale circulation with multiple equilibrium. *J. Atmos. Sci.*, **38**, 2606–2618.

——, 1987: Simple models of the valley-plain circulation. Part I: Minimum resolution model. Part II: Flow resolving model. *Meteor. Atmos. Phys.*, **36**, 231–254.

Ekhart, E., 1939: Mittlere temperaturverhältnisse der alpen und der freien atmosphäre über dem alpenvorland. Teil II: Die temperaturverhältnisse über München und Vergleich mit den alpen. *Meteor. Z.*, **56**, 49–57.

——, 1944: Beiträge zur alpinen meteorologie. [Contributions to alpine meteorology.] *Meteor. Z.*, **61**, 217–231. [English translation: C. D. Whiteman and E. Dreiseitl, 1984: Alpine meteorology: Translations of classic contributions by A. Wagner, E. Ekhart and F. Defant. PNL-5141/ASCOT-84-3. Pacific Northwest Laboratory, Richland, Washington, 121 pp.]

——, 1948: De la structure thermique de l'atmosphere dans la montagne. [On the thermal structure of the mountain atmosphere.] *La Météorologie*, **4**, 3–26. [English translation: C. D. Whiteman, and E. Dreiseitl, 1984: Alpine meteorology: Translations of classic contributions by A. Wagner, E. Ekhart and F. Defant. PNL-5141/ASCOT-84-3. Pacific Northwest Laboratory, Richland, Washington, 121 pp.]

——, 1949: Ber inversionen in den Alpen. [On inversions in the Alps.] *Meteor. Rundsch*, **2**, 153–159.

Elassal, A. A., and V. M. Caruso, 1983: USGS Digital Cartographic Data Standards: Digital Elevation Models. Geological Survey Circular 895-B, 40 pp.

Elder, J., 1965: Laminar free convection in a vertical slot. *J. Fluid Mech.,* **23,** 77–98.

Eliassen, A., and E. Palm, 1960: On the transfer of energy in stationary mountain waves. *Geofys. Publ.,* **22,** 1–23.

Ellison, T., and J. Turner, 1959: Turbulent entrainment in stratified flows. *J. Fluid Mech.,* **6,** 423–448.

Elliot, R. D., and R. W. Shaffer, 1962: The development of quantitative relationships between orographic precipitation and air mass parameters for use in forecasting and cloud seeding evaluation. *J. Appl. Meteor.,* **1,** 218–220.

——, and E. L. Hovind, 1964: The water balance of orographic clouds. *J. Appl. Meteor.,* **3,** 235–239.

Eloranta, E. W., J. M. King and J. A. Weinman, 1975: The determination of wind speeds in the boundary layer by monostatic lidar. *J. Appl. Meteor.,* **14,** 1485–1489.

EPA, 1983: Regional workshops on air quality modeling. Summary report, U.S. Environmental Protection Agency/QAQPS (revised October 1983).

Errico, R., and D. P. Baumhefner, 1987: Predictability experiments using a high resolution limited-area model. *Mon. Wea. Rev.,* **114,** 1625–1641.

Ertel, H., 1944: Die unmoglichkeit einer exakten wetterprognose auf grund synoptischer luftdruckkarten von teilgebieten der erde. [The impossibility of exact weather forecasts based on synoptic pressure charts from limited portions of the earth.] *Meteor. Z.,* **58,** 309–313.

Etling, D., J. Preuss and M. Wamser, 1986: Application of a random walk model to turbulent diffusion in complex terrain. *Atmos. Environ.,* **20,** 741–747.

Fabrick, A., R. Sklarew and J. Wilson, 1977: Point source model evaluation and development study. Report A5-058-87 to California Air Resources Board, by Science Applications Inc.

Faller, A. J., 1966: An experimental study of the instability of the laminar Ekman boundary layer. *J. Fluid Mech.,* **15,** 560–576.

Falvey, H. T., and R. A. Dodge, 1977: Atmospheric simulation using a stratified liquid model. Bureau of Reclamation Report REC-ERC-yy-8, U.S. Dept. of the Interior, 96 pp.

Feigenbaum, M. J., 1979: The onset spectrum of turbulence. *Phys. Letters,* **A74,** 375–410.

Fiedler, F., and T. Prenosil, 1980: Das MESOKLIP-experiment, mesoskaliges klimaprogramm im Oberrheintal. [The MESOKLIP experiment, mesoscale climate program in the Upper Rhine Valley.] Technical Report 1, Wiss. Berichte Meteor. Inst. Univ. Karlsruhe, 107 pp.

Field, J. H., and R. Warden, 1929: A survey of air currents in the Bay of Gibraltar1929-1930. *Geophysics Memoirs,* No. 59, Published by Her Majesty's Stationary Office, U.K., 84 pp. [J. H. Field and R. Warden, 1963: Scientific Paper No. 18, Meteorological Office, London.]

Filliger, P., B. Rickli and H. Wanner, 1987: Slope and valley wind measurements and their comparison with ambient winds and stability. *Extended Abstracts, Fourth Conference on Mountain Meteorology,* Seattle, Amer. Meteor. Soc., 70–75.

Finnigan, J. J., 1988: Airflow over complex terrain. *Flow and Transport in the Natural Environment: Advances and Applications.* W. K. Steffen and O. T. Denmead, Eds., Springer–Verlag.

Foote, G. B., 1984: Influences of gust fronts on the propagation of storms. *Proceedings Ninth International Cloud Physics Conference,* Tallinn, Estonia, Intl. Assoc. Meteor. Atmos. Phys., 419–421.

Fosberg, M. A., 1984: A diagnostic three-dimensional mass and momentum conserving wind model for complex terrain. *Third Conference on Mountain Meteorology,* Portland, Amer. Meteor. Soc.

Frenkiel, J., 1962: Wind profiles over hills (in relation to wind-power utilization). *Quart. J. Roy. Meteor. Soc.,* **88,** 156–169.

——, 1963: Gusts over hills (in relation to wind-power utilization). *Quart. J. Roy. Meteor. Soc.,* **89,** 281–283.

Freytag, C., 1985: MERKUR-Results: Aspects of the temperature field and the energy budget in a large alpine valley during mountain and valley wind. *Contrib. Atmos. Phys.,* **58,** 458–476.

——, 1986: Results from the MERKUR Experiment: Aspects of the momentum budget in a valley during mountain and valley wind. *Meteor. Rundsch.,* **39,** 240–247.

——, 1987: Results from the MERKUR Experiment: Mass budget and vertical motions in a large valley during mountain and valley wind. *Meteorol. Atmos. Phys.,* **37,** 129–140.

——, and B. Hennemuth, Eds., 1979: Hangwindexperiment Innsbruck Oktober 1978—Datensammlung. [Innsbruckslope wind experiment, October 1978—Datavolume.] Wiss. Mitt. 36, Meteorologisches Institut, Universität München, 261 pp.

——, and ——, 1981: DISKUS, Gebirgswindexperiment im Dischmatal—Datensammlung. Teil 1: Sondierungen. [DISKUS, mountain wind experiment in the Dischma Valley—Data volume. Part 1: Soundings.] Wiss. Mitt. 43, Meteorologisches Institut, Universität München, 250 pp.

——, and ——, 1982: DISKUS, Gebirgswindexperiment im Dischmatal—Datensammlung. Teil 2: Bodennahe messungen und flugzeugmessungen. [DISKUS, mountain wind experiment in the Dischma Valley—Data volume. Part 2: Near-ground and airplane measurements.] Wiss. Mitt. 46, Meteorologisches Institut, Universität München, 192 pp.

——, and ——, 1983: MERKUR: Mesoskaliges experiment in Raum Kufstein-Rosenheim. [MERKUR: Mesoscale experiment in the area between Kufstein and Rosenheim.] Wiss. Mitt. 48, Meteorologisches Institut, Universität München, 132 pp.

Frisch, A. S., and S. F. Clifford, 1974: A study of convection capped by a stable layer using Doppler radar and acoustic echo sounders. *J. Atmos. Sci.,* **31,** 1622–1628.

Fritz, R. B., and W. D. Neff, 1989: The use of an optical crosswind sensor to study urban airflows. Preprints, *Sixth Joint Conference on Applications of Air Pollution Meteorology.* Anaheim, Amer. Meteor. Soc., 120–121.

Fujita, T. T., 1988: Rugged Rocky Mountain tornado. *Weatherwise,* **41,** 80–83.

——, 1989: The Teton–Yellowstone tornado of 21 July 1987. *Mon. Wea. Rev.,* **117,** 1913–1940.

——, K. A. Styber and R. A. Brown, 1962: On the mesometeorological field studies near Flagstaff, Arizona. *J. Appl. Meteor.,* **1,** 26–42.

Gallagher, M. W., T. W. Choularton and M. K. Hill, 1988: Some observations of airflow over a large hill of moderate slope. *Bound.-Layer Meteor.,* **42,** 229–256.

Garratt, J. R., R. A. Pielke and W. Miller, 1989: Mesoscale model response to random, surface-based perturbations: A sea breeze experiment. *Bound.-Layer Meteor.,* submitted.

Garrett, A. J., 1980: Orographic cloud over the eastern slopes of Mauna Loa Volcano, Hawaii, related to insolation and wind. *Mon. Wea. Rev.,* **108,** 931–941.

——, and F. G. Smith, 1984: Two-dimensional simulations of drainage winds and diffusion compared to observations. *J. Climate Appl. Meteor.,* **23,** 597–610.

Gebhardt, B., 1988: *Buoyancy Induced Flows and Transport.* Hemisphere Publishing Corporation, 984 pp.

Geiger, R., 1965: *The Climate Near the Ground.* Revised ed., Harvard University Press, 611 pp. [Translated by Scripta Technica, Inc. from the 4th German edition, 1961.]

George, R. L., 1979: Evolution of mesoscale systems over mountainous terrain. Atmos. Sci. Paper 318, Colorado State University, Fort Collins, 160 pp.

Gifford, F. A., 1976: Turbulent diffusion typing schemes: A review. *Nuclear Safety,* **17,** 68–86.

Gill, A., 1966: The boundary-layer regime for convection in a rectangular cavity. *J. Fluid Mech.,* **26,** 515–536.

Gill, A. E., 1982: *Atmosphere–Ocean Dynamics.* Academic Press, 662 pp.

Glass, M., and T. N. Carlson, 1963: The growth characteristics of small cumulus clouds. *J. Atmos. Sci.,* **20,** 397–406.

Gleick, J., 1987: *Chaos: Making a New Science.* Viking Press, 352 pp.

Golding, E. W., 1955: *The Generation of Electricity by Wind Power.* Philosophical Library, 318 pp.

Goldstein, R. J., 1983: *Fluid Mechanics Measurements.* Hemisphere Publishing Corporation, 676 pp.

Gossard, E. E., and W. H. Munk, 1954: On gravity waves in the atmosphere. *J. Meteor.,* **11,** 259–269.

——, and W. H. Hooke, 1975: *Waves in the Atmosphere.* Elsevier, 456 pp.

Graham, N. E., G. H. Taylor, E. L. Hovind, R. L. Petersen, J. E. Cermak and P. C. Sinclair, 1978: An analysis of terrain-induced aerodynamic disturbances near the Kingston Steam Plant, Kingston, Tennessee. North American Weather Consultants Report AQ-78-22, prepared for TVA, 170 pp.

Grant, L. O., and A. M. Kahan, 1974: Weather modification for augmenting orographic precipitation. *Weather and Climate Modification,* W. H. Hess, Ed., John Wiley, 282–317.

Gross, G., H. Vogel and F. Wippermann, 1987: Dispersion over and around a steep obstacle for varying thermal stratifications–numerical simulations. *Atmos. Environ.,* **21,** 483–490.

Grossman, R. L., and D. R. Durran, 1984: Interaction of low-level flow with the western Ghat Mountains and offshore convection in the summer monsoon. *Mon. Wea. Rev.,* **112,** 652–672.

Gudiksen, P. H., 1989: Categorization of nocturnal drainage flows within the Brush Creek valley and the variability of sigma theta in complex terrain. *J. Appl. Meteor.,* **28,** 489–495.

——, and J. J. Walton, 1981: Categorization of nocturnal drainage flows in the Anderson Creek valley. Preprints, *Second Conference on Mountain Meteorology,* Steamboat Springs, CO. Amer. Meteor. Soc., 218–221.

——, and M. H. Dickerson, 1982: ASCOT FY-82. Progress Report UCID-18878-82, ASCOT, Lawrence Livermore National Laboratory.

——, and D. L. Shearer, 1989: The dispersion of atmospheric tracers in nocturnal drainage flows. *J. Appl. Meteor.,* **28,** 602–608.

——, G. J. Ferber, M. M. Fowler, W. L. Eberhard, M. A. Fosberg and W. R. Knuth, 1984: Field studies of transport and dispersion of atmospheric tracers in nocturnal drainage flows. *Atmos. Environ.,* **18,** 713–731.

Haagenson, P. L., 1979: Meteorological and climatological factors affecting Denver air quality. *Atmos. Environ.,* **13,** 79–85.

Hagemeyer, B. C., 1984: An investigation of summertime convection over the Upper Current River Valley of Southwest Missouri. NOAA Tech. Memo, NWS CR-71, 35 pp.

——, 1985: On the observation and modeling of the slope winds of the Upper Current River Valley of southeast Missouri and their relationship to air-mass thunderstorm formation. Tech. Memo. NWS CR-74, NOAA, 40 pp.

Hahn, D. C., 1980: Observed characteristics of turbulence in the atmospheric boundary layer over mountainous terrain. Atmos. Sci. Paper 332, Colorado State University.

Halbsguth, G., M. J. Kerschgens, H. Kraus, G. Meindl and E. Schaller, 1984: Energy fluxes in an alpine valley. *Arch. Meteor. Geophys. Bioklim.,* **A33,** 11–20.

Hales, Jr., J. E., 1972: Surges of maritime tropical air northward over the Gulf of California. *Mon. Wea. Rev.,* **100,** 298–306.

——, 1974: Southwestern United States summer monsoon source—Gulf of Mexico or Pacific Ocean? *J. Appl. Meteor.,* **13,** 331–342.

——, 1975: A severe southwest desert thunderstorm: 19 August 1973. *Mon. Wea. Rev.,* **103,** 344–351.

——, 1977: On the relationship of convective cooling to nocturnal thunderstorms at Phoenix. *Mon. Wea. Rev.,* **105,** 1609–1613.

Hanna, S. R., 1980: Measured σ_y and σ_θ in complex terrain near the TVA Widows Creek, Alabama Steam Plant. *Atmos. Environ.,* **14,** 401–409.

——, G. A. Briggs and R. P. Hosker, 1982: *Handbook on Atmospheric Diffusion.* U.S. DOE Tech. Inf. Center, 102 pp. [DOE/TIC-11223.]

——, B. A. Egan, C. J. Vando and A. J. Curreri, 1984: A complex terrain dispersion model for regulatory applications at the Westvaco Lake Mill. *Atmos. Environ.,* **18,** 685–699.

Hansen, C. A., and J. E. Cermak, 1975: Vortex-containing wakes of surface obstacles. Colorado State University Report CER75-76ACH-JEC16, 163 pp.

Hawkes, H. B., 1947: Mountain and valley winds with special reference to the diurnal mountain winds of the Great Salt Lake Region. Ph.D. dissertation, Ohio State University, 312 pp.

Hawthorne, W. R., and M. E. Martin, 1951: The effect of density gradient and shear on the flow over a hemisphere. *Proc. Roy. Soc. London,* **A232,** 184–195.

Heggli, M. F., and D. W. Reynolds, 1985: Radiometric observations of supercooled liquid water within a split front over the Sierra Nevada. *J. Climate Appl. Meteor.,* **24,** 1258–1261.

Heldt, K., and K. Höschele, 1989: Hang- und bergwinde am Rheintalrand bei Karlsruhe. [Slope and mountain winds at the border of the Rhine Valley near Karlsruhe.] *Meteor. Rundsch.,* **41,** 104–110.

Hennemuth, B., 1985: Temperature field and energy budget of a small alpine valley. *Contrib. Atmos. Phys.,* **58,** 545–559.

——, 1986: Thermal asymmetry and cross-valley circulation in a small alpine valley. *Bound.-Layer Meteor.,* **36,** 371–394.

——, and H. Schmidt, 1985: Wind phenomena in the Dischma Valley during DISKUS. *Arch. Meteor. Geophys. Bioklim.,* **B35,** 361–387.

——, and I. Neureither, 1986: Das feuchtefeld in einem alpinen endtal. [The humidity field in an alpine end valley.] *Meteor. Rundsch.,* **39,** 233–239.

Henz, J. F., 1974: Colorado high plains thunderstorm systems: A descriptive radar-synoptic climatology. Master's thesis, Dept. of Atmos. Sci., Colorado State University, 82 pp.

Hertig, J. A., 1986: Some aspects of exotic simulations. *Proceedings Third International Workshop on Wind and Water Tunnel Modelling of Atmospheric Flow and Dispersion,* Lausanne, Switzerland, 233–247.

——, and P. Liska, 1983: Simulation of regional atmospheric flows on small scale topographical models. *J. Wind Engr. Ind. Aero.,* **15,** 77–89.

Hess, S. L., 1959: *Introduction to Theoretical Meteorology.* Holt, Rinehart and Winston, 362 pp.

Hewson, E. W., 1973: Wind power potential in selected areas of Oregon. Oregon State University Report PUD73-1, 63 pp.

——, and G. C. Gill, 1944: Meteorological investigations in Columbia River Valley Near Trail, B.C. Bur. Mines Bull. 453, U.S. Dept. of the Interior, 23-228.

Hidy, G. M., 1967: Adventures in atmospheric simulation. *Bull. Amer. Meteor. Soc.,* **48,** 143–161.

Hill, F. F., K. A. Browning and M. J. Bader, 1981: Radar and raingauge observations of orographic rain over South Wales. *Quart. J. Roy. Meteor. Soc.,* **107,** 643–670.

Hindman, II, E. E., 1973: Air currents in a mountain valley deduced from the breakup of a stratus deck. *Mon. Wea. Rev.,* **101,** 195–200.

——, and E. J. Wick, 1989: Air motions in the vicinity of Mt. Everest as deduced from Pilatus Porter flights. *Technical Soaring,* in press.

Hinkley, E. D., Ed., 1976: *Laser Monitoring of the Atmosphere.* Springer–Verlag, 380 pp.

Hobbs, P. V., R. C. Easter and A. B. Fraser, 1973: A theoretical study of the flow of air and fallout of solid precipitation over mountainous terrain. Part II: Microphysics. *J. Atmos. Sci.,* **30,** 813–823.

Hoehne, W. E., 1980: Precision of National Weather Service upper air measurements. NOAA Tech. Memo. NWS TED-16, 23 pp.

Hogstrom, A. S., and U. Hogstrom, 1985: The wind regime in coastal areas with special reference to results obtained from the Swedish wind energy program. *Bound.-Layer Meteor.,* **30,** 351–373.

Hoinka, K. P., 1985: A comparison of numerical simulations of hydrostatic flow over mountains and observations. *Mon. Wea. Rev.*, **113,** 719–735.

——, and F. Roesler, 1987: The surface layer on the leeside of the Alps during foehn. *Meteorol. Atmos. Phys.*, **37,** 245–258.

Holmboe, J., and H. Klieforth, 1957: Investigation of mountain lee waves and the airflow over the Sierra Nevada. Final Report, Contract No. AF 19(604)-728, University of California, Los Angeles, 290 pp.

Holton, J. R., 1979: *An Introduction to Dynamic Meteorology.* 2nd ed., Academic Press, 391 pp.

Holzworth, C. G., 1980: The EPA program for dispersion model development for sources in complex terrain. *Second Joint AMS-APCA Conference on Applications of Air Pollution Meteorology,* New Orleans.

Horst, T. W., and J. C. Doran, 1982: Simple nocturnal slope flow data from the Rattlesnake Mountain Site. Technical Report PNL-4406/ASCOT Report 82-5, Pacific Northwest Laboratory, Richland, WA.

——, and ——, 1983: Nocturnal drainage flow on simple slopes. ASCOT Progress Report PNL-SA-11245, Pacific Northwest Laboratory, Richland, WA.

——, and ——, 1986: Nocturnal drainage flow on simple slopes. *Bound.-Layer Meteor.*, **34,** 263–286.

——, and ——, 1988: The turbulence structure of nocturnal slope flow. *J. Atmos. Sci.*, **45,** 605–616.

——, K. J. Allwine and C. D. Whiteman, 1987: A thermal energy budget for nocturnal drainage flow in a simple valley. Preprints, *Fourth Conference on Mountain Meteorology,* Seattle, Amer. Meteor. Soc., 15–19.

Houghton, D. D., and A. Kasahara, 1968: Nonlinear shallow fluid flow over an isolated ridge. *Commun. Pure and Appl. Math.*, **21,** 1–23.

Howard, A. D., and J. L. Walmsley, 1985: Simulation model of isolated dune sculpture by wind. *Proc. Int. Workshop on the Physics of Blown Sand,* Aarhus University, Denmark, in press.

Hoxit, L. R., 1973: Variability of planetary boundary layer winds. Atmos. Sci. Paper 199, Colorado State University, 157 pp.

Huber, A. H., 1988: Distribution of pollutant concentrations downwind of a point-source in the near wake of a building. Ph.D. dissertation, North Carolina State University, 289 pp.

Hunt, J. C. R., 1980: Wind over hills. Survey paper for AMS workshop on the planetary boundary layer, Amer. Meteor. Soc., Boulder, 107-157. [Also J. C. R. Hunt, 1978: Survey paper for NOAA-NSF Workshop, Boulder, Colorado.]

——, 1985: Turbulent diffusion from sources in complex flows. *Ann. Rev. Fluid Mech.*, **17,** 447–485.

——, and P. J. Mulhearn, 1973: Turbulent dispersion from sources near two dimensional obstacles. *J. Fluid Mech.*, **62,** 245–274.

——, and W. H. Snyder, 1980: Experiments on stably and neutrally stratified flow over a model three-dimensional hill. *J. Fluid Mech.*, **96,** 671–704.

——, and K. J. Richards, 1984: Stratified air flow over one or two hills. *Bound.-Layer Meteor.*, **30,** 223–259. [Also in H. Kaplan and N. Dinar, Eds.: *Proc. 29th Oholo Biol. Conf. on Boundary Layer Structure: Modeling and Application to Air Pollution and Wind Energy,* Zichron Ya'Acov, Israel.]

——, W. H. Snyder and R. E. Lawson, Jr., 1978: Flow structure and turbulent diffusion around a three-dimensional hill: Fluid modeling study on effects of stratification. Part I: Flow structure. EPA Report 600/1-78-041 and 600/4-76-047, U.S. Environmental Protection Agency, 96 pp.

——, J. S. Puttock and W. H. Snyder, 1979: Turbulent diffusion from a point source in stratified and neutral flows around a three-dimensional hill. Part I: Diffusion equation analysis. *Atmos. Environ,* **13,** 1227–1239.

——, S. Leibovich and J. L. Lumley, 1981: Prediction method for the dispersal of atmospheric pollutant in complete terrain. Technical Report P85-81-04, Flow Analysis Associates, Ithaca, NY.

——, J. C. Kaimal and E. Gaynor, 1985: Some observations of turbulence structure in stable layers. *Quart. J. Roy. Meteor. Soc,* **114,** 827–858.

——, S. Leibovich and K. J. Richards, 1988a: Turbulent shear flow over hills. *Quart. J. Roy. Meteor. Soc.*, **114,** 1435–1470.

——, K. J. Richards and P. W. M. Brighton, 1988b: Stably stratified flow over low hills. *Quart. J. Roy. Meteor. Soc.*, **114,** 859–886.

——, D. Stretch and R. E. Britter, 1988c: Length scales in stably stratified turbulent flows and their use in turbulence models. *Proc. IMA Conference on Stably Stratified Flows and Dense Gas Dispersion,* Clarendon Press, 285–322.

——, W. Weng and D. J. Carruthers, 1988d: Modelling deposition fluxes on hills. *Proceedings 16th NATO/CCMS International Tech. Meeting on Air Pollution Modelling and its Applications,* Plenum, Lindau, FRG, 623–640.

——, M. R. Raupach, W. Weng and D. J. Carruthers, 1989: Temperature and humidity fields in the airflow over hills, in preparation.

Hunt, T. L., 1947: Weather conditions on Kilimanjaro. *Weather,* **2,** 338–345.

Iribarne, J. V., and W. L. Godson, 1973: *Atmospheric Thermodynamics.* D. Reidel, 222 pp.

Isard, S. A., 1983: Estimating potential direct insolation to Alpine terrain. *Arct. Alp. Res.*, **15,** 77–89.

Ives, R. L., 1964: Winds of the past. *Weatherwise,* **17,** 224–235.

Jackson, P. S., 1979: The influence of local terrain features on the site selection for wind energy generating systems. Boundary Layer Wind Tunnel Report BLWT-1-1979, University of Western Ontario, 83 pp.

——, and J. C. R. Hunt, 1975: Turbulent wind flow over a low hill. *Quart. J. Roy. Meteor. Soc.*, **101,** 929–955.

Jelinek, A., 1937: Beiträge zur mechanik periodischer hangwinde. *Beitr. Phys. Atmos.*, **24,** 60–84.

Jensen, M., 1958: The model-law for phenomena in a natural wind. *Ingenioeren, Int. Ed.*, **2,** 121–128.

Jensen, N. O., E. L. Petersen and I. Troen, 1984: Extrapolation of mean flow statistics with special regard to wind energy applications. WMO Report WCP-86, WMO/TD-No. 15.

Johnson, D. B., 1982: The role of giant and ultragiant aerosol particles in warm rain initiation. *J. Atmos. Sci.*, **39,** 448–460.

Jones, H. L., 1917: *The Geography of Strabo.* Vol. 2, Harvard University Press, 480 pp.

Kaimal, J. C., and J. E. Gaynor, 1983: The Boulder Atmospheric Observatory. *J. Climate Appl. Meteor.*, **22,** 863–880.

——, N. L. Abshire, R. B. Chadwick, M. T. Decker, W. H. Hooke, R. A. Kropfli, W. D. Neff, F. Pasqualucci and P. H. Hildebrand, 1982: Estimating the depth of the daytime convective boundary layer. *J. Appl. Meteor.*, **21,** 1123–1129.

Karr, T. W., and R. L. Wooten, 1976: Summer radar echo distribution around Limon, Colorado. *Mon. Wea. Rev.*, **104,** 728–734.

Katsoulis, B. D., 1988: Some meteorological aspects of air pollution in Athens, Greece. *Meteor. Atmos. Phys.*, **39,** 203–212.

Kelly, R. D., 1988: Asymmetric removal of temperature inversions in a high mountain valley. *J. Climate Appl. Meteor.*, **27,** 664–673.

Kimball, B. A., R. D. Jackson, F. S. Nakayama, S. B. Idso and R. J. Reginato, 1976: Soil-heat flux determination: Temperature gradient method with computed thermal conductivities. *Soil Sci. Soc. Amer. J.*, **40,** 25–28.

Kimura, F., 1986: Formation mechanism of the nocturnal mesoscale vortex in Kanto Plain. *J. Meteor. Soc. Japan*, **64,** 857–869.

King, C. W., 1989: Representativeness of single vertical wind profiles for determining volume fluxes in valleys. *J. Appl. Meteor.*, **28,** 463–466.

Kitabayashi, K. K., M. M. Orgill and J. E. Cermak, 1971: Laboratory simulation of airflow in atmospheric transport-dispersion over Elk Mountain, Wyoming. Colorado State University Report CER70-71KKK-MMO-JEC65, 98 pp.

Kitada, T., K. Igarashi and M. Owada, 1986: Numerical analysis of air pollution in a combined field of land/sea breeze and mountain/valley wind. *J. Climate Appl. Meteor.*, **25,** 767–784.

Klemp, J. B., and D. K. Lilly, 1975: The dynamics of wave induced downslope winds. *J. Atmos. Sci.*, **32,** 320–339.

——, and ——, 1978: Numerical simulation of hydrostatic mountain waves. *J. Atmos. Sci.*, **35,** 78–107.

——, and D. R. Durran, 1987: Numerical modelling of bora winds. *Meteor. Atmos. Phys.*, **36,** 215–227.

Kline, S. J., 1965: *Similitude and Approximation Theory.* McGraw-Hill, 229 pp.

Klitch, M. A., J. F. Weaver, F. P. Kelly and T. H. Vonder Haar, 1985: Convective cloud climatologies constructed from satellite imagery. *Mon. Wea. Rev.*, **113,** 326–337.

Knupp, K. R., 1987: Downdrafts within High Plains cumulonimbi. Part I: General kinematic structure. *J. Atmos. Sci.*, **44,** 987–1008.

Kondo, H., 1984: The difference of the slope wind between day and night. *J. Meteor. Soc. Japan*, **62,** 224–233.

Kondo, J., and T. Sato, 1988: A simple model of drainage flow on a slope. *Bound.-Layer Meteor.*, **43,** 103–123.

——, T. Kuwagata and S. Haginoya, 1989: Heat budget analysis on nocturnal cooling and daytime heating in a basin. *J. Atmos. Sci.*, submitted.

Kropfli, R. A., 1986a: Single Doppler radar measurements of turbulence profiles in the convective boundary layer. *J. Atmos. Oceanic Technol.*, **3,** 305–313.

——, 1986b: Radar probing and measurement of the planetary boundary layer. Part II: Scattering from particulates. *Probing the Atmospheric Boundary Layer,* D. Lenschow, Ed., Amer. Meteor. Soc., 183–199.

——, J. M. Wilczak and W. F. Dabberdt, 1986: Characteristics of terrain-forced eddies in the marine planetary boundary layer from Doppler radar observations and a mesoscale numerical model. Preprints, *23rd Conference on Cloud Physics,* Snowmass, Amer. Meteor. Soc., JP186–JP189.

Kuettner, J. P., Ed.,1982: ALPEX experimental design. Technical report, GARP-ALPEX Report No. 1, WMO, 99 pp.

Kunkel, K. E., E. W. Eloranta and J. A. Weinman, 1980: Remote determination of winds, turbulence spectra and energy dissipation rates in the boundary layer from lidar measurements. *J. Atmos. Sci.*, **37,** 978–985.

Kuo, H. L., 1963: Perturbations of plane Couette flow in stratified fluid and origin of cloud streets. *Phys. Fluids,* **6,** 195–211.

Kuo, J. T., and H. D. Orville, 1973: A radar climatology of summertime convective clouds in the Black Hills. *J. Appl. Meteor.*, **12,** 357–368.

Kyle, T. G., 1987: The energy budget in a valley nocturnal flow. *Tellus,* **A39,** 226–234.

Lamb, D., K. W. Nielson, H. E. Kleiforth and J. Hallett, 1976: Measurements of liquid water content in winter cloud systems over the Sierra Nevada. *J. Appl. Meteor.*, **15,** 763–775.

Lange, R., 1978: ADPIC—A three-dimensional particle-in-cell model for the dispersal of atmospheric pollutants and its comparison to regional tracer studies. *J. Appl. Meteor.*, **17**, 320–329.

Lavery, T. F., A. Bass, D. G. Strimaitis, A. Venkatram, B. R. Greene, P. J. Drivas and B. A. Egan, 1982: EPA Complex Terrain Modeling Program. First Milestone Report—1981 EPA 600/3-82-036, U.S. Environmental Protection Agency.

——, D. G. Strimaitis, A. Venkatram, B. R. Greene, D. DiCristofaro and B. A. Egan, 1983: EPA Complex Terrain Model Development. Third Milestone Report—1983 EPA 600/3-83-101, U.S. Environmental Protection Agency.

——, G. C. Holzworth, F. Schiermeier, W. H. Snyder, L. E. Truppi, B. R. Greene and B. A. Egan, 1986: The full scale plume study: A summary of data collected and phenomena observed. *Fifth Joint Conference on Applications of Air Pollution Meteorology,* Chapel Hill, NC.

Lawrence, R. S., G. R. Ochs and S. F. Clifford, 1972: Use of scintillations to measure average wind across a light beam. *Appl. Opt.*, **11,** 239–243.

Lawson, Jr., R. E., and W. H. Snyder, 1985: Stack heights and locations in complex terrain. Preprints, *Seventh Symposium on Turbulence Diffusion,* Boulder, Amer. Meteor. Soc., 223–226.

——, W. H. Snyder and R. S. Thompson, 1989: Estimation of maximum surface concentrations from sources near complex terrain in neutral flow. *Atmos. Environ.*, **23,** 321–331.

Lee, J. T., D. L. Call, R. E. Lawson, Jr., W. E. Clements and D. E. Hoard, 1988: A video image analysis system for concentration measurements and flow visualization in building wakes. Los Alamos National Laboratory Report LA-UR-88-3853, 63 pp.

Legg, B. J., and M. R. Raupach, 1982: Markov-chain simulations of particle dispersion in inhomogeneous flows: The mean drift velocity induced by a gradient in Eulerian velocity variance. *Bound.-Layer Meteor.*, **24,** 3–13.

Lenschow, D., Ed., 1986: *Probing the Atmospheric Boundary Layer.* Amer. Meteor. Soc., 269 pp.

Lilly, D. K. 1984: Some facets of the predictability problem for atmospheric mesoscales. *AIP Conference Proc. No. 106,* H. C. Wolfe, Ed., 287–294.

——, and J. B. Klemp, 1979: The effects of terrain shape on nonlinear hydrostatic mountain waves. *J. Fluid Mech.*, **95,** 241–261.

——, and D. R. Durran, 1984: Stably stratified moist airflow over mountainous terrain. *Proceedings First Sino-American Workshop Mountain Meteorology,* E. R. Reiter, Z. Baozhen, and Q. Yongfu, Eds., Science Press, Beijing, and Amer. Meteor. Soc., 569–608.

Lin, J. T., and G. J. Binder, 1967: Simulation of mountain lee waves in a wind tunnel. Colorado State University Report CER67-68JTL-GJB24, AD-664-172, 132 pp.

Liu, J. Y., and H. D. Orville, 1969: Numerical modeling of precipitation and cloud shadow effects on mountain induced cumuli. *J. Atmos. Sci.*, **26,** 1283–1298.

Liu, H. T., and J. T. Lin, 1975: Laboratory simulation of plume dispersion from a lead smelter in Glover, Missouri: Neutral and stable atmosphere. Tech. Report 55, Flow Research, Inc., Kent, WA, 61 pp.

Long, R. R., 1953a: A laboratory model resembling the "Bishop-wave" phenomenon. *Bull. Amer. Meteor. Soc.*, **34,** 205–211.

——, 1953b: Some aspects of the flow of stratified fluids. Part I: A theoretical investigation. *Tellus,* **5,** 42–58.

——, 1954: Some aspects of the flow of stratified fluids. Part II: Experiments with a two fluid system. *Tellus,* **6,** 97–115.

——, 1959: A laboratory model of airflow over the Sierra Nevada Mountains. *The Atmosphere and the Sea in Motion: Scientific Contributions to the Rossby Memorial Volume,* Rockefeller Institute Press, 372–380.

Longstaff, T. G., 1923: Meteorological notes from the Mount Everest expedition of 1922. *Quart. J. Roy. Meteor. Soc.*, **49,** 273–276.

Lopez, R. E., and R. L. Holle, 1986: Diurnal and spatial variability of lightning activity in northeastern Colorado and central Florida during the summer. *Mon. Wea. Rev.*, **114,** 1288–1312.

Lorenz, E. N., 1963: Deterministic non-periodic flow. *J. Atmos. Sci.*, **20,** 282–293.

——, 1969: The predictability of a flow which possesses many scales of motion. *Tellus,* **21,** 289–302.

——, 1972: Barotropic instability of Rossby wave motion. *J. Atmos. Sci.*, **29,** 258–264.

——, 1980: Attractor sets and quasi-geostrophic equilibrium. *J. Atmos. Sci.*, **37,** 1685–1699.

Lott, R. A., 1986: Model performance: Plume dispersion over elevated terrain. *Atmos. Environ.*, **20,** 1547–1554.

Lu, N.-P., W. D. Neff and J. C. Kaimal, 1983: Wave and turbulence structure in a disturbed nocturnal inversion. *Bound.-Layer Meteor.*, **26,** 141–155.

Ludlum, F. H., 1956: The structure of rainclouds. *Weather,* **11,** 187–196.

MacCready, Jr., P. B., 1955: High and low elevations as thermal source regions. *Weather,* **10,** 35–40.

Machalek, A., 1974: Inversionsuntersuchungen in einem gebirgstal. [Inversion research in a mountain valley.] *Wetter Leben,* **26,** 157–168.

Maddox, R. A., 1980: Mesoscale convective complexes. *Bull. Amer. Meteor. Soc.*, **61,** 1374–1387.

——, 1981: The structure and life cycle of midlatitude mesoscale convective complexes. Atmos. Sci., Paper

No. 336, Dept. of Atmospheric Science, Colorado State University, Fort Collins, 311 pp.

——, F. Caracena, L. R. Hoxit and C. F. Chappell, 1977: Meteorological aspects of the Big Thompson flash flood of 31 July 1976. Tech. Report ERL 388-APCL 41, NOAA, 83 pp.

——, L. R. Hoxit, C. F. Chappell and F. Caracena, 1978: Comparison of meteorological aspects of the Big Thompson and Rapid City flash floods. *Mon. Wea. Rev.*, **106,** 375–389.

Mahrer, Y., and R. A. Pielke, 1977: The effects of topography on sea and land breezes in a two-dimensional numerical model. *Mon. Wea. Rev.*, **105,** 1151–1162.

Mahrt, L., 1982: Momentum balance of gravity flows. *J. Atmos. Sci.*, **39,** 2701–2711.

Maki, M., and T. Harimaya, 1988: The effect of advection and accumulation of downslope cold air on nocturnal cooling in basins. *J. Meteor. Soc. of Japan,* **66,** 581–597.

——, —— and K. Kikuchi, 1986: Heat budget studies on nocturnal cooling in a basin. *J. Meteor. Soc. of Japan,* **64,** 727–740.

Manins, P. C., and B. L. Sawford, 1979a: Katabatic winds: A field case study. *Quart. J. Roy. Meteor. Soc.*, **105,** 1011–1025.

——, and ——, 1979b: A model of katabatic winds. *J. Atmos. Sci.*, **36,** 619–630.

——, and ——, 1982: Mesoscale observations of upstream blocking. *Quart. J. Roy. Meteor. Soc.*, **108,** 427–434.

Marks, D., and J. Dozier, 1979: A clear-sky longwave radiation model for remote Alpine areas. *Arch. Meteor. Geophys. Bioklim.*, **B27,** 159–187.

Marwitz, J. D., 1972a: Precipitation efficiency of thunderstorms on the High Plains. *J. Rech. Atmos.*, **6,** 367–370.

——, 1972b: The structure and motion of severe hailstorms. Part III: Severely sheared storms. *J. Appl. Meteor.*, **11,** 189–201.

——, 1980: Winter storms over the San Juan Mountains. Part I: Dynamical processes. *J. Appl. Meteor.*, **19,** 913–926.

——, 1983: The kinematics of orographic airflow during Sierra storms. *J. Atmos. Sci.*, **40,** 1218–1227.

——, K. Waight, B. Martner and G. Gordon, 1985: Cloud physics studies in SCPP during 1984–85. Tech. Report Contract No. 2-07-81-V0256, Division of Atmospheric Resource Research, Bureau of Reclamation, U.S. Dept. of the Interior.

Maryon, R. H., J. B. G. Whitlock and G. J. Jenkins, 1986: An analysis of short-range dispersion experiments on the windward slope of an isolated hill. *Atmos. Environ.*, **20,** 2157–2173.

Mason, P. J., 1952: Production of rain and drizzle by coalescence in stratiform clouds. *Quart. J. Roy. Meteor. Soc.*, **78,** 377–386.

——, 1986: Flow over the summit of an isolated hill. *Bound.-Layer Meteor.*, **37,** 385–405.

——, and R. I. Sykes, 1979: Flow over an isolated hill of moderate slope. *Quart. J. Roy. Meteor. Soc.*, **105,** 385–395.

——, and J. C. King, 1985: Measurements and predictions of flow and turbulence over an isolated hill of moderate slope. *Quart. J. Roy. Meteor. Soc.*, **111,** 627–640.

Mathews, T., P. F. Lester and R. B. Hicks, 1984: Sodar observation of chinooks and arctic front passages across the eastern slopes of the Canadian Rockies. *Atmosphere; Atmos. Ocean.*, **22,** 328–342.

May, P. T., R. G. Strauch and K. P. Moran, 1988: The altitude coverage of temperature measurements using RASS with wind profiler radars. *Geophys. Res. Lett.*, **15,** 1381–1384.

——, ——, —— and W. L. Ecklund, 1989a: Temperature sounding by RASS with wind profiler radars: A preliminary study. *IEEE Trans. Geosci. Remote Sens.*, in press.

——, ——, —— and W. D. Neff, 1989b: High resolution weather observations with combined RASS and wind profilers. Preprints, *24th Conference on Radar Meteorology,* Tallahassee, Amer. Meteor. Soc., 746-749.

McAfee, J. R., B. B. Balsley and K. S. Gage, 1989: Momentum flux measurements over mountains: Problems associated with the symmetrical two-beam radar technique. *J. Atmos. Oceanic Technol.*, **6,** 500–508.

McAnelly, R. L., and W. R. Cotton, 1986: Meso (beta)-scale characteristics of an episode of meso (alpha)-scale convective complexes. *Mon. Wea. Rev.*, **114,** 1740–1770.

McCain, J. F., L. R. Hoxit, R. A. Maddox, C. F. Chappell and F. Caracena, 1979: Storm and flood of July 31–August 1, 1976, in the Big Thompson River and Cache la Poudre River basins, Larimer and Weld Counties, Colorado, Part A. Meteorology and hydrology in Big Thompson River and Cache la Poudre River basins. U.S. Geological Survey Professional Paper 1115, U.S. Government Printing Office, Washington, D.C., 152 pp.

McElroy, J. L., and T. B. Smith, 1986: Vertical pollutant distributions and boundary layer structure observed by airborne lidar near the complex Southern California coastline. *Atmos. Environ.*, **20,** 1555–1566.

McEwen, R. B., H. W. Calkins and B. S. Ramey, 1983: USGS digital cartographic data standards: Overview and USGS activities. Geological Survey Circular 895-A, 20 pp.

McKee, T. B., and C. D. Whiteman, 1977: Components

of infrared net radiation in a mountain valley. Atmos. Sci. Paper 279, Colorado State University, 102 pp.

——, and R. D. O'Neal, 1989: The role of valley geometry and energy budget in the formation of nocturnal valley winds. *J. Appl. Meteor.*, **28,** 445–456.

McNaughton, K. G., 1976: Evaporation and advection I: evaporation from extensive homogeneous surfaces. *Quart. J. Roy. Meteor. Soc.*, **102,** 181–191.

McNider, R. T., and R. A. Pielke, 1981: Diurnal boundary layer development over sloping terrain. *J. Atmos. Sci.*, **38,** 2198–2212.

——, S. R. Hanna and R. A. Pielke, 1981: Numerical simulation of the transport and dispersion of pollutants in slope and mountain-valley flows. *Fifth Symposium on Turbulence, Diffusion and Air Pollution,* Amer. Meteor. Soc.

McVehil, G. E., G. R. Ludwig and T. R. Sundaram, 1967: On the feasibility of modeling small scale atmospheric motions. Cornell AeroLab Report ZB-2328-P-1, 178 pp.

Melbourne, W. H., 1977: Development of natural wind models at Monash University. *Proceedings Sixth Australian Hydraulics and Fluid Mechanics Conference,* Adelaide, Australia, 190–194.

Melfi, S. H., D. Whiteman and R. Ferrare, 1989: Observation of atmospheric fronts using Raman lidar moisture measurements. *J. Appl. Meteor.*, **28,** 789–806.

Mendonca, B. G., 1969: Local wind circulation on the slopes of Mauna Loa. *J. Appl. Meteor.*, **8,** 533–541.

Meroney, R. N., 1979: Field verification and laboratory simulation of airflow patterns in complex terrain. *Proc. Fourth Symposium on Turbulence, Diffusion and Air Pollution,* Reno, 592–595.

——, 1981: Physical simulation of dispersion in complex terrain and valley drainage flow situations. *Air Pollution Modeling and Its Application.* C. De Wispelaere, Ed., Plenum Publishing Co., 21 pp.

——, 1986: Guideline for fluid modeling of liquefied natural gas cloud dispersion. Vol II: Technical support document. Gas Research Institute Report GRI86/0102.2, 262 pp. [Colorado State University Report CER84-85RNM50b.]

——, and J. E. Cermak, 1967: Wind tunnel modeling of flow and dispersion over San Nicolas Island. Pacific Missile Range Report PMR-MR-68-1, Point Mugu, CA, 108 pp. [Colorado State University Report CER66-67RNM-JEC44.]

——, and V. A. Sandborn, 1978: Sites for windpower installations: Physical modelling of the influence of hills, ridges and complex terrain on windspeed and turbulence. Final Report CER77-78RNM-VAS50b, Civil Eng. Dept., Colorado State University, 197 pp.

——, J. E. Cermak and B. T. Yang, 1975: Modeling of atmospheric transport and fumigation at shorelines. *Bound.-Layer Meteor.*, **9,** 69–90.

——, A. J. Bowen, D. Lindley and J. R. Pearse, 1978a: Wind characteristics over complex terrain: Laboratory simulation and field measurements at Rakaia Gorge, New Zealand. U.S. Dept. of Energy Report RLO/2438-77/2, 219 pp.

——, V. A. Sandborn, R. J. B. Bouwmeester, H. C. Chien and M. Rider, 1978b: Sites for wind-power installations: Physical modeling of the influence of hills, ridges and complex terrain on wind speed and turbulence. Part I: Executive summary. Tech. Report RLO-2438-78/1, U.S. Dept. of Energy, 91 pp.

Mery, P., 1969: Reproduction en similitude de la diffusion dans la cauche limite atmospherique. *La Houille Blanche,* 327–344.

Moran, K. P., R. G. Strauch, K. B. Earnshaw, D. A. Merritt, B. L. Weber and D. B. Wuertz, 1989: Lower tropospheric wind profiler. Preprints, *24th Conference on Radar Meteorology,* Tallahassee, Amer. Meteor. Soc., 728–731.

Motallebi, N., 1982: Doppler radar observation of the evolution of downdrafts in convective clouds. Atmos. Sci. Paper 355, Colorado State University, 139 pp. [NTIS HC A07/MF A01 CSCL 04B.]

Müller, H., and C. D. Whiteman, 1988: Breakup of a nocturnal temperature inversion in the Dischma Valley during DISKUS. *J. Climate Appl. Meteor.*, **27,** 188–194.

——, R. Reiter, S. Sladkovic and K. Munzert, 1982: Aerologische untersuchungen des tagesperiodischen windsystems im Loisachtal. [Aerological studies of the diurnal wind system in the Loisach Valley.] *Vortrag, 17 Internationale Tagung Alpine Meteorologie,* Berchtesgaden, FRG.

Nappo, C. J., and K. S. Rao, 1987: A model study of pure katabatic flow. *Tellus,* **A39,** 61–71.

——, —— and J. A. Herwehe, 1989: Pollutant transport and diffusion in katabatic flow. *J. Climate Appl. Meteor.*, **28,** 617–625.

Neal, D., 1979: Wind flow and structure over Gebbies Pass, New Zealand: A comparison between wind tunnel simulation and field measurements. Ph.D. dissertation, University of Canterbury, 486 pp.

——, 1983: The influence of model scale on a wind-tunnel simulation of complex terrain. *J. Wind Engr. Ind. Aero.*, **12,** 125–143.

Neff, W. D., 1975: Quantitative evaluation of acoustic echoes from the planetary boundary layer. NOAA Tech. Report ERL 322-WPL, Boulder, 38 pp.

——, 1978: Beamwidth effects on acoustic backscatter in the planetary boundary layer. *J. Appl. Meteor.*, **17,** 1514–1520.

——, 1980: An observational and numerical study of the

atmospheric boundary layer overlying the East Antarctic ice sheet. Ph.D. dissertation, University of Colorado, 300 pp.

——, 1985: Overview of sodar data from FSPS, August 1984 (unpublished draft project report). NOAA Wave Propagation Laboratory, Boulder.

——, 1986: On the use of sodars to study stably stratified flow influenced by terrain. *Atmos. Res.,* **20,** 279–308.

——, 1987: The role of waves and turbulence in complex terrain flows. *Fourth Conference on Mountain Meteorology,* Seattle, Amer. Meteor. Soc., 64–69.

——, 1988: Observations of complex terrain flows using acoustic sounders: Echo interpretation. *Bound.-Layer Meteor.,* **42,** 207–228.

——, 1989a: An overview of the 1986–1987 Denver pollution studies. Preprints, *Sixth Joint Conference on Applications of Air Pollution Meteorology,* Anaheim, Amer. Meteor. Soc., 228–229.

——, 1989b: The ontogony of Denver pollution episodes. Preprints, *Sixth Joint Conference on Applications of Air Pollution Meteorology,* Anaheim, Amer. Meteor. Soc., 230–234.

——, 1989c: Meteorological classifications used in the 1987–1988 Denver Brown Cloud Study. Preprints, *Sixth Joint Conference on Applications of Air Pollution Meteorology.* Anaheim, Amer. Meteor. Soc., 74–77.

——, and C. W. King, 1985: Studies of complex terrain flows using acoustic remote sensors. Wave Propagation Laboratory Internal Report ASCOT 85-1, NOAA/ERL, Boulder, 131 pp.

——, and R. L. Coulter, 1986: Acoustic remote sensing. *Probing the Atmospheric Boundary Layer,* D. Lenschow, Ed., Amer. Meteor. Soc., 201–236.

——, and Z. C. Juan, 1987: Momentum balance and temperature flux with a quasi-steady nocturnal drainage jet. Preprints, *Fourth Conference on Mountain Meteorology,* Seattle, Amer. Meteor. Soc., 30–31.

——, and C. W. King, 1987: Observations of complex terrain flows using acoustic sounders: Experiments, topography and winds. *Bound.-Layer Meteor.,* **40,** 363–392.

——, and ——, 1988: Observations of complex terrain flows using acoustic sounders: Drainage flow structure and evolution. *Bound.-Layer Meteor.,* **43,** 15–41.

——, and ——, 1989: The accumulation and pooling of drainage flows in a large basin. *J. Appl. Meteor.,* **28,** 518–529.

——, —— and H. E. Ramm, 1980: An analysis of acoustic sounder and tethersonde data obtained for the Atmospheric Studies in Complex Terrain (ASCOT) project during 1979. A collection of papers based on drainage wind studies in the Geysers area of Northern California: Part I. M. H. Dickerson, Ed., Lawrence Livermore National Laboratory, Livermore, CA, E1-E27. [UCID-18884, ASCOT-80-7.]

Neiman, P. J., R. M. Hardesty, M. A. Shapiro and R. E. Cupp, 1988: Doppler lidar observations of a downslope windstorm. *Mon. Wea. Rev.,* **116,** 2265–2275.

Neininger, B., 1982: Mesoklimatische Messungen im oberwallis [Mesoclimatic measurements in the Upper Valais.] *Ann. Meteor.,* **N.F., 19,** 105–107.

——, and M. E. Reinhardt, 1986: Meteorological aircraft data set of the "First Himalaya soaring expedition." *DFVLR-Forschungbericht 86-39.* DFVLR, Oberpfaffenhofen, FRG., 149 pp.

Nemoto, S., 1961a: Similarity between natural wind in the atmosphere and model wind in a wind tunnel. Modeling criteria for a local wind. Part I: Theoretical treatment. *Pap. Meteor. Geophys.,* Tokyo, **12(1),** 30–52.

——, 1961b: Similarity between natural wind in the atmosphere and model wind in a wind tunnel. Modeling criteria for a local wind. Part II: Similarity between vertical distribution of mean wind velocity near the ground and that above the model rough surface set in a wind tunnel. *Pap. Meteor. Geophys.,* Tokyo, **12(2),** 117–128.

——, 1961c: Similarity between natural wind in the atmosphere and model wind in a wind tunnel. Modeling criteria for a local wind. Part III: Some examples of experiments. *Pap. Meteor. Geophys.,* Tokyo, **12(2),** 129–154.

——, 1963: Similarity between natural wind in the atmosphere and model wind in a wind tunnel. Modeling criteria for a local wind. Part IV: Supplement to the theoretical treatment. *Pap. Meteor. Geophys.,* Tokyo, **13(2),** 171–195.

Newley, T. M. J., 1985: Turbulent airflow over hills. Ph.D. dissertation, University of Cambridge.

——, H. Pearson and J. C. R. Hunt, 1989: Stably stratified rotating flow through a group of obstacles. In preparation.

Nicholls, J. M., 1973: The airflow over mountains: Research 1958–1972. Tech. Note 127, WMO, 73 pp.

Nickus, U., and I. Vergeiner, 1984: The thermal structure of the Inn Valley atmosphere. *Arch. Meteor. Geophys. Bioklim.,* **A33,** 199–215.

Nicolini, M., J. Paegle and M. L. Altinger, 1987: Numerical simulation of convection and boundary layer convergence. Preprints, *Second Interamerica Congress of Meteorology,* Buenos Aires.

Nieman, P. J., R. M. Hardesty, M. A. Shapiro and R. E. Cupp, 1988: Doppler lidar observations of a downslope windstorm. *Mon. Wea. Rev.,* **116,** 2265–2275.

Nkemdirim, L., and K. Leggat, 1978: The effect of chinook weather on urban heat islands and air pollution. *Water, Air Soil Pollut.,* **9,** 53–67.

Novikov, E. A., 1959: Contributions to the problem of

the predictability of synoptic processes. *Izv. Acad. Sci. USSR, Geophys. Ser.*, **11**, 1721–1724.

Ogawa, Y., P. G. Dosey, K. Uehara and H. Ueda, 1981: A wind tunnel for studying the effects of thermal stratification in the atmosphere. *Atmos. Environ.*, **15**, 807–821.

Ogura, Y., and Yoshizaki, M., 1988: Numerical study of orographic–convective precipitation over the eastern Arabian Sea and the Ghat Mountains during the summer monsoon. *J. Atmos. Sci.*, **45**, 2097–2122.

Olfe, D. B., and R. L. Lee, 1971: Linearized calculations of urban heat island correction effects, AIAA Paper No 71-73, AIAA Ninth Aerospace Science Meeting, New York, 14 pp.

Olyphant, G. A., 1984: Insolation topoclimates and potential ablation in Alpine snow accumulation basins: Front Range, Colorado. *Water Resour. Res.*, **20**, 491–498.

——, 1986a: Longwave radiation in mountainous areas and its influence on the energy balance of Alpine snowfields. *Water Resour. Res.*, **22**, 62–66.

——, 1986b: The components of incoming radiation within a mid-latitude Alpine watershed during the snowmelt season. *Arct. Alp. Res.*, **18**, 163–169.

Ookouchi, Y., M. Segal, R. C. Kessler and R. A. Pielke, 1984: Evaluation of soil moisture effects on the generation and modification of mesoscale circulations. *Mon. Wea. Rev.*, **112**, 2281–2292.

Orgill, M. M., 1981: Atmospheric studies in complex terrain: A planning guide for future studies. Tech. Report PNL-3656/ASCOT-80-4, Pacific Northwest Laboratory, Richland, WA.

——, and R. I. Schreck, 1985: An overview of the ASCOT multi-laboratory field experiments in relation to drainage winds and ambient flow. *Bull. Amer. Meteor. Soc.*, **66**, 1263–1277.

——, J. E. Cermak and L. O. Grant, 1971a: Laboratory simulation and field estimates of atmospheric transport-dispersion over mountainous terrain. Colorado State University Report CER70-71MMO-JEC-LOG40, 302 pp.

——, J. E. Cermak and L. O. Grant, 1971b: Research and development technique for estimating airflow and diffusion parameters related to the atmospheric water resources program. Colorado State University Report CER71-72MMO-JEC-LOG20, 111 pp.

Orville, H. D., 1965: A photogrammetric study of the initiation of cumulus clouds over mountainous terrain. *J. Atmos. Sci.*, **22**, 700–709.

——, 1968: Ambient wind effects on the initiation and development of cumulus clouds over mountains. *J. Atmos. Sci.*, **25**, 385–403.

——, and L. J. Sloan, 1970: A numerical simulation of the life history of a rainstorm. *J. Atmos. Sci.*, **27**, 1148–1159.

Padro, J., 1987: Boundary-layer pollutant concentrations over complex terrain. *Bound.-Layer Meteor.*, **38**, 17–28.

Paegle, J., and D. W. McLawhorn, 1983: Numerical modeling of diurnal convergence oscillations above sloping terrain. *Mon. Wea. Rev.*, **101**, 746–756.

——, and T. Vukicevic, 1987a: On the predictability of low-level flow during ALPEX. *Meteor. Atmos. Physics*, **36**, 45–60.

——, and ——, 1987b: The analysis and forecast sensitivity of low-level mountain flows to input data. Preprints, *Fourth Conference on Mountain Meteorology*, Amer. Meteor. Soc., Seattle, 168–179.

——, J. N. Paegle, M. McCorcle and E. Miller, 1984: Diagnoses and numerical simulation of a low-level jet during ALPEX. *Beitr. Phys. Atmos.*, **57**, 419–430.

Paine, R. J., and B. A. Egan, 1987: User's guide to the Rough Terrain Diffusion Model (RTDM)—Revision 3.20. ERT, 696 Virginia Road, Concord, MA 01742.

——, D. G. Strimaitis, M. G. Dennis, R. J. Yamartino, M. T. Mills and E. M. Insley, 1987: User's guide to the Complex Terrain Dispersion Model. Volume 1: Model description and user instructions. Technical Report 600/8-87-058a, U.S. Environmental Protection Agency.

Palmer, T. N., G. J. Shutts and R. Swinbank, 1986: Attenuation of systematic westerly bias in general circulation and numerical weather prediction models through an orographic wave drag parameterisation. *Quart. J. Roy. Meteor. Soc.*, **112**, 1001–1031.

Pamperin, H., and G. Stilke, 1985: Nächtliche grenzschicht und llj im alpenvorland nahe dem Inntalausgang. [Nocturnal boundary layer and low level jet near the Inn Valley exit.] *Meteor. Rundsch.*, **38**, 145–156.

Panofsky, H. A., and A. A. Townsend, 1964: Change of terrain roughness and the wind profile. *Quart. J. Roy. Meteor. Soc.*, **90**, 147–155.

——, D. Larko, R. Lipschutz, G. Stone, E. F. Bradley, A. J. Bowen and Hojstrup, 1982: Spectra of velocity components over complex terrain. *Quart. J. Roy. Meteor. Soc.*, **108**, 215–230.

Parish, T., 1982: Barrier winds along the Sierra Nevada. *J. Appl. Meteor.*, **21**, 925–930.

Parsons, E. B., and P. V. Hobbs, 1983: The mesoscale and microscale structure and organization of clouds and precipitation in midlatitude cyclones. IX: Some effects of orography on rainbands. *J. Atmos. Sci.*, **40**, 1930–1949.

Pasquill, F., and F. B. Smith, 1983: *Atmospheric Diffusion*, 3rd ed., Ellis Horwood, 437 pp.

Pearse, J. R., 1982: Wind flow over conical hills in a simulated atmospheric boundary layer. *J. Wind Engr. Ind. Aero.*, **10**, 303–313.

Peltier, W. R., and T. L. Clark, 1979: The evolution and stability of finite-amplitude mountain waves. Part II:

Surface wave drag and severe downslope windstorms. *J. Atmos. Sci.,* **36,** 1498–1529.

——, and ——, 1983: Nonlinear mountain waves in two and three spatial dimensions. *Quart. J. Roy. Meteor. Soc.,* **109,** 527–548.

Petersen, R. A., and J. E. Cermak, 1980: Physical modeling of plume dispersion at Alkali and Coal Creek, near Crested Butte, Colorado. Colorado State University Report CER79-80RLP-JEC43, 302 pp.

Peterson, T. C., 1989: The effect of decoupled low-level flow in winter orographic clouds in northern Colorado. Master's thesis, Dept. of Atmospheric Science, Colorado State University, 145 pp.

Petkovsek, Z., 1978a: Relief meteorologically relevant characteristics of basins. *Meteor. Z.,* **28,** 333–340.

——, 1978b: Zones of convergence in local air flow in valleys and basins. *Tagungsbericht, 1. Teil, 15. Internationale Tagung Fuer Alpine Meteorologie,* Grindelwald, 92–96.

——, 1980: Additional relief meteorologically relevant characteristics of basins. *Meteor. Z.,* **30,** 379–382.

Pielke, R. A., 1984: *Mesoscale Numerical Modeling.* Academic Press, 612 pp.

——, and W. R. Cotton, 1977: A mesoscale analysis over South Florida for a high rainfall event. *Mon. Wea. Rev.,* **105,** 343–362.

——, and E. Kennedy, 1980: Mesoscale terrain features. Technical Report UVA-ENV SCI-MESO-1980-1, University of Virginia.

——, R. T. McNider, M. Segal and Y. Mahrer, 1983: The use of a mesoscale numerical model for evaluation of pollutant transport and diffusion in coastal regions and over irregular terrain. *Bull. Amer. Meteor. Soc.,* **64,** 243–250.

Pierce, T. D., and D. B. Turner, 1980: User's guide for MPTER. EPA Tech. Report, EPA 600/8-80-016, U.S. Environmental Protection Agency.

Piliè, R. G., E. J. Mack, W. C. Kocmond, C. W. Rogers and W. J. Eadie, 1975: The life cycle of valley fog. Part I: Micrometeorological characteristics. *J. Appl. Meteor.,* **14,** 347–363.

Plate, E. J., and J. E. Cermak, 1963: Micro-meteorological wind-tunnel facility. Colorado State University Report CER63EJP-JEC9, 65 pp.

——, and C. W. Lin, 1965: The velocity field downstream from a two-dimensional model hill. Part 2. Colorado State University Report CER65EJP-CW41, 59 pp.

Porch, W. M., 1982: Implication of spatial averaging in complex-terrain wind studies. *J. Appl. Meteor.,* **21,** 1258–1265.

——, W. D. Neff and C. W. King, 1988: Comparison of meteorological structure parameters in complex terrain using optical and acoustical techniques. *J. Appl. Optics,* **27,** 2222–2228.

——, S. Barr, W. E. Clements, J. A. Archuleta, A. B. Fernandez, C. W. King, W. D. Neff and R. P. Hosker, 1989a: Smoke flow visualization in a tributary of a deep valley. *Bull. Amer. Meteor. Soc.,* **70,** 30–35.

——, R. B. Fritz, R. L. Coulter and P. H. Gudiksen, 1989b: Tributary, valley and sidewall air flow interactions in a deep valley. *J. Appl. Meteor.,* **28,** 578–589.

Post, M. J., and W. D. Neff, 1986: Doppler lidar measurements of winds in a narrow mountain valley. *Bull. Amer. Meteor. Soc.,* **67,** 274–281.

——, R. A. Richter, R. M. Hardesty, T. R. Lawrence and F. F. Hall, Jr., 1982: National Oceanic and Atmospheric Administration (NOAA) pulsed coherent infrared Doppler lidar: Characteristics and data. *Proc. Soc. Photo-Opt. Instrum. Eng.,* **300,** 60–65.

Prandtl, L., 1942: *Strömungslehre.* Vieweg und Sohn, Braunschweig, 396-399.

Putnam, P. C., 1948: *Power from the Wind.* Van Nostrand Reinhold Company, 224 pp.

Puttock, J. S., and J. C. R. Hunt, 1979: Turbulent diffusion from sources near obstacles with separated wakes. I: An eddy diffusivity model. *Atmos. Environ.,* **13,** 1–13.

Queney, P., 1947: Theory of perturbations in stratified currents with applications to airflow over mountain barriers. University of Chicago Report No. 23, Dept. of Meteorology, University of Chicago Press. [Also described in P. Morel, Ed., 1970: *Dynamic Meteorology,* D. Reidel Publishing Co., 622 pp.]

——, 1948: The problem of air flow over mountain: A summary of theoretical studies. *Bull. Amer. Meteor. Soc.,* **29,** 16–26.

——, G. Corby, N. Gerbier, H. Koschmieder and J. Zierep, 1960: The airflow over mountains. WMO Tech. Note 34., 135 pp.

Rae, W. H., and A. Pope, 1984: *Low-Speed Wind Tunnel Testing.* Wiley Interscience, 534 pp.

Rao, K., and H. Snodgrass, 1981: A nonstationary nocturnal drainage flow model. *Bound.-Layer Meteor.,* **20,** 309–320.

Rao, P. K., 1968: An investigation of the valley-plain wind circulation in the valleys of Vermont. Ph.D. dissertation, New York University. [University Microfilms, Dissertation 69-11837, Ann Arbor, Michigan.]

——, 1970: Theoretical investigation of the change of wind speed along the axis of a valley. *Arch. Meteor. Geophys. Bioklim.,* **A19,** 59–70.

Rasmussen, R. M., P. K. Smolarkiewicz and J. Warner, 1989: On the dynamics of Hawaiian cloud bands: comparison of model results with observations and island climatology. *J. Atmos. Sci.,* **46,** 1589–1608.

Rau, M., and E. J. Plate, 1986: Concept of a thermal stratified boundary layer wind tunnel. *Proceedings Third International Workshop on Wind and Water Tunnel Modelling of Atmospheric Flow and Dispersion,* Lausanne, Switzerland, 213–222.

Rauber, R. M., L. O. Grant, D. Feng and J. B. Snider, 1986: The characteristics and distribution of cloud water over the mountains of northern Colorado during wintertime storms. Part I: Temporal variations. *J. Climate Appl. Meteor.,* **25,** 468–488.

Raymond, D. J., and M. H. Wilkening, 1980: Mountain-induced convection under fair weather conditions. *J. Atmos. Sci.,* **37,** 2693–2706.

——, and M. H. Wilkening, 1982: Flow and mixing in New Mexico mountain cumuli. *J. Atmos. Sci.,* **39,** 2211–2228.

Reap, R. M., 1986: Evaluation of cloud-to-ground lightning data from the western United States for the 1983–84 summer seasons. *J. Climate Appl. Meteor.,* **25,** 785–799.

Reinking, R. F., and J. F. Boatman, 1986: Upslope precipitation events. *Mesoscale Meteorology and Forecasting,* P. S. Ray, Ed., Amer. Meteor. Soc., 437–471.

Reiter, E. R., J. D. Sheaffer, J. E. Bossert, R. C. Fleming, W. E. Clements, J. T. Lee, S. Barr, J. A. Archuleta and D. E. Hoard, 1987: ROMPEX—The Rocky Mountain Peaks Experiment of 1985: Preliminary assessment. *Bull. Amer. Meteor. Soc.,* **68,** 321–328.

Reiter, R., R. Sladkovic and K. Munzert, 1981: Untersuchung des Berg-Tal-Windes und Erfassung des Temperaturprofils während des "Gebingswindexpermentes DISKUS in Dischmatal, August 1980." [Investigation of mountain and valley winds and collection of temperature profiles during the "DISKUS Mountain Wind Experiment of August 1980."] Wiss Mitteil. Nr. 11, Graunhofer-institut für Atmosphärische Umweltforschung, Garmisch-Partenkirchen, FRG.

——, ——, —— and H. Müller, 1982: Erfassung des temperaturund windfeldes während des experimentes "MERKUR" im Inntal bei Radfeld, Tirol, März/April 1982. [Recording of the temperature and wind field during the experiment "MERKUR" in the Inn Valley near Radfeld, Tyrol.] Wiss. Mitteil 13, Fraunhofer-Institut für Atmosphärische Umweltforschung, Garmisch-Partenkirchen, FRG.

——, H. Müller, R. Sladkovic and K. Munzert, 1983: Aerologische untersuchungen der tagesperiodischen gebirgswinde unter besonderer berücksichtigung des windfeldes im talquerschnitt. [Aerological investigations of diurnal mountain winds with special consideration of wind fields in the valley cross section.] *Meteor. Rundsch.,* **36,** 225–242.

——, ——, —— and ——, 1984: Aerologische untersuchungen des tagesperiodischen windsystems im Inntal während MERKUR. [Aerological investigations of diurnal wind systems in the Inn Valley during MERKUR.] *Meteor. Rundsch.,* **37,** 176–190.

Reynolds, D. W., and A. S. Dennis, 1986: A review of the Sierra Cooperative Pilot Project. *Bull. Amer. Meteor. Soc.,* **67,** 513–523.

Reynolds, S. D., T. W. Tesche and L. E. Reid, 1979: An introduction to the SAI Airshed model and its usage. Tech. Report EF78-53R4-EF79-31, SAI, San Rafael, CA.

Rhea, J. O., 1978: Orographic precipitation model for hydrometeorological use. Atmos. Sci. Paper 287, Colorado State University, 198 pp.

Richard, E., P. Mascart and E. C. Nickerson, 1989: On the role of surface friction in downslope windstorms. *J. Appl. Meteor.,* in press.

Rider, M. A., and V. A. Sandborn, 1977: Boundary layer turbulence over two-dimensional hills. Colorado State University Report CER77-78MAR-VAS4.

Roach, W. T., 1976: On the effect of radiative exchange on the growth by condensation of a cloud of fog droplets. *Quart. J. Roy. Meteor. Soc.,* **102,** 361–372.

Robins, A. D., and J. A. Hertig, 1986: Summary. *Proc. Third International Workshop on Wind and Water Tunnel Modelling of Atmospheric Flow and Dispersion,* Lausanne, Switzerland, 1–5.

Rodi, W., 1980: Turbulence models for environmental problems. *Prediction Methods for Turbulent Flows,* W. Kollmann, Ed., McGraw Hill, 259–350.

Ross, D. G., I. N. Smith, P. C. Manins and D. G. Fox, 1988: Diagnostic wind field modeling for complex terrain: Model development and testing. *J. Climate Appl. Meteor.,* **27,** 785–796.

Sasaki, K., H. Kurita, G. R. Carmichael, Y.-S. Chang, K. Murano and H. Ueda, 1988: Behavior of sulfate, nitrate and other pollutants in the long-range transport of air pollution. *Atmos. Environ.,* **22,** 1301–1308.

Sato, T., and J. Kondo, 1988: A simple model of drainage flow in a valley. *Bound.-Layer Meteor.,* **45,** 355–369.

Sauberer, F., and I. Dirmhirn, 1954: Über die entstehung der extremen temperaturminima in der Doline Gstettner-Alm. [On the occurrence of extreme temperature minimums in the Gstettner-Alm Doline.] *Arch. Meteor. Geophys. Bioklim.,* **B5,** 307–326.

——, and ——, 1958: Das Strahlungsklima. [Radiation climate.] Klimatographie von Österreich, Vienna, 13–102.

Saucier, W. J., 1955: *Principles of Meteorological Analysis.* Univ. of Chicago Press, 438 pp.

Schaaf, C. L. B., J. Wurman and R. M. Banta, 1986: Satellite climatology of thunderstorm initiation sites in the Rocky Mountains of Colorado and Northern New Mexico. Technical Report AFGL-TR-86-0075, Air Force Geophysics Laboratory, Hanscom AFB, MA, vi + 35 pp. [NTIS AD-A178052.]

——, J. Wurman and R. M. Banta, 1988: Thunderstorm-producing terrain features. *Bull. Amer. Meteor. Soc.,* **69,** 272–277.

Schlatter, T. W., D. V. Baker and J. F. Hena, 1983: Profiling Colorado's Christmas Eve blizzard. *Weatherwise,* **36,** 60–66.

Schon, J. P., and P. Mery, 1971: A preliminary study of the simulation of neutral atmospheric boundary layer using air injection in a wind tunnel. *Atmos. Environ.,* **5,** 299–312.

Schroeder, T. A., 1977: Meteorological analysis of an Oahu flood. *Mon. Wea. Rev.,* **105,** 458–468.

Schwiesow, R. L., 1986a: A comparative overview of active remote-sensing techniques. *Probing the Atmospheric Boundary Layer,* D. Lenschow, Ed., Amer. Meteor. Soc., 129–138.

——, 1986b: Lidar measurement of boundary-layer variables. *Probing the Atmospheric Boundary Layer,* D. Lenschow, Ed., Amer. Meteor. Soc., 139–162.

Scinocca, J. F., and W. R. Peltier, 1989: Pulsating downslope windstorms. *J. Atmos. Sci.,* 2885–2914.

Scorer, R. S., 1949: Theory of waves in the lee of mountains. *Quart. J. Roy. Meteor. Soc.,* **75,** 41–56.

——, 1953: Theory of airflow over mountains. II: The flow over a ridge. *Quart. J. Roy. Meteor. Soc.,* **79,** 70.

——, 1954: Theory of airflow over mountains. III: Airstream characteristics. *Quart. J. Roy. Meteor. Soc.,* **80,** 417

——, 1955: Theory of airflow over mountains. IV. Separation of flow from the surface. *Quart. J. Roy. Meteor. Soc.,* **81,** 340–350.

——, 1978: *Environmental Aerodynamics,* Ellis Horwood Ltd., 487 pp.

Segal, M., C.-H. Yu and R. A. Pielke, 1988a: Model evaluation of the impact of thermally induced valley circulations in the Lake Powell area on long-range pollutant transport. *J. Air Pollut. Control Assoc.,* **38,** 163–170.

——, R. Avissar, M. C. McCumber and R. A. Pielke, 1988b: Evaluation of vegetation effects on the generation and modification of mesoscale circulations. *J. Atmos. Sci.,* **45,** 2268–2292.

——, R. A. Pielke, R. W. Arritt, M. D. Moran, C.-H. Yu and D. Henderson, 1986: Southern Florida air pollution climatology study and selected episodic impacts. Report for the National Park Service by Colorado State University.

Shapiro, M. A., 1984: Structure and physical processes within meteorological tower measurements from a surface cold front. Analysis of some cloud and frontal events, E. E. Gossard, Ed., NOAA Environmental Research Laboratories, 63–78. [NTIS PB84-179662.]

Sheppard, P. A., 1956: Airflow over mountains. *Quart. J. Roy. Meteor. Soc.,* **82,** 528–529.

Sherman, C. A., 1978: A mass-consistent model for wind fields over complex terrain. *J. Appl. Meteor.,* **17,** 312–319.

Shinn, J. H., R. T. Cederwall, F. J. Gouveia and K. R. Chapman, 1989: Micrometeorology of slope flows in a tributary canyon during the 1984 ASCOT experiment. *J. Appl. Meteor.,* **28,** 569–577.

Silverman, B. A., 1960: The effect of a mountain on convection. *Cumulus Dynamics,* C. A. Anderson, Ed., Pergamon Press, 4–27.

Simpson, J. E., 1987: *Gravity Currents, In the Environment and the Laboratory.* Halsted Press, 244 pp.

Smagorinsky, J., 1969: Problems and promises of deterministic extended range forecasting. *Bull. Amer. Meteor. Soc.,* **50,** 286–311.

Smith, R. B., 1977: The steepening of hydrostatic mountain waves. *J. Atmos. Sci.,* **34,** 1634–1654.

——, 1979: The influence of mountains on the atmosphere. *Advances in Geophysics,* B. Saltzman, Ed., Vol. 21, Academic Press, 87–230.

——, 1980: Linear theory of stratified hydrostatic flow past an isolated mountain. *Tellus,* **32,** 348–364.

——, 1982a: Aerial observations of the Yugoslavian bora. ALPEX preliminary scientific results. Tech. Report GARP No. 7, ICSU, WMO.

——, 1982b: A differential advection model of orographic rain. *Mon. Wea. Rev.,* **110,** 306–309.

——, 1985a: On severe downslope winds. *J. Atmos. Sci.,* **42,** 2597–2603.

——, 1985b: Comments on "Interactions of low-level flow with the western Ghat Mountains and offshore convection in the summer monsoon." *Mon. Wea. Rev.,* **113,** 2176–2177.

——, 1986: ALPEX: Some current issues. *Seminar/Workshop on observation theory and modelling of orographic effects,* Vol. 1. ECMWF, Reading, U.K., 51–56.

——, 1987: Aerial observations of the Yugoslavian bora. *J. Atmos. Sci.,* **44,** 269–297.

——, 1989: Mountain induced stagnation points in hydrostatic flow. *Tellus,* **A41,** 270–274.

——, and Y.-L. Lin, 1983: Orographic rain on the western Ghats. *Proceedings First Sino-American workshop on Mountain Meteorology,* E. R. Reiter, B. Zhu and Y. Qian, Eds., Science Press, Beijing, and Amer. Meteor. Soc., 71–98.

——, and J. Sun, 1987: Generalized hydraulic solutions pertaining to severe downslope winds. *J. Atmos. Sci.,* **44,** 2934–2939.

Smolarkiewicz, P. J., and T. L. Clark, 1985: Numerical simulation of the evolution of a three-dimensional field of cumulus clouds. Part I: Model description, comparison with observation and sensitivity studies. *J. Atmos. Sci.,* **42,** 502–522.

Smolarkiewicz, P. K., and R. Rotunno, 1989: Low Froude number flow past three-dimensional obstacles. Part I: Baroclinically generated lee vortices. *J. Atmos. Sci.,* **46,** 1154–1164.

——, R. M. Rasmussen and T. L. Clark, 1988: On the dynamics of Hawaiian cloud bands: Island forcing. *J. Atmos. Sci.,* **45,** 1872–1905.

Snyder, W. H., 1972: Similarity criteria for the application of fluid models to the study of air pollution meteorology. *Bound.-Layer Meteor.,* **3,** 112–134.

——, 1981: Guidelines for fluid modeling of atmospheric diffusion. EPA Report 600/8-81-009, U.S. Environmental Protection Agency, 185 pp.

——, 1985: Fluid modelling of pollutant transport and diffusion in stable stratified flows over complex terrain. *Ann. Rev. Fluid Mech.,* **17,** 239–266.

——, 1987: Contributions of the fluid modeling facility to EPA's Complex Terrain Model Development Program. EPA Tech. Report 600/3-87-026, U.S. Environmental Protection Agency.

——, 1988: Integration of fluid modeling with complex-terrain field studies and model-development efforts. *Proceedings Eighth Symposium on Turbulence and Diffusion,* San Diego, 189–192.

——, and Y. Ogawa, 1982: Simulation of flow and diffusion over Cinder Cone Butte in a stratified wind tunnel. Data report, National Inst. for Envir. Studies, Tsukaba, Japan.

——, and J. C. R. Hunt, 1984: Turbulent diffusion from a point source in stratified and neutral flows around a three-dimensional hill. Part II: Laboratory measurements of surface concentrations. *Atmos. Environ.,* **18,** 1969–2002.

——, and R. E. Lawson, Jr., 1987: Stable plume dispersion over an isolated hill: Releases above the dividing-streamline height. *Proceedings Third International Symposium on Stratified Flows.* Cal. Inst. Tech., Pasadena, CA.

——, and R. E. Britter, 1987: A wind tunnel study of the flow structure and dispersion from sources upwind of three-dimensional hills. *Atmos. Environ.,* **21,** 735–751.

——, —— and J. C. R. Hunt, 1980: A fluid modeling study of the flow structure and plume impingement on a three-dimensional hill in stably stratified flow. *Proceedings Fifth International Conference on Wind Engineering,* J. E. Cermak, Ed., Vol. 1, Pergamon Press, 319–329.

——, R. S. Thompson, R. E. Eskridge, R. E. Lawson, Jr., I. P. Castro, J. T. Lee, J. C. R. Hunt and Y. Ogawa, 1985: The structure of strongly stratified flow over hills: Dividing-streamline concept. *J. Fluid Mech.,* **152,** 249–288.

Somervell, T. H., 1926: The meteorological results of the Mount Everest expedition. *Quart. J. Roy. Meteor. Soc.,* **52,** 131–136.

Spangler, T. C., 1986: The role of near terrain turbulence in the prediction of ground-level pollutant concentrations in complex terrain. *Atmos. Environ.,* **20,** 861–865.

Sroga, J. T., E. W. Eloranta and T. Barber, 1980: Lidar measurement of wind velocity profiles in the boundary layer. *J. Appl. Meteor.,* **19,** 598–605.

Start, G. E., C. R. Dickson and L. L. Wendell, 1975: Diffusion in a canyon within rough mountainous terrain. *J. Appl. Meteor.,* **14,** 333–346.

Steinacker, R., 1984: Area-height distribution of a valley and its relation to the valley wind. *Contrib. Atmos. Phys.,* **57,** 64–71.

Stone, G. L., and D. E. Hoard, 1989: Low frequency velocity and temperature fluctuations in katabatic valley flows. *J. Appl. Meteor.,* **28,** 477–488.

Stone, R. G., 1934: The history of mountain meteorology in the United States and the Mount Washington Observatory. *Trans. Amer. Geophys. Union,* **15,** 124–133.

Strauch, R. G., D. A. Merritt, K. P. Moran, K. B. Earnshaw and D. Van de Kamp, 1984: The Colorado wind-profiling network. *J. Atmos. Oceanic Technol.,* **1,** 37–49.

——, K. P. Moran, P. T. May, A. J. Bedard and W. L. Ecklund, 1988: RASS temperature sounding techniques. NOAA Tech. Memo. ERL WPL-158, Wave Propagation Laboratory, Boulder, CO, 12 pp.

Strimaitis, D. G., R. J. Paine, B. A. Egan and R. J. Yamartino, 1988: EPA Complex Terrain Model Development. EPA Final Report 600/S3-88-006, U.S. Environmental Protection Agency.

——, A. Venkatram, B. R. Greene, S. Hanna, S. Heisler, T. F. Lavery, A. Bass and B. A. Egan, 1983: EPA Complex Terrain Model Development. Second Milestone Report—1982 EPA 600/3-83-015, U.S. Environmental Protection Agency.

——, D. C. DiChristofaro, B. R. Greene, R. J. Yamartino, A. Venkatram, D. A. Gooden, P. Karamchandani, T. F. Lavery and B. A. Egan, 1986: EPA Complex Terrain Model Development. Fifth Milestone Report 1985, U.S. Environmental Protection Agency.

Strom, G. H., 1952: Wind tunnel techniques used to study influence of building configuration on stack gas dispersal. *Amer. Ind. Hygiene Ass. Quart.,* **13,** 76–80.

Stull, R. B., 1988: *An Introduction to Boundary-Layer Meteorology.* Kluwer Academic Publishers, 666 pp.

Sutera, A., 1980: Stochastic perturbation of a pure convective motion. *J. Atmos. Sci.,* **37,** 245–249.

Tampieri, F., and J. C. R. Hunt, 1985: Two-dimensional stratified fluid flow over valleys: Linear theory and a laboratory investigation. *Bound.-Layer Meteor.,* **32,** 257–279.

Tatarskii, V. I., 1971: *The Effects of the Turbulent Atmosphere on Wave Propagation.* Israel Program for Scientific Translations, U.S. Dept. of Commerce, Nat. Tech. Inform. Serv., Springfield, VA, 472 pp., [Also published by Nauka, Moscow, 1967.]

Taylor, G. I., 1921: Diffusion by continuous movements. *Proc. London Math Soc Ser.,* **2,** 196.

Taylor, P. A., 1977: Some numerical studies of surface boundary layer flow above gentle topography. *Bound.-Layer Meteor.,* **11,** 439–465.

——, and H. W. Teunissen, 1987: The Askervein Hill

project: Overview and background data. *Bound.-Layer Meteor.,* **39,** 15–39.

——, P. J. Mason and E. F. Bradley, 1987: Boundary layer flow over low hills. *Bound.-Layer Meteor.,* **39,** 107–132.

Tennekes, H., and J. L. Lumley, 1973: *A First Course in Turbulence.* MIT Press, 300 pp.

Tesche, T. W., J. L. Haney and R. E. Morris, 1987: Performance evaluation of four grid-based dispersion models in complex terrain. *Atmos. Environ.,* **21,** 233–256.

Teunissen, H. W., 1975: Simulation of the planetary boundary layer in a multiple jet wind tunnel. *Atmos. Environ.,* **9,** 145–174.

——, M. E. Shokr, A. J. Bowen, C. J. Wood and D. W. R. Green, 1987: The Askervein Hill project: Wind-tunnel simulations at three length scales. *Bound.-Layer Meteor.,* **40,** 1–29.

Thompson, P. D., 1957: Uncertainty of initial state as a factor in the predictability of large scale atmospheric flow patterns. *Tellus,* **9,** 275–295.

Thomson, D. J., 1988: Turbulent diffusion. *Proceedings Second European Turbulence Conference,* H. H. Fernholz and H. E. Fiedler, Eds., Technische University, Berlin.

Thorp, J. M., and M. M. Orgill, 1986: A "waterfall" of air. *Weatherwise,* **39,** 319–322.

Todd, C. J., 1964: Aircraft traverses in a growing mountain cumulus cloud. *J. Atmos. Sci.,* **21,** 529–538.

Toth, J. J., and R. H. Johnson, 1985: Summer surface flow characteristics over northeast Colorado. *Mon. Wea. Rev.,* **113,** 1458–1469.

Townsend, A. A., 1956: *The Structure of Turbulent Shear Flow.* Cambridge University Press, 315 pp.

——, 1965: The response of a turbulent boundary layer to abrupt changes in surface conditions. *J. Fluid Mech.,* **22,** 799–822.

——, 1966: The flow in a turbulent boundary layer after a change in surface roughness. *J. Fluid Mech.,* **26,** 255–266.

Tripoli, G. J., and W. R. Cotton, 1989a: Numerical study of an observed orogenic mesoscale convective system. Part I: Simulated genesis and comparison with observations. *Mon. Wea. Rev.,* **117,** 273–304.

——, and ——, 1989b: Numerical study of an observed orogenic mesoscale convective system. Part II: Analysis of governing dynamics. *Mon. Wea. Rev.,* **117,** 305–328.

Turfus, C., 1986: Diffusion from a continuous source near a surface in steady reversing flow. *J. Fluid Mech.,* **172,** 183–209.

——, 1988: Calculating mean concentration from a steady source in recirculating wakes by a particle trajectory method. *Atmos. Environ.,* **22,** 1271–1290.

Turner, D. B., 1979: Atmospheric dispersion modeling: A critical review. *J. Air Pollut. Control Assoc.,* **29,** 502–519.

Turner, J. S., 1973: *Buoyancy Effects in Fluids.* Cambridge University Press, 379 pp.

Ukejurchi, N., H. Sakata, H. Okamoto and Y. Ide, 1967: Study on stack gas diffusion. *Mitsubishi Tech. Bull.,* No. 52, 1–13.

Urfer-Henneberger, C., 1964: Wind- und temperaturverhältnisse an ungestörten schönwettertagen im Dischmatal bei Davos. [Wind and temperature relationships on undisturbed clear weather days in the Dischma Valley near Davos.] *Mitt. Schweiz. Anst. Forstl. Versuchsw.,* **40,** 389–441.

——, 1970: Neure beobachtungen über die entwicklung des schönwetterwindsystems in einem v-förmigen alpental (Dischmatal bei Davos). [New observations of the development of a clear weather wind system in a v-shaped mountain valley] (Dischma Valley near Davos). *Arch. Meteor. Geophys. Bioklim.,* **B18,** 21–42.

Uthe, E. E., W. Viezee, B. M. Morley and J. K. S. Ching, 1985: Airborne lidar tracking of fluorescent tracers for atmospheric transport and diffusion studies. *Bull. Amer. Meteor. Soc.,* **66,** 1255–1262.

Uttal, T., J. B. Snider, R. A. Kropfli and B. W. Orr, 1989: A remote sensing method of measuring atmospheric vapor fluxes: Application to winter mountain storms. *J. Appl. Meteor.,* **28,** in press.

Van der Hoven, I., et al., 1972: Southwest energy study, report of the meteorology work group. Draft report, U.S. Dept. of Interior, 113 pp.

Venkatram, A., D. G. Strimaitis and D. C. DiCristofaro, 1984: A semiempirical model to estimate vertical dispersion of elevated releases in the stable boundary layer. *Atmos. Environ.,* **18,** 923–928.

Vergeiner, I., 1982: Eine energetische theorie der hangwinde. [An energetics theory of slope winds.] *Internationale Tagung Alpine Meteorologie, Tagungsbericht,* 105–107, Berchtesgaden, FRG. [Vergeiner, I., 1982: *Meteor. Atmos. Phys.,* Neue Folge **19,** 189-191.]

——, 1983: Dynamik alpiner windsysteme. [Dynamics of alpine wind systems.] Bericht zum forschungsvorhaben 3556, Fonds zur Forderung der wissenschaftlichen Forschung, Wien, Austria, 129 pp.

——, 1987: An elementary valley wind model. *Meteorol. Atmos. Phys.,* **36,** 255–263.

——, and E. Dreiseitl, 1987: Valley winds and slope winds: Observations and elementary thoughts. *Meteor. Atmos. Phys.,* **36,** 264–286.

——, —— and C. D. Whiteman, 1987: Dynamics of katabatic winds in Colorado's Brush Creek Valley. *J. Atmos. Sci.,* **44,** 148–157.

Verma, S. B., and J. E. Cermak, 1974: Wind-tunnel investigation of mass transfer from soil corrugations. *J. Appl. Meteor.,* **13,** 578–587.

Vukicevic, T., and J. Paegle, 1989: The influence of one-way interacting lateral boundary conditions on predictability of flow in banded numerical models. *Mon. Wea. Rev.,* in press.

Wackter, D. J., and R. J. Londergan, 1984a: Evaluation of complex terrain air quality simulation models. EPA Tech. Report, 450/4-84-017, U.S. Environmental Protection Agency/OAQPA.

——, and ——, 1984b: Operational evaluation of eight complex terrain models for potential use in regulatory applications. *Fourth Joint Conference on Applications of Air Pollution Meteorology,* Amer. Meteor. Soc., J17–J20.

Wagner, A., 1932a: Hangwind—ausgleichsströmung—berg—und talwind. [Slope wind—compensation currents—mountain and valley winds.] *Meteor. Z.,* **49,** 209–217.

——, 1932b: Der tägliche luftdruck- und temperaturgang in der freien atmosphäre und in gebirgstälern. [Daily air pressure and temperature changes in the free atmosphere and in mountain valleys.] *Gerlands Beitr. Geophys.* (Leipzig), **37,** 315–344.

——, 1932c: Neue theorie des berg- und talwindes. *Meteor. Z.,* **49,** 329–341.

——, 1938: Theorie und beobachtung der periodischen gebirgswinde. [Theory and observation of periodic mountain winds.] *Gerlands Beitr. Geophys.* (Leipzig), **52,** 408–449. [English translation: C. D. Whiteman and E. Dreiseitl, 1984: Alpine Meteorology: Translations of Classic Contributions by A. Wagner and E. Ekhart and F. Defant. PNL-5141/ASCOT-84-3. Pacific Northwest Laboratory, Richland, Washington, 121 pp.]

Wakimoto, R. M., and J. L. McElroy, 1986: Lidar observation of elevated pollution layers over Los Angeles. *J. Climate Appl. Meteor.,* **25,** 1583–1599.

Walmsley, J. L., P. A. Taylor and T. Keith, 1986: A simple model of neutrally stratified boundary-layer flow over complex terrain with surface roughness modulations (MS3DJH/3R). *Bound.-Layer Meteor.,* **36,** 157–186.

Watanabe, H., and Y. Ogura, 1987: Effects of orographically forced upstream lifting on mesoscale heavy precipitation: A case study. *J. Atmos. Sci.,* **44,** 661–675.

Watson, J. G., J. C. Chow, L. W. Richards, W. D. Neff, S. R. Anderson, D. L. Dietrich, J. E. Houck and I. Olmez, 1988: The 1987–1988 Metro Denver Brown Cloud Study. Volume III: Data interpretation, 350 pp. [Available from Atmospheric Sciences Center, Denver University, Denver, CO.]

Weickmann, H., 1981: Mechanism of shallow winter-type stratiform cloud systems. Technical Report U.S. Government Printing Office 1982-576-001/1220, NOAA Environmental Research Laboratories, 61 pp.

Weisman, M. L., and J. B. Klemp, 1986: Characteristics of isolated convective storms. *Mesoscale Meteorology and Forecasting,* P. S. Ray, Ed., Amer. Meteor. Soc., 331–358.

Weissbluth, M. J., and W. R. Cotton, 1989: Radiative and nonlinear influences in orographic gravity wave drag. *Mon. Wea. Rev.,* in press.

Weng, W., K. J. Richards and D. J. Carruthers, 1988: Some numerical studies of turbulent wakes over hills. Fernholz, H. H., and H. E. Fiedler, Eds., *Proc. Second European Turbulence Conference,* Berlin.

Wetzel, P. J., 1973: Moisture sources and flow patterns during the northeast Colorado hail season. Master's thesis, Dept. of Atmos. Sci., Colorado State University, 90 pp.

——, W. R. Cotton and R. L. McAnelly, 1983: A long-lived mesoscale convective complex. Part II: Evolution and structure of the nature complex. *Mon. Wea. Rev.,* **111,** 1919–1937.

Whiteman, C. D., 1980: Breakup of temperature inversions in Colorado mountain valleys. Atmos. Sci. Paper 328, Ph.D. dissertation, Colorado State University. [Available from University Microfilms, Ann Arbor, Michigan.]

——, 1982: Breakup of temperature inversions in deep mountain valleys. Part I: Observations. *J. Appl. Meteor.,* **21,** 270–289.

——, 1986: Temperature inversion buildup in Colorado's Eagle Valley. *Meteor. Atmos. Phys.,* **35,** 220–226.

——, 1989: Morning transition tracer experiments in a deep narrow valley. *J. Appl. Meteor.,* **28,** 620–635.

——, and J. G. Whiteman, 1974: An historical climatology of damaging downslope windstorms at Boulder, Colorado. NOAA Tech. Report ERL 336-APCL 35. [Available from the Superintendent of Documents, U.S. Government Printing Office, Washington, D.C.]

——, and T. B. McKee, 1982: Breakup of temperature inversions in deep mountain valleys. Part II: Thermodynamic model. *J. Appl. Meteor.,* **21,** 290–302.

——, and E. Dreiseitl, 1984: Alpine meteorology: Translations of classic contributions by A. Wagner, E. Ekhart and F. Defant. Tech. Report PNL-5141/ASCOT-84-3, Pacific Northwest Laboratory, Richland, WA, 121 pp.

——, and K. J. Allwine, 1986: Extraterrestrial solar radiation on inclined surfaces. *Environ. Software,* **1,** 164–169.

——, and S. Barr, 1986: Atmospheric mass transport by along-valley wind systems in a deep Colorado valley. *J. Climate Appl. Meteor.,* **25,** 1205–1212.

——, K. J. Allwine, M. M. Orgill, L. J. Fritschen and J. R. Simpson, 1989a: Deep valley radiation and surface energy budget microclimates. I: Radiation. *J. Appl. Meteor.,* in press.

——, ——, ——, —— and ——, 1989b: Deep valley radiation and surface energy budget microclimates. II: Energy budget. *J. Appl. Meteor.,* in press.

Wilson, J. W., and C. Mueller, 1987: Thunderstorm nowcasting using doppler radar and soundings. Preprints, *Symposium on Mesoscale Analysis and Forecasting In-*

corporating Nowcasting, Vancouver, European Space Agency, 55–60. [ESA SP-282.]

Wippermann, F., and G. Gross, 1981: On the construction of orographically-influenced wind roses for given distributions of the large-scale wind. *Beitr. Phys. Atmos.,* **54,** 492–501.

Woods, J. D., 1969: On Richardson's number as a criterion for laminar-turbulent transition in the ocean and atmosphere. *Radio Sci.,* **4,** 1289–1298.

Woodcock, A. H., 1975: Anomalous orographic rains of Hawaii. *Mon. Wea. Rev.,* **103,** 334–343.

Wyngaard, J. C., Y. Isumi and S. A. Collins, 1971: Behavior of the refractive index structure parameter near the ground. *J. Opt. Soc. Am.,* **61,** 1646–1650.

——, W. T. Pennell, D. H. Lenschow and M. A. LeMone, 1978: The temperature-humidity covariance budget in the convective boundary layer. *J. Atmos. Sci.,* **35,** 47–58.

Xu, D. P., and J. C. R. Hunt, 1989: The response of the turbulent boundary layer to arbitrary distributed surface roughness. *Quart. J. Roy. Meteor. Soc.,* in press.

Yamada, T., 1983: Simulations of nocturnal drainage flows by a q^2l turbulence closure model. *J. Atmos. Sci.,* **40,** 91–106.

——, 1985: Numerical simulations of valley ventilation and pollutant transport. *Seventh Symposium on Turbulence and Diffusion,* Amer. Meteor. Soc., 312–314.

——, and R. N. Meroney, 1971: Numerical and wind tunnel simulation of response of stratified shear layers to nonhomogeneous surface features. Colorado State University Report CER70-71TY-RNM62 AD-730-953, 215 pp.

——, and S. Bunker, 1988: Development of a nested grid, second moment turbulence closure model and application to the 1982 ASCOT Brush Creek data simulation. *J. Appl. Meteor.,* **27,** 562–578.

——, and ——, 1989: A numerical model study of nocturnal drainage flows with strong wind and temperature gradients. *J. Appl. Meteor.,* **28,** 545–554.

Yocke, M. A., and M. K. Liu, 1978: Modeling wind distributions over complex terrain. Technical Report EF78-78, SAI, Inc., San Rafael, CA.

Yoshino, M. M., 1975: *Climate in a Small Area.* University of Tokyo Press, 549 pp.

Yoshizaki, M., and Y. Ogura, 1988: Two- and three-dimensional modeling studies of the Big Thompson storm. *J. Atmos. Sci.,* **45,** 3700–3722.

Zeman, O., and N. O. Jensen, 1987: Modification of turbulence characteristics in flow over hills. *Quart. J. Roy. Meteor. Soc.,* **113,** 55–80.